Next Generation Batteries

Kiyoshi Kanamura
Editor

Next Generation Batteries

Realization of High Energy Density Rechargeable Batteries

Springer

Editor
Kiyoshi Kanamura
Applied Chemistry Graduate School of Engineering
Tokyo Metropolitan University
Hachioji-shi, Japan

ISBN 978-981-33-6667-1 ISBN 978-981-33-6668-8 (eBook)
https://doi.org/10.1007/978-981-33-6668-8

This Springer imprint is published by the registered company Springer Nature Singapore Pte Ltd.
The registered company address is: 152 Beach Road, #21-01/04 Gateway East, Singapore 189721, Singapore

Foreword

Specially Promoted Research for Innovative Next Generation Batteries of Advanced Low Carbon Technology Research and Development Program (ALCA-SPRING)—A Next Generation Battery Project in Japan

Low-cost, high-performance secondary batteries are essential to establish a sustainable society by reducing CO_2 emissions from vehicles, which account for around 10% of total CO_2 emissions, and stabilizing renewable energy supply. The energy and power densities as well as the cost of lithium-ion battery, which is currently the most popular battery, is, however, not satisfactory and the development of innovative next generation secondary batteries is required.

At the joint committee on the innovative technologies on energy and environment hosted by Executive Directors of the Ministry of Education, Culture, Sports, Science and Technology (MEXT) and the Ministry of Economy, Trade and Industry (METI) in 2012, research on next generation secondary batteries was selected as one of the top priority research areas to be conducted jointly by the two Ministries.

Accordingly, MEXT set up a research project on next generation secondary batteries within the Advanced Low Carbon Technology Research and Development Program (ALCA) of the Japan Science and Technology Agency (JST) in April 2013. The project was named Specially Promoted Research for Innovative Next Generation Batteries (ALCA-SPRING). While researches in the original ALCA programs are conducted by bottom-up approach, ALCA-SPRING was designed to promote research by a top-down approach with the clear goal of realizing practical batteries as it is often the case that even if battery materials show excellent properties under specific conditions set by the researchers, it is not enough for practical use. Initially, four teams were set up to be recruited; namely, 1. All Solid State Battery Team with two sub-teams of sulfide type and oxide type, 2. Metal–air Battery Team, 3. Medium/long-term team, and 4. Long-term Team. Each team leader was requested to form a team with group members not only for the development of materials such as active materials and electrolytes but also for battery integration technology so that each team could undertake a complete study of the respective battery. No other projects not only in Japan but also worldwide had taken the comprehensive approach to carry out battery research from the battery design to battery integration with the developments of materials as well as characterization, evaluation, theoretical calculation with a clear target of realizing novel batteries in mind.

Kohei Uosaki of NIMS was appointed as Program Officer (PO) and team leaders were selected by the PO with the help of Advisory members from the industry (Toyota, Nissan, Toshiba, and Mitsubishi Chemicals) and academia (Osaka University, Yamanashi University, and AIST). Prof. Tatsumisago of Osaka Prefecture University for All Solid State Battery Team and its Sulfide Sub-team, Dr. Takada of NIMS for Oxide Sub-team of All Solid State Battery Team, Dr. Kubo of NIMS for Metal-air Battery Team, Prof. Watanabe of Yokohama National University for Mid/long-term Team, and Prof. Kanamura of Tokyo Metropolitan University for Long-Term Team. As of the kick-off meeting on September 3, 2013, the project consisted of 70 PIs from 40 institutions.

Other unique feature of ALCA-SPRING is the setup of System Research and Strategy Review Team, which discusses the open and close strategies and managing policies for intellectual properties, to maximize the outcome of the project and lead the world overwhelmingly not only in battery technology and also in battery business.

Progress of the projects supported by MEXT and METI and technology exchange among the battery projects are monitored by the Governing Board, which is co-chaired by 2 Directors of MEXT and 2 Directors of METI with experts in battery research such as the leaders of various national projects including ALCA-SPRING as members.

ALCA-SPRING has been receiving full support of the Battery Research Platform, which was established in National Institute for Materials Science (NIMS), National Institute of Advanced Industrial Science and Technology (AIST), and Waseda University to promote research and development on next generation batteries in Japan funded by supplementary budget in FY 2012 with world top facility for advanced characterization and battery fabrication and started full operation from October 2014.

Major restructuring of ALCA-SPRING was made twice based on stage gates held in 2015 and 2017. As a result of the first stage gate evaluation, Mid/long-term Team led by Prof. Watanabe and Long-term Team led by Prof. Kanamura were converted to Li-sulfur Battery with Non-soluble Positive Electrode Team and Next Generation Battery Team, respectively. Metal-air Battery Team became Metal-air Battery Sub-team of Next Generation Battery Team. Acceleration and Promotion for Practical Application Team with Lithium Metal Anode Special Unit and Evaluation, Analysis and Common Materials/technology Group were founded. To support the research by Lithium Metal Anode Special Unit, lithium processing facility was installed in Battery Research Platform in NIMS in 2017 so that metallurgical characteristics can be correlated with electrochemical characteristics.

Restructuring was made again in April 2018 following the second stage gate. A part of Sulfide Sub-team of All Solid State Battery Team was transferred to the SOLiD-EV project, which is developing all solid-state lithium-ion battery with sulfide electrolyte supported by NEDO/METI, because of its significant progress and

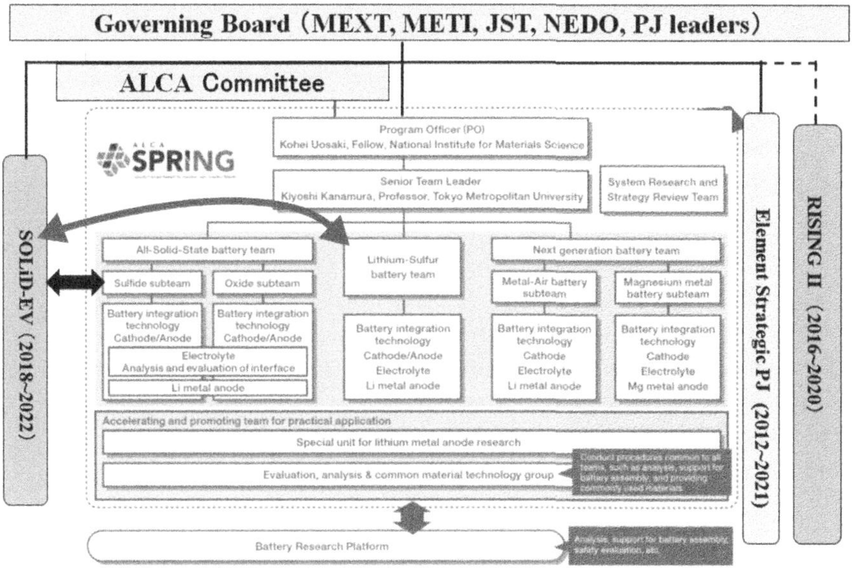

Fig. 1 Structure of ALCA-SPRING and other national battery projects in Japan

successful collaboration with the NEDO project in the first 5 years. Sulfide Sub-team is now concentrating its effort on Li-S battery research and is in close collaboration with the SOLiD-EV project. Members of Oxide Sub-team of All Solid State Battery Team were increased and collaboration between two sub-teams was strengthened. Targets of Li-Sulfur Battery with Non-soluble Positive Electrode Team were more focused. This team is also collaborating with the SOLiD-EV project. Anion Battery Group of Next Generation Battery Team was transferred to another JST project and Metal-air Battery Sub-team started joint development of Li-Air battery with a private company based on its outcome. Figure 1 shows the current structure of ALCA-SPRING and other national projects under the Governing Board in Japan.

Kohei Uosaki
Program Officer of ALCA-SPRING
National Institute for Materials Science
Tsukuba, Japan

Preface

A reduction of global warming gases is now one of the big issues for future human society. The average temperature is increasing year by year according to an increase in carbon dioxide concentration in the atmosphere which comes from human activity. We have to reduce the production of carbon dioxide from our social activity. Our society now strongly depends on fossil fuel, leading to a large production of carbon dioxide. In order to shift from fossil fuel to renewable energy, some new technologies have been developed and already utilized in our society, such as natural energy and bio-mass energy. Another important technology is the electric vehicle which does not release carbon dioxide. In both natural energy and electric vehicle, a storage of electric energy is a key issue. In the case of natural energy, such as solar and wind power energies, a production of electric energy is unstable, depending on season, weather, and day/night time. An effective utilization of natural energy can be realized by the electric energy storage system, which can stabilize an output from natural energy. Therefore, an energy storage system is necessary for the natural energy system. By the way, the electric vehicle is driven by a motor. The rechargeable battery system is necessary, too. A part of future energy society will be operated by natural energy and a part of transportation will shift to an electric vehicle from an internal combustion system. A rechargeable battery system plays an important role in these systems. There are several kinds of rechargeable batteries such as Lead–Acid Battery, Ni-Cd Battery, Ni-metal Hydride Battery, and Lithium-Ion Battery (LIB). LIB has the largest energy density among rechargeable batteries. A small size LIB has been utilized in smartphone and laptop computer. Recently, a large size LIB has been utilized in an energy storage system for natural energy and power source of electric vehicles. However, even LIBs occupy a large space in energy storage systems and electric vehicles. The space for rechargeable battery (or weight of rechargeable battery) should be reduced by using new generation batteries with much higher energy density. So far, there are so many research activities around the world. In Japan, there are several national projects for next generation battery development. ALCA-SPRING (The Advanced Low Carbon Technology Research and Development Program—Specially Promoted Research for Innovative Next Generation Batteries) is one of the big projects supported by Japan Science and Technology. In this project, four kinds of next generation batteries have been researched and developed by a strong collaboration among many researchers. All solid-state batteries, Lithium–Sulfur batteries, Lithium-Air batteries, and Mg metal batteries are the main research topics. In this book, research and development of these next generation batteries are summarized. The development of materials are not only discussed, but the cell performance is investigated. The energy density of cell (real cell with anode, cathode, electrolytes, and others) is estimated by using a laminated cell with relatively higher capacity. This book includes the present status of the ALCA-SPRING project. I hope this book may be useful for researchers who want to develop next generation batteries. I am very glad to understand the aim of this project which is not only material sciences and development of real cell (Battery technology).

Hachioji-shi, Japan
May 2020

Kiyoshi Kanamura

Contents

Lithium-Sulfur Battery

Lithium-Air Battery

Introduction

Importance of Next-Generation Batteries

Kiyoshi Kanamura and Yuto Yamada

Abstract Rechargeable batteries have been utilized in Natural Energy System and Electric Vehicle Applications, in order to reduce the production of carbon dioxide in our society. Lithium-ion battery has been utilized in these applications due to its highest energy density among rechargeable batteries. However, a higher energy density of rechargeable batteries is needed for future energy society. In this section, a background of development of next-generation batteries is introduced to understand the present status of rechargeable lithium-ion battery and next-generation batteries. Especially, some specification and performance of batteries for electric vehicles and electric power plant with natural energy are described. In addition, life cycle assessment of carbon dioxide from four kinds of vehicles, such as gasoline vehicle, diesel vehicle, hybrid vehicle and electric vehicle, is compared and discussed to know a proper direction of next-generation battery development. Moreover, some of the fundamental electrochemical aspects of rechargeable battery are basically discussed to realize real batteries with higher energy density, with other reasonable performance, such as power density, safety and life.

Keywords Carbon dioxide · Natural energy · Electric vehicle · Rechargeable batteries · Next-generation batteries

1 Reduction of Carbon Dioxide

A global warming is a big problem now, which has been caused by huge carbon dioxide production by human activity. This problem results in a rising temperature of atmosphere and abnormal weather. Therefore, reducing carbon dioxide production is an important task for all mankind. There are so many kinds of technologies for reducing the carbon dioxide release rate. Energy technologies, which are independent of fossil fuels, should be developed for future society. One of the possible solutions for new energy system is an introduction of natural energy such as solar energy and

K. Kanamura (✉) · Y. Yamada
Department of Applied Chemistry, Graduate School of Urban Environmental Sciences, Tokyo Metropolitan University, 1-1 Minami-Osawa, Hachioji, Tokyo 192-0397, Japan
e-mail: kanamura@tmu.ac.jp

K. Kanamura (ed.), *Next Generation Batteries*,
https://doi.org/10.1007/978-981-33-6668-8_1

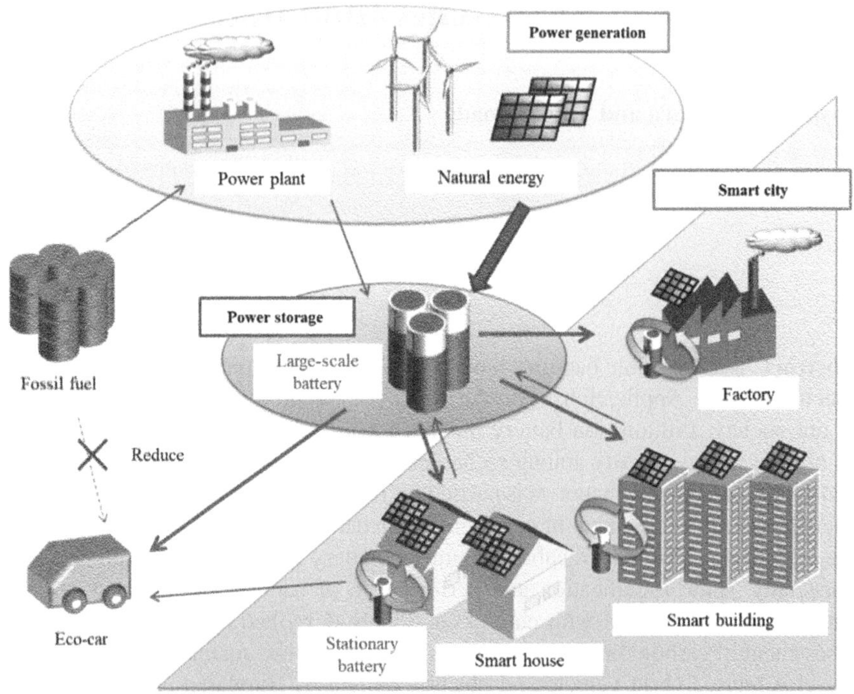

Fig. 1 Schematic illustration of clean energy society with natural energy

wind power energy. Figure 1 shows a schematic illustration of clean energy society with natural energy. The introduction of natural energy has been already started in a small scale. Solar energy or wind power energy is unstable, depending on change in weather and time zone. Such an unstable energy cannot be directly introduced in electric power system network. Therefore, the energy from solar power and wind power should be stored in rechargeable battery before deliver to main gird according to demand from users [1]. This means that a rechargeable battery plays an important role in a smart grid system. So far, traditional batteries, such as lead-acid battery, have been utilized in the smart grid system. However, the power storage system with lead-acid battery needs a vast land. In the small size solar power generation system, lead-acid battery may be applicable. The large-scale smart grid system cannot use lead-acid battery because of a vast land for battery system. Figure 2 shows a photograph of battery system for large-scale solar power plant [2]. The battery system, which is set inside a large size container, is placed in a vast land. When the energy density of rechargeable battery increases, the land used for battery system decreases. Even for the stationary application, the rechargeable battery with higher energy density is strongly demanded.

Another application contributing to reducing carbon dioxide production is an electric vehicle. Figure 3 shows a summary of CO_2 production from various fields

Fig. 2 Photograph of battery system for large scale solar power plant

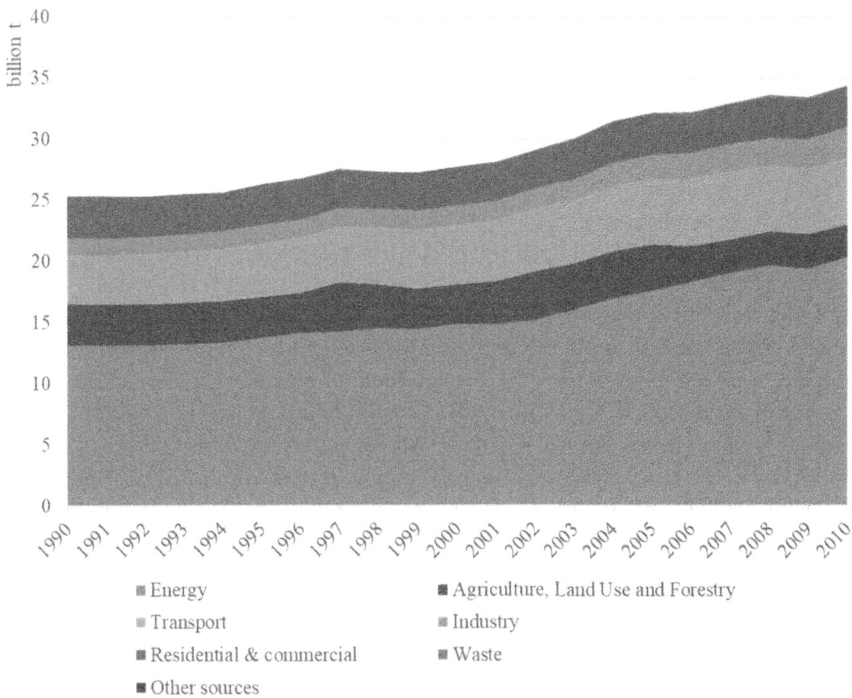

Fig. 3 Summary of CO_2 production from various fields

[3]. The field of transportation produces relatively larger amount of carbon dioxide. Electric vehicles are useful to reduce carbon dioxide production. Automobiles with thermal engines are consuming fossil fuels for long time and release carbon dioxide to air atmosphere. Instead of thermal engine, motor with rechargeable battery has been utilized to automobile, namely electric vehicles (EVs), to reduce the production

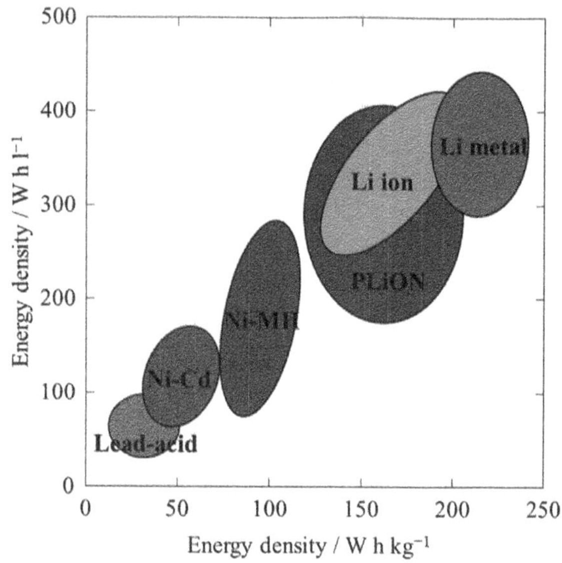

Fig. 4 Comparison of the different batteries

of carbon dioxide. In fact, Lithium-Ion Battery (LIB) has been already utilized in EVs. Tesla car, Leaf (Nissan) and other EVs have been already commercialized. A size of the rechargeable battery (LIB) depends on electric power consumption by EVs. An electric vehicle can travel 8 km per 1 kW h of rechargeable battery energy. The EV with 20 kW h rechargeable battery can travel 160 km. The EV with 80 kW h can travel 640 km. Among existing rechargeable batteries, LIB has the largest energy density, as shown in Fig. 4 [4]. LIB for EVs has 400 W h L^{-1}. The volume of 20 kW h battery is 50 L. The volume of 80 kW h battery is 200 L. The volume of car is around 3000 L. It is not so easy to equip such a large battery (200 L) to automobile. New rechargeable battery with higher energy density is really needed for new EVs. One of the energy density targets is 1000 W h L^{-1}, leading to 40 L battery volume. By the way, a reduction of carbon dioxide production depends on the kind of electricity used in EVs. How can we obtain electricity? When the electricity is produced by thermal power station, EVs do not highly contribute to the reduction of carbon dioxide production. In order to investigate the contribution of EVs to carbon dioxide production, Life Cycle Assessment (LCA) was calculated for various kinds of automobiles. The carbon dioxide production from automobiles can be calculated from a sum of carbon dioxide from production process, battery production process and consumption of gasoline during traveling of car. All data used in this calculation are summarized in Tables 1, 2 and 3 [5–13]. Figure 5 shows the calculated results for LCA. Here, electricity produced from various electric power generation systems is utilized by cars. LCA of gasoline vehicle, diesel vehicle, hybrid vehicle and electric vehicle (20 kW h or 80 kW h battery is installed) was calculated. At the initial stage (zero traveling distance), LCA of EVs is larger than that of gasoline vehicle. This is due to carbon dioxide generation by a battery manufacturing process.

Table 1 The data of CO_2 emission

Item		CO_2 emission(kg)	Reference
Vehicle body	Manufacturing gasoline vehicle (1300 kg)	2824	[5] *1, 2
	Manufacturing gasoline or diesel vehicle (x kg)	2824 × (x/1300)	*3
	Manufacturing HV including battery	5571	[7] *4
	Manufacturing EV without battery (x kg)	2824 × [(x–10.4 × y)/1300]	*5
Battery	Manufacturing 1kWh battery	75	[5] *6
Driving (gasoline)	Consumption of 1L gasoline	2.32	[8]
CO_2 emission by 1kWh Power generation	Current power generation method in Japan	0.54	[9]
	Solar energy	0.038	[10]

*1: Scale ratio of Fig. 3, *2: Vehicle weight is estimated from fuel combustion in car and 2nd page in [6], *3: Calculation assuming that emissions are proportional to vehicle weight, *4: Calculation from the scale ratio, *5: 10.4 = battery weight per 1 kWh, y = battery capacity (kWh), *6: Scale ratio of Fig. 2

Table 2 The data of battery

Battery weight	LIB weight per 1 kWh			Weight(kg)		Reference
				10.4		[11]
Battery for HV	Battery capacity equipped in one HV (kWh)			Guaranteed distance (km)*7		Reference
	0.73			100,000		[12]
EV	Battery capacity equipped in one HV (kWh)	Utilization of battery capacity(%)	Electric efficiency (km/kWh)	Driving distance per one charge (km)	Cycle life (times)	Total driving distance(km)
EV (20kWh)	20	100	9	180	500	90,000
EV (80kWh)	80	100	9	720	500	360,000

*7: The distance promised free repair of battery

During traveling of car, an increase of LCA for EVs is smaller than that of other cars. From a comparison of LCA of gasoline vehicle with those of the EV with 80 kW h battery, the EV exhibits a larger carbon dioxide emission before 90,000 km traveling distance in Japan. If a vehicle is driven 10,000 km per year, the EV does not reduce a carbon dioxide emission during 9 years. If a life of EV is 9 years, the EV increases a carbon dioxide emission. On the other hand, a carbon dioxide

Table 3 The estimation of LIB electric efficiency and cycle life

	Battery capacity(kWh)	Driving distance(km) [13]	Electric efficiency(km/kWh)	Guaranteed distance(km) [13] *8	Cycle life(times)
EV(40kWh)	40	400	10	160,000	400
EV(62kWh)	62	570	9	160,000	280

*8: The limit of driving distance for free repair when battery capacity was below 90% under standard use condition

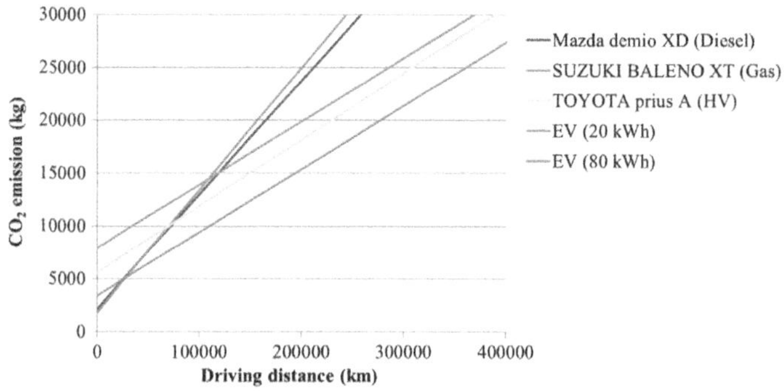

Fig. 5 LCA using current power generation method, without battery exchange

emission can be reduced by EVs, when natural energy is utilized more and more. A battery exchange is needed according to a battery cycle life. By taking into account of battery exchange and utilization of natural energy, Fig. 5 was modified to Fig. 6. In this case, EVs are extremely useful for the reduction of carbon dioxide production. In a sense of suppression of carbon dioxide production by EVs, an introduction of natural energy must be done. Another important point of EVs is a carbon dioxide production at manufacturing process of battery. Even when the energy density of

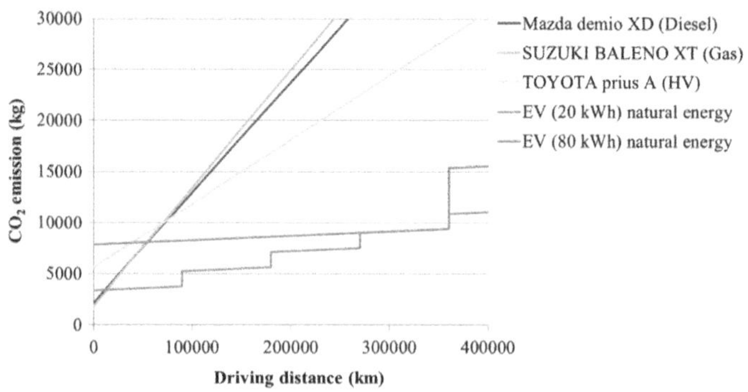

Fig. 6 LCA using solar energy, with battery exchange

rechargeable battery becomes twice larger, a carbon dioxide production from battery manufacturing process does not increase very much. The carbon dioxide emission from the manufacturing can be reduced in a unit of CO_2 amount/Wh. In addition, the production cost of battery is also reduced by higher energy density battery. From these points, the higher energy density of battery should be achieved. In other words, the improvement of energy density of battery is an eternal theme.

2 Energy Density of Battery

Figure 7 shows an essential structure of rechargeable battery consisting of cathode, electrolyte and anode. The longer distance between cathode and anode results in a high resistance of electrolyte part, so that the electrolyte part should be thin. A separator is a key material to reduce the distance between cathode and anode. In general, the separator consists of porous polymer film. Of course, a solid electrolyte system may not need a separator.

By using separator or solid electrolyte, the distance between cathode and anode has to be reduced as possible as we can. The energy density of rechargeable battery is determined by capacity densities of cathode and anode materials. $LiCoO_2$ and graphite are used as cathode and anode materials in LIB, respectively. Non-aqueous electrolyte is used in LIB. Electrochemical reactions taking place in LIB are described as follows.

Fig. 7 Structure of rechargeable battery

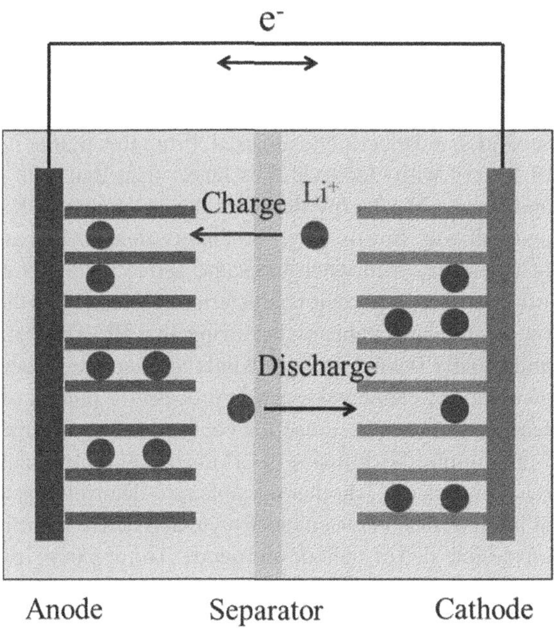

Cathode reaction	$LiCoO_2 \leftrightarrow xLi^+ + xe^- + Li_{1-x}CoO_2$
Anode reaction	$C_6 + xLi^+ + xe^- \leftrightarrow LixC_6$
Total reaction	$LiCoO_2 + C_6 \leftrightarrow Li_{1-x}CoO_2 + Li_xC_6$

When Li^+ ion is extracted from $LiCoO_2$ by $x = 0.5$ during the charging process, the capacity of 140 mA h g^{-1}can be released during the discharge process. Li^+ ion is intercalated into Graphite (one Li^+ ion per 6C) corresponding to 372 mA h g^{-1}. The capacity of 1 g of $LiCoO_2$ is equal to that of 0.38 g of graphite. The total weight of cathode and anode materials is 1.38 g, corresponding to 140 mA h. The battery voltage is 3.7 V. From this estimation, the energy density of this battery can be calculated to be 375 W h kg^{-1}. However, electrolyte, current collector, cell case and other materials used in LIB are not included in this calculation. The real energy density of LIB is estimated at 150 W h kg^{-1}. The energy density of battery is different from the capacity density of active material. This is a very important point for the production of rechargeable battery. Even when active materials have very high capacity density, the battery consisting of these materials does not have high energy density. The energy density of battery strongly depends on both materials and cell structure. The next-generation batteries should be developed with the design of electrode structure and cell configuration. In this way, the above simple estimation only based on active materials misleads the energy density of battery. The volumetric energy density of LIB can be also calculated to be 300 W h L^{-1} from density of LIB. Here, the energy density calculation for LIB with different thickness of electrodes is introduced as an example. Electrode A consists of 100 μm cathode thickness and the 60 μm thickness of anode. Electrode B consists of 50 μm cathode thickness and 30 μm anode thickness. The batteries with Electrode A and Electrode B have the same capacity. Electrodes A and B need current collectors for cathode and anode. Al and Cu foils are used as cathode and anode current collectors in LIB, respectively. Figure 8 shows the schematic illustration of both electrodes. In both cells, the same separator and electrolyte are used. From this figure, it is clear that the energy density of battery with electrode A is larger than that with electrode B. Both gravimetric and volumetric energy densities are increased with increasing thickness of anode and cathode. In this way, the energy density of battery depends on the structure of electrode, even when the same active materials are utilized in the cell. By the way, another important characteristic of battery is the power density. This depends on reaction mechanisms occurring in LIB. The battery reaction is not so simple and usually involves several kinds of elemental reaction process. Figure 9 shows a summary of elemental reaction process for porous cathode or anode. Among these reactions, the slowest reaction becomes a rate-determining step for battery reaction. For example, the diffusion of Li^+ ion in solid active materials is sometimes the slowest reaction process. Another possible rate-determining step is a charge transfer process at the interface between electrolyte and active material. Both reaction steps depend on particle size of cathode and anode. The apparent resistance due to both diffusion in solid matrix and interfacial reaction can be reduced by decreasing particle size (larger surface area). More or less, in practical batteries, the particle size of active materials

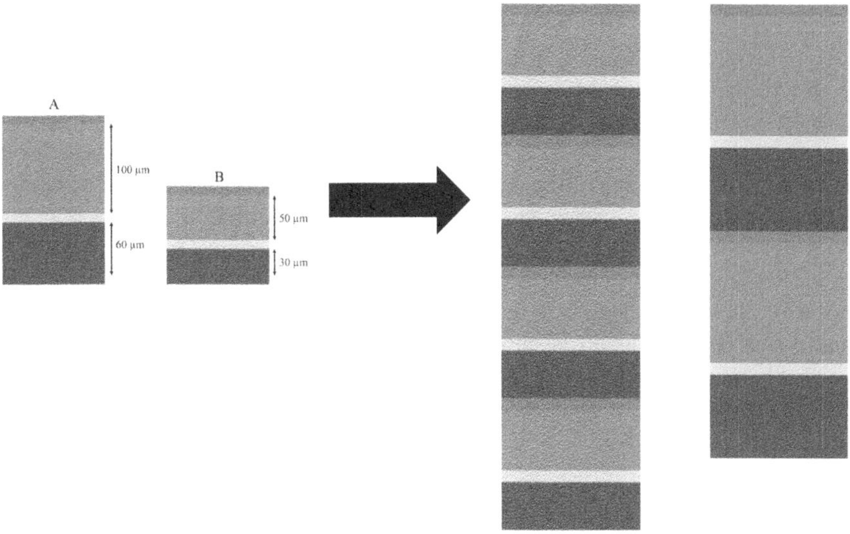

Fig. 8 Schematic illustration of electrodes A and B

Fig. 9 Summary of elemental reaction process for porous electrode

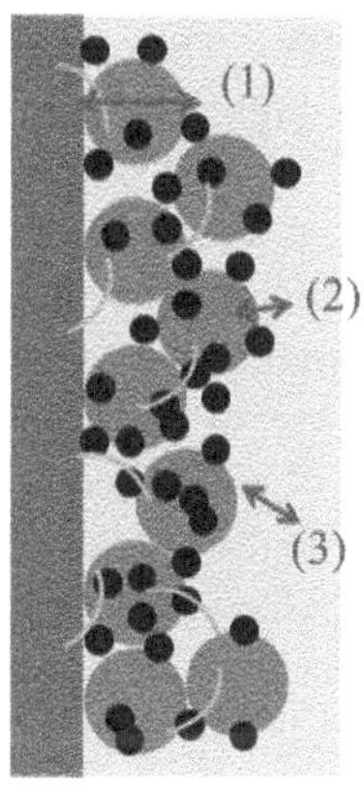

has been already optimized, so that the diffusion in solid matrix and interfacial resistance are not the rate-determining steps. Mostly, the diffusion of Li^+ ion in electrolyte involved in porous electrode is the slowest process, especially at high current discharge or charge [14] This reaction step strongly depends on the thickness of electrodes. The standard thickness of the electrode is 30 ~ 100 μm in LIB. The diffusion of Li^+ ion in electrolyte involved in porous cathode and anode determines the power density of LIB. The elemental reaction process relating to the thickness of electrodes is only a diffusion of Li^+ ion in electrolyte. The rate determining step for both cells in Fig. 8 is the diffusion of Li^+ ion in electrolyte. The diffusion resistance is proportional to the square of the thickness of porous electrode. Therefore, the power density of the cell with electrode A should be much lower than that with electrode B. In simply, the resistance of the cell with electrode A is four times larger compared with electrode B. The energy density of the cell with electrode A is larger than that with electrode B. This is very important point to develop practical battery. The design of electrode and battery must be considered very well. Otherwise, the battery does not work. A simple discussion on active material is not useful. The energy density and power density should be discussed based on the cell with adequate capacity needed by applications. In our ALCA-SPRING project, this point of view has been included to evaluate the materials.

3 Batteries for EVs

There are three kinds of EVs, such as HEV (Hybrid Vehicle), PHEV (Plug in hybrid vehicle) and EV (Electric vehicle). Table 4 shows a summary of characteristics of batteries used in these EVs. The capacity of module battery for HEV is not so large, but its power density is high. The current needed by motors in HEV, PHEV and EV is not so different each other. It only depends on a size of car. In the case of PHEV, vehicles have to travel at least 100 km for one time of battery charge. The module battery for PHEV should have a larger capacity than that of HEV. The power of battery for PHEV is smaller than that of HEV. In the case of EV, vehicles should travel more than 200 km (if possible 500 km). The capacity of battery becomes very large. The power of cell in module battery is so large. In this way, the battery characteristic depends on the kind of EV. The battery design should be optimized for each EV. In ALCA-SPRING project, the battery for EV is a main target, so that the energy density of battery must be increased by developing next-generation batteries, such as Li–Air battery, Li–Sulfur battery, all solid-state battery and Mg battery. These

Table 4 A summary of characteristics of batteries used in these EVs

	Battery capacity (kWh)	Battery weight (kg)	Current value (C)
HEV	5	10	100
PHEV	10	20	50
EV	50	100	10

new batteries may provide higher energy density than LIB. The material science is not only important but also the battery technology is also very critical to realize real battery with high energy density 500 W h kg^{-1} (1000 W h L^{-1}).

References

1. Liu, J., Zhang, J.-G., Yang, Z., Lemmon, J. P., Imhoff, C., Graff, G. L., et al. (2013). *Advanced Functional Materials, 23*(8), 929–946.
2. https://www.tesla.com/energy.
3. https://ourworldindata.org/co2-and-other-greenhouse-gas-emissions.
4. Tarascon, J.-M., & Armand, M. (2001). *Nature, 414*(6861), 359–367.
5. Larcher, D., & Tarascon, J.-M. (2015). *Nat. Chem., 7,* 19–29.
6. http://www.mlit.go.jp/common/001031308.pdf.
7. https://toyota.jp/pages/contents/prius/004_p_001/pdf/spec/prius_ecology_201512.pdf.
8. Saiki, Y., & Nakazawa, M. (1990). *J Japan Soc Air Pollut, 25*(4), 287–293.
9. https://www.jepic.or.jp/data/g08.html.
10. https://www.ene100.jp/zumen/2-1-9.
11. https://www.nies.go.jp/social/traffic/pdf/7-3.pdf.
12. https://toyota.jp/pages/contents/prius/004_p_001/pdf/spec/prius_spec_201512.pdf.
13. https://www3.nissan.co.jp/vehicles/new/leaf/charge/battery.html.
14. Kanamura, K., Yamada, Y., Annaka, K., Nakata, N., & Munakata, H. (2016). *Electrochemistry, 84*(10), 759–765.

Lithium Metal Battery

Rechargeable Lithium Metal Battery

Kiyoshi Kanamura and Yukihiro Nakabayashi

Abstract Li metal anode is now an extremely important anode material for next-generation batteries, such as Li-air, Li-sulfur and Li metal battery. Li metal is necessary to realize higher energy density of these batteries. However, the electrochemical performance of Li metal anode is very low, especially very low cycleability and safety. These problems are related to the lithium metal dendrite formation. In this section, the research history of Li metal anode is reviewed. The surface of Li metal is always covered by some surface layer, which strongly influences the electrochemical behavior of Li metal. The surface state analysis on Li metal has been carried out by using various surface analyses. Recently, the surface modification and creation of artificial layers on Li metal have been reported to improve the cycleability of Li metal anode with a suppression of Li metal dendrite formation. The solid electrolyte is also one possible material to avoid the dendrite formation. In order to realize Li metal anode with high cycleability and safety, various kinds of researches and concepts for the interface between Li metal anode electrolyte (separator).

Keywords Li metal anode · Solid electrolyte interphase · Li metal dendrite · Suppression of li metal dendrite · Artificial surface film

1 Li Metal Anode

Lithium-Ion Battery (LIB) has been commercialized and widely used in various kinds of applications, such as cellar phones, smartphones, laptop computers and other communication devices. In addition, LIB has been utilized in electric vehicles (EVs) and stationary applications (SAs) for smart grid. In these applications, the size of LIB becomes larger. All these devices always need higher energy density of rechargeable battery. Among commercialized rechargeable batteries, LIB has the largest energy density. Rechargeable batteries with higher energy density are really required from various devices, EVs and SAs. The energy density of LIB with lithiated transition

K. Kanamura (✉) · Y. Nakabayashi
Department of Applied Chemistry for Environment, Graduate School of Urban Environmental Sciences, Tokyo Metropolitan University, 1-1 Minami-Osawa, Hachioji, Tokyo 192-0397, Japan
e-mail: kanamura@tmu.ac.jp

K. Kanamura (ed.), *Next Generation Batteries*,
https://doi.org/10.1007/978-981-33-6668-8_2

metal oxide cathode and graphite anode is now near to theoretical values. In order to increase the energy density, new cathode and anode materials have to be developed and utilized in real cells. Ni-based cathodes, such as $LiNi_xMn_yCo_zO_2$ (x + y + z = 1) [1] and $LiAl_{0.05}Ni_{0.8}Co_{0.15}O_2$ [2] have been developed as high capacity cathode materials. Si-based anodes, such as composite between graphite and nanosized Si, have been utilized as high capacity anode materials. However, Si anode has a large volume expansion/shrinkage during discharge and charge processes. This behavior is strongly related to cycle life of anode. Therefore, Si anode itself cannot be utilized to anode material. Usually, the composite between graphite and Si has been utilized in LIB. In this case, the capacity density of composite anode is roughly 500 mA h g^{-1} [3]. This is adequate to improve energy density of LIB. The energy density of LIB with the composite anode could be more than 300 W h kg^{-1}. However, it cannot be higher than 400 W h kg^{-1}. In order to realize the energy density of more than 400 W h kg^{-1}, another anode material is strongly required. Li metal anode has ~4000 mA h g^{-1} capacity density, which is the highest capacity density among various kinds of anode materials. Capacity density of several anode materials is summarized in Table 1 [4]. The capacity density of Li metal is particularly larger than those of other anode materials, while Li metal has a critical problem for practical use, in which unique morphology of Li metal formed during cycling of cell [5]. This morphology has been so-called "dendrite" Li metal. The formation of dendritic Li metal leads to serious chemical reactions between Li metal and electrolyte, resulting in low cyclability and poor safety of cell. The rechargeable Li metal battery has been developed in 1990 and commercialized as power source of cellar phone. Unfortunately, this Li metal battery ignited in use. After this accident, the cell with Li metal anode has not been used for any applications. Recently, new applications, such as smartphone, electric vehicle and stationary application require higher energy density for rechargeable battery, for example 500 W h kg^{-1} and 1000 W h L^{-1}. Li metal anode for next-generation batteries that can exhibit higher energy density will be introduced.

Table 1 Capacity densities and potentials of several anode materials [4]

Anode material	Capacity density/Ah kg^{-1}	Potential/V (versus Li/Li^+)
Carbon	~370	0.1
$Li_4Ti_5O_{12}$	~175	1.5
Conversion electrode	700–900	0.95
Li (Si) alloy	~2100	0.1
Li metal	~3850	0

Kiyoshi Kanamura, Yukihiro Nakabayashi. Tokyo Metropolitan University (2020)

2 SEI and Cyclability of Li Metal Anode

Li metal has been investigated more than 30 years. There are so many literatures, in which morphology changes of Li metal, rechargability of Li metal, surface nature of Li metal and reactions with electrolyte have been investigated. The most serious research subject is the formation of dendritic Li metal during charging process. The electrode potential of Li metal is -3.04 V versus NHE, leading to chemical reaction with components in electrolyte, such as solvent, salt, additive and impurity. Li metal has a surface film, which contains LiOH, Li_2CO_3 and Li_2O. This is the native surface film. The surface film reacts with electrolyte components to form new surface films depending on the kind of electrolyte. The surface film consists of inorganic and organic compounds, which is so-called a solid electrolyte interphase (SEI). The SEI was proposed by Peled [6] and then has been extensively investigated by Aurbach et al. [7] Fig. 1a shows a schematic illustration of native surface film on Li metal anode [8]. The chemical reaction of this surface film was investigated by X-ray photoelectron spectroscopy [8–12]. Electrolytes can penetrate into the native surface film and finally react with Li metal. The reductive reactions of electrolyte take place after immersion of Li metal into electrolyte. On the other hand, when the electrolyte contains a small amount of acid as impurity, such as HF [8–11] and HCl [9], the native surface film reacts with these acid impurities to form LiF or LiCl. The organic solvent, such as propylene carbonate, diethyl carbonate, ethylene carbonate can easily react with Li metal, especially linear carbonate solvent. Even in electrolytes with ether solvent, Li metal reacts with components in electrolyte. There are so many

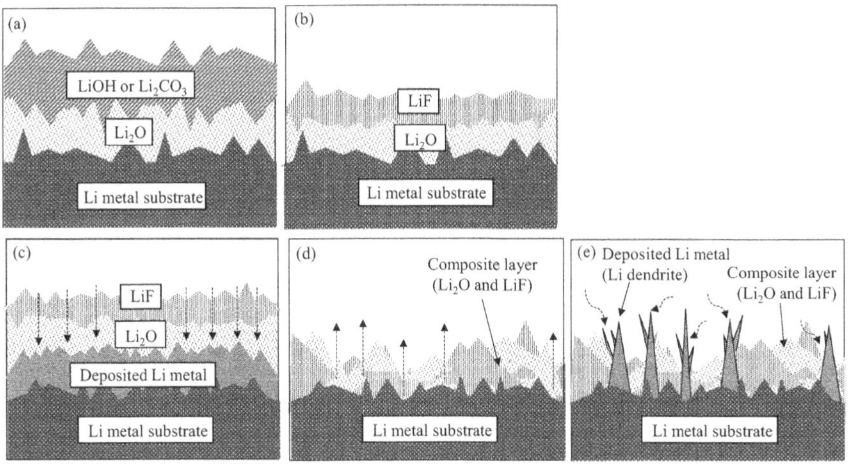

Fig. 1 Schematic illustration of native surface film of oxide (Li_2O), hydroxide (LiOH), carbonate (Li_2CO_3) on Li metal (**a**) and after formation of LiF layer by the reaction with HF in electrolyte (**b**) in addition to the change of Li metal morphology in the processes of charge (**c**) and discharge (**d**) and after repeating charge and discharge cycles (**e**) in electrolytes containing HF; broken arrows indicate penetration passes for electrolyte solution

reports on the chemical reaction between Li metal and electrolytes, mostly focused on the structure and chemical compositions of SEI formed on Li metal anode during discharge and charge processes. Especially, the chemical composition of SEI has been focused mainly. In order to modify the chemical composition of SEI, some additives have been proposed; e.g. inorganic acid such as HF [8–11], inorganic salt such as $LiClO_4$ [7–14], organics such as propylene carbonate [13] and gaseous species such as CO_2 [15]. CO_2 and HF can modify the original SEI to Li_2CO_3 and LiF rich SEI layer. The morphology of lithium metal deposition strongly depends on the modified chemical composition. Unfortunately, all of additive cannot keep a positive effect for Li metal deposition during cycles. XPS spectra (Fig. 2) and SEM images (Fig. 3), which have been measured for the surface of Li metal deposited in the propylene carbonate solution of $LiClO_4$ and a small amount of HF [9], indicating that LiF layer is formed on the surface of deposited Li metal and facilitates to generate current more

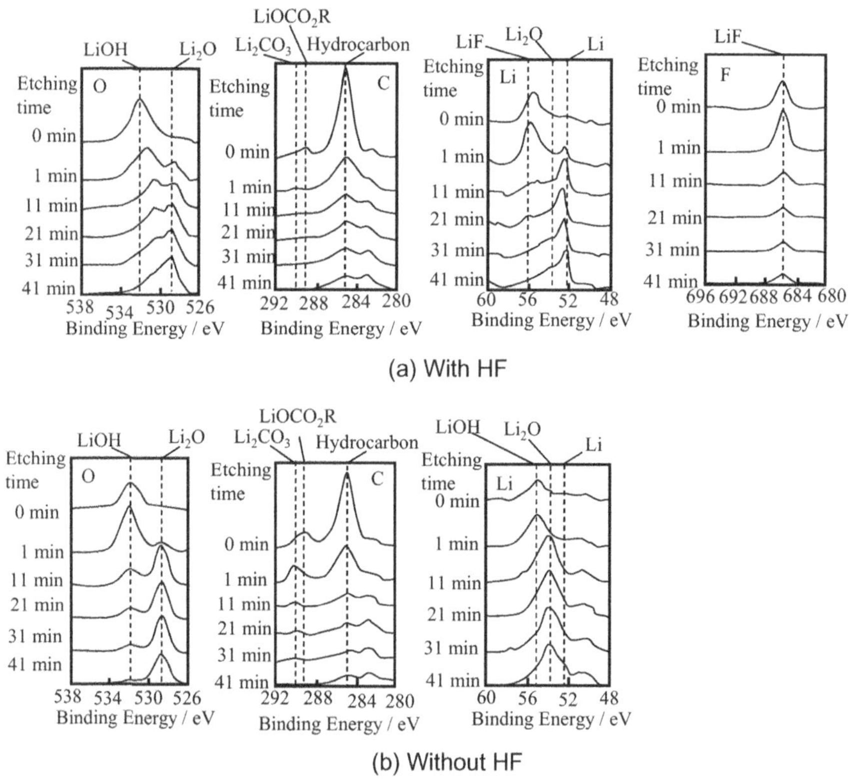

Fig. 2 XPS spectra of the surface of Li metal deposited at 0.2 mA cm^{-2} in the propylene carbonate solution of 1 M $LiClO_4$ with (**a**) or without (**b**) 10 mM HF using three-electrode systems, in which working electrode of Ni metal, counter electrode of Li metal and reference electrode of Li metal were employed; Ar^+ etching times are described also in Fig. 2; XPS spectra were prepared based on Figs. 7 and 8 in [9]

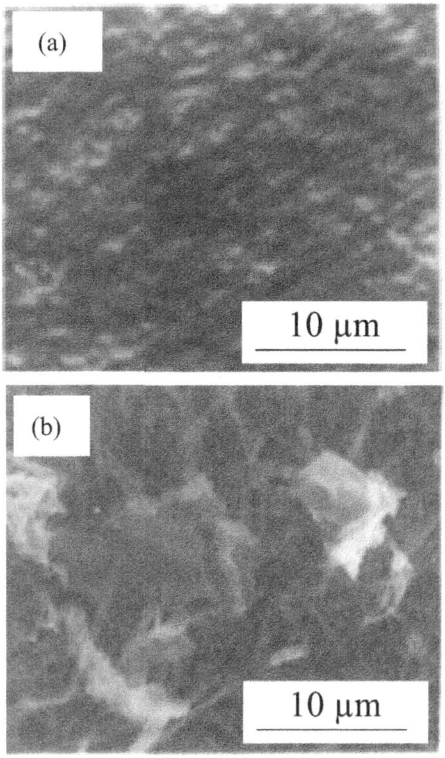

Fig. 3 SEM images of the surface of Li metal deposited at 0.2 mA cm^{-2} and 1.0 C cm^{-2} in the propylene carbonate solution of 1 M $LiClO_4$ with (**a**) or without (**b**) 10 mM HF using three-electrode systems, in which working electrode of Ni metal, counter electrode of Li metal and reference electrode of Li metal were employed; the SEM images were prepared based on Figs. 5 and 6 in [9]

homogenously to suppress the formation of dendritic Li. Formation of LiF layer was also confirmed on the surface of Li metal anode after charge and discharge cycles in propylene carbonate solution of $LiPF_6$ without HF [11]. The electrolyte with $LiPF_6$ has a small amount of HF depending on H_2O content, according to the following chemical reaction of $LiPF_6$ and H_2O.

$$LiPF_6 + H_2O \rightarrow LiF + 2HF + POF_3$$

In the following step, HF can react with the original SEI chemical composition (LiOH, Li_2CO_3 and Li_2O) to form LiF in the new SEI layer; schematic diagram of LiF layer on Li metal anode is shown in Fig. 1b [8]. The deposited lithium metal is not dendrite and has a smooth surface. Similar results have been also conducted by using other additives and electrolytes. However, during the deep discharge and charge cycles (e.g. 5–10 mA h cm^{-2}), the morphology changes to more dendritic and moss-like ones [9]. The SEI layer is one of the important key factors to suppress dendritic or moss-like lithium metal. Unfortunately, the SEI layer is not sufficient to obtain high cyclability and safety of lithium metal anode. The current researches still focus on SEI chemical compositions.

3 Effect of Morphology Change on Cyclability of Li Metal Anode

The change in the morphology of deposited lithium metal affects the cyclability of lithium metal anode. By repeating discharge and charge cycles, the morphology of Li metal gradually changes more dendritic or moss-like [8, 12, 13]. This is attributed to low stability of SEI layer and chemical reaction of Li metal with electrolyte components. Figure 1c–e shows a schematic illustration of Li metal morphology change during the discharge and charge cycles in the electrolyte with HF additive [12]. In the initial deposition of Li metal on Cu, the morphology is very smooth without any dendrite formation. In the initial dissolution of Li metal, the SEI formed on Li metal surface partly breaks down. Even in the dissolution process, new SEI formation on lithium metal takes place, leading to very low Coulombic efficiency. This process is repeated during each discharge and charge cycle. The SEI components are accumulated on lithium metal surface, resulting in non-uniform current distribution for the deposition of lithium metal. Finally, the surface state of lithium metal anode is covered with non-uniform SEI layer, contributing the formation of dendritic lithium metal. In order to realize highly rechargeable lithium metal anode, the chemical reaction of lithium metal with electrolyte during the discharge process should be suppressed by using mechanically highly stable SEI. Therefore, the mechanical stability of SEI layer is one of the important factors to realize non-broken layer, which can protect lithium metal.

In literatures, the protected surface film on lithium metal has been proposed by using inorganic or organic thin solid electrolyte. The mechanical stability of these protect films depends on the kind of solid electrolyte and their thickness. Of course, these films should have a high ionic conductivity. Li_3N [16] and LIPON [17] have been proposed as protect surface films. These films may not be stable when deposited lithium metal is so thick; the capacity of 1 mA h cm^{-2} corresponding to metallic lithium film with 5 μm thick. When using a cathode with 50–100 μm, the cathode capacity is roughly 2–5 mA h cm^{-2}. Correspondingly, the lithium metal anode should have the thickness change of 20–25 μm. In the course of many cycles of lithium metal anode, the thickness change may not be uniform. This morphology change of lithium metal can be illustrated in Fig. 4a [17]. More or less, the surface area of lithium metal increases during many cycles. An ideal morphology change of lithium metal anode can be illustrated in Fig. 4b [17]. In this case, the surface area of lithium anode hardly changes during charge and discharge cycles, and SEI layer forms at only the initial few cycles. When lithium metal is deposited on Cu current collector without Li metal layer and dissolved into bulk electrolyte by using beaker-type electrochemical cell, there is no artificial driving force for uniform deposition of lithium metal in macro scale. It suggests that, in beaker type electrochemical cells, SEI layer produced on lithium metal is not effective to suppress the increment of the surface area of deposited lithium metal, resulting in the formation of dendritic lithium metal. By using laminated or coin type cell, a separator should be employed to avoid an internal short circuit between cathode and anode, in which the anode consists of

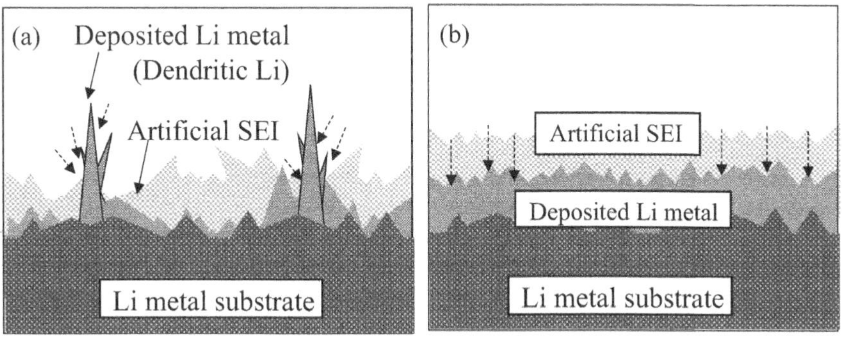

Fig. 4 Schematic illustration of the change in morphology for the surface of Li metal after charge and discharge cycles under non-ideal condition (**a**) or ideal condition (**b**); both non-ideal condition and ideal condition were described as deduced as Li anode modified with solid artificial SEI

lithium metal with SEI layer, separator and liquid electrolyte. The interface between lithium metal and separator has not been investigated at this moment. Most of the researches have been focused on SEI or surface film. New cell design and separator have to be investigated to realize an artificial control of lithium metal deposition with smooth surface.

4 New Electrolyte System

The chemical reaction of Li metal with electrolyte has been extensively investigated to form better SEI layer for Li metal deposition and dissolution. As well as organic solvents, electrolytes react with lithium metal more and less; e.g. $LiPF_6$, $LiClO_4$, $LiBF_4$, LiTFSI (Lithium Bis(trifluoromethanesulfonyl)imide), $LiCF_3SO_3$ and LiBOB (Lithium bis-(oxalato)borate) [18]. By the reaction of electrolytes with lithium metal, surface films are produced on lithium metal. Many researchers have been analyzed surface films and attempted to modify the films to suppress the formation of dendritic or mossy lithium metal in dissolution and deposition cycles. When solvents react with Li metal, some of the organic and inorganic compounds are formed, such as alkyl carbonate, some of polymer (oligomer) and Li_2CO_3. When salts react with Li, inorganic compounds are formed, such as Li_2CO_3, LiF and Li_2O. Therefore, the surface film formed on Li metal surface depends on reactivity of solvent and/or salt with Li metal. Ether solvents are chemically stable even in the contact with lithium metal, while the solvents are easily decomposed by electrochemical oxidation reaction [19, 20]. If a cell has 4.0 V operation potential, the ether solvents may be oxidized easily. Therefore, it is not so easy to use ether solvents for lithium metal battery with 4.0 V cathode versus Li/Li^+. Recently, lithium air battery and lithium sulfur battery have been extensively studied as next-generation batteries with higher energy density than 400 W h kg^{-1} [21–23]. In these cells, lithium metal

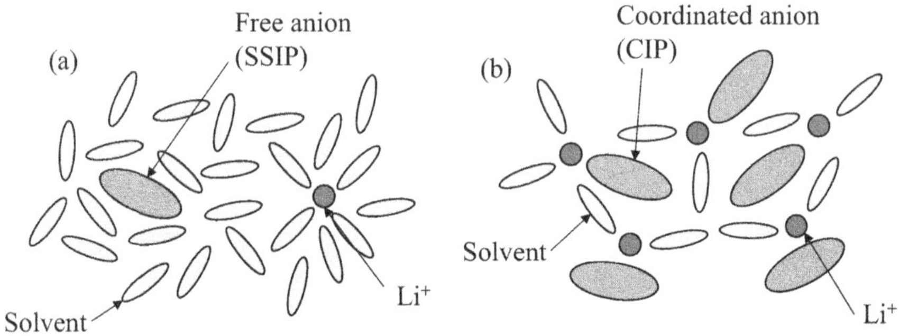

Fig. 5 Schematic illustration of interaction of Li^+ with solvent and anion in electrolyte solution, in which the illustration for dilute electrolyte (**a**) and concentrated electrolyte (**a**) are compared

anode should be used. These cells have a relatively lower operation potential around 2–3 V versus Li/Li^+. In these cells, ether solvents can be utilized as non-reactive solvent against Li metal. In other words, some of next-generation batteries with lower operation potential than 3 V versus Li/Li^+ can be realized with Li metal anode.

Standard organic electrolytes, which have been already used in lithium ion batteries or proposed in primary lithium metal batteries, should be still investigated to find the most suitable electrolyte system. On the other hand, new electrolyte systems have also been suggested by several research groups. Most interesting new electrolyte is highly concentrated one. The molar ratio between solvent and Li^+ ion is almost equal in this new electrolyte. Most of solvents interact with Li^+ ions, as shown in Fig. 5. [24] The strong interaction between solvent molecule and Li^+ ion provides higher stability of solvent. This higher stability of highly concentrated electrolyte has been certified by theoretical calculation. For example, dimethyl carbonate solvent with around 5.0 mol dm^{-3} LiFSI (Lithium bis (fluorosulfonyl)imide) shows high stability against Li metal and cathodes with high operation potential. More detailed discussion will be presented in another chapter.

5 Artificial SEI Formation

HF and CO_2 have been used as additive to modify SEI layer formed on Li metal surface. However, these additives do not exhibit adequate effects to suppress the formation of dendritic Li metal. One of the solutions is the artificial SEI consisting of thin polymer layer. PVdF (PolyVinylidene DiFluoride) [25], PEO (polyethylene oxide) [26] and PAN (polyacrylonitrile) [27] have been used as thin layer cover for Li metal. Some positive effects have been observed and the cyclability of Li metal anode has been reported. The mechanical property of the thin layer cover may be important. When Li metal deposition takes place uniformly at the entire surface of Li metal, the thin layer cover suppresses the dendrite formation. When the current distribution occurs during Li deposition and dissolution processes, the Li metal deposition is not homogenous. This non-uniform current distribution cannot be compensated by the thin layer cover. In the case of the discharge and charge processes with a small discharge and charge capacity, the thin layer cover is very useful but is not so efficient for the discharge and charge cycles with a larger capacity, such as 5–10 mA h cm^{-2}. The mechanical property of the thin layer cover is very important for cyclability of Li metal anode. The researches on the mechanical property of the artificial SEI should be conducted to realize high cyclability of Li metal anode.

6 Additives

Various kinds of additives have been proposed. For example, Cs^{2+} ion is one of the interesting additives to avoid Li metal dendrite [28]. Some cations, which cannot be reduced by Li metal, can adsorb on Li metal surface to modify the surface state of Li metal. More uniform deposition of Li metal may take place in the electrolyte containing Cs^{2+} ion. However, this effect depends on current density of Li metal deposition. At small current density, Cs^{2+} adsorption exhibits an excellent positive effect on Li metal deposition. The formation of dendritic Li metal can be suppressed in the electrolyte with Cs^{2+} ion. At large current, adsorption of Cs^{2+} is not so effective for a suppression of the formation of dendritic Li metal. The surface concentration of Cs^{2+} ion is not enough to realize uniform deposition of Li metal at large current.

Some of electrolyte salts [7–14] and organic compounds [13] have been proposed to improve the rechargability of Li metal anode. Most of these additives react with Li metal to modify the surface film on Li metal. One of the important chemical components is LiF. When the surface film involves LiF, the rechargability of Li metal is improved. However, Li metal can react with electrolyte during discharge process as mentioned above [7–14, 18]. The mechanical property of the surface film containing LiF is also key property for stability of surface film on Li metal during deposition and dissolution cycles. More or less, the surface film is not so mechanically strong. The surface film on Li metal is really important but should be stabilized by other methods. If this problem is solved, Li metal anode will be realized in rechargeable lithium metal battery.

7 New Current Collector

Another important problem is volume change of Li metal anode, as shown in Fig. 6 [29]. As mentioned above, the thickness of Li metal changes during deposition and dissolution cycles. In some cases, an initially rigid bulk of Li metal changes to somehow porous structure, leading to much larger volume change during deposition and dissolution cycles, as shown in the SEM images of porous Li metal after 200 discharge and charge cycles of Li metal battery with NMC cathode using the current

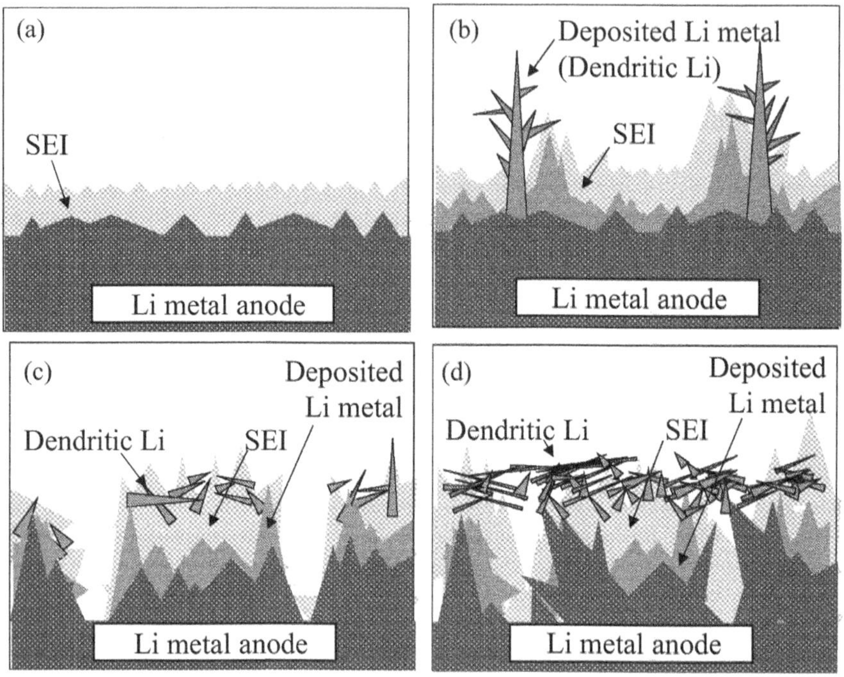

Fig. 6 Schematic illustration of the four steps for volume change of Li metal anode; SEI layer is formed on Li anode by the reaction with electrolyte. (**a**) SEI layer is cracked by expansion of Li anode in the process of Li metal deposition, and dendritic Li metal is generated from the cracks in the layer. (**b**) Dendritic Li metal is released from Li anode in the process of Li metal dissolution and accumulated on the surface of Li anode. (**c**) Compared with the surface of Li anode in the step (**a**), the surface becomes rough after the step of (**c**), as the Li metal is dissolved unevenly. By repeating the steps from (**a**) to (**c**), the surface of Li metal anode becomes rough more, and thick SEI layer and compiled dendritic Li metal cover the surface.(**d**) As a result, Li metal anode become porous, and the volume of the anode is changed

of 0.8 mA cm^{-2} and 2.0 mA cm^{-2} (Fig. 7 [30]). The volume change of Li metal anode leads to a thickness change of cell, which strongly influences the cycleability of cell. In general, the volume change lowers the cyclability of cell. In order to prevent the volume change of Li metal anode, new type of current collectors such as three dimensionally porous copper (3D Cu) foil has been proposed [31, 32]. Lithium metal is electrodeposited inside porous structures on 3D Cu foil without dramatic change of the thickness of Li metal anode in micrometer scale. The electrodeposited lithium metal is dissolved into electrolyte in the process of discharge, and then electrolyte penetrates into the porous structure of Cu foil. The penetration of electrolyte has to be prevented for continuous charge and discharge cycles. Except for 3D Cu foil, carbon fiber cloth has also been proposed to suppress the formation of dendritic lithium metal [33].

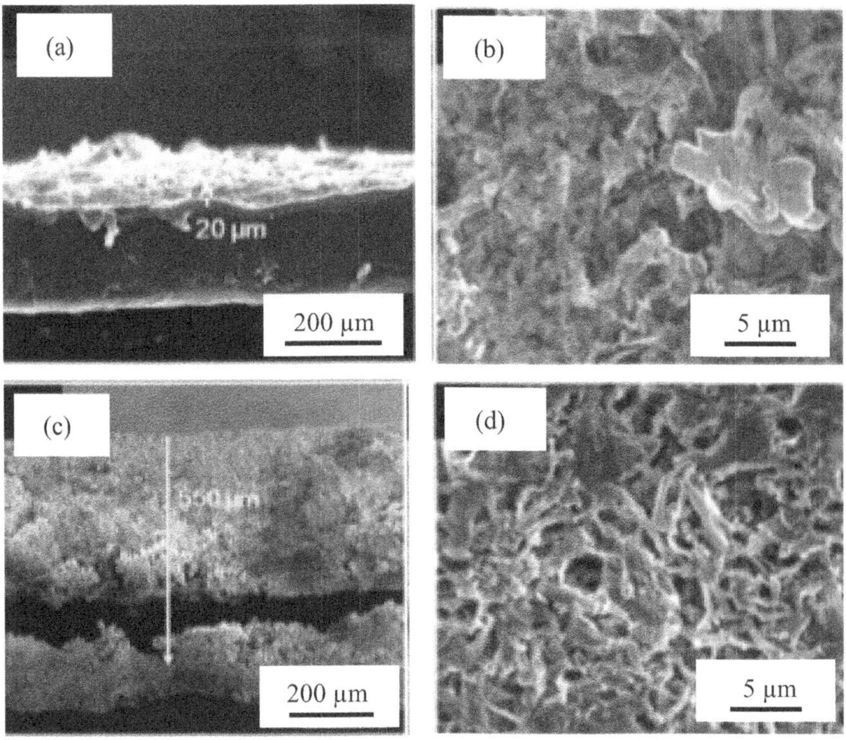

Fig. 7 SEM image of volume change of Li metal anode after 200 charge and discharge cycles in Li||NMC coin cells with the capacity of 3.5 mAh cm^{-2} operating 0.8 mA cm^{-2} (**a**, **b**) and 2.0 mA cm^{-2} (**c**, **d**);the cross sectional images (**a**, **c**) and the top views (**b**, **d**) are prepared from Fig. 5 in [30]

8 Solid Electrolyte

Solid electrolytes can improve essential safety of battery with Li metal anode, due to less chemical activity under the contact with Li metal anode; particularly, $Li_7La_3Zr_2O_{12}$ [34–38] and Li_3PS_4 [39, 40] are more promising due to higher stability among a series of solid electrolytes. For developing lithium metal battery with solid electrolytes, internal short circuit is a critical problem. Figure 8 shows photograph of LLZO solid electrolyte sheet after dissolution and deposition cycles of Li metal; the distinct linear morphology was observed on the surface of LLZO [34]. In addition, Fig. 9 shows the time profile of potentials in charge and discharge cycles in Li/Li symmetric cell; a potential drop toward 0 V (versus Li/Li^+) was observed in Li/Li symmetric cell [34]. The distinct linear morphology and the potential drop indicate that internal short circuit inside LLZO sheet by forming dendritic Li metal in deposition process. Some literatures mentioned that a mechanical stress generated at the interface between Li metal and solid electrolyte sheet is one of the reasons

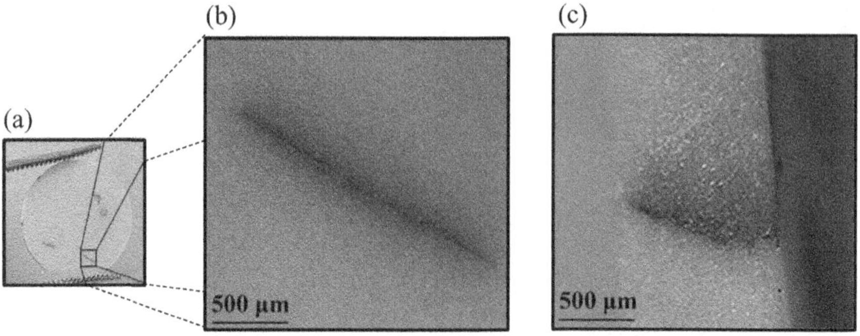

Fig. 8 Photographs of LLZO sheet after charge and discharge cycles in Li/Li symmetric cell, in which the image from top of the crack on LLZO (**a**), the magnified image (**b**) and the cross sectional image of the crack (**c**) are shown here; the photographs were prepared from Fig. 4 [34]

for the internal short circuit [35]. At smaller current, the internal short circuit hardly occurs due to lower mechanical stress. In the practical cells, a few mA cm^{-2} current density is needed to operate the cell for various applications. In order to reduce a generation of mechanical stress, some new ideas have been proposed to suppress the formation of dendritic Li metal. Dendritic Li metal can be formed due to poor contact of Li metal anode with solid electrolyte sheets; poor contact can be confirmed by measuring cross-sectional SEM image of solid electrolytes and Li metal, as shown in Fig. 10 [36]. The poor contact results in inhomogeneity of current at the interface to lead huge mechanical stress inside electrolyte sheets. Electrolyte sheets are cracked by loading huge mechanical stress, and then Li metal is penetrating the cracks inside the sheets in deposition process to generate internal short circuit. Finally, electrolyte sheets are broken into fragments [37]. The poor contact can be derived from low wettability of Li metal to the surface of solid electrolytes. The wettability will be improved by forming interlayer between Li metal and solid electrolyte sheet. Figure 11 exhibits the illustration of the interface between Li metal and solid electrolyte sheets with (Fig. 11a) and without (Fig. 11b) the interlayer to improve the wettability [36]. The interlayer of Au has been investigated for solid electrolytes of metal oxide [38] and metal sulfide [39, 40]. Au interlayer was effective to suppress internal short circuit in solid electrolytes of metal oxide, as shown in time profile of potentials under constant current (Fig. 12) [38], in which Li/Li symmetrical cell with LLZO is employed. Internal short circuit was also suppressed for solid electrolyte of metal sulfide by using Au interlayer, in which Li/Li symmetrical cell with Li_3PS_4 was employed for galvanostatic cycles [39, 40]. Internal short circuit will be critical problem, when Li metal secondary battery with solid electrolytes are operated by using large amount of current for practical use. Therefore, solid electrolytes should be more durable against internal short circuit.

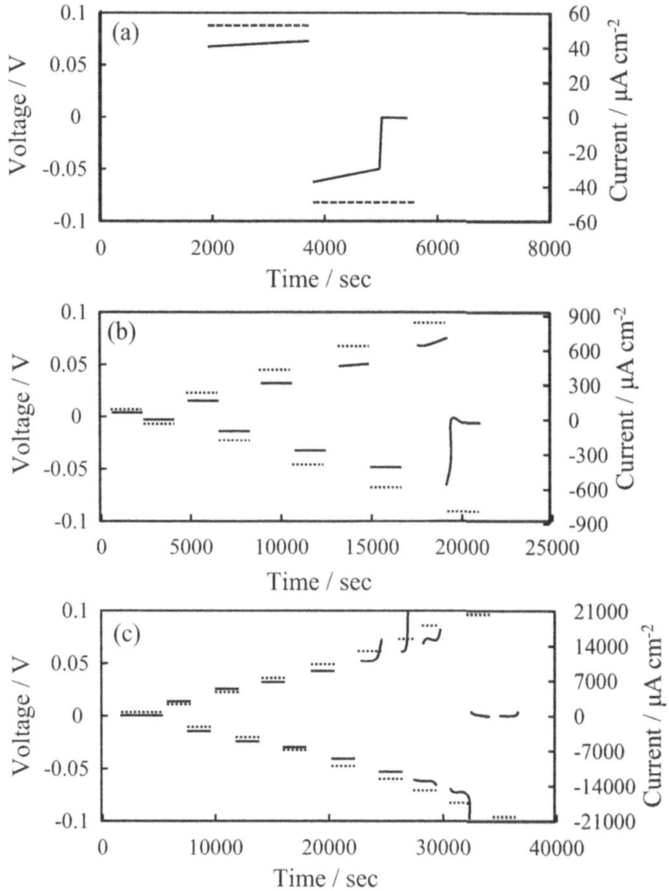

Fig. 9 Time profiles of potentials in charge and discharge cycles in Li/Li symmetric cell using LLZO under the conditions of 30 °C (**a**), 100 °C (**b**) and 160 °C (**c**), in which solid and broken lines indicate the values of voltage and current, respectively; the time profiles were prepared from Fig. 3 in [34]

9 New Separator for Li Metal Anode

There are several kinds of separators for LIB. These separators have been applied to Li metal battery. In most cases, Li metal dendrite was formed during discharge and charge cycles. The SEM images for conventional separators of polyolefin (Fig. 13a) [41] and non-woven cellulose (Fig. 13b) [42] are shown; the porosity of them is around 40%. Both the separators of polyolefin and non-woven cellulose cause internal short circuit facilely due to large sizes of pore and inhomogenous distribution of pores. Figure 14 shows the SEM images of polypropylene separator (Fig. 14a) and

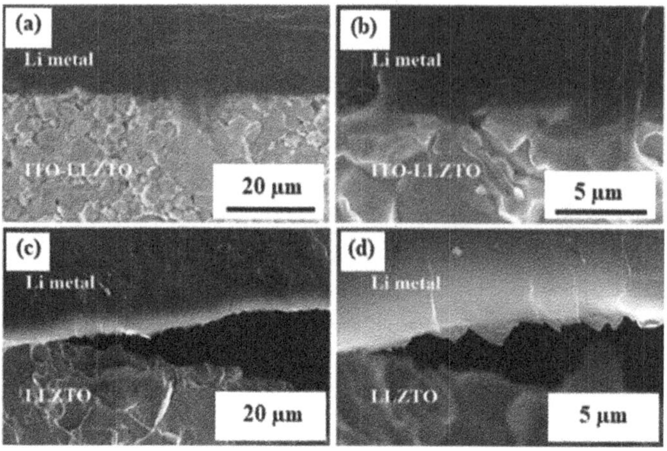

Fig. 10 Cross-sectional SEM images of Li metal and solid electrolyte with (**a**, **b**) or without (**c**, **d**) interlayer, in which a solid electrolyte of $Li_{6.4}La_3Zr_{1.4}Ta_{0.6}O_{12}$ and an interlayer of indium tin oxides are employed as an example; the SEM images were prepared from Fig. 2 in [36]

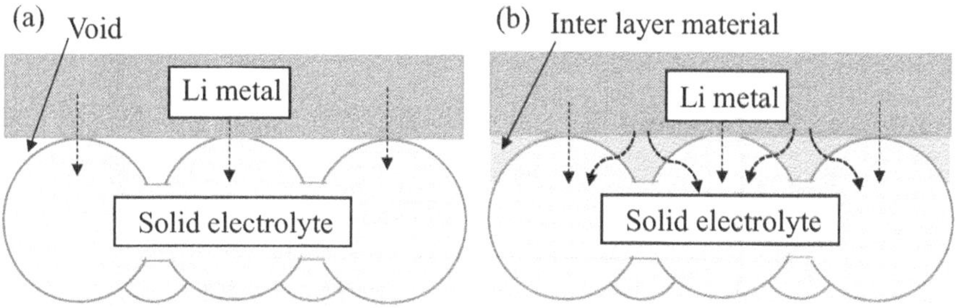

Fig. 11 Schematic the illustration of the interface between Li metal and solid electrolyte sheets with (**a**) and without (**b**) the interlayer

dendritic Li metal deposited in coin cell using polypropylene separator (Fig. 14b), as one of the examples [5].

As a new separator, three dimensionally ordered macroporous (3DOM) polyimide (PI) separator has been proposed by TMU group. Figure 14c shows the SEM image of 3DOM-PI separator. This polymer film consists of macro-sized pores in the range from 100 to 1000 nm, and the length among macro-sized pores in the range from 10 to 200 nm. This membrane has 1.8 tortuosity, which is smaller than those of polyolefin separators (the value is around 4) [5]. This indicates that 3DOM-PI has more uniform pore structure, which can provide more uniform current distribution. In addition, the 3DOM-PI separator with 300 nm size macro-sized pores has smaller connecting pore with 50 nm, so that Li metal dendrite cannot physically penetrate into 3DOM separator. By using this separator, the dendrite formation is suppressed dramatically, leading to prevent internal short circuit of the cell with Li

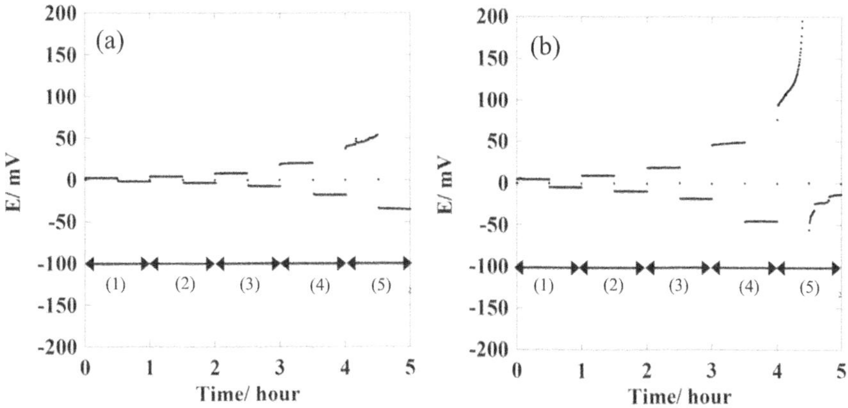

Fig. 12 Time profiles of potentials in charge and discharge cycles in Li metal symmetric cell using Al-doped LLZO (**a**) with and (**b**) without Au interlayer under the condition of 60 °C, in which the time profiles were measured by generating current of at (1) 10 μAcm^{-2}, (2) 20 μAcm^{-2}, (3) 40 μAcm^{-2}, (4) 100 μAcm^{-2}, (5) 200 μAcm^{-2}; the time profiles were prepared from Fig. 6 in [38]

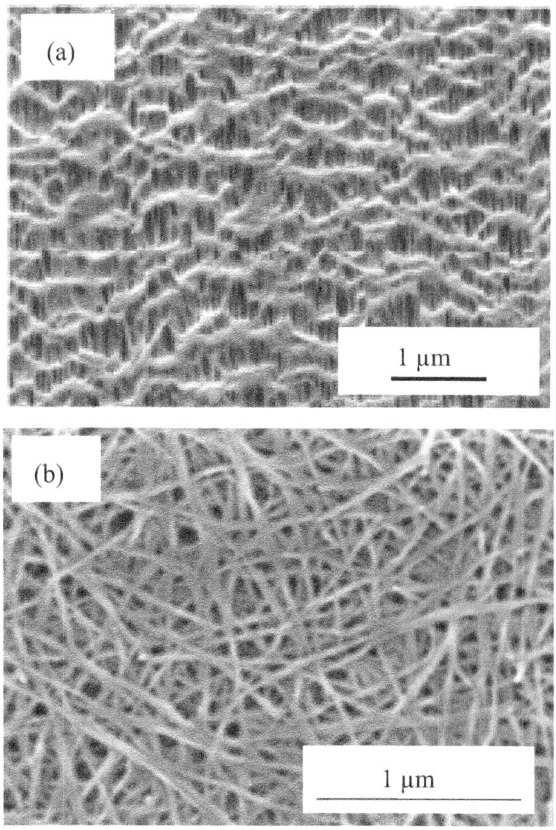

Fig. 13 SEM images for polyolefin separator (**a**) and cellulose separator (**b**); the images were prepared from Fig. 4 in [41] and Fig. 2 in [42]

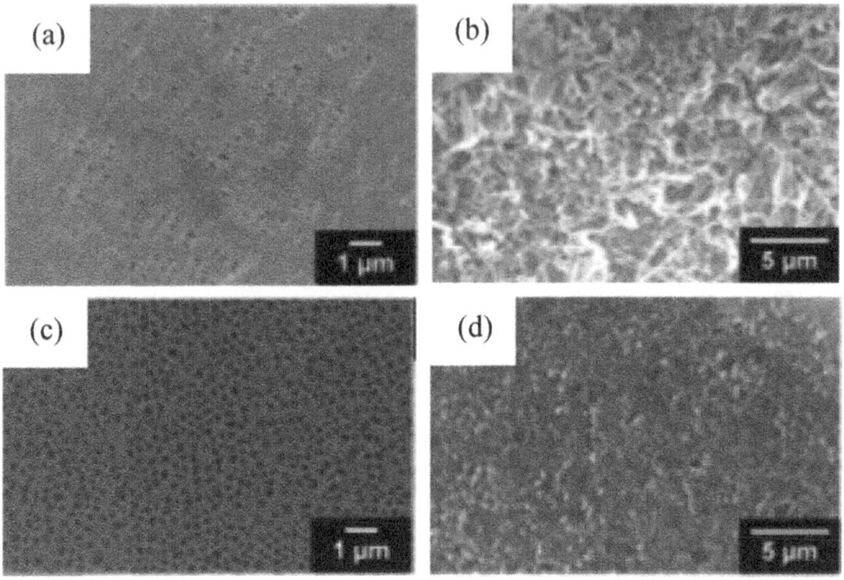

Fig. 14 SEM images of polyolefin (polypropylene) separator (**a**) and 3DOM separator (**c**), Li metal deposited on copper substrate in ethyl carbonate electrolyte of $LiPF_6$ and Li/Cu coin cell using polypropylene (**b**) and 3DOM (**d**) separators; the images were prepared from Figs. 2 and 7 in [5]

metal anode, as shown in Fig. 14d. Figure 15 shows the time profile of potential of Li(600 μm)/Li(100 μm) symmetrical coin cell under a constant current [5]. More than 3000 cycles can be realized by using 3DOM-PI separator. 3DOM separator is one of the possible materials for Li metal secondary battery, while volume expansion and shrinkage is caused in full cells to deplete the performance of charge and

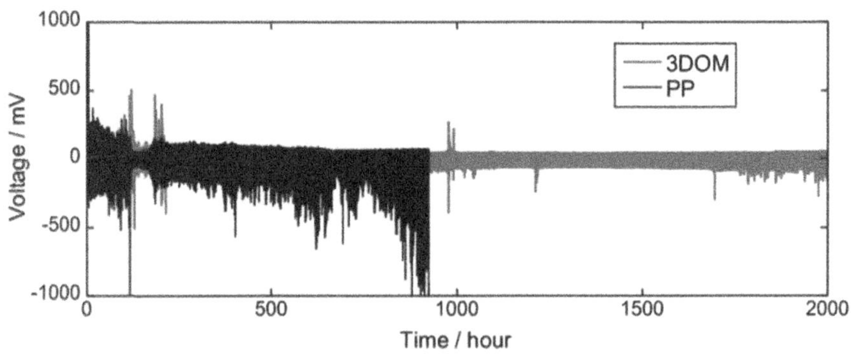

Fig. 15 Time profile of potential during charge and discharge cycles at 10.3 mA cm^{-2} in Li/Li symmetrical coin cell using the separator of polypropylene and 3DOM, in which the pore size of 3DOM was around 300 nm; the time profiles were prepared from [5]

discharge cycles. Therefore, improvement of 3DOM separator or development of new types of separators is required in the future.

10 Summary

Li metal anode has been extensively investigated around world, especially in last several years, there are many publications about improvement of Li metal anode and realization of Li metal battery with oxide cathode, sulfur cathode and air cathode. Unfortunately, still the cyclability of these cells is too low to apply to real applications. The rechargeability of Li mental battery should be improved by using various materials and technologies.

References

1. Yabuuchi, N., & Ohzuku, T. (2003). Novel lithium insertion material of $LiCo_{1/3}Ni_{1/3}Mn_{1/3}O_2$ for advanced lithium-ion batteries. *Journal of Power Sources, 119–121,* 171–174.
2. Chen, C. H., Liu, J., Stoll, M. E., Henriksen, G., Vissers, D. R., & Amine, K. (2004). Aluminum-doped lithium nickel cobalt oxide electrodes for high-power lithium-ion batteries. *Journal of Power Sources, 128,* 278–285.
3. Shen, X., Tian, Z., Fan, R., Shao, L., Zhang, D., Cao, G., et al. (2018). Research progress on silicon/carbon composite anode materials for lithium-ion battery. *J Energy Chem, 27,* 1067–1090.
4. Miao, Y., Hynan, P., Jouanne, A. V., & Yokochi, A. (2019). Current Li-Ion battery technologies in electric vehicles and opportunities for advancements. *Energies, 12,* 1074.
5. Shimizu, Y., & Kanamura, K. (2019). Effect of pore size in three dimensionally ordered macroporous polyimide separator on lithium deposition/dissolution behavior. *Journal of the Electrochemical Society, 166,* A754–A761.
6. Paled, E. (1979). The electrochemical behavior of alkali and alkaline earth metals in nonaqueous battery systems—the solid electrolyte interphase model. *Journal of the Electrochemical Society, 126,* 2047–2051.
7. Aurbach, D., Weissman, I., Schechter, A., & Cohen, H. (1996). X-ray photoelectron spectroscopy studies of lithium surfaces prepared in several important electrolyte solutions. A comparison with previous studies by fourier transform infrared spectroscopy. *Langmuir, 12,* 3991–4007.
8. Kanamura, K., Tamura, H., Shiraishi, S., & Takehara, Z. (1995). XPS analysis of lithium surfaces following immersion in various solvents containing $LiBF_4$. *Journal of the Electrochemical Society, 142,* 340–347.
9. Kanamura, K., Shiraishi, S., & Takehara, Z. (1996). Electrochemical deposition of very smooth lithium using nonaqueous electrolytes containing HF. *Journal of the Electrochemical Society, 143,* 2187–2197.
10. Shiraishi, S., Kanamura, K., & Takehara, Z. (1995). Effect of surface modification using various acids on electrodeposition of lithium. *Journal of Appled Electrochemistry, 25,* 584–591.
11. Shiraishi, S., Kanamura, K., & Takehara, Z.-I. (1999). Influence of initial surface condition of lithium metal anodes on surface modification with HF. *Journal of Appled Electrochemistry, 29,* 867–879.

12. Shiraishi, S., Kanamura, K., & Takehara, Z. (1999). Surface condition changes in lithium metal deposited in nonaqueous electrolyte containing HF by dissolution-deposition cycles. *Journal of the Electrochemical Society, 146,* 1633–1639.
13. Matsuda, S., Nishioka, K., & Nakanishi, S. (2019). High-throughput combinatorial screening of multi-component electrolyte additives to improve the performance of Li metal secondary batteries. *Sci Rep., 9,* 6211.
14. Miao, R., Yang, J., Feng, X., Jia, H., Wang, J., & Nuli, Y. (2014). Novel dual-salts electrolyte solution for dendrite-free lithium-metal based rechargeable batteries with high cycle reversibility. *Journal of Power Sources, 271,* 291–297.
15. Besenhard, J. O., Wagner, M. W., Winter, M., Jannakoudakis, A. D., Jannakoudakis, P. D., & Theodoridou, E. (1993). Inorganic film-forming electrolyte additives improving the cycling behaviour of metallic lithium electrodes and the self-discharge of carbon—lithium electrodes. *Journal of Power Sources, 44,* 413–420.
16. Li, Y., Sun, Y., Pei, A., Chen, K., Vailionis, A., Li, Y., et al. (2018). Robust Pinhole-free Li_3N Solid Electrolyte Grown from Molten Lithium. *ACS Cent Sci, 4,* 97–104.
17. Kozen, A. C., Lin, C.-F., Zhao, O., Lee, S. B., Rubloff, G. W., & Noked, M. (2017). Stabilization of lithium metal anodes by hybrid Artificial solid electrolyte interphase. *Chemistry ofMaterials, 29,* 6298–6307. https://doi.org/10.1021/acs.chemmater.7b01496.
18. Nie, M., & Lucht, B. L. (2014). Role of lithium salt on solid electrolyte interface (SEI) formation and structure in lithium ion batteries. *Journal of the Electrochemical Society, 161,* A1001–A1006.
19. Wang, J., Yamada, Y., Sodeyama, K., Chiang, C. H., Tateyama, Y., & Yamada, A. (2016). Superconcentrated electrolytes for a high-voltage lithium-ion battery. *Nat Commun., 7,* 12032.
20. Yoshida, K., Nakamura, M., Kazue, Y., Tachikawa, N., Tsuzuki, S., Seki, S., et al. (2011). Oxidative-stability enhancement and charge transport mechanism in glyme-lithium salt equimolar complexes. *Journal of the American Chemical Society, 133,* 13121–13129.
21. Jian, Z., Liu, P., Li, F., He, P., Guo, X., Chen, M., et al. (2014). Core–shell-structured CNT@RuO2 composite as a high-performance cathode catalyst for rechargeable Li–O2 Batteries. *Angewandte Chemie Int Ed, 53,* 442–446.
22. Ogasawara, T., Débart, A., Holzapfel, M., Novák, P., & Bruce, P. G. (2006). Rechargeable Li_2O_2 Electrode for Lithium Batteries. *Journal of the American Chemical Society, 128,* 1390–1393.
23. Mizuno, F., Nakanishi, S., Kotani, Y., Yokoishi, S., & Iba, H. (2010). Rechargeable Li-Air batteries with carbonate-based liquid electrolytes. *Electrochemistry, 78,* 403–405.
24. Yamada, Y., & Yamada, A. (2015). Review—superconcentrated electrolytes for lithium batteries. *Journal of the Electrochemical Society, 162,* A2406–A2423.
25. R. Xu, X.-Q. Zhang, X.-B. Cheng, H.-J. Peng, C.-Z. Zhao, C. Yan, J.-Q. Huang, Artificial Soft–Rigid Protective Layer for Dendrite-Free Lithium Metal Anode*Adv. Funct. Mater.,* 28 (2018)no.1705838.
26. Khurana, R., Schaefer, J. L., Archer, L. A., & Coates, G. W. (2014). Suppression of lithium dendrite growth using cross-linked polyethylene/poly(ethylene oxide) electrolytes: A new approach for practical lithium-metal polymer batteries. *Journal of the American Chemical Society, 136,* 7395–7402.
27. Abraham, K. M., & Alamgir, M. (1990). Li + -conductive solid polymer electrolytes with liquid-like conductivity. *Journal of the Electrochemical Society, 137,* 1657–1658.
28. Kubota, K., & Matsumoto, H. (2014). Cation mixtures of alkali metal (Fluorosulfonyl)(trifluoromethylsulfonyl)amide as electrolytes for lithium secondary battery. *Journal of the Electrochemical Society, 161,* A902–A907.
29. Lin, D., Liu, Y., & Cui, Y. (2017). Reviving the lithium metal anode for high-energy batteries. *Nature Nanotechnology, 12,* 194–206.
30. Jiao, S., Zheng, J., Li, Q., Li, X., Engelhard, M. H., Cao, R., et al. (2018). Behavior of lithium metal anodes under various capacity utilization and high current density in lithium metal batteries. *Joule., 2,* 110–124.
31. Yun, Q., He, Y.-B., Lv, W., Zhao, Y., Li, B., Kang, F., et al. (2016). Chemical dealloying derived 3D porous current collector for Li metal anodes. *Advanced Materials, 28,* 6932–6939.

32. Yang, C.-P., Yin, Y.-X., Zhang, S.-F., Li, N.-W., & Guo, Y.-G. (2015). Accommodating lithium into 3D current collectors with a submicron skeleton towards long-life lithium metal anodes. *Nat Commun, 6,* 8058.
33. Zuo, T.-T., Wu, X.-W., Yang, C.-P., Yin, Y.-X., Ye, H., Li, N.-W., et al. (2017). Graphitized carbon fibers as multifunctional 3D current collectors for high areal capacity Li anodes. *Advanced Materials, 29,* 1700389.
34. Sharafi, A., Meyer, H. M., Nanda, J., Wolfenstine, J., & Sakamoto, J. (2016). Characterizing the Li–$Li_7La_3Zr_2O_{12}$ interface stability and kinetics as a function of temperature and current density. *Journal of Power Sources, 302,* 135–139.
35. Li, Q., Yi, T., Wang, X., Pan, H., Quan, B., Liang, T., et al. (2019). *In-situ* visualization of lithium plating in all-solid-state lithium-metal battery. *Nano Energy, 63,* 103895.
36. Lou, J., Wang, G., Xia, Y., Liang, C., Huang, H., Gan, Y., et al. (2020). Achieving efficient and stable interface between metallic lithium and garnet-type solid electrolyte through a thin indium tin oxide interlayer. *Journal of Power Sources, 448,* 227440.
37. Duan, H., Zheng, H., Zhou, Y., Xu, B., & Liu, H. (2018). Stability of garnet-type Li ion conductors: An overview. *Sol St Ion, 318,* 45–53.
38. Wakasugi, J., Munakata, H., & Kanamura, K. (2017). Effect of gold layer on interface resistance between lithium metal anode and $Li_{6.25}Al_{0.25}La_3Zr_2O_{12}$ solid electrolyte. *Journal of the Electrochemical Society, 164,* A1022–A1025.
39. Nagao, M., Hayashi, A., & Tatsumisago, M. (2012). Bulk-type lithium metal secondary battery with indium thin layer at interface between Li electrode and Li_2S-P_2S_5 solid electrolyte. *Electrochemistry, 80,* 734–736.
40. Kato, A., Suyama, M., Hotehama, C., Kowada, H., Sakuda, A., Hayashi, A., et al. (2018). High-temperature performance of all-solid-state lithium-metal batteries having Li/Li3PS4 interfaces modified with Au thin films. *Journal of the Electrochemical Society, 165,* A1950–A1954.
41. Arora, P., & Zhang, Z. J. (2004). Battery separators. *Chemical Reviews, 104,* 4419–4462.
42. Pan, R., Wang, Z., Sun, R., Lindh, J., Edström, K., Strømme, M., et al. (2017). Thickness difference induced pore structure variations in cellulosic separators for lithium-ion batteries. *Cellulose, 24,* 2903–2911.

Concentrated Electrolytes for Lithium Metal Negative Electrodes

Yuki Yamada

Abstract Lithium metal is one of the most promising negative electrodes for high-energy-density rechargeable batteries, but its critical problem is low Coulombic efficiency resulting from the reductive decomposition of an electrolyte thereon, which has hampered its commercial applications. Various electrolytes and additives have been proposed to form a stable interphase between Li metal and electrolyte. Among them, salt-concentrated electrolytes have been most extensively studied, which create a unique anion-derived interphase that can highly stabilize the Li metal/electrolyte interface. The optimization of the salt–solvent combinations as well as the salt concentration enables us to achieve high Coulombic efficiency of over 99%. Herein the unique interphasial properties of concentrated electrolytes, as well as other beneficial interfacial/bulk properties such as extended potential windows, prevented metal corrosion, accelerated electrode reactions, increased transference number and decreased volatility/flammability, are discussed with a focus on their peculiar coordination structures.

Keywords Electrolyte · Concentration · Coordination · Interphase

1 Development of Concentrated Electrolytes

The salt concentration of battery electrolytes has long been fixed to approximately 1 M because the ionic conductivity is usually maximized at this concentration. Increasing the salt concentration over 1 mol dm^{-3} (M, molarity) increases the viscosity, thus compromising the ionic conductivity (Fig. 1a). Hence, such a salt-concentrated region was not extensively studied for battery electrolytes. However, the cation–anion and cation–solvent coordination states are significantly changed over a certain threshold concentration [1, 2]. In conventional 1 M electrolytes, in which the solvent/Li$^+$ ratio is around 10, approximately four solvent molecules are coordinated to one Li$^+$ to dissociate a large portion of Li salts, and remaining solvent

Y. Yamada (✉)
Department of Chemical System Engineering, The University of Tokyo, 7-3-1, Hongo, Bunkyo-Ku, Tokyo 113-8656, Japan
e-mail: y_yamada@chemsys.t.u-tokyo.ac.jp

K. Kanamura (ed.), *Next Generation Batteries*,
https://doi.org/10.1007/978-981-33-6668-8_3

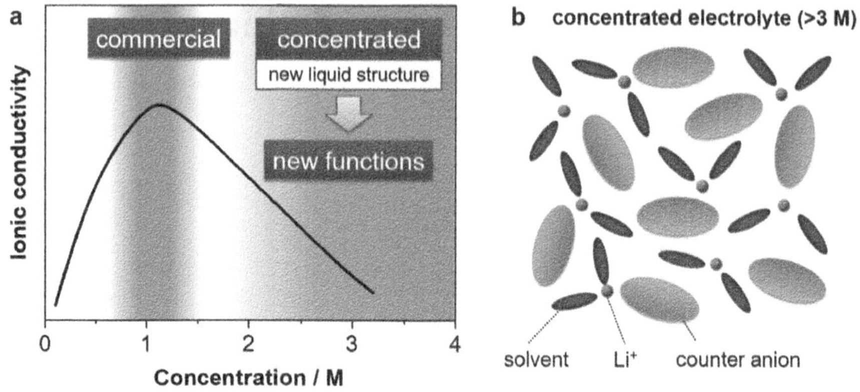

Fig. 1 **a** Typical ionic conductivity curve of a Li-salt organic electrolyte at room temperature. **b** Liquid structure of salt-concentrated electrolytes. Reprinted with permission from [2]. Copyright 2014 American Chemical Society

molecules are present in a free (uncoordinated) state. On the other hand, in concentrated electrolytes, in which the solvent/Li^+ ratio is equal to or lower than two, almost all solvent molecules are incorporated in the solvation shell of Li^+, and there is no free solvent (Fig. 1b). Besides, almost all counter anions are coordinated to Li^+ to form contact ion pairs (CIPs, one Li^+ coordinated to an anion) and even aggregates (AGGs, two or more Li^+ coordinated to an anion). Importantly, such unique coordination states provide the concentrated electrolytes with various unusual functions including (i) extended potential windows [2, 3], (ii) prevented metal corrosion [4, 5], (iii) accelerated electrode reactions [2, 6], (iv) increased Li^+ transference number [7], and (v) decreased volatility and flammability [3, 8]. Owing to these beneficial properties and functions, concentrated electrolytes are attracting increasing attention as a key to developing advanced rechargeable batteries including Li-metal batteries.

2 Li Metal in Concentrated Electrolytes

The use of concentrated electrolytes is effective in utilizing Li metal negative electrodes. Jeong et al. are the pioneer of the research on concentrated battery electrolytes [9], and they also first applied this concept to Li metal electrodes in 2008 [10]. They studied Li plating/stripping in $LiN(SO_2C_2F_5)_2$ (LiBETA or LiBETI)/propylene carbonate (PC) electrolytes to find that the Coulombic efficiency could be remarkably improved at a high salt concentration of 3.27 mol kg^{-1} (m, molality). A similar result was also reported in 2013 on a Li–S battery electrolyte of $LiN(SO_2CF_3)_2$ (LiTFSA or LiTFSI) in 1,3-dioxolane (DOL) and 1,2-dimethoxyethane (DME) [7]. As an extreme case of this strategy, Li metal plating/stripping (though at a low efficiency) in acetonitrile (AN) solvent was also reported in 2014, which is usually reduced rapidly by Li metal at low concentrations [2]. Li metal is also stabilized in dimethyl sulfoxide (DMSO) solvent at a high LiTFSA concentration [11, 12]. Since such solvents as AN and DMSO are reported to be stable against a superoxide radical (i.e., an intermediate of oxygen reduction in nonaqueous media), these concentrated electrolytes will be promising for use in lithium–oxygen batteries. Overall, the use of concentrated LiTFSA or LiBETA salt can generally stabilize Li metal electrodes in a variety of solvents, but the improvement of the Coulombic efficiency was limited to < 90%, which is far from satisfactory for practical applications.

A breakthrough was achieved by the development of concentrated $LiN(SO_2F)_2$ (LiFSA or LiFSI) electrolytes [6, 13]. Qian et al. applied concentrated 4 M LiFSA/DME to Li metal electrodes to demonstrate high Coulombic efficiency of 98.4% on average over 1000 cycles in a Cu|Li half cell [13]. After this report, various concentrated LiFSA electrolytes with a variety of solvents have been studied including ethers [14], carbonates [15], sulfones [16, 17], and fluorinated solvents [18], etc., all of which show high Coulombic efficiencies of Li plating/stripping approaching to or even higher than 99% (Fig. 2). The choice of Li salt is essential to achieve such high Coulombic efficiencies in concentrated electrolytes, and at present, LiFSA is the best salt for use in Li metal batteries.

However, concentrated electrolytes have drawbacks of lower ionic conductivity, higher viscosity, and higher cost (arising from more expensive salts) than conventional dilute electrolytes [19]. To address these issues, a strategy of diluting concentrated electrolytes with specific solvents (a diluent) was first proposed by Dokko et al. in 2013 [20]. The dilution with a low-polar inert solvent (e.g. fluorinated ethers) can retain the original functions of concentrated electrolytes while achieving higher ionic conductivity, lower viscosity, and lower cost. Such a class of electrolytes has local coordination states similar to those of concentrated electrolytes, thus referred to as "localized high-concentration electrolytes" [21]. This strategy was also applied to various LiFSA electrolytes for Li metal electrodes [21–23]. In 2018, Zhang et al. reported a localized high-concentration electrolyte of 1.2 M LiFSA in a mixture of DMC and bis(2,2,2-trifluoroethyl) ether (1:2 by mol) that enables highly reversible

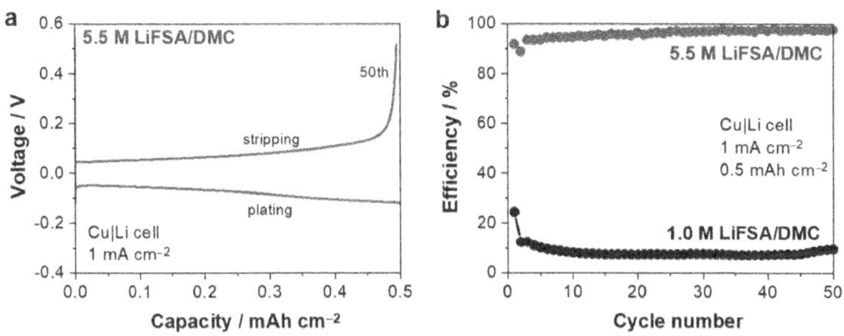

Fig. 2 **a** Charge–discharge voltage curve of a Cu|Li cell with 5.5 M LiFSA/dimethyl carbonate (DMC) in the 50th cycle. Li was plated on the Cu at 1 mA cm^{-2} for 0.5 h (0.5 mAh cm^{-2}) and then stripped up to 0.5 V cut-off. **b** Coulombic efficiencies of Cu|Li cells with 1.0 M and 5.5 M LiFSA/DMC electrolytes

and dendrite-free cycling of Li metal electrodes at a high Coulombic efficiency of well over 99% [21]. Hence, this dilution strategy is a promising way toward practical application of concentrated electrolytes, although there is still large room for electrolyte optimization including the molecular design of the diluent, the selection of the primary solvent and salt, and their mixing ratio.

3 Mechanism of Better Plating/Stripping Reversibility

The better reversibility of Li metal plating/stripping in concentrated electrolytes has been discussed from at least two viewpoints of (i) the formation of better solid electrolyte interphase (SEI) and (ii) the suppression of dendritic deposition. This section overviews the mechanism based on the unique characteristics of concentrated electrolytes.

3.1 SEI Formation in Concentrated Electrolytes

Generally, concentrated electrolytes have higher stability against reduction, thus having wider potential windows than their dilute counterparts. As a visual demonstration of the enhanced reduction stability, the reactivity of Li metal foil (i.e., a strong reducing agent) in dilute (1 M) and concentrated (4.2 M) LiTFSA/AN electrolytes was studied (Fig. 3) [2]. Since AN solvent is prone to reduction, the Li metal was quickly reacted with the AN to dissolve in the dilute electrolyte. On the other hand, no visual change was observed for the Li metal in the concentrated electrolyte, suggesting its enhanced reduction stability even at the low potential of Li metal.

The enhanced reduction stability is primarily owing to the formation of peculiar SEI on Li metal or other low-potential negative electrodes (e.g. graphite) [2, 24]. In most of the dilute electrolytes, solvent molecules are predominantly reduced on such negative electrodes, thus dominating the nature of SEI or resultant reduction

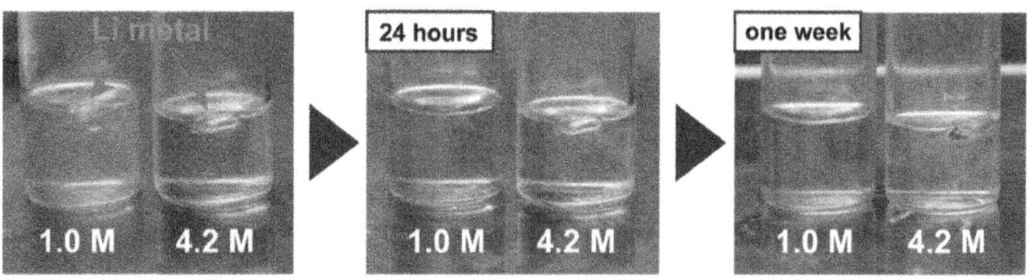

Fig. 3 Reactivity of Li metal foil in dilute 1.0 M and concentrated 4.2 M LiTFSA/AN electrolytes at room temperature. Reprinted with permission from [2]. Copyright 2014 American Chemical Society

stability. If the solvent has poor SEI-forming ability, then it continues to be reduced, as is observed for dilute LiTFSA/AN. On the other hand, in concentrated electrolytes with the aforementioned unique coordination state, the counter anion of Li salt is predominantly reduced on negative electrodes, thus forming a peculiar anion-derived SEI. Hence, the nature of the SEI in concentrated electrolytes is dominated by the Li salt species. Sulfonylamide-based salts such as LiFSA, LiTFSA, and LiBETA form an excellent SEI film rich in F and S elements, which effectively suppresses further electrolyte decomposition, thus rendering high reduction stability to the concentrated electrolytes. Among them, the LiFSA-derived SEI seems to be the best to provide both high stability and low interfacial resistance, which account for the stable and high-rate cycling of Li metal electrodes in concentrated LiFSA electrolytes [13].

The nature of the SEI in concentrated sulfonylamide salt electrolytes has been studied by various methods. Based on transmission electron microscopy (TEM), scanning electron microscopy (SEM), and X-ray photoelectron spectroscopy (XPS), the SEI is thin and compact as compared with those in corresponding dilute electrolytes or conventional ethylene carbonate (EC) electrolytes [7, 10, 13]. XPS analysis shows that the SEI is mainly composed of inorganic compounds such as LiF, Li_2O, and some S-contained substances (e.g. sulfides) [2, 8, 13, 25], which is distinct from conventional EC-based SEI composed of the mixture of organic and inorganic compounds.

The mechanism of forming anion-derived SEI has been discussed based on the coordination states and electronic structures. In concentrated electrolytes over a certain threshold concentration, all counter anions of Li salts are coordinated with Li^+, forming extensive ion pairs (e.g. CIPs and AGGs) [2]. When coordinated with Li^+ (a strong Lewis acid), the anion partially donates its electron to the Li^+, thus being less negatively charged. In this situation, the anion is more prone to reduction. As theoretical evidence, density functional theory-based molecular dynamics (DFT-MD) simulations demonstrate that the lowest unoccupied molecular orbital (LUMO) level of counter anions is shifted downward with increasing salt concentration and finally becomes lower than the LUMO of solvent molecules at a certain high concentration (Fig. 4); hence, the counter anion is energetically the most likely component to receive an electron, thus being reduced [2, 24]. Furthermore, when an excess electron is introduced in the supercell in the DFT-MD simulation, it is indeed received by the ion-paired counter anion, which is then decomposed reductively [24]. This predominant reductive decomposition of the counter anion supported theoretically accounts for the peculiar anion-derived SEI formation in concentrated electrolytes. The anion-derived SEI might be further stabilized at the interface because the dissolution of the SEI components can be suppressed in concentrated electrolytes without a free solvent molecule that readily solvates and dissolves them.

Importantly, the unique local coordination state of concentrated electrolytes is retained even after dilution with a low-polar inert solvent (e.g., fluorinated ethers) [21, 22, 26, 27]. Hence, the anion-derived SEI can be formed similarly with the original concentrated electrolyte, which accounts for stable cycling of Li metal electrodes in localized high-concentration electrolytes [21–23]. This suggests that creating the

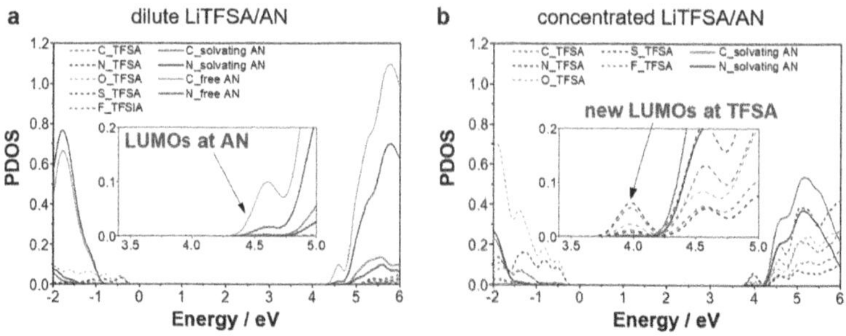

Fig. 4 Projected density of states (PDOS) obtained via DFT-MD simulations on **a** dilute 0.4 M and **b** concentrated 4.2 M LiTFSA/AN electrolytes. Insets show the magnified figures of the lowest energy edge of the conduction band. Reprinted with permission from [2]. Copyright 2014 American Chemical Society

unique local coordination state, rather than just increasing the salt concentration, is essential to achieve better SEI formation for stable Li metal electrodes.

3.2 Morphology of Li Metal in Concentrated Electrolytes

The morphology of plated Li is also important to achieve its stable charge–discharge cycling. In conventional carbonate electrolytes, dendritic Li forms during charge, which not only degrades the cycling performance but also poses a safety risk of penetrating a separator. On the other hand, increasing the salt concentration can effectively suppress the dendritic Li formation [10, 13]. For example, Qian et al. reported that 4 M LiFSA/DME enables smoother Li deposition with nodule-like structure with round-shaped edges (Fig. 5) [13]. As compared with dendritic Li, the nodule-like structure has lower surface area in contact with the electrolyte, thus reducing further electrolyte decomposition that leads to poor Coulombic efficiency.

The mechanism behind the uniform Li deposition is still not clear, but it may be related to (a) the nature of SEI and (b) high Li^+ concentration. The concentrated LiFSA/DME forms a thin and compact SEI, which is not only stable but also highly ion-conductive. Hence, it may enable uniform Li^+ flux from the electrolyte to the electrode surface. Furthermore, the high Li^+ concentration results in high Li^+ flux during deposition, which may circumvent an unfavorable diffusion-controlled condition, thus contributing to uniform Li deposition in concentrated electrolytes.

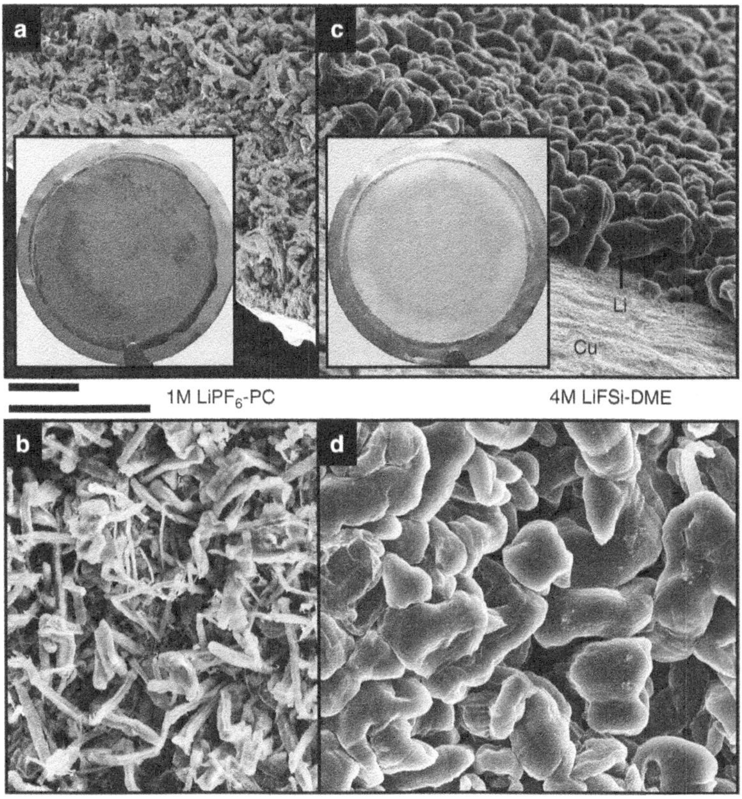

Fig. 5 SEM images of the Li metal plated on Cu in **a**, **b** 1 M $LiPF_6$/PC and **c**, **d** 4 M LiFSA/DME. The scale bar represents 10 μm. Reprinted with permission from [13]. Copyright 2015 Springer Nature

4 Summary and Future Perspectives

Concentrated electrolytes and their derivatives (e.g. localized high-concentration electrolytes) are being extensively studied for use in Li-metal batteries. Concentrated electrolytes have a unique coordination structure of forming extensive ion pairs, which shifts downward the LUMO level of counter anion, thus inducing its reductive decomposition to form an anion-derived SEI on negative electrodes. After optimizing salt–solvent combinations, a high Coulombic efficiency of over 99% has been achieved for Li metal electrodes, which, though still not satisfactory, makes substantial progress toward the commercialization of Li-metal batteries. Furthermore, many other functions have been discovered for concentrated electrolytes, including enhanced oxidation stability, prevented metal corrosion, accelerated electrode reactions, increased Li^+ transference number, and decreased flammability. Hence, concentrated electrolytes might be a major technological breakthrough in

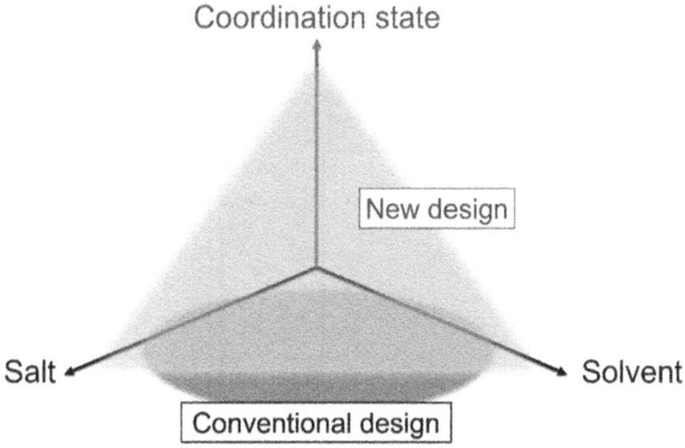

Fig. 6 New electrolyte design manipulating the coordination state as a primary factor in addition to exploring salt/solvent combinations

realizing Li-metal batteries with high voltages, high safety, and quick-charging characters. Finally, the emphasis is placed on the fact that all the beneficial features of concentrated electrolytes arise from their unique coordination structure, as is demonstrated by the concept of dilution with low-polar inert solvents. Hence, manipulating the coordination state of ions and solvent molecules will be a key to further electrolyte development toward advanced Li-metal batteries (Fig. 6).

References

1. Yamada, Y., Takazawa, Y., Miyazaki, K., & Abe, T. (2010). *Journal of Physical Chemistry C, 114,* 11680.
2. Yamada, Y., Furukawa, K., Sodeyama, K., Kikuchi, K., Yaegashi, M., Tateyama, Y., et al. (2014). *Journal of the American Chemical Society, 136,* 5039.
3. Yoshida, K., Nakamura, M., Kazue, Y., Tachikawa, N., Tsuzuki, S., Seki, S., et al. (2011). *Journal of the American Chemical Society, 133,* 13121.
4. Matsumoto, K., Inoue, K., Nakahara, K., Yuge, R., Noguchi, T., & Utsugi, K. (2013). *Journal of Power Sources, 231,* 234.
5. McOwen, D. W., Seo, D. M., Borodin, O., Vatamanu, J., Boyle, P. D., & Henderson, W. A. (2014). *Energy & Environmental Science, 7,* 416.
6. Yamada, Y., Yaegashi, M., Abe, T., & Yamada, A. (2013). *Chemical Communications, 49,* 11194.
7. Suo, L., Hu, Y.-S., Li, H., Armand, M., & Chen, L. (2013). *Nature Communications, 4,* 1481.
8. Wang, J., Yamada, Y., Sodeyama, K., Chiang, C. H. C. H., Tateyama, Y., & Yamada, A. (2016). *Nature Communications, 7,* 12032.
9. Jeong, S. K., Inaba, M., Iriyama, Y., Abe, T., & Ogumi, Z. (2003). *Electrochemical and Solid-State Letters, 6,* A13.
10. Jeong, S. K., Seo, H. Y., Kim, D. H., Han, H. K., Kim, J. G., Lee, Y. B., et al. (2008). *Electrochemistry Communications, 10,* 635.
11. Togasaki, N., Momma, T., & Osaka, T. (2016). *Journal of Power Sources, 307,* 98.
12. Liu, B., Xu, W., Yan, P., Kim, S. T., Engelhard, M. H., Sun, X., et al. (2017). *Advanced Energy Matericals, 7,* 1602605.
13. Qian, J., Henderson, W. A., Xu, W., Bhattacharya, P., Engelhard, M., Borodin, O., et al. (2015). *Nature Communications, 6,* 6362.
14. Liu, P., Ma, Q., Fang, Z., Ma, J., Hu, Y. S., Bin, Z., et al. (2016). *Chinese Physics B, 25,* 7.

15. Fan, X., Chen, L., Ji, X., Deng, T., Hou, S., Chen, J., et al. (2018). *Chem, 4,* 174.
16. Alvarado, J., Schroeder, M. A., Zhang, M., Borodin, O., Gobrogge, E., Olguin, M., et al. (2018). *Materials Today, 21,* 341.
17. Maeyoshi, Y., Ding, D., Kubota, M., Ueda, H., Abe, K., Kanamura, K., et al. (2019). *ACS Applied Materials & Interfaces, 11,* 25833.
18. Suo, L., Xue, W., Gobet, M., Greenbaum, S. G., Wang, C., Chen, Y., et al. (2018). *Proceedings of National Academy of Sciences, 115,* 1156.
19. Yamada, Y., Wang, J., Ko, S., Watanabe, E., & Yamada, A. (2019). *Nature Energy, 4,* 269.
20. Dokko, K., Tachikawa, N., Yamauchi, K., Tsuchiya, M., Yamazaki, A., Takashima, E., et al. (2013). *Journal of the Electrochemical Society, 160,* A1304.
21. Chen, S., Zheng, J., Mei, D., Han, K. S., Engelhard, M. H., Zhao, W., et al. (2018). *Advanced Materials, 30,* 1706102.
22. Ren, X., Chen, S., Lee, H., Mei, D., Engelhard, M. H., Burton, S. D., et al. (2018). *Chem, 4,* 1877.
23. Chen, S., Zheng, J., Yu, L., Ren, X., Engelhard, M. H., Niu, C., et al. (2018). *Joule, 2,* 1548.
24. Sodeyama, K., Yamada, Y., Aikawa, K., Yamada, A., & Tateyama, Y. (2014). *Journal of Physical Chemistry C, 118,* 14091.
25. Wang, J., Yamada, Y., Sodeyama, K., Watanabe, E., Takada, K., Tateyama, Y., et al. (2018). *Nature Energy, 3,* 22.
26. Ueno, K., Murai, J., Ikeda, K., Tsuzuki, S., Tsuchiya, M., Tatara, R., et al. (2016). *Journal of Physical Chemistry C, 120,* 15792.
27. Takada, K., Yamada, Y., & Yamada, A. (2019). *ACS Applied Materials & Interfaces, 11,* 35770.

All-Solid-State Battery with Sulfide Electrolyte

Crystalline Electrolyte

Satoshi Hori, Kota Suzuki, Masaaki Hirayama, and Ryoji Kanno

Abstract Crystalline lithium ion conductors are expected to be employed as solid electrolytes in all-solid-state lithium batteries. Sulfide-based materials are attractive owing to their high ionic conductivity compared to the counterparts of oxides. For the commercialization of solid-state batteries, however, there are still demands for the development of solid electrolytes with various characteristics to satisfy the essential requirements such as high conductivity, low price, and high chemical/electrochemical stability. This chapter introduces $Li_7Ge_3PS_{12}$ (argyrodite) and $Li_{10+\delta}[Sn_ySi_{1-y}]_{1+\delta}P_{2-\delta}S_{12}$ ($Li_{10}GeP_2S_{12}$-type) as representative examples for the recent material search for crystalline lithium ionic conductors.

Keywords Solid electrolyte · Lithium-ion conductor · Material search · Crystalline material

1 Development of Lithium-Ion Conductors

Solid electrolytes are an indispensable part of all-solid-state batteries and, therefore, should meet various requirements including high lithium ionic conductivity and electrochemical stability for increasing energy and power characteristics of cells, as well as thermal and mechanical stability in the air atmosphere for producing a battery package. It is desirable to develop various types of solid electrolytes to identify the most suitable one among a number of potential candidates.

The development of solid electrolytes began from a material search for Li-ion conductors, which has a long history. Prior to the 1990s, there were reports on β-Li aluminum [1, 2], LIthium SuperIonic CONductor (LISICON) [3], Li_3N [4], PEO-based polymer [5], and chalcogenide glass [6, 7]. After phosphorus-sulfide glass

S. Hori · R. Kanno (✉)
All-Solid-State Battery Unit, Institute of Innovative Research, Tokyo Institute of Technology, Yokohama 226-8503, Japan
e-mail: kanno@echem.titech.ac.jp

K. Suzuki · M. Hirayama
Department of Chemical Science and Engineering, School of Materials and Chemical Technology, Tokyo Institute of Technology, Yokohama 226-8502, Japan

K. Kanamura (ed.), *Next Generation Batteries*,
https://doi.org/10.1007/978-981-33-6668-8_4

with superionic conductivity was discovered in the 1990s [8], studies on crystalline sulfide having Li-ion conductivity (described in this chapter) began attracting much attention since the 2000s. Such crystalline materials include thio-LISICONs, having conductivity over 1 mS cm^{-1} at room temperature [9], which led to the discovery of $Li_{10}GeP_2S_{12}$ (LGPS)-related materials in the 2010s [10] that exhibited conductivity in the order of 10 mS cm^{-1}. Currently, there are a number of sulfide crystalline materials such as $Li_7P_3S_{11}$ and argyrodite phases that exhibit Li conductivity that is comparable to those of liquid electrolytes [11–13].

Table 1 summarizes sulfide crystalline electrolytes and shows the crystal structure, chemical composition, and Li conductivity at around room temperature. Some of the sulfides [11–13] show exceptional ionic conductivity, which is comparable to those for liquid electrolytes at room temperature. This class of material maintains high conductivity at low temperatures (i.e. below −30°C) where liquid electrolytes show a significant decrease in conductivity, even at high temperatures (i.e. above 100 °C) in which liquid electrolytes may decompose.

Thio-LISICONs, discovered at the beginning of the 2000s, are a group of sulfides with Li conductivity in the order of 10^{-3}–1 mS cm^{-1} that possess a crystal structure similar to Li_3PO_4. The MS_4 tetrahedral unit is composed of M (= P, Si, Ge, Al, Ga, and so on) coordinated by four sulfur atoms. Arrangements of MS_4 units, which consist of a framework structure, provide a variety of crystal structures such as α-, β-, and γ-Li_3PS_4 [23, 24]. After one member of this group, in which Li ions are widely spread in the crystal, was found to show a high conductivity exceeding 1 mS cm^{-1} [25], researchers began to focus on sulfide crystalline electrolytes as well as amorphous materials that were mainly studied previously. In the 2010s, there have been still reports on new materials belonging to thio-LISICONs [19].

Subsequently, $Li_7P_3S_{11}$ crystal was discovered as glass ceramic [22]. It has a unique crystal structure consisting of a PS_4 unit a and P_2S_7 unit, which correspond

Table 1 Lithium conductivity for sulfide crystalline phases

Structure type	Chemical composition system	Representative chemical composition	Conductivity at around R.T./mS cm^{-1}	References
LGPS	Li–M–P–S–X (M = Ge, Si, Sn; X = Cl, O)	$Li_{10}GeP_2S_{12}$	12	[10]
		$Li_{9.54}Si_{1.74}P_{1.44}S_{11.7}Cl_{0.3}$	25	[11]
		$Li_{9.42}Si_{1.02}P_{2.1}S_{9.96}O_{2.04}$	0.32	[14]
		$Li_{10.35}[Sn_{0.27}Si_{1.08}]P_{1.65}S_{12}$	11	[15]
Argyrodite	Li–M–P–S –X (M = Si, Ge; X = Cl, Br, I)	Li_6PS_5Cl	1.9	[16, 17]
		$Li_{6.6}Ge_{0.6}P_{0.4}S_5I$	18.4	[12]
		$Li_7Ge_3PS_{12}$	0.11	[18]
thio-LISICON	Li–M'–M''–S (M', M'' = P, Si, Ge, Sn, Al, Ga, etc.)	$Li_{4.275}Ge_{0.61}Ga_{0.25}S_4$	6.5×10^{-2}	[9, 19]
		Li_4SnS_4	7.0×10^{-2}	[20]
		β-Li_3PS_4	0.16	[21]
$Li_7P_3S_{11}$	Li–P–S	$Li_7P_3S_{11}$	17	[13, 22]

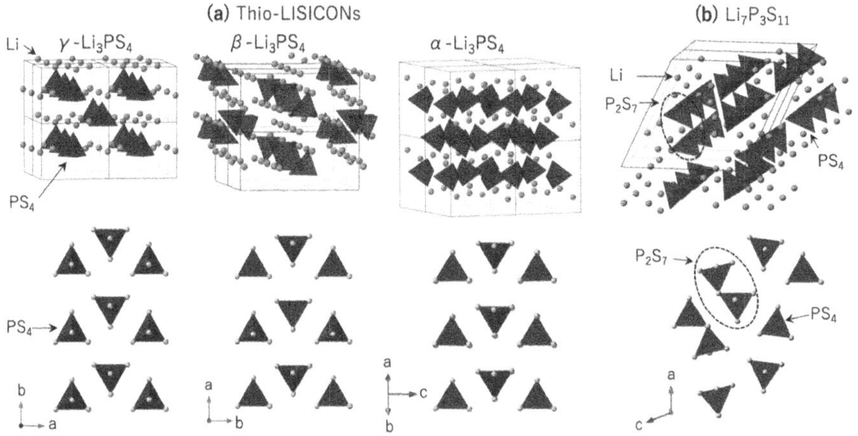

Fig. 1 Crystal structures of **a** thio-LISICONs and **b** $Li_7P_3S_{11}$. Upper side shows lithium sites with PS_4 tetrahedral or P_2S_7 connected-tetrahedral units. For thio-LISICONs, Li_3PS_4 modifications with various PS_4 tetrahedra apex are illustrated

to corner-shared two PS_4 units [26]. This crystal structure is a derivative of the Ag-ion conductor $Ag_7P_3S_{11}$. Derivatives based on $A_7P_3S_{11}$ composition and structure remain undiscovered. Crystal structures for thio-LISICONs and $Li_7P_3S_{11}$ are shown in Fig. 1.

The Li-argyrodite family is a group of solid electrolytes having a crystal structure similar to the mineral Ag_8GeS_6, referred to as argyrodite [16]. This crystal, with cubic symmetry, has a unique feature; there are three types of anion sites, one of which coordinates cation *M'* (= P, Si, Ge, and so on) and forms tetrahedral units while the other two sites do not coordinate or form tetrahedral units. Sulfur atoms occupying one of the latter two sites are preferentially substituted by halogen elements. The Li-argyrodites can be regarded as an analogue to Ag- or Cu- argyrodites, as with the case of $Li_7P_3S_{11}$. Since the first report, novel derivatives are still being discovered [12].

While the abovementioned crystals are analogues of known substances (such as oxides or Cu- or Ag- minerals), $Li_{10}GeP_2S_{12}$ (LGPS) possesses a completely new crystal structure that has previously not been reported. The LGPS crystal was designed by focusing on the variety of MS_4 tetrahedral arrangement, where M represents P, Ge, Si, or Sn. Incorporating a specific elemental and molar ratio for MS_4, the crystal takes unique tetrahedra arrays that are not aligned such that the vertices point in the same direction or in a zig-zag arrangement where the direction alternates. As a result, the array of sulfur atoms transforms from *hcp*-packing, which is taken in thio-LISICONs with the conductivity reaching 1 mS cm^{-1}, to *bcc*-packing, which has been reported as preferable for ion conduction [27], and as a result, LGPS realized exceptionally high Li conductivity over 10 mS cm^{-1}. Since the discovery of the original LGPS, researchers have challenged structure optimization for higher conductivity by chemical composition tuning, which finally led to a conductivity of

25 mS cm^{-1} in the $Li_{9.54}Si_{1.74}P_{1.44}S_{11.7}Cl_{0.3}$ phase [11]. Current research interests focus on the potential for even higher conductivity, as well as the contribution to the Li-ion battery industry.

Among the abovementioned material chemistries, the following sections describe a few examples of recent studies on a member of the argyrodite and LGPS group.

2 $Li_7Ge_3PS_{12}$ Argyrodite Phase

An argyrodite-type novel lithium conductor with Li–Ge–P–S composition was discovered during an exhaustive material search [18]. A material search was conducted based on the formation diagram of the quasi-ternary Li_2S–GeS_2–P_2S_5 system.

Over 30 identical compositions were examined as targets in this diagram. Raw materials (Li_2S, GeS_2, and P_2S_5) were mixed in the appropriate molar ratios under an inert atmosphere. After the mixing process, the obtained powder was pelletized, sealed in a carbon-coated quartz glass tubes at ~ 10 Pa, and heated at 870 °C for 8 h. X-ray diffraction (XRD) patterns of the synthesized powders were collected using an X-ray diffractometer with Cu Kα radiation.

As the XRD pattern in Fig. 2 shows, a monophasic pattern similar to the argyrodite phase was obtained for the composition of $Li_7Ge_3PS_{12}$, which was found to be a new argyrodite phase composition. So far, material searches for the argyrodite phase have been performed on Li_7PS_6-based compositions such as $Li_{7-k}PS_{6-k}X_k$ ($X = $ Cl, Br, I). After the discovery of the $Li_7Ge_3PS_{12}$, a wide variety of argyrodite families were reported [12, 28, 29], in which multi cation and anion combinations, and their composition ratios, were achieved, indicating that a wide solid-solution formation range with argyrodite structure can be expected.

The crystal structure determined by Rietveld refinement using synchrotron XRD data is illustrated in Fig. 3. The novel $Li_7Ge_3PS_{12}$ argyrodite phase with a cubic

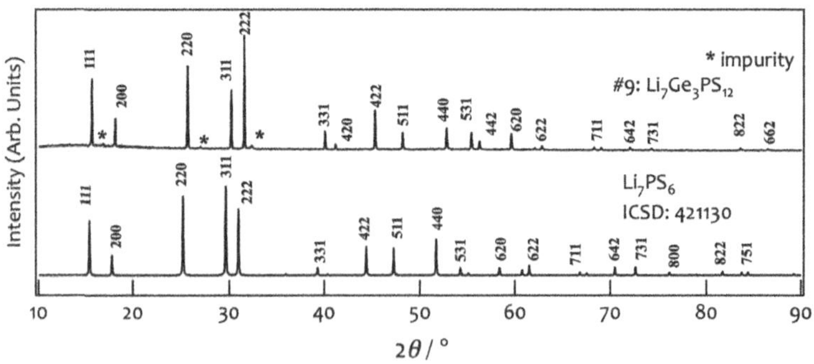

Fig. 2 X-ray diffraction patterns of novel argyrodite-type phase with $Li_7Ge_3PS_{12}$ composition, and of the reported argyrodite-type Li_7PS_6 (ICSD: 421130)

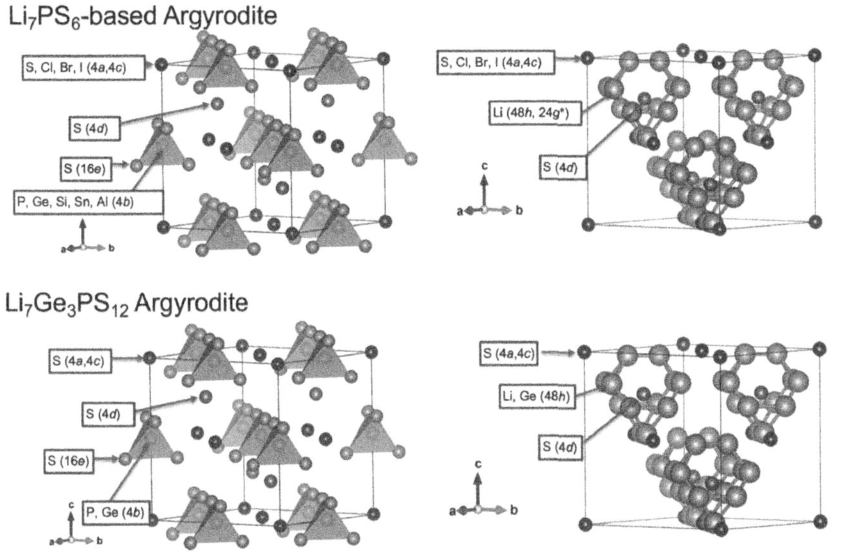

Fig. 3 Crystal structure of the argyrodite-type $Li_7Ge_3PS_{12}$ structure determined by Rietveld analysis (lower panel) together with Li_7PS_6-based argyrodite phase (upper panel). Both phases have a cubic symmetry ($F - 43\ m$, Space group No. 216). For each panel, the left side shows a scaffold composed of tetrahedral units, and the right side represents Li sites as a cluster around the sulfur site. Atoms that can occupy each site are indicated by arrows. For the Li_7PS_6-based phase, possible chemical substitutions reported so far are indicated for each site [12, 16, 28, 29]

symmetry possesses the ordered arrangement of four Isolated (Ge/P)S_4 tetrahedra, which forms the scaffold of the crystal, in contrast to the Li_7PS_6-based argyrodite such as Li_6PS_5X (X = Cl, Br, I) system in which the framework is composed of PS_4 tetrahedra. While 48 h (Wyckoff notation) sites around isolated sulfur (4d) are occupied by Li atoms for Li_7PS_6-based argyrodite, Li and Ge share the 48 h site for the $Li_7Ge_3PS_{12}$ phase.

The conductivity value determined by the AC-impedance method for the new material at 25 °C was 0.11 mS cm^{-1}. The high ionic conductivity could be attributed to the disordered arrangement of lithium ions within the structure which is characteristic of argyrodite-type lithium-ion conductors.

For argyrodite having a solid-solution range reported so far, the highest ionic conductivity over 10 mS cm^{-1} was achieved for the composition of $Li_{6.6}P_{0.4}Ge_{0.6}S_5I$ [12]. The discovery of $Li_7Ge_3PS_{12}$ implies a new insight into structure-conductivity relationships and the possibility of new solid solutions. Further development of elemental combinations and detailed structure-property analysis for the argyrodite-based materials could enhance conductive properties, and other required properties such as chemical/electrochemical stability, cost, and softness.

3 $Li_{10+\delta}[Sn_ySi_{1-y}]_{1+\delta}P_{2-\delta}S_{12}$ with LGPS-Type Structure

The LGPS-type materials have a suitable framework for fast lithium-ion conduction (Fig. 4); the one-dimensional tunnels and/or two-dimensional planes participate to contribute to high ionic diffusion [11, 30, 31]. Several substituents belonging to the LGPS family, e.g., LGPS-type derivatives ($Li_{11}Si_2PS_{12}$, $Li_{10}SiP_2S_{12}$, and $Li_{10}SnP_2S_{12}$) and solid solutions of Li_{10}(Ge, Sn)P_2S_{12} and Li_{10}(Ge, Si)P_2S_{12} were synthesized [32–35]. However, these synthesized derivatives showed lower conductivities than the original LGPS.

The material search was performed focusing on Sn and Si substitution for Ge in the LGPS structure since Sn and Si are inexpensive and abundant elements. Among the LGPS-type solid solutions described by the composition $Li_{10+\delta}M_{1+\delta}P_{2-\delta}S_{12}$, with different Li/*M*/P ratios (*M* = Si, Ge, Sn), the monophasic region of the phase is dependent on the *M* element [34, 36]. A material search was completed for the Li_3PS_4–Li_4SnS_4–Li_4SiS_4 quasi-ternary system to elucidate the compositional region of the LGPS-type phase and improve the ionic conductivity of these solid solutions [15, 37].

Target samples were synthesized by solid-state reactions (550 °C, 24 h). By the phase identification via XRD data, the solid-solution region for the LGPS-type phase in the Li_4SnS_4–Li_4SiS_4–Li_3PS_4 quasi-ternary system is indicated by the green dotted lines in Fig. 5. The P ratio (*x* value) of the LGPS-type single-phase region in each tie line varies with respect to the Sn/Si ratio, i.e., the solid-solution ranges shift to

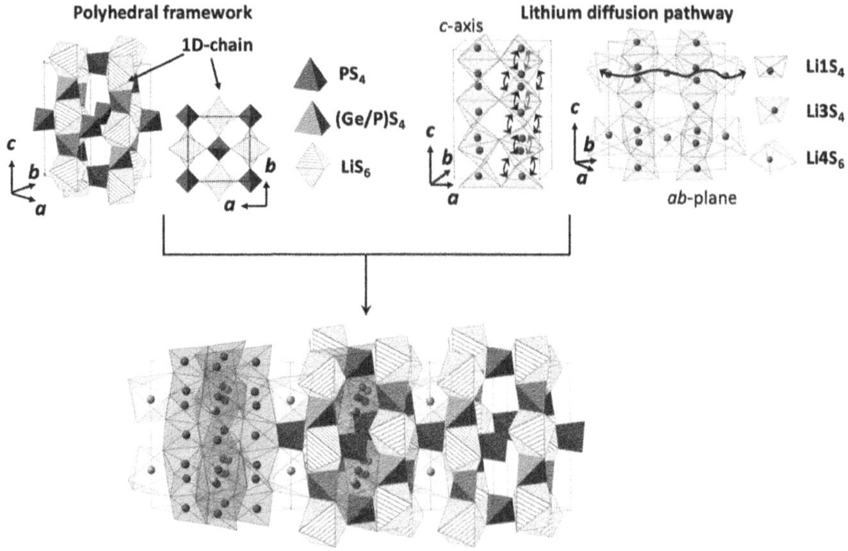

Fig. 4 Crystal structure of the LGPS. Polyhedral framework (upper left) and lithium diffusion pathway (upper right) are illustrated. The merged structure of the framework and lithium units are depicted (bottom). Blue dotted lines depict the size of the single unit cell

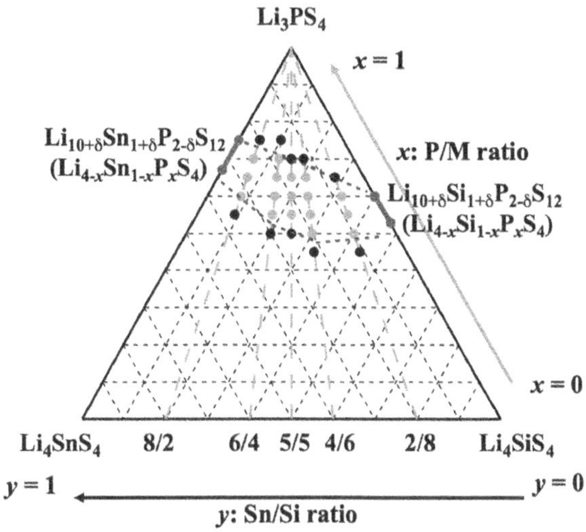

Fig. 5 Quasi-ternary diagram of the Li_3PS_4–Li_4SnS_4–Li_4SiS_4 system. The reported solid-solution ranges of $Li_{10+\delta}Sn_{1+\delta}P_{2-\delta}S_{12}$ ($Li_{4-x}Sn_{1-x}P_xS_4$) and $Li_{10+\delta}Si_{1+\delta}P_{2-\delta}S_{12}$ ($Li_{4-x}Si_{1-x}P_xS_4$) are represented by the green lines [34]. Compositions of the synthesized samples are indicated by black and red points in the diagram. These compositions are described by the formula, $Li_{10+\delta}[Sn_ySi_{1-y}]_{1+\delta}P_{2-\delta}S_{12}$ ($Li_{4-x}[Sn_ySi_{1-y}]_{1-x}P_xS_4$), and the composition variations are indicated by the red lines along the tie lines between Li_3PS_4 and $Li_4[Sn_ySi_{1-y}]S_4$ (the ratio, such as 8/2, represents the ratio of Sn/Si (y/1−y)). The monophasic region is marked between the green dotted lines, and the red points within the region represent the synthesized LGPS-type single phases

a higher P ratio area from Sn/Si = 0/1 to Sn/Si = 1/0. This behavior relates to an increase of the average ionic radius of M^{4+}, which was also confirmed for Si, Ge, and Sn analogues [34], i.e. the solid-solution ranges varied with respect to the higher P content area with the M^{4+} cation in the order of Si, Ge, and Sn. On the other hand, in the examined Si-Sn substitution system, varying the Sn/Si ratio extended the range of the solid solution, which could be due to finely adjusting the average radius of M^{4+} in the $(M/P)S_4$ tetrahedra.

The total ionic conductivities (bulk + grain boundary contributions) of the obtained materials with the LGPS structure were evaluated using the AC-impedance method (see Table 2 for representative samples). All the samples showed high ionic conductivity (over 1 mS cm^{-1}), and relatively high conductive properties were

Table 2 Composition and ionic conductivity of the solid electrolytes synthesized in this study

Sample No.	δ, y, x in $Li_{10+\delta}[Sn_ySi_{1-y}]_{1+\delta}P_{2-\delta}S_{12}$ ($Li_{4-x}[Sn_ySi_{1-y}]_{1-x}P_xS_4$)	σ: pressed pellet (mS cm^{-1})
(i)	$\delta = 0.2, y = 0.8, x = 0.6$	2.50
(ii)	$\delta = 0.5, y = 0.2, x = 0.5$	2.93
(iii)	$\delta = 0.35, y = 0.2, x = 0.55$	3.31 [11.0: sintered pellet]

confirmed by the Si-rich composition. The highest conductivity was observed for the sample with a composition of $Li_{10.35}[Sn_{0.27}Si_{1.08}]P_{1.65}S_{12}$ ($Li_{3.45}[Sn_{0.09}Si_{0.36}]P_{0.55}S_4$) ($y = 0.2$, $\delta = 0.35$, $x = 0.55$). The sintered pellet of this composition exhibited a remarkably high ionic conductivity value of 11 mS cm^{-1}, which is comparable to the original LGPS [10].

For elucidating the structure-conductivity of the $Li_{10+\delta}[Sn_ySi_{1-y}]_{1+\delta}P_{2-\delta}S_{12}$ ($Li_{4-x}[Sn_ySi_{1-y}]_{1-x}P_xS_4$) system, crystal structure analysis was conducted using powder neutron diffraction data.

Three representative LGPS-type materials in the Li–Sn–Si–P–S system were subjected to the structural analysis (see Table 2): (i) a Sn-rich composition with relatively high ionic conductivity, (ii) a Si-rich composition with higher ionic conductivity, and (iii) the Li–Sn–Si–P–S system with the highest ionic conductivity. The absolute value of the ionic conductivity at 298 K increased from sample (i) to (iii). Diffraction data were analyzed by the Rietveld refinement, and were then subjected to the maximum entropy method (MEM) analysis to calculate the nuclear density distributions using crystal structure factors and standard deviations obtained by the Rietveld refinement. The MEM analysis results are illustrated in Fig. 6. Lithium distribution maps were described as a minimum iso-surface level of -0.06 fm Å^{-3} (the negative portion of the scattering amplitude). For all samples, continuous lithium distribution through Li1 and Li3 sites was confirmed, indicating rapid lithium diffusion along the *c*-axis. These observed distributions were comparable to previous results reported both experimentally and computationally [11, 30, 31, 38].

The large atomic displacement parameters of the Li1 and Li3 sites in the Rietveld refinements, which are described elsewhere [37], could correspond to the observed lithium distributions. This unique structural characteristic is also found in superionic conductors such as α-AgI and $Rb_4Cu_{16}I_{7.2}Cl_{12.8}$ [39, 40], which respectively show silver and copper ion conductivities of over 100 mS cm^{-1}. An additional lithium distribution connecting the Li1 and Li4 sites appeared for sample (iii), which showed the highest ionic conductivity. In the case of the lower conductivity samples, the lithium distributions of the 1D chain and Li4 sites were isolated. Therefore, 3D lithium diffusion, which contributes to its high conductivity, may be activated only for sample (iii). Visualized nuclear distribution of lithium by the MEM analysis provided reasonable evidence of high ionic conductivity in the LGPS-type Li–Sn–Si–P–S system. Therefore, the crystal structure analysis is a promising tool for understanding the lithium diffusion phenomena in materials and the design of novel solid ion conductors.

4 Concluding Remarks

Identifying crystalline lithium conductors is strongly desired to meet the various demands required for solid electrolytes employed in solid-state lithium batteries. In this chapter, the discovery of novel sulfides based on crystal structures, known as superionic conductors, is presented. Until the discovery of the $Li_7Ge_3PS_{12}$ phase,

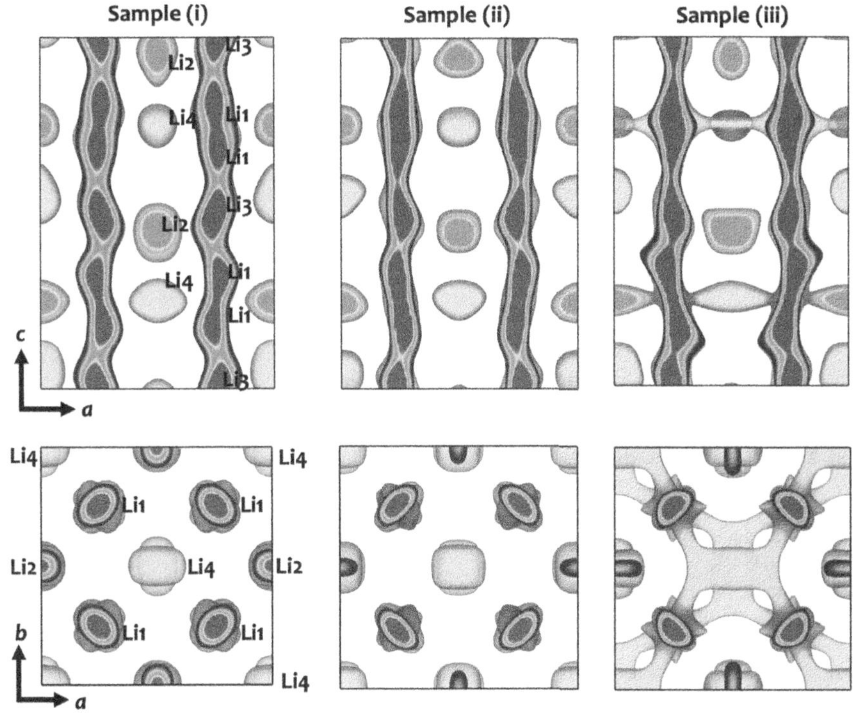

Fig. 6 Nuclear distributions of Li atoms in the LGPS-type $Li_{10+\delta}[Sn_ySi_{1-y}]_{1+\delta}P_{2-\delta}S_{12}$ ($Li_{4-x}[Sn_ySi_{1-y}]_{1-x}P_xS_4$) unit cell at 300 K. The equi-contour surfaces of the lithium nuclear density distribution appear in yellow. Contour maps for slices are shown for (1 0.75 1) (top) and at (1 1 1) (bottom)

argyrodite composition was limited to being Li_7PS_6-based, and anion solid solutions with halogen or selenium were mainly investigated [16, 41]. The study on $Li_7Ge_3PS_{12}$ implied a new class of argyrodite compositions as well as solid solutions with respect to cations such as Ge/P. Indeed, the $Li_{6.6}P_{0.4}Ge_{0.6}S_5I$ argyrodite-type phase [12], which has solid solutions for both cations and anions and exhibits high conductivity, i.e., > 10 mS cm^{-1}, was reported after the discovery of $Li_7Ge_3PS_{12}$. In the case of $Li_{10+\delta}[Sn_ySi_{1-y}]_{1+\delta}P_{2-\delta}S_{12}$, the solid-solution range for the LGPS-type phase was expanded by incorporating multiple kinds of cations such as Sn^{4+}, Si^{4+}, and P^{5+} to the crystal structure, and thereby produced an LGPS-type phase possessing extended Li distribution and excellent ion conductivity over 10 mS cm^{-1}. In both cases of argyrodite- and LGPS-type solid electrolytes, higher conductivity is obtained for a new phase by increasing the chemical species in the crystal structure. This strategy will be effective and practical for future material searches for crystalline ionic conductors.

References

1. Farrington, G. C., Dunn, B. S., & Briant, J. L. (1981). Li^+ and divalent ion conductivity in beta and beta″ alumina. *Solid State Ionics, 3–4*, 405–408.
2. Yung-Fang, Yu Y, & Kummer, J. T. (1967). Ion exchange properties of and rates of ionic diffusion in beta-alumina. *Journal of Inorganic and Nuclear Chemistry, 29*(9), 2453–2475.
3. Hong, H. Y.-P. (1978). Crystal structure and ionic conductivity of $Li_{14}Zn(GeO_4)_4$ and other new Li^+ superionic conductors. *Materials Research Bulletin, 13*(2), 117–124.
4. Lapp, T., Skaarup, S., & Hooper, A. (1983). Ionic conductivity of pure and doped Li_3N. *Solid State Ionics, 11*(2), 97–103.
5. Vincent, C. A. (1987). Polymer electrolytes. *Progress in Solid State Chemistry, 17*(3), 145–261.
6. Kennedy, J. H., Sahami, S., Shea, S. W., & Zhang, Z. (1986). Preparation and conductivity measurements of SiS_2–Li_2S glasses doped with LiBr and LiCl. *Solid State Ionics 18&19, Part 1*, 368–371.
7. Kennedy, J. H., & Yang, Y. (1986). A highly conductive Li^+-Glass system: $(1 - x)$ $(0.4SiS_2$–$0.6Li_2S)$–xLiI. *Journal of the Electrochemical Society, 133*(11), 2437–2438.
8. Zhang, Z., & Kennedy, J. H. (1990). Synthesis and characterization of the B_2S_3–Li_2S, the P_2S_5–Li_2S and the B_2S_3–P_2S_5–Li_2S glass systems. *Solid State Ionics, 38*(3), 217–224.
9. Kanno, R., Hata, T., Kawamoto, Y., & Irie, M. (2000). Synthesis of a new lithium ionic conductor, thio-LISICON–lithium germanium sulfide system. *Solid State Ionics, 130*(1–2), 97–104.
10. Kamaya, N., Homma, K., Yamakawa, Y., Hirayama, M., Kanno, R., Yonemura, M., et al. (2011). A lithium superionic conductor. *Nature Materials, 10*(9), 682–686.
11. Kato, Y., Hori, S., Saito, T., Suzuki, K., Hirayama, M., Mitsui, A., et al. (2016). High-power all-solid-state batteries using sulfide superionic conductors. *Nature Energy, 1*(4), 16030.
12. Kraft, M. A., Ohno, S., Zinkevich, T., Koerver, R., Culver, S. P., Fuchs, T., et al. (2018). Inducing high ionic conductivity in the lithium superionic argyrodites $Li_{6+x}P_{1-x}Ge_xS_5I$ for all-solid-state batteries. *Journal of the American Chemical Society, 140*(47), 16330–16339.
13. Seino, Y., Ota, T., Takada, K., Hayashi, A., & Tatsumisago, M. (2014). A sulphide lithium super ion conductor is superior to liquid ion conductors for use in rechargeable batteries. *Energ Environ Sci, 7*(2), 627–631.
14. Hori, S., Suzuki, K., Hirayama, M., Kato, Y., Kanno, R. (2016). Lithium superionic conductor $Li_{9.42}Si_{1.02}P_{2.1}S_{9.96}O_{2.04}$ with $Li_{10}GeP_2S_{12}$-Type structure in the Li_2S–P_2S_5–SiO_2 pseudoternary system: Synthesis, electrochemical properties, and structure–composition relationships. *Frontiers in Energy Research, 4*(38).
15. Sun, Y., Suzuki, K., Hori, S., Hirayama, M., & Kanno, R. (2017). Superionic conductors: $Li_{10+\delta}[SnySi1{-}y]_{1+\delta}P_{2-\delta}S_{12}$ with a $Li_{10}GeP_2S_{12}$-type structure in the Li_3PS_4–Li_4SnS_4–Li_4SiS_4 quasi-ternary system. *Chemistry of Materials, 29*(14), 5858–5864.
16. Deiseroth, H.-J., Kong, S.-T., Eckert, H., Vannahme, J., Reiner, C., Zaiß, T., et al. (2008). Li_6PS_5X: A class of crystalline Li-rich solids with an unusually high Li^+ mobility. *Angewandte Chemie Int Ed, 47*(4), 755–758.
17. Rao, R. P., & Adams, S. (2011). Studies of lithium argyrodite solid electrolytes for all-solid-state batteries. *Physica Status Solidi (a), 208*(8), 1804–1807.
18. Inoue, Y., Suzuki, K., Matsui, N., Hirayama, M., & Kanno, R. (2017). Synthesis and structure of novel lithium-ion conductor $Li_7Ge_3PS_{12}$. *Journal of Solid State Chemistry, 246*, 334–340.
19. Leube, B. T., Inglis, K. K., Carrington, E. J., Sharp, P. M., Shin, J. F., Neale, A. R., et al. (2018). Lithium transport in $Li_{4.4}M_{0.4}M'_{0.6}S4$ ($M = Al^{3+}$, Ga^{3+}, and $M' = Ge^{4+}$, Sn^{4+}): Combined crystallographic, conductivity, solid state NMR, and computational studies. *Chemistry of Materials, 30*(20), 7183–7200.
20. Kaib, T., Haddadpour, S., Kapitein, M., Bron, P., Schröder, C., Eckert, H., et al. (2012). New lithium chalcogenidotetrelates, LiChT: Synthesis and characterization of the Li^+-conducting Tetralithium ortho-Sulfidostannate Li_4SnS_4. *Chemistry of Materials, 24*(11), 2211–2219.

21. Liu, Z., Fu, W., Payzant, E. A., Yu, X., Wu, Z., Dudney, N. J., et al. (2013). Anomalous high ionic conductivity of nanoporous β-Li_3PS_4. *Journal of the American Chemical Society, 135*(3), 975–978.
22. Mizuno, F., Hayashi, A., Tadanaga, K., Tatsumisago, M. (2005) New, highly ion-conductive crystals precipitated from Li_2S–P_2S_5 Glasses. *Adv Mater, 17*(7), 918–921.
23. Homma, K., Yonemura, M., Kobayashi, T., Nagao, M., Hirayama, M., & Kanno, R. (2011). Crystal structure and phase transitions of the lithium ionic conductor Li_3PS_4. *Solid State Ionics, 182*(1), 53–58.
24. Homma, K., Yonemura, M., Nagao, M., Hirayama, M., Kanno, R. (2010). Crystal structure of high-temperature phase of lithium ionic conductor, Li_3PS_4. *J Phys Soc Jpn, 79*(Suppl. A), 90–93.
25. Kanno, R., & Murayama, M. (2001). Lithium ionic conductor Thio-LISICON: The Li_2S–Ge_2S–P_2S_5 system. *Journal of the Electrochemical Society, 148*(7), A742–A746.
26. Yamane, H., Shibata, M., Shimane, Y., Junke, T., Seino, Y., Adams, S., et al. (2007). Crystal structure of a superionic conductor, $Li_7P_3S_{11}$. *Solid State Ionics, 178*(15–18), 1163–1167.
27. Wang, Y., Richards, W. D., Ong, S. P., Miara, L. J., Kim, J. C., Mo, Y., et al. (2015). Design principles for solid-state lithium superionic conductors. *Nature Materials, 14*(10), 1026–1031.
28. Huang, W., Yoshino, K., Hori, S., Suzuki, K., Yonemura, M., Hirayama, M., et al. (2019). Superionic lithium conductor with a cubic argyrodite-type structure in the Li–Al–Si–S system. *Journal of Solid State Chemistry, 270,* 487–492.
29. Minafra, N., Culver, S. P., Krauskopf, T., Senyshyn, A., & Zeier, W. G. (2018). Effect of Si substitution on the structural and transport properties of superionic Li-argyrodites. *J Mater Chem A, 6*(2), 645–651.
30. Kwon, O., Hirayama, M., Suzuki, K., Kato, Y., Saito, T., Yonemura, M., et al. (2015). Synthesis, structure, and conduction mechanism of the lithium superionic conductor $Li_{10+\delta}Ge_{1+\delta}P_{2-\delta}S_{12}$. *J Mater Chem A, 3*(1), 438–446.
31. Hori, S., Taminato, S., Suzuki, K., Hirayama, M., Kato, Y., & Kanno, R. (2015). Structure-property relationships in lithium superionic conductors having a $Li_{10}GeP_2S_{12}$-type structure. *Acta Crystallographica Section B, 71*(6), 727–736.
32. Kuhn, A., Gerbig, O., Zhu, C., Falkenberg, F., Maier, J., & Lotsch, B. V. (2014). A new ultrafast superionic Li-conductor: ion dynamics in $Li_{11}Si_2PS_{12}$ and comparison with other tetragonal LGPS-type electrolytes. *Phys Chem Phys, 16*(28), 14669–14674.
33. Bron, P., Johansson, S., Zick, K., Schmedt auf der Günne, J., Dehnen, S., & Roling, B. (2013). $Li_{10}SnP_2S_{12}$: An affordable lithium superionic conductor. *Journal of the American Chemical Society, 135*(42), 15694–15697.
34. Hori, S., Suzuki, K., Hirayama, M., Kato, Y., Saito, T., Yonemura, M., et al. (2014). Synthesis, structure, and ionic conductivity of solid solution, $Li_{10+\delta}M_{1+\delta}P_{2-\delta}S_{12}$ (M = Si, Sn). *Faraday Discussions, 176,* 83–94.
35. Kato, Y., Saito, R., Sakano, M., Mitsui, A., Hirayama, M., & Kanno, R. (2014). Synthesis, structure and lithium ionic conductivity of solid solutions of $Li_{10}(Ge_{1-x}M_x)P_2S_{12}$ (M = Si, Sn). *Journal of Power Sources, 271,* 60–64.
36. Hori, S., Kato, M., Suzuki, K., Hirayama, M., & Kato, Y. (2015). Kanno R (2015) Phase diagram of the Li_4GeS_4–Li_3PS_4 quasi-binary system containing the superionic conductor $Li_{10}GeP_2S_{12}$. *Journal of the American Ceramic Society, 98*(10), 3352–3360.
37. Inagaki, M., Suzuki, K., Hori, S., Yoshino, K., Matsui, N., Yonemura, M., Hirayama, M., Kanno, R. Conduction mechanism of $Li_{10}GeP_2S_{12}$-type lithium superionic conductors in a Li–Sn–Si–P–S System. *Chem Mater, 31*(9), 3485–3490.
38. Mo, Y., Ong, S. P., & Ceder, G. (2012). First principles study of the $Li_{10}GeP_2S_{12}$ lithium super ionic conductor material. *Chemistry of Materials, 24*(1), 15–17.
39. Wright, A. F., & Fender, B. E. F. (1977). The structure of superionic compounds by powder neutron diffraction. I. Cation distribution in α-AgI. *Journal of Physics C: Solid State Physics, 10*(13), 2261.
40. Izumi, F., & Momma, K. (2011). Three-dimensional visualization of electron- and nuclear-density distributions in inorganic materials by MEM-based technology. *IOP Conference Series: Materials Science and Engineering, 18*(2), 022001.

41. Kong, S. T., Gun, O., Koch, B., Deiseroth, H. J., Eckert, H., & Reiner, C. (2010). Structural characterisation of the Li argyrodites Li_7PS_6 and Li_7PSe_6 and their solid solutions: Quantification of site preferences by MAS-NMR spectroscopy. *Chemistry, 16*(17), 5138–5147.

Glass Electrolyte

Akitoshi Hayashi

Abstract Sulfide glasses are promising solid electrolytes for all-solid-state batteries because of their excellent conductivity and good formability. These glasses are prepared via mechanochemistry as room temperature process. Crystallization of a metastable phase from glass electrolytes often increases conductivity. Both electrochemical stability and chemical stability are improved by selecting compositions of sulfide electrolytes.

Keywords Glass electrolyte · Sulfide electrolyte · Glass-ceramic · Mechanochemistry

1 Introduction

Sulfide glasses have excellent characteristics as solid electrolytes [1, 2], some of which are as follows: (1) high ionic conductivity in a wide composition range, (2) excellent mechanical properties of good formability and appropriate elastic modulus, and (3) stabilization of metastable phases exhibiting superionic conductivity via the crystallization process. This chapter describes the synthesis and characterization of sulfide glass electrolytes.

2 Synthesis Procedure

To synthesize glass, the melt quenching method is generally used. In this method, a glass raw material is melted at high temperature, following which the melt is rapidly cooled to produce glass. As the lithium concentration in the raw materials increases, crystallization tends to occur when the melt is cooled. Therefore, to obtain glasses with high lithium content, rapid quenching of the melt is vital. Sulfide glasses in

A. Hayashi (✉)
Osaka Prefecture University, Sakai, Osaka, Japan
e-mail: hayashi@chem.osakafu-u.ac.jp

K. Kanamura (ed.), *Next Generation Batteries*,
https://doi.org/10.1007/978-981-33-6668-8_5

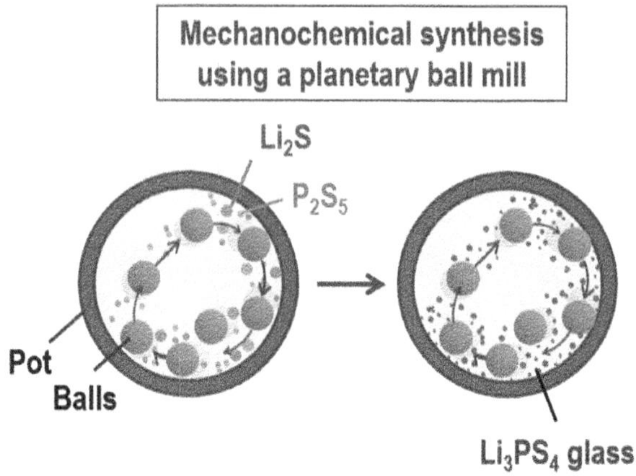

Fig. 1 Schematic of mechanochemical synthesis of Li_3PS_4 glass from sulfide starting materials of Li_2S and P_2S_5 using a planetary ball mill

the binary system Li_2S-P_2S_5 are prepared by the quenching method, but the melting reaction must be carried out in sealed quartz tubes because of the high vapor pressure of sulfide starting materials at high temperatures. Furthermore, glasses with high lithium content can be produced using the mechanochemical method. In this method, the chemical reaction proceeds using mechanical energy. As schematically shown in Fig. 1, Li_2S-P_2S_5 glasses are prepared at room temperature and normal pressure from the mixture of crystalline Li_2S and P_2S_5 powders by mechanochemical synthesis using a planetary ball mill device. It is possible to produce $75Li_2S{\cdot}25P_2S_5$ (mol%, Li_3PS_4) and $75Li_2S{\cdot}25Al_2S_3$ (Li_3AlS_3) glasses with ortho compositions that are generally difficult to synthesize even by rapid quenching.

3 Conductive and Mechanical Properties

The electrical conductivities at 25 °C of sulfide glasses are summarized in Table 1 [3–15]. These conductivities are for a green compact obtained by cold-pressing glass powder synthesized by the mechanochemical method at room temperature. Sulfide glass electrolytes exhibit good formability (or ductility). With an increase in molding pressure, the relative density of green compacts increases gradually because of the decreasing grain boundaries and voids in the compacts. Sulfide glasses are densified by cold pressing without heat treatment; this densification phenomenon is called "room-temperature pressure sintering" [16].

Li_3PS_4 glass has a higher room temperature conductivity (10^{-4} S cm^{-1}) than Li_3AlS_3 glass. Moreover, Li_2S-P_2S_5 glasses, including Li_3PS_4, not only comprise relatively inexpensive elements but also have high ionic conductivity and good deformability. Thus, they are expected to act as a base composition for application in all-solid-state batteries. When LiI or $LiBH_4$ are added to Li_2S-P_2S_5 glasses, the conductivity increases by an order of magnitude, exceeding 10^{-3} S cm^{-1}. These glasses also have an excellent tolerance to reduction by Li metal; therefore, they

Table 1 Room temperature conductivity for sulfide glass, glass-ceramic, and crystal electrolytes

Composition	Conductivity at 25°C ($S\ cm^{-1}$)	Classification	Year	References
$Li_{9.54}Si_{1.74}P_{1.44}S_{11.7}Cl_{0.3}$ (LGPS)	2.5×10^{-2}	Crystal	2016	[3]
$Li_{10}GeP_2S_{12}$ (LGPS)	1.2×10^{-2}	Crystal	2011	[4]
Li_6PS_5Cl (argyrodite)	1.3×10^{-3}	Crystal	2012	[5]
$70Li_2S{\cdot}30P_2S_5$ ($Li_7P_3S_{11}$)	1.7×10^{-2}	Glass-ceramic	2014	[6]
$63Li_2S{\cdot}27P_2S_5{\cdot}10LiBr$	8.4×10^{-3}	Glass-ceramic	2014	[7]
$60Li_2S{\cdot}25P_2S_5{\cdot}10Li_3N$	1.4×10^{-3}	Glass-ceramic	2017	[8]
$80Li_2S{\cdot}20P_2S_5$	1.3×10^{-3}	Glass-ceramic	2006	[9]
$67Li_2S{\cdot}33SnS_2$ (Li_4SnS_4)	1.1×10^{-4}	Glass-ceramic	2018	[10]
$54Li_3PS_4{\cdot}46LiI$	1.8×10^{-3}	Glass	2018	[11]
$50Li_2S{\cdot}17P_2S_5{\cdot}33LiBH_4$	1.6×10^{-3}	Glass	2013	[12]
$75Li_2S{\cdot}25P_2S_5$ (Li_3PS_4)	1.1×10^{-4}	Glass	2001	[13]
$75Li_2S{\cdot}25Al_2S_3$ (Li_3AlS_3)	3.4×10^{-5}	Glass	2004	[14]
$75Li_2S{\cdot}25Sb_2S_5$ (Li_3SbS_4)	1.1×10^{-6}	Glass	2019	[15]

exhibit long-term electrochemical Li plating/stripping cycles [11]. Further details are described in Chap. 13.

The elastic modulus of solid electrolytes is important for maintaining good contact, even when the volume of the active material changes. Young's moduli for densified Li_2S-P_2S_5 glass electrolytes are initially 18–25 GPa, and decrease gradually with the decrease of Li_2S content and the addition of LiI [16–18]. These sulfide glasses have a Young's modulus between those of the oxide glass and the organic polymer. The elastically deformable sulfide electrolyte is expected to function as a buffer according to the volume change of the active material during the charge–discharge cycles.

Table 1 also includes the conductivities for the LGPS crystal ($Li_{9.54}Si_{1.74}P_{1.44}S_{11.7}Cl_{0.3}$), with the highest conductivity of 2.5×10^{-2} $S\ cm^{-1}$, the argyrodite-type crystal Li_6PS_5Cl, and several glass-ceramics, whose conductivities are increased by crystallizing glasses. A metastable phase, which is difficult to obtain by ordinary solid-phase reaction, often precipitates in glass-ceramics, and this crystalline phase has high conductivity, which is thought to have increased conductivity. When the glass at the composition $70Li_2S{\cdot}30P_2S_5$ (mol%) crystallizes at 280 °C, a high temperature phase of $Li_7P_3S_{11}$ with the same composition precipitates, and the resulting glass-ceramic has a high conductivity of 1.7×10^{-2} $S\ cm^{-1}$at 25 °C. Meanwhile, when the heat treatment temperature is 550 °C, the $Li_7P_3S_{11}$ phase disappears, multiple thermodynamic stable phases including $Li_4P_2S_6$ precipitate, and the conductivity decreases by several orders of magnitude. Therefore, the high conductivity of $Li_7P_3S_{11}$ is responsible for increasing the conductivity of glass-ceramics. As another example, Fig. 2 shows the XRD patterns of $67Li_2S{\cdot}33SnS_2$ (Li_4SnS_4) glass obtained by the mechanochemical method and the

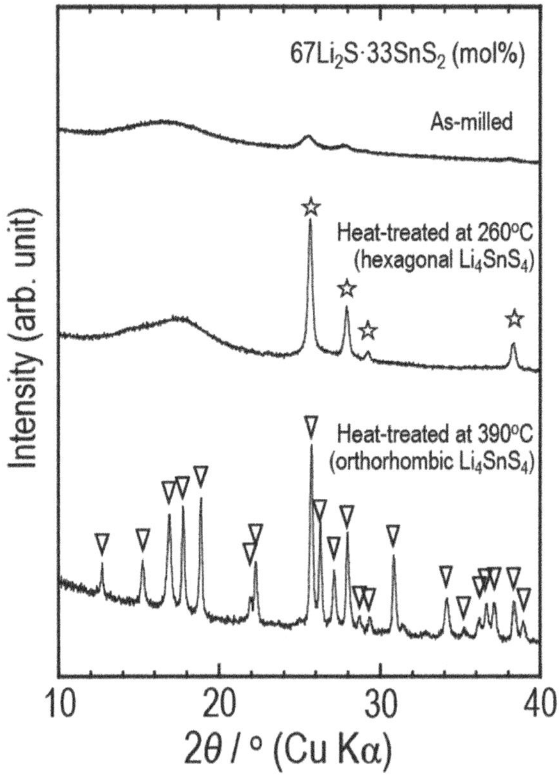

Fig. 2 XRD patterns of $67Li_2S·33SnS_2$ (Li_4SnS_4) glass obtained by the mechanochemical method and the glass-ceramics obtained by subsequent heat treatment at 260 °C or 390 °C

glass-ceramics obtained by the subsequent heat treatment. When the glass is heat-treated at 260 °C for crystallization, a new crystalline phase with hexagonal structure precipitates and the conductivity increases. Furthermore, orthorhombic Li_4SnS_4, a thermodynamically stable phase, precipitates in glass-ceramics heat-treated at a higher temperature of 390 °C. Hexagonal $Li_4SnS_4$is expected to be developed as a new mother structure as an ionic conductor. In addition, in the glass-ceramics obtained by adding Li_3N or LiBr to the Li_2S-P_2S_5 system, a new crystalline phase is precipitated, and the prepared glass-ceramics shows a high conductivity of 10^{-3} S cm^{-1} or more at 25 °C. Considering glass as a precursor for the production of a new metastable phase, researches on the optimization of glass compositions and crystallization processes are expected to develop new solid electrolytes with higher ionic conductivity.

4 Stability Against Moisture

A drawback of sulfide electrolytes is their poor chemical stability in air. To reduce the processing cost of the electrolytes, it is desirable to increase their moisture resistance. Based on our early experiments, the selection of compositions in sulfide electrolytes provides moderate stability in air to sulfide electrolytes. The chemical stability of sulfide glass electrolytes in the binary system Li_2S-P_2S_5 is examined by exposing them to an air atmosphere. The amount of hydrogen sulfide generated from the electrolytes varies depending on the composition and is minimum at the $75Li_2S{\cdot}25P_2S_5$ (Li_3PS_4) composition [19].

When exposed to the atmosphere under more severe conditions with a relative humidity of 70%, Li_3PS_4 glass gradually generates hydrogen sulfide, but in the electrolyte obtained by the introduction of a part of nitrogen or oxygen into the sulfide glass using Li_3N [8] or Li_2O [20], this generation of hydrogen sulfide is suppressed. The $60Li_2S{\cdot}25P_2S_5{\cdot}10Li_3N$ glass-ceramic has a high conductivity of 10^{-3} S cm^{-1}, as listed in Table 1; thus, the glass-ceramic is considered an electrolyte with both high conductivity and moderate chemical stability. Using the preferred M_xO_y(M_xO_y:Fe_2O_3, ZnO, and Bi_2O_3) with a larger negative Gibbs energy change (ΔG) in the reaction with hydrogen sulfide is also effective in decreasing the total amount of hydrogen sulfide [21].

Another approach is the use of sulfide compositions based on the hard and soft acids and bases theory. For instance, Li_4SnS_4 and Li_3SbS_4 are more effective in suppressing hydrogen sulfide than Li_3PS_4, as shown in Fig. 3. In particular, Li_3SbS_4 glass is expected to be an electrolyte with excellent safety because it hardly generates hydrogen sulfide. However, since the room temperature conductivity is as low as 10^{-6} S cm^{-1}, as listed in Table 1, increasing the conductivity is a future challenge.

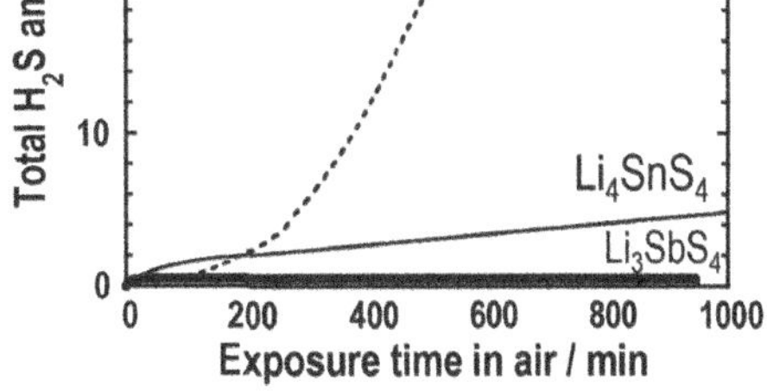

Fig. 3 Total hydrogen sulfide amount generated from Li_3SbS_4, Li_4SnS_4, and Li_3PS_4 glass powders after exposure in air with relative humidity of 70%

5 Concluding Remarks

The mechanochemical synthesis of sulfide glass electrolytes is useful for directly obtaining powder materials and enlarging the glass-forming region to include high lithium-ion content. Sulfide glass-based electrolytes with excellent conductivity, formability, electrochemical stability, and chemical stability were achieved. Further investigations will be conducted on balancing these properties of electrolytes.

References

1. Tatsumisago, M., & Hayashi, A. (2014). *Int J Appl Glass Sci, 5,* 226.
2. Hayashi, A., et al. (2016). *Front Energy Res, 4,* 25.
3. Kato, Y., et al. (2016). *Nat. Energy, 1,* 16030.
4. Kamaya, N., et al. (2011). *Nature Materials, 10,* 682.
5. Boulineau, S., et al. (2012). *Solid State Ionics, 221,* 1.
6. Seino, Y., et al. (2014). *Energy & Environmental Science, 7,* 627.
7. Ujiie, S., et al. (2014). *Mater Renew Sustain Energy, 3,* 18.
8. Fukushima, A., et al. (2017). *Solid State Ionics, 304,* 85.
9. Mizuno, F., et al. (2006). *Solid State Ionics, 177,* 2721.
10. Kanazawa, K., et al. (2018). *Inorganic Chemistry, 57,* 9925.
11. Suyama, M., et al. (2018). *Electrochimica Acta, 286,* 158.
12. Yamauchi, A., et al. (2013). *Journal of Power Sources, 244,* 707.
13. Hayashi, A., et al. (2001). *Journal of the American Ceramic Society, 84,* 477.
14. Hayashi, A., et al. (2004). *Journal of the Ceramic Society of Japan, 112,* S695.
15. Kimura, T., et al. (2019). *Solid State Ionics, 333,* 45.
16. Sakuda, A., et al. (2013). *Sci Rep, 3,* 2261.
17. Kato, A., et al. (2018). *ACS Appl. Energy Mater., 1,* 1002.
18. Kato, A., et al. (2018). *Journal of the Ceramic Society of Japan, 126,* 719.
19. Muramatsu, H., et al. (2011). *Solid State Ionics, 182,* 116.
20. Ohtomo, T., et al. (2013). *Electrochemistry, 81,* 428.
21. Hayashi, A., et al. (2013). *J Mater Chem A, 1,* 6320.

Suspension Process

Huu Huy Phuc Nguyen and Atsunori Matsuda

Abstract Sulfide-based solid electrolytes were conventionally prepared by solid-state reaction methods since long time ago and the introduction of liquid phase synthesis which employing organic solvents as reaction media was just reported recently in 2012 and 2013 [1, 2]. The first two papers described the synthesis of β-Li_3PS_4 using tetrahydrofuran (THF) and thio-LISICON $Li_{3.25}Ge_{0.25}P_{0.75}S_4$ using hydrazine as mediated solvents. The preparation of Li_3PS_4 using THF underwent the *suspension process* whilst hydrazine dissolved all the raw materials and product of the reaction which resulted in a *solution process*. In this section, we will focus on the synthesis of some typical sulfide-based solid electrolytes using *suspension process*.

Keywords Liquid phase · Suspension · Ethyl propionate · $Li_7P_3S_{11}$ · Li_3PS_4 · $Li_7P_2S_8I$

1 Preparation of $Li_7P_3S_{11}$

The suspension of solid electrolyte precursor was prepared by liquid-phase shaking method using ethyl propionate as organic solvent to mediate the reaction between Li_2S and P_2S_5. Li_2S (99.9%) and P_2S_5 (99%) were purchased from Mitsuwa and Merck, respectively, and used without purification. Ethyl propionate (EP), obtained from Aldrich, was dehydrated with 3Å molecular sieves prior to use. In a typical batch, 1 g of Li_2S, P_2S_5 (molar ratio Li_2S:P_2S_5 = 7:3), 4 mm zirconia balls (30 g) and EP (10 mL) were mixed in a 45 mL centrifugation tube (polypropylene, Labcon) and shaken in a dry Ar atmosphere for 5 h at 1500 rpm with an amplitude of about 1 cm. Thus the obtained suspension was evacuated at room temperature for 2 h and then at 190 °C for 1 h at low pressure (using a rotary pump). The as-dried precursor was further heat-treated at 230, 240 or 250 °C to obtain the $Li_7P_3S_{11}$ solid electrolyte.

H. H. P. Nguyen · A. Matsuda (✉)
Toyohashi University of Technology, Toyohashi, Japan
e-mail: matsuda@ee.tut.ac.jp

H. H. P. Nguyen
e-mail: nguyen.huu.huy.phuc.hr@tut.jp

K. Kanamura (ed.), *Next Generation Batteries*,
https://doi.org/10.1007/978-981-33-6668-8_6

Structure of the prepared powder samples was characterized using X-ray diffraction (XRD; Ultima IV, Rigaku), and Raman spectroscopy (NRS-3100, Jasco). Samples were sealed in special holders in an Ar-filled glove box to protect them from exposure to humidity prior to their characterization by XRD and Raman spectroscopy. The temperature dependences of the total conductivity of the prepared samples were investigated using alternating-current impedance spectroscopy (SI 1260, Solatron) from 1 MHz to 10 Hz in a dry Ar flow. The samples for impedance measurements were prepared by uniaxial pressing into pellets of about 1.0 mm in thickness and about 10.0 mm in diameter at a pressure of 70 MPa (at 230, 240 or 250 °C). The prepared pellet was placed in a holder made from PEEK with two stainless steel rods used as blocking electrodes. The cell was then placed in an Ar stream in a glass tube for the temperature dependence measurements. The temperature was gradually increased from room temperature to 130 °C. The sample was held at each temperature for 1 h prior to the impedance measurement.

Figure 1 shows the XRD patterns of raw materials and powder samples obtained at different steps in $Li_7P_3S_{11}$ preparation procedure. The pattern of the powder dried at room temperature exhibited unknown crystal structure without any characteristic of the starting materials. This result indicated that Li_2S and P_2S_5 were completely consumed during the reaction time proposed in this study. Further evacuation of this sample at 190 °C resulted in the formation of amorphous structure due to elimination of ethyl propionate solvent. The samples received after heat treatment at 230 and 240 °C possessed peaks of β-Li_3PS_4 [2]. Upon treating the received powder at 250 °C, $Li_7P_3S_{11}$ was detected together with $Li_4P_2S_6$ and the disappearance of β-Li_3PS_4 [3]. It should be noticed that $Li_7P_3S_{11}$ is composed of Li_3PS_4 and $Li_4P_2S_7$ in 1:1 of molar ratio; thus its formation and disappearance of β-Li_3PS_4 at 250 °C revealed the reaction between β-Li_3PS_4 and amorphous $Li_4P_2S_7$ at this temperature. Furthermore,

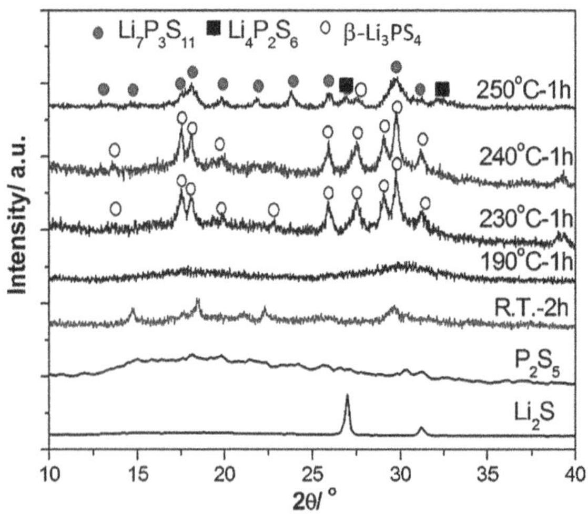

Fig. 1 XRD patterns of raw materials and powder samples obtained at different steps in $Li_7P_3S_{11}$ preparation procedure

the decomposition of $Li_4P_2S_7$ into $Li_4P_2S_6$ and S also took place at this temperature in parallel with the formation of $Li_7P_3S_{11}$.

Figure 2 shows the Raman spectra of raw materials and powder samples formed at different steps in $Li_7P_3S_{11}$ preparation procedure. Raman spectrum of Li_2S exhibited one sharp peak at about 370 cm^{-1} which disappeared in the spectra of the prepared samples. This observation proved that Li_2S was completely expended during the reaction time, which was in agreement with the XRD result. The sample obtained at room temperature had peaks located at 410, 420 and shoulders lying from 2900 to 3000 cm^{-1} indicating the existence of PS_4^{3-}, $P_2S_7^{4-}$ and CH_3 group of ethyl propionate [3, 4]. The samples obtained after drying at 190, 230 and 240 °C showed only peaks of PS_4^{3-}, $P_2S_7^{4-}$ ions; thus ethyl propionate was considered to be eliminated from the samples. Peaks of PS_4^{3-}, $P_2S_7^{4-}$ ions were observed to be slightly shifted as compared with those of the sample being dried at room temperature. This observation might be originated from the interaction among Li^+ ions and ethyl propionate molecules. It should be noticed that both PS_4^{3-} and $P_2S_7^{4-}$ ions were detected by Raman spectroscopy while only β-Li_3PS_4 was observed using XRD. This fact illustrated that $Li_4P_2S_7$ existed in the received samples in the amorphous form. PS_4^{3-}, $P_2S_7^{4-}$ and $P_2S_6^{4-}$ ions were detected in sample being heat-treated at 250 °C, which was also in agreement with the detection of both $Li_7P_3S_{11}$ and $Li_4P_2S_6$ in XRD results.

Figure 3 illustrated the temperature dependence of ionic conductivity of the pelletized samples molded at 240 and 250 °C. Both of the samples exhibited high ionic conductivity at 25 °C, over 1.0×10^{-3} Scm^{-1}; however, those values were still lower than that of glass-ceramic $Li_7P_3S_{11}$ as reported elsewhere. The high ionic conductivity observed from those samples was in comparable with the values obtained from $Li_7P_3S_{11}$ prepared using acetonitrile [5–7]. As a consequence of $Li_7P_3S_{11}$ formation,

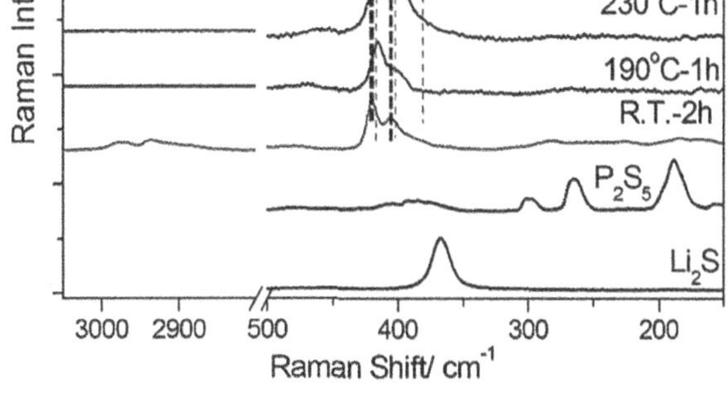

Fig. 2 Raman spectra of raw materials and powder samples obtained at different steps in $Li_7P_3S_{11}$ preparation procedure

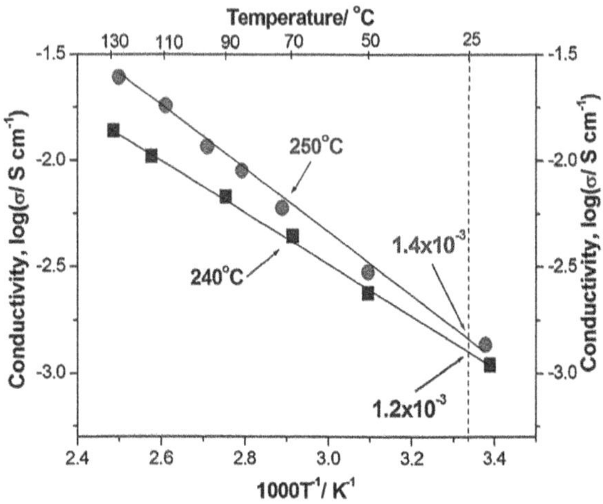

Fig. 3 Temperatures dependence of ionic conductivity of $Li_7P_3S_{11}$ obtained at different temperatures

the conductivity of sample prepared at 250 °C was higher than that of the sample prepared at 240 °C in the measurement temperature range.

2 Preparation of β-Li_3PS_4

The suspension of solid electrolyte precursor was prepared by conventional stirring method using ethyl propionate as organic solvent to mediate the reaction between Li_2S and P_2S_5. Li_2S (99.9%) and P_2S_5 (99%) were purchased from Mitsuwa and Merck, respectively, and used without purification. Ethyl propionate (EP), obtained from Aldrich, was dehydrated with 3Å molecular sieves prior to use. In a typical batch, 1 g of Li_2S, P_2S_5 (molar ratio Li_2S:P_2S_5 = 3:1) and EP (10 mL) were mixed in a 45 mL screw bin and stirred in a dry Ar atmosphere for 6 h at 500 rpm. The thus obtained suspension was evacuated at room temperature for 2 h and then at 170 °C for 1 h at low pressure (using a rotary pump). For comparison, Li_3PS_4 was also prepared by the liquid-phase shaking method as reported elsewhere [8].

Structure of the prepared powder samples was characterized using X-ray diffraction (XRD; Ultima IV, Rigaku). Samples were sealed in special holders in an Ar-filled glove box to protect them from exposure to humidity prior to their characterization by XRD. Morphology of the prepared Li_3PS_4 was observed using scanning electron microscopy (SEM; S4800, Hitachi). The temperature dependences of the total conductivity of the prepared samples were investigated using alternating-current impedance spectroscopy (SI 1260, Solatron) from 1 MHz to 10 Hz in a dry Ar flow. The samples for impedance measurements were prepared by uniaxial pressing into pellets of about 1.0 mm in thickness and about 10.0 mm in diameter at a pressure of 330 MPa (at room temperature). The prepared pellet was placed in a holder made from PEEK with two stainless steel rods used as blocking electrodes. The cell was

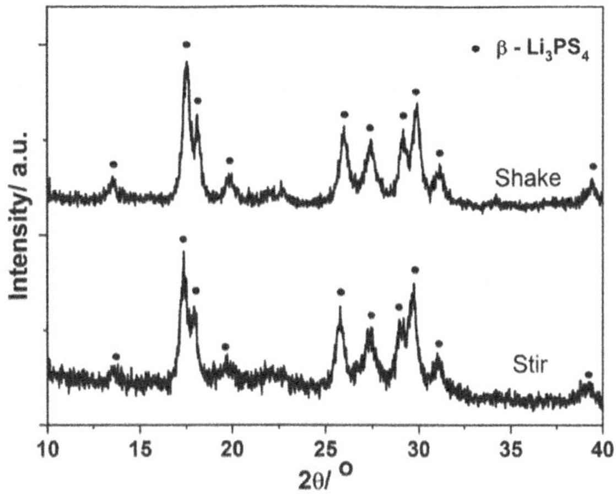

Fig. 4 XRD patterns of Li_3PS_4 prepared by either liquid-phase shaking method or conventional stirring method

then placed in an Ar stream in a glass tube for the temperature dependence measurements. The temperature was gradually increased from room temperature to 150 °C. The sample was held at each temperature for 1 h prior to the impedance measurement.

Figure 4 illustrated the XRD patterns of β-Li_3PS_4 prepared using either conventional stirring method or recently proposed liquid-phase shaking method [2]. The two patterns exhibited resemble features of the beta phase of Li_3PS_4 and no distinct difference was observed. In addition, peaks of Li_2S were undetected in the sample obtained using conventional stirring method. This result proved that β-Li_3PS_4 could also be synthesized by conventional stirring method using ethyl propionate as reaction medium.

Figure 5 shows SEM images of Li_3PS_4 prepared by either conventional stirring method (a) or liquid-phase shaking method (b) and temperatures dependence of ionic conductivities of the synthesized Li_3PS_4 (c). Despite of having same β-Li_3PS_4 structure as pointed out by XRD result, morphologies of the two samples were different. β-Li_3PS_4 prepared by liquid-phase shaking method possessed ribbon-like morphology whilst large flake-shaped β-Li_3PS_4 was received from conventional method. The difference in morphology might be originated from the effect of zirconia balls employed in liquid-phase shaking method. In addition, nearly no difference in temperature dependence of ionic conductivity of the two samples was detected in this study. The ionic conductivity measurement result seemed to be in consistent with XRD results, which revealed the similar structure of both samples. This study also showed that the observed morphology had nearly no impact on the ionic conductivities of the proposed samples.

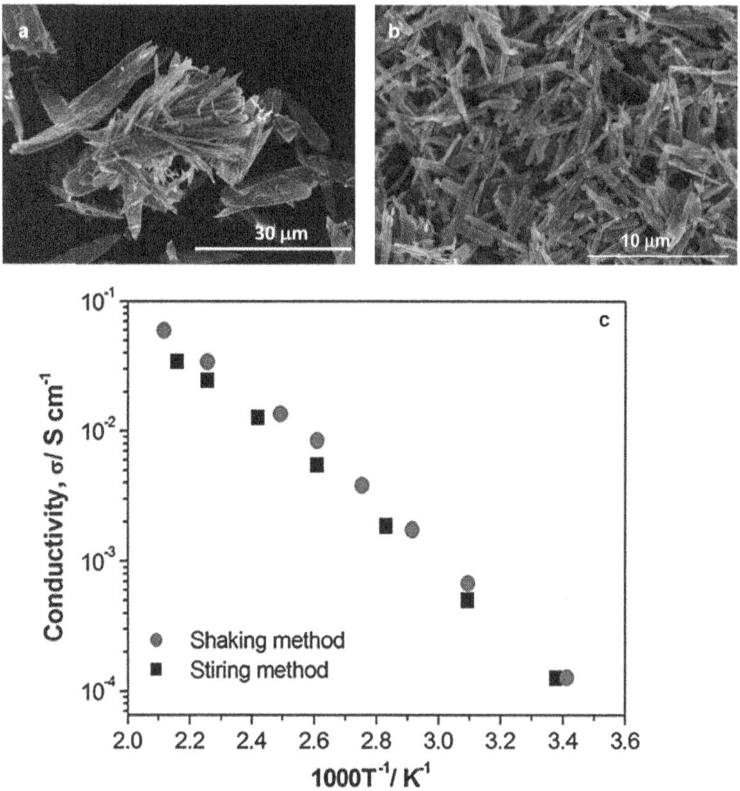

Fig. 5 SEM images of Li_3PS_4 prepared by either conventional stirring method **a** or liquid-phase shaking method **b** and temperatures dependence of ionic conductivities of the synthesized Li_3PS_4 **c**

3 Preparation of $Li_7P_2S_8I$

Ethyl propionate (EP) (99%, Aldrich) was dehydrated by 3Å molecular sieve prior to using. Li_2S (99.9%, Mitsuwa) and P_2S_5 (99%, Merck) were used without any purification process. In a typical trial, 1 g of Li_2S, P_2S_5 (molar ratio of 3:1 to form Li_3PS_4) and LiI (molar ratio $2Li_3PS_4$: 1LiI) were put into a 50 ml screw bin together with 20 ml of EP and seal with tape in a Argon-filled glove box. The prepared mixture was then taken out of glove box and sonicated (US) for 5 h using commercialized ultrasonic cleansing bath (AS ONE VS-100 III) as illustrated in Fig. 6. The resultant mixture was then evacuated at room temperature and 170 °C using a rotary pump to obtain $Li_7P_2S_8I$. Structure of the dried samples was then characterized with XRD (ULTIMA IV, Rigaku) and ionic conductivity was measured using AC Impedance Spectroscopy (Solatron SI 1260).

Figure 7 illustrated the XRD patterns of starting materials (Li_2S, P_2S_5, LiI), β-Li_3PS_4 and $Li_7P_2S_8I$ prepared in this study. No trace of either Li_2S or LiI was detected

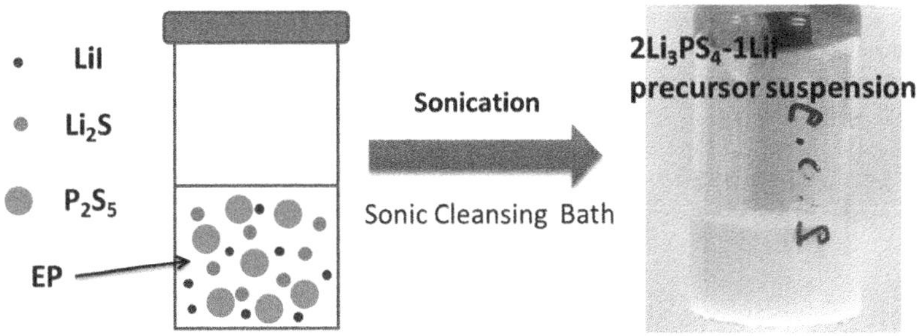

Fig. 6 Schematic described the preparation of $2Li_3PS_4$-LiI precursor suspension

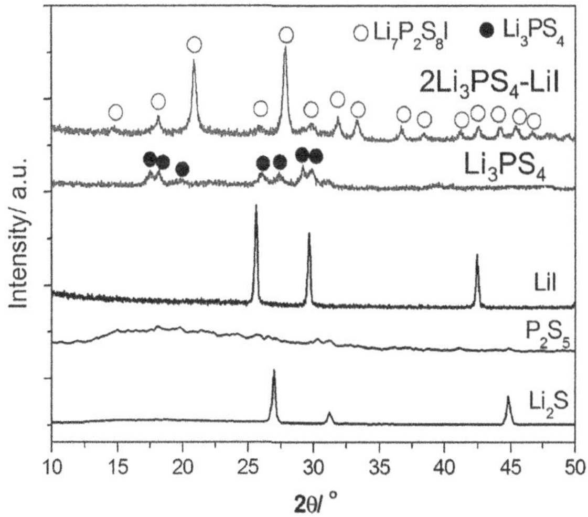

Fig. 7 XRD patterns of starting materials, β-Li_3PS_4 and $Li_7P_2S_8I$ prepared in this study

in XRD pattern and all the observed peaks could be assigned to $Li_7P_2S_8I$ as reported previously [9]. However, peak intensity obtained from this study was much lower than those of reported papers. The reason still remained to be investigated but it might have relation to the properties of solvent employed in this study, ethyl propionate, instead of the reported acetonitrile. In addition, dissolved LiI and precursor of Li_3PS_4 co-existed in the as-received suspension instead on precursor of $Li_7P_2S_8I$. Upon solvent elimination, LiI was introduced into Li_3PS_4 to form $Li_7P_2S_8I$ structure; thus, the maximum molar ratio of LiI that could be inserted into Li_3PS_4 should be depended on solvent structures. Indeed, Li_4PS_4I ($2Li_3PS_4$: 2LiI) was discovered when dimethoxy ethane was employed as solvent instead of acetonitrile; whilst, LiI remaining was observed when Li_4PS_4I was synthesized using either acetonitrile or ethyl propionate and only $Li_7P_2S_8I$ was thus obtained [10]. It also worth noticed that precursor suspension of $Li_7P_2S_8I$ was successfully prepared in just only about 5 h using the commercialized ultrasonic cleansing bath, which was much faster than the reported 24 h [9].

Fig. 8 Temperatures dependence of ionic conductivity of $Li_7P_2S_8I$ obtained in this study

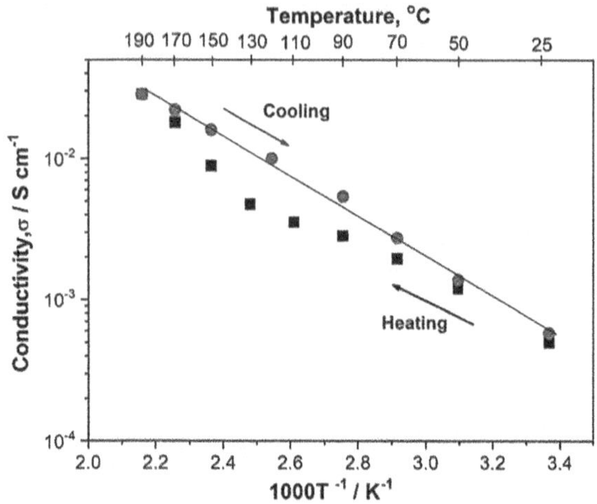

Figure 8 depicted the temperature dependence of ionic conductivity of $Li_7P_2S_8I$ obtained in this study. The conductivity was measured twice: in heating and cooling processes. In the heating process, the conductivity at room temperature (25 °C) was estimated to be about 4.8×10^{-4} Scm^{-1} but the conductivity did not obey Arrhenius law. Upon cooling down from 190 °C, the conductivity at room temperature was slightly increased to 5.5×10^{-4} Scm^{-1}. This observation could be explained as the improvement of total conductivity due to the elimination of the absorbed solvent at high temperature.

References

1. Wang, Y., Liu, Z., Zhu, X., Tang, Y., & Huang, F. (2013). *Journal of Power Sources, 224,* 225–229.
2. Liu, Z., Fu, W., Payzant, E. A., Yu, X., Wu, Z., Dudney, N. J., et al. (2013). *Journal of the American Chemical Society, 135,* 975–978.
3. Mizuno, F., Hayashi, A., Tadanaga, K., & Tatsumisago, M. (2005). *Advanced Materials, 17,* 918–921.
4. Phuc, N. H. H., Morikawa, K., Mitsuhiro, T., Muto, H., & Matsuda, A. (2017). *Ionics, 23,* 2061–2067.
5. Xu, R. C., Xia, X. H., Yao, Z. J., Wang, X. L., Gu, C. D., & Tu, J. P. (2016). *Electrochimica Acta, 219,* 235–240.
6. Yao, X., Liu, D., Wang, C., Long, P., Peng, G., Hu, Y. S., et al. (2016). *Nano Letters, 16,* 7148–7154.
7. Calpa, M., Rosero-Navarro, N. C., Miura, A., & Tadanaga, K. (2017). *RSC Advanced, 7,* 46499–46504.
8. Phuc, N. H. H., Morikawa, K., Totani, M., Muto, H., & Matsuda, A. (2016). *Solid State Ionics, 285,* 2–5.

9. Rangasamy, E., Liu, Z., Gobet, M., Pilar, K., Sahu, G., Zhou, W., et al. (2015). *Journal of the American Chemical Society, 137,* 1384–1387.
10. Sedlmaier, S. J., Indris, S., Dietrich, C., Yavuz, M., Dräger, C., von Seggern, F., et al. (2017). *Chemistry of Materials, 29,* 1830–1835.

Solution Process

Masahiro Tatsumisago and Atsushi Sakuda

Abstract Liquid-phase synthesis processes of sulfide-based solid electrolytes can be classified into two categories: suspension processes and solution processes. In the solution process, the precursors of sulfide-based solid electrolytes form homogenous solutions. This chapter summarizes the preparation of sulfide-based solid electrolytes by the solution process. The development of argyrodite solid electrolytes via precursor solution is also shown as a reprehensive of liquid-phase synthesized sulfide-based solid electrolyte with a high conductivity of more than 10^{-3} S cm^{-1} at 25 °C.

Keywords Liquid-phase synthesis · Solid electrolyte · Sulfide · Argyrodite

1 Preparation of Solid Electrolyte via Homogeneous Solution

As described above, liquid-phase synthesis processes can be categorized into two categories: suspension processes [1–20] and solution processes [21–38].

Suspension processes are conducted by stirring the starting materials in organic solvents in which they have low solubility and the precursor particles of solid electrolytes are suspended in the solvents as shown in Chap. 6 "Suspension Process". In the solution process, the precursors of sulfide-based solid electrolytes form homogenous solutions. This process is mainly conducted via a dissolution-precipitation process in which solid electrolytes prepared by solid-state or mechanochemical techniques are dissolved into solvents. Some papers report the preparation of precursor solutions of sulfide-based solid electrolytes solely by a liquid-phase process. Table 1 summarizes the preparation of sulfide-based solid electrolytes by the solution process. Polar solvents such as amides [22–25], alcohols [21, 26–30, 36–38], and

M. Tatsumisago (✉) · A. Sakuda
Osaka Prefecture University, Sakai, Japan
e-mail: tatsu@osakafu-u.ac.jp

A. Sakuda
e-mail: saku@chem.osakafu-u.ac.jp

K. Kanamura (ed.), *Next Generation Batteries*,
https://doi.org/10.1007/978-981-33-6668-8_7

Table 1 Sulfide-based solid electrolytes prepared via homogeneous precursor solution

Electrolyte	Starting material	Solvent	Pre-treatment	Conductivity/S cm^{-1}	Refs.
$Li_{3.25}Ge_{0.25}P_{0.75}S_4$	Li_2S, GeS_2, P_2S_5	Hydrazine	Solid phase	1.8×10^{-4}	[22]
Li_3PS_4	Li_2S, P_2S_5	NMF	Mechanochemical	2.6×10^{-6}	[23]
Li_3PS_4	Li_2S, P_2S_5	NMF, hexane	-	2.3×10^{-6}	[25]
Amorphous Li_3PS_4	Li_2S, P_2S_5, S	DEGDME	-	2.8×10^{-5}	[33]
Li_6PS_5Cl	Li_2S, P_2S_5, LiCl	EtOH	Mechanochemical	1.9×10^{-4}	[36, 37]
Li_6PS_5Cl	Li_2S, P_2S_5, LiCl	EtOH	Mechanochemical	6.0×10^{-4}	[26]
Li_6PS_5Br	Li_2S, P_2S_5, LiBr	EP + EtOH	-	3.4×10^{-5}	[27]
Li_6PS_5Br	Li_2S, P_2S_5, LiBr	THF + EtOH	-	3.1×10^{-3}	[38]
$Li_6PS_5BH_4$	Li_2S, P_2S_5, $LiBH_4$	THF + EtOH	-	1.3×10^{-4}	[28]
Li_4SnS_4	Li_2S, SnS_2	Water	Solid phase	1.4×10^{-4}	[31]
Li_4SnS_4	Li_2S, SnS_2	MeOH	Solid phase	8.9×10^{-5}	[29]
Amorphous LiI-Li_4SnS_4	Li_2S, SnS_2, LiI	MeOH	Solid phase	4.1×10^{-4}	[29]
Na_3PS_4	Na_2S, P_2S_5	NMF	-	2.6×10^{-6}	[39]
Na_3SbS_4	Na_2S, Sb_2S_3, S	MeOH	Solid phase	2.3×10^{-4}	[21]
Na_3SbS_4	Na_2S, Sb_2S_3, S	Water	Solid phase	2.6×10^{-4}	[21]
Na_3SbS_4	Na_2S, Sb_2S_3, S	Water	-	1.2×10^{-3}	[40]
NaI-Na_3SbS_4	Na_2S, Sb_2S_3, S, NaI	MeOH	Solid phase	7.4×10^{-4}	[30]

(continued)

Table 1 (continued)

Electrolyte	Starting material	Solvent	Pre-treatment	Conductivity/S cm^{-1}	Refs.
Na_3SbS_4-Na_4SnS_4	Na_2S, Sb_2S_3, SnS_2, S	Water	Solid phase	3.0×10^{-4}	[32]

water [31, 32] are generally utilized for the solution process. The first report was made by Wang et al. in 2012 [22]; $Li_{3.25}Ge_{0.25}P_{0.75}S_4$ thin film with a conductivity of 1.82×10^{-4} S cm^{-1} was prepared by dissolving the solid electrolyte in hydrazine to form a homogenous solution and subsequent drying at 240 °C. According to a study by Teragawa and Tatsumisago et al., Li_2S-P_2S_5 solid electrolyte having a conductivity of 2.6×10^{-6} S cm^{-1} at room temperature was prepared using *N*-methylformamide (NMF) [23–25]. Sulfide-based solid electrolytes can be dissolved more stably in basic amide solvents than other solvents. Park and Jung et al. demonstrated that a LiI-Li_4SnS_4 solution was obtained by dissolving Li_4SnS_4 synthesized by the solid-phase method and LiI in methanol [29]. Amorphous LiI-Li_4SnS_4 solid electrolytes with a high conductivity of 4.1×10^{-4} S cm^{-1} were synthesized without side reactions because of the high stability of $SnS_4{}^{4-}$ to oxide ions in nucleophilic solvents.

2 Precursor Solution of Argyrodite Solid Electrolytes

Argyrodite-type solid electrolytes Li_6PS_5X (X = Cl, Br) have been reported to show high conductivity [41–44]. Recently, they were found to be particularly suitable for the solution process [36–38]. Thus, the focus here is on Argyrodite-type solid electrolytes. Some tips on the preparation of Argyrodite-type solid electrolytes are discussed.

Yubuchi et al. produced an ethanol solution of Li_6PS_5Cl by dissolving mechanochemically prepared Li_6PS_5Cl in anhydrous ethanol [36, 37]. Argyrodite Li_6PS_5X electrolytes with high ionic conductivities were also synthesized from Li_2S, P_2S_5, and LiX solely via solution processing using a mixture of tetrahydrofuran and ethanol as the solvent [38]. Scheme 1 shows a schematic illustration of the liquid-phase synthesis of argyrodite-type Li_6PS_5Br and a picture of the obtained precursor solution. The concentrations of Li_6PS_5Br were 5-10 wt%. Figure 1a shows the Raman spectrum for the THF-EtOH precursor solution of Li_6PS_5Br and Li_6PS_5Br solid electrolyte particles prepared by liquid-phase or mechanochemical processes. Raman bands originating from the $PS_4{}^{3-}$ unit were detected at around 420 cm^{-1} in all the samples, indicating that THF and EtOH did not kinetically decompose the $PS_4{}^{3-}$ unit. Figure 1b presents XRD patterns of Li_6PS_5Br solid electrolyte particles prepared with different processes and heat treatment temperatures. All the samples consisted of mainly lithium-ion conducting argyrodite Li_6PS_5Br crystals, along with

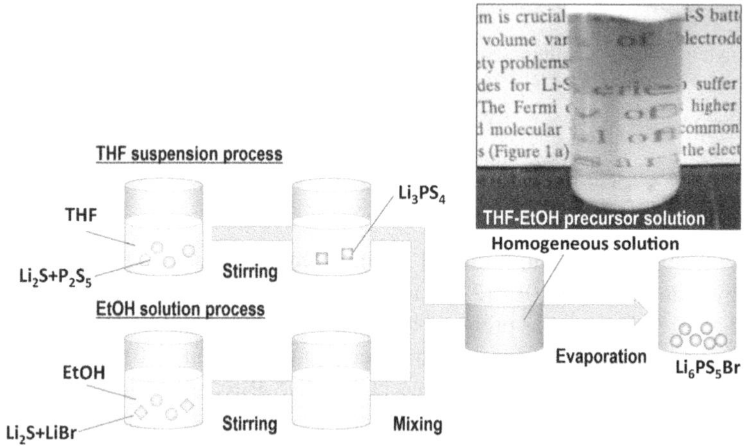

Scheme 1 Schematic illustrations of liquid-phase synthesis of Li_6PS_5Br with both EtOH and THF

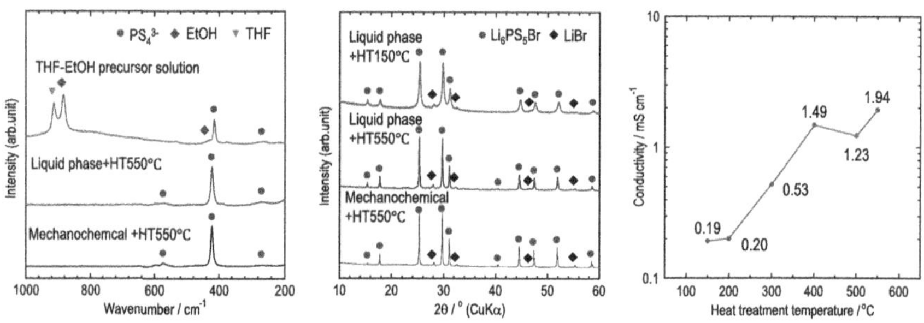

Fig. 1 **a** Raman spectra and **b** XRD patterns of Li_6PS_5Br prepared by liquid-phase process and mechanochemical processes and **c** the conductivities at 25 °C of the green compacts of the argyrodite electrolytes prepared by different heat treatment temperatures

a small amount of LiBr crystals. The crystallinity of Li_6PS_5Br was increased by heat treatment at 550 °C. The Li_6PS_5Br prepared by liquid-phase process and heat treatment at 550°C was confirmed to have the argyrodite structure ($a = 9.9641(2)$ Å, $F\bar{4}3m$, (216)) from the powder XRD pattern and the Rietveld refinement analysis technique. Figure 1c shows the conductivities of green compacts (powder compressed pellets) of the solution-synthesized Li_6PS_5Br. Conductivities of over 10^{-3} S cm^{-1} were achieved by heat treatment at 400-550°C. The sintered body at 550 °C showed a high ionic conductivity of 3.1×10^{-3} S cm^{-1}, which is comparable to the conductivity of Li_6PS_5Br prepared by the solid-phase method. These results show that liquid-phase synthesis is a viable candidate for replacing conventional synthetic techniques. Figure 2 summarizes the temperature dependence of the conductivities of liquid-phase prepared solid electrolytes including both suspension and solution processes.

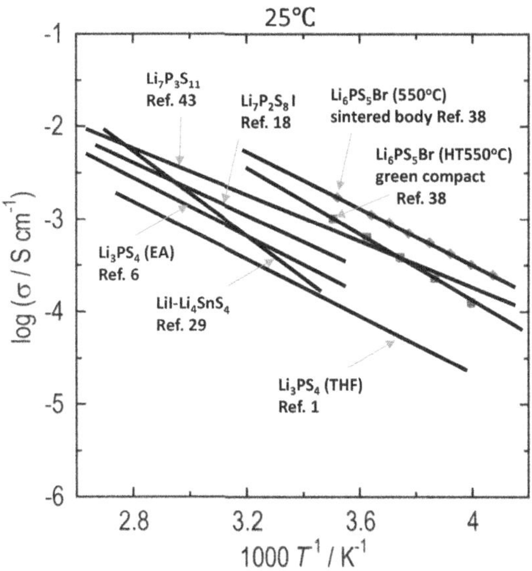

Fig. 2 Arrhenius plots of sulfide-based solid electrolytes prepared by the liquid-phase process. The green compacts were prepared by pressing the obtained powder at room temperature. The sintered body was obtained by heat treatment of the pellet. [1, 6, 18, 29, 38, 45]

Argyrodite Li_6PS_5Br shows higher conductivity than the other solid electrolytes prepared by a liquid-phase process.

Rosero-Navarro et al. reported that the morphology of precipitated particles is controllable by the addition of dispersant [26]. The addition of dispersant produced homogeneous and submicron-sized Li_6PS_5Cl particles while the same conditions without dispersant produced aggregates a few microns in size.

The generation mechanism of the sulfide-based solid electrolytes via solution has not been fully clarified yet. Deeper insight into the mechanism will enable the production of other kinds of sulfide-based solid electrolytes in the future.

References

1. Liu, Z., Fu, W., Payzant, E. A., Yu, X., Wu, Z., Dudney, N. J., et al. (2013). *Journal of the American Chemical Society, 135*, 975–978.
2. Sedlmaier, S. J., Indris, S., Dietrich, C., Yavuz, M., Dräger, C., Von Seggern, F., et al. (2017). *J Chem Mater, 29*(4), 1830–1835.
3. Choi, S., Lee, S., Park, J., Nichols, W. T., & Shin, D. (2018). *Appl Surface Sci., 444*, 10–14.
4. Lim, H.-D., Yue, X., Xing, X., Petrova, V., Gonzalez, M., Liu, H., et al. (2018). *J Mater Chem A, 6*, 7370–7374.
5. Ito, S., Nakakita, M., Aihara, Y., Uehara, T., & Machida, N. (2014). *Journal of Power Sources, 271*, 342–345.
6. Phuc, N. H. H., Morikawa, K., Totani, M., Muto, H., & Matsuda, A. (2016). *Solid State Ionics, 285*, 2–5.
7. Matsuda, A., Muto, H., & Phuc, N. H. H. (2016). *J Jpn Powder Powder Metallurgy, 63*(11), 976–980.

8. Phuc, N. H. H., Totani, M., Morikawa, K., Muto, H., & Matsuda, A. (2016). *Solid State Ionics, 288,* 240–243.
9. Phuc, N. H. H., Morikawa, K., Totani, M., Muto, H., & Matsuda, A. (2017). *Ionics, 23,* 2061–2067.
10. Phuc, N. H. H., Hirahara, E., Morikawa, K., Muto, H., & Matsuda, A. (2017). *Journal of Power Sources, 365,* 7–11.
11. Phuc, N. H. H., Tokuhara, Y., Muto, H., & Matsuda, A. (2017). *Inorganic Chem Front, 4*(10), 1660–1664.
12. Wang, H., Hood, Z. D., Xia, Y., & Liang, C. (2016). *J Mater Chem A, 4,* 8091–8096.
13. Hood, Z. D., Wang, H., Pandian, A. S., Peng, R., Gilroy, K. D., Chi, M., et al. (2018). *Adv Energy Mater, 8*(21), 100014.
14. Calpa, M., Rosero-Navarro, N. C., Miuram, A., & Tadanaga, K. (2017). *RSC Adv, 7*(73), 46499–46504.
15. Xu, R. C., Wang, X. L., Zhang, S. Z., Xia, Y., Xia, X. H., Wu, J. B., et al. (2018). *J Power Sourves, 374,* 107–112.
16. Wang, Y., Lu, D., Bowden, M., Khoury, P. Z. E., Han, K. S., Deng, Z. D., et al. (2018). *Chemistry of Materials, 30,* 990–997.
17. Xu, R. C., Xia, X. H., Wang, Z. L., Gu, C. D., & Tu, J. P. (2016). *Electrochimica Acta, 219,* 235–240.
18. Rangasamy, E., Liu, Z., Gobet, M., Pilar, K., Sahu, G., Zhou, W., et al. (2015). *Journal of the American Chemical Society, 137*(4), 1384–1387.
19. Wan, H., Mwizerwa, J. P., Qi, X., Xu, X., Li, H., Zhang, Q., et al. (2018). *Mater Interfaces, 10,* 12300–12304.
20. Wan, H., Mwizerwa, J. P., Qi, X., Liu, X., Xu, X., Li, H., et al. (2018). *ACS Nano, 12,* 2809–2817.
21. Banerjee, A., Park, K. H., Heo, J. W., Nam, Y. J., Moon, C. K., Oh, S. M., et al. (2016). *Angewandte Chemie Int Ed, 55,* 9634–9638.
22. Wang, Y., Liu, Z., Zhu, X., Tang, Y., & Huang, F. (2013). *Journal of Power Sources, 224,* 225–229.
23. Teragawa, S., Aso, K., Tadanaga, K., Hayashi, A., & Tatsumisago, M. (2013). *Chemistry Letters, 42,* 1435–1437.
24. Teragawa, S., Aso, K., Tadanaga, K., Hayashi, A., & Tatsumisago, M. (2014). *Journal of Power Sources, 248,* 939–942.
25. Teragawa, S., Aso, K., Tadanaga, K., Hayashi, A., & Tatsumisago, M. (2014). *J Mater Chem A, 2,* 5095–5099.
26. Rosero-Navarro, N. C., Miura, A., & Tadanaga, K. (2018). *Journal of Power Sources, 396,* 33–40.
27. Chida, S., Miura, A., Rosero-Navarro, N. C., Higuchi, M., Phuc, N. H. H., Muto, H., et al. (2018). *Ceram. Inter., 44*(1), 742–746.
28. Yubuchi, S., Uematsu, M., Sakuda, A., Hayashi, A., Tatsumisago, M. (2018). *255th ACS National Meeting & Exposition.*
29. Park, K. H., Oh, D. Y., Choi, Y. E., Nam, Y. J., Han, L., Kim, J.-Y., et al. (2016). *Advanced Materials, 288*(9), 1874–1883.
30. Park, K. H., Kim, D. H., Kwak, H., Jung, S. H., Lee, H.-J., Banerjee, A., et al. (2018). *J Mater Chem A, 6,* 17192–17200.
31. Choi, Y. E., Park, K. H., Kim, D. H., Oh, D. Y., Kwak, H. R., Lee, Y.-G., et al. (2017). *Chem Sus Chem, 10*(12), 2605–2611.
32. Heo, J. W., Banerjee, A., Park, K. H., Jung, Y. S., & Hong, S.-T. (2018). *Adv Energy Matet, 8*(11), 1702716.
33. Lim, H.-D., Lim, H.-K., Xing, X., Lee, B.-S., Liu, H., Coaty, C., et al. (2018). *Adv Mater Interfaces, 5,* 1701328.
34. Kim, T. W., Park, K. H., Choi, Y. E., Lee, J. Y., & Jung, Y. S. (2018). *J. Mater. Chem. A, 6,* 840–844.
35. Kim, D. H., Oh, D. Y., Park, K. H., Choi, Y. E., Nam, Y. J., Lee, H. A., et al. (2017). *Nano Letters, 17*(5), 3013–3020.

36. Yubuchi, S., Teragawa, S., Aso, K., Tadanaga, K., Hayashi, A., & Tatsumisago, M. (2015). *Journal of Power Sources, 293,* 941–945.
37. Yubuchi, S., Uematsu, M., Deguchi, M., Hayashi, A., Tatsumisago, M., & Appl, A. C. S. (2018). *Energy Mater, 1*(8), 3622–3629.
38. Yubuchi, S., Uematsu, M., Hotehama, C., Sakuda, A., Hayashi, A., & Tatsumisago, M. (2019). *J Mater Chem A, 7,* 558.
39. Yubuchi, S., Hayashi, A., & Tatsumisago, M., (2015). *Chemistry Letters, 44,* 884–886.
40. Yubuchi, S., Ito, A., Masuzawa, N., Sakuda, A., Hayshi, A. & Tatsumisago, M., (2020). *Journal of Materials Chemistry A, 8,* 1947–1954. https://doi.org/10.1039/C9TA02246E
41. Deiseroth, H.-J., Kong, S.-T., Eckert, H., Vannahme, J., Reiner, C., Zaiß, T., et al. (2008). *Angewandte Chemie Int Ed, 47,* 755–758.
42. Boulineau, S., Courty, M., Tarascon, J.-M., & Viallet, V. (2012). *Solid State Ionics, 221,* 1–5.
43. Deiseroth, H.-J., Maier, J., Weichert, K., Nickel, V., Kong, S.-T., & Reiner, C. (2011). *Anorg Allg Chem, 367,* 1287–1294.
44. Kraft, M. A., Culver, S. P., Calderon, M., Böcher, F., Krauskopf, T., Senyshyn, A., et al. (2017). *Journal of the American Chemical Society, 139*(31), 10909–10918.
45. Yao, X., Liu, D., Wang, C., Long, P., Peng, G., Hu, Y.-S., et al. (2016). *Nano Letters, 16,* 7148–7154.

Wet Chemical Processes for the Preparation of Composite Electrodes in All-Solid-State Lithium Battery

Kiyoharu Tadanaga, Nataly Carolina Rosero-Navarro, and Akira Miura

Abstract The preparation of composite electrodes by using wet chemical processes is briefly reviewed in this chapter. Electrode materials and conductive additive or binder are first dispersed in precursor solutions (dissolution–reprecipitation process) or suspensions (suspension process) of solid electrolytes, and then the solvent is evaporated to form the electrode–electrolyte composite. By using these processes, solid electrolyte covers the surface of active material, and lithium conduction path in the composite electrode can be formed with very small amount of solid electrolyte loading, and thus all-solid-state battery with composite electrode of high loading of solid electrolyte can be constructed. The wet chemical process must be very important for the practical application of the all-solid-state batteries.

Keywords Dissolution–reprecipitation · Suspension · Infiltration · Composite electrode

1 Introduction

To maximize the energy density of all-solid-state batteries, a limited amount of solid electrolytes that still ensures the lithium-ion conduction path in the composite electrode should be used. In addition, the formation of favorable interfaces between electrode materials and solid electrolytes with large contact areas is essential in the construction of all-solid-state batteries. Because surface of a solid is easily covered with liquids, an electrode material coated with a thin layer of a solid electrolyte can be prepared by coating of a liquid precursor of solid electrolyte on the electrode material by a wet chemical process, and subsequent drying and heating. As described in the previous chapter, sulfide-based solid electrolytes have been prepared by wet chemical processes, and preparation of sulfide-based solid electrolytes using wet chemical processes has been applied to the preparation of composite electrodes [1, 2].

K. Tadanaga (✉) · N. C. Rosero-Navarro · A. Miura
Faculty of Engineering, Hokkaido University, Sapporo 060-8628, Hokkaido, Japan
e-mail: tadanaga@eng.hokuda.ac.jp

K. Kanamura (ed.), *Next Generation Batteries*,
https://doi.org/10.1007/978-981-33-6668-8_8

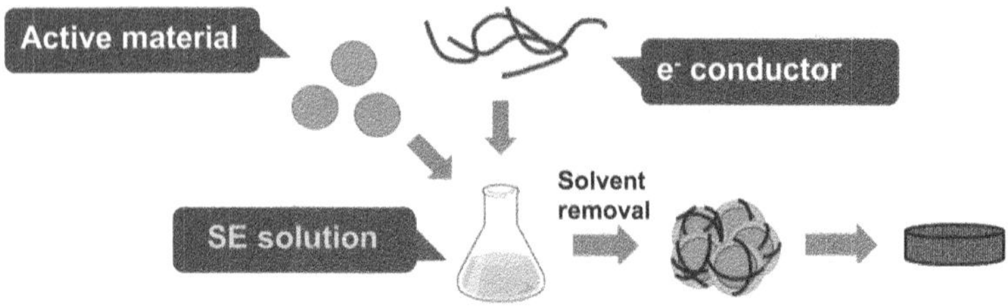

Fig. 1 Preparation process of composite electrodes using a wet chemical process

In this chapter, preparation of composite electrodes by using wet chemical processes is described. In the typical process (Fig. 1), electrode materials and conductive additive or binder are first dispersed in precursor solutions or suspensions of solid electrolytes, and then the solvent is evaporated to form the electrode–electrolyte composite particles. Thin layer of solid electrolyte can be formed on electrode materials by wet chemical processes, and the amount of solid electrolyte in the composite electrode can be minimized using this process. Thus, the energy density of all-solid-state batteries can be increased.

In the preparation of composite electrodes using wet chemical processes, the process is categorized into three processes based on the wet chemical processes for the preparation of sulfide-based solid electrolyte: 1. Dissolution–reprecipitation, 2. Suspension synthesis and 3. Combination of suspension synthesis and dissolution–reprecipitation processes. Table 1 shows some examples of the preparation of composite electrodes using wet chemical processes and the capacities of the cells including the composites [1].

2 Dissolution–Reprecipitation Process

In the Dissolution–Reprecipitation process [1], sulfide-based solid electrolytes are first prepared by a mechanical milling process or solid-state reactions, and the obtained solid electrolyte is dissolved in a solvent to obtain homogenous solution. The homogenous precursor solution can be used for the composite electrodes of both bulk-type and sheet-type all-solid-state batteries.

In the preparation of bulk-type batteries, the active material and sometimes electron conductive additive or binders are dispersed in the precursor solution, and the composite electrodes are obtained after the solvent removal. For the preparation of a sheet-type composite electrode on current corrector, slurry of active material, electron conductive additives and binder is firstly painted on a current corrector. Then, the precursor solution of solid electrolyte is penetrated (infiltrated) into the electrode sheet. Solid electrolyte is precipitated on the surface and gap of the active materials to form composite electrode sheet.

Table 1 Some examples of the preparation of composite electrodes by dissolution–reprecipitation process and suspension processes

Electrolyte	wt%	Active material	wt%	Others	wt%	Solvent	Heating Temp.(°C)	Capacity* (mAh g^{-1})	Refs.
Dissolution–Reprecipitation									
Li_3PS_4	7.5	$LiCoO_2$	92.5	–		NMF	180	32 (bulk)	[3]
Li_6PS_5Cl	7.5	$LiCoO_2$	92.5	–		Ethanol	80	45 (bulk)	[4]
$Li_4SnS_4{\cdot}LiI$	15	$LiCoO_2$	85	–		Methanol	200	135 (bulk)	[5]
Li_4SnS_4	15	$LiCoO_2$	85	–		Water	320	130 (bulk)	[6]
Li_6PS_5Cl	14	NCM_{111}	84	VGCF	2	Ethanol	80	44 (bulk)	[7]
Li_6PS_5Cl	3	$LiCoO_2$	97	Super P	2	Ethanol	180	150 (sheet)	[8]
				PVDF	1				
Li_6PS_5Cl	5	Graphite	95	PVDF	5	Ethanol	180	320 (sheet)	[8]
Li_6PS_5Cl	14	NCM_{111}	84	VGCF	2	EA + Ethanol	150	160 (bulk)	[9]
Li_6PS_5Cl	9	NCM_{111}	89	VGCF	2	ACN + Ethanol	150	115 (bulk)	[10]
				Triton	0.1				
Li_6PS_5Br	36	$Li_4Ti_5O_{12}$	61	CNT	3	THF + Ethanol	150	163 (sheet)	[11]
Li_6PS_5Cl	50	Si	40	Super P PVDF	5 5	Ethanol	180	3246 (sheet)	[12]
Suspension Synthesis									
Li_3PS_4	10	NCM_{111}	90	–		EP	170	130 (bulk)	[13]
$Li_7P_3S_{11}$	50	Co_9S_8	40**	Super P	10	ACN	260	650 (bulk)	[14]
$Li_7P_3S_{11}$	50	Fe_3S_4	45**	Super P	5	ACN	260	1001 (bulk)	[15]

(continued)

Table 1 (continued)

Electrolyte	wt%	Active material	wt%	Others	wt%	Solvent	Heating Temp.(°C)	Capacity* (mAh g^{-1})	Refs.
$Li_7P_3S_{11}$	66	MoS_2	24	AB	10	ACN	250	700 (bulk)	[16]
Li_3PS_4	27.5	NCM_{622}	70	Super C65 NBR	1 1.5	THF	140	153 (sheet)	[17]
Li_3PS_4	47.5	Graphite	50	NBR	2.5	THF	140	338 (sheet)	[17]
Combination of Suspension Synthesis and Dissolution–Reprecipitation									
Li_6PS_5Br	14	NCM_{111}	84	VGCF	2	EP + Ethanol	180	109 (bulk)	[18]
Li_6PS_5Br	10	NCM_{111}	90	–	-	THF + Ethanol	150	120 (bulk)	[19]

AB: acetylene carbon black, NBR: nitrile-butadiene rubber as binder; VGCF: vapor-grown carbon fiber, PVDF: polyvinylidene difluoride, NCM_{111}: $LiNi_{1/3}Co_{1/3}Mn_{1/3}O_2$, NCM_{622}: $LiNi_{0.6}Co_{0.2}Mn_{0.2}O_2$, ACN: acetonitrile, EA: ethylacetate, EP: ethylpropionate, THF: tetrahydrofuran

*Capacity per weight of active material

**Including $Li_7P_3S_{11}$ layer

For example, solid electrolytes with composition of $80Li_2S{\cdot}20P_2S_5$ glass was prepared by ball milling process, and $80Li_2S{\cdot}20P_2S_5$ was dissolved in N-methyl formamide [3]. Then, $LiCoO_2$ particles were coated with sulfide solid electrolyte by using the solution to obtain composite electrode. Li_6PS_5Cl was found to be dissolved in ethanol, and $LiCoO_2$ [4] or $LiNi_{1/3}Co_{1/3}Mn_{1/3}O_2$ (NCM111) [7] were coated by Li_6PS_5Cl using ethanol solution of Li_6PS_5Cl. By the addition of other solvent, such as ethlylacetate or acetonitrile, to ethanol, precipitation process of Li_6PS_5Cl precursor can be controlled, and thus, ethanol with ethyl acetate [9] or acetonitrile [10] was also used of the coating on active materials. Figure 2 shows the SEM image of Li_6PS_5Cl-coated NCM111 obtained by the wet chemical process using mixture of ethanol and ethylacetate as solvent [9]. EDS mapping images of cobalt, sulfur and chlorine are also shown. The coating of a solid electrolyte layer on NCM111 active materials was confirmed by these images [9]. As shown in Fig. 3, bulk-type all-solid-state batteries with 14 wt% of the solid electrolyte in the composite electrode, using the coated particles, exhibited larger charge–discharge capacity (160 mAh g^{-1}) than the composite electrode prepared by a simple mixing of solid electrolyte powder and the active material (90 mAh g^{-1}) [9].

Infiltration of Li_6PS_5Cl-precursor solution into a sheet-type electrode with active material, electron conductive additive, and binder, to form composite electrode sheets has also been reported [8, 11, 12]. In this process, Li_6PS_5Cl precursor was precipitated

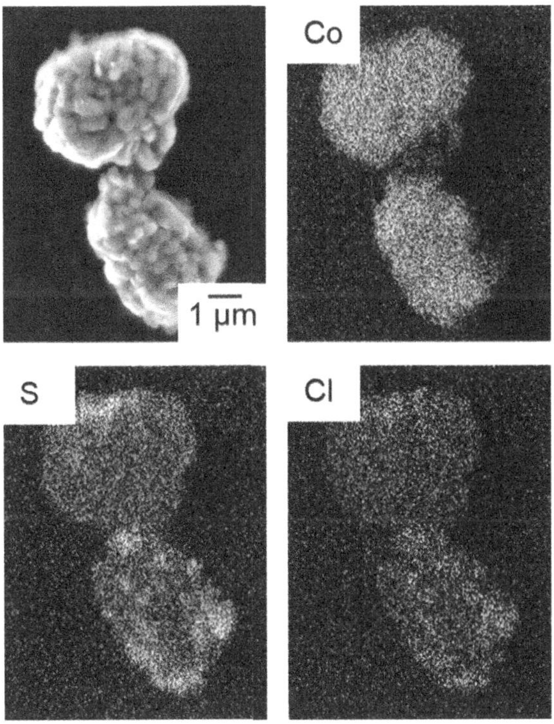

Fig. 2 SEM and EDS images of NCM111 coated with Li_6PS_5Cl using dissolution–reprecipitation process [9]

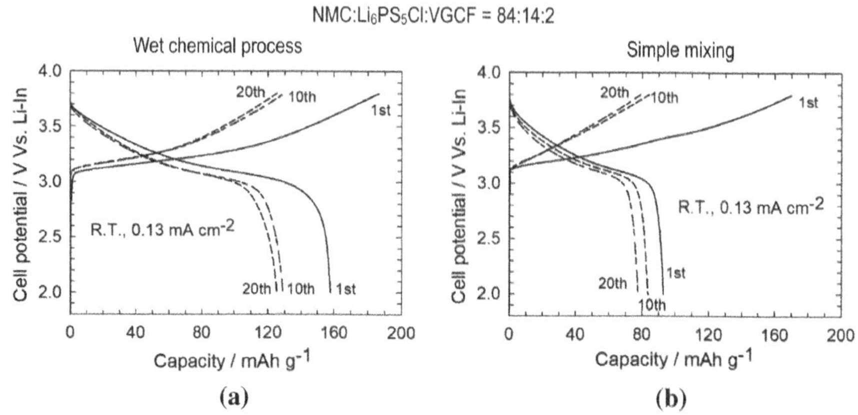

Fig. 3 Charge–discharge curves of all-solid-state lithium batteries using composite electrodes prepared by **a** wet chemical (dissolution–reprecipitation) process and **b** simple mixing process [9]

on the surface and gap of the particles of active material or electron conductive additives, and after drying, heating and adding pressure to the sheet, sheet-type composite electrodes can be obtained.

3 Suspension Syntheses

In the suspension syntheses [1], a precursor suspension of sulfide solid electrolyte is prepared in a solvent with starting materials such as Li_2S and P_2S_5, and then, electrodes and electron conductive additives are dispersed in the precursor suspensions. After evaporation of the solvent, a composite electrode containing active material, electron conductive additives and solid electrolyte is obtained.

For example, the preparation of composite electrode using precursor suspension, in which Li_2S and P_2S_5 were dispersed and shaken with zirconia balls in ethyl propionate, has been reported [12]. After adding 90 wt% NCM111 to the precursor suspension, the suspension was heated at 170 °C under vacuum for 2 h to remove solvent to prepare the composite electrode. The active material particles were well dispersed in the solid electrolyte matrix of the composite electrode, resulting in a high initial discharge capacity of up to 130 mAh g^{-1} in bulk-type all-solid-state batteries [12]. Sheet-type composite electrode was also prepared using suspension synthesis process [16].

4 Combination of Suspension Synthesis and Dissolution–Reprecipitation Process

Our group has proposed a process combining suspension synthesis and dissolution–reprecipitation processes for the formation of composite electrodes with Li_6PS_5Br [17]. In this process, the suspension was first prepared from Li_2S, P_2S_5, LiBr and ethylpropionate by ultrasonication treatment at 60 °C for 1 h. Then, NCM111, vapor grown carbon fiber (VGCF) and ethanol were added to the suspension. By the addition of ethanol, the precursor for the solid electrolyte Li_6PS_5Br was dissolved in the mixed solvent of ethylpropionate and ethanol, and the active material was dispersed in the clear precursor solution. After drying and heating, composite electrodes were obtained. All-solid-state battery with the mass ratio of NCM111: the electrode components (Li, P, S, Br): VGCF = 84: 14: 2 showed the discharge capacity of about 110 mAh g^{-1} at first cycle. On the other hand, the cell using the composite prepared by the mixing process showed lower capacity (55 mAh g^{-1}) than that prepared by the solution process. Yubuchi et al. have also reported the combined process of suspension synthesis and dissolution–reprecipitation process using THF and ethanol [18].

5 Conclusion

Preparation of composite electrodes by using wet chemical processes is briefly reviewed. By using wet chemical processes, favorable electrode–electrolyte interfaces with large contact area are easily formed. In addition, all-solid-state battery with composite electrode of high loading of active material can be prepared, because thin layer of solid electrolyte covers the surface of active material, and lithium conduction path in the composite electrode can be formed with very small amount of solid electrolyte loading. The wet chemical process must be very important for the practical application of the all-solid-state batteries.

References

1. Miura, A., Rosero-Navarro, N. C., Sakuda, A., Tadanaga, K., Phuc, N. H. H., Matsuda, A., et al. (2019). Liquid-phase syntheses of sulfide electrolytes for all-solid-state lithium battery. *Nature Reviews Chemistry, 3,* 189–193.
2. Park, K. H., Bai, Q., Kim, D. H., Oh, D. Y., Zhu, Y. Z., Mo, Y. F., & Jung, Y. S. (2018). Design strategies, practical considerations, and new solution processes of sulfide solid electrolytes for all-solid-state batteries. *Advanced Energy Materials, 8,* 1800035.
3. Teragawa, S., Aso, K., Tadanaga, K., Hayashi, A., & Tatsumisago, M. (2014). Preparation of Li_2S–P_2S_5 solid electrolyte from N-methylformamide solution and application for all-solid-state lithium battery. *Journal of Power Sources, 248,* 939–942.

4. Yubuchi, S., Teragawa, S., Aso, K., Tadanaga, K., Hayashi, A., & Tatsumisago, M. (2015). Preparation of high lithium-ion conducting Li_6PS_5Cl solid electrolyte from ethanol solution for all-solid-state lithium batteries. *Journal of Power Sources, 293,* 941–945.
5. Park, K. H., Oh, D. Y., Choi, Y. E., Nam, Y. J., Han, L. L., Kim, J. Y., et al. (2016). Solution-processable glass LiI-Li_4SnS_4 superionic conductors for all-solid-state Li-ion batteries. *Advanced Materials, 28,* 1874–1883.
6. Choi, Y. E., Park, K. H., Kim, D. H., Oh, D. Y., Kwak, H. R., Lee, Y. G., & Jung, Y. S. (2017). Coatable Li_4SnS_4 solid electrolytes prepared from aqueous solutions for all-solid-state lithium-ion batteries. *Chemsuschem, 10,* 2605–2611.
7. Rosero-Navarro, N. C., Kinoshita, T., Miura, A., Higuchi, M., & Tadanaga, K. (2017). Effect of the binder content on the electrochemical performance of composite cathode using Li_6PS_5Cl precursor solution in an all-solid-state lithium battery. *Ionics, 23,* 1619–1624.
8. Kim, D. H., Oh, D. Y., Park, K. H., Choi, Y. E., Nam, Y. J., Lee, H. A., et al. (2017). Infiltration of solution-processable solid electrolytes into conventional Li-Ion-battery. *Nano Letters, 17,* 3013–3020.
9. Rosero-Navarro, N. C., Miura, A., & Tadanaga, K. (2019). Preparation of lithium ion conductive Li_6PS_5Cl solid electrolyte from solution for the fabrication of composite cathode of all-solid-state lithium battery. *Journal of Sol-Gel Science and Technology, 89,* 303–309.
10. Rosero-Navarro, N. C., Miura, A., & Tadanaga, K. (2018). Composite cathode prepared by argyrodite precursor solution assisted by dispersant agents for bulk-type all-solid-state batteries. *Journal of Power Sources, 396,* 33–40.
11. Yubuchi, S., Nakamura, W., Bibienne, T., Rousselot, S., Taylor, L. W., Pasquali, M., et al. (2019). All-solid-state cells with $Li_4Ti_5O_{12}$/carbon nanotube composite electrodes prepared by infiltration with argyrodite sulfide-based solid electrolytes via liquid-phase processing. *Journal Power Sources, 417,* 125–131.
12. Kim, D. H., Lee, H. A., Song, Y. B., Park, J. W., Lee, S. M., & Jung, Y. S. (2019). Sheet-type Li_6PS_5Cl-infiltrated Si anodes fabricated by solution process for all-solid-state lithium-ion batteries. *Journal of Power Sources, 426,* 143–150.
13. Phuc, N. H. H., Morikawa, K., Mitsuhiro, T., Muto, H., & Matsuda, A. (2017). Synthesis of plate-like Li_3PS_4 solid electrolyte via liquid-phase shaking for all-solid-state lithium batteries. *Ionics, 23,* 2061–2067.
14. Yao, X., Liu, D., Wang, C., Long, P., Peng, G., Hu, Y. S., et al. (2016). High-energy all-solid-state lithium batteries with ultralong cycle life. *Nano Letters, 16,* 7148–7154.
15. Zhang, Q., Mwizerwa, J. P., Wan, H., Cai, L., Xu, X., & Yao, X. (2017). Fe_3S_4@$Li_7P_3S_{11}$ nanocomposites as cathode materials for all-solid-state lithium batteries with improved energy density and low cost. *Journal Materials Chemistry A, 5,* 23919–23925.
16. Xu, R. C., Wang, X. L., Zhang, S. Z., Xia, Y., Xia, X. H., Wu, J. B., & Tu, J. P. (2018). Rational coating of $Li_7P_3S_{11}$ solid electrolyte on MoS_2 electrode for all-solid-state lithium ion batteries. *Journal of Power Sources, 374,* 107–112.
17. Oh, D. Y., Kim, D. H., Jung, S. H., Han, J. G., Choi, N. S., & Jung, Y. S. (2017). Single-step wet-chemical fabrication of sheet-type electrodes from solid-electrolyte precursors for all-solid-state lithium-ion batteries. *Journal Materials Chemistry A, 5,* 20771–20779.
18. Chida, S., Miura, A., Rosero Navarro, N. C., Higuchi, M., Phuc, N. H. H., Muto, H., et al. (2018). Liquid-phase synthesis of Li_6PS_5Br using ultrasonication and application to cathode composite electrodes in all-solid-state batteries. *Ceramics International, 44,* 742–746.
19. Yubuchi, S., Uematsu, M., Hotehama, C., Sakuda, A., Hayashi, A., & Tatsumisago, M. (2019). An argyrodite sulfide-based superionic conductor synthesized by a liquid-phase technique with tetrahydrofuran and ethanol. *Journal Materials Chemistry A, 7,* 558–566.

Dry Coating of Electrode Particle with Solid Electrolyte for Composite Electrode of All-Solid-State Battery

Hideya Nakamura and Satoru Watano

Abstract A dry coating technology to produce composite electrode of an all-solid-state battery has been proposed. First, from the viewpoint of the powder technology, the fundamental concept of the dry coating is introduced. Second, a feasibility study on the dry coating process was conducted by using a cathode active material particle ($LiNi_{1/3}Co_{1/3}Mn_{1/3}O_2$, NCM) and a model material of sulfide solid electrolyte (Na_2SO_4). The results demonstrated that by means of a dry coating process, single NCM particle was uniformly coated with Na_2SO_4 without any breakage and attrition of the NCM particle. Finally, the dry coating was applied to produce the core–shell-type composite particle with an actual sulfide solid electrolyte (Li_3PS_4, LPS). It was demonstrated that the core–shell composite particle, where single particle of NCM is coated with shell layer of LPS (NCM@LPS), is able to be produced even with Li_3PS_4. A composite cathode prepared with NCM@LPS particles exhibited much larger contact area between NCM and LPS, resulting in significant improvement of rate and cycle performances of an all-solid-state half-cell.

Keywords Dry coating · Composite cathode · Solid–solid contact

1 Introduction

For the development of the bulk-type all-solid-state lithium battery, designing the structure of composite electrode consisting of active material (AM), solid electrolyte (SE), and (if necessary) conductive additives is one of the major issues. In the all-solid-state lithium battery, transfer of lithium ions during charging and discharging mainly occurs at the interfacial contact surface between AM and SE. This requires well-designed composite electrode with high contact area between AM particles and SEs.

H. Nakamura (✉) · S. Watano
Department of Chemical Engineering, Graduate School of Engineering, Osaka Prefecture University, Sakai, Japan
e-mail: hnakamura@chemeng.osakafu-u.ac.jp

K. Kanamura (ed.), *Next Generation Batteries*,
https://doi.org/10.1007/978-981-33-6668-8_9

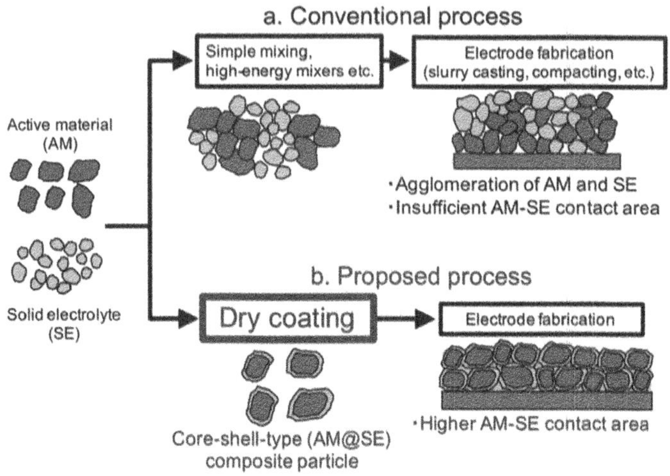

Fig. 1 Proposed dry coating process in this study

However, fabrication of such well-designed electrode structure is very difficult, because the solid electrolyte is a form of solid particulate material and does not spontaneously penetrate into the void space of the composite electrode such as conventional lithium battery with liquid electrolyte. With an increase in a fraction of AM particles in the composite electrode (for higher energy density), fabrication of such well-designed electrode structure with high contact area is much harder. Moreover, AM particles and SE particles are mostly fine powder (ca. smaller than 20 μm), that is generally classified into a group of cohesive and highly agglomerated powder with poor flowability from the viewpoint of the powder technology [1]. This results in particle agglomeration, leading to poor interfacial contact between AM and SE (Fig. 1a). To overcome this issue, a core–shell-type composite particle, in which single AM particle is coated with SE, is considered to be a favorable structure of composite particle (Fig. 1b). If a single AM particle can be coated with SE, the interfacial contact between AM and SE can be formed within a single composite particle. In a composite electrode fabricated with the core–shell composite particles, much higher interfacial contact area between AM and SE can be achieved. However, development of a powder processing technology to produce the core–shell composite particles had been a critical issue.

We here present our effort to create a novel core–shell composite particle for an all-solid-state battery by using a dry coating technology. First, a fundamental concept of the dry coating technology is explained from the viewpoint of the powder technology. Second, fundamental research to investigate the feasibility of the dry coating process was shown. The feasibility study was conducted with Na_2SO_4, which is a model material of sulfide SE. Mechanism of the dry coating of cathode AM particles with SE was also discussed [2, 3]. Third, results on the dry coating with actual sulfide SEs and performance of an all-solid-state half cell prepared with the dry coated particles were presented [4, 5].

2 Fundamental Concept of Dry Coating

Dry coating is a particle processing technology in which the surface of a large core particle is coated with smaller guest particles using solely external mechanical energy applied within a powder processing equipment, without the use of any solvents or other additives such as binder. The basis of the dry coating is the powder mixing technology. Here, a fundamental concept of the dry coating is explained with a typical case of the mixing of a dry binary powder mixture.

Figure 2 shows a schematic of temporal change in the degree of powder mixing. This figure is a logarithmic plot of the standard deviation (σ) of uniformity of two kinds of different powders (black and white powders, shown in Fig. 2) within a powder mixture as a function of the mixing time t, where σ_0 and σ_r are the standard deviations of the powder mixture at the initial state (unmixed powder) and of the powder with perfectly mixed state, respectively. When two kinds of different powders (black and white powders) are mixed, macroscopic exchange of each particle position occurs initially among the particle assemblies. This stage is termed as convective mixing, which occurs at the beginning of the powder mixing. Subsequently, particle assemblies are gradually disintegrated by collision and/or slip between particles, which is mainly induced by the powder shear flow. This results in the microscopic exchange of individual particle, of which the stage is referred as shear mixing. When

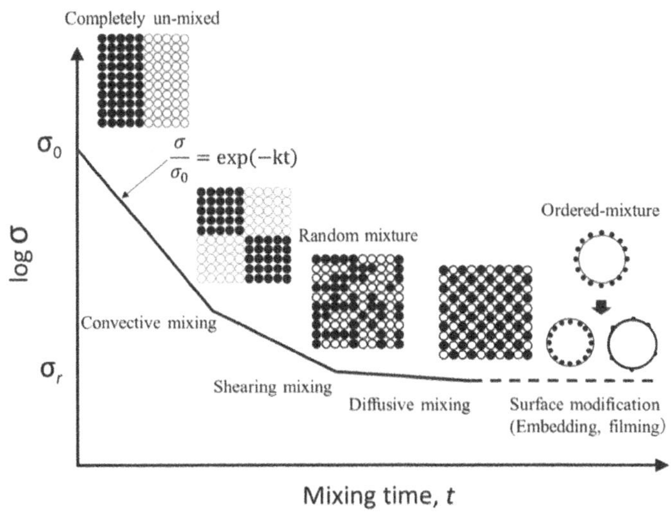

Fig. 2 Schematic of temporal change in the degree of powder mixing

the powder mixing proceeds further, more microscopic exchange of each particle proceeds due to differences in the particle properties (such as size, density, and shape), local inter-particle void space, and slight gradients of translational and rotational velocities between individual particles. This mixing stage is termed as diffusive mixing. Normally, the final mixing state achieved by the diffusive mixing is a random mixture. However, when the two kinds of particles have significant size difference (approximately higher than 10 times), a unique final mixing state can be achieved, i.e., the larger particle is uniformly covered with the smaller particles. This mixing state is termed as an ordered mixture. When additional mechanical energy is added to the ordered mixture, the small guest particles can be immobilized onto the large core particles. This is a fundamental concept of dry coating. To achieve successful dry coating, the following two things are keys: (i) adequate size control of core and guest particles and (ii) an adequate particle processing equipment that can add highly strong mechanical energy to the powder with high transfer efficiency without breakage of the core particles.

3 Feasibility Study of Dry Coating Process Using Model Sulfide SE

Figure 3 demonstrates the particle design concept attempted in this study. A cathode AM particle, which was coated in advance with an oxide solid electrolyte (e.g., $LiNbO_3$) to reduce the interfacial resistance between AM and SE [6], was used as a core particle. We aimed to produce a core–shell type of composite particle in which the surface of the core particle was uniformly coated with SE without breakage and attrition of the core particle.

We first conducted the fundamental study to investigate the feasibility of the dry coating process for producing the core–shell composite particle. As powder samples, $LiNi_{1/3}Co_{1/3}Mn_{1/3}O_2$ (NCM) (ALCA-SPRING) and sodium sulfate (Na_2SO_4) (JIS special grade, Wako Pure Chemical Industries, Ltd.) were used, NCM is a typical positive-electrode active material for LIBs. Figure 4a indicates a SEM image of the NCM particles. Median diameter of the NCM particles was 5.4 μm, measured

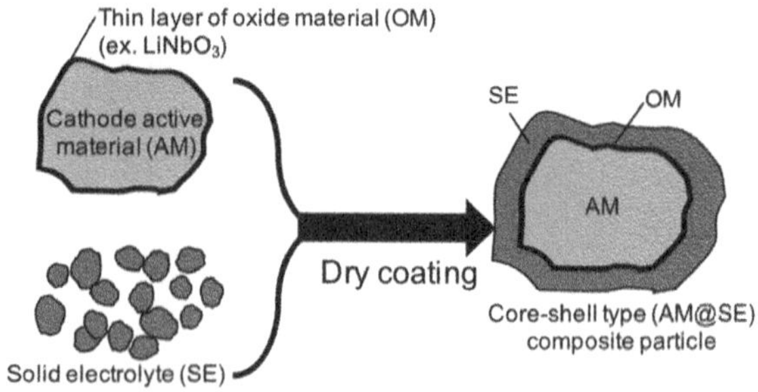

Fig. 3 Particle design was attempted in this study

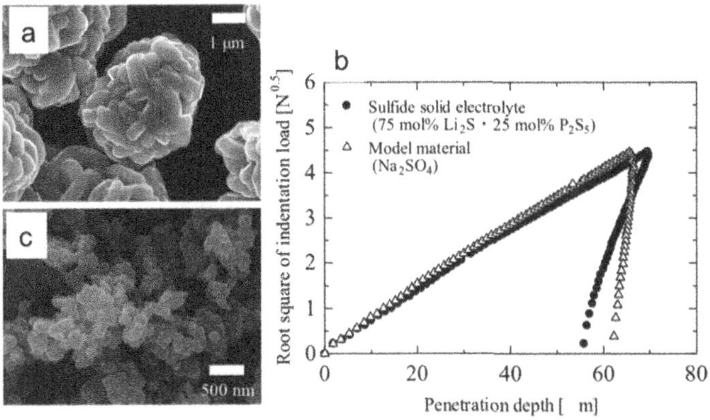

Fig. 4 **a** $LiNi_{1/3}Co_{1/3}Mn_{1/3}O_2$ (NCM) and **b** Na_2SO_4 used in this study. **c** Load-penetration depth curve of sulfide solid electrolyte and model material (Na_2SO_4) measured by indentation test

by a laser diffraction particle size analyzer. NCM is relatively rigid, brittle, and electrically high conductive material, as compared to Na_2SO_4, which was selected as a model material of sulfide SEs [2, 3] because a sufficient amount of sulfide SEs to conduct our feasibility study was commercially unavailable. For selecting a model material of sulfide SEs, we considered that the following properties are important: (i) nature of the room temperature pressure sintering; (ii) mechanical properties such as stiffness and elastoplasticity. It has been reported that sulfide SEs exhibit a unique densification property so-called "room-temperature pressure sintering [7]": when powder of a sulfide SE is compressed, the sulfide SE particles are able to be easily deformed and coalesced even at room temperature. In this study, the model material of sulfide SEs was selected from few inorganic materials, which exhibit the room temperature pressure sintering. We then measured the stiffness and elastoplasticity of the candidate materials by means of an indentation test [3]. As a result, as shown in Fig. 4b, Na_2SO_4 showed the room temperature pressure sintering as well as similar stiffness and elastoplasticity with a sulfide SE [3]. Consequently, Na_2SO_4 was selected and used in this study as a model material of sulfide SEs. The as-received powder of Na_2SO_4 was preliminarily milled by a wet grinding system to enable the powder mixture of NCM and Na_2SO_4 to be nearly an ordered-mixture. Figure 4c shows the milled Na_2SO_4 used in the dry coating. Median diameter of the milled Na_2SO_4particles was 0.95 μm. As a dry coating device, a dry impact-blending device was used (Fig. 5). This device mainly consists of chamber, rotor with blades, and circulation pipe. The dry coating was conducted as follows. NCM and Na_2SO_4 powders were preliminarily mixed by a mortar-and-pestle mixing, and the premixed simple mixture was fed into the dry impact-blending device. The rotor was then rotated and the particles were circulated in the device. During the circulation, the particles are subjected to high impact forces due to particle-to-wall and particle-to-particle collisions. As a result, the surface of the NCM particles is coated with Na_2SO_4 particles. After a predetermined processing time, the discharge valve was

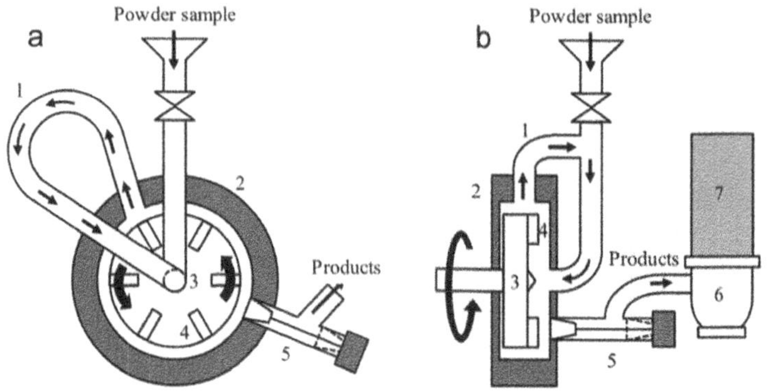

Fig. 5 Schematic of dry impact-blending device used for the dry coating. **a** Front view and **b** side view

opened, and coated particles were collected. To evaluate the performance of the dry coating process, three kinds of powder samples (A, B, and C) were produced by different processes. Powder A was prepared by a vibration mixing using a vibrating mixer (FLX-F60, AS ONE Corp.) for 1 min. Powder B was prepared by mortar-and-pestle mixing for 30 min. These two mixing processes were typical procedures to produce a composite cathode powder before assembling all-solid-state test cells [8, 9]. Powder C was produced by the dry coating process. Rotating speed of the rotor and processing time were set as 16,000 rpm and 5 min, respectively. In all processes, total amount of initial powders was 10 g, and the ratio of NCM to Na_2SO_4 within the initial powders was NCM:Na_2SO_4 = 70:30 wt%.

Figure 6 indicates FESEM and EDX images of powders A, B, and C. In powder A prepared by a vibration mixing (Fig. 6a), surface morphology of the pristine NCM (i.e., rectangular-shaped primary grains) was still observed, indicating that NCM particles were not covered with Na_2SO_4 particles. In the powder B prepared by a mortar-and-pestle mixing (Fig. 6b), NCM particles were not fully covered with Na_2SO_4 particles, although some Na_2SO_4 particles adhered onto NCM particles and NCM particles were partially covered with Na_2SO_4. In the powder C prepared by the dry impact-blending device (Fig. 6c), NCM particles were fully covered with Na_2SO_4 particles. Figure 6d indicates cross-sectional FESEM and EDX images of a single composite particle in the powder C. It was confirmed that NCM particle was coated with continuous layer of Na_2SO_4. Thickness of the coating layer was around 0.5 μm. To investigate damage to the NCM particles by the dry coating, NCM particles after washing out Na_2SO_4 were evaluated. Figure 7a indicates the particle size distributions of NCM particles. There was no change in the size distribution of NCM particles even after the dry coating. Figure 7b shows FESEM image of NCM particle after washing out Na_2SO_4 from the powder C. No significant changes were observed in the surface morphology from the pristine NCM particles. These results confirmed that NCM

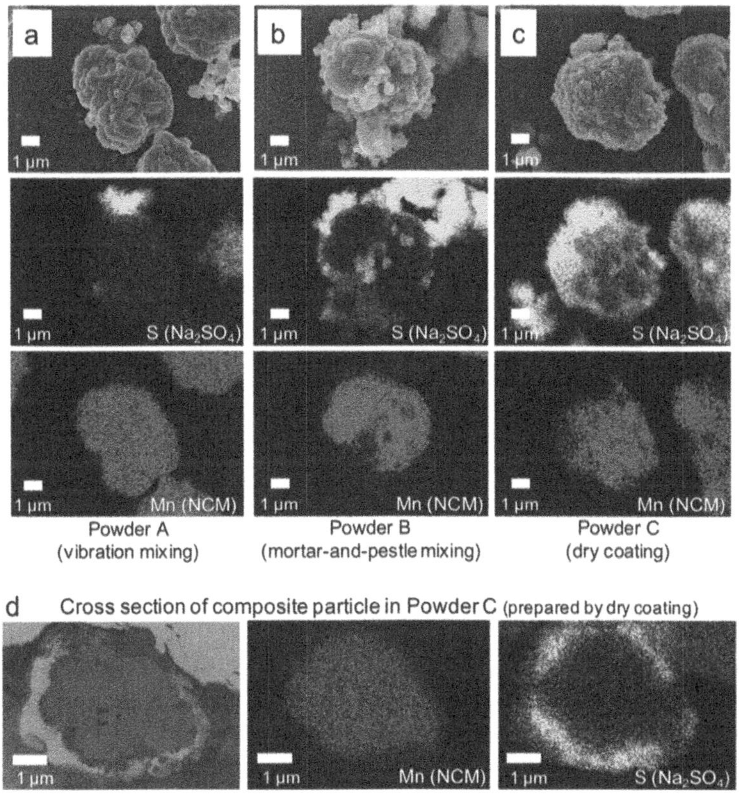

Fig. 6 **a–c** FESEM and EDX images of each powder product. **d** FESEM and EDX images of cross-section of single composite particle prepared by dry coating process

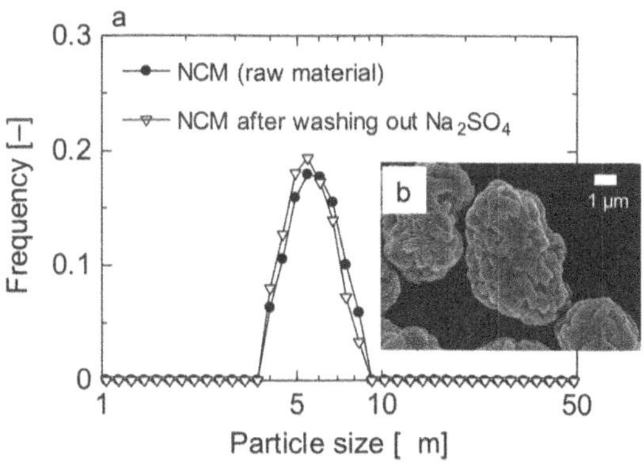

Fig. 7 **a** Particle size distribution of pristine NCM and NCM after washing out Na_2SO_4 from Powder C. **b** SEM image of NCM particles after washing out Na_2SO_4

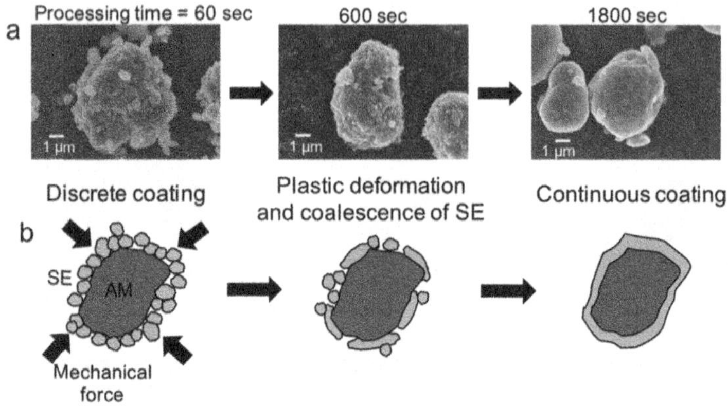

Fig. 8 **a** Dry coated particle at various processing times. **b** Schematic of mechanism of dry-coating in this study

particles (core particles) are not damaged even by mechanical energy applied during the dry coating processing. This is important for all-solid-state batteries because damage to the cathode AM particles, i.e., loss of precoated oxide material layer, can cause an increase in the interfacial resistance and highly negative impact on the battery performance. Consequently, by using the dry coating technology a core–shell composite particle, in which cathode AM particle is uniformly coated with model material of sulfide SE, was successfully produced without physical damage to the AM particles.

From further fundamental investigations, mechanism of the dry coating in this study was revealed as described in Fig. 8. Initially, under an external mechanical force generated in the dry impact-blending device, densification of guest particles occurs on the surface of host particles, resulting in the discrete coating. With an increase in the processing time, i.e., an increase in the mechanical input energy, plastic deformation and coalescence of SE particles on the surface of NCM particle begin to occur. With further increase in the processing time, the plastic deformation and coalescence of SE particles further proceed, resulting in the continuous coating. This is due to significant difference in the mechanical property between cathode AM and SE, i.e., the SE (guest particle) is soft and ductile materials, while the NCM (core particle) is rigid and brittle material [2, 3]. This also indicates that a key property for successful dry coating is a significant difference in mechanical properties between AM and SE. We have also performed parametric study and optimization of the dry coating process [2].

Compressed pellets, mimicking composite cathode of an all-solid-state cell, were then prepared from each powder product. To evaluate the formation of the contact surface between NCM and Na_2SO_4 within the pellet, a DC electrical resistivity of the compressed pellets was measured. Through this measurement, the state of the NCM–Na_2SO_4 interfacial contacts can be evaluated: if NCM is well covered with Na_2SO_4 and the contact surface between NCM and Na_2SO_4 is well formed, the DC

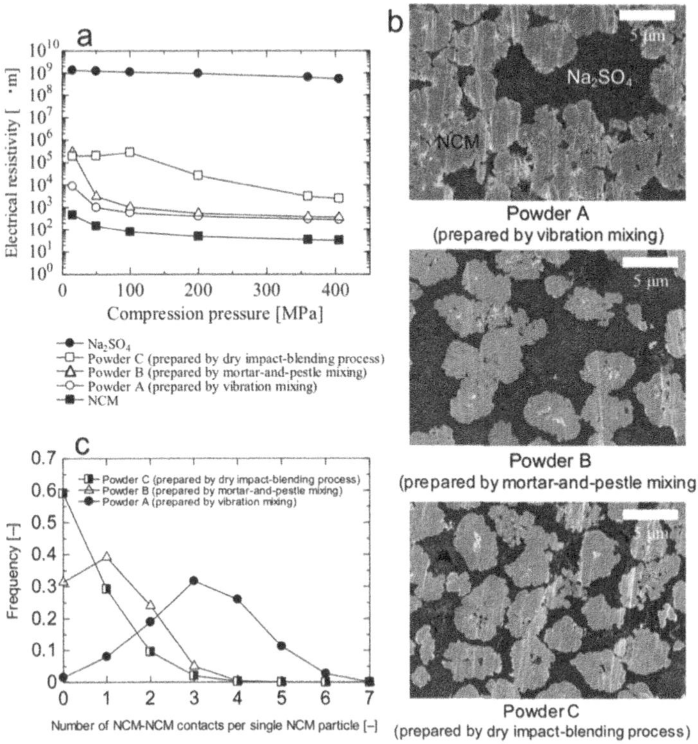

Fig. 9 **a** Electrical resistivity of compressed pellet prepared from each powder. **b** FESEM images of cross-section of compressed pellet (360 MPa) prepared from each powder product. **c** Number of NCM–NCM contacts per single particle of NCM on cross-section of compressed pellets (360 MPa) prepared from each powder product

electrical resistivity becomes higher due to disconnecting NCM-to-NCM percolated network by poorly conductive Na_2SO_4. Figure 9a indicates the electrical resistivity of compressed pellets prepared with various powder products as a function of compression pressure. Since electrical resistivity of Na_2SO_4 was much higher than that of NCM, electrical resistivity of the powders A, B, and C (mixture of Na_2SO_4 and NCM) showed electrical resistivity between Na_2SO_4 and NCM. Electrical resistivity of pellet of powder C (core–shell composite particles) exhibited significantly higher value than those of powders A and B, suggesting that interfacial contacts between NCM and Na_2SO_4 were well formed in the composite cathode prepared from the core–shell composite particles. Figure 9b indicates FESEM images of cross-section of pellets prepared with each powder product. From EDX images, it was found that light gray and dark gray places in FESEM images correspond to NCM and Na_2SO_4, respectively. In the pellet of powder C (core–shell composite particles), however, NCM particles were well dispersed in Na_2SO4 matrix without agglomeration, resulting in the formation of interfacial contacts between NCM and Na_2SO_4. Figure 9c indicates distributions of the number of NCM-NCM contacts per single particle of NCM on cross-section of compressed pellets at different powder products. The number of NCM-NCM contacts in the pellet of powder C exhibited much smaller than the powders A and B. In the pellet of powder C, 60% of NCM particles did not contact with other NCM particles but contacted with Na_2SO_4.

Consequently, the feasibility study demonstrated that by means of a dry coating, single AM particle was uniformly coated with a continuous layer of SE without breakage and attrition of the AM, resulting in an AM@SE core–shell particle. It was also confirmed that the composite cathode prepared with the core–shell particle exhibited high interfacial contacts between AM and SE.

4 Dry Coating of NCM with Li_3PS_4 Sulfide SE and Performance of All-Solid-State Half Cell

Further parametric study of the dry coating process with the model material of sulfide SE (Na_2SO_4) was conducted, and the optimum processing conditions were determined [2]. Under the optimum processing conditions, we attempted to produce a core–shell composite particle with an actual sulfide SE by means of the dry coating process. Performance of an all-solid-state half cell prepared with the core–shell composite particles was then evaluated. As a cathode AM particle, the NCM, which was precoated with 1 wt%-$LiNbO_3$, was used. A glassy Li_3PS_4 (LPS, ALCA-SPRING), which is a typical sulfide SE, was used as a material of shell layer. The as-received LPS powder, which was synthesized by a dry mechanical-milling process, was preliminarily milled using a wet grinding process, and the resultant milled-LPS was used in the dry coating. Number-based median diameter of the milled-LPS was 0.49 μm. The lithium-ion conductivity of the compressed pellet of the milled-LPS was 0.84×10^{-4} S cm^{-1} at 298 K. All dry coating experiments were conducted in an Ar-filled glove box at a low dew-point (<193 K).

Figure 10a indicates FESEM and EDX images of composite particles (NCM:LPS = 87:13 wt%) prepared by the dry coating. A drastic change in the surface morphology was observed, i.e., rectangular-shaped primary grains of the pristine NCM were not observed, but a smooth surface morphology was observed. In the EDX images, both of manganese of NCM and sulfur of LPS were detected from same location. The results demonstrated that by using the dry coating process, the NCM@LPS core–shell composite particle, where a single particle of NCM was uniformly coated with shell layer of LPS, was successfully produced. From FESEM image of the cross-section of a single composite particle, an NCM particle was almost fully coated with continuous LPS layer with 0.1–0.2 μm thickness. We also confirmed that there was no breakage and/or attrition of NCM particles after the dry coating (data were not shown).

All-solid-state half cells were prepared with the composite particles, and efficacy of the core–shell structure on the improvement of electrochemical performance of the all-solid-state cell was investigated. A typical all-solid-state half cell, consisting

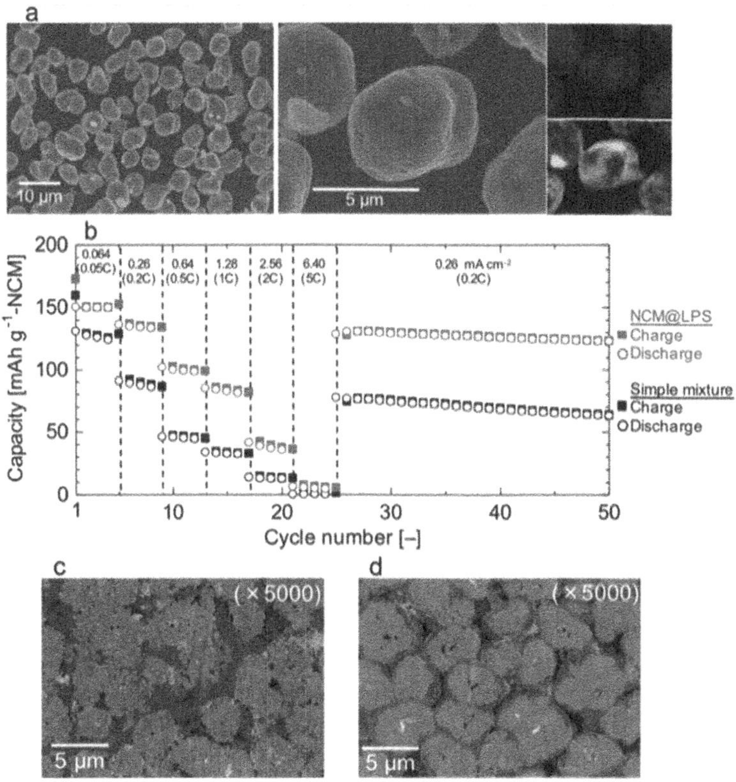

Fig. 10 **a** FESEM and EDX images of dry coated particles prepared with $LiNi_{1/3}Co_{1/3}Mn_{1/3}O_2$ and Li_3PS_4. **b** Rate capability and cycle performance of all-solid-state half cells prepared with core–shell type composite particle (NCM@LPS) and simple mixture at various discharge current density **c** rates). FESEM images of cross-section of composite cathodes prepared with **d** simple mixture and **e** NCM@LPS composite particles

of composite cathode, LPS separator layer, and a counter electrode of Li-In alloy, was fabricated [10]. The composite cathode was prepared solely with the core–shell composite particles without any conductive additives. The half cell was prepared by compressing dry powder of each material. Figure 10b shows electrochemical performance of all-solid-state half cells prepared with the NCM@LPS core–shell composite particles and simple mixture of NCM and LPS obtained by the mortar-and-pestle mixing. For adequate comparison, the mass fraction of NCM and LPS within the simple mixture was set to the same value as that in the NCM@LPS composite particles (i.e., NCM:LPS = 87:13 wt%, in this case). The half cell prepared with the NCM@LPS composite particles exhibited significantly higher charge/discharge capacities as compared to the half cell prepared with the simple mixture. In particular, improvement of the rate capability by dry coating was more pronounced at higher current density. In the half cell prepared with the simple mixture, discharge capacities at the 26th and 50th cycles were 76 mAh g^{-1} and 63 mAh g^{-1}, respectively, i.e.,

83% of the discharge capacity was retained between 26 and 50th cycles. In the cell prepared with the NCM@LPS particles, however, discharge capacities at the 26th and 50th cycles were 131 mAh g^{-1} and 123 mAh g^{-1}, respectively, i.e., 94% of the discharge capacity was retained. These results exhibit that the NCM@LPS particle was also advantageous in terms of capacity retention.

To identify the cause of improving the cell performance by using the NCM@LPS particles, internal structure of the composite cathode was observed. Figure 10c–d shows the cross-section of the composite cathodes. In the composite cathode prepared with the simple mixture, NCM particles (light-gray region) were highly agglomerated, resulting in less interfacial contact with LPS. However, in the composite cathode prepared with the NCM@LPS particles, much higher interfacial contact between NCM and LPS was observed. It was also confirmed that the NCM-to-NCM contacts, i.e., electron transport pathways, were formed in the composite cathode even when the core–shell composite particles were used. It is considered that by compressing the cathode powder bed, the continuous shell layer of LPS was partially broken, resulting in NCM–NCM contacts and the formation of the electron transport pathway.

In summary, it was demonstrated that the dry coating process was applicable to an actual sulfide SE, and the NCM@LPS core–shell composite particle was successfully produced. The rate and cycle performances of an all-solid-state half-cell were significantly improved by using the NCM@LPS composite particles, because a composite cathode with high interfacial contact between NCM and LPS was constructed. Thus, it was concluded that the core–shell composite particle contributes to the structure of the composite cathode, leading to improvement of the performance of an all-solid-state battery.

5 Concluding Remarks

A dry coating technology to produce novel core–shell composite particles for an all-solid-state battery was presented. First, a fundamental concept of the dry coating technology was explained from the viewpoint of dry powder mixing. Key points for successful dry coating were also shown. Second, feasibility study of the dry coating process for producing the core–shell composite particles was conducted by using Na_2SO_4, which is the model material of sulfide SE. It was demonstrated that by means of a dry coating process, single AM particle was uniformly coated with SE without breakage and attrition of the AM, resulting in an AM@SE core–shell particle. Plastic deformation and coalescence of SE particles were promoted by mechanical energy on the rigid NCM particles, resulting in the continuous coating layer of SE. It was also confirmed that the composite cathode prepared with the core–shell composite particle exhibited high interfacial contacts between AM and SE. Finally, the dry coating was performed with an actual sulfide SE (Li_3PS_4). It was demonstrated that the core–shell composite particle is able to be produced even with Li_3PS_4. A composite cathode prepared with NCM@LPS composite particles exhibited higher contact area between NCM and LPS, leading to significant improvement of the rate and cycle performances of an all-solid-state half-cell.

From the viewpoint of industrialization, the developed dry coating process is highly promising. The processing time of the dry coating process is very short (within a few minutes), and the scale-up is relatively easy. Thus, the developed dry coating process is expected to highly contribute to mass production of an all-solid-state battery. From scientific perspective, analysis of the three-dimensional structure of the composite electrode prepared with the core–shell particles can provide important insights into structure–performance relationship of an all-solid-state battery, contributing to further understanding of the mechanism of the all-solid-state battery.

Acknowledgements This study was financially supported by Japan Science and Technology Agency (JST), Advanced Low Carbon Technology Research and Development Program–Specially Promoted Research for Innovative NextGeneration Batteries (ALCA-SPRING). The authors would like to acknowledge Prof. M. Tatsumisago, Prof. A. Hayashi, and Dr. A. Sakuda (Osaka Prefecture University) for supporting FESEM and EDX observations and electrochemical tests. The authors also would express thanks to Prof. A. Matsuda and Prof. H. Muto (Toyohashi University of Technology) for indentation tests.

References

1. Geldart, D. (1973). Types of gas fluidization. *Powder Technology, 7*, 285–292.
2. Kawaguchi, T., Nakamura, H., & Watano, S. (2017). Parametric study of dry coating process of electrode particle with model material of sulfide solid electrolytes for all-solid-state battery. *Powder Technology, 305*, 241–249.
3. Kawaguchi, T., Nakamura, H., & Watano, S. (2018). Dry coating of electrode particle with model particle of sulfide solid electrolytes for all-solid-state secondary battery. *Powder Technology, 323*, 581–587.
4. Watano, S., & Nakamura, H. (2016). *Japanese Patent Application* 2016–162003.
5. Watano, S., & Nakamura, H. (2017). PCT/JP2017/29731.
6. Takada, K., Ohta, N., Zhang, L., Fukuda, K., Sakaguchi, I., Ma, R., et al. (2008). Interfacial modification for high-power solid-state lithium batteries. *Solid State Ionics, 179*, 1333–1337.
7. Sakuda, A., Hayashi, A., & Tatsumisago, M. (2013). Sulfide solid electrolyte with favorable mechanical property for all-solid-state lithium battery. *Science Report, 3*, 2261.
8. Sakuda, A., Hayashi, A., & Tatsumisago, M. (2010). Interfacial observation between $LiCoO_2$ electrode and Li_2S-P_2S_5 solid electrolytes of all-solid-state lithium secondary batteries using transmission electron microscopy. *Chemistry of Materials, 22*, 949–956.
9. Mizuno, F., Hayashi, A., Tadanaga, K., & Tatsumisago, M. (2005). Design of composite positive electrode in all-solid-state secondary batteries with Li_2S-P_2S_5 glass-ceramic electrolytes. *Journal of Power Sources, 146*, 711–714.
10. Sakuda, A., Takeuchi, T., & Kobayashi, H. (2016). Electrode morphology in all-solid-state lithium secondary batteries consisting of $LiNi_{1/3}Co_{1/3}Mn_{1/3}O_2$ and Li_2S-P_2S_5 solid electrolytes. *Solid State Ionics, 285*, 112–117.

Bulk-Type Solid-State LIB

Nobuya Machida

Abstract The all-solid-state batteries have the potential advantages of safety performance for an accident such as thermal shock during utilization. In this chapter, we introduce some studies using the bulk-type solid-state test cells as examples for the investigation of electrode materials and electrode composites. $LiAlO_2$ coating on $LiNi_{1/3}Mn_{1/3}Co_{1/3}O_2$ (NMC) was investigated in order to improve the electrochemical properties of the all-solid-state battery using a sulfide-based solid electrolyte amorphous Li_3PS_4. The $LiAlO_2$ coating on NMC powders was carried out by the use of a sol–gel method. A bulk-type solid-state test cell was constructed with the obtained $LiAlO_2$-coated NMC. The charge–discharge cycling test of the cell revealed that the test cell with $LiAlO_2$-coated NMC showed the capacity larger than 124 $mAhg^{-1}$ even after 400 cycles. Incidentally, the amount of solid electrolyte in the composite electrode limits the total energy density of the battery. In order to reduce the solid electrolyte content in the positive electrode composites, we were conducted on the solid electrolyte coating on the active materials NMC by a wet process. The test cell with the Li_3PS_4-coated NMC showed the capacity of $145mAhg^{-1}$, which was calculated on the base of the NMC weight in the composite electrode. The capacity was larger than that of the non-coated NMC. The all-solid-state battery with the Li_3PS_4-coated NMC had larger specific energy density than that with the non-coated NMC with the same Li_3PS_4 content.

Keywords All-solid-state battery · Lithium-ion battery · $LiAlO_2$ coating · Li_3PS_4 coating

1 Introduction

Toward the goal of realizing sustainable and greener energy scenario, economic and efficient energy storage forms a key challenge. In this context, rechargeable lithium-ion batteries have become a frontrunner device empowering small-scale portable

N. Machida (✉)
Faculty of Science, Konan University, Kobe, Japan
e-mail: machida@konan-u.ac.jp

K. Kanamura (ed.), *Next Generation Batteries*,
https://doi.org/10.1007/978-981-33-6668-8_10

electronic devices to large-scale electric vehicles. The solidification of lithium-ion batteries, in which non-flammable lithium-ion conducting solid electrolytes are used, is very useful for improving the safety and reliability of the lithium-ion batteries [1–4]. The all-solid-state batteries have the potential advantages, no leakage of electrolyte solutions, easy miniaturization such as thin-film batteries, and have safety performance for an accident such as thermal shock during utilization. However, there are no commercially available all-solid-state batteries for practical high-power uses. It indicates that the all-solid-state batteries for high power uses are under the stage of research and development. The all-solid-state test cells for the research and developments are roughly divided into two types, (1) bulk-type solid-state test cell and (2) sheet-type solid-state test cell, according to the purpose and stage of the research and developments.

The advantages of the bulk-type test cells are easy assembling and simple cell constitutions. The bulk-type test cells can be assembled by only successive pressing positive electrode composites, solid electrolytes, and negative electrode composites into a pellet. The bulk-type test cells can be also constituted without additives such as polymer binders, which sometimes cause deterioration in the electrochemical properties of the test cells. The bulk-type solid-state test cells, in consequent, are frequently used in order to investigate the electrochemical properties of newly appeared electrode materials and also newly prepared electrode composites for the all-solid-state batteries. The bulk-type test cells are also used to estimate the appropriate loading amount of active materials for the development of the cells with high energy densities. On the other hand, the sheet-type solid-state test cells are prototype batteries for commercially available batteries. The details on the sheet-type cells are described in Chap. 11. In this chapter, some studies using the bulk-type solid-state test cells are introduced as examples for the investigation of electrode materials and electrode composites.

2 Typical Bulk-Type Solid-State Test Cells and Their Properties

In order to improve the electrochemical properties of the all-solid-state battery using sulfide-based solid electrolytes, many studies have been conducted to form an effective electrode–electrolyte interface, because electrochemical reactions occur through the solid–solid interface between the electrode and solid electrolyte materials, and the interface governs the electrochemical properties of the all-solid-state batteries. Coating of lithium-ion conductive oxides, for example, $Li_4Ti_5O_{12}$ and $LiNbO_3$ on positive electrode active materials such as $LiCoO_2$ and $LiNi_{1/3}Mn_{1/3}Co_{1/3}O_2$, has been reported as an effective modification technique [5–11]. Accordingly, $LiAlO_2$ coating on $LiNi_{1/3}Mn_{1/3}Co_{1/3}O_2$ (NMC) was investigated [11].

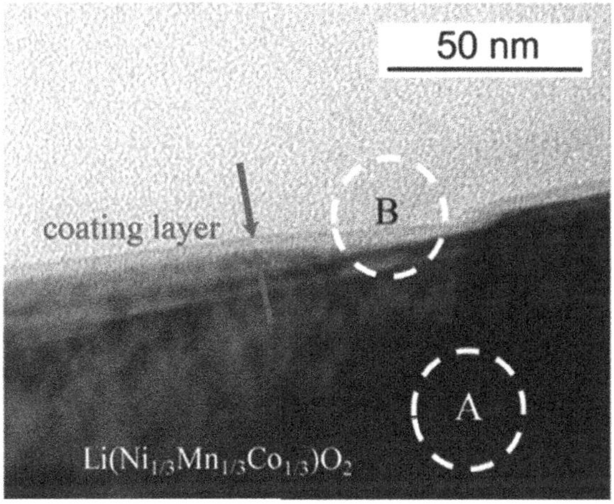

Fig. 1 A TEM image of the 1.0 mol% $LiAlO_2$-coated $LiNi_{1/3}Mn_{1/3}Co_{1/3}O_2$ (NMC) particles annealed at 350 °C. The points for EDX measurements are also shown as dotted circles, A and B, in the figure. Reprinted from Ref. [11], Copyright 2014, with permission from Elsevier

The $LiAlO_2$ coating on NMC powders was prepared by the use of a sol–gel method. The coating sol was prepared from the raw materials, aluminium (III) sec-butoxide, ethyl-acetoacetate, H2O, and 2-propanol. After mixing those raw materials, an appropriate amount of lithium methoxide was added to the mixture, which was stirred continuously at room temperature. NMC powders were added into the coating sol and ultrasonically agitated 40 °C for 30 min. The mixture was concentrated under reduced pressure, and the residual was obtained as a precursor. The $LiAlO_2$-caoted NMC powders were obtained by heat treatment of the precursor at 350 °C. The TEM image of the surface of the 1.0 mol% $LiAlO_2$-coated NMC powder is depicted in Fig. 1. It is clearly seen that the surface of the NMC particle is coated with a homogeneous layer with thickness of 8–15 nm. The TEM observation confirms that the $LiAlO_2$ coating layer is amorphous state. The elemental point analysis using EDX spectroscopy was carried out on the sample. The EDX spectra were measured at positions, A and B, denoted by dotted circles in Fig. 1. The electron beam of about 30 nm in diameter was used for the EDX measurements. The EDX spectrum measured in the surface area B showed a signal of Al element in addition to the signals of Ni, Mn, and Co elements. On the other hand, the signal of Al element cannot be detected in EDX spectrum that has been measured in the bulk area (point A). Those results suggest that the surface of NMC particle has been covered $LiAlO_2$ layer and Al element has not diffused into the inner area of the NMC particles during the heat treatment in the $LiAlO_2$ coating process.

A bulk-type solid-state test cell was constructed with the obtained $LiAlO_2$-coated NMC. A schematic diagram of the bulk-type test cell is shown in Fig. 2. In this case, amorphous Li_3PS_4 was used as a separator of solid electrolytes and $Li_{4.4}Si$ alloy as negative electrode material. The positive electrode of the cell was a mixture of the active material, the $LiAlO_2$-coated NMC (59 wt%), amorphous Li_3PS_4 solid electrolyte (39 wt%) and acetylene black (2 wt%). The cell was prepared by successively

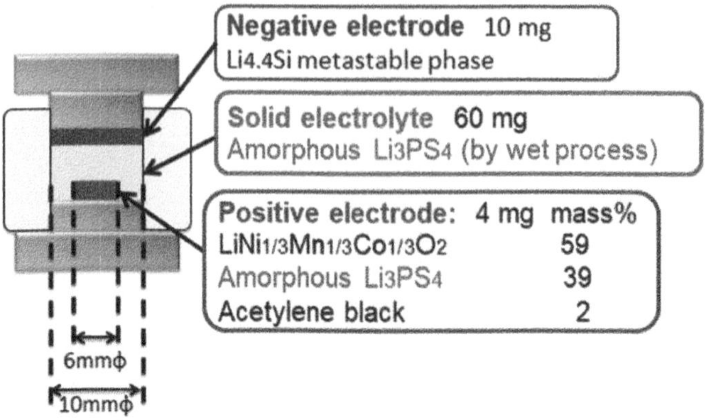

Fig. 2 Bulk-type solid-state cell for electrochemical measurement

pressing the positive electrode mixture, the solid electrolyte powders, and $Li_{4.4}Si$ alloy powders at 380 MPa into a pellet of 10 mm diameter.

The charge–discharge cycling tests of the cells were carried out under a constant current density of 0.1 mA cm^{-2} at 25 °C. In the test, charge–discharge criteria of voltage were set 4.3 and 2.5 V, respectively. Figure 3 shows the charge–discharge curves of the test cell with the 1.0 mol% $LiAlO_2$-coated NMC. The charge–discharge curves of the cell with uncoated NMC are also shown in the figure for comparison. The abscissa of the figure is the capacity calculated on the base of weight of the NMC in the composite electrode. The first discharge capacity of the cell with the uncoated NMC is about 102 $mAhg^{-1}$ and the capacities gradually decreased with an increase in the cycle number. After 100 cycles, the capacity was degraded to

Fig. 3 Charge–discharge curves of the cells $Li_{4.4}Si$ | a-Li_3PS_4 | NMC; **a** the uncoated sample and **b** 1.0 mol% $LiAlO_2$-coated sample. The current density was 0.1 $mAcm^{-2}$. Reprinted from Ref. [11], Copyright 2014, with permission from Elsevier

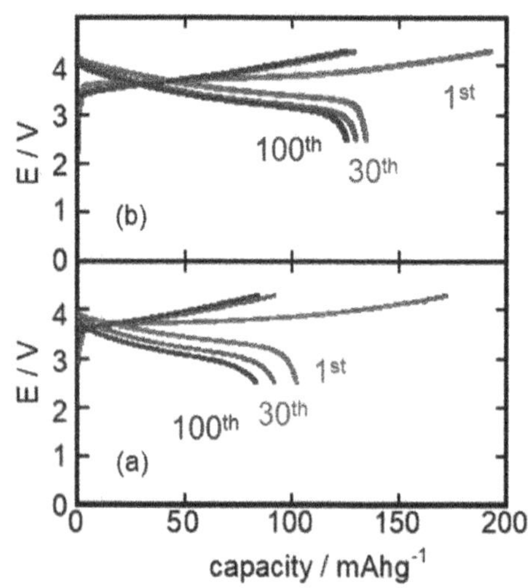

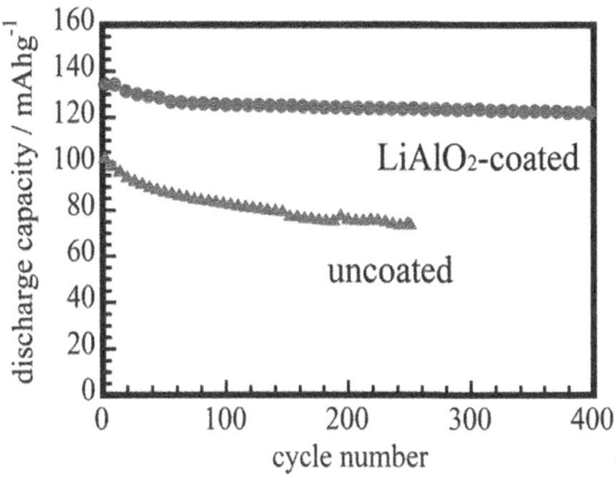

Fig. 4 Cycle performance of the cells $Li_{4.4}Si$ | a-Li_3PS_4 | NMC at the current density of 0.1 $mAcm^{-2}$. The closed triangles denote the cell performance of the uncoated NMC and the closed circles denote the cell performance of the 1.0 mol% $LiAlO_2$-coated NMC. Reprinted from Ref. [11], Copyright 2014, with permission from Elsevier

be 82 $mAhg^{-1}$. On the other hand, the cell with the $LiAlO_2$-coated NMC shows the capacity of 134 $mAhg^{-1}$ for the first cycle. The cell with the $LiAlO_2$-coated NMC keeps the capacity larger than 125 $mAhg^{-1}$ even after 100th charge–discharge cycles. The capacity retention rate after 100 cycles is about 95% for the cell with the $LiAlO_2$-coated NMC.

Figure 4 shows cycle dependence of specific capacities of the bulk-type test cells with the uncoated and the 1.0 mol% $LiAlO_2$-coated NMC as positive electrodes. The first discharge capacity of the cell with the $LiAlO_2$-coated NMC is 134 $mAhg^{-1}$. The discharge capacity is gradually decreasing with an increase in cycle number in the range of 1–50 cycles. After 50 cycles, however, the cell keeps a specific capacity larger than 124 $mAhg^{-1}$ even after 400 cycles. The capacity after 400 cycles is more than 92% of the initial discharge capacity. On the other hand, the cell with the uncoated NMC shows the capacity of 102 $mAhg^{-1}$ for the first discharge. The capacity is continuously decreased with an increase in cycle number and degrades to be 74 $mAhg^{-1}$ after 250 cycles. Those results suggest that the $LiAlO_2$ coating improves the cell performance and the capacity retention for charge–discharge cycling.

Indeed, the $LiAlO_2$ coating is very effective to improve battery performance, but the energy density of the bulk-type solid-state cell is only about 15 $Whkg^{-1}$. The energy density of the cell was calculated from the total weights of the positive electrode composite, the solid electrolyte separator, and the negative electrode materials, excepting the weights of current collectors and the cases. Hence further improvement of the bulk-type test cell is needed to develop practical batteries.

In order to improve the energy density of the test cell and to estimate the limit of the loading amounts of the positive electrode composite, the test cells with a large amount of the electrode composite are examined. The schematic diagram of the test cell is illustrated in Fig. 5a. The weight of the positive electrode composite of the cell is 65 mg, which is about 4.5 times larger than that of the cell as shown in Fig. 2. The total weight of the cell is about 150 mg, excepting the weights of the current

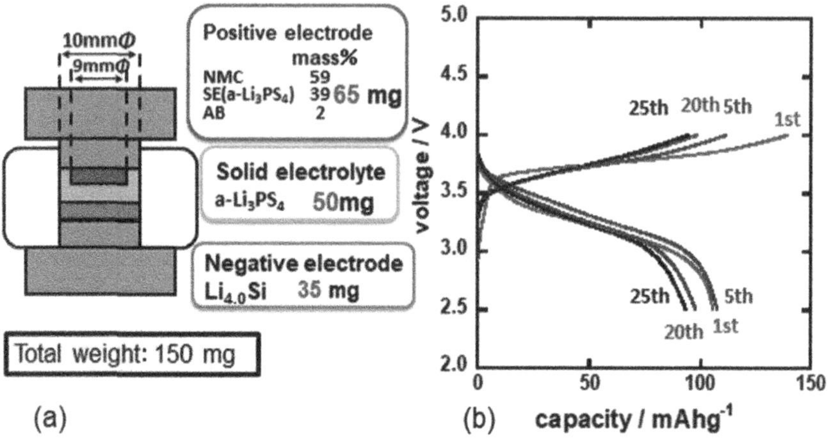

Fig. 5 **a** Schematic diagram of a solid-state test cell for high energy density. **b** Charge–discharge curves of the test cell at the current density of 0.1 $mAcm^{-2}$ at 25 °C

collectors and the cases. The charge–discharge cycle test was carried out under a constant current density of 0.1 $mAcm^{-2}$ at 25 °C. The charge–discharge curves of the cell with the large amount of the positive electrode composite are shown in Fig. 5b. The first discharge capacity of the test cell is about 110 $mAhg^{-1}$ and the energy density of the cell is 90 $Whkg^{-1}$ for the first discharge. The energy density is about six times larger than that of the cell shown in Fig. 2.

The charge–discharge cycle test was performed under several different current densities, 0.1, 0.4, 0.8 $mAcm^{-2}$. The current density was changed every five cycles. After the test with the current density of 0.8 $mAcm^{-2}$, the current density was changed back to the initial current density of 0.1 $mAcm^{-2}$. The cycle dependences of the charge and discharge capacities and typical discharge curves at those current densities are shown in Fig. 6a–b, respectively. The bulk-type solid-state cell with the large amount of the electrode composite, 65 mg, shows good performance even under the current density of 0.8 $mAcm^{-2}$. However, the capacity is decreased to 48 $mAhg^{-1}$ under the current density of 0.8 $mAcm^{-2}$. The deterioration of the capacity under the high current density would be related to the thick layer of the positive electrode. The thickness of the electrode layer is about 500 μm. The electrode layer would be too thick to operate the cell under the high current density. Those results suggest that there is a limit of the loading amount of the positive electrode composite in the bulk-type solid-state cells, and the limit of thickness of the positive electrode layer would be estimated to be 500 μm. Hence, layer-type solid-state cells, which have thin layer structure, are needed to develop solid-state batteries for practical uses.

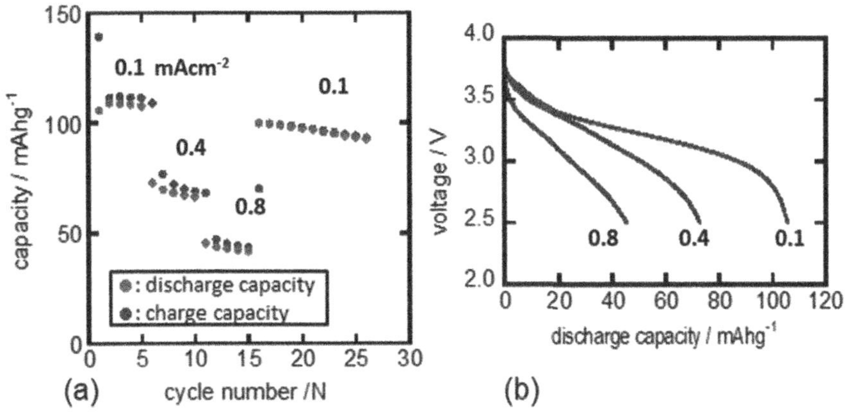

Fig. 6 **a** Cycle dependences of the charge and discharge capacities of the test cell with a large amount of positive electrode composite. **b** Typical discharge curves at the current densities of 0.1, 0.4, and 0.8 mAcm^{-2}

3 Solid Electrolyte-Coated Electrode Materials

The amount of solid electrolyte in the composite electrode is crucial to obtain a low interfacial electrode/electrolyte resistance, however, large amount of the solid electrolyte limits the total energy density of the batteries. In order to reduce the solid electrolyte content in the positive electrode composites and to achieve the good contact between solid electrolyte and positive electrode active materials, many researches were conducted on the solid electrolyte coating on the active materials by the use of dry [12, 13] and wet [14–18] processes. Here, a wet process is introduced as an example of solid electrolyte coating on active materials. Amorphous Li_3PS_4 solid electrolyte shows relatively high lithium-ion conductivity 4×10^{-4} Scm^{-1} at room temperature and a wide electrochemical window in the range of 0–5.0 V versus the Li^+/Li electrode. Hence, the amorphous Li_3PS_4 was selected as the coating solid electrolyte on the $LiNi_{1/3}Mn_{1/3}Co_{1/3}O_2$ (NMC) positive electrode materials.

The Li_3PS_4-coated NMC powder was prepared by the use of a three-step wet process. For the first step, the raw materials, Li_2S (Mitsuwa pure chemicals, 99.9%) and P_2S_5 (Sigma-Aldrich, 99%) were reacted in tetrahydrofuran (THF) for 24 h at room temperature. The mixture was concentrated under reduced pressure at 80 °C and the residual was obtained as a precursor of Li_3PS_4. The obtained precursor contains a little amount of THF. The remained THF causes the transformation of amorphous Li_3PS_4 into β-Li_3PS_4 crystalline phase, when the precursor is heated at high temperature to remove the remained THF. In order to solvent exchange THF for 1,2-diethoxyethane (DEE), in the second step, the precursor of Li_3PS_4 was added into DEE, which was used as a solvent and stirred for 48 h at room temperature. The slurry solution of $Li_3PS_4$with DEE was obtained. In the third step, the $LiNbO_3$-coated NMC powders were added into the slurry and ultrasonically agitated at 40 °C

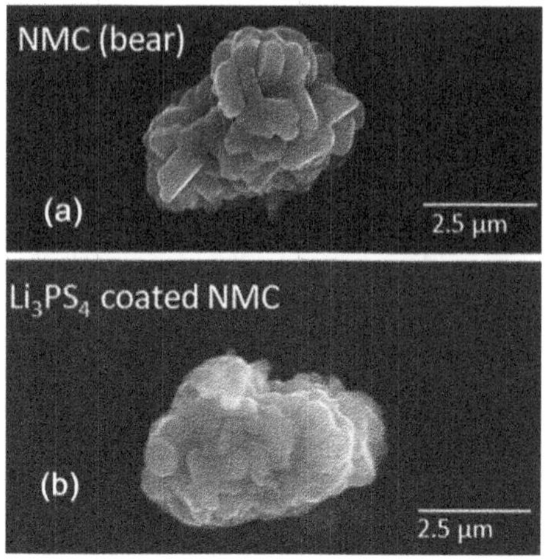

Fig. 7 FE-SEM images of **a** the uncoated $LiNi_{1/3}Mn_{1/3}Co_{1/3}O_2$ particle and **b** the 24.5 wt% Li_3PS_4-coated $LiNi_{1/3}Mn_{1/3}Co_{1/3}O_2$ particle

for 30 min. The $LiNbO_3$ coating is very effective to passivate the interdiffusion of transition metal and sulfur species at the boundary between NMC particle and Li_3PS_4 solid electrolyte. The interdiffusion of the transition metal and sulfur species causes the deterioration of the electrochemical performance of the active materials [9]. The slurry was prepared for the final products to have the Li_3PS_4 content in the range of 0–40 wt%. The mixture was concentrated under reduced pressure, and the residual was obtained as a precursor. The precursor was heated at 160 °C for 2 h and then the amorphous Li_3PS_4-coated NMC powders were obtained.

FE-SEM and energy dispersive X-ray (EDX) measurements were carried out in order to observe morphology of the uncoated NMC and the Li_3PS_4-coated NMC powders. Figure 7 displays FE-SEM images of the NMC particle coated with 24.5 wt% Li_3PS_4. Those NMC powders were heat-treated at 160 °C for 2 h in vacuo. The image of the uncoated NMC powder is also shown in the figure for comparison. For the uncoated NMC sample, as shown in Fig. 7a, the primary particle size is about 0.5–2.0 μm in diameter and those small particles aggregated each other to form spherical secondary particles about 5–8 μm in diameter. In the Li_3PS_4-coated sample, there are no significant differences in the surface morphology and in the particle size in the scale of the SEM observation. The Li_3PS_4-coated particle, however, has smooth surfaces and the surfaces can be attributed to Li_3PS_4.

The EDX mappings of the Li_3PS_4-coated NMC are depicted in Fig. 8. Both P and S atoms of Li_3PS_4 and Mn atom of NMC are detected in the same area. Those results indicate the formation of Li_3PS_4 layer on NMC particles. In addition, whitish particles that are denoted by dotted circles in Fig. 8a are newly observed and stick on the surface of the secondary particles of NMC. The newly appeared particles have smooth surfaces and can be attributed to Li_3PS_4.

Laboratory-scale solid-state cells were constructed with the obtained Li_3PS_4-coated NMC, amorphous Li_3PS_4 as the solid electrolytes, and $Li_{4.4}Si$ alloy as the negative electrode materials. The amorphous sulfide-based electrolyte, Li_3PS_4, was

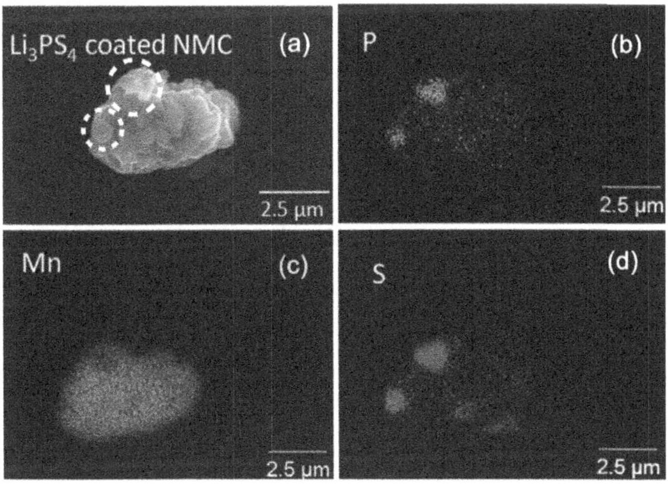

Fig. 8 **a** FE-SEM image of the 24.5 wt% Li_3PS_4-coated NMC particle. The EDX mappings of the 24.5 wt% Li_3PS_4-coated NMC particle for **b** P, **c** Mn, and **d** S atoms obtained from the same field of the SEM image

synthesized by the wet process without the third step already described above. The Li_3PS_4 was used as the electrochemical separator between the positive and negative electrodes. The synthesized Li_3PS_4 shows high lithium-ion conductivity, 2×10^{-4} Scm^{-1} at room temperature. The positive electrode of the test cell was a mixture of the obtained Li_3PS_4-coated NMC (98 wt%) and acetylene black (2 wt%). The test cell was prepared by successively pressing the positive electrode mixture, the Li_3PS_4 solid electrolyte powder, and the $Li_{4.4}Si$ alloy powder at 380 MPa into a pellet of 10 mm diameter. In order to elucidate the effect of the Li_3PS_4 coating on the active materials, the test cell with the composite of the non-coated NMC, Li_3PS_4 which was separately synthesized as solid electrolytes, and acetylene black (2 wt%) was also constructed. In this case, the composite electrode has the same Li_3PS_4 content as the electrode with the Li_3PS_4-coated NMC.

Galvanostatic charge–discharge cycling tests of the cells were performed under a constant current density of 0.1 $mAcm^{-2}$ at 298 K. Figure 9 shows the first charge–discharge curves of the bulk-type test cell with the Li_3PS_4-coated NMC. The positive electrode composite contains 24.5 wt% Li_3PS_4. The first charge–discharge curves of the test cell with the non-coated NMC are also shown in the figure for comparison. The positive electrode with non-coated NMC contains the same amounts of Li_3PS_4 as that with the Li_3PS_4-coated NMC. In the charge process, both test cells with the Li_3PS_4-coated and with the non-coated NMCs show almost same properties. The charge capacities of both test cells are about 160 $mAhg^{-1}$. On the other hand, in the discharge process, the test cell with the Li_3PS_4-coated NMC shows a higher discharge capacity of 140 $mAhg^{-1}$ than the test cell with the non-coated NMC, which shows the discharge capacity of about 120 $mAhg^{-1}$. The discharge process induces the volume shrinkage of NMC particles and the shrinkage separates some portions of the active

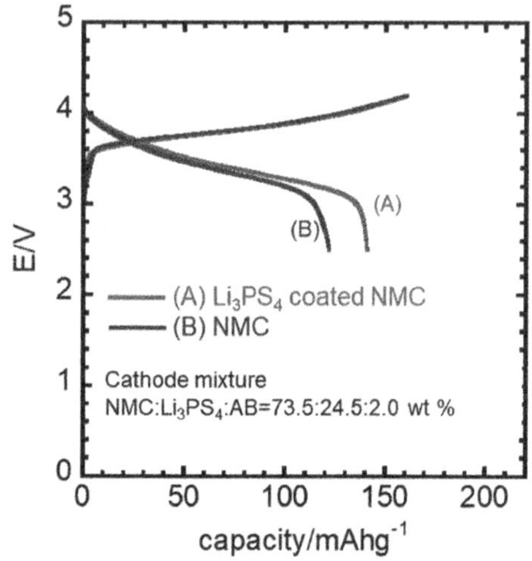

Fig. 9 The first charge–discharge curves of the test cells (**A**) with the 24.5 wt% Li_3PS_4-coated NMC and (**B**) with the uncoated NMC

materials NMC from the solid electrolyte Li_3PS_4 in the composite electrode. The breaking contact between the active materials and solid electrolyte in the electrode composite should cause the decrease in the discharge capacity. Those results indicate that the Li_3PS_4-coated NMC would have the excellent close contactness between the Li_3PS_4 solid electrolyte and the NMC active materials in the positive electrode composite.

Figure 10a shows the Li_3PS_4-content dependence of the discharge capacity of the Li_3PS_4-coated NMC. The discharge capacity is calculated on the base of the weight of NMC contained in the positive electrode composites. The discharge capacity of the

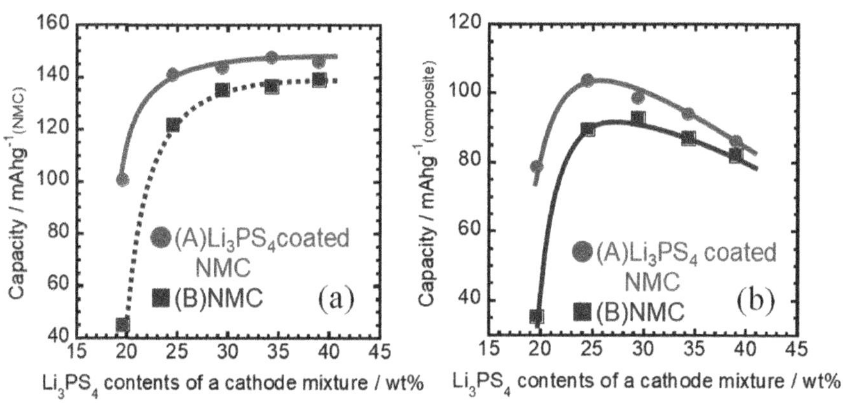

Fig. 10 Li_3PS_4 content dependences of the first discharge capacities of the bulk-type solid-state cells with the uncoated NMC and with the Li_3PS_4-coated NMC; **a** the discharge capacities were calculated on the base of weight of the NMC contained in the positive electrodes and **b** the discharge capacities were calculated on the base of the total weights of the positive electrodes

non-coated NMC is also shown in the figure for comparison. The capacity is increased with an increase in the Li_3PS_4 contents and reaches almost a constant capacity of about 145 $mAhg^{-1}$ for the cell with the Li_3PS_4-coated NMC. The capacities of the Li_3PS_4-coated NMC are larger than those of the non-coated NMC all over the composition range of 19–40 wt%.

The discharge capacity that is calculated on the base of the total weight of the composite electrode is shown in Fig. 10b as a function of the Li_3PS_4 contents. In this case, the discharge capacity shows the maximum of 105 $mAhg^{-1}$ at around the Li_3PS_4 content 25 wt%. The capacity of the non-coated NMC is also shown the maximum at the same composition range. However, the maximum capacity of the non-coated NMC is about 90 $mAhg^{-1}$ and is smaller than that of the Li_3PS_4-coated NMC. Those results suggest that the all-solid-state battery with the Li_3PS_4-coated NMC should have larger specific energy density than that with the non-coated NMC with the same Li_3PS_4 content. Hence, the solid-electrolyte coating on the active materials such as NMC is an effective way to develop all-solid-state lithium-ion batteries that have high energy density.

References

1. Minami, T. (2005). *Solid state ionics for batteries*. Tokyo: Springer.
2. Kamaya, N., Homma, K., Yamakawa, Y., Hirayama, M., Kanno, R., Yonemura, M., et al. (2011). *Nature Materials, 10*, 682–686.
3. Richards, W. D., Miara, L. J., Wang, Y., Kim, J. C., & Ceder, G. (2016). *Chemistry of Materials, 28*, 266–273.
4. Takada, K. (2013). *Acta Materialia, 61*, 759–770.
5. Ohta, N., Takada, K., Zhang, L. Q., Ma, R. Z., Osada, M., & Sasaki, T. (2006). *Advanced Materials, 18*, 2226–2229.
6. Takada, K., Ohta, N., Zhang, L., Fukuda, K., Sakaguchi, I., Ma, R., et al. (2008). *Solid State Ionics, 179*, 1333–1337.
7. Takada, K., Ohta, N., Zhang, L., Xu, X., Hang, B. T., Ohnishi, T., et al. (2012). *Solid State Ionics, 225*, 594–597.
8. Sakuda, A., Kitaura, H., Hayashi, A., Tadanaga, K., & Tatsumisago, M. (2009). *Journal of the Electrochemical Society, 156*, A27–A32.
9. Sakuda, A., Hayashi, A., & Tatsumisago, M. (2010). *Chemistry of Materials, 22*, 949–956.
10. Machida, N., Kashiwagi, J., Naito, N., & Shigematsu, T. (2012). *Solid State Ionics, 225*, 354–358.
11. Okada, K., Machida, N., Naito, N., Shigematsu, T., Ito, S., Fujiki, S., et al. (2014). *Solid State Ionics, 255*, 120–127.
12. Sakuda, A., Hayashi, A., Ohtomo, T., Hama, S., & Tatsumisago, M. (2011). *Journal of Power Sources, 196*, 6735–6741.
13. Ito, Y., Otoyama, M., Hayashi, A., Ohtomo, T., & Tatsumisago, M. (2017). *Journal of Power Sources, 360*, 328–335.
14. Yubuchi, S., Teragawa, S., Aso, K., Tadanaga, K., Hayashi, A., & Tatsumisago, M. (2015). *Journal of Power Sources, 293*, 941–945.
15. Rosero-Navarro, N. C., Kinoshita, T., Miura, A., Higuchi, M., & Tadanaga, K. (2017). *Ionics, 23*, 1619–1624.
16. Phuc, N. H. H., Morikawa, K., Mitsuhiro, T., Muto, H., & Matsuda, A. (2017). *Ionics, 23*, 1–7.

17. Rosero-Navarro, N. C., Miura, A., & Tadanaga, K. (2018). *Journal of Power Sources, 396,* 33–40.
18. Yubuchi, S., Uematsu, M., Hotehama, C., Sakuda, A., Hayashi, A., & Tatsumisago, M. (2019). *Journal of Materials Chemistry A, 7,* 558–566.

Sheet-Type Solid-State LIB

Atsushi Sakuda

Abstract Sheet-type all-solid-state batteries, which consist of a solid electrolyte separator layer and positive and negative electrodes coated on current collectors, have a more practical configuration than pellet-type batteries. In this section, the slurry coating process, which is the most commonly used method for sheet-type all-solid-state batteries, is discussed. The developments of solid electrolyte fine particles, electrode and solid electrolyte sheets, pressing, and confining pressure are outlined.

Keywords All-solid-state batteries · Slurry · Solid electrolyte particles · Electrode sheets

1 Introduction

Sheet-type all-solid-state batteries, which consist of a solid electrolyte separator layer and positive and negative electrodes coated on current collectors, have a more practical configuration than pellet-type batteries. However, reports on sheet-type all-solid-state batteries remain scarce [1–7]. In this chapter, the slurry coating process, which is the most commonly used method for sheet-type all-solid-state batteries, is discussed.

2 Basic Requirements

Cell design is basically the same as with lithium-ion batteries except for the differences derived from using solid electrolytes. Positive and negative electrodes are separated by the solid electrolyte layer. The electrode layers require active materials and sufficient ionic and electronic conductivity. Both ionic and electronic conductivity must be implemented based on the desired current density. Because all materials are

A. Sakuda (✉)
Osaka Prefecture University, Sakai, Japan
e-mail: saku@chem.osakafu-u.ac.jp

K. Kanamura (ed.), *Next Generation Batteries*,
https://doi.org/10.1007/978-981-33-6668-8_11

solid, good contact between the solids, which include active materials, solid electrolytes, conducting additives, and current collectors, must be ensured. In particular, ionic conduction must be ensured by means of homogeneous solid electrolyte distribution. Thus, composite electrodes in which solid electrolytes are mixed into the positive and electrode layer are standard design. Because the solid electrolyte plays a role in maintaining the structure, a binder is not essential in all-solid-state batteries [8]. By contrast, bulk-type all-solid-state cells are constructed using a powder process. Therefore, the use of a binder facilitates cell fabrication [4].

The most important requirement in this process is determining a good combination of solid electrolyte, binder, and solvent to enable slurry coating [4]. The binder should have a low resistance, and appropriate-sized solid electrolyte particles should be used to obtain homogeneous solid electrolyte and electrode sheets.

All-solid-state cells that use inorganic solid electrolytes are less flexible than conventional lithium-ion batteries. Thus, round/flat jelly roll-type cells, which are common for conventional lithium-ion batteries with liquid electrolytes, are difficult to construct, and stacking-type cells are considered to be the most promising configuration for all-solid-state batteries.

3 Solid Electrolyte Fine Particles

To obtain a homogeneous solid electrolyte separator layer, the largest solid electrolyte particles should be at least less than half the target thickness of the solid electrolyte layer. Typical target thickness is 10–30 μm for bulk-type all-solid-state batteries. Thus, the maximum size of the solid electrolyte must be less than at least ten micrometers. To form an effective lithium-ion conducting pathway, using fine solid electrolyte particles that are smaller than active material particles is essential [9]. However, excessive miniaturization makes handling difficult, increases side reactions, and reduces the density of the solid electrolyte. Therefore, examining an appropriate size according to the purpose is necessary.

It is known that sulfide-based solid electrolytes densify easily simply by cold pressing [10]. This is the greatest advantage of sulfide-based solid electrolytes. However, their mechanical property also makes it difficult to obtain small solid electrolyte particles, as their size increases during the pulverization process [11]. Thus, preparing solid electrolytes without secondary particles of several tens of micrometers is relatively difficult. It has been reported that wet ball milling in the presence of dispersants is an effective means of obtaining small particles [12]. Figure 1 shows fine Li_3PS_4 glass powder synthesized from Li_2S and P_2S_5 using dibutyl ether as a dispersant [4, 12]. Solid electrolyte small particles were obtained by using dibutyl ether.

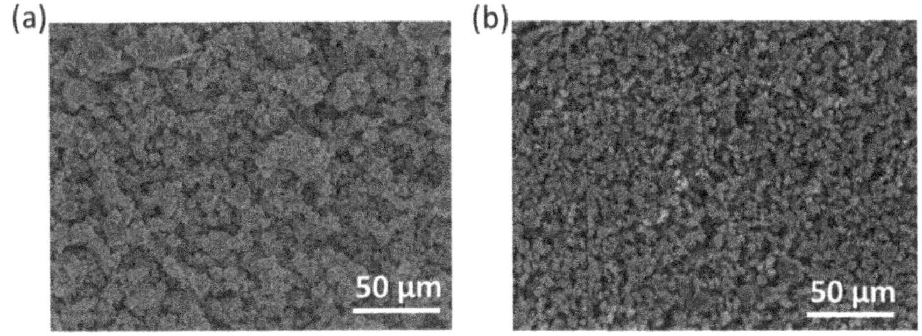

Fig. 1 SEM images of Li_3PS_4 glass powder mechanochemically prepared by wet ball milling using a heptane solvent **a** without and **b** with dibutyl ether as a dispersant

4 Electrode and Solid Electrolyte Sheets

Preparation of electrode and electrolyte slurries is essential to achieve high performance. Figure 2 shows simplified preparation procedures of electrode and electrolyte sheets. The first requirement is to identify the right combination of solid electrolyte, binder, and solvent that does not exhibit an unfavourable side reaction and enables slurry coating. Note that the binder can interfere with the contact between the particles and become a large resistance component [4]. Reducing the binder amount affects the binding force between the current collector foil and electrode layer. Therefore, a device that prevents the peeling of the electrode layer is critical for the current collector foil. Roughening the current collector surface or introducing an additional layer is also effective. For the solvent, using one that has low polarity and that does not easily react with the sulfide-based solid electrolyte is preferable. Viscosity, wettability with particles, and volatility of solvents are also important parameters for preparing a good slurry. Figure 3 is a photograph of a solid electrolyte slurry. That the particles are well dispersed in the solvent is necessary, and this often requires a suitable disperser.

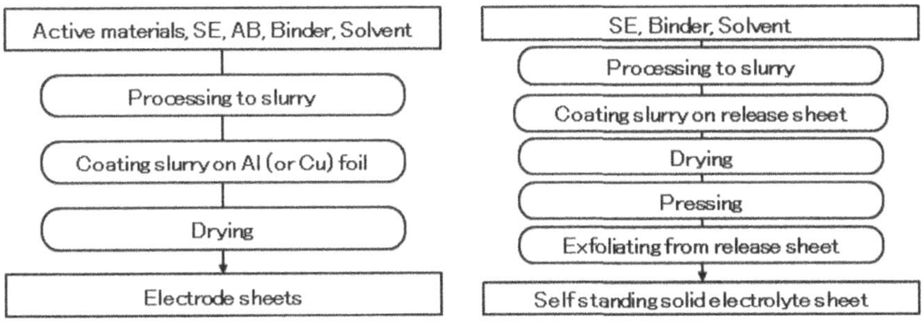

Fig. 2 Procedure for preparing electrode sheets and self-standing solid electrolyte sheets

Fig. 3 Photo of a sulfide solid electrolyte slurry

5 Pressing

Pressing is one of the most important processes in constructing all-solid-state batteries. Figure 4 is an SEM image of the cross-section of a sheet-type all-solid-state test cell constructed by stacking the electrode and electrolyte sheets by applying uniaxial pressing of 370 MPa at room temperature. A relatively dense all-solid-state cell was constructed simply by pressing at room temperature. For mass production, a roll-to-roll process is favourable, and roll pressing is used to press the electrode layer for conventional lithium-ion batteries using liquid electrolytes. However, all-solid-state cells require higher pressure than conventional cells to enable good contact between solid particles. This means using roll pressing is not easy. Development of advanced technology that enables roll pressing for fabrication of all-solid-state cells is anticipated. Furthermore, not only pressing of the electrode sheet but also pressing of the whole cell is required to ensure the positive electrode, solid electrolyte, and negative electrode layers all contact one another. As an alternative, the processes using high-pressure uniaxial pressing and cold (or hot) isostatic pressing have been

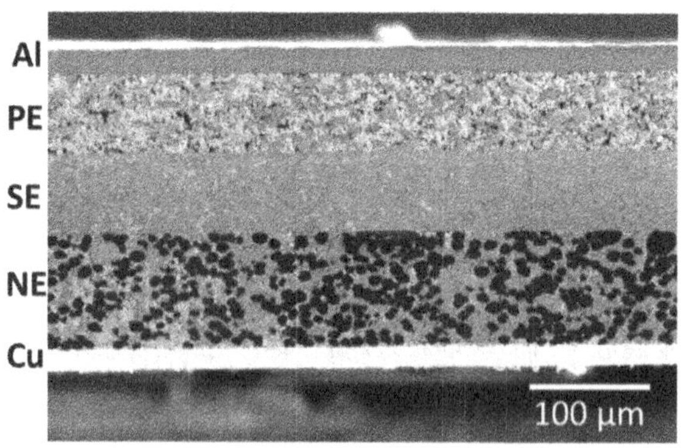

Fig. 4 SEM image of the cross-section of a sheet-type all-solid-state test cell

developed. Development of new materials and processes that decrease the required pressure is desired. The test cell shown in Fig. 4 has a thick solid electrolyte layer and inhomogeneous electrolyte distribution in the electrode layers. Recent progress in the manufacturing of cells has enabled to construct a homogeneous electrode layer that has a thinner electrolyte layer and a less-solid electrolyte.

6 Confining Pressure

All-solid-state cells usually perform well when applied with high confining pressure because the pressure assists the contact between solids. To apply such a high pressure for large-scale all-solid-state batteries, a heavy and large jig is required. The development of external packaging for all-solid-state batteries is critical. In addition, the development of high-temperature-resistant exteriors that take advantage of the high-temperature stability of all-solid-state batteries is expected.

References

1. Inada, T., Kobayashi, T., Sonoyama, N., Yamada, A., Kondo, S., Nagao, M., & Kanno, R. (2009). *Journal of Power Sources, 194,* 1085–1088.
2. Nam, Y. J., Cho, S. J., Oh, D. Y., Lim, J. M., Kim, S. Y., Song, J. H., et al. (2015). *Nano Letters, 15,* 3317–3323.
3. Yamamoto, M., Takahashi, M., Terauchi, Y., Kobayashi, Y., Ikeda, S., & Sakuda, A. (2017). *Journal of the Ceramic Society of Japan, 125,* 391–395.
4. Sakuda, A., Kuratani, K., Yamamoto, M., Takahashi, M., Takeuchi, T., & Kobayashi, H. (2017). *Journal of the Electrochemical Society, 164,* A2474–A2478.
5. Kim, D. H., Oh, D. Y., Park, K. H., Choi, Y. E., Nam, Y. J., Lee, H. A., et al. (2017). *Nano Letters, 17,* 3013–3020.
6. Riphaus, N., Strobl, P., Stiaszny, B., Zinkevich, T., Yavuz, M., Schnell, J., et al. (2018). *Journal of the Electrochemical Society, 165,* A3993–A3999.
7. Park, K. H., Bai, Q., Kim, D. H., Oh, D. Y., Zhu, Y., Mo, Y., & Jung, Y. S. (2018). *Advanced Energy Materials, 8,* 1800035.
8. Yamamoto, M., Terauchi, Y., Sakuda, A., & Takahashi, M. (2018). *Science Reports, 8,* 1212.
9. Sakuda, A., Takeuchi, T., & Kobayashi, H. (2016). *Solid State Ionics, 285,* 112–117.
10. Sakuda, A., Hayashi, A., & Tatsumisago, M. (2013). *Science Reports, 3,* 2261.
11. Sakuda, A. (2018). *Journal of the Ceramic Society of Japan, 126,* 675–683.
12. Sugiura, K., & Ohashi, M. *WO Pat.* 2013/073035.

Sulfur and Sulfide Positive Electrode

Nobuya Machida and Akitoshi Hayashi

Abstract All-solid-state Li/S and Li/Li_2S batteries have received a lot of attention as next-generation lithium-ion batteries, because elemental sulfur and lithium sulfide are attractive positive electrode materials for their high-specific electrochemical capacities (theoretical capacities of S and Li_2S are 1675 and 1167 $mAhg^{-1}$, respectively). However, the all-solid-state cells, Li/S and Li/Li_2S, have suffered from the low utilization of the electrode materials. In this chapter sulfur, and lithium sulfide positive electrode materials that are suitable for all-solid-state lithium batteries are reported. Sulfur–VGCF–Li_3PS_4 composites were prepared by a two-step ball-milling process. Electrochemical properties of the sulfur–VGCF–Li_3PS_4 composite were investigated using an all-solid-state test cell constructed in three layers: a positive electrode layer comprising the sulfur–VGCF–Li_3PS_4 composite (weight fraction 30: 10: 60), a solid electrolyte layer of amorphous Li_3PS_4, and a negative electrode layer consisting of $Li_{4.4}Si$ alloy. The first discharge capacity of the test cell at a current density of 0.1 $mAcm^{-2}$ at 50 °C was about 1530 $mAhg^{-1}$ estimated from the mass of sulfur in the composite electrode. The test cell retained the capacity of more than 1300 $mAhg^{-1}$ up to 50th cycle. On the other hand, for improving the utilization of Li_2S positive electrode, Li_2S–LiI solid solution was prepared and applied as active materials to the all-solid-state batteries. Ionic conductivity of the Li_2S–LiI solid solutions was increased with an increase in the LiI contents and the $80Li_2S \cdot 20LiI$ (mol%) solid solution showed the conductivity of more than 10^{-6} Scm^{-1} at room temperature. The all-solid-state test cell using the $80Li_2S \cdot 20LiI$ solid solution as positive electrode material showed good performance and retained the capacity of 980 $mAhg^{-1}$ at 2C for 2000 cycles without capacity degradation.

Keywords Li/s battery · Lithium-sulfur battery · Lithium-sulfide battery · Li_2S active materials

N. Machida (✉)
Faculty of Science, Konan University, Kobe, Japan
e-mail: machida@konan-u.ac.jp

A. Hayashi
Graduate School of Engineering, Osaka Prefecture University, Sakai, Japan
e-mail: hayashi@chem.osakafu-u.ac.jp

K. Kanamura (ed.), *Next Generation Batteries*,
https://doi.org/10.1007/978-981-33-6668-8_12

1 Introduction

New materials with high electrochemical capacities are required to develop positive electrodes for next-generation lithium-ion batteries, since the transition metal oxides (such as $LiCoO_2$ and $LiMn_2O_4$) used for these materials in conventional lithium-ion batteries have small electrochemical capacities ranging from 120 to 220 $mAhg^{-1}$[1, 2].

Elemental sulfur is an attractive positive electrode material for high-specific-energy lithium-ion batteries due to its high theoretical capacity of 1675 $mAhg^{-1}$. Furthermore, elemental sulfur is inexpensive and nontoxic. However, sulfur cannot be used as a reversible electrode material in lithium-ion batteries with conventional electrolyte solutions [2–6]. The capacity of lithium-sulfur batteries using conventional electrolyte solution rapidly fades with charge–discharge cycling. Capacity loss is attributed to the production of several lithium polysulfides in the battery during sulfur reduction; these lithium polysulfides can dissolve in the electrolyte solution [3, 5, 6]. Inorganic solid electrolytes are effective separators in Li/S batteries because polysulfide ions are not soluble in inorganic solid electrolytes [7–14]. An all-solid-state lithium-ion battery using inorganic solid electrolytes is expected to be one of the most promising devices in which elemental sulfur functions effectively as a rechargeable positive electrode material.

A sulfur positive electrode requires the use of lithium-containing negative electrode because the sulfur positive electrode has no lithium. Lithium sulfide, Li_2S, a Li/S battery discharge reaction product, is a promising positive electrode material, since it can be coupled with negative electrode materials that do not contain lithium.

In this chapter, sulfur and lithium sulfide positive electrode materials that are suitable for all-solid-state lithium batteries are reported.

2 Sulfur Positive Electrode

Sulfur–VGCF–Li_3PS_4 composites were prepared from sulfur (Wako Chemicals, 99%), VGCF (Showa Denko), and amorphous Li_3PS_4 by a two-step ball-milling process using a high energy ball-milling apparatus (Fritch, P-7). To prepare the composite, sulfur (75 wt%) and VGCF (25 wt%) were mixed first in order to provide electronic conductivity for sulfur. Next, amorphous $Li_3PS_4$was milled with the sulfur–VGCF precursor obtained in the first step in order to provide the composites with lithium-ion conductivity. Amorphous Li_3PS_4 (60 wt%) and the precursor (40 wt%) were ball-milled with a rotation rate of 380 rpm and milling time of 20 h. We obtained sulfur–VGCF–Li_3PS_4 composites with ratios of S/VGCF/Li_3PS_4 of 30:10:60 by weight. Electrochemical properties of the obtained sulfur–VGCF–Li_3PS_4 composite were investigated using an all-solid-state test cell constructed in three layers: a positive electrode layer comprising the sulfur-VGCF–Li_3PS_4

composite, a solid electrolyte layer of amorphous Li_3PS_4, and a negative electrode layer consisting of a metastable $Li_{4.4}Si$ alloy. The metastable alloy displays good charge–discharge reversibility and has a wide solid-solution range of $Li_{3.8}Si$ to $Li_{5.2}Si$ [15, 16]. Thus, the metastable $Li_{4.4}Si$ alloy is applicable as a negative electrode material for both types of all-solid-state batteries: those for which charge–discharge cycles begin with charging (lithium insertion into the alloy) or discharging (lithium extraction from the alloy). The cell was prepared by successively pressing the sulfur–VGCF–Li_3PS_4 composite powder, the amorphous Li_3PS_4 solid electrolyte, and the $Li_{4.4}Si$ alloy powder into a pellet 10 mm in diameter at 380 MPa.

Figure 1 shows the first charge–discharge curves of the all-solid-state test cells using the sulfur-VGCF-Li_3PS_4 composite (weight fraction 30:10:60) from 25 to 100 °C. Electrochemical capacity was calculated on the basis of the mass of sulfur in the composite electrode. The charge–discharge test was carried out under a constant current density of 0.1 $mAcm^{-2}$ (24 mAg^{-1}, 0.016 C in C rate) and began with the discharge process. Cut-off voltages of 2.6–0.9 V were used for the test at 25 and 50 °C. At 100 °C, the cut-off voltage range was 2.3–1.2 V. Under a current density of 0.1 $mAcm^{-2}$, the test cells indicate initial discharge and charge capacities greater than 1360 $mAhg^{-1}$, with little regard to operating temperature. Capacities were estimated using the mass of sulfur in the composite electrode. Flatness of the charge–discharge curves, however, improved with increasing operating temperature. These results suggest that the internal resistance of the all-solid-state test cells decreases with increasing operating temperature.

Figure 2a–d shows the first charge–discharge curves for the all-solid-state test cells at a current density of 1.0 $mAcm^{-2}$ (233 mAg^{-1}, 0.14 in C rate) at various temperatures. A cut-off voltage range of 2.6–0.9 V was used for test cells at 25–50 °C, while the range of 1.2–2.3 V was adopted for tests at 100–120 °C. The

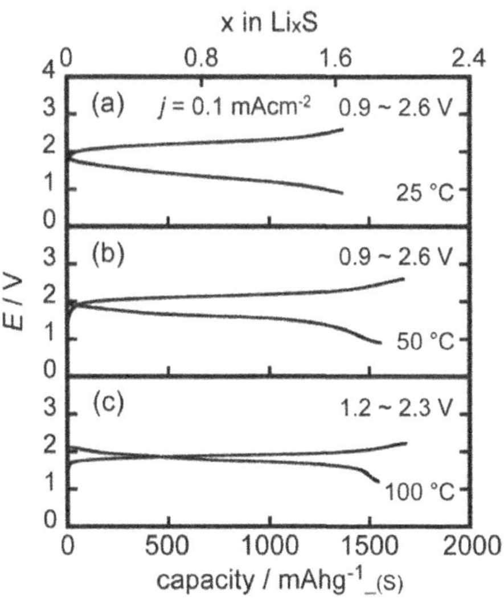

Fig. 1 First charge–discharge curves for the all-solid-state $Li_{4.4}Si$/amorphous Li_3PS_4/S–VGCF–Li_3PS_4 composite test cell at various temperatures; **a** 25 °C, **b** 50 °C, and **c** 100 °C. The charge–discharge test was carried out under a constant current density of 0.1 $mAcm^{-2}$ and began with the discharge process. Capacity was calculated on the basis of the mass of sulfur in the composite electrode

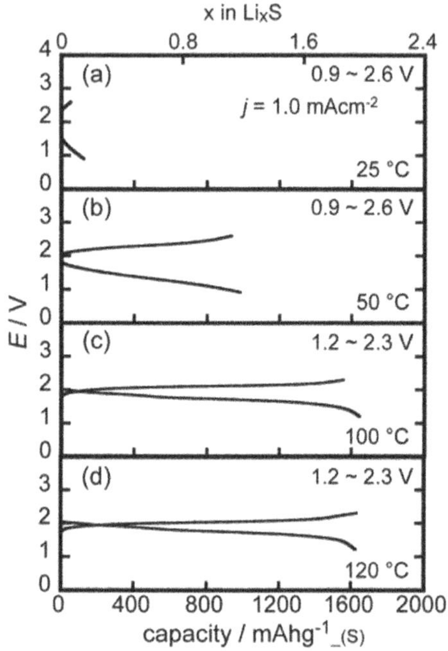

Fig. 2 First charge–discharge curves for the all-solid-state $Li_{4.4}Si$/amorphous Li_3PS_4/S–VGCF–Li_3PS_4 composite test cell, at various temperatures; **a** 25 °C, **b** 50 °C, **c** 100 °C, and **d** 120 °C. Charge–discharge current density was 1.0 mAcm^{-2}. Capacity was calculated on the basis of sulfur mass in the composite electrode

initial discharge capacity of the test cell at 25 °C was less than 200 mAhg^{-1}. The electrochemical capacity of the test cell decreased considerably under a high current density of 1.0 mAcm^{-2} at 25 °C, as shown in Fig. 2a. These results suggest that the all-solid-state test cell with the sulfur–VGCF–Li_3PS_4 composite struggles under high current density at room temperature. However, capacity increased with increasing operating temperature, reaching 1630 mAhg^{-1} at 120 °C, as shown in Fig. 2d. This capacity is approximately 97% of the theoretical capacity for sulfur. The all-solid-state lithium-sulfur test cell can operate at temperatures above 100 °C, where the initial discharge capacity of the cell reaches greater than 1600 mAhg^{-1}, even under a high current density of 1.0 mAcm^{-2}.

Figure 3 shows charge–discharge cycling curves for the all-solid-state cell with the sulfur composite electrode under a constant current density of 0.1 mAcm^{-2} at 50 °C. The cycle test was performed in the cut-off voltage range of 2.6–0.9 V. The 1st, 5th, 10th, 20th, and 50th charge–discharge curves are shown in Fig. 3. The discharge capacity of the first cycle is approximately 1530 mAhg^{-1}, estimated from the mass of sulfur in the composite electrode. Discharge capacity decreases slightly as the cycle number increases, reaching 1350 mAhg^{-1} after the 20th cycle. The capacity remains greater than 1300 mAhg^{-1} up to the 50th cycle. Discharge–charge efficiency is nearly 100% over those cycles.

The charge–discharge cycling test was also carried out at 120 °C under a constant current density of 1.0 mAcm^{-2} in the cut-off voltage range of 2.3–1.2 V. The charge–discharge current density of 1.0 mAcm^{-2} is 10 times larger than that used for

Fig. 3 Charge–discharge cycle curves for the $Li_{4.4}Si$/amorphous Li_3PS_4/S–VGCF–Li_3PS_4 composite test cell. The cycle test was carried out under a constant current density of 0.1 $mAcm^{-2}$ with cut-off voltage range of 0.9–2.6 V at 50 °C

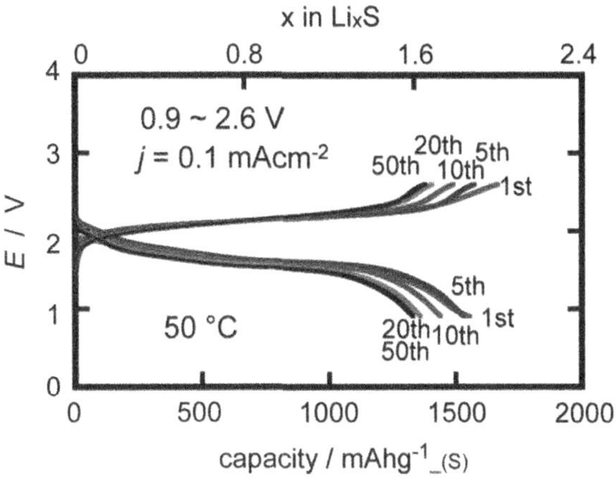

the test at 50 °C, as shown in Fig. 3. Charge–discharge curves for the test cell at 120 °C are shown in Fig. 4. The first discharge capacity is more than 1600mAhg^{-1}; capacity decreased significantly to approximately 1100 mAhg^{-1} by the third cycle. The decrease in capacity after the initial cycle for charge–discharge reactions at high temperature (120 °C) results from large changes in the volume of sulfur during cycling, concomitant with a reaction between S and Li_2S. The melting temperature of sulfur is approximately 120 °C. Thus, the working electrode composite should be softened at that temperature. A change in sulfur volume during cycling would strongly influence the electronic and/or ionic contacts in the composite electrodes.

After initial capacity degradation, however, the test cell retained a high capacity (more than 1000 mAhg^{-1}). This suggests that the all-solid-state lithium-sulfur battery is reversible at 120 °C. Conversely, lithium-sulfur batteries with conventional electrolyte solutions cannot operate at temperatures greater than 100 °C, as the organic solvents used for electrolyte solutions are volatile and flammable.

Fig. 4 Charge–discharge cycle curves for the test cell under a constant current density of 1.0 $mAcm^{-2}$ with cut-off voltage range of 1.2–2.4 V at 120 °C

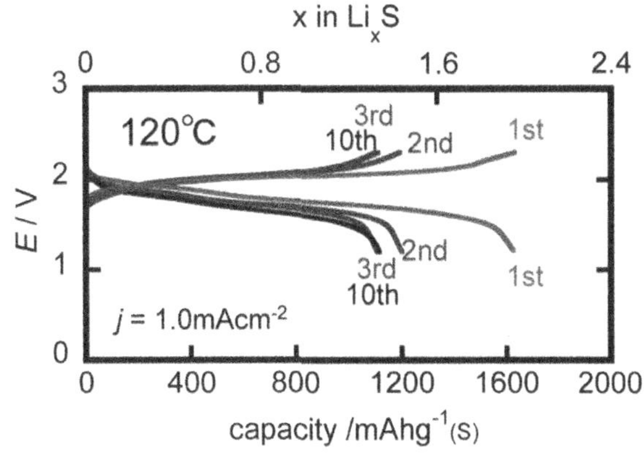

The discharge–charge cycle test was performed under several different current densities: 1.0, 2.0, 3.0, and 4.0 mAcm^{-2}. Current density was changed every 10 cycles. For cycles 1–10, current density was 1.0mAcm^{-2}. After the 10th cycle, the current density was set at 2.0, 3.0, and 4.0 mAcm^{-2} for cycles 11–20, 21–30, and 31–40, respectively. After the 40th cycle, the current density reverted to the initial current density of 1.0 mAcm^{-2}.

Typical discharge curves for the test cell under those current densities are shown in Fig. 5. Even under the high current density of 4.0 mAcm^{-2}, the all-solid-state battery with the sulfur-VGCF-Li_3PS_4 composite exhibits good performance. The average discharge capacities under 1.0, 2.0, 3.0, and 4.0 mAcm^{-2} are 1050, 950, 830, and 700 mAhg^{-1}, respectively. When the current density was reduced to 1.0 mAcm^{-2} after the 40th cycle, the charge–discharge capacity for the test cell returned to greater than 1000 mAhg^{-1}.

The dependence of charge and discharge capacities on cycle number for the sulfur–VGCF–Li_3PS_4 composite at 120 °C are shown in Fig. 6. Cycle number dependence

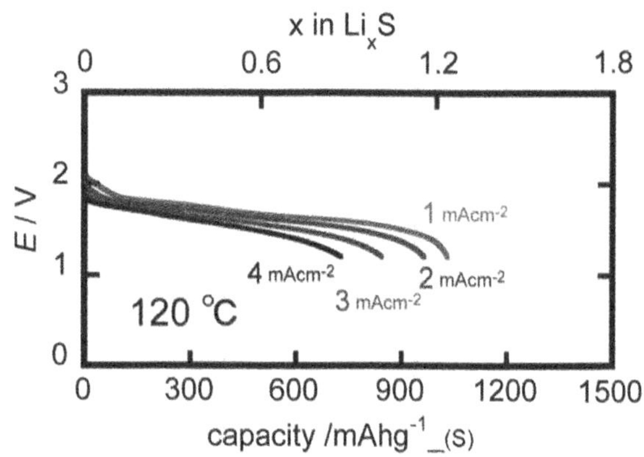

Fig. 5 Typical discharge curves at various current densities: 1.0, 2.0, 3.0, and 4.0 mAcm^{-2}, at 120 °C

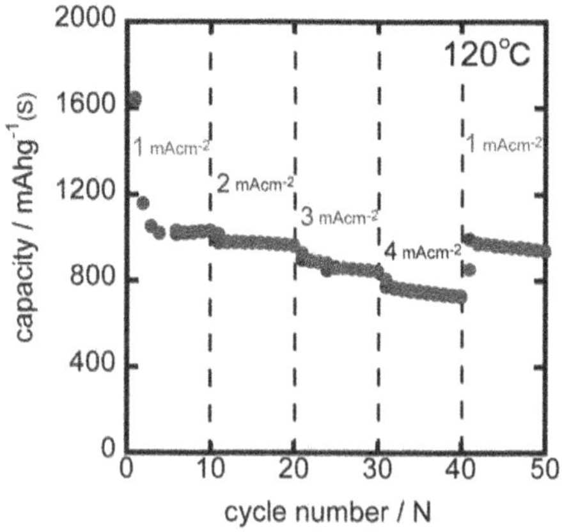

Fig. 6 Cycle number dependence of charge and discharge capacities. The cycle test was carried out with a cut-off voltage range of 1.2–2.4 V at 120 °C. Capacity was calculated on the basis of sulfur mass in the composite electrode in the test cell with $Li_{4.4}Si$/amorphous Li_3PS_4/S–VGCF–Li_3PS_4 composite

of Coulombic efficiency is also shown. Current density was changed every 10 cycles as mentioned above. Even at the high current density of 4.0 $mAcm^{-2}$ (960 mAg^{-1}, 0.64C in C rate), the test cell maintains charge–discharge capacities greater than 700 $mAhg^{-1}$. Charge–discharge efficiency is almost 100% over the cycle test. These results suggest that the all-solid-state cell achieves good charge–discharge cycling performance at 120 °C.

3 Sulfide Positive Electrode

Li_2S, a discharge reaction product of Li/S batteries, is also a promising positive electrode material because it can be coupled with Li-free high capacity negative electrode materials [17–23]. The high specific capacity (1167 $mAhg^{-1}$) of lithium sulfide presents another advantage. However, the lithium-ion and electronic conductivities of Li_2S are very low. Composites comprising Li_2S and lithium halides, such as LiCl, LiBr, and LiI, were investigated as positive electrode materials for all-solid-state lithium batteries [17, 18], because lithium-ion conductivity enhancement is expected for solid solutions of Li_2S–LiX (X = Cl, Br, and I).

Li_2S–LiX (X = Cl, Br, and I) solid solutions were prepared by ball milling using a procedure described elsewhere [17, 18]. X-ray diffraction (XRD) patterns of the obtained Li_2S–LiI samples are shown in Fig. 7a. Silicon powder was added as an internal standard in order to calculate lattice constants for the obtained samples. For all prepared samples, diffraction peaks were indexed by the cubic structure of Li_2S. Peaks attributed to LiI are observed at x = 25. On the other hand, peaks attributed to LiCl and LiBr were not found in compositions up to x = 30 in the Li_2S–LiCl and Li_2S–LiBr systems. The average lattice constants of the samples as a function of composition are shown in Fig. 7b. Lattice constants for the Li_2S–LiCl solid solutions

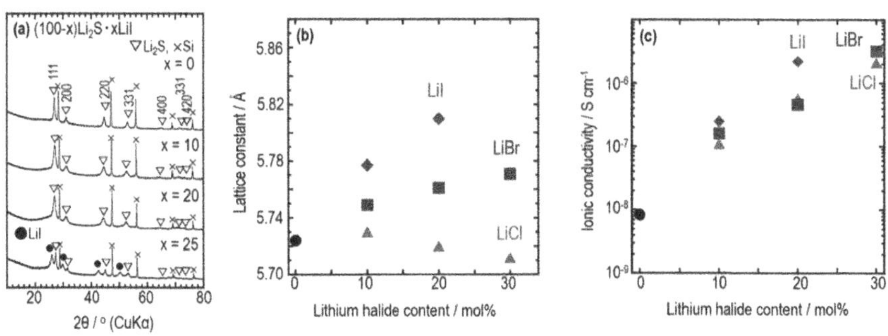

Fig. 7 XRD patterns of prepared samples for **a** (100 − x)Li_2S·xLiI. **b** Lattice constants and **c** ionic conductivities of the prepared solid solutions in the Li_2S–LiCl, Li_2S–LiBr, and Li_2S–LiI systems. From Ref. [17] Copyright © 2017 by John Wiley Sons. Inc. Reprinted by permission of John Wiley & Sons, Inc.

do not differ significantly from those of Li_2S. However, the lattice constants for Li_2S-LiBr and Li_2S-LiI solid solutions increased with increasing lithium halide content. The change in lattice constants for Li_2S-LiI solid solutions is most remarkable in the prepared samples, suggesting that the lattice constants are affected by anionic radius (S^{2-}: 184 pm, Cl^-: 181 pm, Br^-: 195 pm, I^-: 216 pm). Disappearance of the lithium halide diffraction peaks and changes in lattice constants indicate the formation of a Li_2S-based solid solution.

The composition dependence of ionic conductivities of the prepared solid solutions is shown in Fig. 7c. Ionic conductivities of the solid solutions increase with increasing lithium halide content, regardless of the halide. These results suggest that Li^+ ion defects are introduced by charge compensation accompanied with anion substitution. Li_2S–LiI solid solutions show relatively high ionic conductivities compared to other solid solutions with the same lithium halide contents. This is likely due to the larger lattice constants of Li_2S–LiI solid solutions compared to those of other solid solutions.

All-solid-state test cells were constructed with Li_2S–LiX solid solutions. Composite electrodes consisted of Li_2S–LiX (X = Cl, Br, and I), VGCF (Showa Denko), and amorphous Li_3PS_4 solid electrolyte in a ratio of 50:10:40 by weight. A mixture of $70Li_2S{\cdot}30LiCl$, $70Li_2S{\cdot}30LiBr$, or $80Li_2S{\cdot}20LiI$ (mol%) and VGCF (as conductive additive) were mechanically milled at 510 rpm for 10 h using a planetary ball-milling apparatus (Fritch, P-7). The prepared Li_2S–LiX–VGCF powder and solid electrolyte (amorphous Li_3PS_4) were softly milled at 150 rpm for 0.5 h. The obtained composite electrode powders were used directly as working electrodes, while amorphous Li_3PS_4 was used as a solid electrolyte separation layer for the all-solid-state test cells. Li–In alloy was used as the test cell counter electrode.

Charge–discharge curves for the test cells with Li_2S, $70Li_2S{\cdot}30LiCl$, $70Li_2S{\cdot}30LiBr$, and $80Li_2S{\cdot}20LiI$ (mol%) over 50 cycles are shown in Fig. 8a–d, respectively. Test cells were initially charged up to the electric potential of 3.6 V against Li^+/Li electrode. The charge–discharge measurements were controlled with the initial charge capacities. The overpotential was relatively large at the initial cycle and decreased over the first 10 cycles. One plateau is observed during charge–discharge processes; the discharge potential observed at approximately 1.8 V is in accordance with the lithiation potential for small sulfur allotropes, such as S_4, S_3, and S_2 [24, 25]. The conversion reaction of Li_2S/S proceeds in the all-solid-state Li/S cell without forming liquid electrolyte-soluble polysulfides (Li_2S_x, x = 4–8). Cells with Li_2S, $70Li_2S{\cdot}30LiCl$, $70Li_2S{\cdot}30LiBr$, and $80Li_2S{\cdot}20LiI$ exhibit capacities of 500, 600, and 900 mAhg^{-1}, respectively, estimated from the mass of Li_2S in the composite electrode. The cell with $80Li_2S{\cdot}20LiI$ exhibited the highest capacity, suggesting that charged $80Li_2S{\cdot}20LiI$ has a higher ionic conductivity than the other charged solid solutions. In the cell containing $80Li_2S{\cdot}20LiI$, the charge potential gradually decreased with cycling, as shown in Fig. 8d. At the 51st cycle, the cell with $80Li_2S{\cdot}20LiI$ was charged at the cut off voltage of 3.6 V vs. Li^+/Li without the limitation of the initial charge capacity. A reversible capacity was increased to 1000 mAhg^{-1} at the 51st cycle. When the charge–discharge test of the cell with the 3.6 V

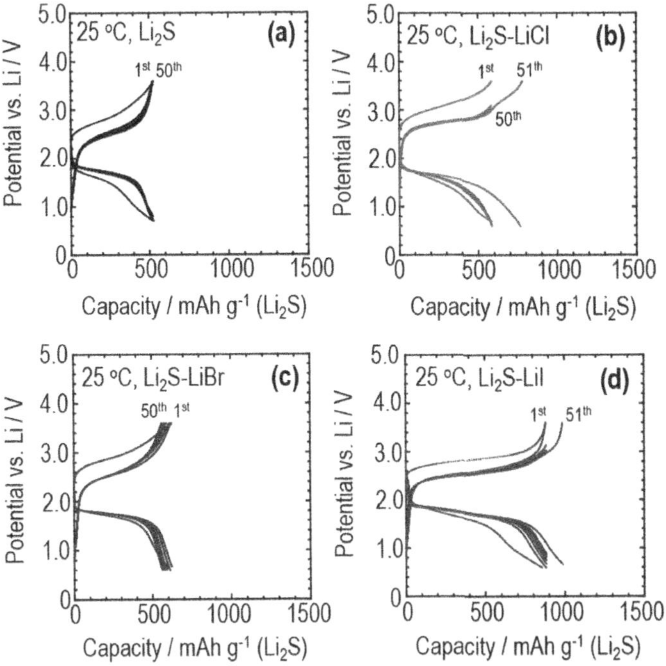

Fig. 8 Charge–discharge curves of the cells with **a** Li_2S, **b** $70Li_2S{\cdot}30LiCl$, **c** $70Li_2S{\cdot}30LiI$, and **d** $80Li_2S{\cdot}20LiI$ electrode materials at a current of 0.5 C (1 C = 1167 $mAhg^{-1}$). From Ref. [17] Copyright © 2017 by John Wiley Sons. Inc. Reprinted by permission of John Wiley & Sons, Inc.

cutoff voltage was performed after the 51st cycle, reversible capacity increased with cycle number, reaching a maximum of 1100 $mAhg^{-1}$.

Rate performance for the $Li_2S/80Li_2S{\cdot}20LiI$ cell at different current densities is shown in Fig. 9a. The cell with $80Li_2S{\cdot}20LiI$ exhibits a capacity of 1130 $mAhg^{-1}$

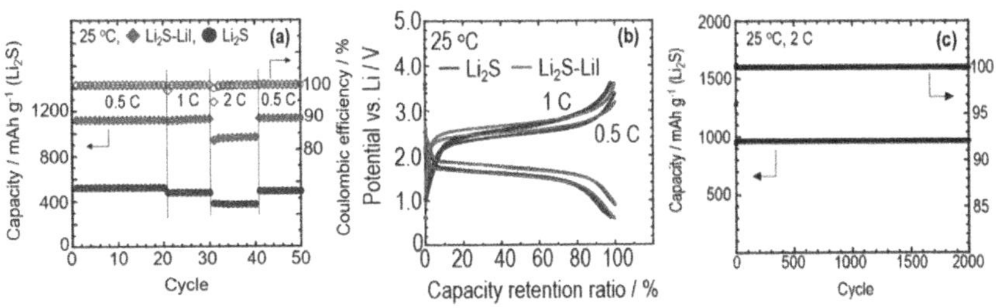

Fig. 9 **a** Rate performance of the all-solid-state cells with Li_2S and $80Li_2S{\cdot}20LiI$ at several current densities and **b** their voltage profiles as a function of capacity retention ratios. **c** Long-term cycle performance of the cell with $80Li_2S{\cdot}20LiI$ electrode materials for 2000 cycles at 2 C. From Ref. [17] Copyright © 2017 by John Wiley Sons. Inc. Reprinted by permission of John Wiley & Sons, Inc.

at 0.5 C and retains its capacity at 1.0 C. Moreover, the cell can deliver a capacity of 970 $mAhg^{-1}$, even at a high current of 2 C. The capacities of the $80Li_2S{\cdot}20LiI$ cell are more than twice those of the Li_2S cell regardless of rate. Charge–discharge profiles for the Li_2S and $80Li_2S{\cdot}20LiI$ cells were compared as a function of capacity retention ratios, as shown in Fig. 9b. Capacity retention ratios were calculated with a discharge capacity of 100% at 0.5 C. Interestingly, the discharge potential profiles do not differ significantly between Li_2S and $80Li_2S{\cdot}20LiI$ at any discharge depth, even though the utilization of $80Li_2S{\cdot}20LiI$ is twice that of Li_2S. During charging, similar overpotentials are observed, though with slightly different potential profiles.

Long-term cycle performance of the $80Li_2S{\cdot}20LiI$ cell at 2 C is shown in Fig. 9c. Remarkably, the $80Li_2S{\cdot}20LiI$ cell displays 100% retention of capacity after 2000th cycle with nearly 100% Coulombic efficiency. In long-term cycle tests (400–1500 cycles), batteries with organic liquid electrolytes demonstrate only 30–80% of their initial capacities after cycling [19–23, 26, 27]. Hence the all-solid-state cell with the $80Li_2S{\cdot}20LiI$ electrode exhibits better cycle performance than those reported for other cells using Li_2S as an active material.

We have demonstrated that all-solid-state cells with $80Li_2S{\cdot}20LiI$ solid solution exhibit a capacity of greater than 1100 $mAhg^{-1}$ at 0.5 C, corresponding 95% utilization. Furthermore, the cell retains a capacity of 980 $mAhg^{-1}$ at 2 C for 2000 cycles. The $80Li_2S{\cdot}20LiI$ cell displays the highest utilization and best cycling performances. These results suggest that the all-solid-state cell with a Li_2S-based solid solution is a promising high energy storage system and provides new insight for improving Li_2S utilization in high-performance Li/S batteries.

References

1. Daniel, C., & Besenhard, J. O. (Eds.). (2011). *Handbook of battery materials 2nd*. Germany: Wiley-VCH and KGaA.
2. Armand, M., & Tarascon, J. M. (2008). *Nature, 451*, 652–657.
3. Mikhaylik, Y. V., & Akridge, J. R. (2004). *Journal of the Electrochemical Society, 151*, A1969–A1976.
4. Zhang, B., Lai, C., Zhou, Z., & Gao, X. P. (2009). *Electrochimica Acta, 54*, 3708–3713.
5. Barghamadi, M., Kapoor, A., & Wen, C. (2013). *Journal of the Electrochemical Society, 160*, A1256–A1263.
6. Zhang, S. S. (2012). *Electrochimica Acta, 70*, 344–348.
7. Machida, N., & Shigematsu, T. (2004). *Chemistry Letters, 33*, 376–377.
8. Machida, N., Kobayashi, K., Nishikawa, Y., & Shigematsu, T. (2004). *Solid State Ionics, 175*, 247–250.
9. Nagao, M., Hayashi, A., & Tatsumisago, M. (2011). *Electrochimica Acta, 56*, 6055–6059.
10. Nagao, M., Hayashi, A., & Tatsumisago, M. (2012). *Electrochemistry Communication, 22*, 177–180.
11. Agostini, M., Aihara, Y., Yamada, T., Scrosati, B., & Hassoun, J. (2013). *Solid State Ionics, 244*, 48–51.
12. Kinoshita, S., Okuda, K., Machida, N., Naito, M., & Shigematsu, T. (2014). *Solid State Ionics, 256*, 97–102.
13. Kinoshita, S., Okuda, K., Machida, N., & Shigematsu, T. (2014). *Journal of Power Sources, 269*, 727–732.

14. Chen, M., & Adams, S. (2015). *Journal of Solid State Electrochemistry, 19,* 697–702.
15. Hashimoto, Y., Machida, N., & Shigematsu, T. (2004). *Solid State Ionics, 175,* 177–180.
16. Tamori, R., Machida, N., & Shigematsu, T. (2001). *Journal of the Japan Society of Powder Metallurgy, 48,* 267–273.
17. Hakari, T., Hayashi, A., & Tatsumisago, M. (2017). *Advanced Sustainable System, 1,* 1700017. https://doi.org/10.1002/adsu.201700017.
18. Hakari, T., Nagao, M., Hayashi, A., & Tatsumisago, M. (2015). *Chemistry Letters, 44,* 1664–1666.
19. Li, C., Zhang, H., Otaegui, L., Singh, G., Armand, M., & Rodriguez, L. M. (2016). *Journal of Power Sources, 326,* 1–5.
20. Takeuchi, T., Kageyama, H., Nakanishi, K., Tabuchi, M., Sakaebe, H., Ohta, T., et al. (2010). *Journal of the Electrochemical Society, 157,* A1196–A1201.
21. Nagao, M., Hayashi, A., & Tatsumisago, M. (2012). *Journal of Materials Chemistry, 22,* 10015–10020.
22. Lin, Z., Liu, Z., Dudney, N. J., & Liang, C. (2013). *ACS Nano, 7,* 2829–2833.
23. Han, F., Yue, J. J., Fan, X., Gao, T., Luo, C., Ma, Z., et al. (2016). *Nano Letters, 16,* 4521–4527.
24. Xin, S., Gu, L., Zhao, N. H., Yin, Y. X., Zhou, L. J., Guo, Y. G., & Wan, L. J. (2012). *Journal of the American Chemical Society, 134,* 18510–18513.
25. Helen, M., Reddy, M. A., Diemant, T., Schindler, U. G., Behm, R. J., Kaiser, U., & Fichtner, M. (2015). *Scientific Reports, 5,* 12146. https://doi.org/10.1038/srep12146.
26. Wang, C., Wang, X., Yang, Y., Kushima, A., Chen, J., Huang, Y., & Li, J. (2015). *Nano Letters, 15,* 1796–1802.
27. Zhang, K., Wang, L., Hu, Z., Cheng, F., & Chen, J. (2014). *Scientific Report, 4,* 6467. https://doi.org/10.1038/srep06467.

Li Negative Electrode

Akitoshi Hayashi

Abstract In order to realize all-solid-state batteries using Li metal as an ultimate negative electrode, interface formation for giving stable Li stripping/plating cycling is important. Modification of electrode–electrolyte interface is effective in improving electrochemical performance by both the insertion of Au buffer layer and the use of Li_3PS_4–LiI with good compatibility to Li metal.

Keywords Lithium metal negative electrode · Sulfide electrolyte · Interface

1 Introduction

An effective use of lithium metal negative electrode is important for realizing all-solid-state Li/S batteries with high energy density. Lithium metal is an ultimate negative electrode because of a large theoretical capacity of 3861 mAh g^{-1} and the lowest electrochemical potential of −3.04 V vs. SHE [1]. However, preventing short-circuit caused by Li whisker is a big challenge. Modification of the interface between solid electrolyte and Li metal electrode is useful for stable Li stripping/plating cycling. This chapter reports interfacial modification from the viewpoint of (1) insertion of metal thin film as a buffer layer and (2) use of sulfide electrolytes with high tolerance to the reduction of Li metal.

2 Interface Modification of Au Thin Film

Theoretical calculations suggest that upon contact with Li metal, Li_3PS_4 glass electrolytes decomposed to Li_2S and Li_3P, which are electronically insulating and thus passivate further decomposition of the electrolyte [2]. In fact, X-ray photoelectron spectroscopy (XPS) reveals that the decomposed layer consisting of Li_2S and Li_3P

A. Hayashi (✉)
Osaka Prefecture University, Sakai, Osaka, Japan
e-mail: hayashi@chem.osakafu-u.ac.jp

K. Kanamura (ed.), *Next Generation Batteries*,
https://doi.org/10.1007/978-981-33-6668-8_13

is formed only in the vicinity of the Li/Li_3PS_4 interface, when vacuum evaporated Li is just deposited on the Li_3PS_4 surface [3]. In electrochemical cells using a liquid electrolyte, the decomposed layer—called the solid electrolyte interphase (SEI)—plays an important role in following Li stripping/plating behavior [4–6]. Therefore, the interface layer should also affect Li stripping/plating stability in all-solid-state Li cells. The insertion of Au thin film as a buffer layer to the Li/Li_3PS_4 interface is effective in enhancing electrochemical cycle performance [7, 8].

Morphology changes of Li metal for all-solid-state symmetrical cells of Li/Li_3PS_4/Li and Li/Au/Li_3PS_4/Au/Li are investigated after galvanostatic Li stripping/plating cycling. Figure 1 shows the SEM images of the Li surfaces and their schematic for both the cells after the initial stripping, the following plating, and after five cycles. Compared to the morphology of Li metal before galvanostatic tests, a contrast caused by the difference of Li thickness is observed after the initial dissolution for the Li/Li_3PS_4/Li cell. Roughness on the surface of the pelletized solid electrolyte, which is prepared simply by powder compaction, may cause inhomogeneous Li stripping at the Li/Li_3PS_4. After the following plating, needle-like and granular deposits are observed. Voids may be formed by non-uniform current distribution, leading to extremely inhomogeneous Li plating such as needle-like deposits. As a result, cycle performance of Li stripping/plating rapidly decreases by the drastic morphology change after the initial cycle. On the other hand, for the cell with the Au thin film, a contrast caused by the difference of Li thickness is not clearly observed.

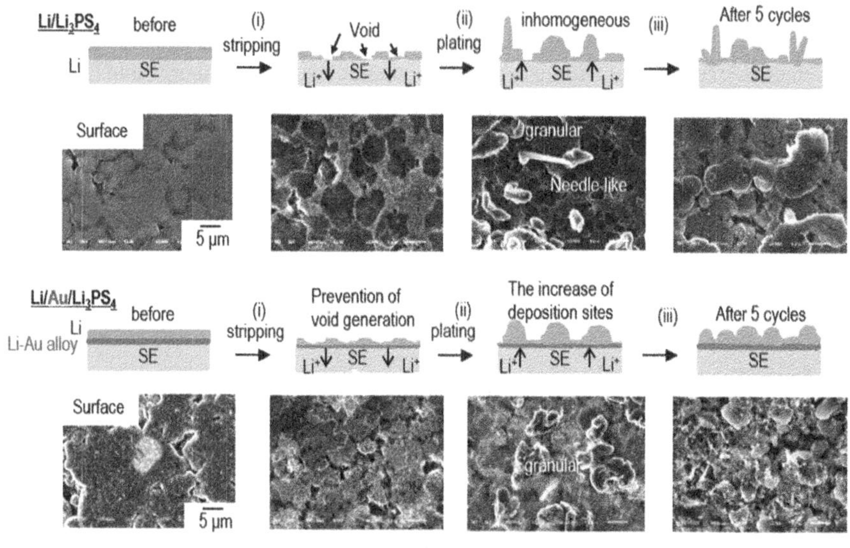

Fig. 1 SEM images of Li surfaces and schematic diagrams of morphology changes for Li metal during galvanostatic cycling tests for the symmetrical Li/Li_3PS_4/Li cell and the Li/Au/Li_3PS_4/Au/Li cell; (i) after the initial stripping, (ii) the following plating, and (iii) after five cycles. Thin films of lithium (3 μm in thickness) and Au (60 nm in thickness) were formed on both the surfaces of the electrolyte pellet by vacuum evaporation. Reprinted from Ref. [3] with permission of Elsevier

In addition, needle-like deposits are few and the similar size of granular deposits is observed after Li plating. The more uniform morphology of Li metal is retained for five cycles. Li–Au alloy formed at the Li/Au/Li_3PS_4 interface prevents void generation during Li stripping. Moreover, the remained Li–Au alloy may increase Li deposition sites, and thus more uniform deposition is observed during cycling. It is noteworthy that more uniform morphology of Li metal improves the reversibility of Li utilization.

Electrochemical performance of all-solid-state symmetric Li cells using sulfide electrolytes is listed in Table 1 [9–15]. The Li stripping/plating performance is defined as the maximum current density at which the cell was cycled without short circuiting. The insertion of Au thin film at the Li_3PS_4 electrolyte/Li metal interface surely enhances the performance at higher areal capacity because the cell without Au film short circuits at one order of magnitude lower areal capacity of 0.64 mA cm^{-2} even at the same operating temperature of 100 °C [8].

3 Use of Sulfide Electrolytes Compatible to Li Metal

Another approach is the use of sulfide electrolytes with high tolerance to the reduction of Li metal. Crystalline lithium halides are promising materials for stabilization of the Li/electrolyte interface, because they are thermodynamically stable against Li metal [2]. The addition of lithium halides to Li_2S–P_2S_5 electrolyte increases ionic conductivity [10] and $Li_7P_2S_8I$ crystal (Li_3PS_4:LiI = 2:1 in molar ratio) shows high electrochemical stability as shown in Table 1 [15]. Therefore, the addition of LiI to Li_3PS_4 glass is expected to improve both the ionic conductivity and the electrochemical stability to Li metal. A glass electrolyte at the composition $54Li_3PS_4 \cdot 46LiI$ (mol%) is mechanochemically prepared and applied to an all-solid-state Li symmetrical cell and long-term galvanostatic cycling test is performed at 100 °C [10]. A constant current density of 1.25 mA cm^{-2} is passed for 4 h per half a cycle (areal capacity of 5 mAh cm^{-2}). As shown in Fig. 2, the cell using $54Li_3PS_4 \cdot 46LiI$ glass electrolyte do not short circuit even after cycling for more than 3400 h (425 cycles). In addition, the voltage profile is stable, and no gradual increase in voltage is observed during cycling. The cell achieves stable Li stripping/plating behavior at a high current density and areal capacity. This result indicates that the $54Li_3PS_4 \cdot 46LiI$ glass shows good electrochemical stability against Li metal without insertion of Au film.

Figure 3 shows cross-sectional back-scattered electron (BSE) images of the interfaces of (a) Li/Li_3PS_4 and (b) Li/$54Li_3PS_4 \cdot 46LiI$ after galvanostatic cycling tests at 100 °C. Crack areas with medium-dark contrast are observed between Li metal and Li_3PS_4 glass in the BSE image, suggesting the existence of reduced components such as Li_2S and Li_3P. On the other hand, compositional contrasts are not observed at the Li/$54Li_3PS_4 \cdot 46LiI$ interface, and no cracks and pores are formed at the interface and good contacts are maintained even after long-term galvanostatic cycling tests for more than 3400 h at 100 °C. X-ray diffraction (XRD) is carried out for the surfaces of Li_3PS_4 and $54Li_3PS_4 \cdot 46LiI$ glass electrolytes, where Li foil is

Table 1 Electrochemical performances of all-solid-state Li symmetric cells with sulfide electrolytes

Electrolyte	Current density/mA cm^{-2}	Areal capacity/mAh cm^{-2}	Cycle number	Temperature /°C	References
Li_3PS_4 glass	0.2	0.1	15	25	[9]
Li_3PS_4 glass	0.64	0.64	5	100	[8]
Li_3PS_4 glass with Au thin film	1.3	6.5	5	100	[8]
$54Li_3PS_4{\cdot}46LiI$ glass	1.25	5	425	100	[10]
$80Li_2S{\cdot}20P_2S_5$ glass–ceramic	0.064	0.192	5	25	[11]
$75Li_2S{\cdot}24P_2S_5{\cdot}1P_2O_5$ glass–ceramics	0.1	0.125	42	25	[12]
$60Li_2S{\cdot}20P_2S_5{\cdot}20LiI$ glass ceramics	0.1	0.1	75	25	[13]
β-Li_3PS_4 crystal	0.75	1.5	24	25	[14]
$Li_7P_2S_8I$ crystal	0.2	0.2	800	25	[15]

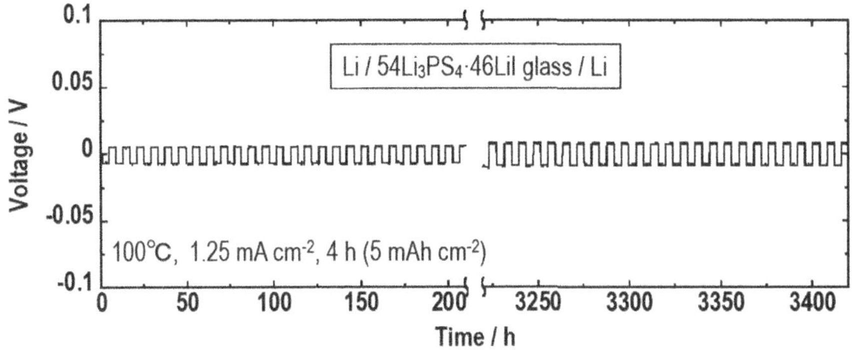

Fig. 2 Long-term galvanostatic cycling tests of Li/54Li_3PS_4·46LiI/Li at 100 °C. The applied current density was 1.25 mA cm^{-2}, and the areal capacity for a half cycle was 5 mAh cm^{-2}. Reprinted from Ref. [10] with permission of Elsevier

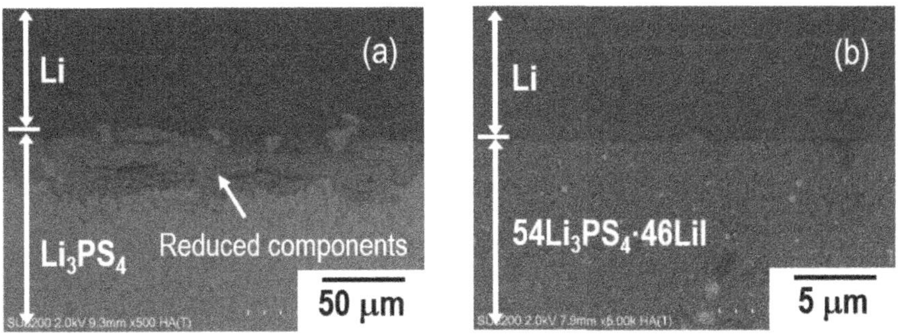

Fig. 3 Cross-sectional back-scattered electron (BSE) images of **a** Li/Li_3PS_4 interface and **b** Li/54Li_3PS_4·46LiI interface after galvanostatic cycling tests at 100 °C. Reprinted from Ref. [10] with permission of Elsevier

detached after galvanostatic cycling tests at 100 °C. Diffraction peaks attributable to crystalline Li_2S are observed on the surface of Li_3PS_4 electrolyte after cycling. The Li_3P component is detected on the surface of Li_3PS_4 by P2p XPS, even though the crystalline size of Li_3P may be too small to detect by XRD; Li_3PS_4 glass at the vicinity of Li metal is thus decomposed to Li_2S and Li_3P, which causes large volume expansion (~150%) during the decomposition ($Li_3PS_4 + 8Li \rightarrow 4Li_2S + Li_3P$). The XRD for 54$Li_3PS_4$·46LiI glass still shows a halo pattern after cycling, suggesting that the glassy state is retained. A side reaction with Li metal is suppressed by adding LiI to Li_3PS_4 glass. Morphology changes of the Li/Li_3PS_4 interface by the formation of crystalline Li_2S with low electronic and ionic conductivity may cause the inhomogeneous distribution of current, which induces the formation of Li whiskers. In contrast, the interface is stabilized by adding LiI to the Li_3PS_4 glass, which effectively prevents the side reaction with Li metal and maintains a close interface during Li stripping/plating cycling.

4 Concluding Remarks

Homogeneous Li stripping/plating behavior at the electrode–electrolyte interface is a key to achieve long-term cyclability of all-solid-state batteries with Li negative electrode. Modification of electrode–electrolyte interface is effective in improving electrochemical performance by both the insertion of Au buffer layer and the use of Li_3PS_4–LiI with good compatibility to Li metal. Continuous studies for finding favourable metal buffer films and sulfide electrolyte with high tolerance to the reduction of Li metal are needed for practical use of Li negative electrode in all-solid-state batteries.

References

1. Tarascon, J. M., & Armand, M. (2001). *Nature, 414,* 359.
2. Zhu, Y., He, X., Mo, Y., & Appl, A. C. S. (2015). *Materials Interfaces, 7,* 23685.
3. Kato, A., Kowada, H., Deguchi, M., Hotehama, C., Hayashi, A., & Tatsumisago, M. (2018). *Solid State Ionics, 322,* 1.
4. Peled, E. (1979). *Journal of the Electrochemical Society, 126,* 2047.
5. Cohen, Y. S., Cohen, Y., & Aurbach, D. (2000). *The Journal of Physical Chemistry B, 104,* 12282.
6. Steiger, J., Kramer, D., & Mönig, R. (2014). *Journal of Power Sources, 261,* 112.
7. Kato, A., Hayashi, A., & Tatsumisago, M. (2016). *Journal of Power Sources, 309,* 27.
8. Kato, A., Suyama, M., Hotehama, C., Kowada, H., Sakuda, A., Hayashi, A., & Tatsumisago, M. (2018). *Journal of the Electrochemical Society, 165,* 1950.
9. Agostini, M., Aihara, Y., Yamada, T., Scrosati, B., & Hassoun, J. (2013). *Solid State Ionics, 244,* 48.
10. Suyama, M., Kato, A., Sakuda, A., Hayashi, A., & Tatsumisago, M. (2018). *Electrochimica Acta, 286,* 158.
11. Nagao, M., Hayashi, A., & Tatsumisago, M. (2012). *Electrochemistry Communications, 22,* 177.
12. Tao, Y., Chen, S., Liu, D., Peng, G., Yao, X., & Xu, X. (2016). *Journal of the Electrochemical Society, 163,* A96.
13. Choi, S., Lee, S., Yu, J., Doh, C. H., & Ha, Y. C. (2017). *ECS Transactions, 77,* 65.
14. Wang, H., Hood, Z. D., Xia, Y., & Liang, C. (2016). *Journal of Materials Chemistry A, 4,* 8091.
15. Rangasamy, E., Liu, Z., Gobet, M., Pilar, K., Sahu, G., Zhou, W., et al. (2015). *Journal of the American Chemical Society, 137,* 1384.

TEM Analyses

Shigeo Mori

Abstract Because the sulfide-based solid electrolytes are sensitively damaged by electron beams during the transmission electron microscopy (TEM) observation, nanoscale microstructures of Li_2S-P_2S_5 glass electrolytes have not been revealed so far. In this chapter, we describe nanoscale microstructures, crystallization processes, and thermal stabilities of Li_2S-P_2S_5 glass electrolytes revealed by using high-resolution electron microscopy, electron diffraction, and hollow-cone dark-field imaging methods. In order to visualize the spatial distribution of the amorphous and crystalline regions in the Li_2S-P_2S_5 glass electrolytes, a TEM observation method called the hollow-cone and conventional dark-field imaging method was applied. We would emphasize that TEM should be a powerful tool to analyze nanostructures and amorphous structures by combining hollow-cone dark-field imaging, nanobeam electron diffraction, and high-resolution electron microscopy.

Keywords Transmission electron microscopy · Nanostructure · Sulfide-based solid electrolytes · Crystallization process · Glass ceramics

1 Introduction

Transmission Electron Microscope (TEM) is widely used as a powerful method for evaluating microstructures inside materials at the atomic level [1–4]. There are several TEM observation methods. In the bright-field imaging method, the contrast is due to Bragg diffraction of electrons by the crystal, atomic number, and sample thickness. In addition to observing crystal grains, alloy phases, precipitates, and thin-film stacking, it is also possible to observe dislocations and stacking faults. On the contrary, in a dark-field image that is formed using only diffracted waves, only specific crystal regions can be imaged, which is useful for confirming precipitates and crystal grain size. Furthermore, high-resolution lattice images can be observed using interference between transmitted waves and diffracted waves. Also, from the electron diffraction

S. Mori (✉)
Department of Materials Science, Osaka Prefecture University, Sakai, Japan
e-mail: mori@mtr.osakafu-u.ac.jp

K. Kanamura (ed.), *Next Generation Batteries*,
https://doi.org/10.1007/978-981-33-6668-8_14

pattern, it is possible to investigate the crystal states of amorphous, single crystal, polycrystalline, etc. and determine the crystal structure.

A bulk type all-solid-state battery is manufactured by laminating fine particles. Since a large amount of the electrode active material can be introduced into the electrode layer, there is an advantage that the battery capacity can be greatly improved. For example, when paying attention to the positive electrode layer, the positive electrode layer is composed of a solid electrolyte and a positive electrode active material that carries a lithium ion conduction path. Therefore, in order to put an all-solid battery into practical use, it is important to develop a solid electrolyte exhibiting high ion conductivity and an electrode material having a high capacity and to construct an undisturbed electrode–electrolyte solid interface [5–8].

A sulfide-based electrolyte, which is one of the inorganic solid electrolytes, has higher ionic conductivity than an oxide-based electrolyte and is electrochemically stable in a wide potential range, so it is expected to be applied to all-solid-state batteries. Among them, the Li_2S-P_2S_5 glass electrolyte has the property of improving ion conductivity when it is heat-treated and crystallized. Therefore, in order to realize high ionic conductivity, it is extremely important to elucidate the morphology and connection of crystalline regions in the amorphous state and in the amorphous matrix. Therefore, we have focused on sulfide-based glass electrolytes, and directly observed the amorphous state in addition to the thermal stability evaluation of all-solid-state batteries using TEM. We have developed TEM observation techniques that can observe the crystallization behavior of glass and glass ceramics [9–11].

A crystallized glass is generally called a glass ceramic, but the actual state of the amorphous state including the crystal has not been elucidated so far. One of the major reasons is the air instability of the sulfide-based solid electrolyte. There has never been an experimental environment for evaluating the sulfide-based solid electrolyte without exposing them to the atmosphere, and it was extremely difficult to directly observe the nanostructures of the sulfide-based solid electrolyte. Therefore, an experimental environment for the TEM observation was established first. Furthermore, in order to visualize the spatial distribution of the amorphous and crystalline regions, a TEM observation method called the hollow-cone and conventional dark-field imaging method was also established [8–13].

2 TEM Experimental Methods

All procedures from the thin TEM sample preparation to TEM observation have to be done inside the glove box, because sulfur-based solid electrolytes break down when exposed to the atmosphere. We prepared the TEM thin samples mounted on an amorphous carbon film supported by a Cu grid and subsequently attached it on a double-tilt vacuum transfer TEM holder (Gatan model 648). Both DF imaging and electron diffraction experiments were carried out to elucidate the crystallization processes of the $75Li_2S$-$25P_2S_5$ glasses in the temperature range from 20 to 450 °C. In order to observe the precipitated crystallites in the amorphous matrix, we developed a TEM observation technique referred to as "hollow-cone dark-field TEM imaging" [14, 15]. This imaging technique is very effective to observe nano-sized crystallites. Technically, the incident electron beam tilted at a fixed angle is circulated with respect to the optical axis of the objective lens. Therefore, the DF image was taken by using the particular diffraction spots in the ED pattern. Figure 1 shows the optical system realizing the hollow-cone bright-field TEM imaging method. The parallel electron beams are irradiated on the specimen with the inclination angles in X and Y directions controlled by using the beam deflector system placed above the specimen. The circulating electron beam is illuminated in all azimuthal directions around the optical axis. In the hollow-cone bright- and dark-field TEM images, we can observe the nano-sized crystallites in the amorphous matrix by integrating conventional dark-field images due to a particular Bragg-diffraction spot.

A computer program called "Process Diffraction" was used to identify the precipitated crystalline phase from the ED patterns. This program can transform digital ED

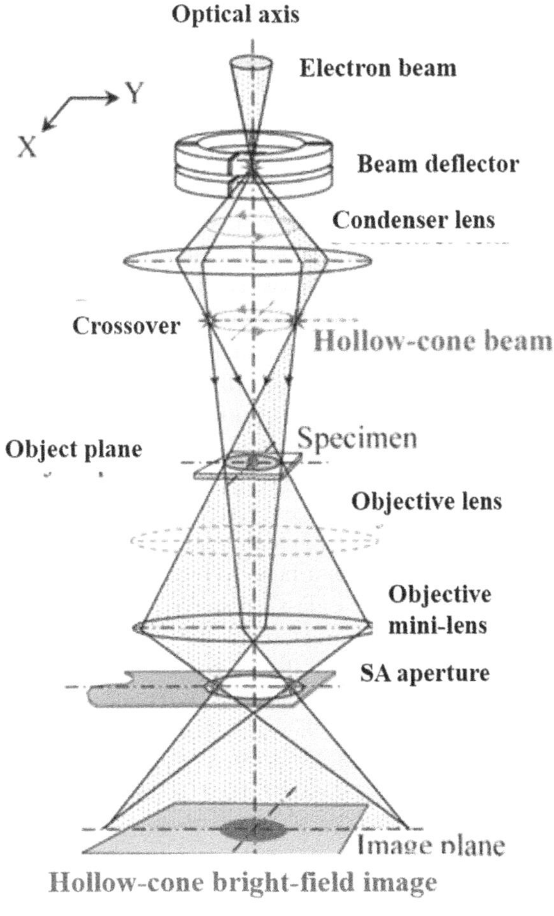

Fig. 1 The optical system realizing the hollow-cone bright-field TEM imaging method

patterns into integral intensities as a function of the scattering angle, and the intensity profile similar to a usual XRD pattern was reproduced [16–18]. Estimating the degree of the crystallization is important to understand the crystallization behaviors of the $75Li_2S{\cdot}25P_2S_5$ glasses. Thus, we tried to evaluate the degree of the crystallization from the intensity of the Bragg and halo-ring diffraction spots in the ED patterns. Note that the degree of crystallization should be defined as the ratio of crystalline regions in an amorphous matrix. When the intensities of diffraction spots of as-prepared glass and crystallized glass samples are, respectively, written as $I_{\text{as-prepared glass}}$ and $I_{\text{crystallized glass}}$, the crystallization degree (x) is given by Eq. (1) as below [19],

$$x = \frac{I_{\text{crystallized glass}} - I_{\text{as-prepared glass}}}{I_{\text{crystallized glass}}} \times 100. \tag{1}$$

3 TEM Analysis of Sulfide-Based Solid Electrolytes

3.1 *Microstructures in Li_2S-P_2S_5 Glass Ceramics*

The microstructure of the $80Li_2S{\cdot}20P_2S_5$ glass ceramics obtained by heat-treatment at 240 °C is shown in Fig. 2. The inset of Fig. 2 shows a low-magnified bright-field image, which shows the morphology of $80Li_2S{\cdot}20P_2S_5$ glass ceramics. As indicated by small arrows in the inset, speckled contrasts due to the aggregation of crystalline nanoparticles could be observed. To get some information on crystalline nanoparticles, we took the high-resolution TEM image from the area indicated by the circle in the inset. We succeeded in taking high-resolution lattice images due to crystalline nanoparticles in the $80Li_2S{\cdot}20P_2S_5$ glass ceramics. In the high-resolution lattice image, lattice fringes were clearly observed in Fig. 2. The average size of crystalline nanoparticles is approximately 5 nm and distributed randomly in an amorphous matrix. One of the characteristic features of the crystalline nanoparticles is that some crystalline nanoparticles are connected to each other, as indicated by dotted circles. This indicates that precipitation and connection of crystalline nanoparticles in an amorphous matrix contributes to high ionic conductivity of more than 10^{-3} S cm^{-1} in the $80Li_2S{\cdot}20P_2S_5$ glass–ceramics.

Here, the microstructure of the $75Li_2S{\cdot}25P_2S_5$ glass ceramics is described. In order to elucidate the microstructure and crystallization behavior in the $75Li_2S{\cdot}25P_2S_5$ glasses, we carried out high-resolution TEM observation and *in-situ* TEM observation in the heating process from room temperature in the $75Li_2S{\cdot}25P_2S_5$ glasses. Figures 3a–b show DF and HR images obtained from the $75Li_2S{\cdot}25P_2S_5$ glasses at room temperature. The corresponding ED pattern is shown in the inset of Fig. 3a. Typical halo rings in the ED pattern indicate that the $75Li_2S{\cdot}25P_2S_5$ glasses

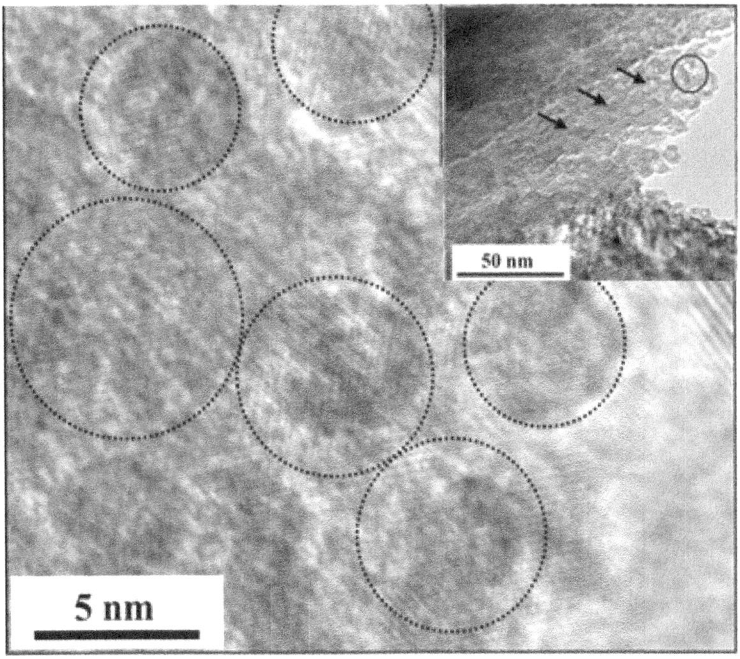

Fig. 2 High-resolution lattice image of the $80Li_2S{\cdot}20P_2S_5$ glass ceramics. In the inset, a low-magnified image is shown

are characterized by an amorphous state. In the DF image of Fig. 3a, however, crystallized regions with bright contrasts should originate from the crystalline nanoparticles with an average size of approximately 5 to 20 nm. In addition, crystalline nanoparticles were distributed randomly over the sample without connecting. In the high-resolution lattice image shown in Fig. 3b, furthermore, lattice fringes due to crystalline nanoparticles with an average size of approximately 5 nm are randomly distributed in an amorphous matrix, as indicated by dotted circles. This clearly indicated that the amorphous state in the $75Li_2S{\cdot}25P_2S_5$ glasses coexisted with crystalline nanoparticles with an average size of approximately 5 nm.

3.2 *Crystallization Process in the Li_2S-P_2S_5 Glass*

Here, the crystallization processes of $75Li_2S{\cdot}25P_2S_5$ glass were described, which were revealed by in-situ TEM observation in the heating processes from room temperature. From the XRD profiles of $75Li_2S{\cdot}25P_2S_5$ samples obtained after heat-treatment at various temperatures, the as-prepared glass and the samples obtained after heat-treatment from 120 to 180 °C are characterized by an amorphous state. On the other hand, it was revealed that the samples after heat-treatment above 200 °C show a

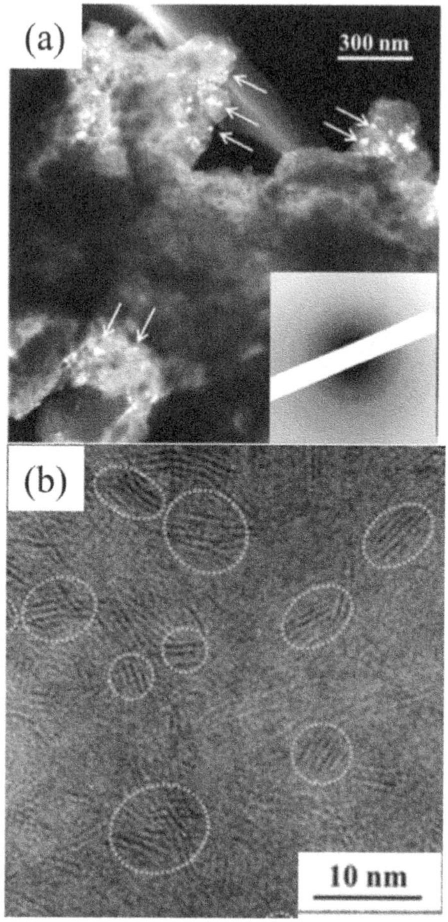

Fig. 3 Microstructures of the $75Li_2S{\cdot}25P_2S_5$ glass. **a** hollow-cone dark-field image and **b** high-resolution lattice image of the $75Li_2S{\cdot}25P_2S_5$ glass. In the inset of (**a**), the ED pattern is shown

single β-Li_3PS_4 crystalline phase [20]. The ionic conductivity of $75Li_2S{\cdot}25P_2S_5$ glass is shown by diamonds in Fig. 4. In the temperature range between 20 and 120 °C, the ionic conductivity gradually increases from 5.9×10^{-4} S/cm to 7.6×10^{-4} S/cm. Subsequently, the high ionic conductivity with approximately 7.5×10^{-4} S/cm is maintained until the temperature reached 180 °C. On the other hand, above 180 °C, the ionic conductivity drastically dropped to about 1.8×10^{-4} S/cm. To elucidate the temperature change of the ionic conductivity, *in-situ* TEM observation was conducted in the heating process from room temperature. As a result, it is revealed that the drastic decrease of the ionic conductivity should originate from the appearance of the nanocrystallites of β-Li_3PS_4 in an amorphous matrix. Figure 5 displays the change of the ED patterns as a function of the temperature, which was obtained from the $75Li_2S{\cdot}25P_2S_5$ glass. Note that the ED patterns in Fig. 5a–f were taken at room temperature, 170 °C, 190 °C, 200 °C, 230 °C, and 250 °C, respectively. The ED pattern at room temperature shows a halo pattern indicating the amorphous

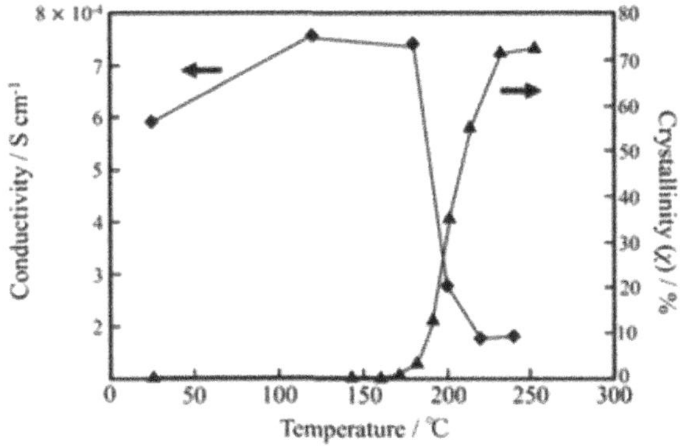

Fig. 4 Changes of the ionic conductivity and the degree of the crystallization of $75Li_2S{\cdot}25P_2S_5$ glass as a function of the temperature. Diamonds and triangles indicate ionic conductivity and the degree of the crystallization evaluated from the ED patterns, respectively

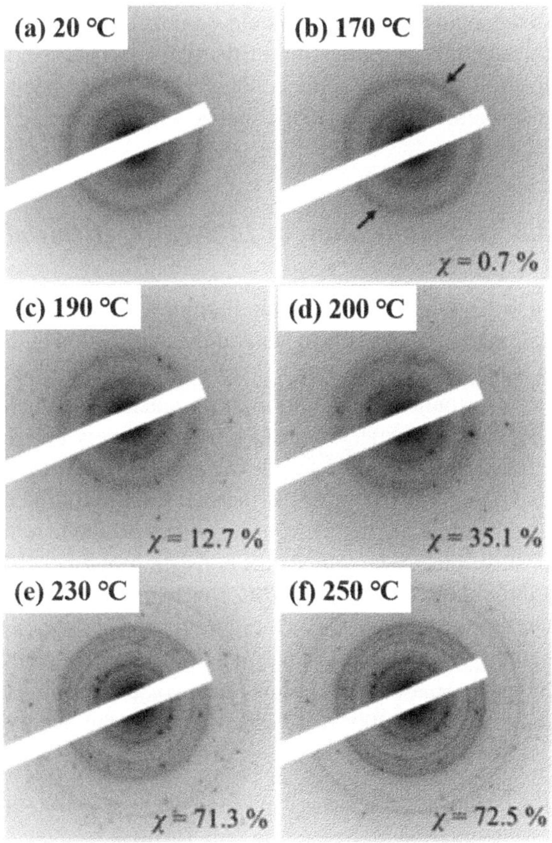

Fig. 5 Changes of the ED patterns as a function of the temperature, obtained from the same area of the sample upon heating from 20 to 250 °C

structure. When the sample was heated from room temperature, diffraction spots due to the nanocrystallites of β-Li_3PS_4 start to appear at around 180 °C. Subsequently, the crystallization into β-Li_3PS_4 crystallites from the amorphous structure proceeded significantly above 200 °C. The ED patterns above 200 °C display Bragg-diffraction spots due to the β-Li_3PS_4 crystallites. We calculated the degrees of crystallinity at each temperature based on equation, (1) and the degrees of the crystallization were described in the bottom of Fig. 5a–f. The degree of the crystallization was evaluated from the ED patterns using the same evaluation method using the Eq. (1) [19]. The degrees of the crystallization are shown by the triangles in Fig. 4. As evident in Fig. 4, no change in the degrees of the crystallization was found with increasing the temperature from room temperature. Subsequently, when the temperature was increased, around 180 °C, the degree of the crystallization drastically increased, whereas the ionic conductivity abruptly decreased. This suggests that the decrease of the ionic conductivity should stem from the formation of the β-Li_3PS_4 crystallites through the crystallization of the glasses. This reciprocal relation between crystallization degree and ionic conductivity has also been proved by the differential pair distribution function (PDF) analysis of high-energy XRD [21].

3.3 *Crystallization of LPS-NMC Electrode Composites*

In order to put all solid-state lithium ion secondary batteries to practical use, it is essential to suppress heat generation and ignition of the batteries and to ensure safety. It is necessary to develop battery materials that can withstand the temperature of the use environment. In addition, heat generation can also be a factor that deteriorates the life of the battery. For this reason, it is very important to evaluate the exothermic behavior of positive electrode composites and elucidate the factors of the exothermic reaction. In this session, the thermal stability of the electrode composite was evaluated by applying the TEM observation technique to the electrode composites. Here, positive electrode composites consisting of Li_2S-P_2S_5 (LPS) glass electrolyte and $LiNi_{1/3}Mn_{1/3}Co_{1/3}O_2$ (NMC) were investigated, which exhibit good long-term cycle characteristics [7, 8]. Figure 6 shows the thermal behavior of the NMC-LPS cathode composite after charging by differential scanning calorimetry. As the temperature is increased at a constant rate, two exothermic reactions occur in the temperature range of 300 to 400 °C. In order to clarify the factors of these exothermic reactions,

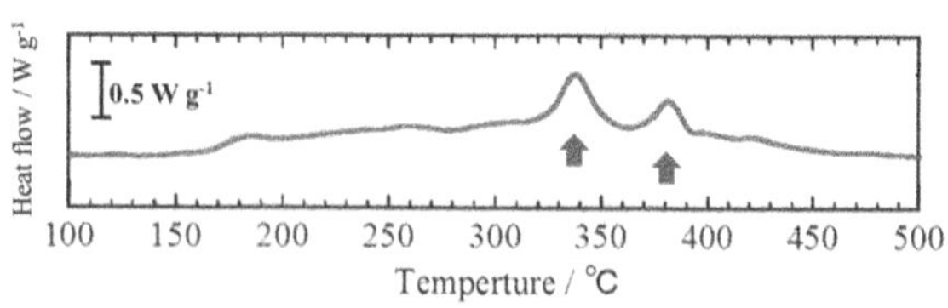

Fig. 6 Differential scanning calorimetry result of NMC-LPS cathode composites after charging

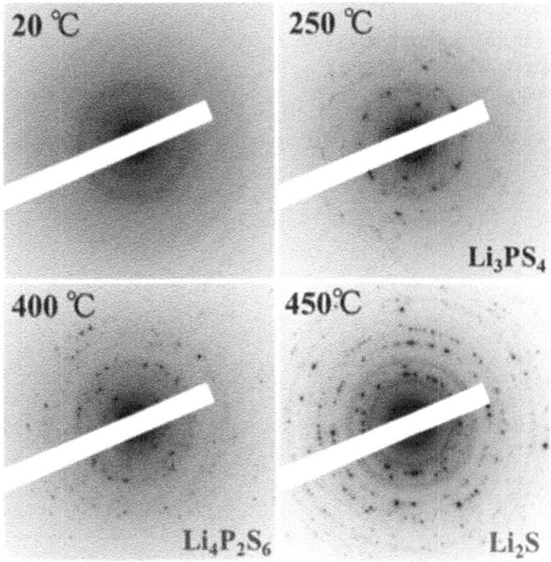

Fig. 7 Changes of the ED patterns as a function of the temperature obtained from the LPS region of NMC-LPS cathode composite after charging

the positive electrode composite was heated inside TEM using a heating TEM holder, and the structural changes in the heating process were observed.

We examined the structural change in the heating process of the NMC-LPS cathode composite after charging and found that there was no structural change in the active material NMC and crystallization occurred in the electrolyte LPS. Here, we described the difference of crystallization behavior between the LPS region with interfacial contact with NMC and the LPS region without interfacial contact. Figure 7 shows the temperature change of the ED patterns obtained from the LPS region near the NMC/LPS interface in the NMC-LPS positive electrode composite after charging. First, at room temperature (20°°C), a halo pattern indicating an amorphous state is observed. When heating the sample from room temperature, Bragg peaks due to the nanoparticles start to appear at about 250°°C and the number of the Bragg peaks increase slowly up to about 350°°C. When the temperature was further increased, a clear Debye-ring having a strong intensity appeared at 400°°C. From the ED patterns at each temperature, we tried to identify the precipitated crystal phases consisting of the nanoparticles. It was found that Li_3PS_4 was precipitated at 200–350°°C and $Li_4P_2S_6$ and Li_2S were precipitated at 400°°C and above. On the other hand, Fig. 8 shows the temperature change of an ED pattern obtained from the LPS region, where there is no interface contact with the NMC in the NMC-LPS positive electrode composite. A halo pattern indicating an amorphous state is observed at room temperature. When the temperature is increased from room temperature, Bragg peaks due to the nanoparticles of Li_3PS_4 start to appear at about 250°°C. On heating the samples further, it was found that only Li_3PS_4 was precipitated in the LPS region having no interfacial contact with NMC. Thus, it was revealed that the crystallization behavior of LPS greatly depends on the presence or absence

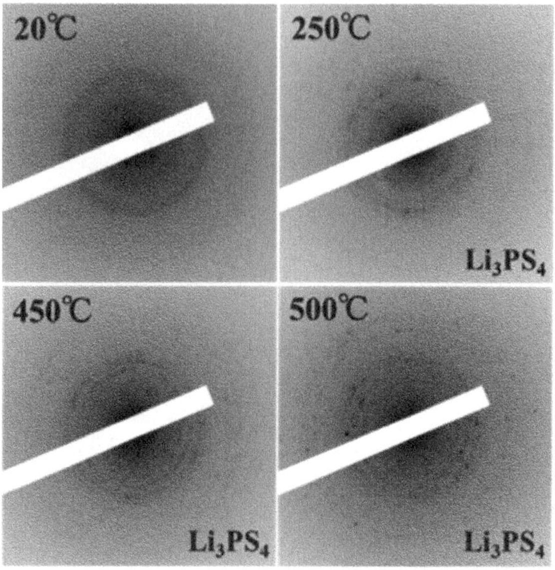

Fig. 8 Temperature change of the ED patterns in the LPS region without interface contact with NMC

of interfacial contact with the active material NMC. Figure 9 shows the temperature changes of microstructures of the LPS/NMC composites. From the temperature change of the microstructures, it is found that the crystallization region drastically increases around 400°°C. As described above, in the LPS electrolyte of the NMC-LPS positive electrode composite, the crystallization of Li_3PS_4 greatly progresses in the high-temperature region where the exothermic reaction exists, accompanying the precipitation of $Li_4P_2S_6$ and Li_2S.

4 Summary

In this chapter, the TEM analysis of the microstructures of the sulfide-based solid electrolyte Li_2S-P_2S_5, which is sensitively damaged by electron beams, was described. In particular, we reported the results obtained using the conventional TEM observation method such as electron diffraction, bright-field imaging, hollow-cone dark-field imaging, and high-resolution imaging experiments.

For better understanding the factors of the exothermic reaction, first-principal calculation of the crystallization energy should be required. In addition, it is expected that the crystallization behavior of LPS greatly depends on the interface contact with NMC. In the amorphous region, the bonding state of atoms and clusters in the local region should have some influence on the crystallization behavior of LPS. We believe that TEM should be a powerful tool to analyze the nanoparticles and the amorphous structure by combining hollow-cone dark-field imaging, nanobeam electron diffraction, and pair distribution function methods.

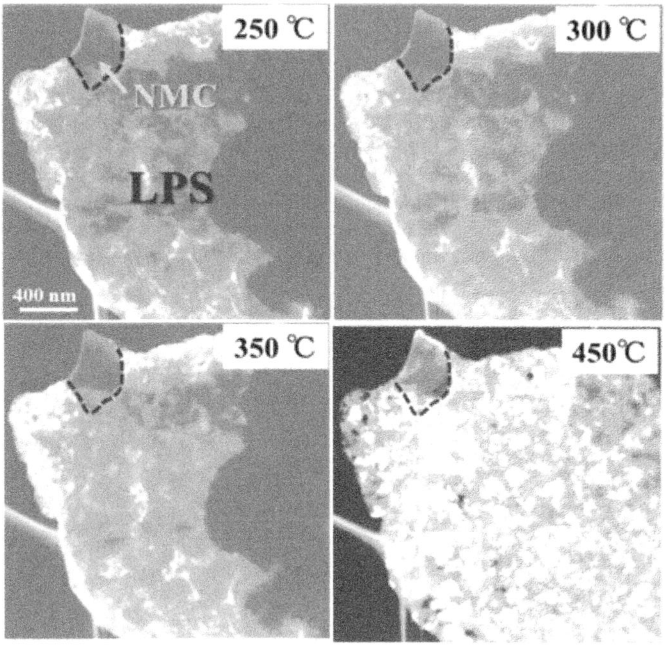

Fig. 9 A series of hollow-cone dark-field images showing the crystallization process of LPS in NMC-LPS cathode composite in the heating process from room temperature to 450 °C

References

1. Williams, D. B., & Carter, C. B. (1986). *Transmission electron microscopy*. New York: Plenum Press.
2. De Drael, M. (2012). *Introduction to conventional transmission*. Cambridge University Press.
3. Reimer, L., & Kohl, H. (2008). *Transmission electron microscopy; physics of image formation (Springer series of optical sciences)*, Springer.
4. Hirsh, P. B., Howie, A., Nicholson, R. B., Pashley, D. W., & Whelan, M. J. (1977). *Electron microscopy of thin crystals* (2nd ed.). Huntington, New York: Krieger.
5. Minami, T., Tatsumisago, M., Wakihara, M., Iwakura, C., Kohjiya, S., & Tanaka, I. (2005). *Solid state ionics for batteries*. Tokyo: Springer.
6. Hayashi, A., Minami, K., Ujiie, S., & Tatsumisago, M. (2010). *Journal Non-Crystalline Solids, 356,* 2670–2673.
7. Mizuno, F., Hayashi, A., Tadanaga, K., & Tatsumisago, M. (2005). *Advanced Materials, 17,* 918–921.
8. Kato, Y., Hori, S., Saito, T., Suzuki, K., Hirayama, M., Mitsui, A., et al. (2016). *Nature Energy, 1,* 16030.
9. Tsukasaki, H., Mori, S., Morimoto, H., Hayashi, A., & Tatsumisago, M. (2017). *Scientific Reports, 7*(4142), 1–7.
10. Tsukasaki, H., Mori, S., Shiotani, S., Yamamura, H., & Iba, H. (2018). *Solid State Ionics, 317,* 122–126.
11. Tsukasaki, H., Mori, Y., Otoyama, M., Yubuchi, S., Asano, T., Tanaka, Y., et al. (2018). *Scientific Reports, 8,* 6214.
12. Tsukasaki, H., Otoyama, M., Mori, Y., Mori, S., Morimoto, H., Hayashi, A., & Tatsumisago, M. (2017). *Journal of Power Sources, 367,* 42–48.

13. Tsukasaki, H., Fukuda, W., Morimoto, H., Arai, T., Mori, S., Hayashi, A., & Tatsumisago, M. (2018). *Scientific Reports, 8,* 15613.
14. Kunath, W., Zemlin, F., & Weiss, K. (1985). *Ultramicroscopy, 16,* 123.
15. Taya, M., Ikuta, T., & Takai, Y. (2008). *Optik, 119,* 153.
16. Lábár, J. L. (2008). Electron diffraction based analysis of phase fractions and texture in nanocrystalline thin films, part I: Principles. *Microscopy and Microanalysis, 14,* 287–295.
17. Lábár, J. L. (2009). Electron diffraction based analysis of phase fractions and texture in nanocrystalline thin films, Part II: Implementation. *Microscopy and Microanalysis, 15,* 20–29.
18. Lábár, J. L., et al. (2012). Electron diffraction based analysis of phase fractions and texture in nanocrystalline thin films, part III: Application examples. *Microscopy and Microanalysis, 18,* 406–420.
19. Tsukasaki, H., Mori, S., Shiotani, S., Yamamura, H., & Iba, H. (2017). *Journal of Power Sources, 369,* 57–64.
20. Homma, K., et al. (2011). *Solid State Ionics, 182,* 53.
21. Shiotani, S., Ohara, K., Tsukasaki, H., Mori, S., & Kanno, R. (2017). *Scientific Reports, 7,* 6972.

XAFS Analysis

Kentaro Yamamoto and Yoshiharu Uchimoto

Abstract All-solid-state batteries using nonflammable inorganic solid electrolytes are expected to have both high energy density and excellent safety and to have higher power density than current lithium-ion batteries. However, the actual power density of the all-solid-state batteries is often smaller than expected, due to factors such as the resistance at the electrode/solid-state electrolyte interface and/or the dispersion of the active material and solid-state electrolyte in the composite electrode. In order to improve the power density of the all-solid-state batteries, analytical methods with several space and time scales are necessary for clarification about the specific polarization factors of the all-solid-state batteries. In this chapter, we introduce the application of X-ray absorption fine structure (XAFS) analysis to the analysis of an interfacial structure between the cathode and solid electrolyte, and a reaction distribution in composite materials. Depth-resolved XAFS analysis proves that the electronic structure of the active material changed and a reactive layer formed at the interface between the active material and the solid electrolyte during charging and discharging, which leads to the increase of resistance. Two-dimensional XAFS analysis proves that the ionic conduction resistance increases the farther away from the solid electrolyte layer because of the large bending of the ionic path in the heterogeneous electrode, leading to large polarization. These XAFS analyses are powerful tools to understand the phenomena inside all-solid-state batteries and improve the performances.

Keywords All-solid-state batteries · XAFS analysis · Cathode/electrolyte interface · Reaction distribution

K. Yamamoto (✉) · Y. Uchimoto
Graduate School of Human and Environmental Studies, Kyoto University, Kyoto 606-8501, Japan
e-mail: yamamoto.kentaro.4e@kyoto-u.ac.jp

Y. Uchimoto
e-mail: uchimoto.yoshiharu.2n@kyoto-u.ac.jp

K. Kanamura (ed.), *Next Generation Batteries*,
https://doi.org/10.1007/978-981-33-6668-8_15

1 Introduction

All-solid-state battery using a nonflammable inorganic solid electrolyte instead of a flammable organic liquid electrolyte as the electrolyte can achieve both high energy density and excellent safety. In recent years, an inorganic solid electrolyte material having properties equivalent to those of an organic electrolyte has been developed [1]. In the organic liquid electrolyte, the anion concentration polarization occurs due to a low transport number of lithium ions, whereas concentration polarization does not occur even when charging and discharging at a high current density because the inorganic solid electrolyte has a lithium-ion transport number of 1. Therefore, the rate performance of all-solid-state batteries can be higher than that of current lithium-ion batteries [2]. However, there are many cases where the output characteristics of an all-solid-state battery are smaller than expected. This is because the actual performance of the storage battery is affected not only by the properties of the material but also by the following factors in the reaction element process. In the practical application of the all-solid-state battery, it is required to quantify the polarization factor, and it can be said that it is necessary to develop various operando analyzes to elucidate the polarization factor of the all-solid-state battery.

The all-solid-state battery, like the conventional liquid lithium-ion secondary battery, has a single cell of the order of cm, and the thickness of the mixture electrode is considered to be several tens to 100 μm. On the other hand, the size of the active material is on the order of several tens nm to several μm, and the reaction at the electrode–electrolyte interface proceeds in the active material/solid electrolyte interface phase formed on the order of several nm. Figure 1 shows the phenomenon that constitutes the reaction of the all-solid-state battery composite electrode. For the cathode and anode of an all-solid battery, a mixture comprising a solid electrolyte, cathode and anode active materials, a conductive material, and a binder is used, and has a complicated three-dimensional structure. At the active material/electrolyte interface, insertion and extraction reactions of lithium ions occur. In the active material, reorganization of the lattice occurs during charge/discharge reactions, and the accompanying phase change progresses. In the reaction of an all-solid-state battery, multiple phenomena are intertwined and in a non-equilibrium state. Ex situ analysis after disassembling battery cell, which has been used in conventional battery research, does not provide a clear picture of the determining factors of the reaction rate and degradation mechanism. In order to these phenomena, analytical methods under battery operating condition (operando) are necessary. The strong transmissivity of synchrotron radiation X-rays and the high temporal/spatial resolution make it possible to measure the temporal and spatial hierarchical structure of an all-solid-state battery in situ as the battery reaction progresses. The polarization characteristic of the solid-state battery is often caused by the resistance component in the active material/solid electrolyte interface phase or the dispersed state of the active material and the solid electrolyte in the mixture. The following describes operand analysis techniques to clarify these phenomena.

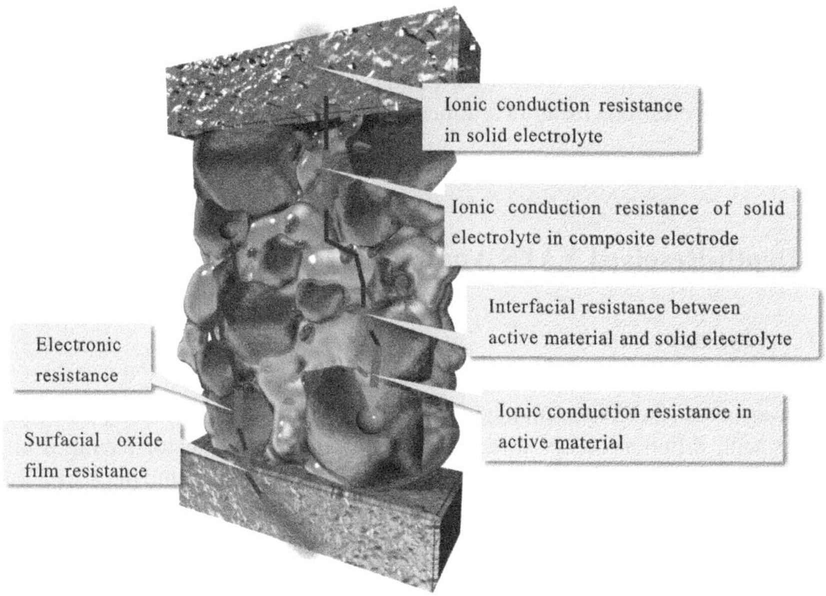

Fig. 1 Phenomena that constitute the reaction of the all-solid-state battery mixture electrode

2 XAFS Analysis

There are many techniques for in situ measurement of lithium-ion secondary batteries using synchrotron radiation X-rays. Typical examples are X-ray absorption spectroscopy (XAS) or X-ray absorption fine structure (XAFS). These techniques are effective for analyzing the cause of polarization in the all-solid-state battery. By examining the X-ray absorption rate, the X-rays, and electrons emitted from the substance caused by irradiation with X-ray, information inside the substance can be obtained. For example, when a material is irradiated with X-ray while changing the energy of X-ray, a steep X-ray absorption phenomenon called an absorption edge occurs at a specific energy corresponding to the contained element. This is a phenomenon in which electrons in the inner core level are excited by X-rays and transit to unoccupied orbitals and vacant energy bands beyond the Fermi level. By analyzing the absorption edge spectrum, the coordination structure and the electronic structure of the absorbing atom can be obtained. This technique is XAFS. In particular, the spectrum near the absorption edge is called X-ray absorption near-edge structure (XANES), while the spectrum observed in the high energy region within about 1 keV from the absorption edge is called Extended X-ray absorption fine structures (EXAFS), respectively. The former can provide information on the electronic state of the absorbing atom, and the latter can provide information on the coordination structure around the absorbing atom. There are excellent books on the principles of XAFS and the analytical theory of EXAFS, so please refer to them [3, 4]. Here, mainly, in a sulfide solid electrolyte system with high lithium-ion conductivity, (1) structural analysis of the electrode–sulfide solid electrolyte interface using depth-resolved X-ray absorption spectroscopy and (2) the reaction distribution analysis in the composite electrode using two-dimensional X-ray absorption spectroscopy will be described.

3 Depth-Resolved XAFS Analysis

3.1 Abstract

Depth-resolved XAFS measurement is a method of separately detecting fluorescent X-rays with different escape directions by introducing a detector having a large number of X-ray detection channels on a two-dimensional plane into the XAFS fluorescence method. For samples with high smoothness and flatness such as thin-film samples, it is possible to detect changes in electron and local structure in the depth direction by separately detecting fluorescent X-rays with different escape angles to the sample surface. This method has been used to analyze the positive electrode/electrolyte interface structure of lithium-ion secondary batteries [5] and the positive electrode/electrolyte interface structure of all-solid-state batteries [6]. Here, the principle of the depth-resolved XAFS method and an example applied to the observation of the structure of the positive electrode/sulfide solid electrolyte interface of an all-solid-state battery [7] will be described.

3.2 Principle

As described above, this measurement detects fluorescent X-rays using a two-dimensional detector (PILATAS). The PILATAS detector is a two-dimensional hybrid pixel X-ray detector that combines a pn diode array sensor and a complementary metal oxide semiconductor (CMOS) readout integrated circuit on a two-dimensional plane. The hybrid pixel can detect a single photon and is composed of a preamplifier (preamplifier), a comparator (wave height discrimination circuit), and a counter circuit. The preamplifier amplifies the charge generated in the sensor by the incoming X-rays. When the charge exceeds a preset threshold, the comparator generates a digital signal. This mechanism allows digital storage and readout of the number of detected X-rays per pixel without dark current or readout noise [8]. The PILATUS 100 K used in this experiment has 487×195 square pixels with a side of 172 μm, and the total area where X-rays can be detected is 84×34 mm^2. The pixel features a 20-bit counter, which can measure up to 220 photons per irradiation [9].

When a sample is irradiated with X-ray, vacancies are created in the inner shell. When electrons fall from the outer shell into the vacancy, fluorescent X-rays corresponding to the absorbed X-rays are emitted. As shown in Fig. 2, the emitted fluorescent X-rays have different distances to escape from the sample depending on the emitted angle. When the distance is large, the fluorescent X-rays are largely absorbed by the sample and cannot reach the detector. In Fig. 2, the detection channel of PILATUS having a small channel number corresponds to a position for detecting fluorescent X-rays emitted at a low angle. Fluorescent X-rays emitted from a shallow position on the sample surface reach the low-angle channel of PILATUS. On the other hand, fluorescent X-rays emitted from a deep position on the sample surface does not reach the low-angle channel because the X-rays have a large amount of self-absorption by the sample. As a result, in the channel on the low-angle side, only the fluorescent X-rays emitted in a region near the sample surface are detected. On the other hand, the detection element of PILATUS having a large channel number in Fig. 2 corresponds to a position for detecting fluorescent X-rays emitted at a high angle. When the emission angle of the fluorescent X-rays is large, the influence of self-absorption is reduced, so that even fluorescent X-rays emitted from a deep region of the sample can be detected. Therefore, in the high-angle channel, information including not only the surface of the sample but also a deep portion is detected. As the Pilatus number increases, the information in a deep portion is added, and the ratio of information on the sample surface decreases. In this way, the depth-resolved XAFS method enabled observation of changes in the electron and local structures in the depth direction at a resolution of 3 to 4 nm by separately detecting X-ray fluorescence with different escape angles.

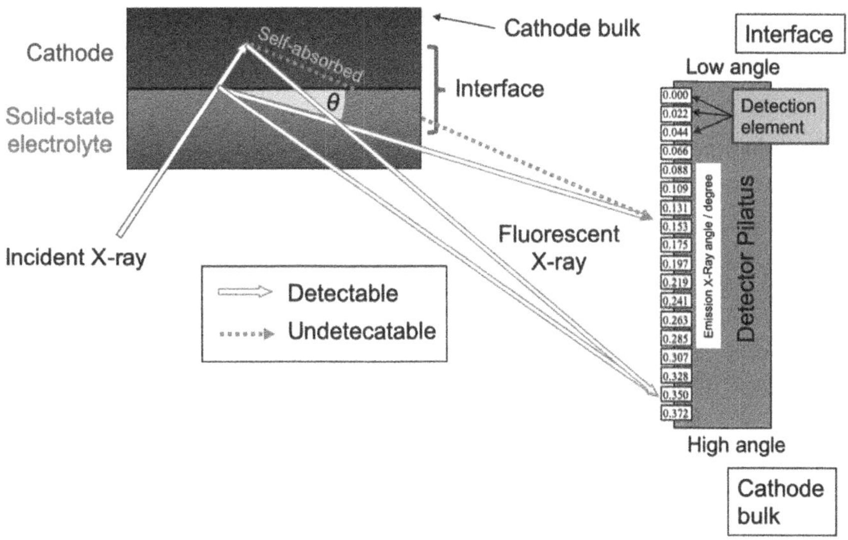

Fig. 2 The principle of depth-resolved XAFS

3.3 Measurement Example

We show an example, of which a Li_3PO_4 intermediate layer was introduced at the solid–solid interface between $LiCoO_2$, a typical cathode material, and Li_2S-P_2S_5, a sulfide solid electrolyte, and the reaction mechanism at the solid–solid interface was clarified by electrochemical measurement and depth-resolved XAFS measurement. An example is shown [7]. Depth-resolved XAFS measurement was performed with BL37XU of SPring-8, and the absorption spectrum was performed in the range including the absorption edge energy of the Co-*K* shell. The size of the incident X-ray beam is 1.0 mm wide × 0.4 mm long, and the distance between the PILATUS detector and the sample is 45 cm. Regarding the notation of data, since the distance from the sample to the detector is 45 cm, the angle between the detector and the thin film is calculated from the channel number of the detected element, and the data is described based on the angle of the emitted X-ray.

Figure 3 shows the absorption edge energy positions (corresponding to the oxidation state of Co) obtained by depth-resolved XAFS for (a) $LiCoO_2$/Li_2S-P_2S_5 interface without the Li_3PO_4 intermediate layer and for (b) $LiCoO_2$/Li_2S-P_2S_5 interface with the Li_3PO_4 intermediate layer. When the Li_3PO_4 intermediate layer was not introduced, the energy of the $LiCoO_2$/Li_2S-P_2S_5 interface was shifted to a lower energy than that of the bulk $LiCoO_2$, indicating that Co was reduced. This difference is more pronounced after the cycle, meaning that the thickness of the reaction layer increased. On the other hand, the Li_3PO_4 interface modification reduces the reduction of Co at the $LiCoO_2$/Li_2S-P_2S_5 interface, and the growth of the reaction phase hardly progresses even after the cycle.

4 Two-Dimensional XAFS Analysis

4.1 Abstract

X-ray absorption spectroscopy is suitable for reaction distribution analysis because it is possible to estimate the valence of an electrode active material using the redox of a transition metal. By devising a light source and a detector, imaging with high spatial resolution can be achieved. In the two-dimensional XAFS, a large beam is applied to the sample and the transmitted light intensity is detected by a two-dimensional detector in order to perform the measurement in a region as wide as possible. By introducing a two-dimensional detector, it is possible to measure the electronic state of a specific element existing at a position of interest and to map the progress of the electrode reaction at each position. Here, we explain the principle of two-dimensional X-ray absorption spectroscopy (two-dimensional XAFS) and an example of analyzing the reaction distribution in the positive electrode of an all-solid-state battery as a method for analyzing the reaction distribution in the composite electrode [10–12].

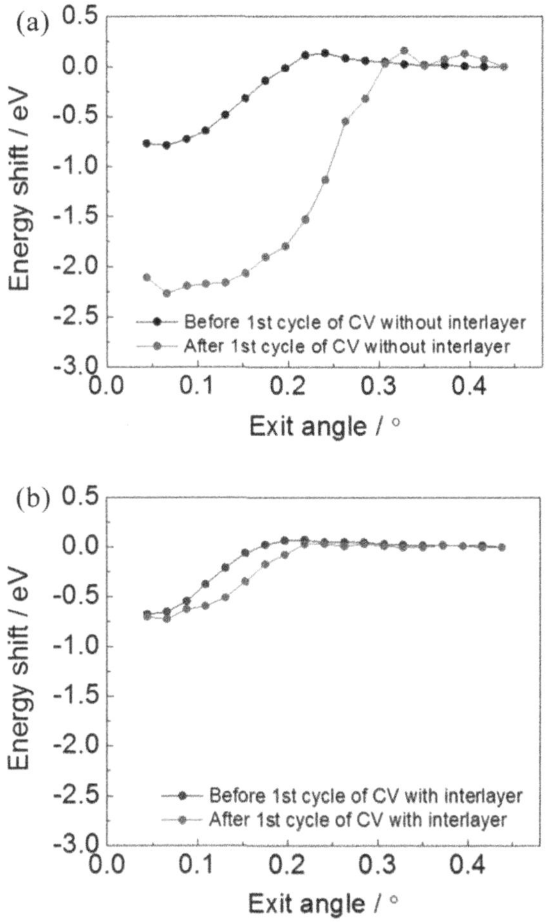

Fig. 3 Energy shift at half of the normalized absorbance of XANES spectra from different exit angles for the thin-film electrodes (**a**) without and (**b**) with the interlayer before and after the first cycle of CV measurement. Reprinted with permission from ref [7]. Copyright 2018 Elsevier

4.2 *Principle*

Figure 4a shows a schematic diagram of the two-dimensional XAFS method. The sample to be measured is sliced and placed in the X-ray beam path. The incident X-ray passes through the slit and the ion chamber for measuring I_0 and reaches the sample. The transmitted X-rays passing through the sample are detected by a two-dimensional detector having a large number of X-ray detecting channels in two-dimensional directions. Figure 4b shows the principle of the two-dimensional detector. A large number of detection channels differentiate the sample irradiated with X-rays and record a transmission XAFS spectrum of one detection element and one minute range. Thereafter, the absorption edge energy value of the XANES spectrum obtained by each detection channel is extracted by calculation. Two-dimensional map images can be obtained by assigning different colors to different absorption edge

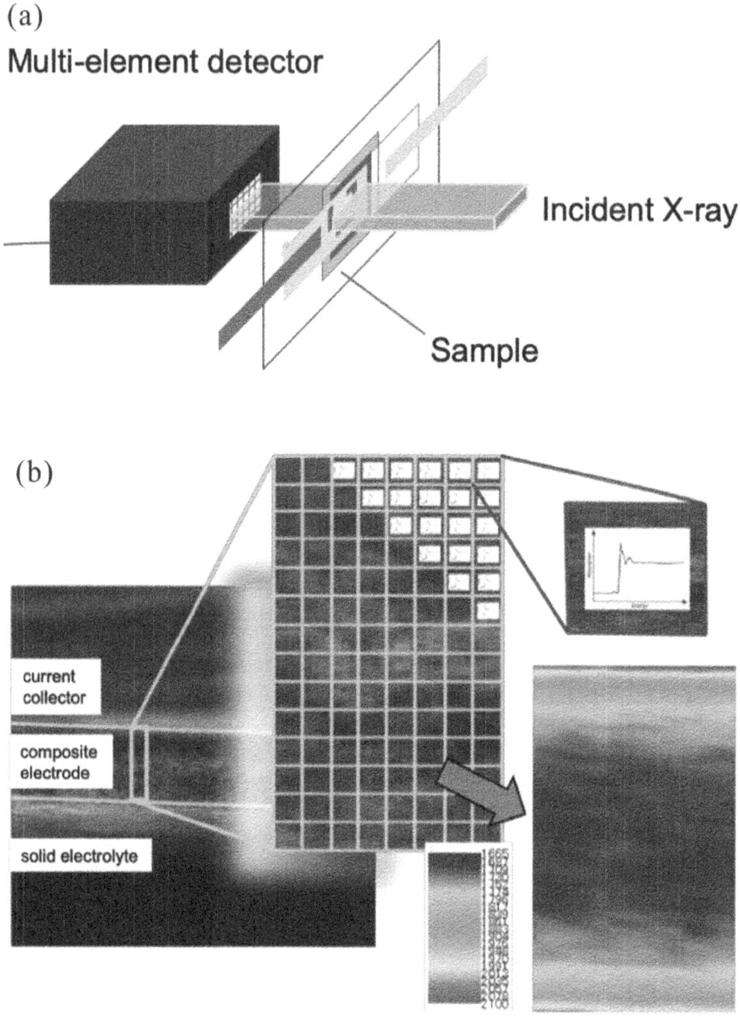

Fig. 4 **a** Schematic diagram of 2D XAFS measurement. **b** Principle diagram of 2D detector

energies and color-coding them by energy. The two-dimensional detector consists of a complementary metal oxide semiconductor (CMOS) image sensor, a scintillator (LuAG: Ce +, 10 μm thick), and a magnifying glass. The transmitted light is converted into visible light by the scintillator. The light converted to visible light is magnified five times by a magnifying glass and guided to an image sensor. The image sensor is composed of 2048 × 2048 CMOS elements and has a pixel size of 6.5 μm × 6.5 μm. Therefore, the field of view is 13.312 mm × 13.312 mm at the maximum, and the resolution in an ideal case is 1.3 μm × 1.3 μm.

4.3 Measurement Example

An example in which two-dimensional XAFS is applied to a mixture electrode of a sheet-type all-solid-state battery is described. The mixture electrode of the sheet-type all-solid-state battery has a complicated three-dimensional structure consisting of an active material, a solid electrolyte, a conductive additive, and a binder, similar to a lithium-ion secondary battery using a liquid electrolyte. An example of the two-dimensional XAFS applied to an all-solid-state battery consisting of a composite electrode using different binders and discussing the correlation between the dispersion state of the electrode and the heterogeneous reaction phenomenon is presented [10].

For the mixture electrode, $LiNi_{1/3}Co_{1/3}Mn_{1/3}O_2$ (NCM), $75Li_2S$-$25P_2S_5$ (SE), acetylene black (AB), and a binder are mixed in a weight ratio of NCM: SE: AB: binder = 70: 30: 3: 3. It was produced by doing. Styrene-butadiene-styrene rubber (SBR) and Styrene-ethylene-butylene-styrene (SEBS) were used as binders, and anisole and heptane were used as solvents during slurry preparation. The cell shown in Fig. 5 was used for the measurement of two-dimensional XAFS. As shown in Fig. 5, a titanium foil having no lithium-ion conductivity is inserted in a part between the positive electrode and the separator (electrolyte layer). In the region covered with the titanium foil, the movement of lithium ions in the electrode is limited only to the in-plane direction of the electrode (the coordinate axis in the drawing indicates the XY plane). Here, this region is called an ion blocking region. Further, it is a portion where the titanium foil is not inserted, where the positive electrode layer and the separator layer directly face each other. This portion operates as a normal battery and is called a non-blocking region. All the movement of the ions in the ion blocking region follows

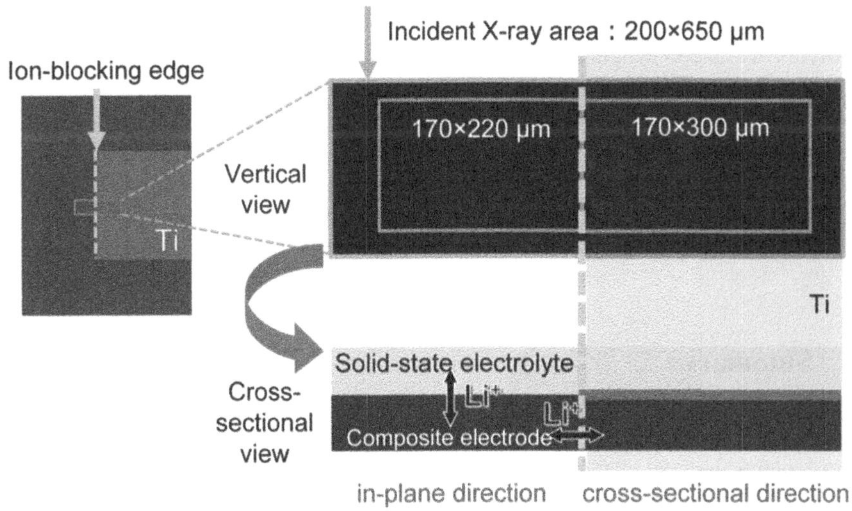

Fig. 5 Schematic illustration of the ion-blocking electrode for 2D- XAS measurements.

the ion movement (potential gradient) generated in the adjacent non-blocking region. For example, during charging, in the ion blocking region, lithium ions are extracted from the boundary with the non-blocking region, and an electrode reaction occurs. Thus, it can be said that the ion blocking region simulates the reaction behavior in the electrode thickness direction. By performing two-dimensional XAFS measurement in the non-blocking region and the ion-blocking region, the reaction distribution in the electrode surface and the electrode cross section can be observed. Two-dimensional XAFS measurement was performed on the Ni *K*-edge of the NCM electrode using a two-dimensional detector (spatial resolution 1.3 μm), and one spectrum was obtained every 20 min. The measurement was performed continuously during charging at the 1/40 C rate.

Figure 6 shows the in-plane X-ray transmission image of the mixture electrode using SBR and SEBS and the results of two-dimensional XAFS at a state of charge of 3% to 18%. The dark part of the X-ray transmission image indicates that the NCM, which is the active material, is present in a large amount because the X-rays are strongly absorbed. It was confirmed that the active material was dispersed nonuniformly in the mixture electrode using SEBS rather than SBR. The two-dimensional XAFS shows that the reaction progresses more uniformly in the SBR cell than in the SEBS cell. This means that the reaction distribution in the SBR cell is less likely to occur than in the SEBS cell. In order to quantitatively discuss the reaction distribution in the cross-sectional direction, the reaction amount was estimated from the energy shift of XANES, and the result plotted against the distance in the cross-sectional direction is shown in Fig. 7. The reaction distribution of the SBR cell having a uniform structure extends to the deep part of the electrode. In order to find the diffusion coefficient in the mixture electrode, fitting was performed using a diffusion equation that draws one-dimensional diffusion. The line in Fig. 7 is a fitting curve. The apparent diffusion coefficient was determined to be $7.5 \times 10^{-11}/\text{cm}^2\text{s}^{-1}$ in the SBR cell and $3.9 \times 10^{-11}/\text{cm}^2\text{s}^{-1}$ in the SEBS cell. Also, the diffusion coefficient of solid electrolyte $75Li_2S$-$25P_2S_5$ glass is reported to be $7.8 \times 10^{-10}/\text{cm}^2\text{s}^{-1}$ [13]. The electrode showed a small value compared with the diffusion coefficient of the solid electrolyte bulk, and the SEBS electrode having a nonuniform structure showed a smaller value than the SBR electrode. The ion mobility depends on the shape (tortuosity) of the ion path. In the case of an inhomogeneous structure, because the conductive phase is divided and the tortuosity is large, the ionic conduction resistance becomes large in the deep region in the electrode, resulting in a large ion potential gradient.

5 Summary

In order to improve the performance and durability of an all-solid-state battery, it is important to understand the chemical and physical conditions of the various materials used under operating conditions, and in situ measurement using synchrotron radiation is also required. This is one of the useful techniques. In this chapter, we have focused on in situ measurements on all-solid-state batteries that we have performed so far. In

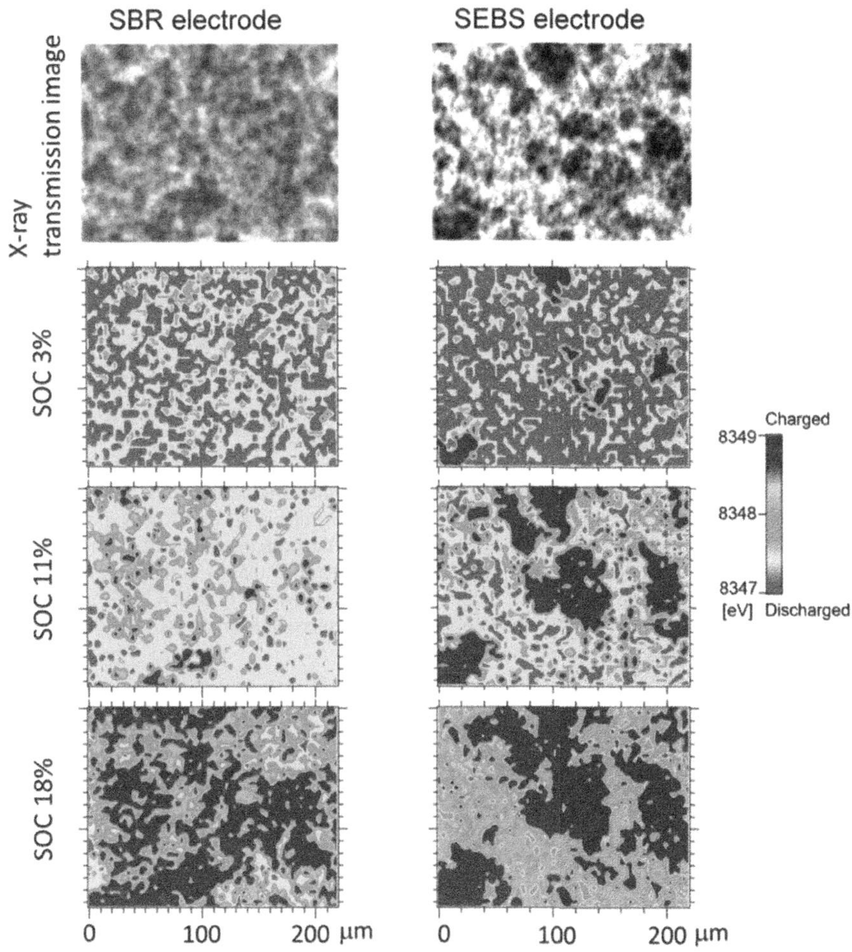

Fig. 6 X-ray transmission images and reacted area mapping results for different SOCs from 3% to 18% in the in-plane direction for the SBR electrode (left side) and SEBS electrode (right side). Reprinted with permission from ref [10]. Copyright 2019 American Chemical Society

the case of an all-solid-state battery, since the electrolyte is solid, a polarization factor unique to a solid-state battery is involved in the formation and dispersion of the interface between the positive and negative electrode active materials/solid electrolyte. It becomes increasingly important to clarify the factors by the operand measurement and realize the battery characteristics such as the high-rate characteristics inherent to the solid-state battery.

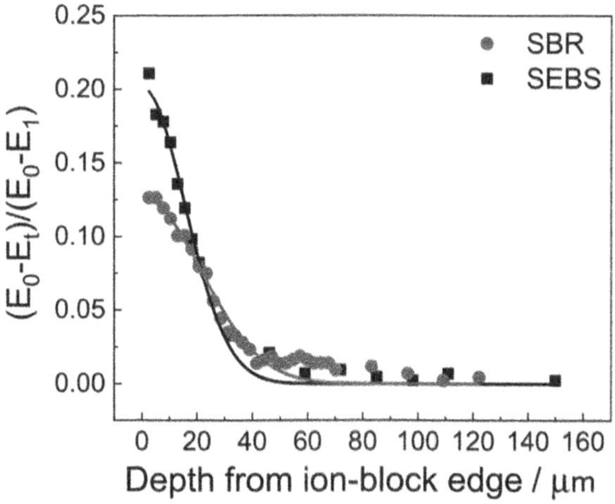

Fig. 7 Reaction amount estimated from XANES spectra of the SBR electrode and SEBS electrode as a function of depth from the ion-block edge. E_0 is the peak top energy of XANES spectra for Ni *K*-edge of $Li_xNi_{1/3}Co_{1/3}Mn_{1/3}O_2$ with Li content of 1. E1 is the peak top energy of XANES spectra for the Ni-*K* edge of $Li_xNi_{1/3}Co_{1/3}Mn_{1/3}O_2$ with a Li content of 0.5. E_t is the peak top energy of average XANES spectra for the Ni *K*-edge of $Li_xNi_{1/3}Co_{1/3}Mn_{1/3}O_2$ in SOC 21% at different position depths from the ion-block edge. Reprinted with permission from ref [10]. Copyright 2019 American Chemical Society

References

1. Kamaya, N., et al. (2011). *Nature Materials, 10,* 682.
2. Kato, Y., et al. (2016). *Nature Energy, 1,* 16030.
3. Calvin, S. (2012). *XAFS for everyone*. CRC.
4. Stohr, J. (1992). *NEXAFS spectroscopy*. Berlin: Springer.
5. Takamatsu, D., et al. (2011). *The Journal of Physical Chemistry Letters, 2,* 2511.
6. Okumura, Y., et al. (2011). *Journal of Materials Chemistry, 21,* 10051.
7. Chen, K., et al. (2018). *Solid State Ionics, 327,* 150.
8. Paul Scherrer Institut, *SLS Detector Group'website*. http://pilatus.web.psi.ch/index.htm.
9. Bech, M., et al. (2008). *Applied Radiation and Isotopes, 66,* 474.
10. Chen, K., et al. (2019). *Journal of Physical Chemistry C, 123,* 3292.
11. Nakamura, T., et al. (2017). *Journal of Physical Chemistry C, 121,* 2118.
12. Nakamura, T., et al. (2019). *Batteries & Supercaps, 2,* 1.
13. Baba, T., & Kawamura, Y. (2016). *Frontiers in Energy Research, 4,* 22.

Characterization of Cathode/Sulfide Electrolyte Interface Using a Thin-Film Model Battery

Masaaki Hirayama, Kota Suzuki, Ryoji Kanno, Takuya Masuda, and Kazuhisa Tamura

Abstract Interfacial phenomena of oxide cathode/sulfide electrolytes are crucial for the improvement of energy density and cycle stability of all-solid-state batteries. Thin-film batteries provide a simple reaction field for mechanistic studies. We fabricated a model film battery consisted of $LiCoO_2$ cathode, $LiNbO_3$ buffer layer, amorphous Li_3PS_4 electrolyte, and Li metal anode by physical vapor deposition. Structural changes at the cathode-side interface were investigated by in situ X-ray diffraction (XRD) and X-ray photoelectron spectroscopy (XPS) using Ar etching. The $LiCoO_2$ lattice showed a reversible change under high voltage operation with the upper cutoff voltage of 4.5 V. A cathode electrolyte interphase (CEI) layer was formed at the interface by decomposition of $LiNbO_3$ and Li_3PS_4. The ionic and electronic conductivities of the CEI layer could be crucial to improve the reaction resistance and the cycle stability of the oxide cathode/sulfide electrolyte interface.

Keywords $LiCoO_2$ cathode · Li_3PS_4 electrolyte · Interfacial reaction · In situ X-ray diffraction · X-ray photoelectron spectroscopy

M. Hirayama (✉) · K. Suzuki
Department of Chemical Science and Engineering, School of Materials and Chemical Technology, Tokyo Institute of Technology, Yokohama 226-8502, Japan
e-mail: hirayama@echem.titech.ac.jp

R. Kanno
All-Solid-State Battery Unit, Institute of Innovative Research, Tokyo Institute of Technology, Yokohama 226-8503, Japan

T. Masuda
Research Center for Advanced Materials and Characterization, National Institute for Materials Science, Ibaraki 305-0044, Japan

K. Tamura
Synchrotron Radiation Research Center, Japan Atomic Energy Agency, Hyogo 679-5148, Japan

K. Kanamura (ed.), *Next Generation Batteries*,
https://doi.org/10.1007/978-981-33-6668-8_16

1 Toward High Voltage Operation of All-Solid-State Batteries

High voltage operation has attracted significant research attention to improve the energy density and power density of next generation lithium batteries. Liquid-type lithium batteries have some issues in the reliability of high voltage operation due to side reactions at cathode/liquid electrolyte interfaces [1]: (i) oxidative decomposition of electrolyte species occurs in high voltage regions above 4.5 V, accompanied with the gas generation, (ii) delithiated phases of the cathode are electrochemically instable, which leads to the dissolution of transition metal species into the electrolyte, (iii) The decomposition products peel off from the cathode surface, and the cathode/electrolyte interfaces are newly formed, which causes further side reactions. All-solid-state lithium batteries using oxide electrolytes are one of the promising candidates owing to superior chemical and electrochemical stability and negligible conductivity of transition metal ions of the oxide electrolytes. In thin-film battery systems, stable high voltage operation has been demonstrated using a spinel-type or olivine-type high-potential cathode with Li_3PO_4-based electrolytes [2]. In contrast, there are few reports on high voltage operation of high-potential cathodes with the sulfide electrolytes [3], as sulfide solid electrolytes have been predicted to thermodynamically decompose in the low voltage region below 4.0 V [4] However, the decomposition product could play a role in solid electrolyte and suppressing further decomposition, that is well known as a solid electrolyte interphase (SEI) formed at graphite anode/organic electrolyte interfaces in liquid-type lithium batteries [5]. A comprehensive understanding of the interfacial phenomena is crucial to design the cathode/sulfide electrode interfaces exhibiting high rate capability and high cycle stability.

Bulk-type all-solid-state batteries generally include composite electrodes consisting of active material, solid electrolyte, and conductive additives [6]. The composite electrodes include many parameters such as particle size/shape, ratios of the components and voids, impurities, and their nonuniformity, which makes it difficult to analyze the interfacial phenomena. Furthermore, there are very few experimental techniques that allow direct observation of the electrode/solid electrolyte interface during the electrochemical reactions. Thin-film batteries offer several advantages for mechanistic studies on the interfacial phenomena: (i) The two-dimensional interface without binders and conductive additives provides a simple reaction field. (ii) The nanometers thick electrode shows very small changes in volume during lithium (de)intercalation, and the reaction area keeps constant. (iii) As the total thickness of the film battery is from sub micrometers to several micrometers, X-rays can penetrate through the film battery. In situ XRDs can be applied to detect the crystal structure changes of the cathode material during electrochemical cycling [1a, 7]. However, there are few reports on the application of sulfide electrolytes to thin-film batteries due to the difficulty of fabricating the films [8].

In this study, we investigate the electrochemical properties and structural changes of a $LiCoO_2$ cathode/$LiNbO_3$ buffer layer/Li_3PS_4 electrolyte interface under high voltage operation. Thin-film batteries of $LiCoO_2$ cathode/$LiNbO_3$ buffer layer/Li_3PS_4 electrolyte/Li metal anode were fabricated on Au/Al_2O_3 substrates by physical vapor deposition. Constant current charge-discharge measurements and AC impedance analyses were conducted to clarify electrochemical properties. Structural changes collected by in situ XRD and XPS using Ar etching provided comprehensive information on interfacial phenomena at oxide cathode/sulfide electrolytes [8b].

2 Fabrication of Sulfide-Type Thin-Film Batteries

Figure 1 illustrates Li/Li_3PS_4/$LiNbO_3$/$LiCoO_2$ thin-film battery fabricated on Au-coated Al_2O_3 substrate. $LiCoO_2$ films were deposited at room temperature using a radio-frequency magnetron sputter system with a target of $LiCoO_2$ bonded onto a copper plate [8b, 9]. The $LiCoO_2$ films were post-annealed at 725 °C under an O_2 flow for crystallization [9]. XRD, Raman spectroscopy, and SEM confirmed that the $LiCoO_2$ films were composed of single-phase domains of layer rocksalt type $LiCoO_2$. The $LiNbO_3$ and Li_3PS_4 films were synthesized at room temperature by pulsed laser deposition (PLD) with a KrF excimer laser. The $LiNbO_3$ layer was introduced to suppress the resistive layer that generally forms at the $LiCoO_2$/Li_3PS_4 interface [10]. For the Li_3PS_4 films, a 14% Li_2S-excess $Li_{3.42}PS_{4.21}$ target was used to compensate for the loss of Li and S during the PLD process. The P-shaped Li metal anode and Au current collector were deposited by thermal evaporation of metallic Li and Au, and they were in contact with the electrochemical device through needle-type probes. The electrochemical reaction area was 9 mm^2. Table 1 summarizes synthetic methods and thicknesses of each component in Li/Li_3PS_4/$LiNbO_3$/$LiCoO_2$ thin-film batteries

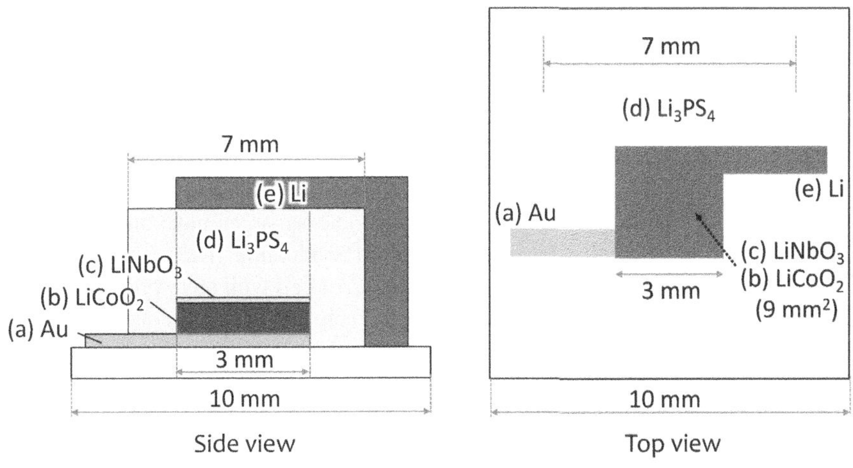

Fig. 1 Schematics of the cell design for Li/Li_3PS_4/$LiNbO_3$/$LiCoO_2$ thin-film batteries fabricated on Au-coated Al_2O_3 substrate

Table 1 Synthetic methods and thicknesses of each component in $Li/Li_3PS_4/LiNbO_3/LiCoO_2$ thin-film batteries fabricated on Au-coated Al_2O_3 substrate

Component	Material	Synthetic method	Thickness
e. Anode	Li metal	Vacuum vapor deposition	~1 μm
d. Solid electrolyte	Amorphous Li_3PS_4	Pulsed laser deposition	2.0–2.5 μm
c. Buffer layer	Amorphous $LiNbO_3$	Pulsed laser deposition	7–10 nm
b. Cathode	Polycrystalline $LiCoO_2$	Radio frequency sputtering	180–200 nm
a. Current collector	Au	Vacuum vapor deposition	~100 nm

fabricated on Au-coated Al_2O_3 substrate. Thicknesses of each layer were estimated by X-ray reflectometry and/or a stylus surface profiler.

Prior to the fabrication of thin-film batteries, the structure and the electrochemical properties of the Li_3PS_4 films were investigated as shown in Fig. 2. The P 2*p* and S 2*p* peaks of the Li_3PS_4 film are observed at similar positions to those of the Li_3PS_4 pellet (Fig. 2a). Furthermore, the film was amorphous in nature and consisted of the PS_4^{3-} units, confirmed by XRD and Raman spectroscopy [8b]. The Li_3PS_4 film was successfully synthesized by PLD with the Li_2S-excess target. From AC impedance measurement, the Li_3PS_4 film showed a resistance component that corresponds to lithium-ion diffusion in the grains. The resistance between the grains for Li diffusion could be negligible due to the close contact between the grains in the dense film. The lithium-ion conductivity was 5.3×10^{-4} S cm^{-1} at room temperature, and the activation energy was 45 kJ mol^{-1} (Fig. 2b). The ionic conductivity was twice the value of the previously reported for amorphous Li_2S-P_2S_5 films (7.9×10^{-5} S cm^{-1} with an activation energy of 43 kJ mol^{-1}) [11]. Considering that the ion conductivity depends on the carrier mobility and the concentration of the active carrier, the Li_3PS_4 film synthesized using the Li-excess target may contain a higher number of lithium ions that participate in ionic conduction. The current response measurement was conducted using a $Au/Li_3PS_4/Au$ cell with an applied voltage of DC 1 V to determine the transport number of Li^+ (Fig. 2c). The DC conductivity was 3.6×10^{-10} S cm^{-1}, which was much lower than the conductivity measured by AC impedance, which confirms that the amorphous Li_3PS_4 film is a pure ionic conductor. The cyclic voltammogram of a $Li/Li_3PS_4/Au$ cell showed no oxidation and reduction peaks except for the deposition/dissolution of lithium at around 0 V (Fig. 2d). The Li_3PS_4 film was electrochemically stable between 5 V and 0 V. The above results

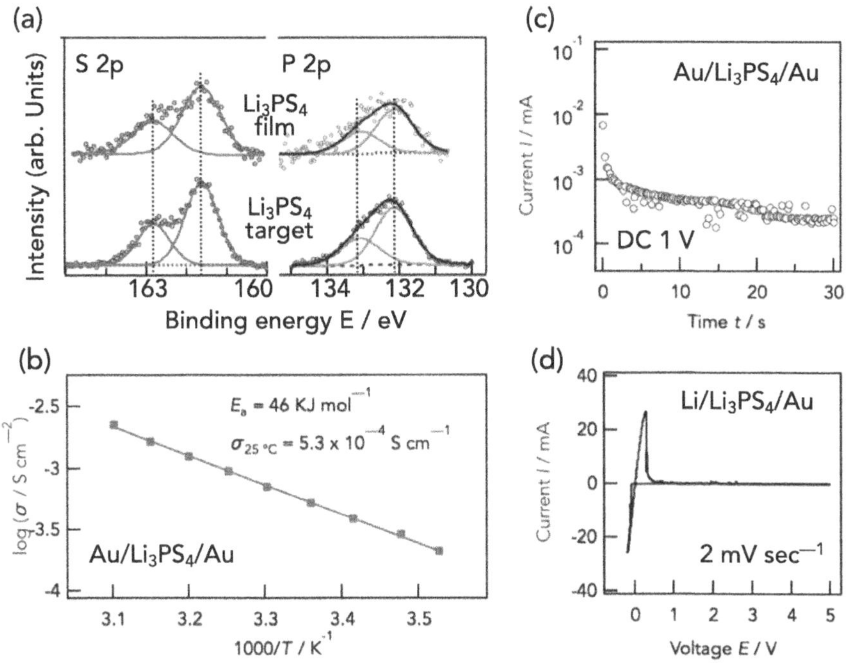

Fig. 2 **a** S 2p and P 2p XPS spectra of a Li_3PS_4 film and target. **b** Arrhenius plot of AC impedance results and **c** DC current response of a Au/Li_3PS_4 film/Au symmetric cell. **d** Cyclic voltammogram of a Li/Li_3PS_4 film/Au cell [8b. © 2016 The American Ceramic Society]

confirm that the amorphous Li_3PS_4 film has the high lithium ion conductivity and sufficiently wide potential windows for application as a solid electrolyte in all-solid-state batteries.

3 High Voltage Operation of $LiCoO_2$ with Sulfide Electrolyte

Constant current charge-discharge measurements were conducted for a thin-film battery of Li/Li_3PS_4/$LiNbO_3$/$LiCoO_2$ with different upper cutoff voltages of 4.2 V and 4.5 V. Figure 3a shows the charge-discharge curves collected with the cutoff voltages of 4.2 V and 2.5 V. The plateau regions are evident at around 3.9 V, which corresponds to the two-phase reaction of rhombohedral $LiCoO_2$ and $Li_{1-x}CoO_2$ ($0 \leq x \leq 0.3$) [1a]. The first charge and discharge capacities were 122 and 112 mAh g^{-1}, respectively. The irreversible capacity of 10 mAh g^{-1} could correspond to side reactions such as the formation of a cathode electrolyte interphase (CEI) layer. The discharge capacity at the 20th cycle was 108 mAh g^{-1} with a Coulomb efficiency of 99.6%. Furthermore, no lowering of the average discharge voltage was observed.

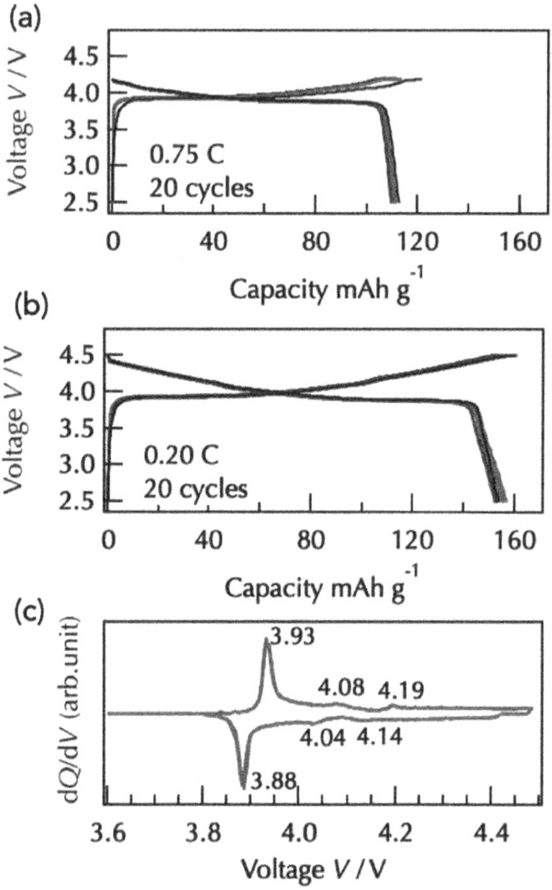

Fig. 3 **a**, **b** Charge-discharge curves of a thin-film battery of $Li/Li_3PS_4/LiNbO_3/LiCoO_2$ operated at different voltage ranges of 4.2–2.5 V and 4.5–2.5 V. **c** the initial dQ/dV curve of $Li/Li_3PS_4/LiNbO_3/LiCoO_2$ operated at a voltage range of 4.5–2.5 V

After the charge-discharge cycling between 4.2 V and 2.5 V, the upper cutoff voltage was changed to 4.5 V (Fig. 3b). The initial charge and discharge capacities were 161 and 153 mAh g^{-1}, respectively, which might be associated with side reactions that proceeded at the charging process between 4.2 V and 4.5 V. During the following 20 cycles, the discharge capacity kept remained over 150 mAh g^{-1} without lowering of the average discharge voltage, which is similar to that observed for the charge-discharge cycles between 4.2 V and 2.5 V. The dQ/dV curve at the 20th cycle clearly showed the oxidation and reduction peaks with a small peak separation of about 50 mV, which indicates the reversible phase transition between rhombohedral and monoclinic $Li_{1-x}CoO_2$ [1a] (Fig. 3c). These results reveal that the reversible lithium intercalation proceeds at the $LiCoO_2/LiNbO_3/Li_3PS_4$ interface under the high voltage operation with the upper cutoff voltage of 4.5 V. The $LiNbO_3$ layer could act as a buffer to facilitate lithium diffusion at high voltage regions, similar to the case of 4 V-class cathode/sulfide electrolyte interfaces.

The crystal structure changes of the $LiCoO_2$ film cathode during electrochemical reactions were investigated by in situ synchrotron XRD. The $Li/Li_3PS_4/LiNbO_3/LiCoO_2$ film battery fabricated on the Au/Al_2O_3 substrate was set in a spectro-electrochemical cell in an Ar-filled glovebox. The cell was kept in a vacuum during the XRD measurements. The XRD patterns were collected using a κ-type six-circle diffractometer installed at the bending-magnet beamline BL14B1 of the synchrotron beam facilities SPring-8 in Japan [7a, 12]. An X-ray wavelength of 0.82,518 Å (15 keV) was employed. Figure 4 presents in situ XRD patterns for the 003 diffraction peaks of the $LiCoO_2$ film electrode before cycling, at the first charged (4.1, 4.3, and 4.5 V) and discharged states (4.3, 4.1, 3.9, and 2.5 V), and after discharge to 2.5 V at 5th and 30th cycles. After the initial charge to 4.1 V, two diffraction peaks were observed at lower angles than the original 003 peak. This corresponds to the lattice expansion along the *c* axis which is induced by Coulombic repulsion between oxide layers due to lithium deintercalation from the $LiCoO_2$ lattice [13]. The diffraction peak shifted to higher angles at charge from 4.1 V to 4.5 V, which indicated the lattice contraction of $Li_{1-x}CoO_2$, accompanied by lithium deintercalation. From previous XRD studies on $LiCoO_2$, the $Li_{1-x}CoO_2$

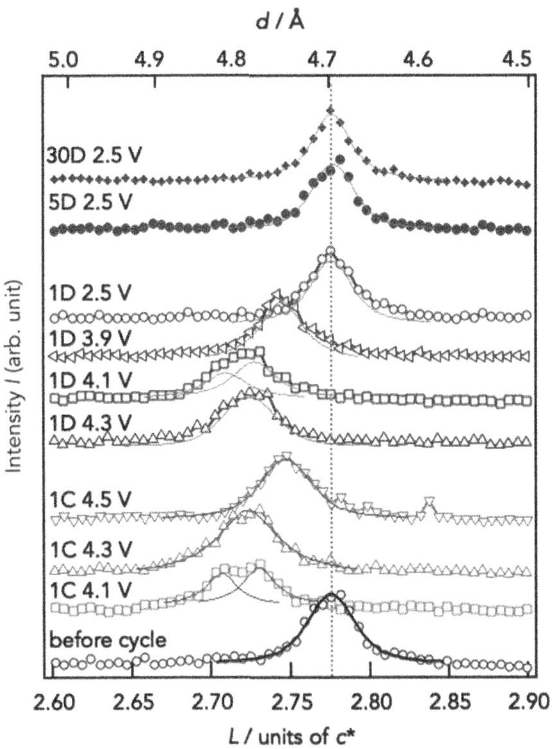

Fig. 4 In situ synchrotron XRD patterns of $LiCoO_2$ 003 peak during lithium (de)intercalation in a thin-film battery of $Li/Li_3PS_4/LiNbO_3/LiCoO_2$

lattice begins to contract along the *c* axis when the amount of lithium deintercalation x is over 0.6 [1a]. Thus, the lattice contraction confirmed the formation of a highly delithiated $Li_{1-x}CoO_2$ phase. The diffraction peak reversibly shifted to lower angles at discharge 4.3 V and to higher angles at discharge from 4.1 to 2.5 V. The $LiCoO_2$ 003 peak at discharge 2.5 V showed no significant changes in the position and the intensity with the cycle number. The in situ XRDs revealed that the $LiCoO_2$ works as a high capacity cathode with high cycle stability under high voltage operation in the sulfide-type all-solid-state batteries.

4 Interfacial Reactions of $LiCoO_2/LiNbO_3/Li_3PS_4$

Charge transfer resistances of lithium intercalation at the $LiCoO_2/LiNbO_3/Li_3PS_4$ interface were investigated by electrochemical impedance spectroscopy. To examine the effects of the high voltage operation, impedance measurements were performed at the discharge 4.2 V from different charged states at 4.3, 4.4, and 4.5 V. Figure 5 depicts the discharge curves collected with different upper cutoff voltages and the Nyquist plots of impedance spectra measured at discharge 4.2 V. The thin-film battery showed two semicircles, and their frequency values corresponding to the apex of the semicircle were ~10^5 Hz and ~10 Hz, respectively. The semicircle observed at the high-frequency regions could be identified to the deposition/dissolution of lithium at the anode-side interface, because it showed no changes with the voltages. The semicircle observed at the low-frequency regions showed voltage dependence, which corresponded to the charge transfer resistance at the cathode side. After charged to 4.3 V, the charge transfer resistance at discharge 4.2 V was analyzed to be 467 Ω cm^2.

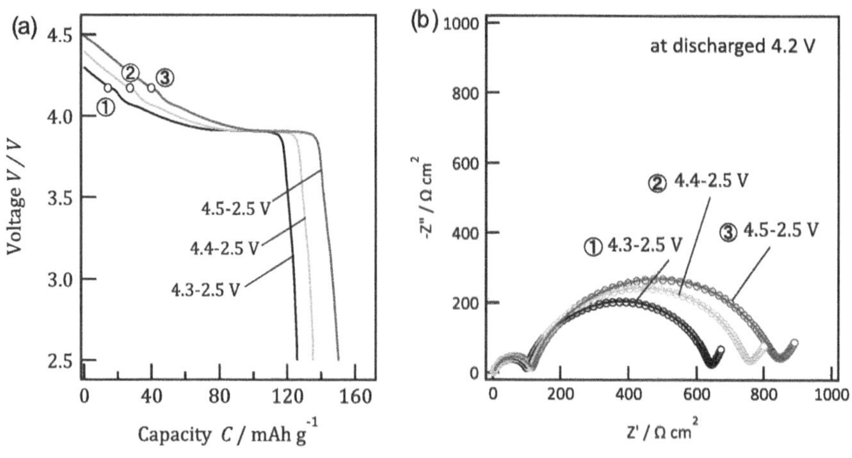

Fig. 5 **a** Discharge curves of a thin-film battery of $Li/Li_3PS_4/LiNbO_3/LiCoO_2$ from different charged states of 4.3, 4.4, and 4.5 V. **b** Nyquist plots of impedance spectra measured after discharged to 4.2 V from different charged states

After the film battery was charged to 4.4 V and 4.5 V, those values slightly increased to 559 and 620 Ω cm^2, respectively. We can speculate that the oxidative decomposition of the $LiNbO_3$ and/or Li_3PS_4 layers is promoted at higher voltage regions, and the decomposition products with poor electric conductivity make the charge transfer resistance higher. Considering that the reversible charge-discharge reactions proceeded under high voltage operation at 4.5 V, the decomposition products might suppress further decomposition.

The electrochemical impedance spectroscopy suggested the formation of a resistive interface layer. The changes in chemical bonding conditions at the $LiCoO_2/LiNbO_3/Li_3PS_4$ interface were investigated using an XPS apparatus with an Al-Kα X-ray source and Ar ion sputtering (2 kV), installed at National Institute for Materials Science (NIMS) Battery Research Platform. The as-grown and cycled $Li/Li_3PS_4/LiNbO_3/LiCoO_2$ cells were sealed in a transfer vessel in an Ar-filled glove box and transferred to the XPS chamber to minimize any possible side reactions with species in the air. The Li metal and Li_3PS_4 layers were removed by Ar etching to observe the $LiCoO_2/LiNbO_3/Li_3PS_4$ interface. Etching rates of Li, Li_3PS_4, and $LiCoO_2$ were estimated to be 11–13, 10–12, and ~6 nm min^{-1}, respectively, from the preliminary etching test. The etching time was set to 0 min when the signals of Nb species were first observed, which corresponds to the interface. The XPS spectra were collected at a pass energy of 46.95 eV and a take-off angle of 60°. In order to avoid the surface charging, charge compensation was performed using an electron neutralizer and an ion gun. The P 2*p* and S 2*p* XPS spectra showed no significant changes with the etching time due to a large amount of residuals of etching species of the micrometer scale Li_3PS_4 layer. Spectral fitting was carried out by using the fitting program, CasaXPS. The binding energy scale was calibrated with metallic $Au4f_{7/2}$ (84.0 eV).

Figure 6 shows XPS spectra in the Nb 3*d*, Co $2p_{3/2}$, and O 1 *s* regions of the as-grown and cycled $LiCoO_2/LiNbO_3/Li_3PS_4$ interfaces, etched for different durations from 0 min to 30 min. At 0 min, Nb $3d_{3/2}$ and $3d_{5/2}$ peaks are observed for the as-grown and cycled samples, which indicates that the observed area corresponds to the interfacial regions (Fig. 6a). The positions of the Nb 3*d* peaks after the second cycle shifted to higher binding energies. The chemical bonding conditions around Nb ions changed during the cycles. No signals in the Co $2p_{3/2}$ region were observed at the as-grown interface (0 min), whereas two Co $2p_{3/2}$ peaks were observed at 779.1 eV and 781.5 eV for the 10 and 30 min-etched samples (Fig. 6b). The 779.1 eV peak is attributable to Co^{3+} in the $LiCoO_2$ lattice [14], and the 781.5 eV peak can be identified as Co^{4+} in Li-deficient $Li_{1-x}Co^{III/IV}O_2$ [15] and/or Co—O bonding in simple oxides such as CoO and Co_3O_4 [16]. From the spectrum fitting, the higher intensity ratio of the 779.1 eV to 781.5 eV peaks was observed for the 10 min sample than that observed for the 30 min sample. Near the interface, the composition and/or structure changed from the stoichiometric $LiCo^{III}O_2$. After the second cycle, the 781.5 eV peak was obviously observed for the 0 min and 10 min samples (Fig. 6c). Furthermore, the XP peak in the O 1 s region appeared at 531.0 eV at 0 min which is attributed to the Co^{II}—O and Nb—O bonds. The existence of both Nb and Co species at the interface suggests the formation of decomposition products of the interfacial elements during

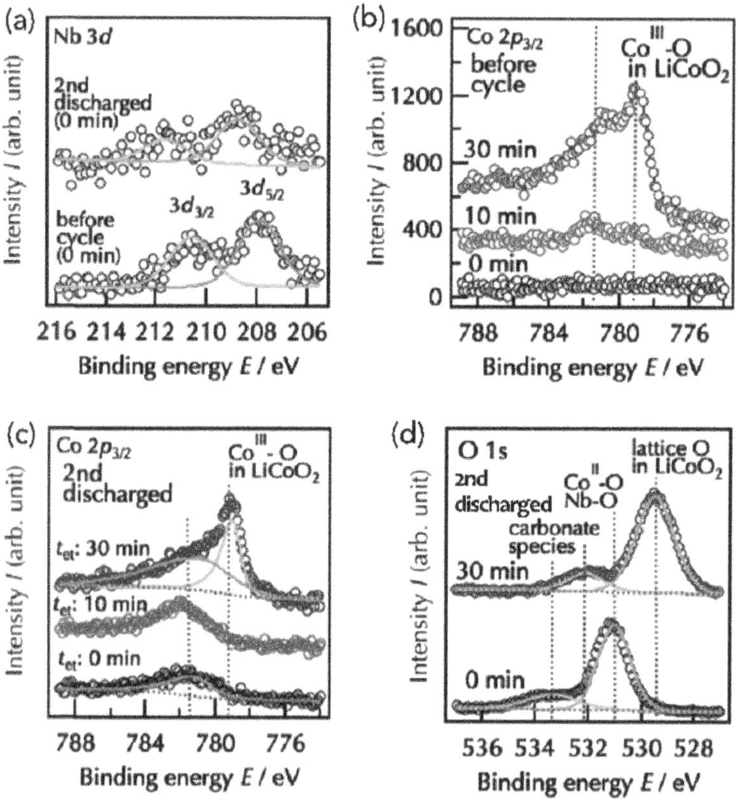

Fig. 6 **a** Nb 3d, **b**, **c** Co $2p_{3/2}$, and **d** O 1 s XPS spectra of thin-film batteries of $Li/Li_3PS_4/LiNbO_3/LiCoO_2$ before and after charge-discharge cycles between 4.6 V and 2.5 V. The etching time when signals of Nb were first observed was normalized to 0 min

the initial cycles. The Co^{3+}—O peaks mainly were observed in the Co $2p_{3/2}$ and O 1 *s* regions after the 30 min etching. These results indicate the reversible lattice change in the $LiCoO_2$ bulk, which is consistent with the in situ XRD results. We can speculate that the decomposition products can play a role in a cathode electrolyte interphase (CEI) layer with lithium ion conductivity that can suppress the further decomposition of interfacial species. However, the electrochemical impedance spectroscopy suggested that the CEI layer formed at high voltage regions increased the charge transfer resistance of the lithium intercalation. Our findings clarify the importance for controlling the composition and structure of the CEI layer to improve the rate capability of the cathode/sulfide electrolyte interface under high voltage operation.

5 Concluding Remarks

Thin-film batteries of $Li/Li_3PS_4/LiNbO_3/LiCoO_2$ have been fabricated on Au/Al_2O_3 substrates by physical vapor deposition. These batteries provide a simple reaction field for electrochemical lithium intercalation at the oxide cathode/sulfide electrolyte interface. The $Li/Li_3PS_4/LiNbO_3/LiCoO_2$ battery delivers the charge-discharge capacities of over 150 mAh g^{-1} in the voltage range of 4.5 V and 2.5 V, followed by highly reversible charging-discharging for 20 cycles without lowering of the average discharge voltage. In situ synchrotron XRDs confirm the formation of a highly delithiated $Li_{1-x}CoO_2$ phase ($x \geq 0.6$) and the reversible lattice changes. This represents the experimental demonstration that the stable lithium intercalation proceeds at the oxide cathode/sulfide electrolyte interface under high voltage operation. In the view of intercalation reaction rate, the charge transfer resistance at discharg 4.2 V increases with the upper cutoff voltages, revealed from AC impedance spectroscopy. XPS analyses with Ar ion etching detect the formation of an interfacial layer by decomposition of interfacial species. The interfacial layer can play a role in poor lithium conductive CEI that suppresses further decomposition of interfacial species. However, the CEI components suggested by XPS could not be identified as compounds that are predicted from first-principle calculations [4a, 17]. The development of in situ experimental technique to detect interfacial phases is of considerable importance for controlling the composition and structure of the CEI layer to improve the rate capability of cathode/sulfide electrolyte interface under high voltage operation.

References

1. a) Chen, Z., Dahn, J. R. (2004). *Electrochimica Acta, 49*, 1079–1090; b) Kajiyama, A., Masaki, R., Wakiyama, T., Matsumoto, K., Yoda, A., Inada, T., Yokota, H., Kanno, R. (2015). *Journal of The Electrochemical Society, 162*, A1516-A1522.
2. Kuwata, N., Kudo, S., Matsuda, Y., & Kawamura, J. (2014). *Solid State Ionics, 262,* 165–169.
3. a) Oh, G., Hirayama, M., Kwon, O., Suzuki, K., Kanno, R. (2016). *Chemistry of Materials, 28*, 2634–2640; b) Oh, G., Hirayama, M., Kwon, O., Suzuki, K., Kanno, R. (2016). *Solid State Ionics, 288*, 244–247.
4. a) Richards, W. D., Miara, L. J., Wang, Y., Kim, J. C., Ceder, G. (2016). *Chemistry of Materials, 28*, 266–273; b) Zhu, Y., He, X., Mo, Y. (2015). *ACS Applied Materials & Interfaces, 7*, 23685–23693.
5. a) Yazami, R. (1999). *Electrochimica Acta 45*, 87–97; b) Yazami, R., Reynier, Y. F. (2002). *Electrochimica Acta 47*, 1217–1223.
6. a) Kamaya, N., Homma, K., Yamakawa, Y., Hirayama, M., Kanno, R., Yonemura, M., Kamiyama, T., Kato, Y., Hama, S., Kawamoto, K., Mitsui, A. (2011). *Nature Materials, 10*, 682–686; b) Li, W. J., Hirayama, M., Suzuki, K., Kanno, R. (2016). *Solid State Ionics 285*, 136–142; c) Kato, Y., Shiotani, S., Morita, K., Suzuki, K., Hirayama, M., Kanno, R. (2018). *The Journal of Physical Chemistry Letters, 9*, 607-613.
7. a) Hirayama, M., Ido, H., Kim, K., Cho, W., Tamura, K., Mizuki, J. I., Kanno, R. (2010). *Journal of the American Chemical Society, 132*, 15268–15276; b) Taminato, S., Hirayama,

M., Suzuki, K., Tamura, K., Minato, T., Arai, H., Uchimoto, Y., Ogumi, Z., Kanno, R. (2016). *Journal of Power Sources, 307*, 599-603.
8. a) Ogawa, M., Kanda, R., Yoshida, K., Uemura, T., Harada, K. *Journal of Power Sources, 205*, 487–490; b) Quan, Z., Hirayama, M., Sato, D., Zheng, Y., Yano, T. -A., Hara, K., Suzuki, K., Hara, M., Kanno, R. (2017). *Journal of the American Ceramic Society, 100*, 746-753.
9. Hata, J.-I., Hirayama, M., Suzuki, K., Dupré, N., Guyomard, D., & Kanno, R. (2019). *Batteries & Supercaps, 2,* 454–463.
10. Ohta, N., Takada, K., Sakaguchi, I., Zhang, L., Ma, R., Fukuda, K., et al. (2007). *Electrochemistry Communications, 9,* 1486–1490.
11. Sakuda, A., Hayashi, A., Hama, S., & Tatsumisago, M. (2010). *Journal of the American Ceramic Society, 93,* 765–768.
12. Hirayama, M., Sonoyama, N., Abe, T., Minoura, M., Ito, M., Mori, D., et al. (2007). *Journal of Power Sources, 168,* 493–500.
13. Amatucci, G. G. (1996). *Journal of the Electrochemical Society, 143,* 1114.
14. Cherkashinin, G., Nikolowski, K., Ehrenberg, H., Jacke, S., Dimesso, L., & Jaegermann, W. (2012). *Physical Chemistry Chemical Physics, 14,* 12321–12331.
15. Dupin, J. C., Gonbeau, D., Benqlilou-Moudden, H., Vinatier, P., & Levasseur, A. (2001). *Thin Solid Films, 384,* 23–32.
16. Wang, J., Yang, S., Liu, X., Ren, S., Guan, F., & Chen, M. (2004). *Applied Surface Science, 221,* 272–280.
17. Xiao, Y., Miara, L. J., Wang, Y., & Ceder, G. (2019). *Joule, 3,* 1252–1275.

All-Solid-State Battery with Oxide-Based Electrolytes

Solid-State Batteries with Oxide-Based Electrolytes

Kazunori Takada

Abstract Lithium-ion batteries are the most widespread rechargeable batteries. However, there remain some problems to be solved, most of which originate from the organic solvent electrolyte, especially when they are applied to electric vehicles and stationary energy storage. Solid electrolytes are expected to provide a solution to the problems, and practical performance has been achieved in solid-state batteries with sulfide electrolytes. Now, oxide-type batteries are regarded as the next target.

Keywords Solid-state battery · Solid electrolyte · Oxide

Studies on solid-state batteries have been initiated by the discovery of fast ionic conduction around room temperature in α-AgI [1], and several solid-state batteries were prototyped around 1950 [2]. The advantages of solid-state batteries over liquid electrolyte systems that motivated the studies at that time are (1) absence of electrolyte leakage from sealed batteries, (2) absence of problems relating to vaporization of liquid electrolytes, (3) absence of phase transitions at low temperature, and (4) ease of miniaturization and multiple shape factors [3]. Although ionic conductivity of α-AgI is higher than 1 S cm^{-1}, it is transformed to less conductive β or γ-phase below 147 °C. Studies aiming at fast ionic conduction at room temperature have provided some solid electrolytes that are as conductive as liquids at room temperature in Ag^+ or Cu^+ ion-conductive systems [4, 5]. They have made solid-state batteries operatable at room temperature [6, 7], and the application of intercalation chemistry to electrical energy storage [8] has made them rechargeable [9, 10].

On the other hand, the first solid-state battery has been put into practical use in 1972 [11]. The Li/I_2-PVP battery for a cardiac pacemaker implanted in a patient was not only the first commercialized solid-state battery but also the first lithium battery in practical use. However, these batteries are very specific. The solid-state batteries with Ag^+ or Cu^+ ion conductors are very low in the cell voltage due to the high ionization energy of Ag or Cu. The Li/I_2-PVP batteries are low in the current

K. Takada (✉)
National Institute for Materials Science, Tsukuba, Japan
e-mail: takada.kazunori@nims.go.jp

K. Kanamura (ed.), *Next Generation Batteries*,
https://doi.org/10.1007/978-981-33-6668-8_17

drain due to the low ionic conductivity of LiI, which is formed in situ between Li and I_2-PVP and acts as the solid electrolyte.

In the early 1990s, two types of rechargeable batteries have been commercialized: one is nickel/metal hydride battery, and the other is a lithium-ion battery, and the latter has strengthened the motivation to solid-state batteries. Because organic solvent electrolytes employed in lithium-ion batteries are combustible substances, they have never been free from safety issues, which has focused our attention on nonflammability of solid electrolytes. Moreover, the safety issues are more serious in large batteries that are strongly required today for electric vehicles, stationary energy storage, etc., to realize a low-carbon society. Because the increasing battery size increases the amount of combustible electrolyte and lowers heat radiation, nonflammability of solid electrolytes that provide a fundamental solution to the safety issues is attracting more attention than ever.

Demands for large batteries have attracted our attention to another shortcoming of lithium-ion batteries. Although the useful life of several years may be enough for consumer use, a 10-year lifetime is necessary for the batteries in vehicles, and stationary batteries must stand more than 20-year use. One of the major factors governing the lifetime is side reactions, e.g., electrolyte decomposition. Lithium-ion batteries generate high cell voltage for their high energy by combining highly-oxidizing cathodes and highly-reducing anodes, which tend to lead to electrochemical decomposition of the electrolytes. Solid electrolytes provide high durability to such batteries. Mobile species in solid electrolytes are only lithium ions at room temperature, and there are no other species that diffuse to the electrode surface and undergo electrochemical oxidation or reduction. As a result, solid-state batteries exhibit a very long cycle life, as demonstrated by a thin-film system [12].

In addition, the employment of solid electrolytes is expected to increase energy densities, although electrochemical energy is stored not in electrolytes but in active materials. For instance, they enable us to use electrode materials that increase the energy densities but do not work stably in liquid electrolytes. One example is transition-metal oxides showing high electrode potentials around 5 V vs. Li^+/Li [13]. Although they raise the cell voltage and thus energy densities, the high-voltage cathodes do not work stably in liquid electrolytes, because the high potentials lead to oxidative decomposition of the electrolytes. On the other hand, reductive decomposition has not been allowing us to use high-capacity anodes including lithium alloys [14].

On the other hand, electrolyte decomposition does not proceed continuously in solid electrolytes due to the absence of diffusive reactants, as mentioned above. That is, because solid electrolytes are kinetically stable in a wide potential range [15], high-voltage cathodes [16] and high-capacity anodes [17] become available in solid electrolytes. In addition, even if such active materials are not employed, there is a room for increasing energy density.

Another advantage of the high energy density is that solid electrolytes simplify the battery system. It is well recognized that the nonflammability of solid electrolytes enables us to simplify the safeguards, which lithium-ion batteries should be equipped with. In addition, solid electrolytes are expected to simplify module structure in large battery systems. For example, tens of cells should be connected in series in a battery module to generate high voltage for vehicle application. Although the corresponding number of battery cases are necessary in liquid systems, the serial connection is achievable in solid systems only by stacking bipolar electrodes and electrolyte layers alternately in a single battery case, which reduces the weight and volume of battery cases and increases the gravimetric and volumetric energy density of the battery module [18].

It may be astonishing that high power density has been anticipated in solid-state system, because solid-state batteries have been suffering from their low power densities. Although it has the most serious shortcoming of solid-state batteries, and achieving practical power density is a critical target in their development, some rate-determining steps in liquid systems do not work in solid-state systems. Because anions as well as lithium ions are mobile in liquid electrolytes, battery operation at high current density brings about severe concentration gradient in the electrolyte, leading to diffusion limitation [19, 20]. In contrast, solid-state batteries are unlikely to reach such a situation, because the negative charge in solid electrolytes is immobilized on their framework.

Another rate-determining step in liquid electrolyte systems that is absent in solid systems is desolvation. Lithium ions are solvated in liquid electrolytes and thus should be desolvated before intercalating into the interlayers of active materials because the solvated ions are too large to intercalate into the interlayers. The desolvation step needs high activation energy and thus limits the kinetics in electrode reactions [21, 22]. Of course, lithium ions intercalate to the active materials without such desolvation in solid systems. Because solid-state batteries are free from these rate-determining steps, they generate higher current drains than liquid systems, when the solid electrolyte has comparable conductivity to liquid electrolytes. However, the potential performance has been revealed only in solid-state batteries with sulfide-based solid electrolytes [23].

Concentration gradient hardly takes place, and electrode reactions are not accompanied by desolvation in solid-state batteries; however, high performance is never expected without fast ionic conduction in solid electrolytes. The highest ionic conductivity has reached 10^{-2} S cm^{-1} among sulfide-based electrolytes [24, 25], and they have provided sulfide-type solid-state batteries with higher performance than lithium-ion batteries with liquid electrolytes [23]. However, sulfide-based electrolytes are hygroscopic and easily decomposed even in the atmosphere to generate H_2S gas [26]. Thus, the battery production needs an inert atmosphere, which lowers the productivity and raises the production cost, and the toxicity of the H_2S is of concern. Although replacement of sulfide-based electrolytes with chemically-stable oxide-based one is strongly required, the replacement lowers the battery performance extremely. The highest ionic conductivities that have been observed among oxide electrolytes [27–29] are of the order of 10^{-3} S cm^{-1}, which are lower than those in sulfide systems by one order of magnitude; however, it is not the main obstacle for the oxide-type batteries to achieve practical performance.

The electric conductivity of organic solvent electrolytes used in the current lithium-ion batteries is of the order of 10^{-2} S cm^{-1}, which is also higher than the highest conductivities of oxide electrolytes by one order of magnitude. However, their transport number for lithium ion is 0.5 in the organic electrolytes at the highest [30]. It means that lithium-ion conductivities of organic solvent electrolytes are of the order of 10^{-3} S cm^{-1}. Since only the movement of lithium ions operates lithium-ion batteries, it can be concluded that oxide-based solid electrolytes have enough conductivities to make performance in the oxide-type solid-state batteries comparable to that of lithium-ion batteries. However, the battery performance of oxide systems is very poor, and solid-state batteries employing oxide-based electrolytes are limited almost only to thin-film systems [31].

The predominant difference of oxide solid electrolytes from sulfides, which results in the poor performance of oxide-type batteries, is in their mechanical properties. Sulfide-based solid electrolytes are relatively soft [32] and thus can be formed into a dense body easily, for instance, only by cold pressing. Electrolyte particles are connected to each other very tightly in the compacted body, and lithium ions move from particle to particle without being disturbed [33]. On the other hand, oxide-based electrolytes are hard and brittle, and it is impossible to connect the electrolyte particles only by pressing.

The sintering process is usually employed to make dense ceramic, and oxide-electrolyte particles can be connected to each other by sintering. However, the grain boundary resistance is still high even in the sintered pellets [28, 34], which hinders the solid-state batteries from generating practical power density. A simple way to reduce the grain boundary resistance is, indeed, raising the sintering temperature. For example, increasing sintering temperature from 1350 to 1450 °C reduces the grain boundary resistance in a perovskite-type solid electrolyte significantly [35]. However, electrolytes are contacted to active materials in solid-state batteries, and the increasing temperature promotes mutual diffusion between the active material and solid electrolyte. It forms impurity phase at the interface [36], and the high resistance of the impurity phase lowers the battery performance.

Low battery performance of oxide systems originates from difficulties in connecting the battery materials. High-temperature sintering is necessary to reduce the grain boundary resistance in highly-conductive oxide electrolytes, whereas it forms a resistive reaction layer at the electrode/electrolyte interface. Our research team in ALCA-SPRING project is tackling this problem to realize oxide-type solid-state batteries.

References

1. Tubandt, C., & Lorenz, E. (1914). Molekularzustand und elektrisches leitvermögen kristallisierter salze. *Zeitschrift für Physikalische Chemie, 24,* 513–543.
2. Brodd, R. J., Kozawa, A., & Kordesch, K. V. (1978). 75th anniversary review Series: Primary Batteries 1951–1976. *Journal of the Electrochemical Society, 125,* 271C–283C.
3. Owens, B. B., & Skarstad, P. M. (1992). Ambient temperature solid state batteries. *Solid State Ionics, 53–56,* 665–672.

4. Owens, B. B., & Argue, G. R. (1967). High-conductivity solid electrolytes: MAg_4I_5. *Science, 157,* 308–310.
5. Takahashi, T., Yamamoto, O., Yamada, S., & Hayashi, S. (1979). Solid-state ionics: High copper ion conductivity of the system CuCl-CuI-RbCl. *Journal of the Electrochemical Society, 126,* 1654–1658.
6. Owens, B. B., & Bottleberghe, J. R. (1993). Twenty year storage test of $Ag/RbAg_4I_5/I_2$ solid state batteries. *Solid State Ionics, 62,* 243–249.
7. Takahashi, T., & Yamamoto, O. (1977). Solid-state ionics—solid electrolyte cells with copper ion conductors. *Journal of Appled Electrochemistry, 7,* 37–43.
8. Whittingham, M. S. (1976). Electrical energy storage and intercalation chemistry. *Science, 192,* 1126–1127.
9. Takada, K., Kanbara, T., Yamamura, Y., & Kondo, S. (1990). Rechargeable solid-state batteries with silver ion conductors. *Solid State Ionics, 40*(41), 988–992.
10. Kanno, R., Takeda, Y., Ohya, M., & Yamamoto, O. (1987). Rechargeable all solid-state cell with high copper ion conductor and copper chevrel phase. *Materials Research Bulletin, 22,* 1283–1290.
11. Greatbatch, W., Lee, J. H., Mathias, W., Eldridge, M., Moser, J. R., & Schneider, A. A. (1971). The solid-state lithium battery: A new improved chemical power source for implantable cardiac pacemakers. *IEEE Transactions on Biomedical Engineering BME, 18,* 317–324.
12. Wang, B., Bates, J. B., Hart, F. X., Sales, B. C., Zuhr, R. A., & Robertson, J. D. (1996). Characterization of thin-film rechargeable lithium batteries with lithium cobalt oxide cathodes. *Journal of the Electrochemical Society, 143,* 3203–3213.
13. Goodenough, J. B., & Kim, Y. (2010). Challenges for rechargeable Li batteries. *Chemistry of Materials, 22,* 587–603.
14. Wu, H., & Cui, Y. (2012). Designing nanostructured Si anodes for high energy lithium ion batteries. *Nano Today, 7,* 414–429.
15. Zhu, Y., He, X., & Mo, Y. (2015). Origin of outstanding stability in the lithium solid electrolyte materials: insights from thermodynamic analyses based on first-principles calculations. *ACS Applied Materials & Interfaces, 7,* 23685–23693.
16. Li, J., Ma, C., Chi, M., Liang, C., & Dudney, N. J. (2015). Solid electrolyte: the key for high-voltage lithium batteries. *Advanced Energy Materials, 5,* 1401408.
17. Sakabe, J., Ohta, N., Ohnishi, T., Mitsuishi, K., & Takada, K. (2018). Porous amorphous silicon film anodes for high-capacity and stable all-solid-state lithium batteries. *Communications Chemistry, 1,* 24.
18. Kato, Y., Kawamoto, K., Kanno, R., & Hirayama, M. (2012). Discharge performance of all-solid-state battery using a lithium superionic conductor $Li_{10}GeP_2S_{12}$. *Electrochemistry, 80,* 749–751.
19. Fuller, T. F., Doyle, M., & Newman, J. (1994). Simulation and optimization of the dual lithiumion insertion cell. *Journal of the Electrochemical Society, 141,* 1–10.
20. Diederichsen, K. M., McShane, E. J., & McCloskey, B. D. (2017). Promising routes to a high Li^+ transference number electrolyte for lithium ion batteries. *ACS Energy Letters, 2,* 2563–2575.
21. Abe, T., Fukuda, H., Iriyama, Y., & Ogumi, Z. (2004). Solvated Li-ion transfer at interface between graphite and electrolyte. *Journal of the Electrochemical Society, 151,* A1120–A1123.
22. Xu, K., Cresce, A. V. W. (2012). Li^+-solvation/desolvation dictates interphasial processes on graphitic anode in Li ion cells. *Journal of Materials Research, 27,* 2327–2341.
23. Kato, Y., Hori, S., Saito, T., Suzuki, K., Hirayama, M., Mitsui, A., et al. (2016). High-power all-solid-state batteries using sulfide superionic conductors. *Nature Energy, 1,* 16030.
24. Kamaya, N., Homma, K., Yamakawa, Y., Hirayama, M., Kanno, R., Yonemura, M., et al. (2011). A lithium superionic conductor. *Nature Materials, 10,* 682–686.
25. Seino, Y., Ota, T., Takada, K., Hayashi, A., & Tatsumisago, M. (2014). A sulphide lithium super ion conductor is superior to liquid ion conductors for use in rechargeable batteries. *Energy & Environmental Science, 7,* 627–631.
26. Muramatsu, H., Hayashi, A., Ohtomo, T., Hama, S., & Tatsumisago, M. (2011). Structural change of $Li_2S–P_2S_5$ sulfide solid electrolytes in the atmosphere. *Solid State Ionics, 182,* 116–119.

27. Aono, H., Sugimoto, E., Sadaoka, Y., Imanaka, N., & Adachi, G. (1989). Ionic conductivity of the lithium titanium phosphate ($Li_{1+X}M_XTi_{2-X}(PO_4)_3$, M = Al, Sc, Y, and La) systems. *Journal of the Electrochemical Society, 136,* 590–591.
28. Inaguma, Y., Liquan, C., Itoh, M., Nakamura, T., Uchida, T., Ikuta, H., et al. (1993). High ionic conductivity in lithium lanthanum titanate. *Solid State Communications, 86,* 689–693.
29. Li, Y., Han, J.-T., Wang, C.-A., Xie, H., & Goodenough, J. B. (2012). Optimizing Li^+ conductivity in a garnet framework. *Journal of Materials Chemistry, 22,* 15357–15361.
30. Capiglia, C., Saito, Y., Kageyama, H., Mustarelli, P., Iwamoto, T., Tabuchi, T., et al. (1999). ^{7}Li and ^{19}F diffusion coefficients and thermal properties of nonaqueous electrolyte solutions for rechargeable lithium batteries. *Journal of Power Sources, 81–82,* 859–862.
31. Bates, J. B., Dudney, N. J., Neudecker, B., Ueda, A., & Evans, C. D. (2000). Thin-film lithium and lithium-ion batteries. *Solid State Ionics, 135,* 33–45.
32. Sakuda, A., Hayashi, A., Takigawa, Y., Higashi, K., & Tatsumisago, M. (2013). Evaluation of elastic modulus of $Li_2S–P_2S_5$ glassy solid electrolyte by ultrasonic sound velocity measurement and compression test. *Journal of the Ceramic Society of Japan, 121,* 946–949.
33. Takada, K., & Kondo, S. (1998). Lithium ion conductive glass and its application to solid state batteries. *Ionics, 4,* 42–47.
34. Aono, H., Sugimoto, E., Sadaoka, Y., Imanaka, N., & Adachi, G. (1991). Electrical property and sinterability of $LiTi_2(PO_4)_3$ mixed with lithium salt (Li_3PO_4 or Li_3BO_3). *Solid State Ionics, 47,* 257–264.
35. Inaguma, Y., & Nakashima, M. (2013). A rechargeable lithiumair battery using a lithium ion conducting lanthanum lithium titanate ceramics as an electrolyte separator. *Journal of Power Sources, 228,* 250–255.
36. Kobayashi, Y., Takeuchi, T., Tabuchi, M., Ado, K., & Kageyama, H. (1999). Densification of $LiTi_2(PO_4)_3$-based solid electrolytes by spark-plasma-sintering. *Journal of Power Sources, 81–82,* 853–858.

Perovskite-Type Lithium-Ion Solid Electrolytes

Yoshiyuki Inaguma

Abstract Perovskite-type compounds with the chemical formula of ABX_3 exhibit versatile functionalities because they can accommodate various elements and tolerate deficiency and distortion. Among them, A-site deficient perovskite-type oxides containing lithium (Li) such as $La_{2/3-x}Li_{3x}TiO_3$ show fast Li-ion conductivity attributable to the percolation-controlled diffusion of Li ions in the vicinity of A-site via vacancies. In this chapter, the crystal structure, the ionic diffusion, and their correlation in perovskite-type Li-ion solid electrolytes are described in connection with the second-order Jahn–Teller effect, which is based on the covalent bond between the B ion and the oxygen. Moreover, the two important properties in perovskite-type Li-ion solid electrolytes for application to all-solid-state batteries, (1) the ionic conductivity at grain boundaries and (2) the electrochemical stability, are mentioned. The governing factors of the electrochemical stability are then elucidated in terms of the electronic structure.

Keywords Perovskite · Lithium-ion conductivity · Activation energy · Vacancy · Site percolation · Second-order Jahn-Teller(SOJT) effect · Electronegativity · Electrochemical stability

1 Introduction

Lithium (Li) all-solid-state batteries (SSB) are attracting attention from the viewpoint of safety, reliability, and high energy density [1]. In the replacement of current Li-ion batteries using liquid electrolytes with Li SSB, the development of solid electrolytes (SE) is a key issue. Among SE, oxide-based SE are expected to be chemically stable, durable, and easy to handle compared to sulfide-based SE. However, the ionic conductivity of oxide-based SE is more than one order smaller than that of sulfide-based SE [2–4]. Therefore oxide-based SE exhibiting high ionic conductivity are desired.

Y. Inaguma (✉)
Department of Chemistry, Faculty of Science, Gakushuin University, Tokyo, Japan
e-mail: yoshiyuki.inaguma@gakushuin.ac.jp

K. Kanamura (ed.), *Next Generation Batteries*,
https://doi.org/10.1007/978-981-33-6668-8_18

Among oxide-based SE, I focus description on perovskite(Pv)-type Li-ion SE [5–7]. Studies on Pv-type Li-ion SE originate from the reports on the Li-ion conductivities of $Ln_{1/3-x}Li_x(Nb_{1-x}Ti_x)O_3$ (Ln = La, Nd) by Latie et al. in 1984 [8], and $La_{2/3-x}Li_{3x}TiO_3$ (LLTO) by Belous et al. in 1987 [9]. Later, it has been recognized again since LLTO was reported to exhibit a Li-ion conductivity of 10^{-3} Scm^{-1} at room temperature in 1993 [10–13], and many studies have been performed. As will be described later, Pv-type Li-ion SE generally has a high reduction potential, which is attributable to the constituent elements and the chemical bonding. Therefore, electrode materials such as Li metal with a negative (low) standard electrode potential, $E°$ of −3.045 V versus SHE, cannot be applied as negative electrodes, which is a disadvantage in the application of Pv-type Li-ion SE to SSB. On the other hand, since the structure is relatively simple and the Li-ion conductivity is high, understanding the correlation between the structure and the ionic conductivity and clarification of the governing factors of the electrochemical stability would give us important guidelines for the development of new SE.

In this chapter, I describe the crystal structure, the chemical bonding, and the bulk ionic conduction of Pv-type Li-ion SE and their relations, and also mention the ionic conductivity at grain boundaries and the electrochemical stability which are important factors for application to SSB. The first half of this chapter was rewritten based on a part of my review article written by Japanese [14]. The schematics of crystal structures shown in this chapter were drawn using VESTA [15].

2 Structure, Chemical Bond, and Ionic Conductivity of Perovskite-Type Li-Ion Solid Electrolytes

2.1 Crystal Structure of Perovskite-Type Compounds

Since Pv-type compounds with a general formula ABX_3 accommodate various ions, a huge number of Pv-type compounds exhibiting various functionality such as electrical conductivity, dielectric properties, magnetism, and optical properties have been reported [16]. The versatile functionalities of Pv-type compounds, called "functional treasures," come from the slight structural changes as well as the difference in the constituent ions.

In the Pv-type structure, as shown in Fig. 1, the BX_6 octahedra share their corners, and the A ions exist in the interstice. In an ideal cubic Pv-type compound, the BX_6 octahedra share their corners so that the B-X-B angle is 180°, and the A ion exists in the center of the interstice (Fig. 1a), but most Pv-type compounds have distorted structures as shown in Fig. 1b, where the BX_6 octahedra rotate (tilt) and the cations displace from their ideal positions [17–23]. In general, the tilt occurs when the tolerance factor t of Pv defined by Eq. (1) [24] is smaller than unity, which corresponds to the ideal cubic Pv structure.

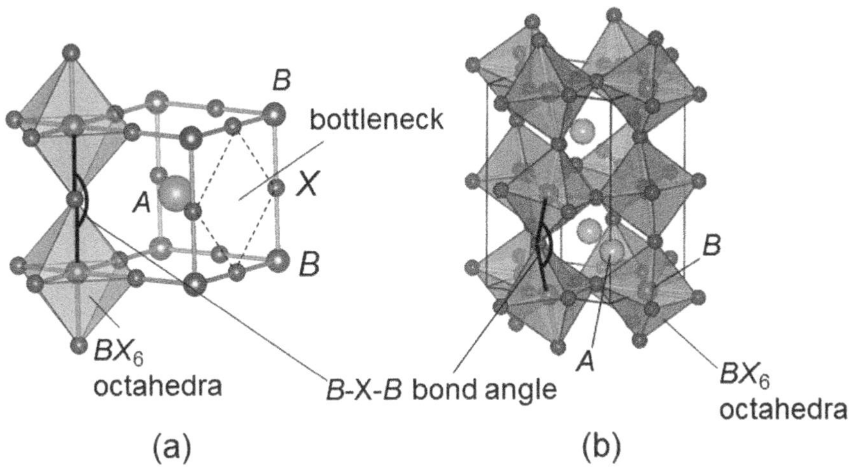

Fig. 1 Perovskite-type structure

$$t = \frac{r_A + r_X}{\sqrt{2}(r_B + r_X)} \quad (1)$$

Here, r_A, r_B, and r_X are the ionic radii of *A*, *B*, and *X* ions, respectively. For example, in Pv-type oxides, when the B-O-B angle becomes smaller than 180° accompanied by the tilt of the BO_6 octahedra, the overlap between the oxygen 2*p* orbitals and the *d* or *s* orbitals of the B ions changes, resulting in differences in the physical properties such as electrical conductivity and magnetism.

2.2 *A-Site Deficient Perovskite-Type Li-Ion Solid Electrolytes and the Li Position*

In Pv-type compounds, a large number of defects at the *A*, *B*, and the anion site can be tolerated. Among deficient Pv-type compounds, in some A-site deficient Pv-type oxides such as $La_{2/3}TiO_3$ [25], $La_{1/3}NbO_3$ [26, 27], and $La_{1/3}TaO_3$ [26], the layers with different A-site occupancies are alternately arranged along the *c*-axis. In $La_{1/3}TaO_3$, as shown in Fig. 2a, the layer where the La ion does not exist and the layer where La ions occupy with a probability of 2/3 are alternately repeated. The common feature of these compounds is that the B-site contains d^0 ions with d^0 electron configuration such as Ti^{4+}, Nb^{5+}, and Ta^{5+}. King et al. [28] pointed out that the ordered arrangement of the A-site vacancies is based on a second-order Jahn–Teller (SOJT) effect or pseudo-Jahn–Teller effect [29–33]. The distortion due to the SOJT effect is attributable to the orbital interaction between the empty $d\pi$ orbital of the d^0 ion and the 2*p* orbital of the oxygen ion. A distorted structure is stabilized by occupying electrons into the bonding orbital between the empty $d\pi$ orbital of the

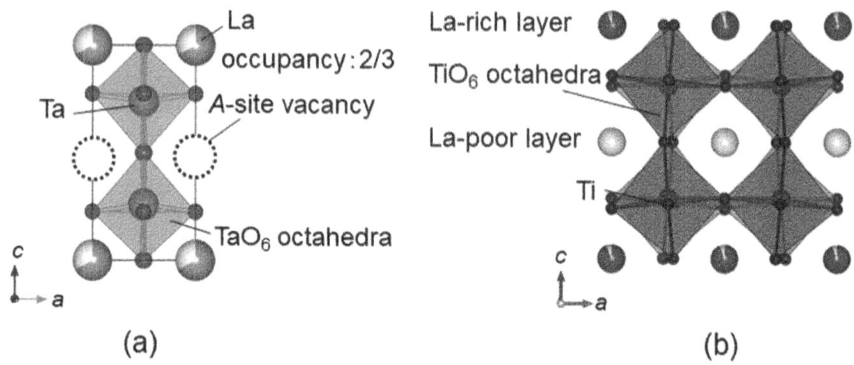

Fig. 2 Crystal structures of **a** $La_{1/3}TaO_3$, **b** $La_{2/3-x}Li_{3x}TiO_3 (x = 0.05)$

d^0 ion and the 2*p* orbital of the oxygen, which has lower energy than that of the original 2*p* orbital of the oxygen. The SOJT effect supports the shift of Ta^{5+} ions from the center of the TaO_6 octahedra toward the La-free layer as seen in Fig. 2a, which also minimizes the electrostatic (Coulomb) repulsion between La^{3+} and Ta^{5+} ions. Therefore, the ordered arrangement of the A-site vacancies occurs in concert with the SOJT effect and the electrostatic energy mandate.

All the Pv-type Li-ion SE also have A-site vacancies, and the Li-ion diffusion occurs via these vacancies. Among Pv-type Li-ion SE, LLTO exhibits an ordered arrangement of A-site vacancies along the *c*-axis as the same as $La_{1/3}TaO_3$ does. Figure 2b shows the structure of LLTO ($x = 0.05$) [34] in which the Ti^{4+} ions displace to the La-poor layer with a low La occupancy so as to reduce the electrostatic repulsion between the La^{3+} and the Ti^{4+} ions. On the other hand, though Na_xWO_3 and one of the Pv-type Li-ion solid electrolytes, $(Sr_{0.5+x}Li_{0.5-2x})(Ti_{0.5}Ta_{0.5})O_3$ [35, 36], are also A-site deficient Pv-type oxides, the ordered arrangement of the vacancies was not observed by X-ray diffraction experiments. Even in such a case, we can expect that the A-site vacancies can be stabilized by the displacement of the SOJT active ions toward the vacancies, and the formation of the vacancies and the SOJT effect are considered to occur in a concerted manner. Since all the known Pv-type Li-ion SE contain d^0 ions such as Ti^{4+}, Nb^{5+}, Ta^{5+}, and W^{6+} at the B-site position, the A-site vacancies are stabilized by the SOJT effect based on the interaction between the empty $d\pi$ orbital of the d^0 ion and the oxygen 2*p* orbital.

In Pv-type Li-ion SE, the Li^+ ions, which diffuse via the A-site vacancies, are located at different positions from the skeletal A ions such as La^{3+} ion. According to neutron diffraction experiments, the position surrounded by four oxygens (the position surrounded by the dotted line in Fig. 1a) is the stable site [37, 38]. Actually, the Li^+ ions exist in the vicinity of the position surrounded by four oxygen atoms in the A-site. It is thought that a Li ion is located at various positions in the vicinity of the A-site so as to mitigate the electrostatic repulsion between cations, since the skeletal A ions such as La^{3+} ion and the other Li^+ ions are present in the adjacent A-site lattice.

2.3 *Ionic Diffusion Mechanism in Perovskite-Type Li-Ion Solid Electrolytes*

2.3.1 Vacancy Mechanism and Site Percolation

A Li^+ ion resides near the A-site in Pv-type Li-ion SE, and the position changes depending on the positions of the other Li^+ ions and the A ions around the Li^+ ion. As mentioned above, since the Li ions diffuse based on the vacancy diffusion mechanism via the A-site vacancies, the ionic conductivity depends on the concentration of the Li^+ ions and the vacancies. When Li ions diffuse from one lattice to the next, if there are skeletal ions such as La^{3+} ions in the adjacent lattices, they cannot hop into the next lattice due to the electrostatic repulsion between the cations. Considering only the A-site (the A-site sublattice), since the lattice is a simple lattice, the diffusion of Li ions is controlled by a site percolation in a simple lattice [39]. When Li^+ ions and vacancies percolate the entire system, a dc ionic conduction would be observed. Therefore, the ionic conductivity depends on the sum of the concentration of Li ions and vacancies. The composition dependence of ionic conductivity in some Pv-type Li-ion SE suggests that the Li-ion diffusion is based on the percolation-controlled vacancy mechanism [35, 40]. In fact, the skeleton ions, the Li ions, and the vacancies are not necessarily in three-dimensionally and completely disordered arrangement, which would make the site percolation more complicated than simple lattice percolation [40, 41].

2.3.2 Activation States in Ionic Diffusion: Bottlenecks and Activation Energy

When the Li ions move to the adjacent lattices in Pv-type Li-ion SE, they pass through the interstice surrounded by four oxygens (the part surrounded by the dotted line in Fig. 1a). If the interstice is narrow, this position becomes the bottleneck of ion diffusion, *i.e.*, the Li ion at the position is in activated state because it receives the electron repulsion by oxygen.

In $Ln_{2/3-x}Li_{3x}TiO_3$ (Ln = La, Pr, Nd, Sm), the activation energy E_a for ionic conduction increases in the order of *Ln*, La, Pr, Nd, and Sm, and the ionic conductivity decreases [12]. When considering simply, as the lattice volume decreased, the bottleneck size also decreased, resulting in a decrease in the ionic conductivity. But in reality, neutron diffraction experiments confirmed that the degree of tilt of BO_6 octahedra increased and the size of interstice decreased accordingly, as the *Ln* changed from La to Pr, Nd [42]. This means that the interstice surrounded by four oxygens is narrow for the diffusion of Li ions, that is, this interstice is a bottleneck of diffusion and is consistent with the fact that E_a increases as *Ln* changes from La to Pr, and Nd. Figure 3 shows the relationship between the cubic root of the volume for the primitive lattice (Pv parameter), a_p and E_a in the Pv-type Li-ion SE. As seen in Fig. 3, the data are divided into several groups depending on the difference in the B ions.

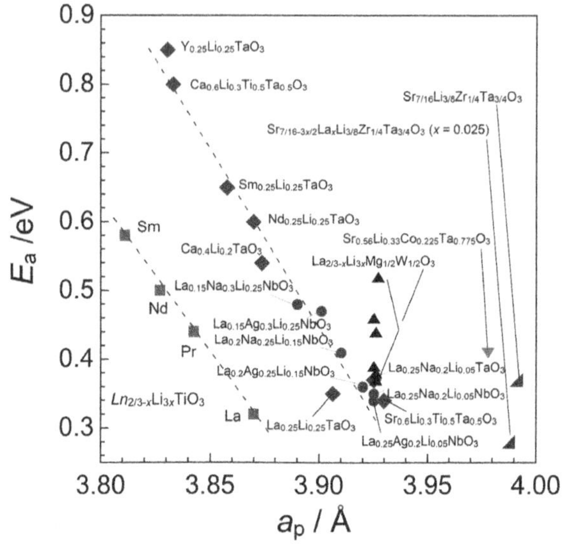

Fig. 3 Relationship between the cubic root of the volume for the primitive lattice (the perovskite parameter), a_p and the activation energy, E_a in perovskite-type Li-ion solid electrolytes

The compounds containing only Ti, the compounds containing only Nb or Ta, or the compounds containing Ti and Nb and Ta, the compounds containing $Mg_{1/2}W_{1/2}$, the compound containing Co, and the compounds containing Zr belong to different groups. In recent years, it has been found that considering the tilt of BO_6 octahedra, the data follow one trend though there are some exceptions [43]. From this, it can be considered that the bottleneck size depends on the tilt of the BO_6 octahedra, which affects the activation energy E_a for diffusion.

As mentioned above, the experimental results suggest that the activation state of ion diffusion depends on the bottleneck size surrounded by four oxygens. On the other hand, the first principle calculation on LLTO [44–49] shows that the position of the Li ion in the activated state is not in the bottleneck surrounded by four oxygen atoms but inside the A-site. Therefore, there still remains room for discussion about the predominant factor of the activation energy in Pv-type Li-ion SE.

3 Ionic Conduction at the Grain Boundary and Electrochemical Stability of Perovskite-Type Li-Ion Solid Electrolytes

3.1 Ionic Conduction at Grain Boundaries

In the previous sections, the fundamental aspects related to the interior (bulk) part of the crystal have been mainly described. Considering the use of polycrystalline samples as SE for SSB, it is essential to improve the ionic diffusion in the grain

Table 1 Bulk (σ_{bulk}), total (σ_{total}) Li-ion conductivity and the ratio of grain boundary resistance to bulk resistance, R_{gb}/R_b at room temperature in several perovskite-type Li-ion solid electrolytes

Compounds	σ_{bulk}/Scm^{-1}	σ_{total}/Scm^{-1}	R_{gb}/R_b	Temp.	References
$La_{2/3-x}Li_{3x}TiO_3$	1×10^{-3}	7×10^{-5}	13.3	300 K	[10]
$La_{2/3-x}Li_{3x}TiO_3$	1.6×10^{-3}	5.2×10^{-4}	2.1	300 K	[50]
$Sr_{0.5}Li_{0.5}Ti_{0.5}Ta_{0.5}O_3$	5.5×10^{-4}	1.2×10^{-4}	3.6	300 K	[35]
$Li_{3/8}Sr_{7/16}Zr_{1/4}Ta_{3/4}O_3$	2×10^{-4}	8×10^{-5}	1.5	303 K	[51]
$Li_{3/8}Sr_{7/16}Zr_{1/4}Ta_{3/4}O_3$	2.8×10^{-4}	2.0×10^{-4}	0.4	300 K	[52]
$Li_{3/8}Sr_{7/16-3x/2}La_xZr_{1/4}Ta_{3/4}O_3$ (x = 0.025)	1.3×10^{-3}	3.3×10^{-4}	2.9	303 K	[53]
$Li_{3/8}Sr_{7/16}Hf_{1/4}Ta_{3/4}O_3$	3.8×10^{-4}	1.7×10^{-4}	1.2	298 K	[54]

boundary in order to reduce the internal resistance of SSB (the ion diffusion at the interface with the active materials is also important but not mentioned in this chapter).

Table 1 shows the bulk (grain) and the total Li-ion conductivity of several Pv-type Li-ion SE at room temperature. As seen in Table 1, for many compounds, the grain boundary resistance R_{gb} is greater than the bulk resistance R_b. In the initial research on LLTO, the bulk ionic conductivity was 1×10^{-3} Scm^{-1}, whereas the total conductivity of the sample was 7 x 10^{-5} Scm^{-1}, resulting in the ratio of R_b/R_{gb} of about 1/10 [10]. Later, it was found that by increasing the sintering temperature and promoting grain growth, the total conductivity was improved to 5×10^{-4} Scm^{-1} [50]. In this case, the grain boundary resistance is equivalent to the bulk resistance. Considering the volume occupied by the grain boundary with respect to the entire sample, the grain boundary resistance contributes greatly, and there is still room for improvement. Moreover, when Sr^{2+} ion is included as the A-site ion, the value of R_{gb}/R_b is smaller than that of LLTO. The difference between the compounds is thought to depend on the substances present at the grain boundaries, but the cause of the grain boundary resistance has not been fully clarified yet. The elimination of grain boundary resistance based on the elucidation of the cause is a key issue for application.

3.2 *Electrochemical Stability as Solid Electrolytes of Batteries*

When considering the application of SE for SSB, the electrochemical stability (potential window) to the active material as well as ionic conductivity is an important factor. Currently reported Pv-type Li-ion SE have the disadvantage that they are unstable to Li metal. For example, LLTO has the reduction potential of 1.7 V vs. Li/Li$^+$ and is reduced by Li metal, resulting in electronic conductivity [10, 55, 56] (Fig. 4a).

Pv-type Li-ion SE containing Ti^{4+}($3d^0$), Nb^{5+}($4d^0$), Ta^{5+}($5d^0$), and W^{6+}($5d^0$) with the nd^0 electron configuration as B ions are semiconductors or insulators from the

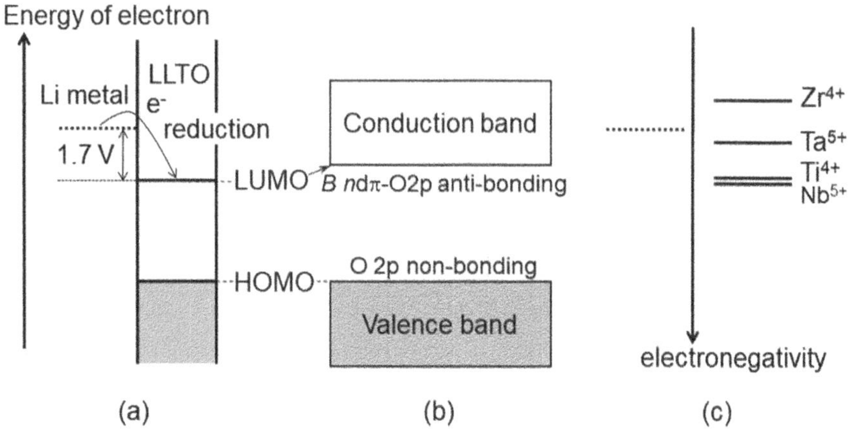

Fig. 4 Schematic diagrams of reduction of LLTO with Li metal (**a**), the band structure of perovskite-type Li-ion solid electrolytes (**b**), and the relationship of the magnitude of electronegativity for each B ion (**c**)

viewpoint of the theory of electron in solid. In these compounds, the upper part of the valence band mainly consists of oxygen nonbonding orbital, and the bottom part of the conduction band mainly consists of the antibonding orbital between the B ion *nd* orbital and the oxygen 2*p* orbital. The valence band is occupied, while the conduction band is empty (Fig. 4b). Therefore, the LUMO (Lowest unoccupied molecular orbital) corresponds to the bottom of the conduction band.

When the electron energy level of Li metal is greater than the LUMO (when the potential with respect to Li is positive), the compound would undergo a reduction (Fig. 4a). As mentioned above, the bottom part of the conduction band mainly consists of antibonding orbital between B *nd* orbital and O 2*p* orbital, therefore the LUMO level is determined by the level of d orbital of B ions (electronegativity) and the bandwidth depending on the magnitude of the orbital interaction between B *n*d and O 2*p*. Since the top of the valence band mainly consists of a nonbonding orbital of oxygen, assuming that the upper level of the valence band does not depend on the material, the lower level of the conduction band can be predicted from the band gap. For example, the band gaps of Pv-type oxides, $CaTiO_3$, $La_{2/3-x}Li_{3x}TiO_3$ ($x = 0.12$), $La_{1/2}Na_{1/2}TiO_3$, $SrTiO_3$, $La_{1/3}NbO_3$, and $La_{1/3}TaO_3$ are 3.56 eV [57], 3.37, 3.29 eV [57], 3.26 eV [57], 3.22 eV [58], and 3.87 eV [58], respectively. Focusing our attention on B ions, the level at the bottom of the conduction band increases in the order of Nb < Ti < Ta, which corresponds to the order of electronegativity (Nb > Ti > Ta) (Fig. 4c). In addition, when comparing Ti compounds, the band gap decreases in the order of $CaTiO_3$, $La_{2/3-x}Li_{3x}TiO_3$ ($x = 0.12$), $La_{1/2}Na_{1/2}TiO_3$, and $SrTiO_3$. This is consistent with that the interaction between the M *n*d and the O 2*p* orbitals becomes stronger, resulting in the increase in the bandwidth, *i.e.*, decrease in band gap as the Ti–O–Ti angle approaches 180° from $CaTiO_3$ to $SrTiO_3$.

Table 2 Reduction potential of perovskite-type Li-ion solid electrolytes

Compounds	*B* ion	Potential (V versus Li/Li^+)	References (Remark)
$La_{2/3-x}Li_{3x}TiO_3$	Ti	1.6–1.7	[55, 56]
$La_{1/3}NbO_3$	Nb	3.27	[59] (Li insertion)
$Nd_{1/3}NbO_3$	Nb	3.10	[59] (Li insertion)
$La_{1/3-x}Li_{3x}NbO_3$	Nb	2.0	[60]
(Sr,Li)(Ti,Ta)O_3	Ti, Ta	1.5	Private communication
$Sr_{7/16}$ $Li_{3/8}Zr_{1/4}Ta_{3/4}O_3$	Zr, Ta	1.0	[51]
$Sr_{7/16-3x/2}La_x$ $Li_{3/8}Zr_{1/4}Ta_{3/4}O_3$ ($x = 0.025$)	Zr, Ta	1.3	[53]
$Li_{3/8}Sr_{7/16}Hf_{1/4}Ta_{3/4}O_3$	Hf, Ta	1.4	[54]

Table 2 shows the reduction potentials reported for some Pv-type Li-ion SE. As seen in Table 2, the reduction potential increases as the electronegativity of B ions increases (Nb > Ti > Ta). In addition, when the compounds include two or more types of B ions, the B ion with a greater electronegativity governs the reduction potential. Therefore, if a Pv-type Li-ion SE containing only $Sc^{3+}(3d^0)$, $Zr^{4+}(4d^0)$, and $Hf^{4+}(5d^0)$ with smaller electronegativity can be obtained, the reduction potential would decrease and the LUMO level would be higher than the energy level of Li, which would satisfy the electrochemical stability against Li metal. For example, $La_{2/3-x}Li_{3x}ZrO_3$, in which the Ti^{4+} ions of LLTO are replaced with Zr^{4+} ions, is a candidate for this purpose, but no Pv-type phase with this composition has been obtained. Therefore, at present, when using Pv-type Li-ion SE for SSB, it is necessary to develop SSB using a negative electrode other than Li metal.

Though I mentioned the reduction side on the electrochemical stability, the oxidation side is still unknown. However, the potential window on the oxidation side is expected to be wide (the oxidation potential is high). In that case, even if a negative electrode with a high reduction potential relative to Li metal is selected, the use of a positive electrode with a higher potential than known positive electrodes would make it possible to develop batteries with high energy density, which can compensate for the drawback of high reduction potential.

4 Summary and Outlook

In this chapter, the crystal structure, the bulk ionic conductivity and their relationship of Pv-type Li-ion SE, and the ionic conductivity at grain boundaries and the electrochemical stability, which are important for the application, were described. The SOJT effect plays an important role in stabilizing the A-site deficiency, and may promote ionic diffusion [14] though it is not mentioned in this chapter. In the future, the relationship between the SOJT effect and ionic diffusion will be understood more deeply. Moreover, Pv-type Li-ion SE exhibit high reduction potential due to the high electronegativity of B ions such as Ti^{4+} and Nb^{5+}, and when applied to SSB, Li metal cannot be used as the negative electrode. This is a major drawback in the application. In order to achieve reduction resistance against Li metal, it is necessary to select B ion with low electronegativity. In addition, the developments of SSB using a negative electrode other than Li metal and a high potential positive electrode are also desired.

References

1. Tatsumisago, M., Nagao, M., & Hayashi, A. (2013). Recent development of sulfide solid electrolytes and interfacial modification for all-solid-state rechargeable lithium batteries. *Journal of Asian Ceramic Societies, 1*(1), 17–25. https://doi.org/10.1016/j.jascer.2013.03.005.
2. Kanno, R., & Murayama, M. (2001). Lithium ionic conductor Thio-LISICON: The Li_2S -GeS_2 - P_2S_5 system. *Journal of the Electrochemical Society, 148*(7), A742–A746. https://doi.org/10.1149/1.1379028.
3. Kamaya, N., Homma, K., Yamakawa, Y., Hirayama, M., Kanno, R., Yonemura, M., et al. (2011). A lithium superionic conductor. *Nature Mater, 10*(9), 682–686. https://doi.org/10.1038/nmat3066.
4. Mizuno, F., Hayashi, A., Tadanaga, K., & Tatsumisago, M. (2005). New, Highly ion-conductive crystals precipitated from Li_2S–P_2S_5 glasses. *Advanced Materials, 17*(7), 918–921. https://doi.org/10.1002/adma.200401286.
5. Stramare, S., Thangadurai, V., & Weppner, W. (2003). Lithium lanthanum titanates: A review. *Chemistry of Materials, 15*(21), 3974–3990. https://doi.org/10.1021/cm0300516.
6. Inaguma, Y. (2006). Fast percolative diffusion in lithium ion-conducting perovskite-type oxides. Journal of the Ceramic Society of Japan, *114* (Dec):1103–1110. https://doi.org/10.2109/jcersj.114.1103.
7. Bohnke, O. (2008). The fast lithium-ion conducting oxides $Li_{3x}La_{2/3-x}TiO_3$ from fundamentals to application. *Solid State Ionics, 179*(1–6), 9–15. https://doi.org/10.1016/j.ssi.2007.12.022.
8. Latie, L., Villeneuve, G., Conte, D., & Le Flem, G. (1984). Ionic conductivity of oxides with general formula $Li_xLn_{1/3}Nb_{1-x}Ti_xO_3$ (Ln = La, Nd). *Journal of Solid State Chemistry, 51*(3), 293–299. https://doi.org/10.1016/0022-4596(84)90345-1.
9. Belous, A. G., Novitskaya, G. N., Polyanetskaya, S. V., Gornikov, Y. I. (1987). Investigating complex oxides of the lanthanum lithium titanate ($La_{2/3-x}Li_{3x}TiO_3$) composition ([English translation: *Inorganic Materials, 23*, 412–415.]). Izv Akad Nauk SSSR, *Inorganic Materials, 23*(3), 470–472.
10. Inaguma, Y., Liquan, C., Itoh, M., Nakamura, T., Uchida, T., Ikuta, H., et al. (1993). High ionic conductivity in lithium lanthanum titanate. *Solid State Communications, 86*(10), 689–693. https://doi.org/10.1016/0038-1098(93)90841-A.
11. Inaguma, Y., Chen, L., Itoh, M., Nakamura, T. (1994). Candidate compounds with perovskite structure for high lithium ionic conductivity. *Solid State Ionics*, 70–71(Part 1), 196–202. https://doi.org/10.1016/0167-2738(94)90309-3.
12. Itoh, M., Inaguma, Y., Jung, W. –H., Chen, L., Nakamura, T. (1994). High lithium ion conductivity in the perovskite-type compounds $Ln_{1/2}Li_{1/2}TiO_3$(Ln = La, Pr, Nd, Sm). *Solid State Ionics, 70–71* (Part 1), 203–207. https://doi.org/10.1016/0167-2738(94)90310-7.
13. Kawai, H., Kuwano, J. (1994). Lithium ion conductivity of A-site deficient perovskite solid solution $La_{0.67-x}Li_{3x}TiO_3$. *Journal of the Electrochemical Society, 141*(7), L78–L79. https://doi.org/10.1149/1.2055043.
14. Inaguma, Y. (2016). A review of recent research on perovskite-type lithium ion-conducting oxides. *Nihon Kessho Gakkaishi, 58*(2), 62–72. https://doi.org/10.5940/jcrsj.58.62.
15. Momma, K., & Izumi, F. (2011). VESTA 3 for three-dimensional visualization of crystal, volumetric and morphology data. *Journal of Applied Crystallography, 44*(6), 1272–1276. https://doi.org/10.1107/S0021889811038970.

16. Mitchell, R. H. (2002). *Pervoskites: Modern and Ancient*. Almaz Press.
17. Glazer, A. (1972). The classification of tilted octahedra in perovskites. *Acta Crystallographica. Section B, Structural Science, 28*(11), 3384–3392. https://doi.org/10.1107/S0567740872007976.
18. Glazer, A. (1975). Simple ways of determining perovskite structures. *Acta Crystallographica, 31*(6), 756–762. https://doi.org/10.1107/S0567739475001635.
19. Woodward, P. (1997). Octahedral tilting in perovskites. *I. Geometrical Considerations. Acta Crystallogr B, 53*(1), 32–43. https://doi.org/10.1107/S0108768196010713.
20. Woodward, P. (1997). Octahedral tilting in perovskites. II. structure stabilizing forces. Acta Crystallogr B, *53*(1), 44–66. https://doi.org/10.1107/s0108768196012050.
21. Howard, C. J., & Stokes, H. T. (1998). Group-theoretical analysis of octahedral tilting in perovskites. *Acta Crystallographica. Section B, Structural Science, 54*(6), 782–789. https://doi.org/10.1107/S0108768198004200.
22. Howard, C. J., Stokes, H. T. (2002). Group-theoretical analysis of octahedral tilting in perovskites. Erratum. *Acta Crystallographica Section B, 58* (3 Part 2), 565. https://doi.org/10.1107/s010876810200890x.
23. Aleksandrov, K. S. (1976). The sequences of structural phase transitions in perovskites. *Ferroelectrics, 14*(1), 801–805. https://doi.org/10.1080/00150197608237799.
24. Goldschmidt, V. M. (1926). Die Gesetze der Krystallochemie. *Naturwissenschaften, 14*(21), 477–485. https://doi.org/10.1007/BF01507527.
25. Abe, M., & Uchino, K. (1974). X-ray study of the deficient perovskite $La_{2/3}TiO_3$. *Materials Research Bulletin, 9*(2), 147–155. https://doi.org/10.1016/0025-5408(74)90194-9.
26. Iyer, P. N., Smith, A. J. (1967). Double oxides containing niobium, tantalum, or protactinium. III. Systems involving the rare earths. *Acta Crystallographica, 23*(5), 740–746. https://doi.org/10.1107/s0365110x67003639.
27. Kennedy, B. J., Howard, C. J., Kubota, Y., & Kato, K. (2004). Phase transition behaviour in the A-site deficient perovskite oxide $La_{1/3}NbO_3$. *Journal of Solid State Chemistry, 177*(12), 4552–4556. https://doi.org/10.1016/j.jssc.2004.08.047.
28. King, G., & Woodward, P. M. (2010). Cation ordering in perovskites. *Journal of Materials Chemistry, 20*(28), 5785–5796. https://doi.org/10.1039/B926757C.
29. Bersuker, I. B. (2013). Pseudo-Jahn–Teller effect—A two-state paradigm in formation, deformation, and transformation of molecular systems and solids. *Chemical Reviews, 113*(3), 1351–1390. https://doi.org/10.1021/cr300279n.
30. Bersuker, I. B. (1966). On the origin of ferroelectricity in perovskite-type crystals. *Physics Letters, 20*(6), 589–590. https://doi.org/10.1016/0031-9163(66)91127-9.
31. Wheeler R.A., Whangbo H., Hughbanks T., Hoffmann R., Burdett K., Albright A. (1986) Symmetric vs. asymmetric linear M-X-M linkages in molecules, polymers, and extended networks. *Journal of the American Chemical Society, 108*(9), 2222–2236. https://doi.org/10.1021/ja00269a018.
32. Kunz, M., & Brown, I. D. (1995). Out-of-center distortions around octahedrally coordinated d^0 transition metals. *Journal of Solid State Chemistry, 115*(2), 395–406. https://doi.org/10.1006/jssc.1995.1150.
33. Halasyamani, P. S., & Poeppelmeier, K. R. (1998). Noncentrosymmetric oxides. *Chemistry of Materials, 10*(10), 2753–2769. https://doi.org/10.1021/cm980140w.
34. Inaguma, Y., Katsumata, T., Itoh, M., Morii, Y. (2002). Crystal structure of a lithium ion-conducting perovskite $La_{2/3-x}Li_{3x}TiO_3$ ($x = 0.05$). *Journal of Solid State Chemistry, 166*(1), 67–72. https://doi.org/10.1006/jssc.2002.9560.
35. Inaguma, Y., Matsui, Y., Shan, Y.-J., Itoh, M., & Nakamura, T. (1995). Lithium ion conductivity in the perovskite-type $LiTaO_3$-$SrTiO_3$ solid solution. *Solid State Ionics, 79*, 91–97. https://doi.org/10.1016/0167-2738(95)00036-6.
36. Inaguma Y, Seo A, Katsumata T (2004) Synthesis and lithium ion conductivity of cubic deficient perovskites $Sr_{0.5+x}Li_{0.5-2x}\square_x Ti_{0.5}Ta_{0.5}O_3$ and the La-doped compounds. *Solid State Ionics, 174*(1-4), 19–26. https://doi.org/10.1016/j.ssi.2004.06.013.

37. Alonso, J. A., Sanz, J., Santamaría, J., León, C., Várez, A., Fernández-Díaz, M. T. (2000). On the location of Li^+ cations in the fast Li-cation conductor $La_{0.5}Li_{0.5}TiO_3$ perovskite. *Angewandte Chemie International Edition, 39*(3), 619–621. https://doi.org/10.1002/(sici)1521-3773(20000204)39:3%3c619::aid-anie619%3e3.0.co;2-o.
38. Yashima, M., Itoh, M., Inaguma, Y., Morii, Y. (2005). Crystal structure and diffusion path in the fast lithium-ion conductor $La_{0.62}Li_{0.16}TiO_3$. *Journal of the American Chemical Society, 127*(10), 3491–3495. https://doi.org/10.1021/ja0449224.
39. Stauffer, D. (1992). *Introduction to percolation theory*. London: Taylor & Francis.
40. Inaguma, Y., & Itoh, M. (1996). Influences of carrier concentration and site percolation on lithium ion conductivity in perovskite-type oxides. *Solid State Ionics, 86–88*(Pt. 1), 257–260. https://doi.org/10.1016/0167-2738(96)00100-2.
41. Katsumata, T., Matsui, Y., Inaguma, Y., Itoh, M. (1996). Influence of site percolation and local distortion on lithium ion conductivity in perovskite-type oxides $La_{0.55}Li_{0.35-x}K_xTiO_3$ and $La_{0.55}Li_{0.35}TiO_3$-KMO_3 (M = Nb and Ta). *Solid State Ionics, 86–88*(PART 1), 165–169. https://doi.org/10.1016/0167-2738(96)00116-6.
42. Skakle, J. M. S, Mather, G. C., Morales, M., Smith, R. I., West, A. R. (1995). Crystal structure of the Li^+ ion-conducting phases, $Li_{0.5-3x}Re_{0.5+x}TiO_3$: RE = Pr, Nd; $x \approx 0.05$. *Journal of Materials Chemistry, 5* (11)1807–1808. https://doi.org/10.1039/jm9950501807.
43. Inaguma, Y., Katsumata, T., Mori, D. (2010). Predominant factor of activation energy for ionic conductivity in perovskite-type lithium ion-conducting oxides. Journal of the Physical Society of Japan, *79*(Suppl.A), 69–71. https://doi.org/10.1143/jpsjs.79sa.69.
44. Catti, M. (2007). First-principles modeling of lithium ordering in the LLTO ($Li_xLa_{2/3-x/3}TiO_3$) superionic conductor. *Chemistry of Materials, 19*(16), 3963–3972. https://doi.org/10.1021/cm0709469.
45. Catti, M. (2008). Ion mobility pathways of the Li^+ conductor $Li_{0.125}La_{0.625}TiO_3$ by Ab Initio simulations. *The Journal of Physical Chemistry C, 112*(29), 11068–11074. https://doi.org/10.1021/jp803345y.
46. Catti, M. (2011). Short-range order and Li^+ ion diffusion mechanisms in $Li_5La_9\square_2(TiO_3)_{16}$ (LLTO). *Solid State Ionics, 183*(1), 1–6. https://doi.org/10.1016/j.ssi.2010.12.016.
47. Nakayama, M., Shirasawa, A., & Saito, T. (2009). Arrangement of La and vacancies in $La_{2/3}TiO_3$ predicted by first-principles density functional calculation with cluster expansion and Monte Carlo simulation. *Journal of the Ceramic Society of Japan, 117*(1368), 911–916. https://doi.org/10.2109/jcersj2.117.911.
48. Kim, D. H., Jeong, Y. C., Seo, H. I., & Kim, Y. C. (2012). Lithium ion migration pathways in $Li_{3x}La_{2/3-x}\square_{1/3-2x}TiO_3$. *Ceramic International, 38*(SUPPL. 1), S467–S470. https://doi.org/10.1016/j.ceramint.2011.05.041.
49. Moriwake, H., Gao, X., Kuwabara, A., Fisher, C. A. J., Kimura, T., Ikuhara, Y. H., et al. (2015). Domain boundaries and their influence on Li migration in solid-state electrolyte (La, Li)TiO_3. *Journal of Power Sources, 276*, 203–207. https://doi.org/10.1016/j.jpowsour.2014.11.139.
50. Inaguma, Y., & Nakashima, M. (2013). A rechargeable lithium–air battery using a lithium ion-conducting lanthanum lithium titanate ceramics as an electrolyte separator. *Journal of Power Sources, 228*, 250–255. https://doi.org/10.1016/j.jpowsour.2012.11.098.
51. Chen, C. H., Xie, S., Sperling, E., Yang, A. S., Henriksen, G., & Amine, K. (2004). Stable lithium-ion conducting perovskite lithium–strontium–tantalum–zirconium–oxide system. *Solid State Ionics, 167*(3), 263–272. https://doi.org/10.1016/j.ssi.2004.01.008.
52. Kimura, K., Wagatsuma, K., Tojo, T., Inada, R., & Sakurai, Y. (2016). Effect of composition on lithium-ion conductivity for perovskite-type lithium–strontium–tantalum–zirconium-oxide solid electrolytes. *Ceramic International, 42*(4), 5546–5552. https://doi.org/10.1016/j.ceramint.2015.12.133.
53. Lu, J., & Li, Y. (2018). Conductivity and stability of $Li_{3/8}Sr_{7/16-3x/2}La_xZr_{1/4}Ta_{3/4}O_3$ superionic solid electrolytes. *Electrochimica Acta, 282*, 409–415. https://doi.org/10.1016/j.electacta.2018.06.085.
54. Huang, B., Xu, B., Li, Y., Zhou, W., You, Y., Zhong, S., et al. (2016). Li-Ion Conduction and Stability of Perovskite $Li_{3/8}Sr_{7/16}Hf_{1/4}Ta_{3/4}O_3$. *ACS Applied Materials & Interfaces, 8*(23), 14552–14557. https://doi.org/10.1021/acsami.6b03070.

55. Shan, Y. J., Chen, L., Inaguma, Y., Itoh, M., & Nakamura, T. (1995). Oxide cathode with perovskite structure for rechargeable lithium batteries. *Journal of Power Sources, 54*(2), 397–402. https://doi.org/10.1016/0378-7753(94)02110-O.
56. Birke, P., Scharner, S., Huggins, R. A., Weppner, W. (1997). Electrolytic stability limit and rapid lithium insertion in the fast-Ion-conducting $Li_{0.29}La_{0.57}TiO_3$ perovskite-type compound. *Journal of the Electrochemical Society, 144* (6), L167-L169. https://doi.org/10.1149/1.1837713.
57. Inaguma, Y., Tsuchiya, T., & Katsumata, T. (2007). Systematic study of photoluminescence upon band gap excitation in perovskite-type titanates $R_{1/2}Na_{1/2}TiO_3$: Pr (R = La, Gd, Lu, and Y). *Journal of Solid State Chemistry, 180*(5), 1678–1685. https://doi.org/10.1016/j.jssc.2007.03.017.
58. Inaguma, Y., Muronoi, T., Sano, K., Tsuchiya, T., Mori, Y., Katsumata, T., et al. (2011). An approach to control of band gap energy and photoluminescence upon band gap excitation in Pr^{3+}-doped perovskites $La_{1/3}MO_3$ (M = Nb, Ta): Pr^{3+}. *Inorganic Chemistry, 50*(12), 5389–5395. https://doi.org/10.1021/ic101955v.
59. Nadiri, A., Le Flem, G., & Delmas, C. (1988). Lithium intercalation in $Ln_{1/3}NbO_3$ perovskite-type phases (Ln = La, Nd). *Journal of Solid State Chemistry, 73*(2), 338–347. https://doi.org/10.1016/0022-4596(88)90118-1.
60. Nakayama, M., Imaki, K., Ikuta, H., Uchimoto, Y., Wakihara, M. (2002). Electrochemical lithium insertion for perovskite oxides of $Li_yLa_{(1-y)/3}NbO_3$ ($y = 0, 0.1, 0.25$). *The Journal of Physical Chemistry B, 106*(25), 6437–6441. https://doi.org/10.1021/jp0258659.

Garnet-Type Lithium Ion Conducting Oxides: $Li_7La_3Zr_2O_{12}$ and Its Chemical Derivatives

Junji Akimoto, Naoki Hamao, and Kunimitsu Kataoka

Abstract Recent advancements of garnet-type lithium ion conducting oxide materials such as $Li_7La_3Zr_2O_{12}$ and its chemical derivative $Li_{6.5}La_3Zr_{1.5}Ta_{0.5}O_{12}$ are reviewed. A relationship between chemical composition and conductivity is revealed in the $Li_{7-x}La_3Zr_{2-x}Ta_xO_{12}$ solid solution system. Small single-crystal samples were synthesized by a flux method so as to demonstrate the structural properties of Al-doped $Li_7La_3Zr_2O_{12}$. We also grew large single-crystal samples of $Li_{6.5}La_3Zr_{1.5}Ta_{0.5}O_{12}$ by floating zone melting method and determined the precise crystal structure using single-crystal neutron diffraction data. The lithium ion conductive properties were investigated by electrochemical measurement and NMR spectroscopy using single-crystal specimens. Novel low-temperature synthetic techniques so as to produce tetragonal $Li_7La_3Zr_2O_{12}$ and cubic $Li_{6.5}La_3Zr_{1.5}Ta_{0.5}O_{12}$ fine powders were developed.

Keywords Garnet · Crystal structure · Crystal growth · Flux method · Floating zone melting method

1 Introduction

All-solid-state Li-ion batteries (LIB) using solid electrolyte attract attention as next-generation batteries without flammable organic liquid electrolytes since the Li-ion batteries require improved safety and high energy density [1–3]. Because the characteristics of the solid electrolyte largely dominate the battery performance, many researchers have performed efforts to explore the excellent lithium ion conductors for the development of all-solid-state batteries. The solid electrolyte for all-solid-state LIB requires high lithium ion conductivity, high density, chemical stability, and electrochemical wide potential windows. Among the various kinds of solid electrolytes, oxide-based solid electrolytes show high chemical stability. Thus, a variety of oxide-based solid electrolytes showing good lithium ion conductivity

J. Akimoto (✉) · N. Hamao · K. Kataoka
National Institute of Advanced Industrial Science and Technology (AIST), Tsukuba, Japan
e-mail: j.akimoto@aist.go.jp

K. Kanamura (ed.), *Next Generation Batteries*,
https://doi.org/10.1007/978-981-33-6668-8_19

have been investigated, including the NASICON-type $Li_{1+x}Al_xTi_{2-x}(PO_4)_3$ [4], LISICON-type $Li_{2+2x}Zn_{1-x}GeO_4$ [5], perovskite-type $La_{2/3-x}Li_{3x}TiO_3$ [6], garnet-type $Li_7La_3Zr_2O_{12}$ [7], and glassy solid electrolyte materials such as in the Li_3BO_3-Li_4SiO_4 system. Among them, the garnet-type $Li_7La_3Zr_2O_{12}$ and its chemical derivatives are one of the most promising candidates because of its excellent comprehensively electrochemical performance including the electrochemical stability against lithium metal [7]. In addition, the bulk lithium ion conductivity of the Ga-doped $Li_7La_3Zr_2O_{12}$ was reported to be ~10^{-3} S cm^{-1} at room temperature, the value of which is highest among the oxide-based materials [8].

In this chapter, we review the crystal structural properties of two kinds of cubic forms and tetragonal form for $Li_7La_3Zr_2O_{12}$ from a viewpoint of lithium ion arrangement. In addition, a relationship between the crystal structure and electrochemical properties in the solid solution system $Li_{7-x}La_3Zr_{2-x}Ta_xO_{12}$ by using polycrystalline samples prepared by solid-state reaction is discussed. The conductive properties for the three kinds of $Li_{6.5}La_3M_{1.5}Ta_{0.5}O_{12}$ (*M*: Zr, Hf, Sn) are demonstrated. Small single-crystal samples were synthesized by a flux method so as to demonstrate the structural properties of Al-doped $Li_7La_3Zr_2O_{12}$. In addition, large single-crystal samples of $Li_{6.5}La_3Zr_{1.5}Ta_{0.5}O_{12}$ were recently grown by the floating zone melting method, and the precise crystal structure was revealed using single-crystal neutron diffraction data. The lithium ion conductive properties were investigated by electrochemical and NMR spectroscopy using the single-crystal specimens. Finally, novel low-temperature synthetic techniques so as to produce the $Li_7La_3Zr_2O_{12}$ and $Li_{6.5}La_3Zr_{1.5}Ta_{0.5}O_{12}$ fine powders are reviewed.

2 Crystal Structure of Garnet-Type Lithium Ion Conducting Oxides

The crystal structure of $Li_7La_3Zr_2O_{12}$ at room temperature belongs to the tetragonal symmetry, space group $I4_1/acd$ [9]. On the other hand, the cubic symmetry is stable only above 600 °C [10]. The high-temperature cubic garnet structure stabilizes by the substitution of Al and Ga at the Li sites, and/or by the substitution of Nb and Ta at the Zr site. Because crystal structural details such as Li–Li distance, Li–O polyhedral volume, and Li arrangement together with defect content are very important indicators to reveal the good Li-ion conduction mechanism in $Li_7La_3Zr_2O_{12}$ and its derivatives, we mention here the crystal structure of cubic Al-doped $Li_7La_3Zr_2O_{12}$ in comparison with the tetragonal structure, because the Li-ion conductivity of tetragonal $Li_7La_3Zr_2O_{12}$ is lower by two orders of magnitude than that of cubic Al-doped $Li_7La_3Zr_2O_{12}$.

In both cubic and tetragonal structures [9, 11], the garnet framework structure is composed of dodecahedral LaO_8 and octahedral ZrO_6. On the other hand, the Li atoms occupy two types of crystallographic sites in the interstices of the framework structure in the cubic phase. As shown in Fig. 1, the Li1 and Li2 atoms are located

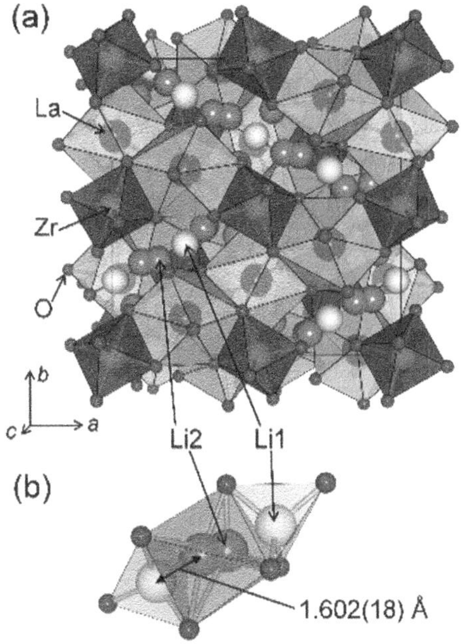

Fig. 1 **a** Crystal structure of cubic $Li_7La_3Zr_2O_{12}$. **b** Coordination polyhedra around the Li1 and Li2 sites in cubic $Li_7La_3Zr_2O_{12}$ [11]

in the tetrahedral 24*d* site and distorted octahedral 96*h* site, respectively. The Li2 site is vacant in the ideal garnet structure such as in yttrium iron garnet $Y_3Fe_5O_{12}$ (YIG). Because the Li2 atom is situated at two equivalent positions in the distorted octahedron as a positional disorder in the lithium distribution, the local coordination environment of the Li2 atom may be considered to be near fourfold LiO_4 coordination. The disordering and partial occupation of the Li atoms at the Li2 site were reported as a key role of Li-ion conduction in cubic Al-doped $Li_7La_3Zr_2O_{12}$ [11].

From a structural viewpoint, the Li-ion migration pathway should correspond to the Li atomic arrangement in the structure. As shown in Fig. 1, cubic $Li_7La_3Zr_2O_{12}$ shows the complicated Li atomic arrangement in the interstices of the garnet-type framework structure. However, by focusing attention only on the Li atoms in the structure, the basic unit of the arrangement can be simply drawn as a loop constructed by the Li1 and Li2 sites (Fig. 2). This loop links to another one, where only the Li1 site is shared by two loops as a junction, and a three-dimensional network of the Li-ion migration pathway is formed in the structure, as shown in Fig. 3.

In the loop structure, the tetrahedral $Li1O_4$ and distorted octahedral Li_2O_6 share with the face, which results in the very short Li–Li distance in this migration pathway (Fig. 1). This is a significant feature of the cubic garnet-type Li-ion conductors and may be related to the good Li-ion conductive properties compared to the other compounds.

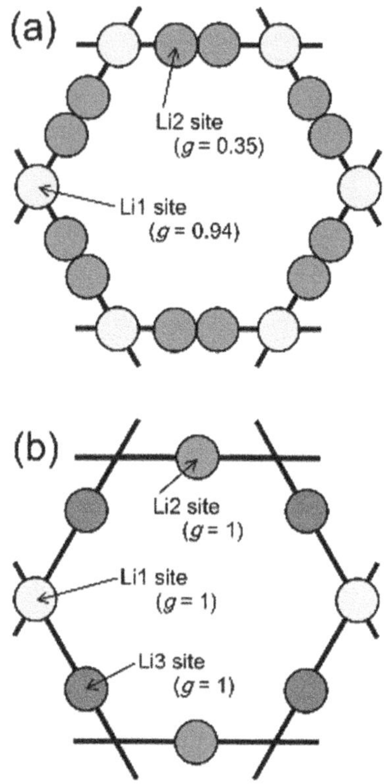

Fig. 2 The loop structures constructed by Li arrangement in **a** cubic and **b** tetragonal $Li_7La_3Zr_2O_{12}$ [11]

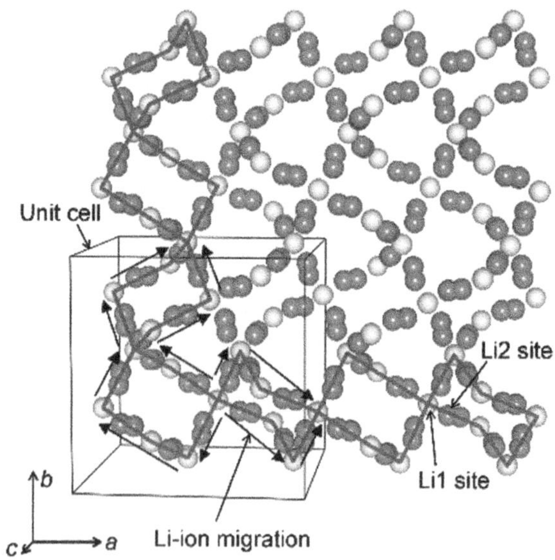

Fig. 3 Three-dimensional network structure of Li arrangement in cubic $Li_7La_3Zr_2O_{12}$ [11]

In the case of Ga-doped $Li_7La_3Zr_2O_{12}$, the Ga occupation in the loop structure causes lowering the cubic symmetry from *Ia*-3*d* to *I*-43*d* and changes the Li arrangement (Fig. 4) [8]. Unfortunately, the origin for high Li-ion conductivity in Ga-doped $Li_7La_3Zr_2O_{12}$ has not been revealed yet.

On the other hand, tetragonal $Li_7La_3Zr_2O_{12}$ shows a complete ordering of the Li atoms [9]. Namely, tetragonal $Li_7La_3Zr_2O_{12}$ has two tetrahedral sites (Li1 site and vacancy), and two distorted octahedral sites (Li2 and Li3 sites) in the loop structure (Fig. 2). The Li1, Li2, and Li3 sites are fully occupied by the Li atoms. Accordingly, the Li–Li distances over 2.5 Å are longer than those in a cubic structure.

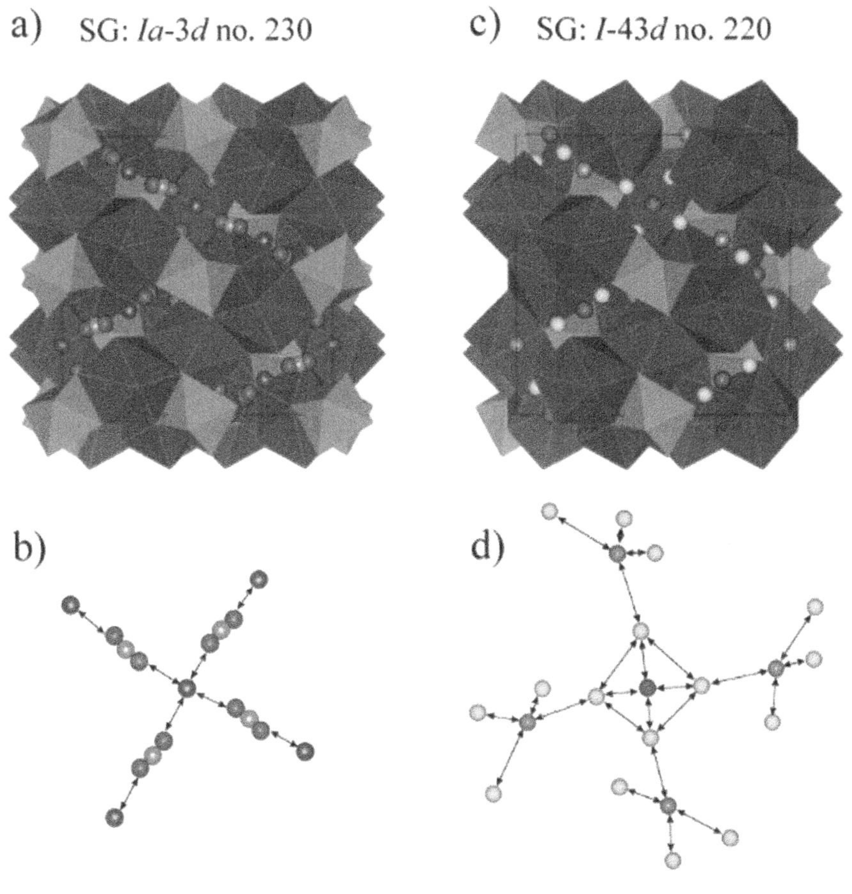

Fig. 4 **a** Crystal structure of cubic Al-doped $Li_7La_3Zr_2O_{12}$ with space group *Ia*-3*d*. The corresponding Li-ion diffusion pathway is shown in (**b**). **c** Crystal structure of cubic Ga-doped $Li_7La_3Zr_2O_{12}$ with space group *I*4-3*d*. The corresponding Li-ion diffusion pathway is shown in (**d**) [8]

3 Synthesis of Polycrystalline $Li_{7-x}La_3Zr_{2-x}Ta_xO_{12}$ by Solid-State Reaction

In order to increase the Li-ion conductivity of the garnet-type $Li_7La_3Zr_2O_{12}$, a large number of studies on the chemical substitution using various cation species have been reported. Among them, $Li_{7-x}La_3Zr_{2-x}Ta_xO_{12}$ with the cubic garnet-related-type structure shows the highest Li-ion conductivity (~ 10^{-4} S cm at 300 K) [12–14]. For this reason, the chemical and structural properties of $Li_{7-x}La_3Zr_{2-x}Ta_xO_{12}$ have been widely investigated in the literature. However, it is difficult to prepare high-quality $Li_{7-x}La_3Zr_{2-x}Ta_xO_{12}$ samples because aluminum is easily contaminated from the crucible material during the high-temperature heating. Recently, the garnet-type Al-free $Li_{7-x}La_3Zr_{2-x}Ta_xO_{12}$ samples were synthesized by some research groups [15, 16]. For example, Wang et al. reported the existence of two phases of tetragonal and cubic structures in the compositional range of $0.1 < x < 0.4$ for the Al-free $Li_{7-x}La_3Zr_{2-x}Ta_xO_{12}$ samples [16]. However, the effect of only Ta substitution on the structural property and the Li-ion conductivity has not been clarified yet.

Recently, the garnet-type Al-free $Li_{7-x}La_3Zr_{2-x}Ta_xO_{12}$ ($0 \leq x \leq 0.6$) samples were synthesized by the conventional solid-state synthesis method under Al-free conditions [17]. In addition, the influence of Ta substitution on the crystal phase formation and the Li-ion conductive properties was demonstrated [17]. The starting materials used were Li_2CO_3, La_2O_3, ZrO_2, and Ta_2O_5 in this method. An excess Li source was added to compensate for the volatilization of lithium during the high-temperature calcination. These materials were mixed by ball milling and calcined at 1127 K in air. Then, the pressed pellets were sintered at 1373 K in the air using YSZ crucibles. At the sintering step, the pellets were covered with the mother powder of the same composition to prevent the Li loss.

Figure 5 shows the lattice parameters of the $Li_{7-x}La_3Zr_{2-x}Ta_xO_{12}$ ($0 \leq x \leq 0.6$) samples. The XRD patterns for $Li_7La_3Zr_2O_{12}$ ($x = 0$) and $Li_{6.4}La_3Zr_{1.4}Ta_{0.6}O_{12}$ ($x = 0.6$) samples were assigned to be single phases of tetragonal (space group: $I4_1/acd$) and cubic (space group: Ia-$3d$) structures, respectively. On the other hand, the intermediate compositional samples of $Li_{7-x}La_3Zr_{2-x}Ta_xO_{12}$ ($0.2 \leq x \leq 0.5$) showed a coexistence of both the tetragonal and cubic phases.

All of the Al-free $Li_{7-x}La_3Zr_{2-x}Ta_xO_{12}$ ($0 \leq x \leq 0.6$) samples exhibit relatively high conductivity of ~ 10^{-4} S cm^{-1} at room temperature, and the $Li_{6.5}La_3Zr_{1.5}Ta_{0.5}O_{12}$ ($x = 0.5$) sample shows the highest Li-ion conductivity of 8.4×10^{-4} S cm^{-1} at room temperature. In order to clarify the relationship between the Li-ion conductivity and the Li-ion arrangement, the crystal structure analysis of $Li_{7-x}La_3Zr_{2-x}Ta_xO_{12}$ ($0 \leq x \leq 0.6$) was performed by Rietveld analysis using powder X-ray diffraction data. Figure 6 shows the loop structure of the Li arrangement for the $x = 0.2$ and $x = 0.5$ samples. The Li2 atom at 96h site is gradually shifted together with increasing Ta-content from $x = 0.2$ to 0.5, resulting in the shorter Li–Li distance in the loop structure of the cubic garnet-type framework structure. Accordingly, the optimum Li arrangement in the loop structure can be established in the $x = 0.5$ composition from a viewpoint of Li conductivity [17].

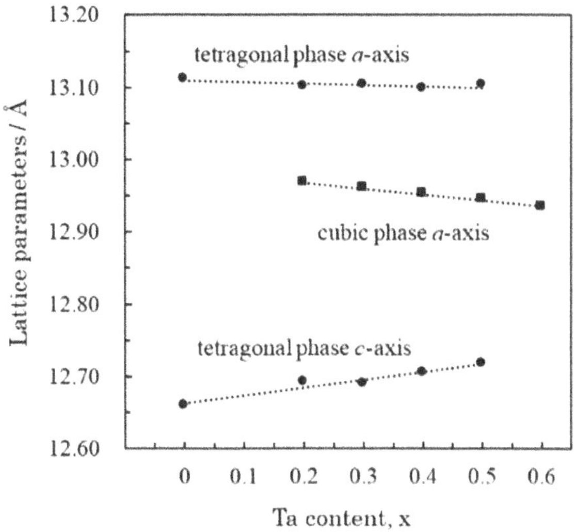

Fig. 5 Lattice parameters of the Al-free $Li_{7-x}La_3Zr_{2-x}Ta_xO_{12}$ ($0 \leq x \leq 0.6$) samples [17]

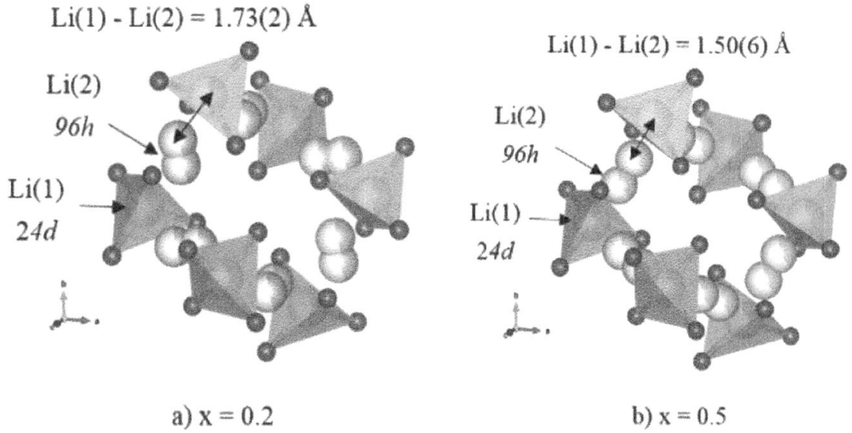

Fig. 6 The coordination polyhedral around Li sites for $Li_{7-x}La_3Zr_{2-x}Ta_xO_{12}$ with **a** $x = 0.2$ and **b** $x = 0.5$ samples [17]

4 Synthesis of Polycrystalline $Li_{6.5}La_3M_{1.5}Ta_{0.5}O_{12}$ (*M*: Zr, Hf, Sn) by Solid-State Reaction

Tantalum is one of the most effective dopants for the garnet-type oxide, and the optimum Ta-content is nearly 0.5 per unit formula, as previously demonstrated in Ref. [17]. Therefore, we focused on the Ta-content of 0.5 for garnet-type oxide having the cubic structure. It is well known that $Li_7La_3Sn_2O_{12}$ [18] and $Li_7La_3Hf_2O_{12}$ [19]

also belong to the tetragonal structure such as $Li_7La_3Zr_2O_{12}$. However, the optimized dopant for $Li_7La_3Sn_2O_{12}$ and $Li_7La_3Hf_2O_{12}$ have not been examined yet including the amount of substitution. It can be expected that the Li-ion conductivity of $Li_7La_3Sn_2O_{12}$ and $Li_7La_3Hf_2O_{12}$ is also improved by Ta substitution for Sn and/or Hf site. Furthermore, the chemical stability of garnet-related-type $Li_7La_3M_2O_{12}$ (*M*: Zr, Hf, Sn) to carbonate and hydroxide was recently reported by first-principle calculation using DFT calculation [20]. Kang et al. reported that $Li_7La_3Sn_2O_{12}$ showed higher chemical stability than the other two materials [20]. Therefore, $Li_7La_3Sn_2O_{12}$ could be expected for solid electrolyte as all-solid-state Li-ion batteries. However, Li-ion conductive properties and crystal structure parameters of $Li_7La_3Sn_2O_{12}$ which has a cubic system by cation substitution have not been clarified yet.

Recently, the garnet-type oxide $Li_{6.5}La_3Hf_{1.5}Ta_{0.5}O_{12}$ and $Li_{6.5}La_3Sn_{1.5}Ta_{0.5}O_{12}$ samples were prepared by conventional solid-state reaction method using HfO_2 and SnO_2 as starting materials according to the reported procedure as previously described in Ref. [21]. Figure 7 shows the powder X-ray diffraction patterns of $Li_{6.5}La_3Hf_{1.5}Ta_{0.5}O_{12}$ and $Li_{6.5}La_3Sn_{1.5}Ta_{0.5}O_{12}$. All diffraction peaks can be attributed to a single phase of cubic garnet-type structure having the space group of *Ia*-3*d*. Figure 8 shows the cross-sectional morphology of $Li_{6.5}La_3Hf_{1.5}Ta_{0.5}O_{12}$ and $Li_{6.5}La_3Sn_{1.5}Ta_{0.5}O_{12}$ after sintering at 1150 °C. These samples have a high relative density above 90%. The small and homogeneous grain size distributions are observed in the $Li_{6.5}La_3Hf_{1.5}Ta_{0.5}O_{12}$ sample. All of the Ta substituted samples exhibit a relatively high conductivity of ~10^{-4} S cm^{-1} at room temperature, and the activation energies of $Li_{6.5}La_3Hf_{1.5}Ta_{0.5}O_{12}$ and $Li_{6.5}La_3Sn_{1.5}Ta_{0.5}O_{12}$, which were determined from the Arrhenius plots, are $E_a = 0.40$ and 0.45 eV, respectively (Fig. 9). The value of $Li_{6.5}La_3Hf_{1.5}Ta_{0.5}O_{12}$ is smaller than that (0.43 eV) of $Li_{6.5}La_3Zr_{1.5}Ta_{0.5}O_{12}$.

The crystal structure was analyzed by the Rietveld method using powder X-ray diffraction data [21]. The unit cell volumes and octahedral (*M*, Ta)O_6 volumes for $Li_{6.5}La_3M_{1.5}Ta_{0.5}O_{12}$ (*M*: Zr, Hf, Sn) decrease together with decreasing ionic radii

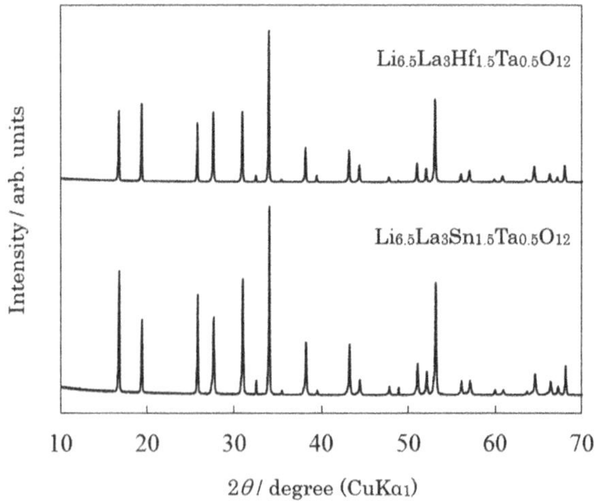

Fig. 7 Powder X-ray diffraction patterns of $Li_{6.5}La_3Hf_{1.5}Ta_{0.5}O_{12}$ and $Li_{6.5}La_3Sn_{1.5}Ta_{0.5}O_{12}$ [21]

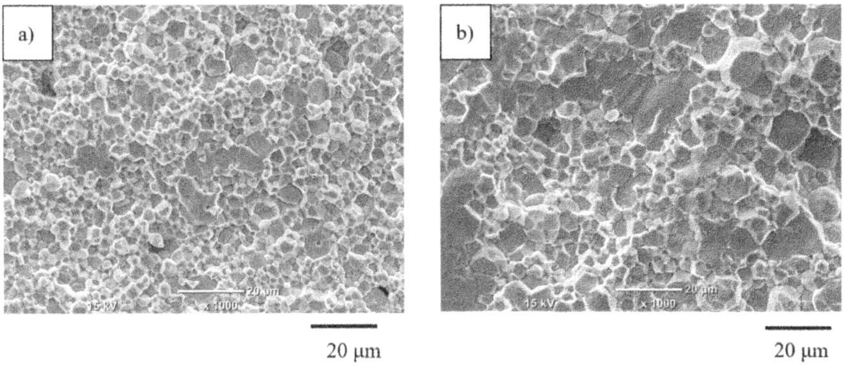

Fig. 8 SEM images of the cross-sectional morphology of **a** $Li_{6.5}La_3Hf_{1.5}Ta_{0.5}O_{12}$ and **b** $Li_{6.5}La_3Sn_{1.5}Ta_{0.5}O_{12}$ [21]

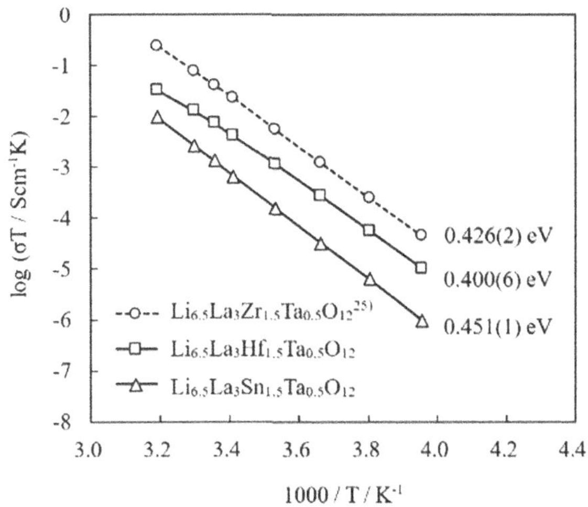

Fig. 9 Arrhenius plots of total conductivities of $Li_{6.5}La_3M_{1.5}Ta_{0.5}O_{12}$ (*M*: Zr, Hf, Sn) [21]

of M^{4+} cations (0.72 Å for Zr^{4+}, 0.71 Å for Hf^{4+}, and 0.69 Å for Sn^{4+}). However, the polyhedral volumes of $Li1O_4$ and Li_2O_6 in $Li_{6.5}La_3Hf_{1.5}Ta_{0.5}O_{12}$ are the largest among these $Li_{6.5}La_3M_{1.5}Ta_{0.5}O_{12}$ compounds. It has been suggested that the bottleneck for Li-ion diffusion from Li 24*d* site to Li 96*h* site is the face of shared edges of neighboring Li site in the garnet-type structure. Therefore, Li-ion conductivity is affected by the polyhedral volume of $Li1O_4$. Since $Li_{6.5}La_3Hf_{1.5}Ta_{0.5}O_{12}$ has a larger polyhedral volume of $Li1O_4$ and $Li2O_6$ in the crystal structure, the Li-ion conductive space is wider than $Li_{6.5}La_3Hf_{1.5}Ta_{0.5}O_{12}$, which diffuse from tetrahedral site to octahedral site. This fact may suggest the lowest activation energy observed in the $Li_{6.5}La_3Hf_{1.5}Ta_{0.5}O_{12}$ sample. From a viewpoint of the Li–O polyhedral volume in the unit cell, it is concluded that $Li_{6.5}La_3Hf_{1.5}Ta_{0.5}O_{12}$ has a most suitable Li-ion environment among the $Li_{6.5}La_3M_{1.5}Ta_{0.5}O_{12}$ (M: Zr, Hf, Sn) compounds from a viewpoint of Li conductivity [21].

5 Low-Temperature Synthesis of Tetragonal $Li_7La_3Zr_2O_{12}$ by Co-Precipitation Method

The particle size and morphology are important for the production of a thin film and sintering body for solid electrolyte in all-solid-state battery. For this reason, low-temperature synthetic techniques such as the sol-gel method and the co-precipitation method have been applied so as to produce fine particle samples for garnet-type materials [22–25]. In fact, the uniform and small particle sample of the garnet-type materials was reported by several low-temperature synthetic techniques. For example, Kokal et al. reported the synthesis of tetragonal $Li_7La_3Zr_2O_{12}$ powder prepared by the modified Pechini sol-gel method [22]. The grain size of their powder sample was reported to be ranged from 500 nm to 2 μm.

A similar feature was recently reported in garnet-related-type $Li_{6.75}La_3Zr_{1.75}Nb_{0.25}O_{12}$ prepared by the co-precipitation method [26]. The primary particle size prepared by the co-precipitation method was reported to be 50–100 nm for the $Li_{6.75}La_3Zr_{1.75}Nb_{0.25}O_{12}$ sample, which was much smaller than that by the solid-state synthetic method. Therefore, it is considered that the control of the particle morphology for the garnet materials was affected by the particle size of the precursor.

Recently, we synthesized a tetragonal garnet-type $Li_7La_3Zr_2O_{12}$ sample by the co-precipitation method using LiOH solution, in order to control the particle size and morphology of the $Li_7La_3Zr_2O_{12}$ powder for the first time [27]. Stoichiometric amounts of $LaCl_3 \cdot 7H_2O$ and $ZrOCl_2 \cdot 8H_2O$ were dissolved in distilled water. Then, the LiOH solution was added slowly into the mixed solution prepared above until the solution pH became 8–9. The precipitated precursor, which was composed of lanthanum hydroxide and zirconium hydroxide, was evaporated and washed with distilled water and was dried at 80 °C overnight. Next, in order to obtain the $Li_7La_3Zr_2O_{12}$ sample, stoichiometric $LiOH \cdot H_2O$ was added into the ground precursor. Finally, the mixture was calcined at 950 °C for 10 h in the air using an alumina crucible. The particle morphology of the obtained $Li_7La_3Zr_2O_{12}$ sample was elliptical in shape and the primary grain size was smaller than the conventional solid-state method, as shown in Fig. 10. The result of the particle size distribution measurement shows that the average secondary particle size was approximately 9.66 μm, even after calcination at 950 °C with a relatively narrow distribution [27].

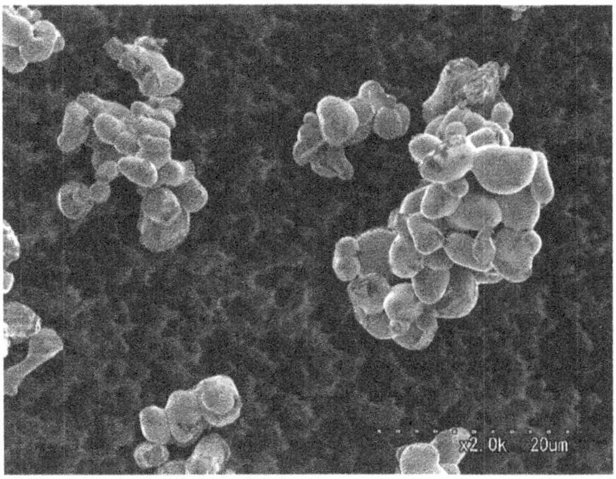

Fig. 10 FE-SEM images of the particle morphology of $Li_7La_3Zr_2O_{12}$ prepared by co-precipitation method [27]

6 Low-Temperature Synthesis of $Li_{6.5}La_3Zr_{1.5}Ta_{0.5}O_{12}$ Using Precursor Oxides

The garnet-type $Li_{6.5}La_3Zr_{1.5}Ta_{0.5}O_{12}$ samples having high Li-ion conductivity have been synthesized by conventional solid-state reaction. However, a high temperature above 1100 °C is needed to prepare a single-phase $Li_{6.5}La_3Zr_{1.5}Ta_{0.5}O_{12}$ sample by this synthetic method. From a viewpoint of the interface between solid electrolyte and electrode materials in all-solid-state battery, lowering the synthetic temperature is important so as to avoid the reaction between these materials at the interface. In the case of $Li_7La_3Zr_2O_{12}$, various low-temperature synthetic methods using sol-gel and/or co-precipitation processes [22–25] or some precursor materials such as pyrochlore-type $La_2Zr_2O_7$ [28, 29] have been developed to avoid Li loss during high-temperature treatment. However, $Li_{6.5}La_3Zr_{1.5}Ta_{0.5}O_{12}$ samples have not been synthesized yet by such low-temperature methods.

Recently, we successfully synthesized a garnet-type $Li_{6.5}La_3Zr_{1.5}Ta_{0.5}O_{12}$ polycrystalline sample at a relatively low temperature of 700 °C using pyrochlore-type $La_2Zr_2O_7$ and weberite-type La_3TaO_7 as precursor oxides [30]. In this synthetic procedure, the precursor materials of pyrochlore-type $La_2Zr_2O_7$ and weberite-type La_3TaO_7 were first synthesized via conventional solid-state reactions using La_2O_3, ZrO_2, and Ta_2O_5 as starting materials. Stoichiometric amounts of these oxides were ball-milled using zirconia balls in ethanol for 2 h. Then, the obtained mixtures were calcined between 1200 and 1300 °C for 12 h in the air using an intermediate grinding. Next, the stoichiometric amount of $LiOH{\cdot}H_2O$ and the obtained precursor materials were mixed according to the following Eq. (1):

$$\begin{aligned} &6.5\,LiOH \cdot H_2O + 0.75\,La_2Zr_2O_7 \\ &\quad + 0.5\,La_3TaO_7 \rightarrow Li_{6.5}La_3Zr_{1.5}Ta_{0.5}O_{12} + 9.75\,H_2O \end{aligned} \tag{1}$$

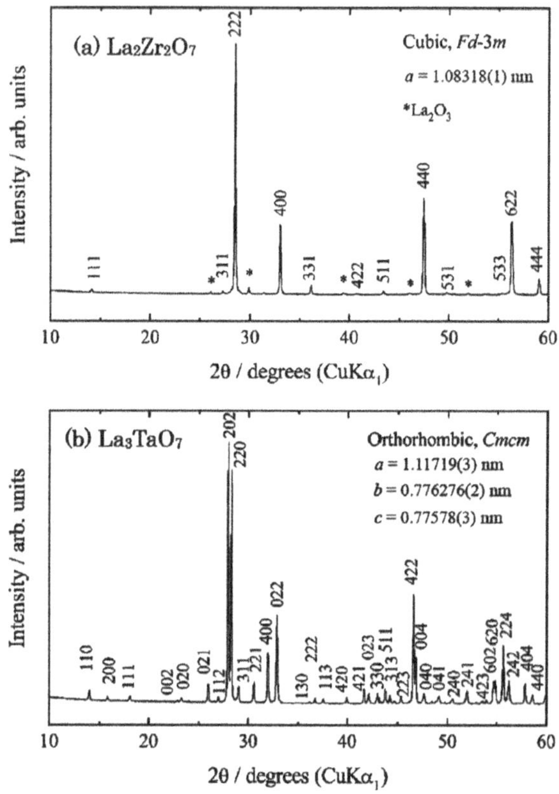

Fig. 11 Powder XRD patterns of **a** pyrochlore-type $La_2Zr_2O_7$ and **b** weberite-type La_3TaO_7 samples [31]

Then, the mixture was calcined at 700 °C for 12 h in the air using a high-purity alumina crucible.

Figure 11 shows the XRD patterns of pyrochlore-type $La_2Zr_2O_7$ and weberite-type La_3TaO_7 as precursor oxides, respectively. The main peaks of both the samples are well-indexed to the structural data of cubic pyrochlore-type $La_2Zr_2O_7$ (space group: *Fd*-3*m*) [31] and orthorhombic weberite-type La_3TaO_7 (space group: *Cmcm*) [32], respectively. Figure 12 shows the XRD pattern of the garnet-type $Li_{6.5}La_3Zr_{1.5}Ta_{0.5}O_{12}$ prepared using the present precursor oxides. All of the observed peaks can be indexed to the cubic garnet-type structure (space group: *Ia*-3*d*), and no impurity phase was observed. The lattice parameter was determined to be $a = 12.9577(1)$ Å by the least-squares method. Accordingly, the garnet-type $Li_{6.5}La_3Zr_{1.5}Ta_{0.5}O_{12}$ was successfully synthesized using $La_2Zr_2O_7$ and La_3TaO_7 as precursor materials at a relatively low temperature of 700 °C. The primary particle size of the present $Li_{6.5}La_3Zr_{1.5}Ta_{0.5}O_{12}$ sample was about 2 μm, as shown in Fig. 13, which is considerably smaller than that prepared by conventional solid-state reaction at higher temperatures. Thermo-gravimetry differential thermal analysis data confirmed that the reaction between $LiOH{\cdot}H_2O$ and precursor oxides occurred above 500 °C after the dehydration reaction of $LiOH{\cdot}H_2O$ [30].

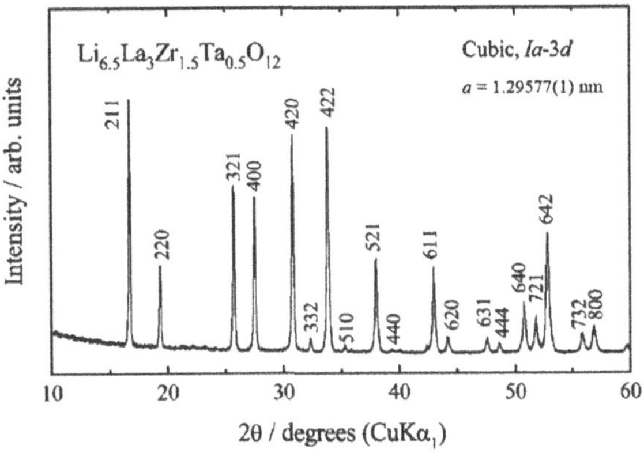

Fig. 12 Powder XRD pattern of garnet-type $Li_{6.5}La_3Zr_{1.5}Ta_{0.5}O_{12}$ sample prepared using the precursor oxides [31]

Fig. 13 SEM image of garnet-type $Li_{6.5}La_3Zr_{1.5}Ta_{0.5}O_{12}$ sample prepared using the precursor oxides [31]

7 Synthesis of al-Doped $Li_7La_3Zr_2O_{12}$ Single Crystals by a Flux Method

Single-crystal specimens have highly been desired to clarify the physical and electrochemical properties. Especially, the intrinsic bulk and/or anisotropic conductive properties can be measured using the single-crystal specimens, as we have previously demonstrated in electrical conductivity measurements using $LiCoO_2$ and Li_xCoO_2 single crystals [33–35] and in electrochemical measurements of the chemical diffusion coefficients of Li-ion in the spinel-type $LiMn_2O_4$ [36, 37]. The single-crystal samples of such lithium transition oxides have been synthesized by a flux method at relatively lower temperatures, because of the high volatility of lithium oxide components at high temperatures.

Single-crystal samples of garnet-type $Li_7La_3Zr_2O_{12}$ with the garnet-type structure were successfully synthesized by a self-flux method using some lithium salts at high temperatures [9, 11, 38]. The tetragonal form of $Li_7La_3Zr_2O_{12}$ crystals was prepared

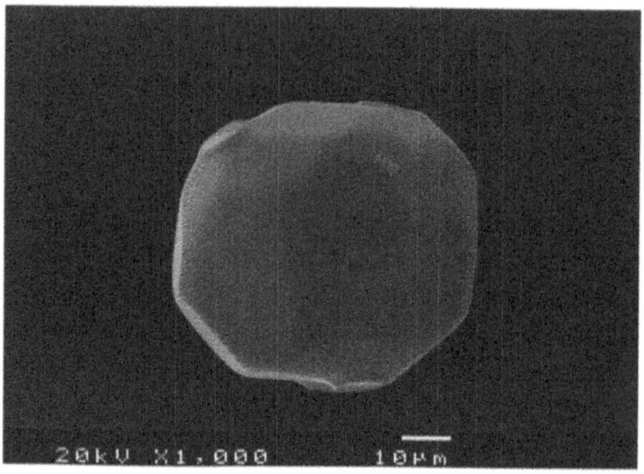

Fig. 14 SEM micrograph of a single crystal of tetragonal $Li_7La_3Zr_2O_{12}$ [9]

using Li_2CO_3 as a flux material at 1313 K in a gold crucible [9]. The crystals were separated from the solidified melts by immersing the gold crucible in hot water. Figure 14 shows the typical photograph of tetragonal $Li_7La_3Zr_2O_{12}$ crystals. The grown crystals are colorless and transparent; they also had a spherical shape with a maximum diameter of 0.05 mm in size.

The cubic $Li_7La_3Zr_2O_{12}$ was first synthesized by high-temperature heating at 1250 °C, and its crystal structure was determined using single-crystal X-ray diffraction data [11]. However, the obtained crystal size was 0.05 mm in diameter which was too small to evaluate the physical properties of the cubic $Li_7La_3Zr_2O_{12}$.

Single crystals of cubic Al-doped $Li_7La_3Zr_2O_{12}$ having larger size were synthesized by a self-flux method using $LiNO_3$ as a starting component at 1150 °C [38]. The starting materials of La_2O_3 and ZrO_2 were first well-mixed in a nominal molar ratio of La: Zr = 3: 2. These were further mixed by excess $LiNO_3$ as self-flux material. The powder mixture was charged into a high-purity alumina crucible, and the temperature was elevated to 1150 °C in air, and then held for 4 h. The obtained crystals were separated from the solidified flux materials by immersing the alumina crucible in hot water.

Figure 15 shows the obtained single crystals of Al-doped $Li_7La_3Zr_2O_{12}$ grown at 1150 °C using $LiNO_3$ as a self-flux material in a starting molar ratio of Li: La: Zr = 20: 3: 2. The crystals were colorless and transparent and had a polyhedral shape with a maximum dimension of $0.15 \times 0.10 \times 0.10$ mm^3. The well-formed crystal morphology with facets can be observed in the SEM photograph [38].

The lattice parameter was determined by the powder XRD data of the pulverized single crystals to be a = 13.00(3) Å. This value was slightly larger than those in the previous reports for Al-doped $Li_7La_3Zr_2O_{12}$ [7, 11]. This fact may suggest that the Li^+/H^+ ion exchange reaction has proceeded in the washing procedure using hot water. We should treat carefully the grown crystals from the solidified flux material in the case of the flux method.

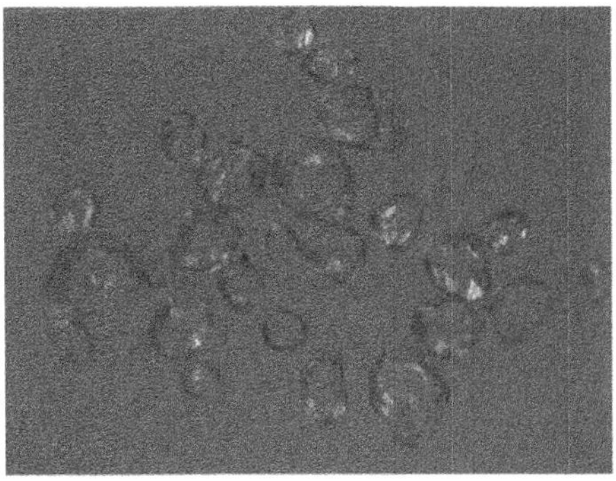

Fig. 15 Single crystals of cubic $Li_7La_3Zr_2O_{12}$ grown at 1150 °C using $LiNO_3$ as a self-flux material

8 Crystal Growth of $Li_{6.5}La_3Zr_{1.5}Ta_{0.5}O_{12}$ Single Crystals by Melt Growth Technique

The centimeter-sized large single crystals of garnet-type oxide materials were recently synthesized from the high-temperature melt. The melt growth technique is usually used for the industrial single-crystal growth such as semiconducting Si, $LiNbO_3$, and $LiTaO_3$. The first report for the lithium ion conducting garnet-type oxides from the melt was single-crystal growth of $Li_5La_3Ta_2O_{12}$ [39], while the experimental details for crystal growth were not reported. Recently, single crystals of $Li_6La_3ZrTaO_{12}$ were grown by the Czochralski (Cz) method, and NMR measurement using a single-crystal sample was reported [40]. Furthermore, single crystals of $Li_{7-x}La_3Zr_{2-x}Nb_xO_{12}$ and $Li_{7-x}La_3Zr_{2-x}Ta_xO_{12}$ were grown by using the floating zone (FZ) melting method [41–43]. From these reports, melt growth techniques such as Cz and FZ method are thought to be useful and suitable for crystal growth of the garnet-type lithium ion conducting oxides.

In the case of FZ method, the sample preparation was reported to be carried out in two steps [42]. The first step is the preparation of the raw materials by a conventional solid-state reaction. A mixture of Li_2CO_3, La_2O_3, ZrO_2, and Ta_2O_5 was heated at 1123 K in the air. In this case, 20% excess composition for the lithium-content was employed so as to prevent the compositional deviation due to the volatilization of lithium during the crystal growth. The sample was reground and isostatically pressed to form a cylindrical shape. The formed rods were subsequently sintered at 1423 K for 4 h in air. The second step is the crystal growth using an optical image furnace equipped with four halogen lamps.

Centimeter-sized single-crystal rods of $Li_{6.5}La_3Zr_{1.5}Ta_{0.5}O_{12}$ were grown by floating zone melting technique [42, 43]. The typical size of the single-crystal rod was 8 mm in diameter and 70 mm in length, as shown in Fig. 16. $Li_{6.5}La_3Zr_{1.5}Ta_{0.5}O_{12}$ crystallizes in a cubic structure with an *Ia*-3*d* space group and the lattice parameter $a = 12.9455$ (6) Å [42]. From the results of structure refinement using single-crystal neutron diffraction data [42], occupation sites for lithium atoms were determined to be two crystallographic sites: the distorted tetrahedral 96*h* site and the distorted

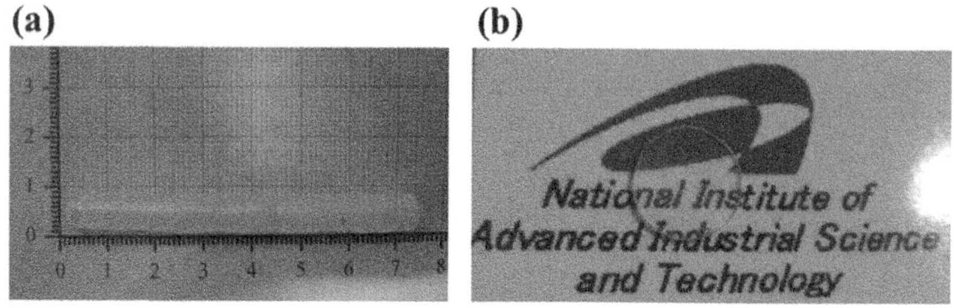

Fig. 16 **a** As-grown $Li_{6.5}La_3Zr_{1.5}Ta_{0.5}O_{12}$ single-crystal rod and **b** a polished $Li_{6.5}La_3Zr_{1.5}Ta_{0.5}O_{12}$ single-crystal plate [42]

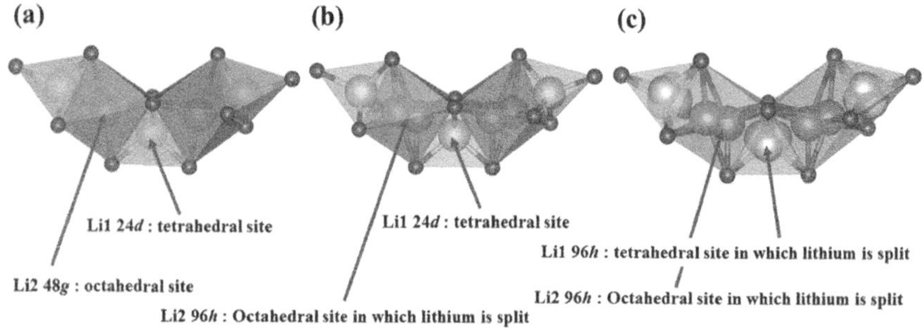

Fig. 17 Three kinds of Li arrangement models; **a** completely ordering model in tetragonal $Li_7La_3Zr_2O_{12}$, **b** partly disordering model as previously reported, and **c** completely disordering model in $Li_{6.5}La_3Zr_{1.5}Ta_{0.5}O_{12}$ determined using single-crystal neutron diffraction data [42]

octahedral 96*h* site (Fig. 17). The total Li-ion conductivity of $Li_{6.5}La_3Zr_{1.5}Ta_{0.5}O_{12}$ was determined to be 1.27×10^{-3} S cm^{-1} at 298 K by AC impedance measurements (Fig. 18). This value is the highest among the reported garnet-type materials, and it is concluded that the bulk nature of conductivity can be utilized as total conductivity in the case of the single-crystal solid electrolyte because there is no grain boundary in the single-crystal solid. Using NMR spectroscopy [42, 44–46], we determined the Li diffusion coefficient to be 1.57×10^{13} m^2 s^{-1} at 298 K and 7.96×10^{13} m^2 s^{-1} at 333 K [42]. These values related to Li-ion migration are higher than those reported for polycrystalline samples.

9 Concluding Remarks

In this chapter, the recent advancements of the garnet-type lithium ion conducting oxide materials including not only large-sized single-crystal growth but also small particle synthesized at lower temperatures has been reviewed. The intrinsic bulk nature of the conductive properties and the precise structural properties have been

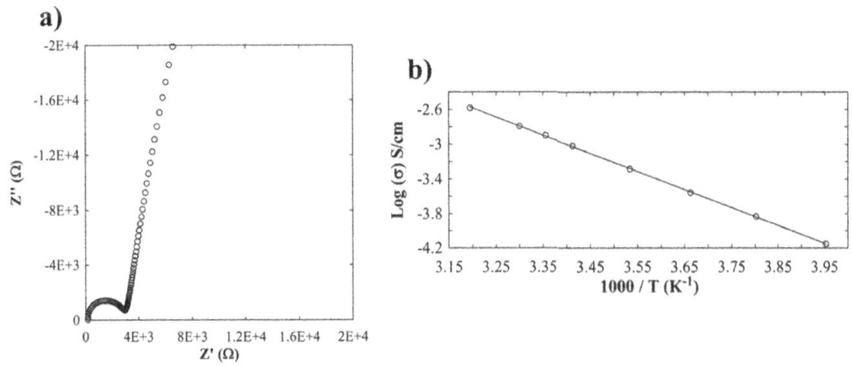

Fig. 18 **a** AC impedance Nyquist plot of the $Li_{6.5}La_3Zr_{1.5}Ta_{0.5}O_{12}$ single-crystal plate at 298 K. **b** Temperature dependence of the Li-ion conductivity for the $Li_{6.5}La_3Zr_{1.5}Ta_{0.5}O_{12}$ single-crystal plate [42]

revealed using the single-crystal samples for the first time. We believe the advantage of single-crystal solid electrolyte for the battery application from viewpoints of low resistance of separator and high critical current density against the lithium penetration phenomena. We are now trying to improve the lithium ion conductivity of the garnet materials by additional chemical modifications and optimization of the local Li arrangement in the loop structure constructing by the garnet-type framework structure.

Acknowledgements The authors thank Dr. Hiroshi Nagata and Dr. Norihito Kijima of AIST for the electrochemical measurements and valuable discussions. A part of this study was financially supported by the Advanced Low Carbon Technology Research and Development Program (ALCA-SPRING) from Japan Science and Technology Agency (JST).

References

1. Goodenough, J. B., & Kim, Y. (2010). *Chemistry of Materials, 22,* 587–603.
2. Armand, M., & Tarascon, J. M. (2008). *Nature, 451,* 652–657.
3. Janek, J., & Zeier, W. G. (2016). *Nature Energy, 1,* 16041.
4. Hong, H. Y. –P. (1978). *Materials Research Bulletin, 13,* 117–124.
5. Aono, H., Sugimoto, E., Sadaoka, Y., Imanaka, N., & Adachi, G. (1989). *Journal of the Electrochemical Society, 136,* 590–591.
6. Inaguma, Y., Liquan, C., Itoh, M., Nakamura, T., Uchida, T., Ikuta, H., et al. (1993). *Solid State Communications, 86,* 689–693.
7. Murugan, R., Thangadurai, V., & Weppner, W. (2007). *Angewandte Chemie International Edition, 46,* 7778–7781.
8. Rettenwander, D., Redhammer, G., Preishuber-Pflügl, F., Cheng, L., Miara, L., Wagner, R., et al. (2016). *Chemistry of Materials, 28,* 2384–2392.
9. Awaka, J., Kijima, N., Hayakawa, H., & Akimoto, J. (2009). *Journal of Solid State Chemistry, 182,* 2046–2052.

10. Matsui, M., Takahashi, K., Sakamoto, K., Hirano, A., Takeda, Y., Yamamoto, O., et al. (2014). *Dalton Transactions, 43,* 1019–1024.
11. Awaka, J., Takashima, A., Kataoka, K., Kijima, N., Idemoto, Y., & Akimoto, J. (2011). *Chemistry Letters, 40,* 60–62.
12. Li, Y., Wang, C., Xie, H., Cheng, J., & Goodenough, J. B. (2011). *Electrochemistry Communications, 13,* 1289–1292.
13. Allen, J. L., Wolfenstine, J., Rangasamy, E., & Sakamoto, J. (2012). *Journal of Power Sources, 206,* 315–319.
14. Buschmann, H., Berendts, S., Mogwitz, B., & Janek, J. (2012). *Journal of Power Sources, 206,* 236–244.
15. Thompson, T., Wolfenstine, J., Allen, J. L., Johannes, M., Huq, A., David, I. N., et al. (2014). *Journal of Materials Chemistry A, 2,* 13431–13436.
16. Wang, Y., & Lai, W. (2015). *Journal of Power Sources, 275,* 612–620.
17. Hamao, N., Kataoka, K., Kijima, N., & Akimoto, J. (2016). *Journal of the Ceramic Society of Japan, 124,* 678–683.
18. Percival, J., Kendrick, E., Smith, R. I., & Slater, P. R. (2009). *Dalton Transactions, 26,* 5177–5181.
19. Awaka, J., Kijima, N., Kataoka, K., Hayakawa, H., Ohshima, K., & Akimoto, J. (2010). *Journal of Solid State Chemistry, 183,* 180–185.
20. Kang, S. G., & Sholl, D. S. (2014). *Journal of Physical Chemistry C, 118,* 17402–17406.
21. Hamao, N., Kataoka, K., & Akimoto, J. (2017). *Journal of the Ceramic Society of Japan, 125,* 272–275.
22. Kokal, I., Somer, M., Notten, P. H. L., & Hintzen, H. T. (2011). *Solid State Ionics, 185,* 42–46.
23. Janani, N., Ramakumar, S., Dhivya, L., Deviannapoorani, C., Saranya, K., Murugan, R. (2011). *Ionics, 17,* 575–580.
24. Toda, S., Ishiguro, K., Shimonishi, A., Hirano, A., Takeda, Y., Yamamoto, O., et al. (2013). *Solid State Ionics, 233,* 102–106.
25. Takano, R., Tadanaga, K., Hayashi, A., & Tatsumisago, N. (2014). *Solid State Ionics, 255,* 104–107.
26. Imagawa, H., Ohta, S., Kihira, Y., & Asaoka, T. (2014). *Solid State Ionics, 262,* 609–612.
27. Hamao, N., & Akimoto, J. (2015). *Chemistry Letters, 44,* 970–972.
28. Deviannapoorani, C., Ramakumar, S., Janani, N., & Murugan, R. (2015). *Solid State Ionics, 283,* 123–130.
29. Kimura, T., Yamada, Y., Yamamoto, K., Matsuda, T., Nomura, H., & Hirayama, T. (2017). *Journal of the American Ceramic Society, 100,* 1313–1319.
30. Hamao, N., & Akimoto, J. (2019). *Journal of the Ceramic Society of Japan, 127,* 374–477.
31. Tabira, Y., Withers, R. L., Yamada, T., & Ishizawa, N. (2001). *Zeitschrift für Kristallographie, 216,* 92–98.
32. Wakeshima, M., Nishimine, H., & Hinatsu, Y. (2004). *Journal of Physics: Condensed Matter, 16,* 4103–4120.
33. Akimoto, J., Gotoh, Y., & Oosawa, Y. (1998). *Journal of Solid State Chemistry, 141,* 298–302.
34. Takahashi, Y., Gotoh, Y., Akimoto, J., Mizuta, S., Tokiwa, K., & Watanabe, T. (2002). *Journal of Solid State Chemistry, 164,* 1–4.
35. Takahashi, Y., Kijima, N., Dokko, K., Nishizawa, M., Uchida, I., & Akimoto, J. (2007). *Journal of Solid State Chemistry, 180,* 313–321.
36. Akimoto, J., Takahashi, Y., Gotoh, Y., & Mizuta, S. (2000). *Chemistry of Materials, 12,* 3246–3248.
37. Dokko, K., Nishizawa, M., Mohamedi, M., Umeda, M., Uchida, I., Akimoto, J., et al. (2001). *Electrochemical and Solid-State Letters, 4,* A151–A153.
38. Awaka, J., Takashima, A., Hayakawa, H., Kijima, N., Idemoto, Y., & Akimoto, J. (2011). *Key Engineering Materials, 485,* 99–102.
39. Hyooma, H., & Hayashi, K. (1988). *Materials Research Bulletin, 23,* 1399–1407.
40. Stanje, B., Rettenwander, D., Breuer, S., Uitz, M., Berendts, S., Lerch, M., et al. (2017). *Annalen der Physik, 529,* 1700140.

41. Kataoka, K., Nagata, H., & Akimoto, J. (2018). *Scientific Reports, 8,* 9965.
42. Kataoka, K., & Akimoto, J. (2018). *ChemElectroChem, 5,* 2551–2557.
43. Kataoka, K., & Akimoto, J. (2019). *Journal of the Ceramic Society of Japan, 127,* 521–526.
44. Dorai, A., Kuwata, N., Takekawa, R., Kawamura, J., Kataoka, K., & Akimoto, J. (2018). *Solid State Ionics, 327,* 18–26.
45. Hayamizu, K., Terada, Y., Kataoka, K., & Akimoto, J. (2019). *The Journal of Chemical Physics—AIP Publishing—Scitation, 150,* 194502.
46. Hayamizu, K., Terada, Y., Kataoka, K., Akimoto, J., & Haishi, T. (2019). *Physical Chemistry Chemical Physics: PCCP, 21,* 23589–23597.

Powder-Process-Based Fabrication of Oxide-Based Bulk-Type All-Solid-State Batteries

Toyoki Okumura

Abstract High-rate operation of oxide-based bulk-type all-solid-state batterie (ASSB) is achieved not only by the development of novel oxide electrolyte (OE) with high bulk ionic conductivity but also by the formable processes of ion-transfer-suited interfaces by powder sintering. Unlike for sulfides and hydrides, the ionic conduction of oxides is bottlenecked by that at grain-boundaries (GBs), that are at the electrolyte/electrolyte homo-interfaces. Moreover, desirable electrode/electrolyte heterointerfaces were difficult to be fabricated by powder sintering because most electrode materials easily engage in thermochemical reactions with the electrolyte. This chapter focuses on the interface formation techniques based on the usage of OE powder.

Keywords All-solid-state Li-ion battery · Solid-state electrolyte · Li-ion conductor · Liquid-phase sintering · Spark-plasma sintering

1 Introduction

The non-flammability, durability, and non-toxicity of oxides make them promising ionic conductors for safe high-capacity ASSBs, the successful realization of which requires the development of OEs with high bulk ionic conductivities close to those of organic liquid electrolytes ($\sigma > 10^{-2}$ S cm^{-1}). Although several OEs (perovskite-type and garnet-type oxides) with conductivities of up to 10^{-3} S cm^{-1} [1] have been reported to date, higher conductivities are needed for ASSB operation at current rates above 1 C, as exemplified by the successful high-rate operation of an ASSB with a 10^{-2} S cm^{-1} sulfide-based electrolyte [2]. Although such high-rate performance is difficult to realize for OEs, they can be used to examine the practical operation of corresponding ASSBs at low rates. As high-ionic-conductivity OEs do not warrant the enhanced rate performance of the corresponding ASSBs, these batteries face the problem of (dis)charge difficulty at room temperature even when operated at

T. Okumura (✉)
Research Institute of Electrochemical Energy, National Institute of Advanced Industrial Science and Technology (AIST), Midorigaoka 1-8-31, Ikeda, Osaka 563-8577, Japan
e-mail: toyoki-okumura@aist.go.jp

K. Kanamura (ed.), *Next Generation Batteries*,
https://doi.org/10.1007/978-981-33-6668-8_20

low rates, which is attributed to the difficulty of interface formation between oxide particles during powder sintering.

The desired interfaces can be produced by vapor deposition, during which active material (AM) and OE source vapor phases are deposited to afford thin oxide films with an atomic-scale smooth plating [3]. This technique has contributed to the commercialization of thin-film (thickness ≈ several μm) batteries with capacities of < 1 mAh used to power small electric devices such as those intended for medical use. On the other hand, bulk-type ASSBs with thicker (tens to hundreds of μm) electrode layers are needed to address the growing demand for higher-capacity batteries with capacities of > 10 mAh [Fig. 1a] used to power next-generation internet-of-things devices. Such thick electrode layers can be prepared by powder sintering, which, however, does not allow one to easily form ion-transfer-suited GBs between OEs [Fig. 1b] and AM-OE heterointerfaces [Fig. 1c]. In this chapter, we introduce methods of obtaining ion-transfer-suitable interfaces in bulk-type ASSBs by powder processes, namely the preparation of (a) OE/OE interfaces with low GB resistance [Fig. 1b, Sect. 2], and (b) impurity-free AM/OE interfaces by powder sintering [Fig. 1c, Sect. 3]. The performances of the thus prepared ASSBs are presented and discussed in detail.

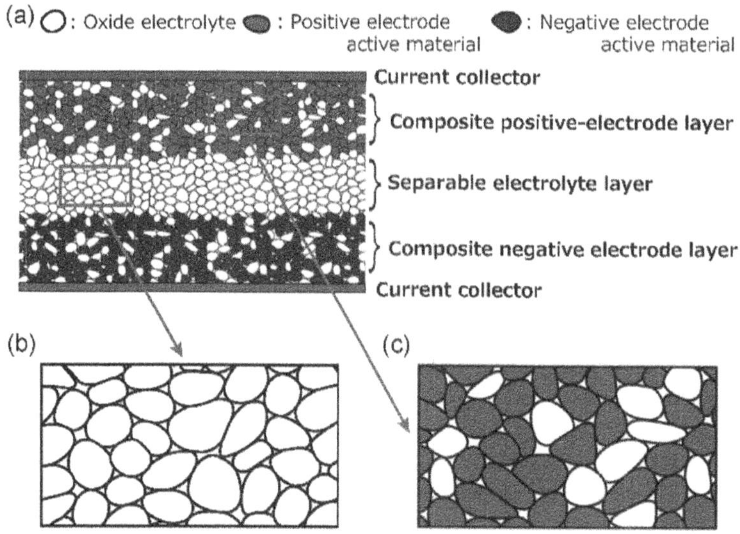

Fig. 1 **a** Schematic cross-sectional view of an OE-based bulk-type ASSB, **b** zoomed-in view of the OE/OE interface commonly denoted as GB, **c** zoomed-in view of the oxide AM/OE interface

2 Fabrication of Low-GB-Resistance Interfaces in OE Compacts by Powder Sintering

The history of inorganic ionic conductor development shows that GB resistance strongly contributes to the total conductivity of OEs and thus reduces the rate performance of corresponding bulk-type ASSBs [1]. AC impedance measurements are commonly used to probe the ionic conductivity of OE sintered compacts, and the obtained complex impedance plots typically feature two semicircular segments (one at high and one at low frequency) as well as a low-frequency spike [4]. The high-frequency semicircle represents bulk ionic resistance (R_{bulk}), while the low-frequency one represents GB ionic resistance (R_{GB}), and the low-frequency spike corresponds to the capacitance between the two blocking electrodes, as presented in Fig. 2a. Bode plots of the imaginary parts of impedance (Z'') [Fig. 2b] and modulus (M'') [Fig. 2c] are also used to determine ionic resistance and (the reciprocal of) capacitance, respectively. AC impedance analysis of γ-Li_3PO_4-type $Li_{2+2x}Zn_{1-x}GeO_4$ (Li superionic conductor, LISICON) showed that R_{GB} varied from sample to sample at the same composition, lying between 0.5 R_{bulk} and 8 R_{bulk} [4]. Moreover, some garnet-type OE sintered compacts exhibit sufficiently low R_{GB} [5], although its origin is still not clear. Thus, R_{GB} observed after powder sintering is difficult to understand, and, hence, to control, which makes it hard to achieve R_{GB} repeatability and minimization in OE sintered compacts for enhancing the rate performance of bulk-type ASSBs. Herein, we examined the effect of sintering conditions on the R_{GB} of OEs, especially of γ-Li_3PO_4-type $Li_{3.5}Ge_{0.75}S_{0.25}O_4$, to understand how GB resistance in OE-based bulk-type ASSBs prepared by powder sintering can be minimized [6].

2.1 *Effect of Sintering Temperature/Time on GB Resistance*

Figure 3 shows the Bode plots for $Li_{3.5}Ge_{0.75}S_{0.25}O_4$ compacts (relative density >90%) sintered at various temperatures and/or different times in a flow of O_2 [6].

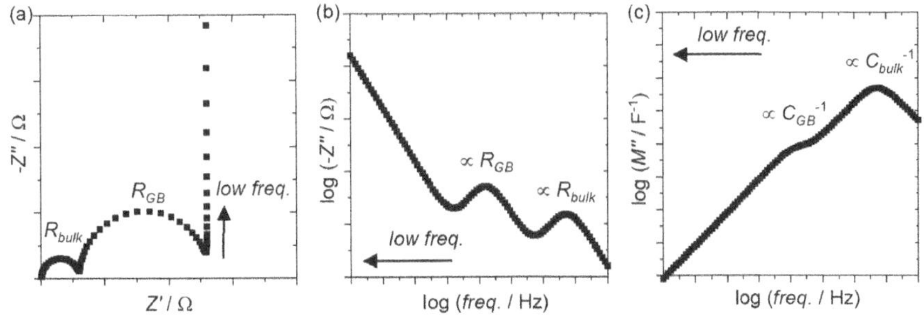

Fig. 2 **a** Typical Nyquist plots as well as Bode plots of **b** Z'' and (c) M'' for OE sintered compacts sandwiched between blocking electrodes

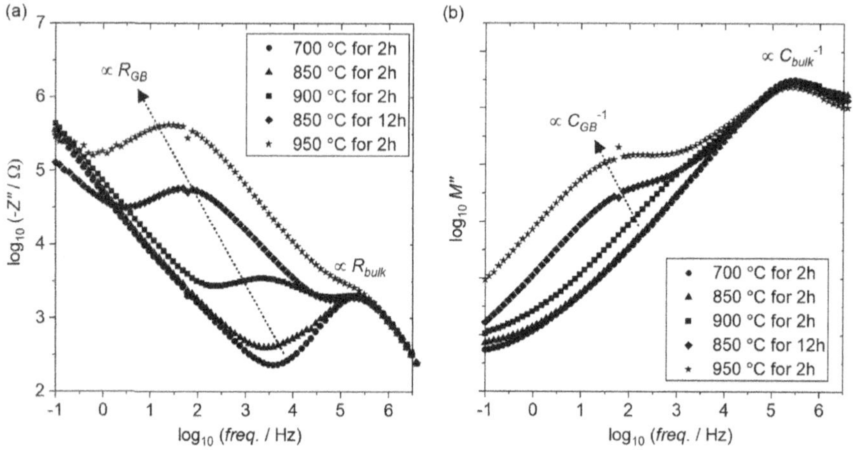

Fig. 3 Bode plots of **a** Z" and **b** M" obtained for Au/$Li_{3.5}Ge_{0.75}S_{0.25}O_4$ sintered compact/Au samples operated at 25 °C. Reproduced with permission [6]. Copyright 2018, American Chemical Society

Crests at ~0.5 MHz observed for both Z" and M" were ascribed to contributions of the bulk, R_{bulk} and C_{bulk}^{-1}, which overlapped under various sintering conditions, i.e., were constant in the chosen sintering temperature/time region. Note that the low relative density observed for sintering at a lower temperature of 600 °C resulted in an aggravation of apparent R_{bulk}. Figure 3a and b also shows additional crests at low frequencies (1 kHz–10 Hz) for compacts sintered above 900 °C and for longer than 12 h, which reflected GB contributions, R_{GB} and C_{GB}^{-1}. These parameters drastically increased after sintering above 900 °C and for a prolonged time, whereas no significant change was observed after sintering at 700–850 °C for 2 h. To explain this behavior, we examined the microstructures of sintered $Li_{3.5}Ge_{0.75}S_{0.25}O_4$ compacts, as shown in Fig. 4 [6]. Although the size of $Li_{3.5}Ge_{0.75}S_{0.25}O_4$ grains significantly increased when the sintering temperature was raised from 700 to 850 °C, the change of R_{GB} was insignificant, i.e., the GB area decrease caused by grain growth did not reduce R_{GB} (and also C_{GB}^{-1}) in this LISICON-type electrolyte. This behavior was in stark contrast to that previously reported for (Li, La)TiO_3, a perovskite-type electrolyte, in which case grain growth induced by high-temperature sintering effectively reduced R_{GB} [7]. The above results demonstrate that no universal method of GB resistance control is applicable to all OEs, which poses the question of why sintering at elevated temperature and/or for prolonged time affects the aggravation of R_{GB} and the increase of C_{GB}^{-1} in LISICON-type electrolytes? Energy-dispersive X-ray spectroscopy (EDX) revealed the localization of S between grains in the compact sintered at an elevated temperature of 950 °C, as shown in Fig. 4d and e [6], i.e., high-temperature sintering resulted in residual Li–S–O phase accumulation after liquid-phase (LP) sintering, in agreement with the biphasic separation of γ-Li_3PO_4-type $Li_{3.5}Ge_{0.75}S_{0.25}O_4$ into [γ-Li_3PO_4-type phase + Li–S–O LP] observed at high temperature for the Li_4GeO_4–Li_2SO_4 pseudo-binary system [8]. The residual phase growth at GBs significantly

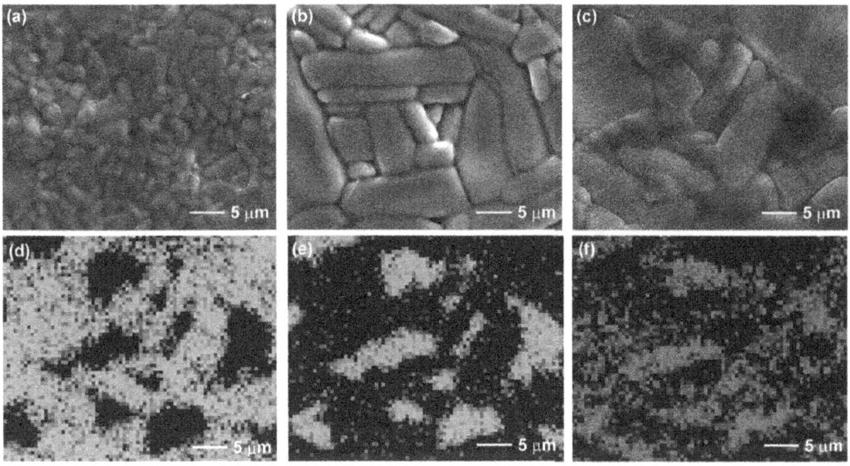

Fig. 4 SEM images of the surfaces of $Li_{3.5}Ge_{0.75}S_{0.25}O_4$ compacts sintered at **a** 700 °C for 2 h, **b** 850 °C for 2 h, and **c** 950 °C for 2 h. EDX mappings of **d** Ge, **e** S, and **f** O for the sample in (c). Reproduced with permission [6]. Copyright 2018, American Chemical Society

contributed to the increase of C_{GB}^{-1} and the aggravation of R_{GB}. Notably, high densification is often required for the production of solid-state devices, as well-connected solid/solid interfaces (e.g., those produced by LP sintering) are likely to enhance the physical properties of ceramics. In the case of OEs, the thickness of the GB residual phase after LP sintering should be carefully controlled to obtain sintered compacts with low R_{GB} and assemble bulk-type ASSBs with high-rate performance [6].

2.2 *Effect of Sintering Atmosphere on GB Resistance*

Oxides are usually described as air-stable electrolytes, which is an advantage for ASSB manufacturing. However, perovskite-type (Li, La)TiO_3 was reported to be an air-sensitive material with Li-ion transport properties strongly influenced by the content/localization of the ion conduction-blocking Li_2CO_3 secondary phase formed during sintering in the presence of moisture [9]. A similar sensitivity to air was also reported for the surface of garnet-type $Li_7La_3Zr_2O_{12}$ [10]. Herein, we describe the effect of sintering atmosphere on the R_{GB} of LISICON-type electrolytes [6]. Figure 5a shows the Bode plots of Z" for $Li_{3.5}Ge_{0.75}S_{0.25}O_4$ compacts sintered in a flow of O_2 and in the air [6], revealing that the compact sintered in the air featured a higher crest at low frequency, which confirmed that R_{GB} is sensitive to the sintering atmosphere. Figure 5b and c shows depth-dependent C 1 s X-ray photoelectron spectra of compacts sintered under different atmospheres. The strong peak A at 290 eV corresponded to the photoemission of CO_3^{2-} [6], and peak B at 284 eV was ascribed to the photoemission of absorbed hydrocarbons in the XPS vacuum

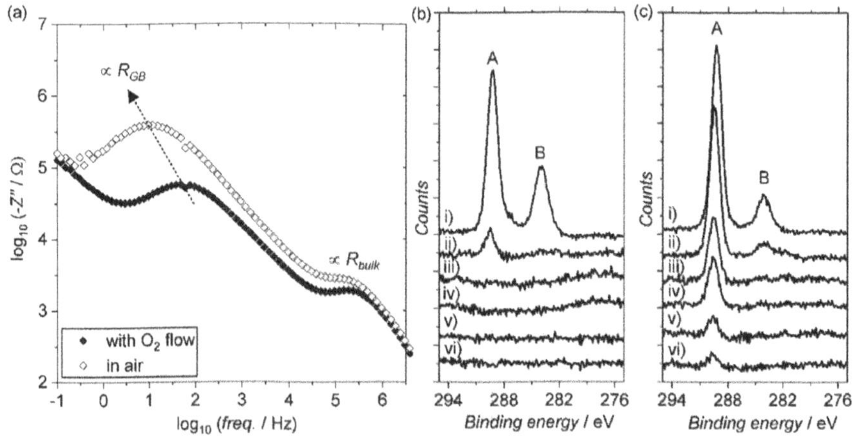

Fig. 5 **a** Bode plots of Z'' for Au/$Li_{3.5}Ge_{0.75}S_{0.25}O_4$ sintered compact/Au samples operated at 25 °C. The compacts were sintered at 850 °C for 12 h in a flow of O_2 (filled diamonds) or in the air (empty diamonds). Depth-dependent C 1 s spectra of $Li_{3.5}Ge_{0.75}S_{0.25}O_4$ powder compacts sintered at 850 °C for 12 h in (**b**) a flow of O_2 and (**c**) air. Etching times of (i) 0.1, (ii) 1, (iii) 4, (iv) 6, (v) 10, and (vi) 30 min were used. Reproduced with permission [6]. Copyright 2018, American Chemical Society

chamber. In the case of the compact sintered in a flow of O_2, the signal of CO_3^{2-} disappeared upon etching, which indicated that no Li_2CO_3 was formed on GBs inside the compact. On the other hand, the formation of Li_2CO_3 extended deep into the compact sintered in air. Thus, sintering in air resulted in increased R_{GB}, and sintering atmosphere control was concluded to be important for suppressing the formation of the Li_2CO_3 secondary phase. Unfortunately, OE surfaces are often sensitive to air at close-to-sintering temperatures, and therefore, the use of air-sintered OEs should be avoided for assembling bulk-type ASS-LIBs in such cases [6].

2.3 *Use of Spark-Plasma Sintering (SPS) to Minimize GB Resistance*

SPS relies on the application of microscopic electrical discharge between ceramic (or metal) powders under pressure, with the employed setup shown in Fig. 6a. The thus obtained samples exhibit a microstructure [Fig. 6c] markedly different from that observed after conventional furnace sintering [Fig. 6b]. As sintering is achieved at a relatively low temperature and in a short time, SPS not only affords dense pellets but also suppresses grain growth and is therefore suited for the production of high-strength ferroelectric ceramics with favorable GB properties [11]. Figure 7 shows the Nyquist plots of OE compacts prepared by SPS and conventional furnace sintering

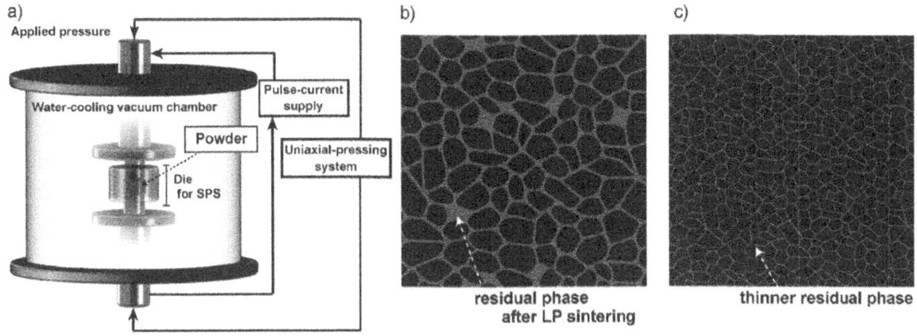

Fig. 6 Schematic illustrations of (**a**) SPS equipment and microstructures observed for OE powder compacts sintered (**b**) in a conventional furnace in O_2 flow and (**c**) by SPS

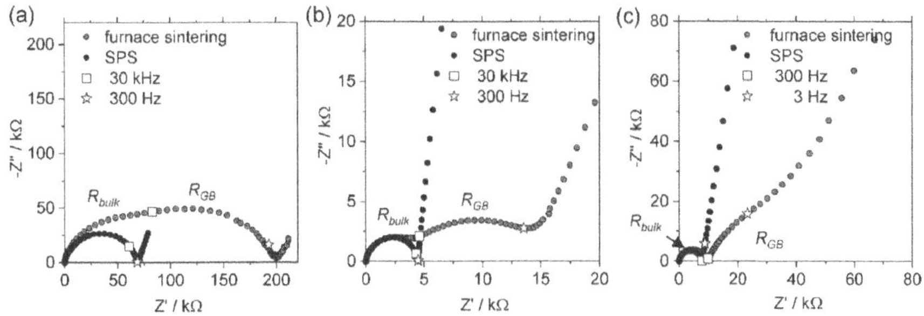

Fig. 7 Nyquist plots of Au/oxide electrolyte sintered compact/Au samples obtained at 25 °C in an Ar-filled cell for **a** $Li_{2.2}C_{0.8}B_{0.2}O_3$, **b** $Li_{3.5}Ge_{0.75}S_{0.25}O_4$, and **c** $Li_{3.75}Ge_{0.75}P_{0.25}O_4$. Reproduced with permission [6]. Copyright 2018, American Chemical Society

[6], revealing that the former featured decreased GB resistances. The microstructural analysis demonstrated that GB resistance minimization resulted in a decrease of residual phase thickness because of the rapid nature of the SPS process, and the expansion of GB area due to grain growth suppression was shown to insignificantly contribute to GB resistance. Thus, as a rapid interface-connecting process, SPS was concluded to be suitable for the production of GB resistance-free electrolyte compacts, and, hence, for the assembly of bulk-type ASSBs with high-rate performance [6].

3 AM/OE Interface Fabrication by Powder Co-Sintering

Composite electrode layers prepared from AM/OE mixtures are commonly used to assemble bulk-type ASSBs with increased electrochemical interface area [Fig. 1c]. Powder sintering allows one to obtain AM–OE connections sufficient for interface formation and prevents the formation of undesirable interfacial impurities during

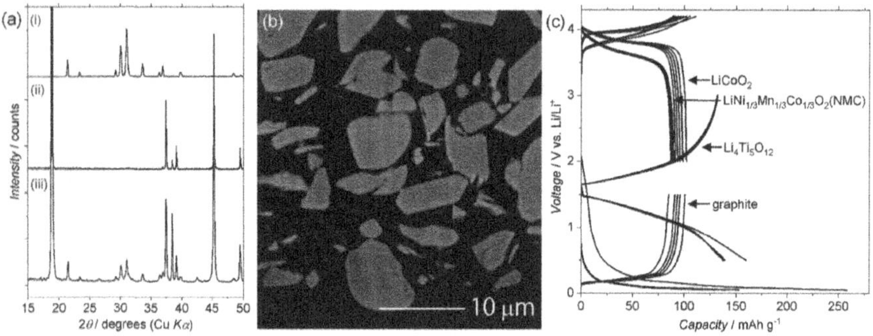

Fig. 8 **a** Powder XRD patterns of (i) the $Li_{2.2}C_{0.8}B_{0.2}O_3$ electrolyte, (ii) the $LiCoO_2$ active material, and (**c**) the $LiCoO_2$–$Li_{2.2}C_{0.8}B_{0.2}O_3$ composite electrode after SPS-based assembly at 450 °C for 5 min. **b** Cross-sectional SEM image of the polished $LiCoO_2$ (light gray particles)–$Li_{2.2}C_{0.8}B_{0.2}O_3$ (dark gray area) composite electrode after SPS-based assembly at 450 °C for 5 min. **c** Electrochemical charge-discharge profiles of ASSBs (AM–$Li_{2.2}C_{0.8}B_{0.2}O_3$ composite electrode|$Li_{2.2}C_{0.8}B_{0.2}O_3$ electrolyte|dry-polymer electrolyte|Li metal) recorded for five cycles at a constant current density of 64 μA cm^{-2} and a temperature of 60 °C. [11]

sintering. However, the high sintering temperature (>1000 °C) required for the densification of OEs such as garnets, perovskites, and NASICONs complicates the fabrication of ASSBs, as most AMs decompose at such temperatures [12]. Ohta et al. used Li_3BO_3 as a composite OE component and successfully operated the corresponding ASSB (Au|$LiCoO_2$–Li_3BO_3 composite electrode|garnet-electrolyte separator|Li metal) [13]. In this case, the low melting point of the Li_3BO_3 electrolyte promoted the formation of a favorable interface between $LiCoO_2$ and Li_3BO_3 [13]. We also assembled an ASSB with $Li_{2.2}C_{0.8}B_{0.2}O_3$, which has the highest conductivity of 2.1×10^{-6} S cm^{-1} at 30 °C in the Li_2CO_3-Li_3BO_3 pseudo-binary system [11, 14]. Figure 8a shows the X-ray diffraction (XRD) pattern and the microstructure of a $LiCoO_2$–$Li_{2.2}C_{0.8}B_{0.2}O_3$ composite electrode prepared by 450-°C SPS [11], revealing that no peaks other than those of $Li_{2.2}C_{0.8}B_{0.2}O_3$ and $LiCoO_2$ were observed. Furthermore, a well-defined interface between $LiCoO_2$ (light gray) and $Li_{2.2}C_{0.8}B_{0.2}O_3$ (dark gray) without any voids was observed at the $LiCoO_2$–$Li_{2.2}C_{0.8}B_{0.2}O_3$ composite electrolyte layer [Fig. 8b]. This means that a well-defined interface between $LiCoO_2$ and $Li_{2.2}C_{0.8}B_{0.2}O_3$ without traces of impurities was prepared by powder sintering, as the melting point of $Li_{2.2}C_{0.8}B_{0.2}O_3$ was low enough to preclude its reaction with $LiCoO_2$ [11]. The low reactivity of $Li_{2.2}C_{0.8}B_{0.2}O_3$ during co-sintering also affected combinations with other AMs. The use of $Li_{2.2}C_{0.8}B_{0.2}O_3$ as an ASSB OE allowed the operation of typical charge-discharge profiles of various well-known AMs [Fig. 8c]. Moreover, differences between the initial cycling performances of various AMs were observed. The large electrochemical window of OEs was expected to inhibit electrochemical decomposition reactions and thus result in good cyclability of OE-based ASSBs. However, gradual capacity degradation was observed when $LiCoO_2$ or graphite was used as ASSB AMs. The microstructure of a cycled ASSB based on $LiCoO_2$ featured significant fracturing at interfaces between

$LiCoO_2$ and $Li_{2.2}C_{0.8}B_{0.2}O_3$ [14], indicating that repetitive AM grain expansion-contraction during cycling can result in degradation. Actually, the ASSB based on $Li_4Ti_5O_{12}$ exhibited less pronounced capacity degradation than that based on $LiCoO_2$, because $Li_4Ti_5O_{12}$ behaves as a zero-strain AM during charge-discharge. Although $LiNi_{1/3}Mn_{1/3}Co_{1/3}O_2$ (NMC) is not a zero-strain AM, the cyclability of an NMC-based ASSB was superior to that of the ASSB with $LiCoO_2$. Note that as both AMs had similar particle sizes of ~10 μm, similar AM grain expansion-contraction was expected to occur in NMC grains during charge-discharge. Herein, we focused on particle morphology differences between $LiCoO_2$ and NMC, demonstrating that the former particles were probably single grains, while particles of the latter probably comprised aggregates of single grains. Although the particle morphologies of ceramics depend on synthesis conditions, those of generally commercialized $LiCoO_2$ and NMC exhibit similar tendencies. Therefore, expansion-contraction in each NMC grain has a small effect on the total expansion-contraction of a given NMC particle. Although the effect of expansion-contraction on the durability of solid-state electrochemical devices has been well studied [15], degradation is especially serious in bulk-type ASSB containing brittle OEs. Thus, the development of AM particles with less strain and/or specific morphology is important for bulk-type ASSB design.

As mentioned above, low-temperature-sinterable $Li_{2.2}C_{0.8}B_{0.2}O_3$ is well suited for the assembly of bulk-type ASSBs and their practical performance enhancement. However, the low conductivity of Li_2CO_3-Li_3BO_3 (~10^{-6} S cm^{-1}) complicates the high-rate operation of these batteries. As described above, rate performance can be enhanced by the combined use of high-ionic-conductivity OEs (e.g., garnet-type ones) and heterointerface-suitable OEs (e.g., Li_3BO_3) at the right places in bulk-type ASSBs [13]. Nonetheless, the conductivity of the latter OEs should be further enhanced to achieve suitable high-rate performance. Among the various OEs, LISICON-based ones co-sintered with layered oxides such as $LiCoO_2$ are suited for this purpose [Fig. 9a] [16]. The ionic conductivities of germanium-based LISICONs exceed 10^{-5} S cm^{-1}, allowing for superior rate performance [Fig. 9b]. Therefore, further search for highly conductive and co-sinterable OEs is important for enhancing the rate performance of bulk-type ASSBs.

4 Conclusions

In this chapter, we introduced interface formation techniques based on the usage of OE powder to assemble bulk-type ASSBs, namely a) control of OE sintering behavior to reduce GB resistance, and b) selection of suitable OEs to prevent interfacial side reactions during AM/OE co-sintering, focusing on interface formability over OE conductivity. The presented data imply that the combined use of highly conductive electrolytes, as presented in the previous chapter, is required to realize bulk-type ASSBs with good high-rate capability. The combination of highly conductive electrolyte development and interfacial solutions presented in this chapter is expected to aid the design of next-generation oxide-based ASSBs.

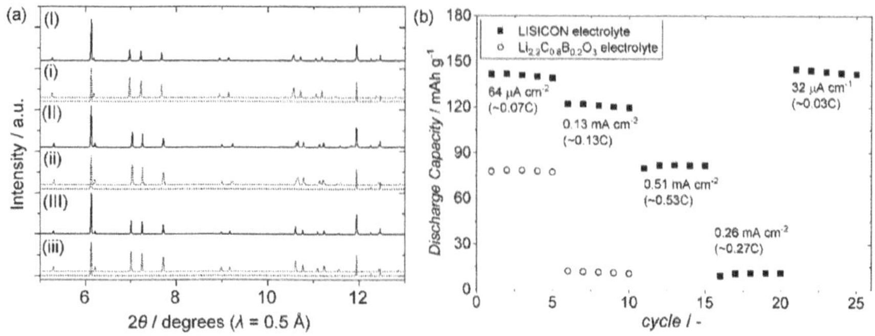

Fig. 9 **a** Synchrotron powder XRD patterns of $LiCoO_2$–LISICON composite electrodes after co-sintering: (I) $LiCoO_2$–$Li_{3.5}Ge_{0.5}V_{0.5}O_4$, (II) $LiCoO_2$–$Li_{3.75}Ge_{0.75}P_{0.25}O_4$, and (III) $LiCoO_2$–$Li_{3.75}Ge_{0.75}S_{0.25}O_4$. Original patterns of (i) $LiCoO_2$ (black dashed line) and $Li_{3.5}Ge_{0.5}V_{0.5}O_4$ (gray dashed line), (ii) $LiCoO_2$ (black dashed line) and $Li_{3.75}Ge_{0.75}P_{0.25}O_4$ (gray dashed line), and (iii) $LiCoO_2$ (black dashed line) and $Li_{3.75}Ge_{0.75}S_{0.25}O_4$ (gray dashed line) are also shown. **b** Discharge capacity retentions of ASSBs at different current densities/60 °C (NMC-OE composite electrode|OE|dry-polymer electrolyte|Li metal) for $Li_{3.5}Ge_{0.5}V_{0.5}O_4$ (filled squares) and $Li_{2.2}C_{0.8}B_{0.2}O_3$ (empty circles) as OEs

References

1. Takada, K. (2018). *Journal of Power Sources, 394,* 74.
2. Kato, Y., Hori, S., Saito, T., Suzuki, K., Hirayama, M., Matsui, A., et al. (2016). *Nature Energy, 1,* 1.
3. Bates, J. B., Dudney, N. J., Neudecker, B., Ueda, A., & Evans, C. D. (2000). *Solid State Ionics, 135,* 33.
4. Bruce, P. G., & West, A. R. (1983). *Journal of the Electrochemical Society, 130,* 662.
5. for example, Thangadurai, V., Weppner, W. (2005). *Journal of the American Ceramic Society, 88,* 411.
6. Okumura, T., Taminato, S., Takeuchi, T., Kobayashi, H., & Appl, A. C. S. (2018). *Energy Materials, 1,* 11.
7. Inaguma, Y., & Nakashima, M. (2013). *Journal of Power Sources, 228,* 250.
8. Dissanayake, M. A. K. L., Gunawardane, R. P., West, A. R., Senadeera, G. K. R., Bandaranayake, P. W. S. K., & Careem, M. A. (1993). *Solid State Ionics, 62,* 217.
9. Aguesse, F., López del Amo, J., Roddatis, V., Aguadero, A., & Kilner, J. A. (2014). *Advanced Materials Interfaces, 1,* 1300143.
10. Cheng, L., Crumlin, E. J., Chen, W., Qiao, R., Hou, H., Lux, S. F., et al. (2014). *Physical Chemistry Chemical Physics: PCCP, 16,* 18294.
11. Okumura, T., Takeuchi, T., & Kobayashi, H. (2017). *Journal of the Ceramic Society of Japan, 125,* 276.
12. He, Y., Lu, C., Liu, S., Zheng, W., & Luo, J. (2019). *Advanced Materials Interfaces, 9,* 1901810.
13. Ohta, S., Komagata, S., Seki, J., Saeki, T., Morishita, S., & Asaoka, T. (2013). *Journal of Power Sources, 238,* 53.
14. Okumura, T., Takeuchi, T., & Kobayashi, H. (2016). *Solid State Ionics, 288,* 248.
15. Julien, C. (1990). *Materials Science and Engineering B, 6,* 9.
16. Taminato, S., Okumura, T., Takeuchi, T., & Kobayashi, H. (2018). *Solid State Ionics, 326,* 52.

Lithium Chloroboracite $Li_4B_4M_3O_{12}Cl$ (M = Al, Ga): Glass-Ceramic Synthesis and Application to Solid-State Rechargeable Lithium Batteries

Koichi Kajihara, Naoto Tezuka, Yuta Okawa, Mayu Saito, Mao Shoji, Jungo Wakasugi, Hirokazu Munakata, and Kiyoshi Kanamura

Abstract The nearly single-phase and almost fully-crystallized lithium-ion-conducting glass-ceramics of cubic lithium chloroboracites, $Li_{4+x}B_7O_{12+x/2}Cl$ ($x = 0$–1), were found to be formed by melt-quench-crystallization method using the Li_2O–B_2O_3–LiCl ternary system. The extension of this method to the Li_2O–B_2O_3–M_2O_3–LiCl (M = Al, Ga) quaternary system enabled to form glass-ceramics containing a new cubic lithium chloroboracite, $Li_4B_4M_3O_{12}Cl$ (M = Al, Ga), as the main phase. They are the first example of boracites with substituted boron sites and formed by fully replacing the tetrahedral BO_4 units in the parent compound, $Li_4B_7O_{12}Cl$, with AlO_4 or GaO_4 units. The conductivity of $Li_4B_4M_3O_{12}Cl$-based glass-ceramics at room temperature was $\sim 1 \times 10^{-5}$ S cm^{-1}, an order of magnitude larger than the conductivity of $Li_4B_7O_{12}Cl$-based glass-ceramics. The $Li_4B_4Al_3O_{12}Cl$-based glass-ceramics exhibited a Li^+ ion transport number of ~1, wide electrochemical window between 0 and 6 V versus Li/Li^+, and stability in contact with Li metal. A solid-state rechargeable lithium cell using the $Li_4B_4Al_3O_{12}Cl$-based glass-ceramic solid electrolyte worked at 60 °C, using $LiCoO_2$-based composite cathode containing a small amount of ionic liquid electrolyte and Li metal anode.

Keywords Lithium boracite · $Li_4B_4M_3O_{12}Cl$ (M = Al, Ga) · Glass-ceramics · Solid electrolyte · Rechargeable lithium battery

1 Introduction

Boracite is a mineral with the formula $Mg_3B_7O_{13}Cl$, but it also refers to a group of crystalline compounds with the general formula $A_3B_7O_{13}X$, containing the divalent cations of A and monovalent anions of X [1–4]. Although the room temperature phases of boracites were often distorted, they commonly transform into cubic phases at high temperatures [2–4]. There also exists a lithium variant of cubic boracite,

K. Kajihara (✉) · N. Tezuka · Y. Okawa · M. Saito · M. Shoji · J. Wakasugi · H. Munakata · K. Kanamura
Department of Applied Chemistry for Environment, Graduate School of Urban Environmental Sciences, Tokyo Metropolitan University, 1-1 Minami-Osawa, Hachioji, Tokyo 192-0397, Japan
e-mail: kkaji@tmu.ac.jp

K. Kanamura (ed.), *Next Generation Batteries*,
https://doi.org/10.1007/978-981-33-6668-8_21

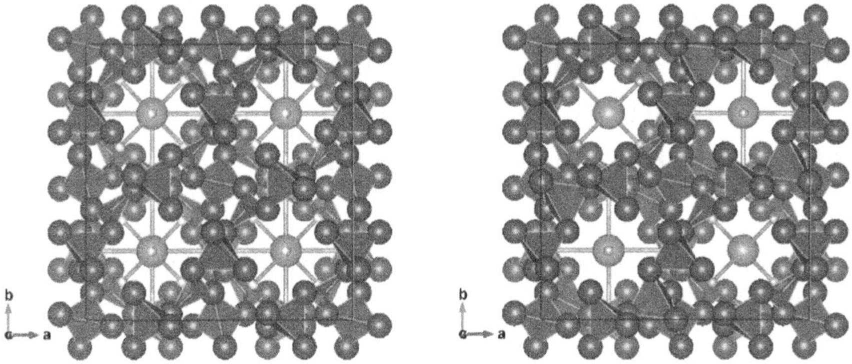

Fig. 1 Unit cell structures of $Li_4B_7O_{12}Cl$ ($x = 0$, space group $F\overline{4}3c$) (left) and $Li_5B_7O_{12.5}Cl$ ($x = 1$, space group $F23$) (right) drawn with VESTA [17]. Red, large pale-green, and small pale-green spheres denote O, Cl, and Li atoms, respectively, and BO_3 and BO_4 units are drawn with green triangles and tetrahedra, respectively. Among eight subcells in the unit cell, only four front-side ones are shown for clarity. Reprinted with permission from Ref. [16]. Copyright 2017 The Ceramic Society of Japan

$Li_{4+x}B_7O_{12+x/2}X$ [5–16]. Figure 1 shows the unit cell structures of the end members $x = 0$ ($Li_4B_7O_{12}Cl$, space group $F\overline{4}3c$) and 1 ($Li_5B_7O_{12.5}Cl$, space group $F23$). The B–O framework of $Li_{4+x}B_7O_{12+x/2}X$ consists of corner-shared triangular BO_3 and tetrahedral BO_4 units. Each cubic octant of the unit cell accommodates an X^- ion coordinated by 10 Li sites in $Li_4B_7O_{12}Cl$ and 4 or 6 Li sites in $Li_5B_7O_{12.5}Cl$. These compounds are lithium-ion conducting and the conductivity is the highest at $x = 0$ and $X = Cl$ (~10^{-7}–10^{-6} S cm^{-1} at room temperature [8–11, 16]). The conductivity of $Li_5B_7O_{12.5}Cl$ is much lower than that of $Li_4B_7O_{12}Cl$ because of lower symmetry and resultant ordering of Li. Lithium chloroboracites have been synthesized by heat treatment in a sealed glass tube [5, 7, 11], hydrothermal methods [6, 10, 13], and sol–gel methods [14]. However, these compounds have attracted less attention recently probably because of the low conductivity and synthetic difficulties. In addition, although boracites exhibit a wide variety of element combinations, the replacement of B in the B–O framework with other cations has not been reported.

2 Synthesis of Glass-Ceramics [16]

The compositions of $Li_{4+x}B_7O_{12+x/2}Cl$ ($x = 0$–1) are located in the glass-forming region of the Li_2O–B_2O_3–LiCl ternary system [18]. This fact prompted us to synthesize the glass-ceramics of this compound through the crystallization of precursor glasses. Powders of Li_2CO_3, B_2O_3, and LiCl were mixed in the Li_2CO_3:B_2O_3:LiCl molar ratio of x_1:7:2x_2. The mixture was melted in a platinum crucible covered with an alumina crucible for 30 min at 1000 °C. The melt was poured onto a stainless steel plate held at room temperature and pressed with another stainless steel plate.

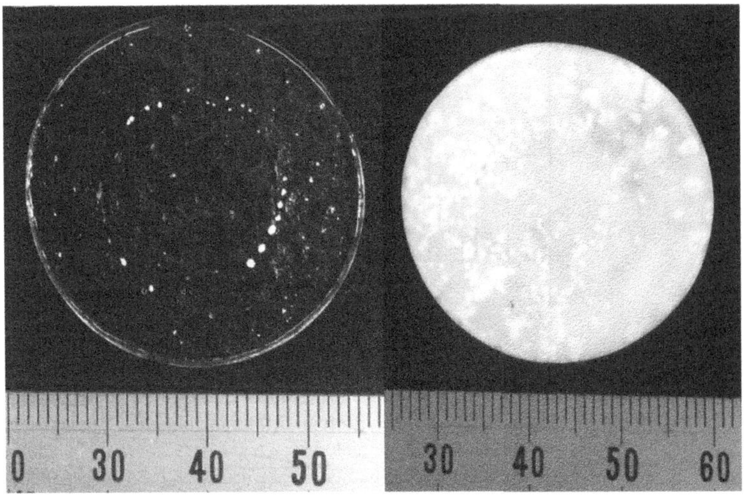

Fig. 2 Photographs of a precursor glass prepared at $(x_1, x_2) = (3.0, 1.4)$ (left) and glass-ceramic derived from the precursor glass by heat treatment for 3 h at 600 °C (right). Reprinted with permission from Ref. [16]. Copyright 2017 The Ceramic Society of Japan

The precursor glasses were then placed in the furnace preheated between 500 and 800 °C and crystallized.

Figure 2 shows a precursor glass prepared at $(x_1, x_2) = (3.0, 1.4)$ and its glass-ceramic obtained by heat treatment for 3 h at 600 °C. The precursor glass was amorphous. The heat treatment maintained the sample's shape whereas altered its appearance from transparent to opaque. Upon heating $Li_2B_4O_7$ was initially formed at ~500 °C and subsequent heating at or above ~600 °C generated $Li_4B_7O_{12}Cl$ from $Li_2B_4O_7$ and residual amorphous phase. The formation of lithium-rich phase $Li_5B_7O_{12.5}Cl$ ($x = 1$) can be detected by the presence of the 111 reflection, which is forbidden in $Li_4B_7O_{12}Cl$, and was insignificant at $x_1 + x_2 \lesssim 4.5$.

Figure 3 shows the result of Rietveld refinement using RIETAN-FP [19] for the powder XRD pattern of a glass-ceramic prepared at $(x_1, x_2) = (2.8, 1.6)$. To enhance the decomposition of $Li_2B_4O_7$, the amount of LiCl (x_2) was increased while maintaining $x_1 + x_2$ at 4.4. The degree of crystallinity and weight fraction of $Li_4B_7O_{12}Cl$ were ~0.96 and ~0.98, respectively. The ac conductivity of this sample was ~4.6×10^{-4} S cm^{-1} at 200 °C and the activation energy was ~0.52 eV.

3 Cubic Lithium Chloroboracite with Substituted Boron Sites, $Li_4B_4M_3O_{12}Cl$ (M = Al, Ga) [20]

Four-coordinated trivalent cations, Al^{3+} (ionic radius: 0.53 Å [21]) and Ga^{3+} (0.61 Å [21]) ions, were selected as the dopants to replace B^{3+} ions in the BO_4 units (0.25 Å [21]) in $Li_4B_7O_{12}Cl$. The melt-quenching-crystallization method developed in Sect. 2

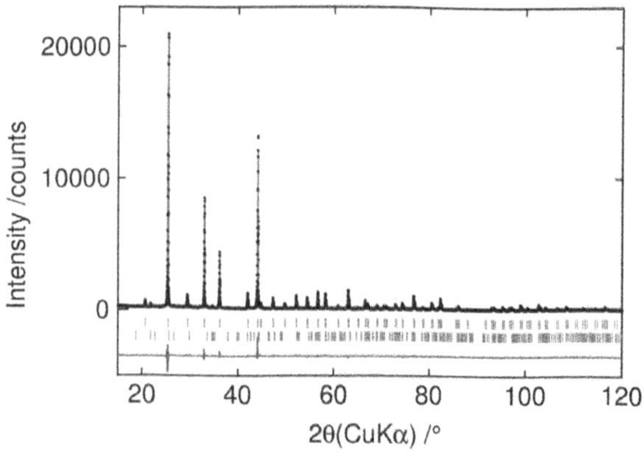

Fig. 3 Observed (dots), calculated (line), and difference (bottom line) XRD patterns of glass-ceramic prepared at $(x_1, x_2) = (2.8, 1.6)$ and heat treated for 3 h at 600 °C. Vertical bars indicate the positions of allowed Bragg reflections of contributing phases, $Li_4B_7O_{12}Cl$ (upper) and $Li_2B_4O_7$ (lower). Reprinted with permission from Ref. [16]. Copyright 2017 The Ceramic Society of Japan

was therefore extended to Li_2O–B_2O_3–M_2O_3–LiCl quaternary system. Powders of Li_2CO_3, B_2O_3, γ-Al_2O_3 or β-Ga_2O_3, and LiCl were mixed in the Li_2CO_3:B_2O_3:M_2O_3 (M = Al, Ga):LiCl molar ratio of 3:7(1−y):7y:2.8 or 3.2. They were melted for 30 or 60 min at 1000–1300 °C, quenched, and crystallized for 3 h at 600 °C.

Figure 4 shows the powder XRD patterns of glass-ceramics. In all samples crystallinity was high. The 222 reflection of the undoped ($y = 0$) sample was observed at ~25.4°. In Al- or Ga-doped samples two additional peaks were seen at lower diffraction angles. The left peak was located at $2\theta \simeq 23.8$ and ~23.4° for the Al- and Ga-doped samples, respectively, and their positions were nearly independent of y. These peaks were most intense at $y = 0.43$ (= 3/7). Rietveld analysis confirmed that the left peak originates from $Li_4B_4M_3O_{12}Cl$ (space group $F\bar{4}3c$), formed by the complete replacement of B in BO_4 units in $Li_4B_7O_{12}Cl$ with Al or Ga. The weight fractions of $Li_4B_4M_3O_{12}Cl$ were ~0.87 and ~0.76 for the Al- and Ga-doped samples, respectively. The lattice parameters of $Li_4B_4Al_3O_{12}Cl$ and $Li_4B_4Ga_3O_{12}Cl$ were ~12.9 and ~13.1 Å, respectively, and they are boracites with the currently largest known unit cells. The middle peak gradually shifted to lower angles with an increase in y and was attributed to $Li_4B_4(B,M)_3O_{12}Cl$, the solid solution between $Li_4B_7O_{12}Cl$ and $Li_4B_4M_3O_{12}Cl$. The solid solution peak became weak at $y \gtrsim 0.2$, indicating its decomposition into $Li_4B_7O_{12}Cl$ and $Li_4B_4M_3O_{12}Cl$, possibly because of strain originating from the coexistence of BO_4 and MO_4 units in the (B, M)–O framework. These observations indicate that $Li_4B_4M_3O_{12}Cl$ is thermodynamically more stable than the solid solution. It is also noteworthy that $Li_4B_4M_3O_{12}Cl$ directly precipitated from the precursor glass below 500 °C, at much lower temperature than the formation temperature of $Li_4B_7O_{12}Cl$ (~600 °C).

The Al- and Ga-doped samples exhibited the highest conductivity at $y = 0.43$ and 0.30, respectively. Figure 5 shows Arrhenius plots of ac conductivity of these samples measured with Au electrodes. The activation energy of conductivity was ~0.36 and

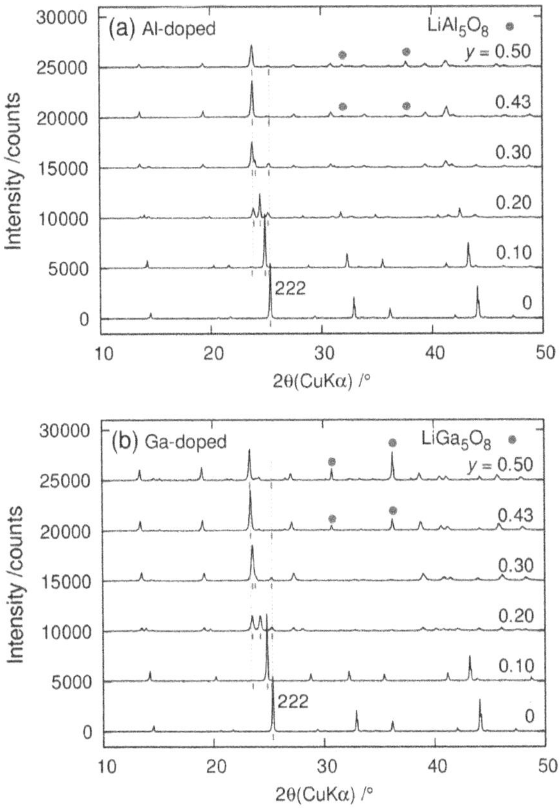

Fig. 4 Powder XRD patterns of Al-doped (**a**) and Ga-doped (**b**) glass-ceramics. Vertical bars indicate the positions of the 222 reflections of cubic boracite phases in each sample. Dotted lines indicate the positions of the 222 reflections in samples prepared at $y = 0$ (undoped sample) and 0.43. Reprinted with permission from Ref. [20]. Copyright 2017 The Chemical Society of Japan

~0.38 eV for the Al- and Ga-doped samples, respectively. Their conductivities at room temperature were $\sim 1\times10^{-5}$ S cm^{-1} and were an order of magnitude larger than that of the undoped ($y = 0$) sample [16]. The conductivities of the Al- or Ga-doped samples were even higher than that of single crystals of $Li_4B_7O_{12}Cl$ and $Li_4B_7O_{12}Cl_{0.68}Br_{0.32}$ [10] below ~150 °C.

Figure 6 shows Nyquist plots and dc polarization profile (inset) measured with Li–Au electrodes at 30 °C for an Al-doped sample prepared at $y = 0.43$. The semicircles observed at high- and low-frequency regions were attributed to the bulk and charge-transfer resistances, respectively. The latter component was significantly reduced by the alloying of Li and Au at 120 °C [22] prior to the dc polarization measurement. The dc polarization measurement was carried out with a constant current of 2 μA and polarity change every 30 min. The dc resistance calculated using Ohm's law was $\sim 3.7\times10^4$ Ω. It was equal to the total resistance evaluated from the low-frequency intercept of the Nyquist plot at the Z' axis after the dc polarization measurement

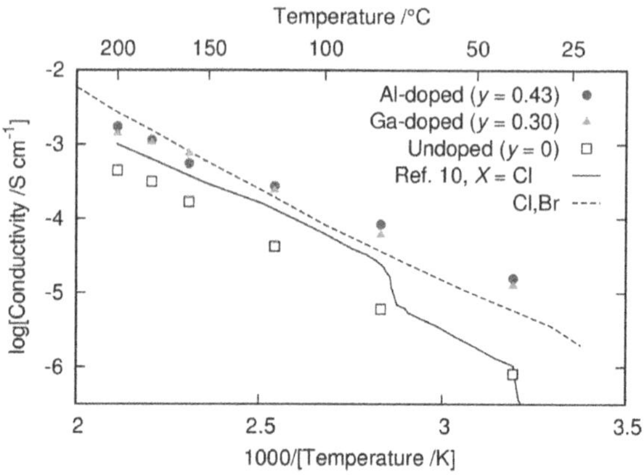

Fig. 5 Arrhenius plots of the ac conductivities of Al-doped ($y = 0.43$), Ga-doped ($y = 0.30$), and undoped ($y = 0$) glass-ceramics. Data reported for single crystals of $Li_4B_7O_{12}Cl$ (solid line) and $Li_4B_7O_{12}Cl_{0.68}Br_{0.32}$ (dashed line) [10] are also shown. After ref. [20]

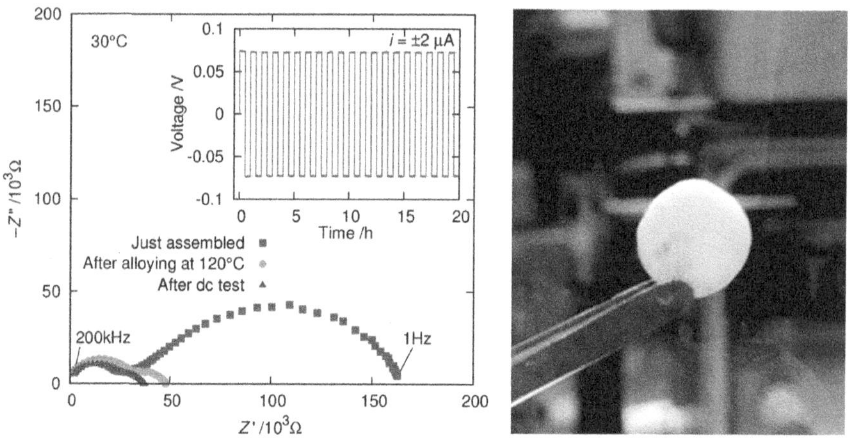

Fig. 6 (Left) Nyquist plots and dc polarization profile (inset) recorded with Li–Au electrodes at 30 °C for an Al-doped glass-ceramic prepared at $y = 0.43$. (Right) Appearance of an Al-doped glass-ceramic prepared at $y = 0.43$ after electrochemical measurements with Li metal electrodes. Reprinted with permission from Ref. [20]. Copyright 2017 The Chemical Society of Japan

(~3.7×10^4 Ω). The absence of transient components in the dc polarization profile and the equal dc and ac resistances indicate that the transport number of Li^+ ions is ~1. Figure 6 also shows that the appearance of the Al-doped sample was unchanged after measurements with Li electrodes. In the cyclic voltammetry performed with Au working and Li–Au alloy counter/reference electrodes, the Al-doped sample exhibited a wide electrochemical window up to 6 V versus Li/Li^+. In contrast, the Ga-doped sample was unstable and dark spots appeared after measurements.

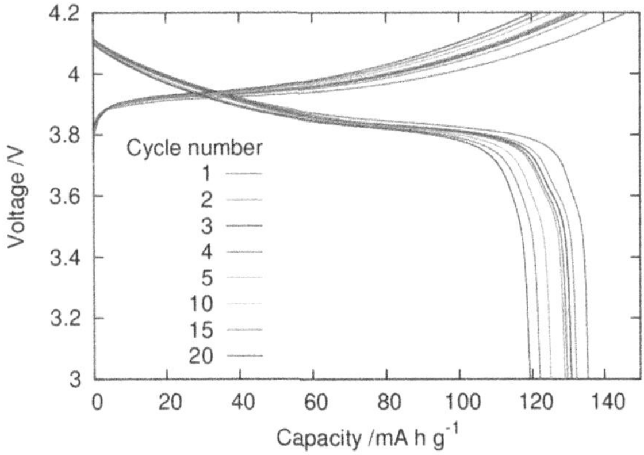

Fig. 7 Charge/discharge curves of a solid-state cell at 60 °C fabricated using an Al-doped glass-ceramic prepared at $y = 0.43$

4 Solid-State Rechargeable Lithium Battery [23]

A solid-state lithium battery was fabricated in a 2032-type coin cell using the $Li_4B_4Al_3O_{12}Cl$-based glass-ceramic solid electrolyte (thickness: ~1 mm), $LiCoO_2$-based composite cathode containing a small amount of ionic liquid electrolyte (1 mol dm^{-3} Li[TFSA]/[emim][TFSA] [24]), and Li–Au alloy anode. Figure 7 shows charge/discharge curves recorded at 60 °C with a current density of 10 μA cm^{-2}. The discharge capacity of the first cycle was comparable to the practical capacity of $LiCoO_2$ (140 mAh g^{-1}). Capacity retention after 20 cycles were ~88%, and average Coulombic efficiency between 5 and 20 cycles was ~99.3%. These results confirmed the compatibility of $Li_4B_4Al_3O_{12}Cl$-based oxychloride glass-ceramic solid electrolytes with solid-state rechargeable lithium batteries.

Acknowledgement This work was supported by ALCA-SPRING project of Japan Science and Technology Agency (JST) and the Murata Science Foundation.

References

1. Schmid, H. (1965). *Journal of Physics and Chemistry of Solids, 26*, 973.
2. Nelmes, R. J. (1974). *Journal of Physics C: Solid State Physics, 7*, 3840.
3. Knorr, K., Peters, L., Winkler, B., Milman, V., & Castellanos-Guzman, A. G. (2007). *Journal of Physics: Condensed Matter, 19*, 275207.
4. Iliev, M. N., & Schmid, H. (2014). *Journal of Raman Spectroscopy, 45*, 267.
5. Levasseur, A., Fouassier, C., & Hagenmuller, P. (1971). *Materials Research Bulletin, 6*, 15.
6. Jeitschko, W., & Bither, T. A. (1972). *Zeitschrift für Naturforschung B: Chemical Science, 27*, 1423.
7. Levasseur, M. A., Lloyd, D. J., Fouassier, C., & Hagenmuller, P. (1973). *Journal of Solid State Chemistry, 8*, 318.
8. Réau, J.-M., Levasseur, A., Magniez, G., & Calès, B. (1976). *Materials Research Bulletin, 11*, 1087.

9. Shannon, R. D., Taylor, B. E., English, A. D., & Berzins, T. (1977). *Electrochimica Acta, 22,* 783.
10. Jeitschko, W., Bither, T. A., & Bierstedt, P. E. (1977). *Acta Crystallographica. Section B, Structural Science, 33,* 2767.
11. Calès, B., Levasseur, A., Fouassier, C., Réau, J. M., & Hagenmuller, P. (1977). *Solid State Communications, 24,* 323.
12. Vlasse, M., Levasseur, A., & Hagenmuller, P. (1981). *Solid State Ionics, 2,* 33.
13. Byrappa, K., & Shekar, K. V. K. (1993). *Journal of Materials Research, 8,* 864.
14. Nagase, T., Sakane, K., & Wada, H. (1998). *Journal of Sol-Gel Science and Technology, 13,* 223.
15. Sorokin, N. I. (2015). *Physics of Solid State, 57,* 314.
16. Tezuka, N., Okawa, Y., Kajihara, K., & Kanamura, K. (2017). *Journal of the Ceramic Society of Japan, 125,* 348.
17. Momma, K., & Izumi, F. (2011). *Journal of Applied Crystallography, 44,* 1272.
18. Soppe, W., Aldenkamp, F., & Hartog, H. W. (1987). *Journal of Non-Crystalline Solids, 91,* 351.
19. Izumi, F., & Momma, K. (2007). *Solid State Phenomena, 130,* 15.
20. Kajihara, K., Tezuka, N., Shoji, M., Wakasugi, J., Munakata, H., & Kanamura, K. (2017). *Bulletin of the Chemical Society of Japan, 90,* 1279.
21. Shannon, R. D. (1976). *Acta Crystallographica, 32,* 751.
22. Wakasugi, J., Munakata, H., & Kanamura, K. (2017). *Journal of the Electrochemical Society, 164,* A1022.
23. Saito, M., Shoji, M., Kizuki, Y., Munakata, H., Kanamura, K., & Kajihara, K, *Unpublished*.
24. Shoji, M., Munakata, H., & Kanamura, K. (2018). *Proceedings of the 31th Fall Meeting of the Ceramic Society of Japan* 1V23.

Operando Analysis of All-Solid-State Lithium Ion Batteries by Using Synchrotron X-ray

Koji Amezawa and Yuta Kimura

Abstract In all-solid-state lithium ion batteries (ASSLIBs), smooth and homogeneous reactions in the electrodes are required to achieve sufficient performance. In this chapter, a novel *operando* analytical technique using synchrotron X-ray, which enables to directly observe reaction distributions in battery composite electrodes of ASSLIBs, is introduced. Computed tomography X-ray absorption near edge structure (CT-XANES) was developed to three-dimensionally visualize the reaction distributions in composite electrodes. The technique was applied to investigate the reaction in the composite cathode of the model ASSLIB, $LiCoO_2$ (LCO)-$Li_{2.2}C_{0.8}B_{0.2}O_3$ (LCBO)|LCBO|poly (ethylene oxide) (PEO)-based polymer|Li. Based on the obtained results, the origin of the reaction distribution formation was discussed. It was found that the primary rate-limiting process was the delay of the ion transport through active material particles in the investigated composite cathode. It was demonstrated that CT-XANES can be a very powerful tool in understanding the factors responsible for reactions and ion/electron transports in ASSLIBs.

Keywords X-ray tomography · X-ray absorption near edge structure spectroscopy · *Operando* analysis

1 Introduction

Synchrotron X-ray has a variety of unique properties, such as extremely high brightness and coherency, energy tunability, high polarization, and short pulse, which enable nondestructive and high position/time-resolved analysis of materials without any specific environments. Particularly in recent years, synchrotron X-ray analysis is often applied for *operando* analysis in the field of battery research. In the last decade, our group also has extensively developed various synchrotron X-ray *operando* analytical techniques for investigating lithium ion batteries [1–3, 23]. In this chapter, as typical examples of such investigations, *operando* visualization of reactions in all-solid-state lithium ion batteries using synchrotron X-ray is described.

K. Amezawa (✉) · Y. Kimura
Institute of Multidisciplinary Research for Advanced Materials, Tohoku University, Sendai, Japan
e-mail: koji.amezawa.b3@tohoku.ac.jp

K. Kanamura (ed.), *Next Generation Batteries*,
https://doi.org/10.1007/978-981-33-6668-8_22

2 Reactions in All-Solid-State Lithium Ion Batteries (ASSLIBs)

Bulk-type all-solid-state lithium ion batteries (ASSLIBs) are expected as new-generation batteries that can achieve safety operation, high output power, and high capacity. In this type of batteries, smooth and homogeneous reactions in the composite electrodes are required. In a composite electrode consisting of electrode active material and electrolyte particles, it is considered that the reaction kinetics is influenced by various mass transport processes, such as transport of ion in the electrolyte particles, transports of ion and electron in the active material particle, and charge transfer at the electrode/electrolyte interface. If one or some of these mass transport processes would be very slow, ion or electron could not be sufficiently supplied to a part of active materials, resulting in the deterioration of the capacity and output power. Moreover, such an inhomogeneous reaction in a composite electrode may cause overcharge or overdischarge of the active material, and lead to fatal degradation of the durability and the safety. Therefore, in order to develop batteries having sufficient performance (rate capability, output power, durability, and reliability, etc.), it is necessary to clarify reasons for the reaction distribution formation and to optimize the electrodes and operation conditions.

Many researches have been devoted to reveal the reaction distribution in composite electrodes of lithium ion batteries [4–17]. A variety of analytical techniques were applied to experimentally evaluate the reaction distribution in composite electrodes, particularly in those of conventional lithium ion batteries using liquid electrolytes [4–15]. However, in some of previous reports, the reaction distribution was investigated with electrodes removed from dismantled batteries because of the limitation of analytical techniques [4, 14, 16]. Such an ex situ observation may not be appropriate to truly observe the reaction distribution because it may change from moment to moment even under open circuit conditions. In addition, the most of the analyses employed in previous reports was limited to one- [4–7] or two-dimensional [8–15] analysis in spite of the reaction distribution three-dimensionally formed in practical electrodes.

From the backgrounds mentioned above, we have been developing *operando* analytical techniques to directly observe reaction distributions in battery composite electrodes. In particular, the developed techniques are recently being applied to visualize reaction distributions in composite electrodes of ASSLIBs. In the following sections, recent researches to investigate reaction distributions in ASSLIB composite cathodes by using computed tomography X-ray absorption near edge structure (CT-XANES) are introduced. To our best knowledge, this is the first work which performed *operando* and three-dimensional observation of reaction distributions in ASSLIB composite electrodes.

3 CT-XANES Measurements of ASSLIBs

CT-XANES is an analytical technique to combine three-dimensional computing tomography imaging with X-ray absorption near edge structure measurement [18]. This technique enables us to nondestructively evaluate three-dimensional distribution of chemical and electronic states of specimen for relatively large observation area within reasonably high position and time resolutions. By taking these advantages of CT-XANES into account, we applied this technique to the *operando* analysis of reaction distribution in ASSLIB electrodes. In our measurements, $LiCoO_2$ (LCO)-$Li_{2.2}C_{0.8}B_{0.2}O_3$ (LCBO)|LCBO|poly (ethylene oxide) (PEO)-based polymer|Li, which was developed by Okumura et al. [19], was chosen as an example of ASSLIBs, and the reaction distribution in its composite cathode was investigated. Schematic illustration and picture of the investigated ASSLIB cell and the cross-sectional image of its composite cathode were given in Fig. 1.

CT-XANES measurements were carried out at the undulator beamline BL37XU, SPring-8, Japan. Figure 2 shows experimental setup of CT-XANES measurements. Synchrotron X-ray monochromated by a double crystal spectrometer of Si (111) was delivered to the specimen, and the transmission X-ray was converted to the visible light by passing a single crystal scintillator of Ce:LuAG (Ce:$Lu_3Al_5O_{12}$). The visible light image was recorded by a low-noise sCMOS camera (ORCA-Flash 4.0, Hamamatsu photonics K.K., Japan) after being magnified by a 10 × objective lens. The pixel size and the field of view of the camera were 0.51 μm and *ca.* 1 × 1 mm^2, respectively.

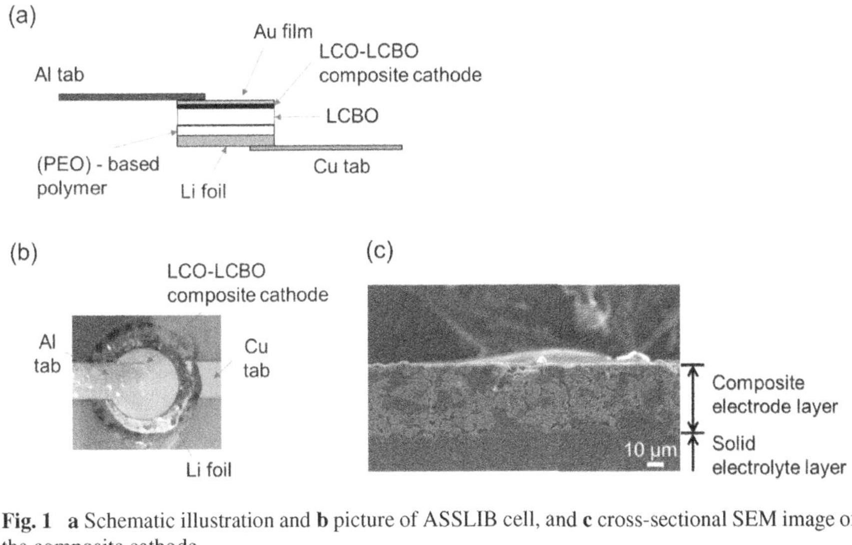

Fig. 1 **a** Schematic illustration and **b** picture of ASSLIB cell, and **c** cross-sectional SEM image of the composite cathode

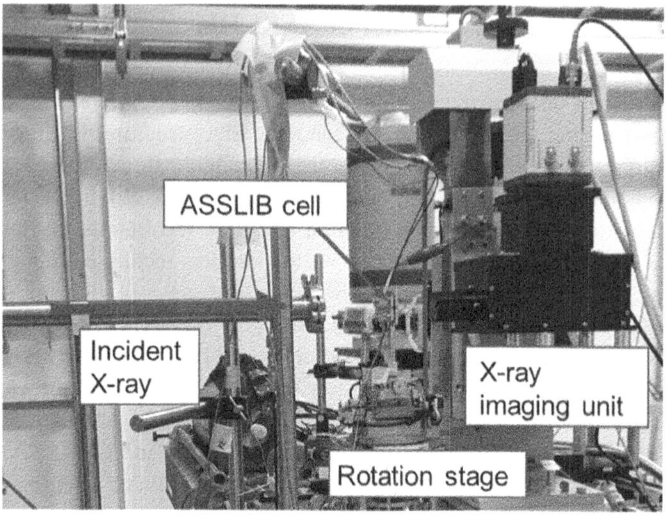

Fig. 2 Experimental setup of CT-XAFS measurements

As seen in Fig. 1, the investigated ASSLIB cell was flat-shaped. Thus, it was impossible to take transmission images of the cell at all projection angles. Therefore, the angle-limited CT technique [18] was employed. In our measurements, the specimen angle was varied from −65 to 65° with a step of 0.2° with respect to the direction of the incident X-ray. An X-ray projected transmission image (I) was taken with an exposure time of 20 ms at each specimen angle. The X-ray image without the cell (I_0) and the dark image of the sCMOS camera without the X-ray irradiation (I_d) were also obtained. The two-dimensional absorption coefficient images (μ_t) were obtained from the X-ray projected transmission images (I), the incident X-rays images (I_0) and the dark image (I_d) from the Beer's law.

$$\mu_t = -\ln\left(\frac{I - I_d}{I_0 - I_d}\right) \tag{1}$$

By correcting the obtained μ_t images, the three-dimensional μ_t image was reconstructed. The correction and the reconstruction were carried out using the software developed by Matsui et al. [18]. This series of CT measurement was repeated while scanning the energy of incident X-ray near Co K-edge to detect the change in the chemical/electronic state of Co ion in LCO (active material) in the investigated composite cathode. The times needed for the single CT and CT-XANES measurements were approximately 20 s and 25 min, respectively, in our measurements.

As shown in Fig. 3, from the three-dimensional μ_t images acquired at different X-ray energies near the absorption edge, XANES spectra can be obtained for each position by plotting absorption coefficient (μ_t) as a function of X-ray energy. In our

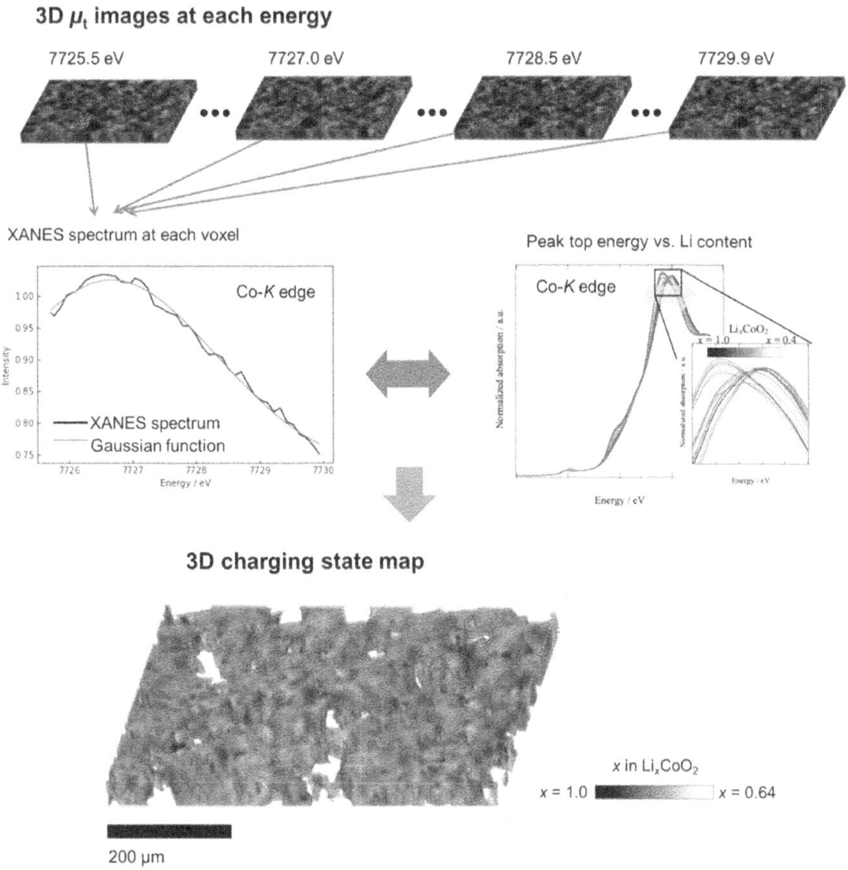

Fig. 3 Schematic illustration of the acquisition processes for three-dimensional charging state map

analyses, in order to reasonably reduce the experimental noise of the μ_t signal, the region of the composite electrode layer was cropped and each of 6 × 6 × 6 voxels was binned into 1 voxel. The size of the three-dimensional μ_t images was 630 × 517 × 46 μm^3 (207 × 170 × 15 voxels) and the voxel size was *ca.* 3.1 μm. As discussed in the previous studies [1, 2, 15], in the case of Li_xCoO_2 (LCO), there is a one-to-one relation between the peak top energy in the Co-*K* edge XANES spectra and the Li content (x in Li_xCoO_2) except for the composition ranges around $x = 0.5$ and $0.9 < x < 1.0$. By taking this relation into account, the Li content x, *i.e.*, the state of charge (SOC), was determined from the peak top energy of the XANES spectrum.

4 *Operando* Observation of Reaction Distribution in an ASSLIB Composite Electrode

By applying CT-XANES technique described in the previous section, *operando* observation of three-dimensional reaction distribution in the ASSLIB composite cathode was performed. The ASSLIB cell was charged/discharged with a current of 100 μA and the cut-off voltage of 2.0–4.35 V. Figure 4a represents the charge/discharge curve. CT-XANES measurements were repeated at every 30 min or 1 h during the charge/discharge. The observed capacities were 57 and 52 mAh/g for charging and discharging, respectively. From these capacity values, the Li contents (x in Li_xCoO_2) after charging and discharging were estimated to be $x = 0.79$ and 0.98, respectively. If the cell is charged to the rated capacity (100 mAh/g in this work) as intended, the average Li content should be $x = 0.64$. But the Li content evaluated from the charge curve ($x = 0.79$) was much larger than this value. This indicated that some of active materials in the composite electrode were not fully charged and an inhomogeneous reaction distribution was formed during charging. On the other hand, most of active materials seemed to be fully discharged after discharging, because the Li content was recovered nearly to the original value ($x = 1.0$).

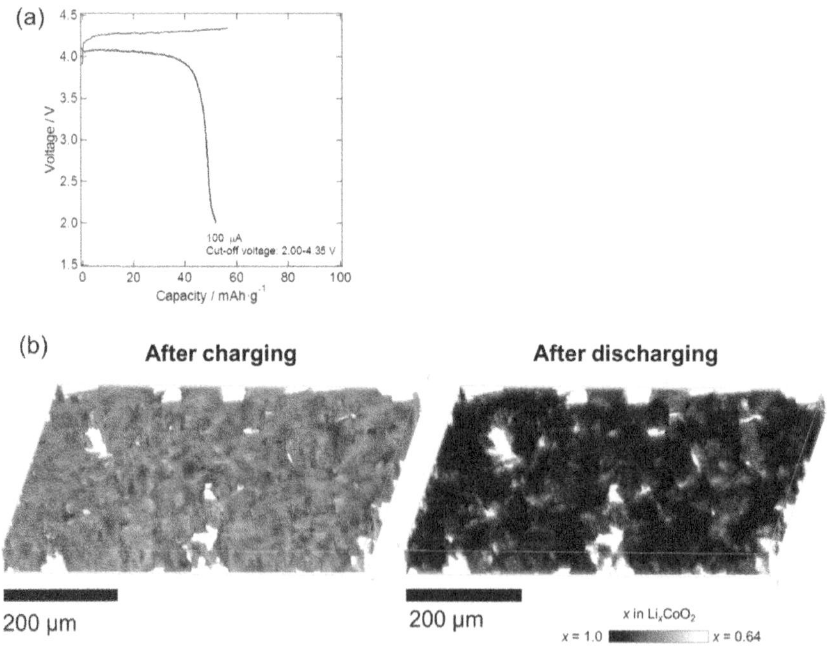

Fig. 4 **a** Charge/discharge curve of the ASSLIB cell and **b** three-dimensional charging state maps after charging and discharging. The black and white regions indicate discharged ($x = 1.0$) and charged ($x = 0.64$) active material regions, respectively

Figure 4b shows three-dimensional reaction distribution maps obtained from CT-XANES measurements. Because of the page limitation, we here show the results after charging and discharging. The colored and transparent areas correspond to the active material and the solid electrolyte/void, respectively. The white/black areas indicate the fully charged/discharged areas. The Li content in Fig. 4b was determined by the shift of the peak top energy in Co *K*-edge XANES spectra after charging and discharging with relative to the peak top energy before charging. The average peak top energy was 1.2 eV higher after charging while it was almost the same after discharging, when it was compared with one for the pristine state (before charging). These shifts of the peak top energy corresponded to the Li contents of $x = 0.77$ and 1.0, respectively, which were in good agreement with those estimated from the charge/discharge capacities in Fig. 4a, i.e., $x = 0.79$ and 0.98. This demonstrated that CT-XANES technique can evaluate the reaction distribution in an ASSLIB composite electrode three-dimensionally and quantitatively.

From the results of CT-XANES measurements, the state of charge (the Li content) can be estimated for any desired positions and times. Figure 5 presents the existence ratio of LCO active material area having the Li content of $0.9 < x \leq 1.0$, $0.8 < x \leq 0.9$, $0.7 < x \leq 0.8$, $0.64 < x \leq 0.7$, and $x \leq 0.64$, which were obtained from the reaction distribution map by CT-XANES after charging. It was found that 12% of LCO had higher Li content than 0.64 (fully charged state) while 88% of active material was not fully charged. The ratio of LCO having extremely high Li content ($0.9 < x \leq 1.0$) was very small and was only 3%. This means that there were few LCO particles which were disconnected from the conduction path of electrons and thus remained uncharged. In other words, the main reason for the low charge capacity of the investigated ASSLIB cell was not due to the electric isolation of active material particles.

From the three-dimensional reaction distribution map, two-dimensional charging state maps can be extracted at any desired cross-sections. Representative results

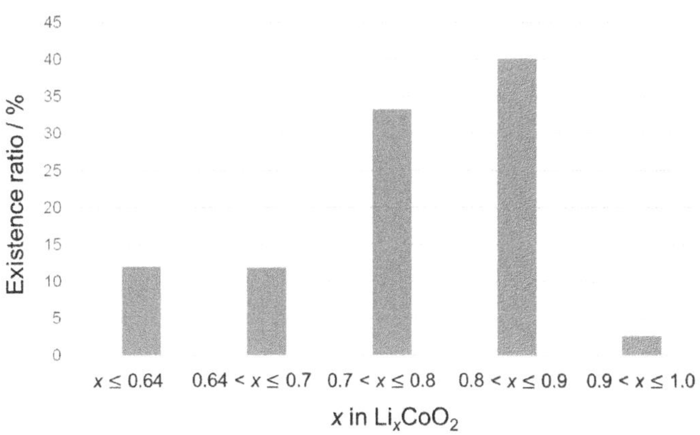

Fig. 5 Ratio of LCO with the Li content (x in Li_xCoO_2) of $0.9 < x \leq 1.0$, $0.8 < x \leq 0.9$, $0.7 < x \leq 0.8$, $0.64 < x \leq 0.7$, and $x \leq 0.64$ in the three-dimensional charging state map after charging

in the thickness and the in-plane directions were shown in Fig. 6a and b, respectively. These figures clearly showed that the charge/discharge reaction in homogeneously progressed both in the thickness and the in-plane directions. A similar two-dimensional charging state map can be obtained by other analytical techniques such as two-dimensional XANES, microscope Raman spectroscopy and so on. However, a two-dimensional charging state map obtained by these two-dimensional techniques is different in principle from one obtained by CT-XANES technique. In order to clearly demonstrate this difference, a two-dimensional charging map in the in-plane direction, which was obtained by two-dimensional XANES measurements of the composite electrode after charging, is presented in Fig. 6c. This map corresponds to the two-dimensional absorption coefficient images at the projection angle of 0°. It is obviously seen that the charging state map acquired by two-dimensional XANES (Fig. 6c) was very different from one by CT-XANES (Fig. 6b). The former seemed more homogeneous than the latter in spite of the observation of the same area. The

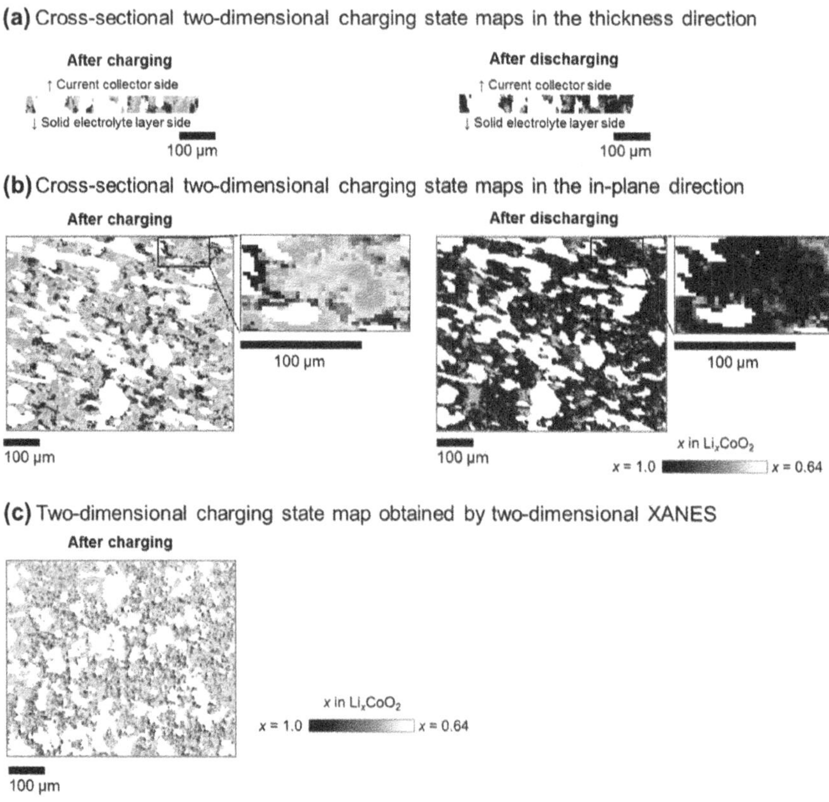

Fig. 6 Representative cross-sectional two-dimensional charging state maps during charging and discharging **a** in the thickness direction and **b** in the in-plane direction. **c** Two-dimensional charging state map after charging obtained by two-dimensional XANES. The black and white regions indicate discharged ($x = 1.0$) and charged ($x = 0.64$) active material regions, respectively

inhomogeneous reaction distribution seen in Fig. 6b was hardly found in Fig. 6b. This is because the absorption signal in two-dimensional XANES measurements was integrated and averaged in the X-ray transmission direction (the thickness direction in the case of Fig. 6c). In this respect, the two-dimensional information obtained by two-dimensional XANES is not always accurate enough and three-dimensional analytical technique like CT-XANES is necessary for the precise analysis of local charging state distribution.

5 Origin of Inhomogeneous Reaction in an ASSLIB Composite Electrode

In the previous sections, it was demonstrated that CT-XANES enables *operando* and three-dimensional evaluation of the charging state for a relatively large area with sufficient resolutions of position and time. Therefore, CT-XANES can be a powerful tool when the local distribution of charging state in an ASSLIB composite electrode is discussed. Taking the advantages of this novel analytical technique, we here discuss the origin of inhomogeneous reaction in the ASSLIB composite electrode based on the obtained three-dimensional charging state maps.

In an ASSLIB composite electrode, a reaction distribution can be formed due to the delay of mass transport processes. Typically, four kinds of mass transport processes are considered to be involved in the reaction of an ASSLIB composite electrode; (a) ion transport through solid electrolyte particles, (b) electron transport through active material particles, (c) ion transport through active material particles, and (d) charge transfer at solid electrolyte/active material interfaces. If the process (a) is rate-limiting, the potential drop due to the ion transport becomes larger with the distance from the solid electrolyte layer. Then, for charging, the reaction preferentially progresses near the solid electrolyte layer, as shown in Fig. 7a. When the process (b) is slow, the reaction preferentially occurs near the current collector (Fig. 7b). In these two cases, the charging state monotonically changes along the thickness direction. On the other hand, when the mass transport (c) is a rate-limiting process, the distribution of the charging state appears within active material particles, as given in Fig. 7c, since the potential drop occurs mainly within active material particles. If the process (d) is rate-limiting, the inhomogeneous reaction distribution is not formed, but the charging state is uniformly low (Fig. 7d).

As seen in Fig. 6a and b, the Li content after charging was inhomogeneous according to location of active material. However, in the thickness direction (Fig. 6a), the charging state is appeared to be random, and there seemed to be no monotonical change in the charging state in this direction. Thus, we can conclude that neither ion transport through solid electrolyte particles (Fig. 7a) nor electron transport through active material particles (Fig. 7b) were rate-limiting in this composite cathode under the applied charge/discharge conditions. On the other hand, in the in-plane direction (Fig. 6b), relatively higher charged regions (grey-white regions) tended to exist

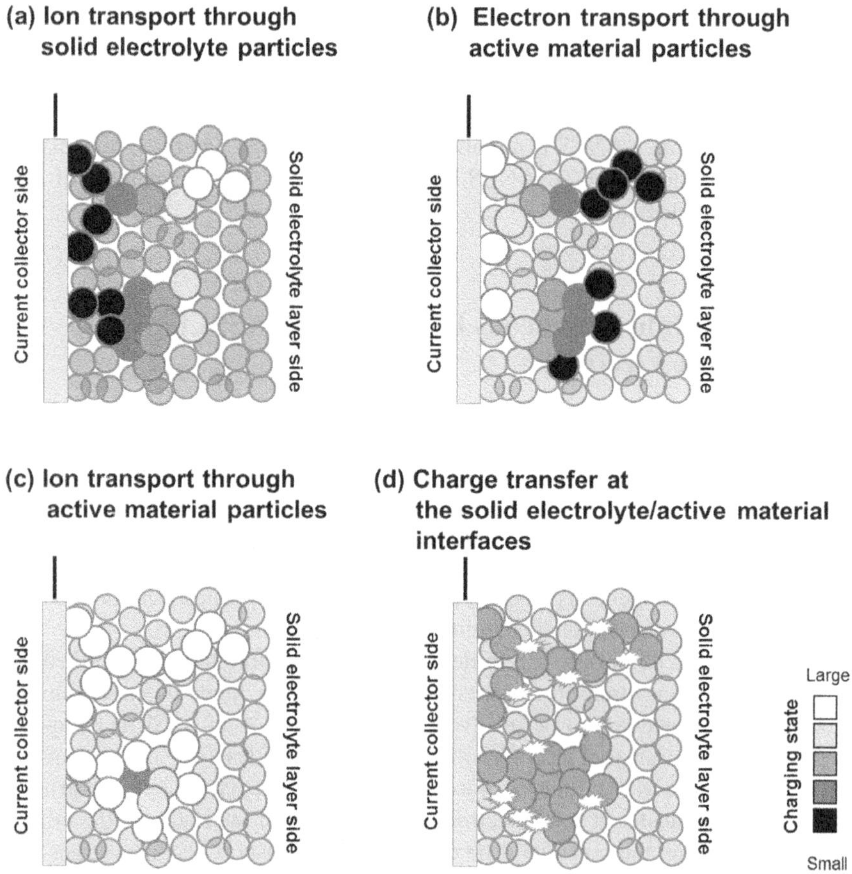

Fig. 7 Schematic illustration of the expected reaction distributions which are formed in an ASSLIB composite electrode, when the rate-limiting process is **a** ion transport through solid electrolyte particles, **b** electron transport through active material particles, **c** ion transport through active material particles, and **d** charge transfer at solid electrolyte/active material interfaces, respectively

near the outside of active material particles, which was thought to be near the solid electrolyte. Contrarily, relatively lower charged regions (black regions) tended to be present inside the active material regions. From these results, we can infer that the ion transport through active material particles is slow in the composite electrode.

Considering that the average size of the LCO active material particles used in this work was about 10 μm, it was supposed that the inhomogeneity of charging state in Fig. 6b did not occur within one active material particle but within aggregated active material particles. Actually, Li ion diffusion in a single LCO particle is known to be fast [20]. It is therefore assumed that the slow ion transport through active material particles is caused by the large interfacial resistance between LCO particles. One of the plausible reasons for this is the small contact area between LCO particles. Since LCO has relatively high Young's modulus and does not easily plastically deform

[21], it is difficult to form large contact areas between LCO particles. In addition, the anisotropic nature of Li ion diffusion in the layered oxide [22] probably makes it more difficult to form good ionic conduction paths at the interfaces.

In conventional LIBs with liquid electrolyte, almost all active material particles are immersed in the liquid electrolyte. Thus, ion transport through active material particles cannot be a serious problem, since active material particles can receive/release Li ions easily from/into the surrounding liquid electrolyte. The formation of reaction distribution due to the delay of the ion transport through active material particles is one of the inherent and unique problems for ASSLIBs.

Considering the origin of the reaction distribution formation identified above, in the investigated ASSLIB, an effective way to improve the charging and discharging capacities is to more uniformly disperse active material particles in the composite cathode, while maintaining the conduction path of electrons. In addition, the results tell us that the total capacity can be enhanced also by increasing the thickness of the composite cathode layer, since the capacity was not restricted by the ion transport through solid electrolyte or the electron transport through active material. In this way, CT-XANES measurements can indicate us directions to improve performance of ASSLIBs.

6 Summary

In this chapter, recent researches on *operando* analysis of all-solid-state lithium ion batteries (ASSLIBs) by using synchrotron X-ray were introduced. By applying CT-XANES, *operando* and three-dimensional evaluation of reaction distribution in an ASSLIB composite electrode was performed. As a result, it was found that the primary rate-limiting process was the delay of the ion transport through active material particles in the investigated ASSLIB composite cathode. As described here, analytical techniques using synchrotron X-ray like CT-XANES can be very powerful tools in understanding the factors responsible for reactions and ion/electron transports in batteries.

References

1. Fakkao, M., Nakamura, T., Kimura, Y., Tsuruta, K., Tamenori, Y., Amezawa, K. (2017). *20th Meeting of Japanese XAFS Society*, P-16, Himeji
2. Fakkao, M., Chiba, K., Kimura, Y., Nakamura, T., Okumura, T., Nitta, K., et al. (2017). *Journal of the Ceramic Society Japan, 125,* 299–302.
3. Kimura, Y., Tomura, A., Fakkao, M., Nakamura, T., Ishiguro, N., Sekizawa, O., Nitta, K., Uruga, T., Okumura, T., Tada, M., Uchimoto, Y., & Amezawa, K. (2020) *Journal of Physical Chemistry Letters, 11,* 3629–3636.
4. Liu, J., Kunz, M., Chen, K., Tamura, N., & Richardson, T. J. (2010). *The Journal of Physical Chemistry Letters, 1,* 2120–2123.

5. Strobridge, F. C., Orvananos, B., Croft, M., Yu, H.-C., Robert, R., Liu, H., et al. (2015). *Chemistry of Materials, 27,* 2374–2386.
6. Kitada, K., Murayama, H., Fukuda, K., Arai, H., Uchimoto, Y., Ogumi, Z., & Matsubara, E. (2016). *Journal of Power Sources, 301,* 11–17.
7. Pietsch, P., Hess, M., Ludwig, W., Eller, J., & Wood, V. (2016). *Scientific Report, 6,* 27994.
8. Harris, S. J., Timmons, A., Baker, D. R., & Monroe, C. (2010). *Chemical Physics Letters, 485,* 265–274.
9. Nishi, T., Nakai, H., & Kita, A. (2013). *Journal of the Electrochemical Society, 160,* A1785–A1788.
10. Katayama, M., Sumiwaka, K., Miyahara, R., Yamashige, H., Arai, H., Uchimoto, Y., et al. (2014). *Journal of Power Sources, 269,* 994–999.
11. Li, Y., El Gabaly, F., Ferguson, T. R., Smith, R. B., Bartelt, N. C., Sugar, J. D., Fenton, K. R., Cogswell, D. A., David Kilcoyne, A. L., Tyliszczak, T., Bazant, M. Z., Chueh, W. C. (2014). *Nature Materials 13,* 1149–1156.
12. Nowack, L., Grolimund, D., Samson, V., Marone, F., & Wood, V. (2016). *Scientific Report, 6,* 21479.
13. Orikasa, Y., Gogyo, Y., Yamashige, H., Katayama, M., Chen, K., Mori, T., et al. (2016). *Scientific Report, 6,* 26382.
14. Tanida, H., Yamashige, H., Orikasa, Y., Gogyo, Y., Uchimoto, Y., & Ogumi, Z. (2016). *Journal of Physical Chemistry C, 120,* 4739–4743.
15. Nakamura, T., Chiba, K., Fakkao, M., Kimura, Y., Nitta, K., Terada, Y., et al. (2019). *Batteries Supercaps, 2,* 688–694.
16. Otoyama, M., Ito, Y., Hayashi, A., & Tatsumisago, M. (2016). *Chemistry Letters, 45,* 810–812.
17. Chen, K., Shinjo, S., Sakuda, A., Yamamoto, K., Uchiyama, T., Kuratani, K., et al. (2019). *Journal of Physical Chemistry C, 123,* 3292–3298.
18. Matsui, H., Ishiguro, N., Uruga, T., Sekizawa, O., Higashi, K., Maejima, N., & Tada, M. (2017). *Angewandte Chemie, 129,* 9499–9503.
19. Okumura, T., Takeuchi, T., & Kobayashi, H. (2016). *Solid State Ionics, 288,* 248–252.
20. Dokko, K., Nakata, N., & Kanamura, K. (2009). *Journal of Power Sources, 189,* 783–785.
21. Qu, M., Woodford, W. H., Maloney, J. M., Carter, W. C., Chiang, Y.-M., & Van Vliet, K. J. (2012). *Advanced Energy Materials, 2,* 940–944.
22. Xie, J., Imanishi, N., Matsumura, T., Hirano, A., Takeda, Y., & Yamamoto, O. (2008). *Solid State Ionics, 179,* 362–370.
23. Kimura, Y., Fakkao, M., Nakamura, T., Okumura, T., Ishiguro, N., Sekizawa, O., Nitta, K., Uruga, T., Tada, M., Uchimoto, Y., & Amezawa, K. (2020). *ACS Applied Energy Materials, 3,* 7782–7793.

First-Principles Simulations on Battery Materials

Takahisa Ohno

Abstract One essential issue to be solved for the development of solid-state lithium ion battery is the high ionic interface resistance between battery materials, especially between the electrodes and solid electrolytes. First-principles calculations are able to analyze the properties of the electrode/solid-electrolyte interfaces including the dynamics of lithium ions on the atomic scale, which is difficult to measure experimentally, so are expected to provide insights into the origin of the high interface resistance. In this article, our recent first-principles studies on the cathode/electrolyte interfaces are presented. We have demonstrated the formation of lithium-depleted layer at interface based on the first-principles molecular dynamics, which explains reasonably the experimental findings of the high interface resistance between the oxide cathodes and sulfide electrolytes and its reduction by the insertion of oxide buffer layers. Besides, the first-principles analysis on the lithium diffusion within the oxide electrolytes is also presented, which has clarified the diffusion obstacles such as bottleneck structures.

Keywords First-principles · Interface · Lithium ion diffusion

1 Introduction

All solid-state lithium ion batteries (LIBs) are expected to be the next-generation energy storage devices because of their high energy–density and good safety performance [1–3]. There are, however, still some issues to be solved for practical applications of solid-state LIBs. Essential one is the low power density, which comes mainly from the lower ionic conductivity of solid electrolytes than organic liquid electrolytes. During battery operation, lithium ions transport between cathode and anode materials through the solid electrolytes, passing across the electrode/electrolyte interfaces. It is highly desirable to improve the lithium ionic conductivity within the solid electrolytes alone and that across the interfaces between electrodes and electrolytes.

T. Ohno (✉)
International Center for Materials Nanoarchitectonics (MANA), National Institute for Materials Science (NIMS), Tsukuba, Japan
e-mail: OHNO.Takahisa@nims.go.jp

K. Kanamura (ed.), *Next Generation Batteries*,
https://doi.org/10.1007/978-981-33-6668-8_23

Nowadays, two types of solid electrolytes are considered for solid-state LIB, which are sulfide and oxide electrolytes. Sulfide electrolytes usually exhibit higher ionic conductivities than oxide counterparts, because of the higher polarizability of sulfide ions [4]. Several sulfide electrolytes with high ionic conductivity comparable to those of liquid systems have been developed by recent intensive studies [5]. As for oxide electrolytes, however, the ionic conductivities are about one order of magnitude lower than those of sulfide electrolytes and further developments are highly demanded.

Even if a solid electrolyte itself shows sufficient ionic conductivity, if the interface with electrodes exhibits high interface resistance, this delays the insertion and extraction of lithium ions across the interface during battery operation and the power density decreases. Sulfide electrolytes show high ionic conductivity as mentioned above, and they are soft and deformable enough to form good interface connection with electrodes only by compression without high temperature sintering. In spite of these advantages, the interfaces between sulfide electrolytes and cathodes show high ionic resistance. It has been reported that the insertion of buffer layers at the cathode/sulfide-electrolyte interfaces reduces the resistance [6–9], but the origin of the reduction has not been clarified yet. On the other hand, oxide electrolytes have the advantage of stability under ambient atmosphere over sulfide electrolytes. However, oxide electrolytes are hard materials and high temperature sintering is inevitable to connect ionic conduction with electrodes, which is in strong contrast to sulfide electrolytes. Such heat treatment promotes mutual diffusion at the interfaces which forms harmful interface products and increases the interface resistance [10]. Methods to suppress the formation of resistive layers and make good interface connections have been intensively explored now.

Understanding the properties of cathode/solid-electrolyte interfaces is indispensable to reducing the interface resistances. The interface resistance comes from a narrow region of several nanometers near the interface [11] and it is difficult to measure such a narrow and buried interface on the atomic scale experimentally. First-principles calculations based on the density functional formalism [12, 13], which describes the electronic properties and ionic dynamics with high accuracy, are expected to make good contribution toward elucidation of the interface properties in solid-state batteries. In this article, we have presented our recent activities based on the first-principles calculations on the interface properties between cathodes and solid electrolytes [14–20] and the ionic conductivity in oxide electrolytes [21].

2 Cathode/Electrolyte Interfaces

2.1 High Interface Resistance

As described above, the interfaces between cathodes and sulfide electrolytes show high ionic resistance. Some mechanisms have been proposed for the origin of the high ionic interface resistance. One is the formation of a space-charge layer at the interface [2, 6, 8, 11]. The high potential of the cathode forms a lithium-depleted layer on the electrolyte side of the interface, which is highly resistive due to the absence of the charge carrier. Based on this space-charge layer model, the insertion of insulating oxide buffer layers such as $LiNbO_3$ and $Li_4Ti_5O_{12}$ at the interfaces has been proposed as a promising method to reduce the interface resistance [6, 8]. The other is the formation of a defective layer induced by mutual diffusion and chemical reactions at the interface [22]. At the interface between $LiCoO_2$ and the sulfide electrolytes, the inter-diffusion of Co ions and the formation of a defective have been reported from TEM experiments [10].

2.2 Calculation Methods

Recently, we have investigated the interface properties between cathodes and solid electrolytes using the first-principles calculations based on the density functional theory [14–18]. The first-principles calculations provide atomistic insights into the origin of the high ionic interface resistance. For cathode materials $LiFePO_4$ is used. As for solid electrolytes Li_3PO_4 and Li_3PS_4 are adopted as oxide and sulfide-based ones, respectively, in order to compare cathode/oxide electrolyte and cathode/sulfide electrolyte interfaces. These adopted materials are typical battery materials used for the solid-state LIBs.

The first-principles calculations have been performed for these interface systems in both discharged and charged states of batteries. Most of computational studies are limited only to the interfaces at the fully discharged state. In reality, the interface phenomena such as the formation of space-charge layers and defective layers, which may cause the high interface resistance, often occur at the charging processes. In order to elucidate the origin of the high interface resistance, computational analysis is necessary not only for the discharged state but also for the charged state. Based on the obtained results of the atomic structures, the electronic properties, and the ionic dynamics at the cathode/electrolyte interfaces, we have discussed the formation of lithium-depleted layers and defective layers and the effect of the buffer layers in comparison with the oxide and sulfide electrolytes.

The methods of first-principles calculations are presented in Sumita [14]. Geometry optimization and electronic structure analysis have been performed based on the density functional theory (DFT) implemented in CP2K program [23]. The Goedecker, Teter, and Hutter (GTH) pseudopotentials [24] are selected for the PBE functional [25] with the local spin density approximation (LSD). Total energies are calculated at the Γ point in the super cell approach. The hybrid Gaussian (MOLOPT DZVP) and plane wave (500 Ry for cutoff energy) basis set [26] was used, where valence pseudo wavefunctions are expanded in Gaussian-type orbitals and the density is represented in a plane wave auxiliary basis. The +U strategy is applied to include the electronic correlation within the d orbital of Fe, in which the value of the effective U is set to 4.3 eV as suggested in the previous research [27].

2.3 Interface Structures

The first step of the calculations is to build up the reasonable atomic structures of the interfaces. As for the cathode/oxide-electrolyte interface, the $LiFePO_4/Li_3PO_4$ interface, which corresponds to a fully discharged state, has been constructed between $LiFePO_4$ (010) and γ-Li_3PO_4 (100) planes by first-principles structure optimization [14, 15]. The $LiFePO_4$ (010) surface is the most dominant on $LiFePO_4$ nano-size crystals and regarded as active for Li ion intercalation/de-intercalation. The stoichiometric coherent $LiFePO_4/Li_3PO_4$ interface can be built up, which is composed of a $(1 \times 3 \times 2)$ $LiFePO_4$ (010) surface slab including 24 $LiFePO_4$ units and a $(3 \times 1 \times 2)$ Li_3PO_4 (100) surface slab including 24 Li_3PO_4 units as shown in Fig. 1.

For the sulfide electrolyte Li_3PS_4, on the other hand, the lattice mismatch between $LiFePO_4$ and Li_3PS_4 is too large to form a coherent interface between them. It is probable that Li_3PS_4 becomes amorphous near the interface with $LiFePO_4$ since Li_3PS_4 is more deformable than $LiFePO_4$. From this consideration, we have built up the interface between the $LiFePO_4$ (010) surface and amorphous Li_3PS_4 by first-principles molecular dynamics simulations (FP-MD), as shown in Fig. 2 [16]. The amorphous Li_3PS_4 is constructed from 16 Li_3PS_4 units randomly stacked on the (010) plane of a $(1 \times 3 \times 2)$ supercell of $LiFePO_4$. This interface structure is stabilized and equilibrated by FP-MD simulations during 40,000 MD steps (40 ps) using a constant-temperature and constant-pressure (NPT) ensemble at the temperature of

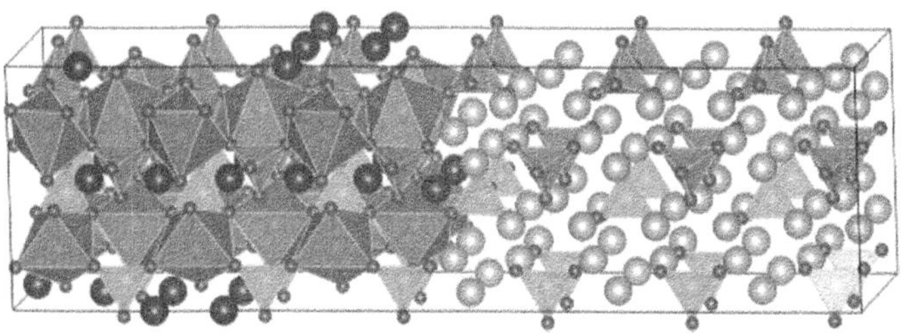

Fig. 1 Atomic structure of the coherent $LiFePO_4$ (010)/Li_3PO_4 (010) interface. The crystal structures are drawn by a computer program, VESTA Ref. [28]

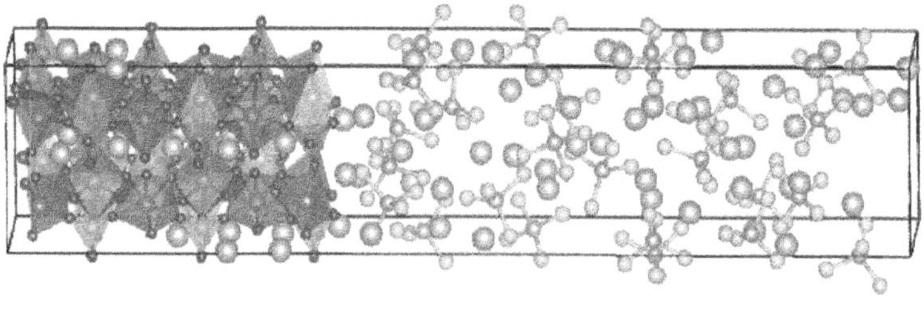

Fig. 2 Atomic structure of the $LiFePO_4$ (010)/Li_3PS_4 interface, which is obtained by averaging each atomic position over the FP-MD trajectories for 40 ps

400 K and the pressure of 1.0 atom. After the equilibration of the initial interface structure, additional FP-MD simulations for 40 ps have been performed to obtain the interface properties.

There are four five-coordinated Fe atoms at the $LiFePO_4$ (010) surface and four one-coordinated O atoms of PO_4 anions at the Li_3PO_4 (100) surface. These active Fe and O atoms are bonded to each other and thus there are no under-coordinated elements at the $LiFePO_4/Li_3PO_4$ interface, which may contribute to the stability of this interface. Actually, it is found that the $LiFePO_4/Li_3PO_4$ interface is stable during FP-MD simulations even at a high temperature of 1500 K [14]. As for the $LiFePO_4/Li_3PS_4$ interface, the active surface Fe atoms are not fully terminated by adsorption of PS_4 anions because of the steric repulsion between bulky PS_4 anions. However, the $LiFePO_4/Li_3PS_4$ interface is stable during the FP-MD simulations for 40 ps at a low temperature of 400 K [16].

2.4 Discharged States

Figures 3 and 4 plot the contour maps of density of states (DOS) for the $LiFePO_4/Li_3PO_4$ and $LiFePO_4/Li_3PS_4$ interfaces. The DOS plot of the $LiFePO_4/Li_3PO_4$ interface [15] is obtained for the optimized interface structure, whereas the DOS plot of the $LiFePO_4/Li_3PS_4$ interface [16] is for the averaged structure during the FP-MD simulations for 40 ps in which the atomic movement is not so significant in the fully discharged state. These DOS plots provide general features of the electronic structures of the cathode/electrolyte interfaces.

First of all, the band gaps of the cathode and electrolyte materials can be estimated from these DOS plots, since the middle layers of the cathode and electrolyte regions have almost the same local density of states (LDOS) as the respective bulk materials and thus are regarded as the bulk regions. The band gap of $LiFePO_4$ is estimated to be 3.7 eV from the energy difference between the valence band maximum (VBM) and the conduction band minimum (CBM) in the bulk region, both of which stem from the 3d orbitals of Fe atom. For the electrolyte materials, the band gaps are calculated to be 5.9 eV for Li_3PO_4 and 2.7 eV for Li_3PS_4. The VBMs of Li_3PO_4 and Li_3PS_4 are mainly composed of the 2p orbitals of O atom and the 3p orbitals of S atom, respectively.

The band offsets at the cathode/electrolyte interfaces are also estimated from the energy differences between the VBM and CBM of the cathode and the electrolyte materials in each bulk regions. Since the charge reaction in cathode materials is ideally the lithium deintercalation accompanied by the extraction of electrons from the VBM states of the cathodes, the valence band offset at the cathode/electrolyte interface significantly affects the charge reaction.

For the $LiFePO_4/Li_3PO_4$ interface, the VBM of the $LiFePO_4$ is located 1.2 eV above the VBM of the Li_3PO_4, which indicates the charge reaction of $LiFePO_4$ ideally proceeds, that is, the charging electrons are extracted from the Fe-3d orbitals

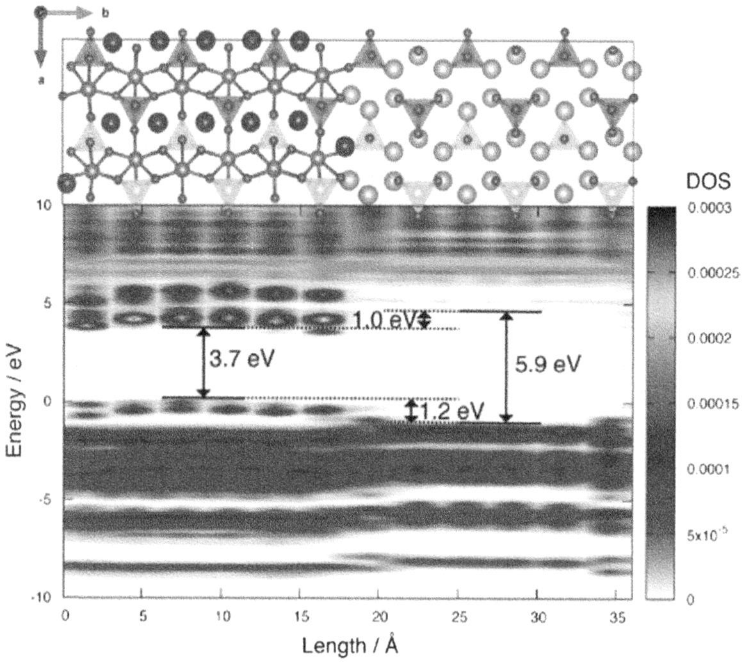

Fig. 3 Contour map of density of states (DOS) of the coherent $LiFePO_4$ (010)/Li_3PO_4 (010) interface. Fermi energy of the whole system is set to zero. Reprinted with permission from Ref. [18], copyright © 2017, American Chemical Society

to give the electromotive force for lithium deintercalation from $LiFePO_4$. For the $LiFePO_4$/Li_3PS_4 interfaces, on the other hand, the VBM of the Li_3PS_4 is located 0.7 eV higher than the VBM of the $LiFePO_4$. This is because the S-3p orbitals in Li_3PS_4 is about 2.0 eV higher than the O-2p orbitals in Li_3PO_4. Therefore, electrons will be extracted not from the $LiFePO_4$ cathode but from the Li_3PS_4 electrolyte at the beginning of the charge reaction, and at the same time lithium ions will be extracted from the Li_3PS_4 to keep charge neutrality. In this way the obtained band offset indicates that the Li_3PO_4 electrolyte is electrochemically stable whereas Li_3PS_4 is not stable against the charged $LiFePO_4$ cathode.

2.5 Charged States and Lithium Depletion

In order to verify this suggestion and reveal the origin of the high interface resistance between the oxide cathode and sulfide electrolyte, we have investigated the $FePO_4$/Li_3PO_4 and $FePO_4$/Li_3PS_4 interfaces in fully charged states [16]. The initial structures of the fully charged state interfaces are obtained by removing all lithium

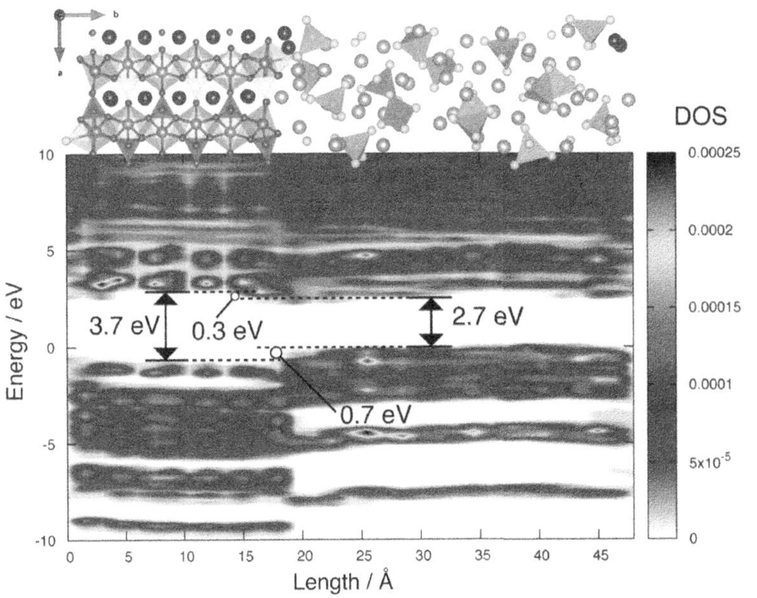

Fig. 4 Contour map of density of states (DOS) of the $LiFePO_4$ (010)/Li_3PS_4 interface for the averaged structure during the FP-MD simulations for 40 ps. Reprinted with permission from Ref. [16], copyright © 2016, American Chemical Society

atoms from the $LiFePO_4$ cathode and the dynamics of these interfaces are examined from FP-MD simulations at the temperature of 400 K for about 200 ps.

For the sulfide electrolyte, the charged-state $FePO_4$/Li_3PS_4 interface is found to become unstable at a relatively early stage of the FP-MD simulation. Figure 5 shows the contour map of DOS of the $FePO_4$/Li_3PS_4 interface at the early stage [16]. It is evident that both a lithium ion and an electron are transferred from the Li_3PS_4 side to the $FePO_4$ side. The transferred electron reduces one Fe^{3+} ion to form a localized level (Fe^{2+}) in the band gap of $FePO_4$, whereas a PS_4 anion in the Li_3PS_4 side is oxidized to generate a hole state. In addition to the electron transfer, one lithium ion is transferred from the Li_3PS_4 side to the $FePO_4$ side. The $FePO_4$/Li_3PS_4 interface gains about 0.3 eV with this transformation, which is nearly identical to the difference in the calculated lithium chemical potentials between the bulk $FePO_4$ (3.59 eV) and Li_3PS_4 (3.26 eV). The FP-MD simulation finally transfers a significant amount of lithium ions and electrons residing in the interface region of the Li_3PS_4 side to the $FePO_4$ side. As a result, the $FePO_4$ side becomes Li_{1-x} $FePO_4$ by accommodating lithium ions, and the interface region of the Li_3PS_4 side becomes low in the lithium ion concentration as shown in Fig. 6 [16, 19]. The Li_3PS_4 side is oxidized and the oxidized Li_3PS_4 tends to undergo a structural transformation and some aggregates such as $(PS_4)_2$ and S–S bonds are observed [16, 18]. In this way, the FP-MD simulation demonstrates clearly the instability of the $FePO_4$/Li_3PS_4 interface and shows that a lithium-depleted layer is formed at the Li_{1-x} $FePO_4$/Li_3PS_4 interface, which is

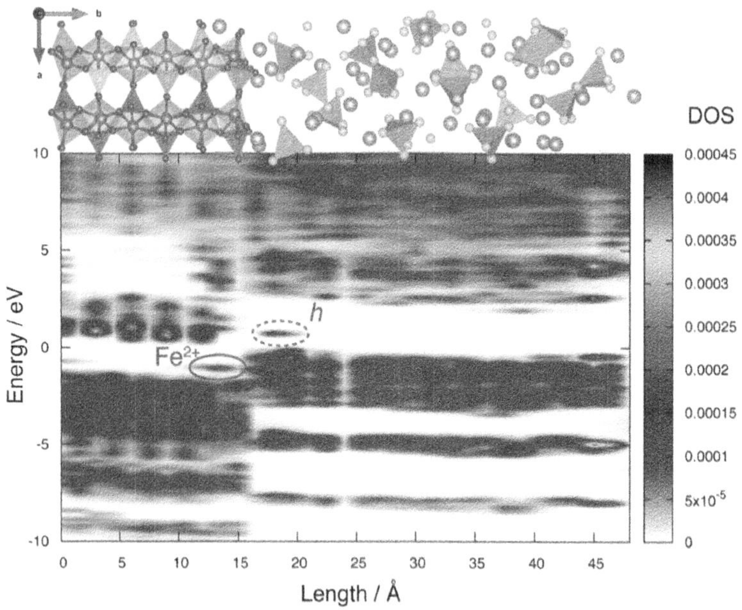

Fig. 5 Contour map of density of states (DOS) of the $FePO_4$ (010)/Li_3PS_4 interface for the snapshot structure at the early stage (about 40 ps) of the FP-MD simulations, where both a Li ion and an electron are transferred from the Li_3PS_4 side to the $FePO_4$ side. Reprinted with permission from Ref. [16], copyright © 2016, American Chemical Society

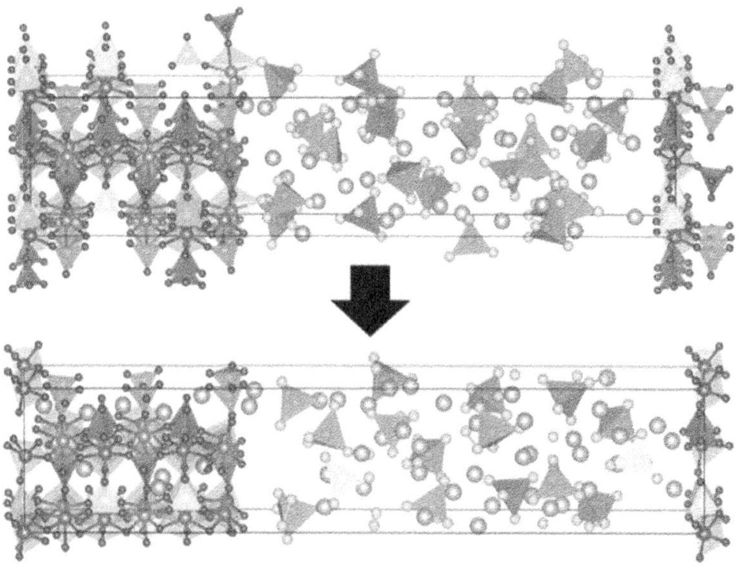

Fig. 6 Atomic structure of the $FePO_4$ (010)/Li_3PS_4 interface before (**a**) and after (**b**) FP-MD simulations

consistent with the proposed model from the experimental results [6]. The lithium-depleted layer would make the interface highly resistive due to the absence of charge-carrying lithium ions.

The $FePO_4/Li_3PO_4$ interface, on the other hand, does not show any instability. The FP-MD simulation changes negligibly the initial structure of the $FePO_4/Li_3PO_4$ interface as shown in Fig. 8 [16, 19]. The Li_3PO_4 side keeps its original structure and lithium depletion is not observed at the interface region of the Li_3PO_4 side. Figure 7 shows the contour map of DOS of the $FePO_4/Li_3PO_4$ interface [16]. Although the VBM of the oxide Li_3PO_4 side is higher than that of the $FePO_4$ side, similar to the sulfide Li_3PS_4 case shown in Fig. 5, the oxidation of Li_3PO_4 by $FePO_4$ does not occur during the FP-MD simulation, since the acceptor level (Fe^{3+}) of $FePO_4$ is located too high in energy to remove electrons from the Li_3PO_4 side. Furthermore, it is found that even if one lithium ion is intentionally transferred from the Li_3PO_4 side to the $FePO_4$ side, the transferred lithium ion migrates back to the Li_3PO_4 side during FP-MD simulations. Thus, both lithium ions and electrons are preferentially removed from the $LiFePO_4$ cathode, not from the Li_3PO_4 electrolyte even at the beginning of charging. The lithium depletion does not occur at the $FePO_4/Li_3PO_4$ interface, in contrast to the $FePO_4/Li_3PS_4$ interface.

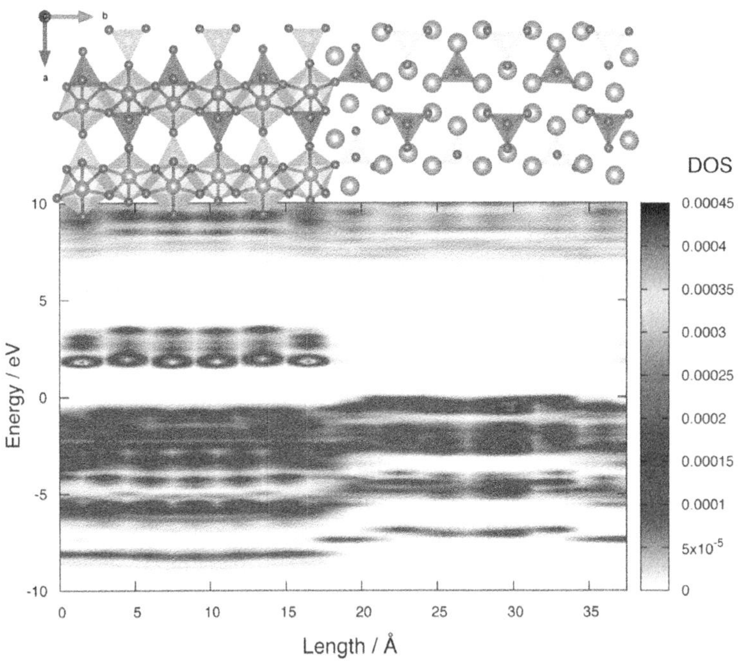

Fig. 7 Contour map of density of states (DOS) of the $FePO_4$ (010)/Li_3PO_4 (010) interface. Reprinted with permission from Ref. [16], copyright © 2016, American Chemical Society

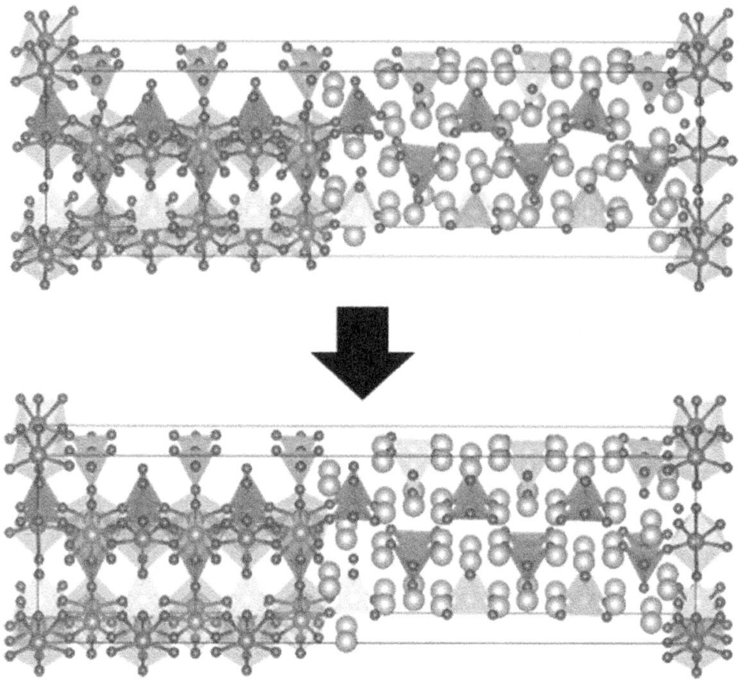

Fig. 8 Atomic structure of the $FePO_4$ (010)/Li_3PO_4 (010) interface before (**a**) and after (**b**) FP-MD simulations

2.6 Discussions

(a) **Scenario of charging**

Based on the obtained results we infer what happens at the $LiFePO_4$/Li_3PS_4 interface during charging as shown in Fig. 9 [16]. At the beginning of charging [Fig. 9b], even if electrons are removed from the cathode side and acceptor Fe^{3+} levels are formed, the Fe^{3+} levels accept electrons immediately from the sulfide side near the interface, which is due to the band alignment at the interface. This means that electrons are preferentially extracted not from the oxide cathode but from the sulfide electrolyte at the beginning of charging. As a result, the cathode side remains neutral and the sulfide side becomes charged. Then, the lithium ions in the sulfide side migrate toward the anode side under the applied electric field. The sulfide side near the interface works as an active material at the beginning of charging, which leads to the formation of the lithium-depleted layer. Actually, recent experiments have reported that Li_3PS_4 is oxidized and works as an active material in solid-state battery systems [10]. When the lithium-depleted layer grows thick on the sulfide side during charging, the electron transfer from the sulfide side to the cathode side becomes suppressed and finally electrons and lithium ions begin to be removed from the cathode side instead of the sulfide

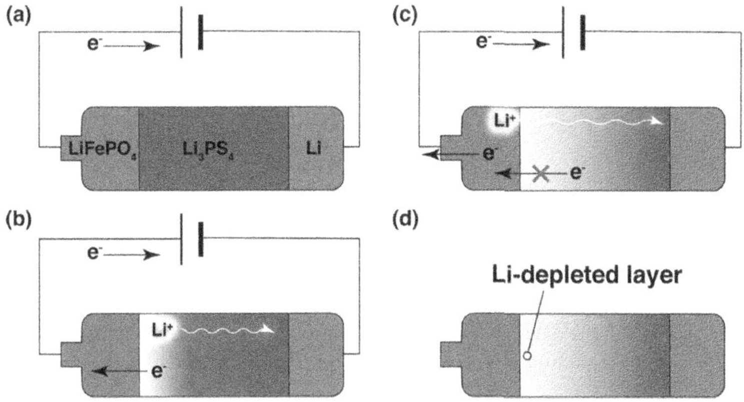

Fig. 9 Schematic diagram of charging process of the Li ion batteries with the cathode/sulfide electrolyte interface: **a** the initial fully discharged state, **b** the state at the beginning of charging, **c** the state when the charging proceeds, and **d** the final state where the Li-depleted layer is formed. Reprinted with permission from Ref. [16], copyright © 2016, American Chemical Society

side [Fig. 9c]. The migration of lithium ions from the cathode side to the anode side may not compensate the lithium-depleted layer in the sulfide, otherwise the compensated sulfide side would donate electrons to the acceptor Fe^{3+} levels again according to the same scenario at the beginning of charging. The thickness of the lithium-depleted layer may depend on the diffusion of lithium ions and electrons in the lithium-depleted sulfide.

(b) **Effects of buffer layer**

As discussed above, the electron transfer from the sulfide electrolyte to the oxide cathode at the interface, that is, the oxidation of the sulfide electrolyte during charging leads to the formation of lithium-depleted layer, which is suggested to result in high interface resistance. Therefore, in order to reduce the interface resistance, the electron transfer from the sulfide electrolyte should be suppressed. The oxide electrolytes like Li_3PO_4 usually have lower VBM relative to the sulfide electrolytes and even to the cathode materials. Therefore, when the oxide layers are inserted at the interfaces, they may act as the buffer layers to suppress the oxidation of the sulfide electrolytes and reduce the interface resistance as shown in Fig. 10 [16]. Actually, recent experiments have found the reduction of the interface resistance by interposing oxide electrolytes [6, 8]. The present findings from the first-principles simulations support the validity of the buffer layer insertion at the interface.

(c) **Other cathode materials**

Although the $LiFePO_4/Li_3PS_4$ interface has been investigated here, other interfaces between oxide cathodes and sulfide solid electrolytes such as $LiCoO_2/Li_{3.25}Ge_{0.25}P_{0.75}S_4$ interface are also examined experimentally. For most of those interfaces, the VBMs of oxide cathodes which are mainly composed of the 3*d* orbitals of transition metals are lower in energy than the

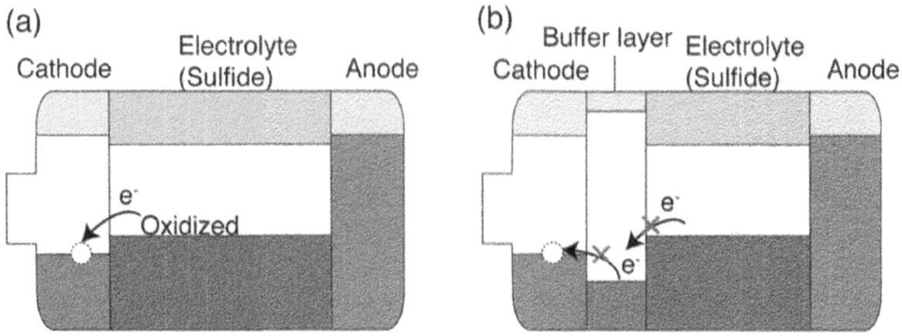

Fig. 10 Schematic diagram of the band lineups of **a** the cathode/sulfide-electrolyte/Li metal system and **b** the system with the oxide buffer layer inserted at the cathode/sulfide-electrolyte interface. Reprinted with permission from Ref. [16], copyright © 2016, American Chemical Society

VBMs of sulfide electrolytes composed of the S-3*p* orbitals and also the Li chemical potentials of oxide cathodes are lower than those of sulfide electrolytes, which are similar to the $LiFePO_4/Li_3PS_4$ interface. As discussed above, these features lead to the instability of the oxide-cathode/sulfide-electrolyte interfaces. Therefore, the present charging picture given from the $LiFePO_4/Li_3PS_4$ interface will be valid for most of the interfaces between oxide cathodes and sulfide electrolytes. We have performed a preliminary FP-MD simulation for the $LiCoO_2$/electrolytes interface. Although the interface spin-state of $LiCoO_2$ is complicated, the manifold spin-states may not affect the band alignment at the interface significantly [17]. As expected, the $Li_{0.5}CoO_2/Li_3PS_4$ interface in a charged state exhibits the instability shown in Fig. 11, such as the transfer of lithium ions from the Li_3PS_4 to the $Li_{0.5}CoO_2$ sides and the formation of S–S bonds in the Li_3PS_4 side. On the other hand, the $Li_{0.5}CoO_2/Li_3PO_4$ interface does not show any instability during FP-MD simulations. The recent experiment has reported that the lithium-depletion effect is negligible at the interface between $LiCoO_2$ and $Li_3PO_{4-x}N_x$ fabricated in all-in-vacuum process [29], which suggests that Li_3PO_4 is not oxidized by $LiCoO_2$ in consistent with the FP-MD simulations.

(d) **Other mechanisms**

The discussions so far have revealed that the high interface resistance and the effects of the buffer layer between the oxide cathodes and the sulfide electrolytes can be explained by the formation of lithium-depleted layer in the sulfide side of the interface. On the other hand, another mechanism has been proposed for the high interface resistance and its reduction [22], which suggests the formation of defective layers induced by mutual diffusion and chemical reactions at the interface. In fact, it has been reported that transition metals and anions as well as lithium ions diffuse across the interface between $LiCoO_2$ and sulfide electrolytes to form a defective layer mainly composed of Co and S atoms [10]. The defective layer is suggested to be the origin of the high interface resistance. The insertion of a buffer layer at the cathode/electrolyte interface suppresses the mutual diffusion and the formation of defective layers to reduce the interface resistance. As

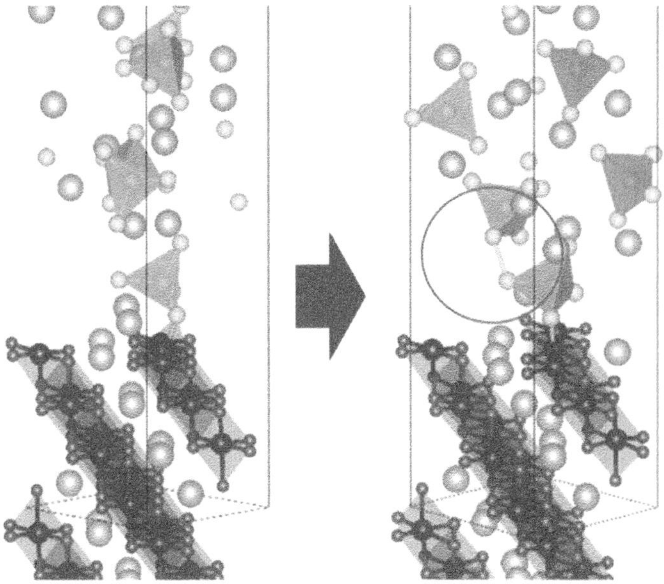

Fig. 11 Atomic structure of the $Li_{0.5}CoO_2/Li_3PS_4$ interface before (**a**) and after (**b**) FP-MD simulation for 100 ps at 400 K

shown in Fig. 6, although a significant amount of lithium ions is transferred from the Li_3PS_4 side to the $FePO_4$ side, the flamework structure does not change significantly without any mutual diffusion at the $FePO_4/Li_3PS_4$ interface. These findings are obtained from the FP-MD simulations at the temperature of 400 K for about 200 ps, which is long enough to examine the kinetics of light diffusive ions like lithium ions. However, the FP-MD simulations for much longer time might be required to examine the mutual diffusion of heavy transition metals and anions and the formation of defective layers. We have performed a preliminary FP-MD simulation at a higher temperature of 700 K to accelerate the mutual diffusion and found a slight indication of mutual diffusion of Fe and S ions. At the moment it is premature to deny the formation of defective layers from the present FP-MD simulations.

(e) **Kinetic and Energetics**

Based on the first-principles simulations, the stability of the cathode/electrolyte interfaces is discussed in terms of energetics as well as kinetics. From the kinetic point of view, the evolution of the interface structures such as the formation of lithium-depleted layers is examined by FP-MD simulations as shown above. From the viewpoint of energetics, on the other hand, possible reactions of electrolytes in contact with electrodes are evaluated from energy calculations. If some reactions are calculated to be exothermic, the interfaces are suggested to be inherently thermodynamically unstable. In reality, however, metastable structures are often formed with the help of kinetic barriers which inhibit phase transitions to lower energy states [30, 31]. The kinetic barriers are probably related to the energetic and physical confinement of electrons within each of the electrode and electrolyte regions, which make them unavailable to take part in the interface reactions, and also related to the slow dynamics of constituent elements of the reactions. It may be reasonable to consider that the buffer layer provides a kinetic barrier by confining electrons in each of the cathode and electrolyte regions and impedes the availability of electrons required for the electrolyte reactions. However, whether interface reactions involving heavy ions actually occur or not to form defective layers is hard to determine only from first-principles simulations. In addition to the atomistic analysis by FP-MD simulations, other techniques such as the continuum modeling like the phase-filed method would be useful to investigate these interface problems.

(f) **Oxide electrolytes**

The interfaces between cathodes and oxide electrolytes are expected to exhibit high ionic conductivity since lithium depletions will not be formed at the interfaces, as shown above. In fact, the $LiCoO_2/Li_3PO_{4-x}N_x$ interface fabricated in all-in-vacuum process is reported to show significantly low resistance [29]. In practical processes, however, high temperature sintering is employed to connect ionic conduction between cathodes and oxide electrolytes, which leads to the formation of resistive layers at the interfaces. For example, garnet-type $Li_7La_3Zr_2O_{12}$ (LLZO) has received much attention as a next-generation oxide electrolyte because of its high lithium ion conductivity at room temperature and sufficient chemical stability against lithium metal [32]. However, $Li_7La_3Zr_2O_{12}$ particles are hard and have limited physical contacts with $LiCoO_2$ ones. The insertion of plastic lithium ion conductors like Li_3BO_3 between $LiCoO_2$ and $Li_7La_3Zr_2O_{12}$ particles is expected to improve the physical contact between them only by low-temperature processes [32, 33]. The inserted Li_3BO_3 is considered to act as both a solid electrolyte and a bonding material. When Li_3BO_3 is inserted at the $LiCoO_2/Li_7La_3Zr_2O_{12}$ interface, besides the good bonding between $LiCoO_2$ and $Li_7La_3Zr_2O_{12}$ via Li_3BO_3, the high ionic conductivity at the interfaces with Li_3BO_3 is also an important issue. Figure 12 shows the atomic trajectory near the $Li_7La_3Zr_2O_{12}/Li_3BO_3$ interface which is obtained from the preliminary FP-MD simulations at the temperature of 1500 K during 20 ps. It is found that Li ions diffuse across the interface and the inserted Li_3BO_3 does not impede the ionic conductivity, which is consistent with the experimental results [34]. The insertion of such plastic lithium ion conductors is thought to be a promising way to make good interface connection without high temperature processes.

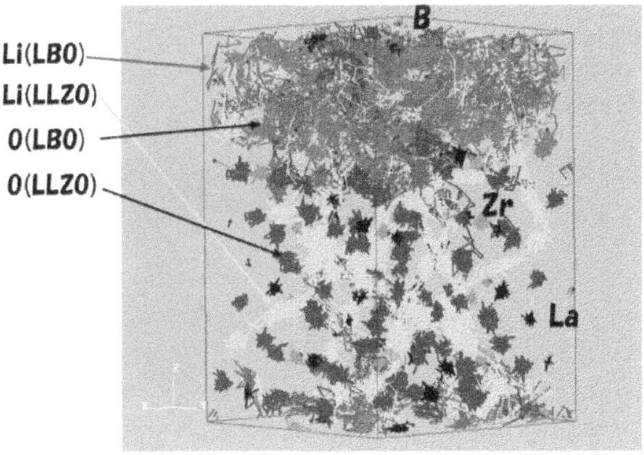

Fig. 12 Trajectory of atoms near the $Li_7La_3Zr_2O_{12}/Li_3BO_3$ interface

3 Lithium Ion Conductors

3.1 Oxide Solid Electrolytes

There is an increasing demand for solid-state batteries with high energy density and good safety performance in a wide range of applications from portable electronics to electric vehicles. Oxide solid electrolytes are more stable under ambient atmosphere with moisture than sulfide electrolytes and have great advantages in terms of safety, which is strongly required in applications to transportation vehicles. However, oxide electrolytes have some difficulties including the low ionic conductivities across themselves and across the interfaces with cathodes, as mentioned above. Another is concerned with electrochemical stability, that is, oxide electrolytes are often reduced by lithium containing anodes. Lithium metal is expected to be used as an anode material since it is the most electropositive and has high energy capacity. The electrochemical stability against lithium metal is demanded in order to meet the requirement of high energy density. Therefore, oxide solid electrolytes with both high ionic conductivity and high electrochemical stability are required for applications.

The electrochemical stability of electrolytes against Li metal depends on the compositions and structures [35]. The oxide electrolytes containing high electronegative Ti^{4+} ions such as $La_{2/3-x}Li_{3x}TiO_3$ [36] is known to be reduced by lithium metal. It is suggested that the electrochemical stability of electrolytes is improved by decreasing electronegativities of cation components, for example, from Ti^{4+} and Sb^{5+} to Ta^{5+} and Zr^{4+} [35]. Based on this concept, Inaguma group has synthesized various kinds of oxide solid electrolytes and tried to enhance their ionic conductivities by substitution of constituent elements [21]. We have investigated the mechanism of lithium ion diffusion of these materials by first-principles calculations for further enhancement of ionic conductivity.

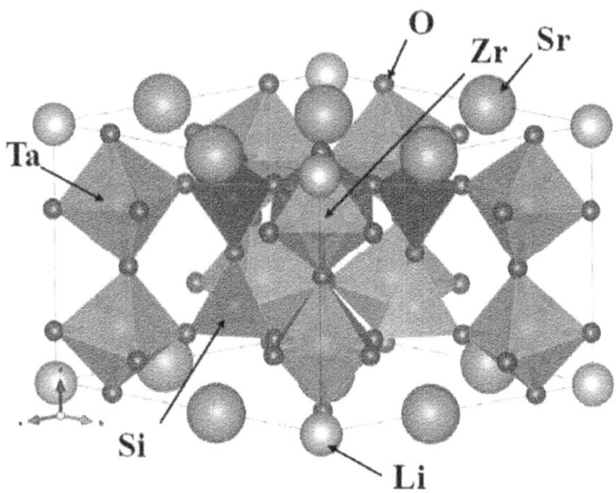

Fig. 13 Crustal structure of the host $Sr_3Ta_5ZrSi_4O_{26}$, in which Li atom is put at the 1a site

3.2 *Calculation Methods*

Recently, Inaguma group has successfully synthesized new compounds of $Sr_3Li_xTa_{6-x}Zr_xSi_4O_{26}$ using $Sr_3Ta_6Si_4O_{26}$ as a host material by element substitution and found that these compounds show considerable lithium ion conductivity [21]. The lithium ion diffusion for the material of $x = 1.0$, that is, $Sr_3LiTa_5ZrSi_4O_{26}$ has been examined by the first-principles calculations (Inaguma 2016). We have employed the projector augmented planed wave method (PAW-VASP) [37] with the generalized gradient approximation (GGA-PBE) [38]. The cut-off energy of 550 eV is used and the Monkhorst–Pack special k points set $4 \times 4 \times 4$ (36 K points) is employed for the Brillouin zone integration.

To construct the crystal structure of $Sr_3LiTa_5ZrSi_4O_{26}$, one Zr ion is substituted for one of six Ta ions and then single Li ion is put at the available sties of 1a, 1b, and 3g in the host $Sr_3Ta_6Si_4O_{26}$ structure shown in Fig. 13, and finally all ions are relaxed so as to search the lowest-energy position of Li ion. The activation energy of lithium ion diffusion is estimated using the nudged elastic band (NEB) method [39, 40]. To ascertain the lithium ion diffusion path, the FP-MD simulations are carried out with using Γ point sampling and the time step of 1 fs.

3.3 *Results*

Table 1 lists the total energies for various Li ion sites in $Sr_3LiTa_5ZrSi_4O_{26}$. The total energy of Li ion at the 1a site is set as zero. The 3g sites are distinguished by the nearest neighbor cations. The 3g-1 and 3g-2 sites are surrounded by Ta ions and Zr ion, whereas the 3g-3 site is surrounded by only Ta ions. Since there are little differences in energy between the 1a and 3g sites, the Li ions are likely to distributed into these sites. There are several possible diffusion paths of Li ions. One is the path

Table 1 The total energy of $Sr_3LiTa_5ZrSi_4O_{26}$ for various Li sites

Li-site	1a	3 g-1	3 g-2	3 g-3	1b
E (eV)	0	0.013	0.014	0.094	0.560

from the 1a to 1b site, which is parallel to the c-axis. The others are from the 1b to 3g site and from the 3g to other 3g site, both of which are vertical to the c-axis. Schematics of the diffusion paths parallel to and vertical to the c-axis are shown in Fig. 14 a and b, respectively. All of these paths include the bottlenecks formed by TO_6 and ZrO_6 octahedra and SiO_4 tetrahedra.

Judging from the estimation of the sizes of these bottlenecks, Li ion is unlikely to diffuse along the c-axis since the vacant space formed by three connected TO_6 or ZrO_6 octahedrons act as a small bottleneck for Li-ion diffusion. In fact, the activation energy of this diffusion path is calculated to be 2.97 eV by using the NEB method as shown in Fig. 15a. On the other hand, the activation energy of the path from the 3g to other 3g sites is estimated to be 0.69 eV as shown in Fig. 15b, which is in

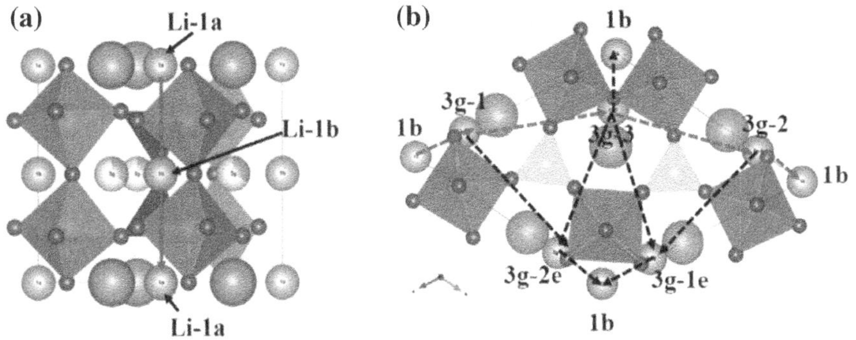

Fig. 14 Schematics of the Li-ion diffusion paths in $Sr_3LiTa_5ZrSi_4O_{26}$. Diffusion paths along the c-axis (**a**) and in the ab plane (**b**) are shown

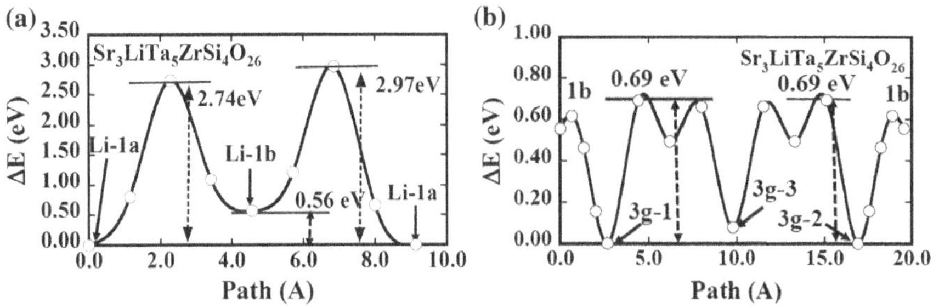

Fig. 15 Relative barrier height energies (ΔE) of Li diffusion in $Sr_3LiTa_5ZrSi_4O_{26}$ between 1a sites through 1b site along the c-axis (**a**) and Li diffusion in the ab plane between 1b sites through 3 g-1, 3 g-3, and 3 g-2 site (**b**). Both diffusion paths are shown as the red arrowed lines in Fig. 14

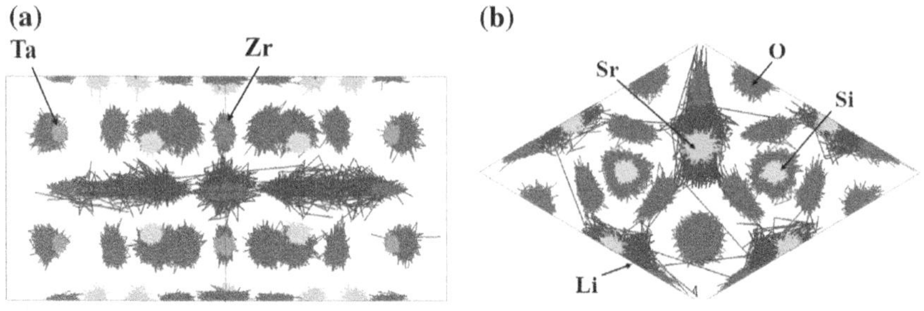

Fig. 16 Atomic trajectory of $Sr_3LiTa_5ZrSi_4O_{26}$ from the [001] (**a**) and [110] (**b**) directions, which are obtained from the FP-MD simulations at the temperature of 1400 K during 1980 ps. The initial position of Li is located at 1b site

good agreement with the experimental value of 0.76 eV. Therefore, the diffusion path between the 3g sites shown in Fig. 14b probably dominates the lithium ion conductivity. Figure 16 plots the atomic trajectory of $Sr_3LiTa_5ZrSi_4O_{26}$ which are obtained from the FP-MD simulations at the temperature of 1400 K during 1980 ps. It is found that Li ions spend almost time at the 3 g sites and diffuse two dimensionally within the Sr-ion deficient layer.

There are two maximum points along the dominant path between the 3g sites for lithium ion diffusion. One maximum point is attributed to the bottleneck formed by the O ions surrounding the diffusing Li ion, whereas the other comes from the large polyhedron around the 3g site where the Li ion lacks the structural stabilization from chemical bonding with the surrounding O ions. The oversized interstice around the 3g site acts as the diffusion barrier. These results suggest that ionic conductivity could not be improved by just expanding the bottlenecks in the Sr-ion deficient layer. Further improvement of ionic conductivity might be attained by forming new three-dimensional diffusion paths.

4 Closing Remarks

In this article, it is reported that first-principles calculations provide instructive insights into the lithium diffusion in the battery materials, that is, the diffusion across the cathode/solid-electrolyte interfaces as well as that in the oxide electrolytes themselves. As for the interface diffusion, the high interface resistance between oxide cathodes and sulfide electrolytes and its reduction by the insertion of oxide buffer layers is explained reasonably by the formation of lithium-depleted layer in the sulfide side of the interface. About the lithium diffusion of oxide electrolytes, the diffusion obstacles such as bottleneck structures can be clarified to improve the conductivity from the first-principles calculations. In this way, first-principles simulations are powerful tools to investigate the bulk and interface properties of battery materials and are expected to contribute to the practical application of solid-state LIB in cooperation with experiments.

References

1. Tarascon, J.-M., Armand, M. (2001). Issues and challenges facing rechargeable lithium batteries. *Nature 414*, 359–367.
2. Hayashi, A., Tatsumisago, M. (2012). Invited paper: Recent development of bulk-type solid-state rechargeable lithium batteries with sulfide glass-ceramic electrolytes. *Electronic Materials Letters, 8*, 199–207.
3. Takada, K. (2013). Progress and prospective of solid-state lithium batteries. *Acta Materials 61*, 759–770.
4. Zheng, N., Bu, X., & Feng, P. (2003). Synthetic design of crystalline inorganic chalcogenides exhibiting fast-ion conductivity. *Nature, 426*, 428–432.
5. Kamaya, N., Homma, K., Yamakawa, Y., Hirayama, M., Kanno, R., Yonemura, M., Kamiyama, T., Kato, Y., Hama, S., Kawamoto, K., Mitsui, A. (2011) A lithium superionic conductor. *Nature Materials 10*, 682–686.
6. Takada, K., Ohta, N., Fukuda, K., Sakaguchi, I., Ma, R., Osada, M., Sasaki T. (2008) Interfacial modification for high-power solid-state lithium batteries. *Solid State Ionics 179*, 1333–1337.
7. Takada, K., Ohta, N., Zhang, L., Cu, X., Hang, B. T., Ohnishi, T. et al. (2012). Interfacial phenomena in solid-state lithium battery with sulfide solid electrolyte. *Solid State Ionics 225*, 594–597.
8. Ohta, N., Takada, K., Zhang, L., Ma, R., Osada, M., Sasaki, T. (2006) Enhancement of the high-rate capability of solid-state lithium batteries by nanoscale interfacial modification. *Advanced Materials 18*, 2226–2229.
9. Ohta, N., Takada, K., Sakaguchi, I., Zhang, L., Ma, R., Fukuda, K. et al. (2007). $LiNbO_3$-Coated $LiCoO_2$ as cathode material for all solid-state lithium secondary batteries. *Electrochemistry Communications 9*, 1486–1490.
10. Hakari, T., Nagao, M., Hayashi, A., Tatsumisago. T. (2015). All-solid-state lithium batteries with Li_3PS_4 glass as active material. *Journal Power Sources 293*, 721–725.
11. Takada, K. (2013). Interfacial nanoarchitectonics for solid-state lithium batteries. *Langmuir, 29*, 7538–7541.
12. Hohenberg, P., & Kohn, W. (1964). *Physical Review B, 136*, 864–871.
13. Kohn, W., & Sham, L. J. (1965). *Physical Review A, 140*, 1133–1138.
14. Sumita, M., Tanaka, Y., Ikeda, M., & Ohno, T. (2015). Theoretically designed Li_3PO_4(100)/$LiFePO_4$(010) coherent electrolyte/cathode interface for all solid-state Li ion secondary batteries. *The Journal of Physical Chemistry C* 119, 14–22.
15. Sumita, M., Tanaka, Y., Ikeda, M., & Ohno, T. (2016). Theoretical insight into charging process in a Li_3PO_4(100)/$LiFePO_4$(010) coherent interface system. *Solid State Ionics 285*, 59–65.
16. Sumita, M., Tanaka, Y., Ikeda, M., Ohno, T. (2016). Charged and discharged states of cathode/sulfide electrolyte interfaces in all-solid-state lithium ion batteries. *The Journal of Physical Chemistry C 120*, 13332–13339.
17. Sumita, M., & Ohno, T. (2016). Multi-spin-state at Li_3PO_4/$LiCoO_2$(104) inter-face. *Physical Chemistry Chemical Physics 18*, 4316–4319.
18. Sumita, M., Tanaka, Y., Ohno, T. (2017) Possible polymerization of PS_4 at a Li_3PS_4/$FePO_4$ interface with reduction of the $FePO_4$ phase. *The Journal of Physical Chemistry C 121*, 9698–9704.
19. Takada, K., & Ohno, T. (2016). Experimental and computational approaches to interfacial resistance in solid-state batteries. *Frontiers Energy in Research, 4*, 10.
20. Takada, K., Ohno, T., Ohta, N., Ohnishi, T., & Tanaka, Y. (2018). Positive and negative aspects of interfaces in solid-state batteries. *ACS Energy Letters, 3*, 98–103.
21. Aimi, A., Inaguma, Y., Kubota, M., Mori, D., Katsumata, T., Ikeda, M., Ohno, T. (2016). Synthesis structure and ionic conductivities of novel Li-ion conductor $A_3Li_xTa_{6-x}Zr_xSi_4O_{26}$ (A = Sr and Ba). *Solid State Ionics 285*, 19–28.
22. Sakuda, A., Hayashi, A., Tatsumisago. M. (2010). Interfacial observation between $LiCoO_2$ electrode and Li_2S–P_2S_5 solid electrolytes of all-solid-state lithium secondary batteries using transmission electron microscopy. *Chemistry of Materials 22*, 949–956.

23. The CP2K developers group. https://www.cp2k.org/ (accessed December 2014).
24. Goedecker, S., Teter, M., & Hutter, J. (1996). Separable dual-space Gaussian pseudopotentials. *Physical Review B: Condensed Matter Material Physical, 1996*(54), 1703–1710.
25. Perdew, J. P., Burke, K., & Ernzerhof, M. (1996). Generalized gradient approximation made simple. *Physical Review Letters, 77,* 3865–3868.
26. Lippert, G., Hutter, J., & Parrinello, M. (1997). A hybrid gaussian and plane wave density functional scheme. *Molecular Physics, 92,* 477–487.
27. Zhou, F., Kang, K., Maxisch, T., Ceder, G., & Morgan, D. (2004). The electronic structure and band gap of $LiFePO_4$ and $LiMnPO_4$. *Solid State Communications, 132,* 181–186.
28. Momma, K., & Izumi, F. (2010). *J. Appl. Crystallogr. 41,* 658.
29. Haruta, M., Shiraki, S., Suzuki, T., Kumatani, A., Ohsawa, T., Takagi, Y., et al. (2015). Negligible, "Negative space-charge layer effects" at oxide-electrolyte/electrode interfaces of thin-film batteries. *Nano Letters, 15,* 1498–1502.
30. Zhu, Y., He, X., & Mo, Y. (2015). Origin of outstanding stability in the lithium solid electrolyte materials: insights from thermodynamic analyses based on first-principles calculations. *ACS Applied Materials & Interfaces, 7,* 23685–23693.
31. Zhu, Y., He, X., & Mo, Y. (2016). First principles study on electrochemical and chemical stability of solid electrolyte-electrode interfaces in all-solid-state Li-ion batteries. *Journal of Materials Chemistry A, 4,* 3253–3266.
32. Ohta, S., Komagata, S., Seki, J., Saeki, T., Morishita, S., & Asaoka, T. (2013). All-solid-state lithium ion battery using garnet-type oxide and Li_3BO_3 solid electrolytes fabricated by screen-printing. *Journal of Power Sources, 238,* 53–56.
33. Park, K., Yu, B. C., Jung, J. W., Li, Y., Zhou, W., Gao, H., et al. (2016). Electrochemical nature of the cathode interface for a solid-state lithium-ion battery: interface between $LiCoO_2$ and garnet-$Li_7La_3Zr_2O_{12}$. *Chemistry of Materials, 28,* 8051–8059.
34. Tadanaga, K., Takano, R., Ichinose, T., Mori, S., Hayashi, A., & Tatsumisago, M. (2013). Low temperature synthesis of highly ion conductive $Li_7La_3Zr_2O_{12}$–Li_3BO_3 composites. *Electrochemistry Communications, 33,* 51–54.
35. Nakayama, M., Kotobuki, M., Munakata, H., Nogami, M., & Kanamura, K. (2012). *Physical Chemistry Chemical Physics: PCCP, 14,* 10008–10014.
36. Inaguma, Y., Liquan, C., Itoh, M., Nakamura, T., Uchida, T., Ikuta, H., & Wakihara, M. (1993). *Solid State Communications, 86,* 689.
37. Kresse, G., & Furthmüller, J. (1996). *Computational Materials Science, 6,* 15.
38. Perdew, J. P. (1985). *Physical Review Letters, 55,* 1665.
39. Henkelman, G., & Jónsson, H. (2000). *the Journal of Chemical Physics, 113,* 9978.
40. Ikeda, M., Yamasaki, T., & Kaneta, C. (2010). *Journal of Physics: Condensed Matter, 22,* 384214.

Lithium-Sulfur Battery

Outline of Li–S Battery Project

Masayoshi Watanabe

Abstract In this project, Li–S batteries have been investigated by a team comprising 12 principal investigators. In particular, we focused on developing Li–S batteries with sparingly solvating electrolytes for cathode active materials. Prior to the start of our project, the dissolution of the discharge products of sulfur (lithium polysulfides: Li_2S_x, $x = 8 \sim 2$) was considered to be inevitable, which resulted in a decrease in the amount of cathode active materials and served as a redox shuttle. We thermodynamically suppressed the dissolution of Li_2S_x by the development of two different materials. One of these are the sparingly solvating electrolytes, which have intrinsically low solubility toward Li_2S_x. The other materials are the microporous carbon substrates in which sulfur is encapsulated, where the direct contact between electrolyte solvent and Li_2S_x is not possible and therefore, Li_2S_x cannot be dissolved into electrolyte. We set a target performance for two different Li–S battery systems. One is the Li–S battery using metallic lithium anode, where the target performance is 300 Wh/kg for energy density. The other includes graphite/Si–Li_2S battery using Li_2S cathode, where the target performance is 200 Wh/kg for energy density. In this part, the state-of-the-art of our project is briefly interpreted.

Keywords Lithium-sulfur battery (Li–S battery) · Sparingly solvating electrolyte · Microporous carbon · Silicon (Si) anode · Lithium sulfide (Li_2S) cathode

It is an obligation of the current society, particularly of the developed countries, to create a sustainable society, where consumption of fossil carbon resources is significantly suppressed and carbon neutral environment (the generation and consumption of CO_2 is balanced) is achieved. For countries with meager natural resources such as Japan, it is urgent to create such society, and the endeavor to achieve it is expected to lead us toward enhanced industrial and social excellence on an international scale. The development of next-generation secondary batteries is one of the promising technologies, which can help us in achieving a low carbon society. Energy density,

M. Watanabe (✉)
Institute of Advanced Sciences, Yokohama National University, Hodogaya-ku, Yokohama, Kanagawa, Japan
e-mail: mwatanab@ynu.ac.jp

K. Kanamura (ed.), *Next Generation Batteries*,
https://doi.org/10.1007/978-981-33-6668-8_24

which is the main parameter that indicates the performance of secondary batteries, significantly depends on charge storage capacity of the anode and cathode materials. In lithium-ion batteries (LIBs), which have the highest energy densities among commercialized secondary batteries, the contribution of anode (graphite) capacity is greater than that of cathode. Therefore, the enhancement of the battery performance has been mainly realized by the development of cathode materials. Consequently, the cathode capacity is approaching the anode capacity, and the energy density of LIBs is approaching its upper limit (~250 Wh/kg). For the next-generation secondary batteries (known as "*beyond LIBs*") having higher energy density than LIBs, it is essential to use higher capacity anode and cathode materials than those utilized in LIBs. Additionally, the cost, natural abundance, and environmental friendliness of the electrode and electrolyte materials should also be considered. Typical examples of such "*beyond LIBs*" include lithium-sulfur (Li–S) battery, multi-valent metal battery (typically Mg battery), and lithium-air (Li–O_2) battery, which are expected to have higher theoretical energy densities than that of LIB.

In this project, abbreviated as ALCA-SPRING and supported by Japan Science and Technology Agency (JST), Li–S batteries have been investigated by a team comprising 12 principal investigators. In particular, we focused on developing Li–S batteries with sparingly solvating electrolytes for cathode active materials. Prior to the start of our project, the dissolution of the discharge products of sulfur (lithium polysulfides: Li_2S_x, $x = 8 \sim 2$) was considered to be inevitable, which resulted in a decrease in the amount of cathode active materials and served as a redox shuttle. The redox shuttle effect is caused by the dissolution of Li_2S_x, diffusion of Li_2S_x to Li anode, and chemical reduction of Li_2S_x (x becomes smaller) at the Li anode during discharge, followed by the back-diffusion of the reduced Li_2S_x to the cathode and its electrochemical oxidation during charge. Such phenomena lead to deteriorated cyclability and low coulombic efficiency of the Li–S battery. To prevent the redox shuttle, typically, $LiNO_3$ was added to the electrolytes. $LiNO_3$ is chemically reduced at the Li anode to form a solid electrolyte interphase (SEI; Li_2O and Li_3N are the main constituents), which prohibits the direct chemical reduction of Li_2S_x (Fig. 1a) [1]. For the suppression of the dissolution of Li_2S_x, the physisorption and/or chemisorption of Li_2S_x to nano-structured materials such as porous carbons and graphene oxide have been attempted. However, SEI formation suffers from the consumption of $LiNO_3$ during charge/discharge cycles due to the breakage of SEI and the active Li surface formation followed by the reaction with $LiNO_3$. The physisorption and/or chemisorption of Li_2S_x is a kinetic trapping phenomenon, and the long-term suppression of the dissolution appears to be difficult.

In this project, we thermodynamically suppressed the dissolution of Li_2S_x (Fig. 1b) [1] by the development of two different materials. One of these are the sparingly solvating electrolytes, which have intrinsically low solubility toward Li_2S_x. The other materials are the microporous carbon substrates in which sulfur is encapsulated, where the direct contact between electrolyte solvent and Li_2S_x is not possible and therefore, Li_2S_x cannot be dissolved into electrolyte.

We also set a target performance for two different Li–S battery systems. One is the Li–S battery (Li|Li_2S_x sparingly solvating electrolyte|S) using metallic lithium

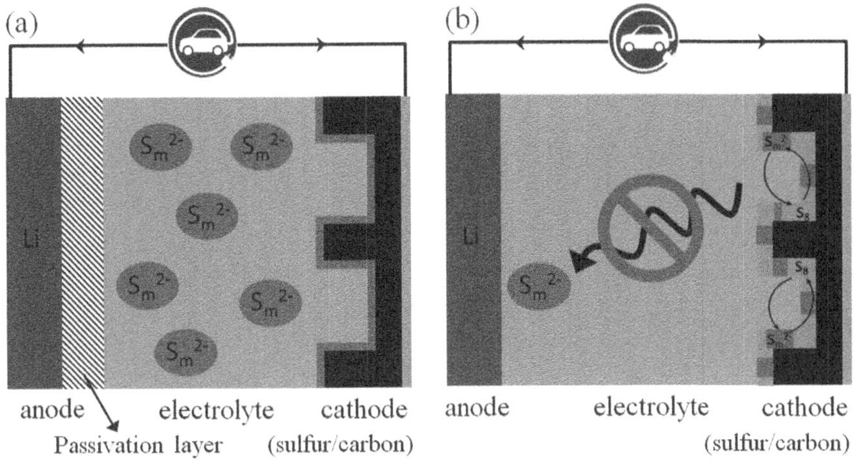

Fig. 1 General liquid-electrolyte-based approaches to problem of Li_2S_x dissolution in Li–S batteries. Reprinted with permission from Ref. [1]. Copyright 2015, Wiley

anode, where the target performance is 300 Wh/kg for energy density, 8/C rate for rate capability, and 500 cycles for cyclability. The other includes graphite/Si–Li_2S battery (Graphite/Si|Li_2S_x sparingly solvating electrolyte|Li_2S) using Li_2S cathode, where the target performance is 200 Wh/kg for energy density, 8/C rate for rate capability, and 500 cycles for cyclability. Hereafter, details of the project studies are described by the principal investigators.

Reference

1. Zhang, S., Ueno, K., Dokko, K., Watanabe, M. (2015). Recent advances in electrolytes for lithium–sulfur batteries. Advanced Functional Materials 1500117

Fundamental Properties and Solubility Toward Cathode Active Materials

Kazuhide Ueno

Abstract Many challenges including insulating properties of active materials, sluggish electrochemical reactions, and dissolution of the sulfur species into electrolyte remain unresolved for practical application of Li–S batteries. Since electrolyte properties are essentially involved in the above serious issues, development of electrolytes is a key to establishing these batteries. Sparingly solvating electrolytes that can thermodynamically suppress the dissolution of the sulfur species are an emerging class of electrolytes used in Li–S batteries. In this chapter, glyme-based solvate ionic liquids were discussed as an example of the sparingly solvating electrolytes. Saturated solubility of lithium polysulfides (Li_2S_m) in various types of electrolytes ranging from molecular solvents to aprotic and solvate ionic liquids was also determined to study factors that govern the dissolution of the sulfur species into electrolyte. The effects of electrolyte properties on the cell performance were also discussed.

Keywords Solubility · Polysulfides · Electrolytes

1 Introduction

The basic principle of lithium-sulfur (Li–S) batteries was proposed in the 1960s, and extensive research in this field has been performed to date. However, it has not been put into practical use due to various inherent issues, such as (i) low electronic conductivity of the active materials, (ii) the volume change observed in the active material during the charge and discharge process, (iii) dissolution of lithium polysulfide (Li_2S_m), which is a reaction intermediate of sulfur-based active materials, into the electrolyte solution, and (iv) dendrite formation on the metallic lithium anode [1] Among them, the dissolution of Li_2S_m into the electrolyte is the most serious problem specific to Li–S batteries.

K. Ueno (✉)
Department of Chemistry and Life Science, Yokohama National University, Yokohama, Japan
e-mail: ueno-kazuhide-rc@ynu.ac.jp

K. Kanamura (ed.), *Next Generation Batteries*,
https://doi.org/10.1007/978-981-33-6668-8_25

Recently, various parameters necessary for realizing a Li–S battery with high energy density (300 Wh kg^{-1} or more) have been investigated in detail [2]. Particularly, the importance of reducing the ratio of the electrolyte to the amount of sulfur active species (E/S [μL/mg] ratio) and increasing the amount of active sulfur-based material per unit area (i.e., sulfur loading) in the cathode have been recognized in the Li–S battery research community. However, it has been reported that high capacity is not obtained in a Li–S cell using ether-based organic electrolytes with a small E/S ratio in which Li_2S_m is saturated under lean electrolyte conditions [3]. It is also not easy to form a thick film of sulfur cathode while maintaining an efficient electron/ion conduction pathway in a sulfur cathode containing a large amount of insulating sulfur; therefore, achieving a high charge/discharge capacity with a thick film electrode is also an issue to be addressed. In this context, the properties of the electrolyte are closely relevant to the dissolution issue of Li_2S_m in the electrolyte and the challenge of reducing the E/S ratio.

Recent studies have shown the efficient suppression of Li_2S_m dissolution in certain highly concentrated electrolytes [4, 5] and molten Li salt solvates (solvate ionic liquids, SILs) [6] due to a combination of effects, including the scarcity (low activity) of non-coordinating "free" solvents capable of dissolving Li_2S_m and the common ion effect arising from intrinsically high Li salt concentrations. Certain aprotic ionic liquids have also been found to show limited Li_2S_m solubility to a similar level to these highly concentrated electrolytes due to their weak coordinating properties toward ionic materials [7, 8]. Therefore, undesired side reactions caused by the redox shuttle effect and a loss of the active materials from the cathode can be largely mitigated in electrolytes with low Li_2S_m solubility. This enables stable charge/discharge cycling with high Coulombic efficiency in Li–S cells. Such "sparingly solvating" liquid electrolytes are now considered as a prospective alternative to ether-based organic electrolytes toward achieving long-lived and high-energy density practical Li–S batteries under lean electrolyte conditions [9]. In this chapter, we introduce the fundamental properties and solubility toward cathode active materials of sparingly solvating liquid electrolytes based on ionic liquids and molten Li salt solvates.

2 Conventional Liquid Electrolytes for Li–S Batteries

Before introducing the properties of the emerging electrolytes with low Li_2S_m solubility, previous research efforts on liquid electrolytes used for Li–S batteries are briefly summarized here. In order to promote the dissociation of Li salts, polar solvents are generally used in the organic electrolyte solution of lithium secondary batteries. Various aprotic polar solvents have been studied during the early stages of research on Li–S battery electrolytes [10]. However, it was found that the nucleophilic polysulfide anion ($S_m{}^{2-}$) undergoes a side reaction with carbonate solvents, such as ethylene carbonate and diethyl carbonate, which are the main components of established lithium-ion battery electrolytes [11]. From the viewpoint of their stability toward metallic lithium anodes, ether solvents are commonly used as the main solvent in Li–S battery electrolytes [12]. In addition, S_m^{2-} was also found to react irreversibly with some of anions of the Li salt, such as BF_4^-, PF_6^-, and bis(fluorosulfonyl)amide ($[FSA]^-$), which are often used in lithium secondary batteries [8, 12, 13]. Therefore, there is a limitation on the kind of Li salt that can be used for Li–S batteries as well as the solvent type. Chemically stable and highly dissociative, lithium bis(trifluoromethanesulfonyl) amide (LiTFSA) is generally used as the Li salt in Li–S battery electrolytes. Although organic electrolyte solutions containing LiTFSA suffer from corrosion of the aluminum current collector at ~3.8 V vs Li/Li^+, Li–S batteries have no problem concerned with Al corrosion since the operating voltage of Li–S batteries is ~2 V. An organic electrolyte solution based on mixed ethers, 1,3-dioxolane/dimethoxyethane (DOL/DME, 1:1 v/v), containing 1 M LiTFSA and ca. 0.1 M $LiNO_3$ additive is used as the standard electrolyte for Li–S batteries.

3 Fundamental Properties of Solvate Ionic Liquids

Some stoichiometric mixtures of molecular solvents (or ligands) and Li salts can form low-melting Li salt solvates. The molten state of these Li salt solvates can be categorized into a subclass of ionic liquids, SILs, if they are comprised of discrete complex cations and their counter anions in the liquid state [14] The physicochemical properties and Li ion solvation structure of mixtures of oligoether solvents (glymes) and Li salts ($[Li(glyme)_n]X$) have been thoroughly studied as one example of molten Li salt solvates that have potential to be categorized as SILs [15, 16].

The stability of $[Li(glyme)_n]^+$ complex cations in the liquid state can be readily evaluated using self-diffusion coefficient measurements of glyme (D_G) and Li ions (D_{Li}) using pulsed-filed-gradient (PFG) NMR spectroscopy [17, 18]. The diffusivity ratio of D_G/D_{Li} is depicted in Fig. 1 for [Li(G3)]X and [Li(G4)]X with different counter anions (Fig. 1a) and $[Li(glyme)_n]X$ with different chain lengths (Fig. 1b), respectively. The D_G/D_{Li} ratio is unity (i.e., $D_G \approx D_{Li}$) for $[Li(G2)_{4/3}][TFSA]$, [Li(G3)][TFSA], [Li(G4)][TFSA], and [Li(G4)][BETA], suggesting the joint diffusion of the longer glyme molecules and Li ions in the form of complex cations. This supports the formation of SILs consisting of long-lived $[Li(glyme)_n]$ complex cations and the counter anion, X^-, where free glyme molecules are scarcely present. In contrast, [Li(G3)]X and [Li(G4)]X containing other anions ($[TFA]^-$, NO_3^-, $[OTf]^-$, BF_4^-, and ClO_4^-), $[Li(THF)_4][TFSA]$ and $[Li(G1)_2][TFSA]$ show D_G/D_{Li} values > 1, indicating either a frequent ligand exchange between the unstable complex cations and counter anions or the presence of free solvent molecules in the molten state. Thus, these molten Li salt solvates should not be classified as SILs. The D_G/D_{Li} values increase upon increasing the Lewis basicity of the anions ($ClO_4^- < BF_4^- < [OTf]^- < NO_3^- < [TFA]^-$) for [Li(G3)]X and [Li(G4)]X, respectively. This suggests that the Li ion–solvent interaction becomes weaker upon increasing the interaction between the Li ions and Lewis basic anion X^-. The D_G/D_{Li} value also increases upon decreasing the number of oxygen atoms in the solvent even with a constant molar ratio of Li ion and ligand oxygen atoms ($[O]/[Li^+]$) of 4 or 5. Therefore, the stability

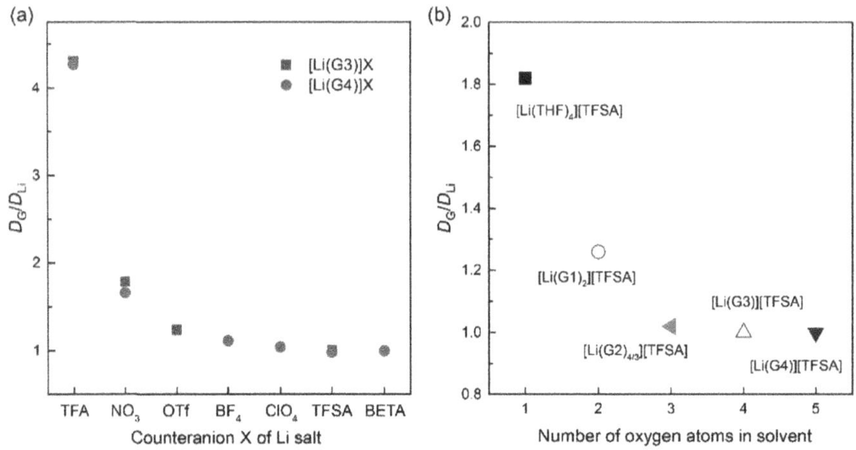

Fig. 1 The diffusivity ratio of solvent and Li ion (D_G/D_{Li}) for **a** [Li(G3)]X and [Li(G4)]X and **b** [Li(glyme)$_n$][TFSA]. TFA: trifluoroacetate, OTf: trifluoromethanesulfonate, TFSA: bis(trifluoromethanesulfonyl)amide, BETA: bis(pentafluoroethylsulfonyl)amide, THF: tetrahydrofuran, G1: monoglyme, G2: diglyme, G3: triglyme, G4: tetraglyme

of the complex cations also depends on the number of coordination sites in the solvent (i.e., chelate effect). Consequently, molten glyme-Li salt solvates are not necessarily classified as SILs, and [Li(glyme)$_n$]X can be divided into two distinct liquid states, SILs or concentrated solutions. The formation of stable complex cations depends on the competition between the glyme molecules and counter anions that interact with the Li ions, and this is only attainable when combined with weakly coordinating, bulky anions (such as perfluorinated sulfonylamide anions), and multidentate ligands (such as G2, G3, and G4). Indeed, the fraction of free glyme molecules estimated using Raman spectroscopy was only a few percent for SILs, whereas it exceeds 10% for other [Li(glyme)$_n$]X [19]. It was also found that the absence of the free glyme molecules in [Li(glyme)$_n$]X has a significant association with enhanced thermal and electrochemical properties for SILs among [Li(glyme)$_n$]X. Such "sparingly solvating" liquid electrolytes are now considered as prospective alternatives to ether-based organic electrolytes toward achieving long-lived and high-energy density practical Li–S batteries under lean electrolyte conditions.

4 Solubility Toward Cathode Active Materials

An efficient way to impede the dissolution of Li_2S_m via the physical confinement of molecular sulfur (S_8) in mesoporous carbon (CMK-3) was proposed in 2009 [20]. This kinetic trapping of S_8/Li_2S_m using mesoporous carbon as the host material has become very popular and has been studied intensively as the main research topic for tackling the Li_2S_m dissolution issue [21]. On the other hand, thermodynamic control of Li_2S_m dissolution using liquid electrolytes has also been recognized as an effective strategy, in addition to the use of solid-state inorganic or polymer electrolytes that can inherently block the dissolution of Li_2S_m [10]. Here, we show examples of liquid electrolytes that can achieve low solubility of Li_2S_m.

S_8 is very soluble in non-polar solvents such as carbon disulfide and toluene, whereas the final discharge product of Li_2S is only soluble in highly polar solvents such as water and alcohols. The intermediate products of Li_2S_m are expected to dissolve in a wide range of low-polarity to high-polarity solvents depending on their polysulfide chain length. Therefore, it is difficult to find an appropriate solvent that does not dissolve all the sulfur-based active material components. In fact, the solubility of the most soluble Li_2S_m (m = 8) species is > 6000 mM S (value converted per sulfur atom) in ether-based organic electrolytes, such as 1 M LiTFSA DOL/DME [7]. However, IL-based electrolytes and highly concentrated electrolytes were found to significantly suppress the dissolution of Li_2S_m, and the poor solubility of Li_2S_m was reported to be relevant to good cyclability and high coulombic efficiency in Li–S batteries. Consequently, the solubility of S8, Li_2S_m, and Li_2S in various liquid electrolytes was systematically investigated and it has been revealed that all sulfur species are poorly soluble in certain ILs and highly concentrated electrolytes [6, 8, 13, 22] As seen in Fig. 2, the solubility of Li_2S_m (m = 8) is as low as 100 mM S or less in aprotic ILs and molten Li salt solvate (SILs) with weakly coordinating anions such as $[TFSA]^-$ and $[BETA]^-$. Upon comparison of the Li_2S_m solubility

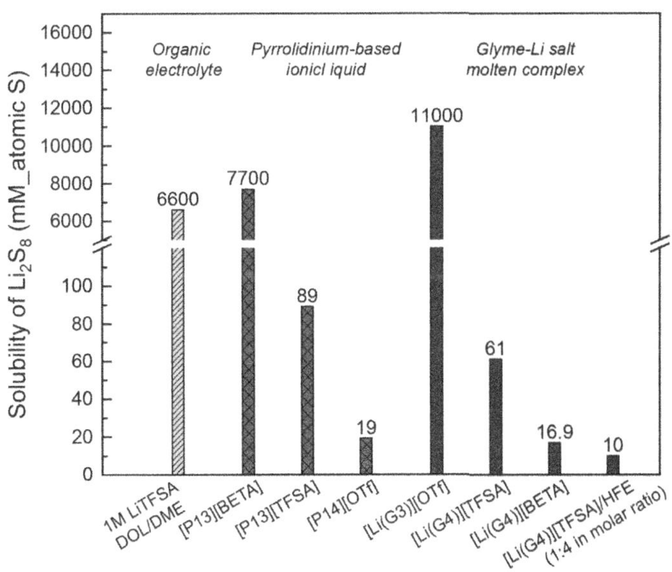

Fig. 2 Solubility of Li_2S_m (m = 8) in various liquid electrolytes, P13: *N*-methyl-*N*-propyl pyrrolidinium, P14: *N*-butyl-*N*-methyl pyrrolidinium, BETA: bis(pentafluoroethylsulfonyl)amide, TFSA: bis(trifluoromethanesulfonyl)amide, OTf: trifluoromethanesulfonate, G3: triglyme, G4: tetraglyme, HFE: 1,1,2,2–tetrafluoroethyl 2,2,3,3–tetrafluoropropyl ether

in 1 M LiTFSA in DOL/DME and [Li(G4)][TFSA], the low solubility of Li_2S_m can be simply interpreted by the common ion effect at first glance since the Li salt concentration reaches near saturation limit for [Li(G4)][TFSA]. On the other hand, the solubility of Li_2S_m (m = 8) in [P14][OTf] and [Li(G3)][OTf] is higher than that in 1 M LiTFSA in DOL/DME. Obviously, the anionic structure of the liquid electrolyte greatly affects the solubility of Li_2S_m in aprotic ILs and molten Li salt solvates. The high solubility despite the absence of a molecular solvent in [P14][OTf] is due to the strong coordination of Li_2S_m by Lewis basic $[OTf]^-$. Even though the Li salt concentration of [Li(G3)][OTf] (c_{Li} = 3.9 M) is much higher than an ether-based organic electrolyte, a non-negligible amount of the free glyme molecules is present and these free glyme molecules and Lewis basic $[OTf]^-$ anions can solubilize a large amount of Li_2S_m. From the above results, there is a correlation between the solubility of Li_2S_m and the Lewis basicity or donor properties of the liquid electrolyte; the coordination of the solvent and/or anion to Li ions in Li_2S_m seems to be a crucial factor for the solubility of Li_2S_m.

The addition of a non-coordinating diluent such as a hydrofluoroether (HFE) to SILs was found to further reduce the solubility of Li_2S_m (m = 8) to ~10 mM_S [6]. Furthermore, dilution with HFE resulted in enhanced ionic conductivity on account of the remarkable decrease in the viscosity of the SILs. These SIL-based electrolytes and highly concentrated electrolytes that enable the thermodynamic suppression of Li_2S_m dissolution are key materials, which can be used as sparingly solvating liquid electrolytes to achieve long-term cycling and high-energy density in practical Li–S batteries.

5 The Effect of Sparingly Solvating Electrolytes on Li–S Battery Performance

In Li–S batteries using ether-based organic electrolytes, Li_2S_m formed in the discharge process is highly solubilized and the charge/discharge reaction proceeds via the dissolution/precipitation chemistry of Li_2S_m via solid phase (S_8)–liquid phase (Li_2S_m, $m \geq 3$)–solid phase (Li_2S_2, Li_2S) [1]. Parasitic reactions between the dissolved Li_2S_m and metallic lithium anode (i.e., redox shuttle effect) and self-discharge are suppressed by a solid electrolyte interphase (SEI) film formed on the lithium electrode surface (Fig. 3a). It has been reported that the $LiNO_3$ additive and dissolved Li_2S_m are reductively decomposed to form an effective SEI containing Li_2O, LiN_xO_y, Li_2S, Li_2S_2, $Li_2S_2O_3$, and Li_2SO_4 on the lithium electrode surface [23]. However, the SEI film may not be stable over repeated plating/stripping reactions of the lithium anode, and the SEI-forming additives are consumed and finally depleted after long-term charge/discharge cycling.

On the other hand, in Li–S batteries using sparingly solvating electrolytes, all the electrochemical reactions on the sulfur cathode basically occur in the solid state because Li_2S_m does not dissolve into the electrolyte (Fig. 3b). A Li–S battery using

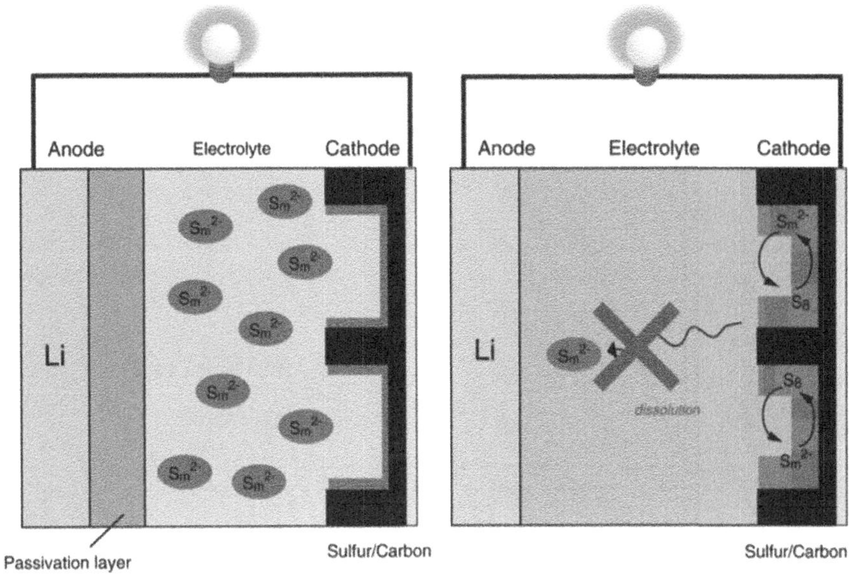

Fig. 3 Liquid-electrolyte-based Li–S cells using **a** ether-based organic electrolytes and **b** sparingly solvating electrolytes

[Li(G4)][TFSA] demonstrated stable charge/discharge cycling over 800 cycles with a high coulombic efficiency of >98% [24]. However, the capacity retention rate was ~60% after 600 cycles, and the capacity decay was presumably due to deterioration of the electron conducting pathway associated with the repeated volume change of the active material in the solid phase during charge/discharge cycling. Li–S batteries using SIL electrolytes also faces poor rate capability when compared to ether-based organic electrolytes because of the sluggish solid-phase reaction of sulfur-based active materials with low electron conductivity.

It has been recognized that the use of sparingly solvating electrolytes is extremely important in improving the energy density via increasing the amount of active sulfur and reducing the amount of electrolyte [9, 25]. Figure 4 shows the relationship between the E/S ratio and the fraction of potentially soluble sulfur active material as Li_2S_m (m = 8) in the cathode, where the fraction can be calculated from the Li_2S_m solubility. In the DOL/DME organic electrolyte, all sulfur species in the cathode can dissolve as Li_2S_m (m = 8) at an E/S ratio of > 5, but Li_2S_m becomes saturated at an E/S ratio of < 5. Therefore, when the E/S ratio is small, the dissolution of a large amount of Li_2S_m results in an increase in the viscosity of the electrolyte and a decrease in the ion conductivity. In fact, the relationship between the E/S ratio and discharge capacity has been investigated and the discharge capacity of the cell using the DOL/DME organic electrolyte was found to decrease from 1100 mAh g^{-1} at E/S = 7 to 700 mAh g^{-1} at E/S = 5, and 200 mAh g^{-1} at E/S = 3 [26]. On the other hand, the influence of the E/S ratio on the amount of dissolved Li_2S_m is

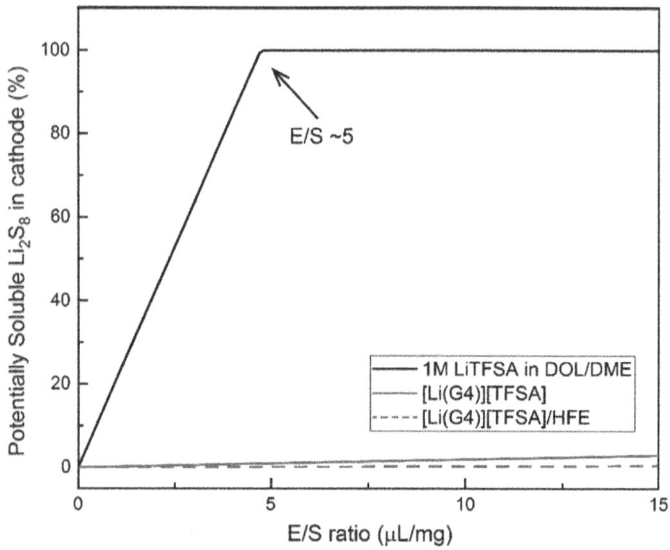

Fig. 4 Relationship between the fraction of potentially soluble Li_2S_m (m = 8) in sulfur cathode and E/S ratio

small with sparingly solvating electrolytes such as SILs, and most of the sulfur-based active materials remain as a solid in the cathode. Indeed, it has been reported that the discharge capacity of a Li–S cell using sparingly solvating electrolytes is less affected by the E/S ratio [25].

Improving the reversibility of the lithium metal electrode while suppressing the dendritic growth of lithium metal is also an important issue to consider when improving the cycle life. In this respect, highly concentrated electrolytes have been reported to exhibit higher reversibility in the lithium plating/stripping reactions when compared with common organic electrolytes [27]. Thus, sparingly solvating electrolytes based on highly concentrated electrolytes and SILs have proven useful when addressing not only the Li_2S_m dissolution issues but also the challenge regarding the dendritic growth of the metallic lithium anode.

6 Conclusions

In this chapter, we have introduced the recent research findings on the thermodynamic suppression of Li_2S_m dissolution using liquid electrolytes for practical Li–S batteries. Among battery systems using electrode active materials that cause a dissolution/precipitation reaction, many of the commercially available batteries are operated by suppressing the solubility of the electrode active material in the electrolyte solution, whereas there are few examples of practical batteries in which a large amount of electrode active material is solubilized in the electrolyte solution.

For example, in lead acid batteries, the extremely low solubility of $PbSO_4$ leads to stable charge and discharge cycling. Also, in Na–$NiCl_2$ batteries, the solubility of the $NiCl_2$ active material is very small in the molten salt electrolyte. Thus, from the historical background of practical batteries, the use of a sparingly solvating electrolyte is now considered a reasonable approach to increase the energy density of Li–S batteries with prolonged charge/discharge cycling life.

References

1. Manthiram, A., Fu, Y. Z., Chung, S. H., Zu, C. X., & Su, Y. S. (2014). *Chemical Reviews, 114,* 11751–11787.
2. Eroglu, D., Zavadil, K. R., & Gallagher, K. G. (2015). *Journal of the Electrochemical Society, 162,* A982–A990.
3. Hagen, M., Hanselmann, D., Ahlbrecht, K., Maça, R., Gerber, D., & Tübke, J. (2015). *Advanced Energy Material, 5,* 1401986.
4. Suo, L., Hu, Y.-S., Li, H., Armand, M., & Chen, L. (2013). *Nature Communication, 4,* 1481.
5. Shin, E. S., Kim, K., Oh, S. H., & Cho, W. I. (2013). *Chemical Communications, 49,* 2004–2006.
6. Dokko, K., Tachikawa, N., Yamauchi, K., Tsuchiya, M., Yamazaki, A., Takashima, E., et al. (2013). *Journal of the Electrochemical Society, 160,* A1304–A1310.
7. Park, J.-W., Yamauchi, K., Takashima, E., Tachikawa, N., Ueno, K., Dokko, K., & Watanabe, M. (2013). *Journal of Physical Chemistry C, 117,* 4431–4440.
8. Park, J.-W., Ueno, K., Tachikawa, N., Dokko, K., & Watanabe, M. (2013). *Journal of Physical Chemistry C, 117,* 20531–20541.
9. Cheng, L., Curtiss, L. A., Zavadil, K. R., Gewirth, A. A., Shao, Y., & Gallagher, K. G. (2016). *ACS Energy Letters, 1,* 503–509.
10. Zhang, S., Ueno, K., Dokko, K., & Watanabe, M. (2015). *Advance Energy Material, 5,* 1500117.
11. Gao, J., Lowe, M. A., Kiya, Y., & Abruña, H. D. (2011). *Journal of Physical Chemistry C, 115,* 25132–25137.
12. Zhang, S. S. (2013). *Journal of Power Sources, 231,* 153–162.
13. Ueno, K., Park, J. W., Yamazaki, A., Mandai, T., Tachikawa, N., Dokko, K., & Watanabe, M. (2013). *Journal of Physical Chemistry C, 117,* 20509–20516.
14. Austen Angell, C., Ansari, Y., Zhao, Z. (2012). *Faraday Discussions, 154,* 9–27.
15. Mandai, T., Yoshida, K., Ueno, K., Dokko, K., & Watanabe, M. (2014). *Physical Chemistry Chemical Physics: PCCP, 16,* 8761–8772.
16. Watanabe, M., Dokko, K., Ueno, K., & Thomas, M. L. (2018). *Bulletin. Chemical Society of Japan, 91,* 1660–1682.
17. Ueno, K., Yoshida, K., Tsuchiya, M., Tachikawa, N., Dokko, K., & Watanabe, M. (2012). *The Journal of Physical Chemistry B, 116,* 11323–11331.
18. Zhang, C., Ueno, K., Yamazaki, A., Yoshida, K., Moon, H., Mandai, T., et al. (2014). *The Journal of Physical Chemistry B, 118,* 5144–5153.
19. Ueno, K., Tatara, R., Tsuzuki, S., Saito, S., Doi, H., Yoshida, K., et al. (2015). *Physical Chemistry Chemical Physics: PCCP, 17,* 8248–8257.
20. Ji, X., Lee, K. T., & Nazar, L. F. (2009). *Nature Materials, 8,* 500.
21. Eftekhari, A., & Kim, D.-W. (2017). *Journal of Materials Chemistry A, 5,* 17734–17776.
22. Zhang, C., Yamazaki, A., Murai, J., Park, J.-W., Mandai, T., Ueno, K., et al. (2014). *Journal of Physical Chemistry C, 118,* 17362–17373.
23. Xiong, S., Xie, K., Diao, Y., & Hong, X. (2014). *Journal of Power Sources, 246,* 840–845.
24. Seki, S., Serizawa, N., Takei, K., Umebayashi, Y., Tsuzuki, S., & Watanabe, M. (2017). *Electrochemistry, 85,* 680–682.

25. Matsumae, Y., Obata, K., Ando, A., Yanagi, M., Kamei, Y., Ueno, K., & Watanabe, M. (2019). *Electrochemistry*. https://doi.org/10.5796/electrochemistry.19-00021.
26. Hagen, M., Fanz, P., & Tübke, J. (2014). *Journal of Power Sources, 264,* 30–34.
27. Qian, J., Henderson, W. A., Xu, W., Bhattacharya, P., Engelhard, M., Borodin, O., & Zhang, J.-G. (2015). *Nature Communications, 6,* 6362.

Thermodynamic and Structural Aspects of Solvate Ionic Liquid Formation

Yasuhiro Umebayashi, Nana Arai, and Hikari Watanabe

Abstract Development of the electrolyte solution plays a key role in realization for lithium-sulfur (Li–S) battery. The lithium ion local structure and the liquid structures of an electrolyte solution have direct influence on the battery performance because of the electromotive force, the charging/discharging speed and the life time. The activity for the free solvent directly contributes to the electrode potential in the concentrated electrolyte such as the solvate ionic liquid. Lithium-glyme solvate ionic liquid consists of only the glyme-solvated lithium ion and its counter ion; similarly a super-concentrated electrolyte solution contains no full solvated Li^+ ion. However, it is unclear whether both electrolyte solutions should be regarded the same. From a thermodynamic point of view, the activity and/or the activity coefficient for the solvent is useful for the distinction among them. In this chapter, the liquid structures and the thermodynamic parameters like an activity coefficient are described for solvate ionic liquid.

Keywords Solvate ionic liquid · Liquid structure · Activity · Lithium ion local structure

Y. Umebayashi (✉)
Institute of Science and Technology, Niigata University, Niigata, Japan
e-mail: yumescc@chem.sc.niigata-u.ac.jp

N. Arai · H. Watanabe
Graduate School of Science and Technology, Niigata University, Niigata, Japan

N. Arai
Inorganic Material Group Material Analysis Team, Energy & Functional Materials Research Laboratory, Sumitomo Chemical Co., Ltd., Tokyo, Japan

H. Watanabe
Department of Pure and Applied Chemistry, Faculty of Science and Technology, Tokyo University of Science, Tokyo, Japan

K. Kanamura (ed.), *Next Generation Batteries*,
https://doi.org/10.1007/978-981-33-6668-8_26

1 Introduction

Angell et al. proposed solvate or chelate ionic liquids as one of the categories for room-temperature ionic liquids composed of solvate ion and a counter ion [1]. On the other hand, molten salts of hydrated metal ions are well-known as hydrate melts. For example, a lot of nitrate salts of hydrated metal ions yield a low melting point, such as the melting point is about 30 °C for $LiNO_3 \cdot 4H_2O$. Recently, Suo et al. reported highly concentrated $LiN(SO_2CF_3)_2$ aqueous solution of 22 mol kg^{-1} ($LiN(SO_2CF_3)_2$:H_2O = 1:2.3) is a liquid at an ambient temperature, and is available as an electrolyte for the aqueous Li-ion secondary batteries of 2.0 V [2]. In addition, highly concentrated Li-salt solutions attracted much attention as the Li^+ conducting electrolyte solutions for the next-generation lithium secondary batteries [3–5]. Such highly concentrated salts solutions are currently called as "water-in-salt" or "super-concentrated electrolyte solution".

Here, we have a simple question: Are solvate ionic liquids and super-concentrated electrolyte solution the same, or different? We should consider whether the glyme—Li salt equimolar molten mixtures can indeed be regarded as solvate ionic liquids or are super-concentrated solutions of the Li salts in the glyme solvents [6]. The ionicity $\Lambda_{imp}/\Lambda_{NMR}$ is convenient for the discussion, where Λ_{imp} and Λ_{NMR} represent the molar conductivity measured by the ac impedance method and that evaluated based on the self-diffusion coefficients for the component cation and anion measured by the PFG-NMR technique using the Nernst—Einstein equation, respectively [7].

As aforementioned in Chapter "Fundamental Properties and Solubility Toward Cathode Active Materials", Ueno et al. clearly proved that the ratio of self-diffusion coefficients D_G/D_{Li} and D_X/D_{Li} for [Li(glyme)]X depends on ionicity, where D_G, D_{Li} and D_X stand for the self-diffusion coefficient for glyme, Li^+ and anion X, respectively, as seen in Fig. 1 [8]. According to the figure, the [Li(glyme)]X can be divided broadly into two classes based on the D_G/D_{Li} ratio. The D_G/D_{Li} ratio is almost unity when the $\Lambda_{imp}/\Lambda_{NMR}$ is greater than 0.4, which strongly suggests the formation of [Li(glyme)] complex cations in the liquid state. These mixtures can be regarded as lithium "solvate ionic liquids". On the contrary, when the $\Lambda_{imp}/\Lambda_{NMR}$ is smaller than 0.4, the glymes can diffuse faster as the solvent among the cations, anions and solvents components in the solution is similar to non-aqueous electrolyte solutions. Hence, it can be said that $\Lambda_{imp}/\Lambda_{NMR} = 0.4$ is the dividing value for whether [Li(glyme)]X are "solvate ionic liquids" or "super-concentrated electrolyte solutions". It should be noted that "free" glyme is practically undetectable, and all the glyme molecules participate in complex formation at $\Lambda_{imp}/\Lambda_{NMR} > 0.4$, while uncoordinated glyme would exist even in the presence of an equivalent amount of the Li salt at $\Lambda_{imp}/\Lambda_{NMR} < 0.4$. Ionicity is based on the transport properties such as a molar ionic conductivity and a self-diffusion coefficient. In that sense, it can be a dynamic measure for the definition of the solvate ionic liquids. From structural and thermodynamic point of views, we can discuss on the static definition. In this session, we survey structural and thermodynamic definition of the solvate ionic liquids.

2 Structural Aspects

One of the clearest evidence is crystal structure for the glyme coordination to the Li^+ to form $[Li(glyme)]^+$ complex cation. Henderson et al. have reported various crystal structures of the glyme-Li salts. For triglyme, glyme:LiX crystals (X: $CF_3SO_3^-$, BF_4^-, ClO_4^- and AsF_6^-) were isolated and structurally characterized [9]. Among them, almost all of Li^+ forms the contact ion-pair (CIP) and aggregates with these anions (AGG). Despite the differences in anion symmetry and structure, all of the solvates have the same form of structure. Each five-coordinate Li^+ cation is coordinated by four oxygen atoms (two each from two different triglyme molecules) and one donor atom from a single anion. The triglyme molecules are coordinated to two different Li^+ cations, resulting in linear, polymeric chains. On the other hand, BPh_4^- forms a new form of glyme:LiX complex, (triglyme):$LiBPh_4$ = 1:4, in which the five-coordinate Li^+ are coordinated solely by the triglyme oxygen atoms. One of the oxygen atoms, however, uses its second electron lone pair to coordinate a second Li^+ cation. Dimeric solvates thus form from two of these in which the cations are each coordinated to oxygen atoms from two triglyme molecules. In addition, polymeric chains form between the cations and anions, but two separate forms of cation coordination sites exist in the $(triglyme)_{2/3}$:$LiCF_3SO_3$ crystal. In the first site, the five-coordinate Li^+ cations are coordinated by three oxygen atoms from a single triglyme molecule and two anion oxygen atoms (one each from two anions). The fourth oxygen atom of the triglyme molecule is coordinated to another cation. The second site consists of four-coordinate Li^+ cations coordinated by two of the triglyme terminal oxygen atoms and two anion oxygen atoms (one each from two anions). TFSA $[(CF_3SO_2)_2 N^-)$ and BETI $((C_2F_5SO_2)_2N^-]$ anions yield rather different crystals from aforementioned anions [10]. In particular, the triglyme:LiBETI crystal consists of a monomeric $[Li(triglyme)]^+$ cation adopting a crown-ether(12-crown-4)-like conformation, and a BETI anion. In this complex, Li^+ is coordinated to four oxygen atoms from a single triglyme molecule and one oxygen atom from the BETI anion. The crystal structure of the low-melting triglyme:LiTFSA was reported as having a similar coordination structure to that consisting of BETI, with monomeric $[Li(triglyme)]^+$ cations. However, Li^+ is six-fold coordinated, with four oxygen atoms from a single triglyme molecule and two oxygen atoms from a single TFSA anion.

For tetraglyme, crystal structures are reported for X = AsF_6^-, BF_4^- and $CF_3CO_2^-$. The tetraglyme:$LiAsF_6$ crystal consists of double-helix dimers in which two six-coordinate Li^+ cations are coordinated by two tetraglyme molecules [11]. Each cation is coordinated by three oxygen atoms from each tetraglyme molecule with the central ethylene oxide units of each tetraglyme molecule coordinated to both cations. The $(tetraglyme)_{1/2}$:$LiBF_4$ crystal has two crystallographically different Li^+ cation coordination sites. In the first coordination site, half of the cations are coordinated by the tetraglyme molecules as $[(tetraglyme)\text{-}Li]_2^{2+}$ double-helix dimers very similar to those found in tetraglyme:$LiAsF_6$, but with a slightly different symmetry. In the second coordination site, the other half of the Li^+ cations are coordinated by the BF_4^- anions forming linear ionic chains. These cations have tetrahedral coordination by four fluorine atoms (one each from four anions). Each of the BF_4^- anions is coordinated to two Li^+ cations. One of the BF_4^- anions is disordered over three positions of equal occupancy. The anions thus form linear $(–(BF_4^-)_2–Li^+)_n$ chains arranged between the $[(\text{tetraglyme})Li]_2^{2+}$ double-helix dimers. $(\text{Tetraglyme})_{2/5}:LiCF_3CO_2$ crystal also has two different Li^+ cation coordination sites. The Li^+ cations, however, have five-fold coordination by three oxygen atoms in the ethyleneoxides and two anions. One anion is coordinated to both cations through a single anion oxygen donor atom. The second oxygen donor atom of the anion is coordinated to a cation in the second site. These anions are thus coordinated to three different Li^+ cations. The second anion coordinating each cation in the first site also coordinates a cation in the second site and these anions are coordinated to two different Li^+ cations. The four-coordinate Li^+ cations in the second site are tetrahedrally coordinated by four anions and serve as bridges between the tetraglyme-Li^+ cation solvates.

Single crystal structure analysis is useful for considering the Li^+ solvation structure in the solvate ionic liquids. However, it should be noted that the translational motion is the most characteristic for liquids comparing with the solids, and also that the solvent and/or anion exchange reactions generally can occur only in solutions. Hence, neutron/X-ray structure analyses in liquids/solutions is important and indispensable for the discussion. Atkin et al. showed bulk liquid structures of equimolar mixtures of tetraglyme with LiTFSA or $LiNO_3$ as the representative "good" and "poor" solvate ionic liquids by means of the neutron diffraction (scattering) technique with the empirical potential-structure refinement (EPSR) analysis [12]. EPSR analysis adequately reproduced the observed neutron structure factors so that the pair correlation functions $g(r)$ and the atomic coordination numbers were evaluated and discussed. As clearly indicated with the coordination numbers, the Li^+ prefers the tetraglyme oxygen atoms to those of the TFSA anions in the tetraglyme:LiTFSA equimolar mixture; the coordination numbers are 2.27, 1.72 and 3.99 for the oxygen atoms in the anion, that in tetraglyme and the sum of them, respectively, while the reverse is the case for tetraglyme:$LiNO_3$; 0.95, 4.39 and 5.32, respectively. The relative coordination power follows the order of $TFSA^- <$ tetraglyme $< NO_3^-$ based on their Gutmann's donor numbers of 22.5, 69.4 and 87.9 kJ mol^{-1} for $TFSA^-$, tetraglyme and NO_3^-, respectively. Therefore, Lewis basicity can be used to indicate whether a given glyme–lithium salt combination will form a "good" or a "poor" solvate ionic liquid.

Although the structural parameters are not obtained such as the atomic distance and the atomic coordination numbers, Raman spectroscopy has an advantage against the neutron/X-ray techniques; it yields the solvation number, which is the number of directly coordinated solvent molecules to the central metal ion. Figure 1 displays a fingerprint region for the triglyme-based solvate ionic liquids. As shown in Fig. 1a, neat triglyme exhibits three Raman bands at 810, 835 and 850 cm^{-1}, while Raman spectra for the 1:1 mixture of the lithium salt have different spectral features strongly depending on the anions. Though the anion dependence looks rather complicated, it can be noticed that the peaks at around 860–880 cm^{-1} are the sharpest and the most intense in this region for the anions of weakly coordinating power, whilst those at around 800–860 cm^{-1} are the largest than those appeared in the higher frequency

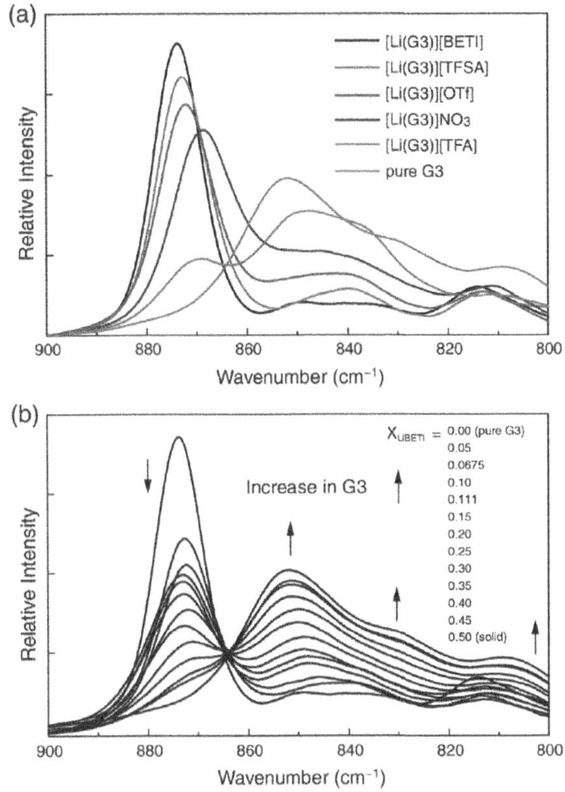

Fig. 1 **a** Raman spectra for the 1:1 mixtures of the lithium salts ($BETI^-$, $TFSA^-$, OTf^-, NO_3^-, TFA^-) and neat triglyme and **b** LiBETI concentration dependence in Raman spectra. Reproduced from Ref. [13] with permission from the PCCP Owner Societies

region for the strongly coordinating anions. Comparing with the observed Raman spectra for the crystals and those theoretically predicted for the isolated glyme-solvate Li^+ in vacuum, the Raman band appearing at around 860–880 cm^{-1} can reasonably be attributed to the triglyme molecule coordinated to the Li^+. This is also clearly shown in the Li-salt concentration dependence in Raman spectra. As shown in Fig. 1b, Raman bands intensity at 810, 835 and 850 cm^{-1} decreased, while that at 873 cm^{-1} increased as the increase of the concentration for the Li salt of weakly coordinating anion. The solvation number can be evaluated with the quantitative analysis of the Li salt concentration dependence on the Raman band intensity; 1.13 and 0.87 for tri- and tetraglyme, respectively. These values strongly suggest that the Li is practically solvated by one glyme molecule in the good solvate ionic liquids.

Neutron total scattering with the $^{6/7}Li$ isotopic substitution technique is quite powerful for the Li^+ local structure analysis in solutions. This technique has been applied to the Li^+ local structure in an equimolar mixture of LiTFSA and tetraglyme [14]. Figure 2 displays the proposed Li^+ local structure model with the experiments. The Li^+–O (tetraglyme) distances are spread such as 2.06, 2.09, 2.13, 1.93 and 2.24 Å in the proposed Li^+ local structure model. The average distance of 2.05 Å except the longest one is significantly larger than values of 1.95–1.97 Å usually found in

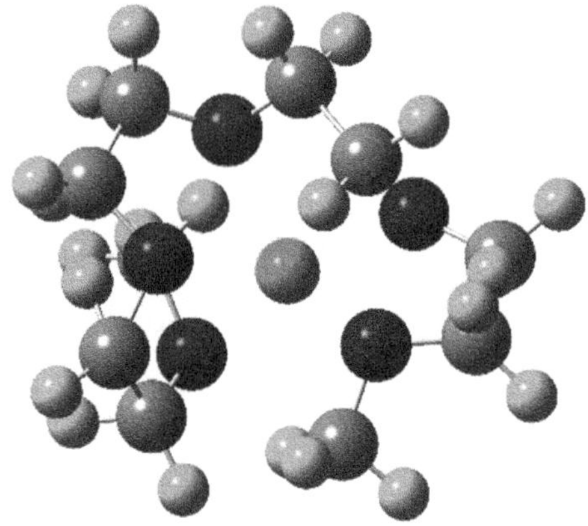

Fig. 2 Li^+ local structure model proposed with neutron total scattering experiments

aqueous solutions. On the other hand, the distances are similar to the values for cyclic and acyclic carbonates (2.04–2.08 Å) and for the polyethyleneoxide solutions (2.07–2.1 Å). Clearly, the longest distance of 2.24 Å is significantly larger than the others. According to the MD simulations, on average one of the terminal oxygen atoms in the glyme molecules is not fully coordinated to the Li^+, which is reasonably consistent with the rather longer 2.24 Å distance for the Li^+–O proposed by the neutron total scattering experiments. Thus, the Li^+ is not simultaneously coordinated by all five oxygen atoms of the tetraglyme molecule (deficient 5-coordination), but only to four of them (actual 4-coordination). The solvate cation is considerably distorted, which can be ascribed to the limited phase space of the ethyleneoxide chain and competition for the coordination sites from the TFSA anion. The coordination numbers are listed in Table 1 along with various methods.

Table 1 Coordination number for Li···O_x atom–atom pairs in [Li(G4)][TFSA] and [Li(G4)][NO_3]

	[Li(G4)TFSA]						[Li(G4)NO_3]	
	Neutron		MD				Neutron	MD
T/K	298	298	303	303	298	503	298	303
References	[12][a]	[14][b]	[15]	[16]	[17]	[18]	[12]	[15]
Li···O_t	1				1.41		0.49	
Li···O_m	0.94				1.76		0.35	
Li···O_c	0.33				0.95		0.09	
Li···O_{anion}	1.72		1.9	0.5	1.06	0.6	4.39	5.66
Li···O_{G4}	2.27		3.86	4.5	4.12	4.2	0.93	0.33
Li···$O_{anion+G4}$	3.99		5.76	5	5.18	4.8	5.32	5.99
Li···G4		0.95(1)						
Li···TFSA		1.4(5)						

x = c, m, t: the center, the middle and the terminal ether oxygen atoms in G4, respectively
Achieved by EPSR simulation[a], $^{6/7}$Li isotopic substitution technique[b]

Co-solvent of lower viscosity is often used to decrease viscosity of the electrolyte solutions. For this purpose, 1,1,2,2-tetrafluoroethyl 2,2,3,3-tetrafluoropropyl ether, hereafter abbreviated as HF in this section, is used for the Li–S secondary batteries. The HFE dilution effect has been investigated by means of high-energy X-ray total scattering technique, Raman spectroscopy and MD simulations [17]. With diluting solvate ionic liquid by the HFE, HFE hardly coordinates to the Li^+, so that [Li^+(glyme)] complex cation could be practically maintained in the HFE diluted solvate ionic liquids. Isotopic substitution Raman spectra (ISRS) clearly exhibits this speculation. Figure 3 shows Raman spectra for ^{7}LiTFSA-tetraglyme solvate ionic liquid (a); that for the neat HFE (b); and the ISRS for the HFE diluted solvate ionic liquid (c) at the frequency range of 630–800 cm^{-1}. The ISRS is defined as the difference spectra between the ^{6}Li enriched and the natural abundance samples of exactly the same components except the isotopic ratio of the Li^+. Figure 3c evidently shows with the HFE dilution that HFE never coordinates to the Li^+ and the SSIP and the CIP are slightly perturbed due to the variation of the surroundings.

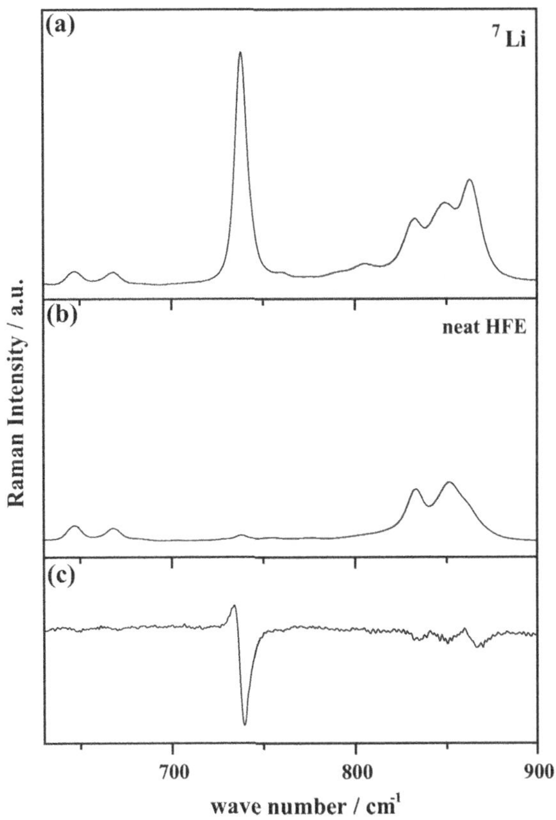

Fig. 3 Raman spectrum for the [*Li(G4)][TFSA]:HFE = 1:4 solution composed of natural abundance Li^+ (**a**), that for neat HFE (**b**) and **c** the isotropic substitution Raman spectrum for [*Li(G4)][TFSA]:HFE = 1:4 solution

3 Thermodynamic Aspects

To yield further insight into the glyme-type solvate ionic liquids, Raman spectra were measured for 1:1 mixtures of a series of glymes [19]. Figure 4a shows Raman spectra at 353 K in the frequency range of $720 \leq \nu/\mathrm{cm}^{-1} \leq 780$. At a glance, the peak position shifted toward the lower frequency side as lengthening the ethylene oxide chain, suggesting that the longer chain glyme more stabilizes Li^+ in free energy. These spectra can be successfully deconvoluted with two *pseudo*-Voigt peak functions. The lower and higher frequency components can be ascribable to the TFSA in the solvent separated ion-pair (SSIP) and the contact ion-pair (CIP), respectively. If these ion-pair formations are in equilibrium, the apparent equilibrium constant $K = [\mathrm{CIP}]/[\mathrm{SSIP}]$, practically a concentration quotient, can be defined and evaluated by using the Raman band area I_{SSIP} and I_{CIP} for the SSIP and the CIP, respectively, taking the Raman scattering factors into consideration.

$$\underset{(\mathrm{SSIP})}{\left[\mathrm{Li}^{+}(\mathrm{glymes})\right] \cdot \mathrm{TFSA}^{-}} \rightleftarrows \underset{(\mathrm{CIP})}{\left[\mathrm{Li}^{+}(\mathrm{glymes})\left(\mathrm{TFSA}^{-}\right)\right]} \quad (1)$$

Thus, the free energy levels can be estimated based on the simple relationship $\Delta G° = -RT\ln K$. Here, it should be noted that the Raman bands assigned to the glymes and TFSA coordinated to the Li^+ are practically, regardless, in the peak position of the glymes. This indicates the Li^+ local structure in the CIP is practically the same. Hence, we assume that the free energy level for the CIP is independent of the glymes so that

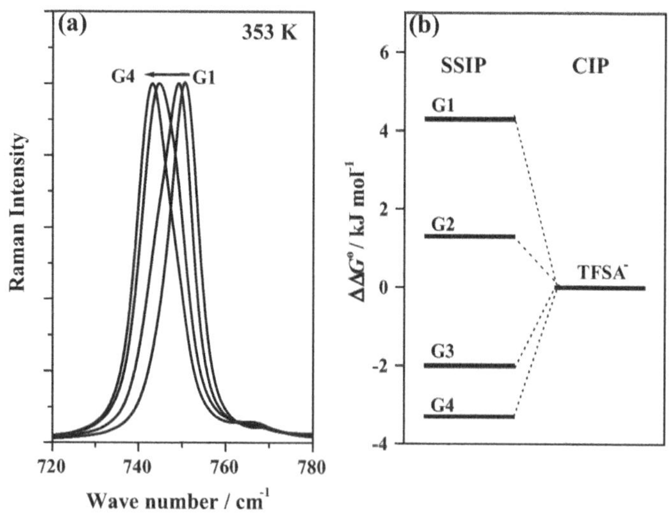

Fig. 4 **a** Raman spectra at 353 K in the frequency range of $720 \leq \nu/\mathrm{cm}^{-1} \leq 780$. **b** The free energy levels $\Delta\Delta G° = -RT\ln\mathrm{K}$ as the reference state of the CIP. Reprinted with permission from Ref. [19]. Copyright 2019 by American Physical Society

it can be considered as the reference state. Thus, the estimated free energy levels are shown in Fig. 4b. As clearly shown in this figure, the free energy difference is rather small, at most about 10 kJ mol^{-1}; among them, [Li(triglyme)]$^+$ and [Li(tetraglyme)]$^+$ have lower relative Gibbs free energy levels than that for the CIP. Furthermore, the temperature dependence of Raman spectra and the corresponding van't Hoff relationship yield more detailed CIP formation thermodynamics. Figure 5a and b exhibits temperature dependence of the Raman spectra of the TFSA and their van't Hoff plots, respectively. As shown in Fig. 5a, when elevating temperature, Raman bands were broadened, and thus peak intensity lowered. However, the Raman intensity at around 730–744 cm^{-1} increased, yet that at 750 cm^{-1} decreased, respectively, suggesting that the CIP formation is evidently exothermic. Thus the evaluated enthalpy $\Delta H°$ is about −5 kJ mol^{-1} for the CIP formation, which is reasonable due to the stronger Coulombic interaction of the Li$^+$ with the anion TFSA rather than electrically neutral glymes. Ab initio calculations agree well with the experiments; the tetraglyme CIP is about 17 kJ mol^{-1} more stable than the SSIP in the isolated gas phase [20].

The other, probably more important, thermodynamic aspect of the solvate ionic liquids can be acquired from the free glyme concentration in the solvate ionic liquid. As mentioned above, a free glyme concentration is practically negligible for the equimolar mixture of tri/tetraglyme with LiFSA. However, significant free glyme existence is suggested for the lithium salts of more strongly coordinating anions such as NO_3^-, CF_3COO^-, BF_4^- and ClO_4^-. By utilizing Raman spectra, we can evaluate the free glyme concentration in such mixtures. Normalized Raman intensity

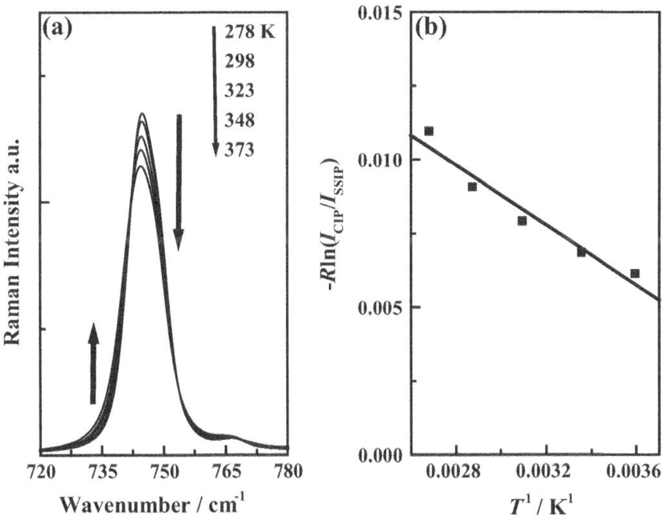

Fig. 5 **a** Temperature dependence of the TFSA Raman bands from 278 to 373 K for [Li(triglyme)][TFSA] SIL in the frequency range of 720 ≤ ν/cm^{-1} ≤ 780 and **b** the corresponding van't Hoff plots. Reprinted with permission from Ref. [19]. Copyright 2019 by American Physical Society

at ν cm^{-1} $I(\nu)$ can be expressed as $I(\nu) = J(\nu) \cdot c_i$, where $J(\nu)$ and c_i represent the Raman scattering factor and the concentration for the species *i*, respectively. On the other hand, the mass balance equation holds species *i* as $c_T = \Sigma c_i$, where c_T stands for the total concentration for the species *i*. Here, we assumed the free or the Li^+ coordinating, hereafter we call them as *unbound* and *bound*, respectively, glyme molecules just exist in the mixture; the two-state approximation can be held. Under this assumption, we obtain simple equation as follows:

$$c_{ub} = \{I(\upsilon) - J_b(\upsilon) \cdot c_T\}/\{J_{ub}(\upsilon) - J_b(\upsilon)\} \quad (2)$$

where c_{ub}, $J_{ub}(\nu)$ and $J_b(\nu)$ represent the concentration and the Raman scattering factor for the *unbound* and that for the *bound*, respectively. For the peak position of 873 cm^{-1} assigned to the *bound* glyme, the free glyme concentrations were evaluated for the equimolar mixtures of the glymes with various lithium salts and shown in Fig. 6. As seen, the percentage of the free glyme is rather small for TFSA and $BETI^-$based solvate ionic liquids, while more than 20% glyme molecules exist as a free in the solvate ionic liquids composed of the lithium salts of more strongly coordinating anions [13].

Recently, we extend single-point analysis of the Raman intensity to the multivariable analysis [21]. Taking into account the solvation number of glymes *n*, we obtain the following equation:

$$I(\upsilon)/c_T = J_{ub}(\upsilon) + \{J_b(\upsilon) - J_{ub}(\upsilon)\} \cdot n \cdot c_{Li}/c_T \quad (3)$$

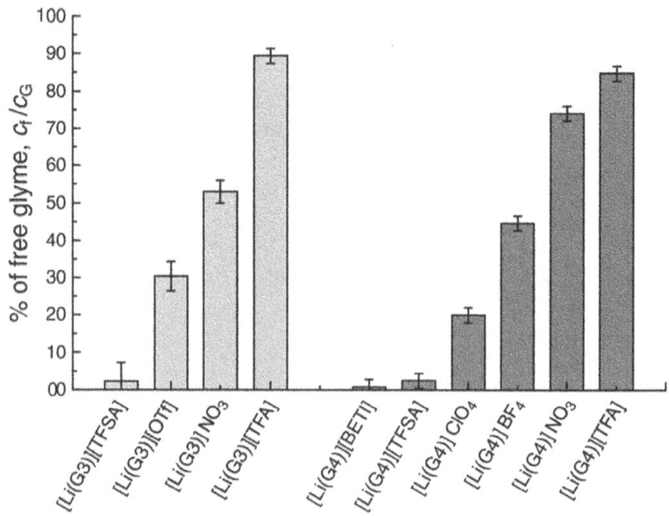

Fig. 6 Free glyme concentrations evaluated for equimolar mixtures of the glymes (G3 and G4) with various lithium salts. Reproduced from Ref. [13] with permission from the PCCP Owner Societies

where c_{Li} is a lithium salt concentration. The $I(\nu)$ versus c_{Li}/c_T plots at various n yield straight lines with an inflection point, whose slopes and intercepts correspond to $J_{ub}(\nu)$ and $\{J_b(\nu) - J_{ub}(\nu)\} \cdot n$, respectively. For glyme solvate ionic liquids, $n = 1$, so that the slopes and intercepts directly give $J_{ub}(\nu)$ and $J_b(\nu)$ as a function of ν; the Raman spectra for *unbound* and *bound* glymes of one moles. Once $J_{ub}(\nu)$ and $J_b(\nu)$ are obtained, then the concentration for the *unbound* and *bound* glymes could be parameters in the following equation as variables $J_{ub}(\nu)$ and $J_b(\nu)$:

$$I(\upsilon) = J_{ub}(\upsilon) \cdot [ub] + J_b(\upsilon) \cdot [b] \tag{4}$$

First values of $J_{ub}(\nu)$ and $J_b(\nu)$ should contain systematic errors, and thus they should be approximate ones. Therefore, more valid values can be guessed by using adequate peak functions: the *pseudo*-Voigt functions. In this procedure negative values in $J_{ub}(\nu)/J_b(\nu)$ can be avoided. Thus, simple linear regression analysis yields [ub] and [b]—the formation distribution functions. Similarly, the formation distribution functions can be properly approximated by using polynomials to avoid negative values, then $J_{ub}(\nu)$ and $J_b(\nu)$ can be refined again. These two linear least squares analyses should be complemental so that they have to be repeated with each other. Hence, these analyses were named as the complemental least squares analysis (CLSA). Figure 7 exhibits the results applied to the LiTFSA-triglyme solvate ionic liquid. Free glymes concentrations evaluated by various methods are listed in Table 2.

From strict thermodynamic viewpoint, an activity or an activity coefficient is more suitable for definition of the solvate ionic liquids. For this purpose, vapor pressure measurements are useful. Chemical potential as a function of vapor pressure for the gas phase, which should contain solely a glyme component, in equilibrium with a given solvate ionic liquid is identical to that as a function of activity in the equilibrium

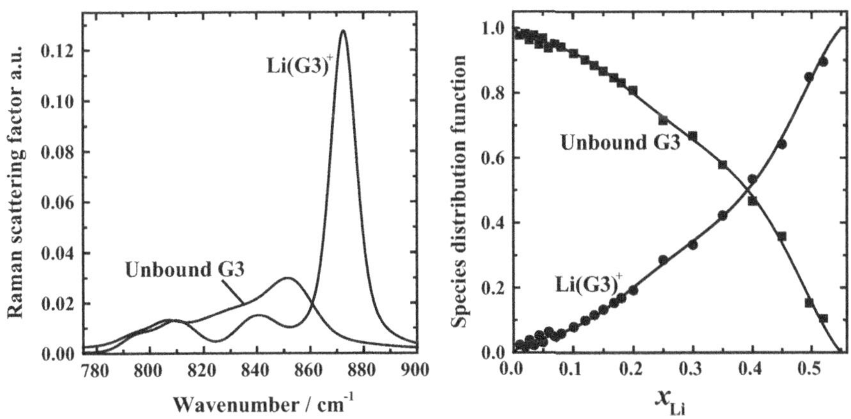

Fig. 7 Raman scattering factors **a** and species distribution functions **b** for mole fraction of the triglyme in $[LiTFSA]_x[G3]_{(x-1)}$ ($x = 0$–0.50) systems. Reprinted with the permission from Ref. [21]. Copyright 2020 American Chemical Society

Table 2 Free glyme concentration for [Li(triglyme)][X] (X = TFSA, NO_3, TfO, TFA) evaluated by various experimental techniques and MD simulations

	T/K	Li(G3)TFSA	Li(G3)NO_3	Li(G3)TfO	Li(G3)TFA	References
Raman	303	2.3	53	30	89	[13]
Raman[a]	298	17	27			
Raman[b]	298	5.3	10.5			
Neutron[c]	298	5	37			[12]
Electrochemical	303	0.057		53		[13]
MD simulations	303	5	89		41	[15]

[a]CLSA, [b]MCR-ALS and [c]EPSR

liquid phase. We roughly estimated the vapor pressure and the activity/activity coefficient for the representative solvate ionic liquids at ambient temperature with a simple pressure dependence of the boiling point and the Clausius–Clapeyron relationship. Figure 8 shows the activity coefficients for the representative solvate ionic liquids and the composition dependence of LiTFSA and triglyme mixtures around 1:1. For comparison, data for the water-in-basalt of LiTFSA + LiBETI:H_2O = 1:2 is also shown in the figure [22]. As shown in the figure, the activity coefficients are greater than 0.01 for the solvate ionic liquids consisting of NO_3^- and CF_3COO^-, while that is less than 0.01 for the TFSA is based on whose activity coefficient dramatically decreases as increasing LiTFSA composition at around the mole fraction of 0.5. Hence, activity coefficient < 0.01 could be a thermodynamic criterion for the solvate ionic liquids, not the concentrated electrolyte solution.

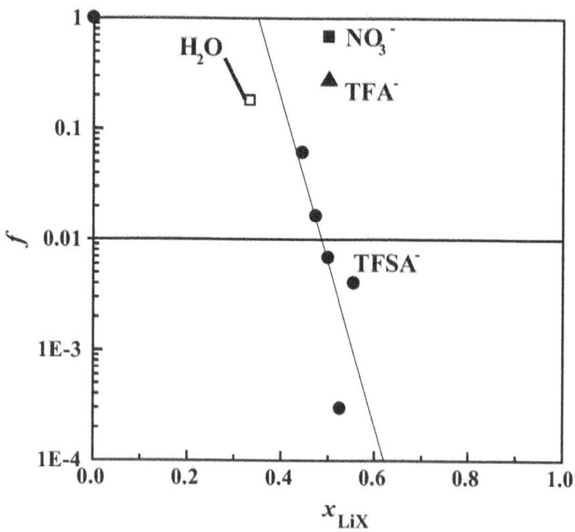

Fig. 8 Plots of the activity coefficients against the Li salt mole fractions (filled circle, triangle and square are $(LiTFSA)_x(G3)$ $x = 0.42 - 0.56$, $(LiTFA)_{0.5}(G3)$ and $(LiNO_3)_{0.5}(G3)$ and open square is $Li(TFSA)_{0.7}(BETI)_{0.3}$ $2H_2O$) taken from Ref. [22]. Reprinted with the permission from Ref. [21]. Copyright 2020 American Chemical Society

4 Conclusion

In this chapter, we surveyed structural and thermodynamic aspects of the solvate ionic liquids. With the triglyme and tetraglyme of a longer ethyleneoxide chain, at most four ether oxygen atoms coordinate to the Li^+ in solvate ionic liquids in a distorted manner. The anions of weakly coordinating ability such as TFSA and BETI, the direct coordination scarcely occur in the equimolar mixtures of triglyme/tetraglyme, and their lithium salt. On the other hand, contact ion-pairing and/or aggregation is obviously found both in crystals and solvate ionic liquids. Free glyme concentration in solvate ionic liquids can be evaluated by means of quantitative analyses of Raman spectra to yield clear indication whether the solvate ionic liquids or ordinary concentrated solution. Activity/activity coefficient of the glyme in the equimolar mixture with lithium salt yields strict thermodynamic criterion of the above inquiry.

References

1. Austen Angell, C., Ansari, Y., Zhao, Z. (2012). *Faraday discussions, 154*, 9–27.
2. Suo, L., Borodin, O., Gao, T., Olguin, M., Ho, J., Fan, X., et al. (2015). *Science, 350*(6263), 938–943.
3. Yamada, Y., Furukawa, K., Sodeyama, K., Kikuchi, K., Yaegashi, M., Tateyama, Y., & Yamada, A. (2014). *Journal of the American Chemical Society, 13*(136), 5039–5046.
4. Tatara, R., Kwabi, D. G., Batcho, T. P., Tulodziecki, M., Watanabe, K., Kwon, H. M., Thomas, M. L., Ueno, K., Thompson, C. V., Dokko, K., Shao-Horn, Y., Watanabe, M. (2017). *The Journal of Physical Chemistry C, 121*, 9162–9172.
5. Fujii, K., Wakamatsu, H., Todorov, Y., Yoshimoto, N., & Morita, M. (2016). *The Journal of Physical Chemistry C, 120,* 17196–17204.
6. Mandai, T., Yoshida, K., Ueno, K., Dokko, K., & Watanabe, M. (2014). *Physical Chemistry Chemical Physics: PCCP, 16,* 8761–8772.
7. Ueno, K., Tokuda, H., & Watanabe, M. (2010). *Physical Chemistry Chemical Physics: PCCP, 12,* 1649–1658.
8. Ueno, K., Yoshida, K., Tsuchiya, M., Tachikawa, N., Dokko, K., & Watanabe, M. (2012). *The Journal of Physical Chemistry B, 116,* 11323–11331.
9. Henderson, W. A., Brooks, N. R., Brennessel, W. W., & Young, V. G. (2003). Jr. *Chemistry of Materials, 15,* 4679–4684.
10. Henderson, W. A., McKenna, F., Khan, M. A., Brooks, N. R., Young, V. G., Jr., & Frech, R. (2005). *Chemistry of Materials, 17*(9), 2284–2289.
11. Henderson, W. A., Brooks, N. R., & Young, V. G. (2003). Jr. *Chemistry of Materials, 15,* 4685–4690.
12. Murphy, T., Callear, S. K., Yepuri, N., Shimizu, K., Watanabe, M., Canongia Lopes, J. N., Darwish, T., Warrf, G. G., Atkin, R. (2016). *Physical Chemistry Chemical Phys*ics, *18,* 17224–17236.
13. Ueno, K., Tatara, R., Tsuzuki, S., Saito, S., Doi, H., Yoshida, K., et al. (2015). *Physical Chemistry Chemical Physics: PCCP, 17,* 8248–8257.
14. Saito, S., Watanabe, H., Hayashi, Y., Matsugami, M., Tsuzuki, S., Seki, S., Canongia Lopes, J. N., Atkin, R., Ueno, K., Dokko, K., Watanabe, M., Kameda, Y., Umebayashi, Y. (2016). *The Journal of Physical Chemistry Letters, 7,* 2832–2837.
15. Shimizu, K., Freitas, A. A., Atkin, R., Warr, G. G., FitzGerald, P. A., Doi, H., Saito, S., Ueno, K., Umebayashi, Y., Watanabe, M., Canongia Lopes, J. N. (2015). *Physical Chemistry Chemical Physics, 17,* 22321–22335.

16. Tsuzuki, S., Shinoda, W., Matsugami, M., Umebayashi, Y., Ueno, K., Mandai, T., et al. (2015). *Physical Chemistry Chemical Physics: PCCP, 17,* 126–129.
17. Saito, S., Watanabe, H., Ueno, K., Mandai, T., Seki, S., Tsuzuki, S., et al. (2016). *The Journal of Physical Chemistry B, 120,* 3378–3387.
18. Shinoda, W., Hatanaka, Y., Hirakawa, M., Okazaki, S., Tsuzuki, S., Ueno, K., & Watanabe, M. (2018). *The Journal of Chemical Physics, 148,* 193809.
19. Arai, N., Watanabe, H., Yamaguchi, T., Seki, S., Ueno, K., Dokko, K., et al. (2019). *Journal of Physical Chemistry C, 123,* 30228–30233.
20. Tsuzuki, S., Shinoda, W., Seki, S., Umebayashi, Y., Yoshida, K., Dokko, K., & Watanabe, M. (2013). *ChemPhysChem, 14,* 1993–2001.
21. Arai, N., Watanabe, H., Nozaki, E., Seki, S., Tsuzuki, S., Ueno, K., Dokko, K., Watanabe, M., Kameda, Y., Umebayashi, Y. *The Journal Physical Chemistry Letters* (2020).
22. Yamada, Y., Usui, K., Sodeyama, K., Ko, S., Tateyama, Y., & Yamada, A. (2016). *Nature Energy, 1*(10), 16129.

Properties and Dynamics by Computer Simulation

Seiji Tsuzuki and Wataru Shinoda

Abstract Stable structures of $[Li(G_n)]^+$ complexes (G_n: CH_3–(–O–CH_2–CH_2–)$_n$–O–CH_3, n = 1–4), their stabilization energies by the formation of complexes (ΔE_{form}) and the binding energies of $[Li(G_n)]^+$ complexes (n = 3, 4) with $[TFSA]^-$ (ΔE_{bind}) were studied by ab initio molecular orbital calculations. The magnitude of ΔE_{form} increases significantly as n becomes larger. The ΔE_{form} calculated for the $[Li(G_3)]^+$ and $[Li(G_4)]^+$ complexes were −95.6 and −107.7 kcal/mol, respectively. The ΔE_{bind} for the $[Li(G_3)]^+$ (−82.8 kcal/mol) is greater than that for the $[Li(G_4)]^+$ (−70.0 kcal/mol), which coincides with the faster diffusion of ions in the equimolar mixture of G_4 and Li[TFSA] compared with the equimolar mixture of G_3 and Li[TFSA]. Molecular dynamics simulations of the equimolar mixtures show that the coordination structures of Li^+ change depending on the length of the glyme. The potential mean force between Li^+ and glyme was evaluated as a function of distance. The lower energy minimum for G_4 indicates the higher stability of $[Li(G_4)]^+$ complex. The coordination number of G_3 oxygen to the Li^+ shows that multiple glymes bind with Li^+ in the intermediate state of glyme exchange. The measurements of the rate of exchange dynamics in the equimolar mixtures show long lifetime of the $[Li(G_n)]^+$ complexes (n = 3, 4).

Keywords Solvate ionic liquid · Solvation energy · Molecular dynamics simulation · Ion exchange · Li-glyme complex

---

S. Tsuzuki (✉)
Research Center for Computational Design of Advanced Functional Materials (CD-FMat), National Institute of Advanced Industrial Science and Technology (AIST), 1-1-1 Umezono, Tsukuba 305-8568, Ibaraki, Japan
e-mail: s.tsuzuki@aist.go.jp

W. Shinoda
Department of Materials Chemistry, Nagoya University, Furo-cho, Chikusa-ku, Nagoya 464-8603, Japan

K. Kanamura (ed.), *Next Generation Batteries*,
https://doi.org/10.1007/978-981-33-6668-8_27

1 Introduction

The physicochemical properties of electrolyte are important for developing high performance batteries. Low viscosity and high ionic conductivity are essential for improving battery capacity. The intermolecular interactions among ions and solvent molecules in electrolyte control the liquid structure and transport properties. Unfortunately, however, it is not easy to reveal the molecular-level liquid structures and intermolecular interactions only by experimental measurements. Recently, computational methods are becoming powerful tools for studying the liquid structures and intermolecular interactions. We have studied the intermolecular interactions of $[Li(Gn)]^+$ complexes (n = 3 and 4, CH_3–(–O–CH_2CH_2–)$_n$–O–CH_3) by ab initio molecular orbital calculations and studied the liquid structures and transport properties of electrolytes composed of glymes and Li[TFSA] by molecular dynamics simulations.

2 $[Li(glyme)]^+$ Complexes

The optimized stable structures of $[Li(G3)]^+$ and $[Li(G4)]^+$ complexes are shown in Fig. 1. The four oxygen atoms of G3 and five oxygen atoms of G4 have contact with Li^+ in the stable structures. The average Li···O distances in the $[Li(G3)]^+$ and $[Li(G4)]^+$ complexes are 2.001 and 2.051 Å, respectively [1]. The short Li···O distances suggest the existence of strong attraction between Li^+ and oxygen atoms. The stabilization energies (ΔE_{form}) calculated for the two complexes (the stabilization energies by the formation of the complexes from isolated species) are −95.6 and −107.7 kcal/mol, respectively [1]. Apparently, the strong attraction between Li^+ and glymes is the cause for the formation of stable $[Li(glyme)]^+$ complexes in the equimolar mixtures of glymes (G3 and G4) and Li[TFSA].

The ΔE_{form} calculated for the Li^+ complexes with dimethyl ether (DME) and glymes (Gn, n = 1–4) is summarized in Table 1. The size of ΔE_{form} increases as

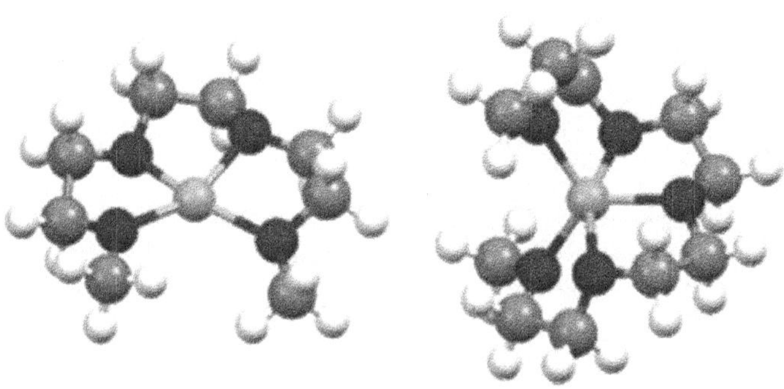

Fig. 1 HF/6-311G** level optimized structures for the $[Li(G3)]^+$ and $[Li(G4)]^+$ complexes

Table 1 Interaction energies for Li^+ complexes with dimethyl ether and glymes[a]

	n[b]	ΔE_{int}[c]	ΔE_{def}[d]	ΔE_{form}[e]	ΔE_{es}[f]	ΔE_{ind}[g]	ΔE_{other}[h]
$[Li(DME)]^+$	1	−38.5	0.7	−37.8	−31.4	−21.3	14.2
$[Li(G1)]^+$	2	−65.2	4.2	−61.0	−58.6	−40.3	33.7
$[Li(G2)]^+$	3	−87.3	7.8	−79.5	−76.8	−55.7	45.1
$[Li(G3)]^+$	4	−106.0	10.4	−95.6	−90.2	−64.8	49.1
$[Li(G4)]^+$	5	−119.7	12.0	−107.7	−103.2	−76.0	59.4

[a]Energy in kcal/mol. [b]Number of oxygen atoms which have contact with Li^+. [c]Interaction energies calculated for complexes at the MP2/6-311G** level. [d]The increase of energy of glyme by the deformation of geometry associated with the complex formation. [e]Stabilization energy by the formation of complex from isolated species. Sum of ΔE_{int} and ΔE_{def}. [f]Electrostatic energy. [g]Induction energy. [h]$\Delta E_{other} = \Delta E_{int} - \Delta E_{es} - \Delta E_{ind}$. ΔE_{other} is mainly exchange-repulsion and dispersion energies

the number of interacting oxygen atoms increases. The ΔE_{form} is the sum of the calculated intermolecular interaction energy (ΔE_{int}) between Li^+ and glyme and the deformation energy (ΔE_{def}), which is the increase of the energy of glyme associated with the complex formation [1]. The electrostatic and induction energies (ΔE_{es} and E_{ind}) were calculated using distributed multipole analysis [1–4] as summarized in Table 1, which shows that the electrostatic and induction interactions are the sources of the strong attraction. Although the electrostatic interactions (Coulombic interaction between static charges of Li^+ and glyme) are stronger than the induction interactions (attraction by induced polarization of glyme), the contribution of the induction interactions to the attraction is significant.

The HOMO energy levels calculated for isolated glymes and those in the $[Li(glyme)]^+$ and [Li(glyme)][TFSA] complexes are summarized in Table 2. The HOMO energy levels calculated for the glymes in the $[Li(gylme)]^+$ complexes are significantly lower than those for isolated glymes. Although the HOMO energy levels for the glymes in the[Li(glyme)][TFSA] complexes are higher than those for the $[Li(gylme)]^+$ complexes, they are still substantially lower than the HOMO energy levels for isolated glymes, which suggests that the interaction with Li^+ lowers the HOMO energy level of glymes and improves the oxidation stability of glymes in the mixtures [5].

Table 2 HOMO energy levels calculated for isolated glymes and glymes in $[Li(glyme)]^+$ and [Li(glyme)][TFSA] complexes

	eV
G3 (all trans)	−11.45
$[Li(G3)]^+$	−15.51
[Li(G3)][TFSA]	−12.10
G4 (all trans)	−11.46
$[Li(G4)]^+$	−14.90
[Li(G4)][TFSA]	−11.80

3 [Li(glyme)][TFSA] Complexes

The optimized stable structures of [Li(G3)][TFSA] and [Li(G4)][TFSA] complexes (TFSA is bis(trifluoromethylsulfonyl) amide) are shown in Fig. 2. Not only oxygen atoms of glyme but also oxygen atoms of $[TFSA]^-$ have contact with Li^+ in the complexes. The ΔE_{form} calculated for the complexes is summarized in Table 3. The ΔE_{form} for the [Li(G3)][TFSA] and [Li(G4)][TFSA] complexes are nearly identical (−178.4 and −177.7 kcal/mol, respectively). The ΔE_{bind} is the difference between the ΔE_{form} for the [Li(glyme)][TFSA] complex and that for the $[Li(glyme)]^+$ complex, which corresponds to the binding energy between $[Li(glyme)]^+$ complex and $[TFSA]^-$ anion. The G3 complex has larger (more negative) ΔE_{bind} compared to G4, which shows that the interaction between $[Li(G3)]^+$ complex and $[TFSA]^-$ is stronger than that between $[Li(G4)]^+$ complex and $[TFSA]^-$ [1].

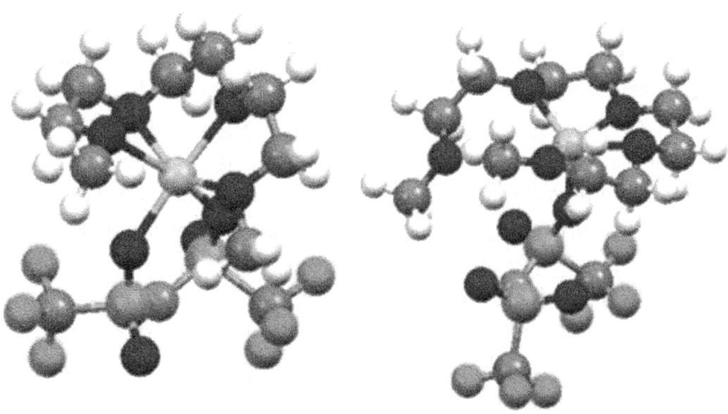

Fig. 2 HF/6-311G** level optimized structures for the [Li(G3)][TFSA] and [Li(G4)][TFSA] complexes

Table 3 Stabilization energies by the formation of $[Li(glyme)]^+$ and [Li(glyme)][TFSA] complexes[a]

Glyme	ΔE_{form}[b]		ΔE_{bind}[c]
	$[Li(glyme)]^+$	[Li(glyme)][TFSA]	
G3	−95.6	−178.4	−82.8
G4	−107.7	−177.7	−70.0

[a]Energy in kcal/mol. [b]Stabilization energy by the formation of complex from isolated species. [c]Difference between the ΔE_{form} for the [Li(glyme)][TFSA] complex and that for the $[Li(glyme)]^+$ complex. ΔE_{bind} corresponds to the binding energy between the $[Li(glyme)]^+$ complex and $[TFSA]^-$ anion

4 Liquid Structures and Transport Properties of Equimolar Mixtures of Glymes and Li[TFSA]

The molecular dynamics simulations of equimolar mixtures of glyme (G3 and G4) and Li[TFSA] were carried out. Radial distribution function between the Li^+ and oxygen atoms of the glymes and that between the Li^+ and oxygen atoms of the $[TFSA]^-$ anions is shown in Fig. 3. The sharp peaks at 2 Å show the short contact between Li^+ and the oxygen atoms of glyme and $[TFSA]^-$ in the mixtures. The peak height of Li–O(TFSA) in the mixture of G4 and Li[TFSA] is lower than that in the mixture of G3 and Li[TFSA], which shows that the increase of the glyme chain length decreases the contact of oxygen atoms of $[TFSA]^-$ with Li^+. The accumulated coordination number of Li^+ in the mixtures (Fig. 4) also shows that the contact of oxygen atoms of $[TFSA]^-$ with Li^+ decreases by the increase of the glyme chain length. The accumulated coordination number shows that all four oxygen atoms of G3 and one oxygen atom of $[TFSA]^-$ have contact with Li^+ in the equimolar mixture of G3 and Li[TFSA], while the coordination number of oxygen atoms of G4 is about 4.5 and that of oxygen atoms of $[TFSA]^-$ is 0.5 in the equimolar mixture of G4 and Li[TFSA] [6].

Self-diffusion coefficients of glyme and ions in the equimolar mixtures were calculated. The calculated self-diffusion coefficients are compared with the experimental values as shown in Table 4. The ratio of self-diffusion coefficients calculated for glyme and Li^+ (D_{glyme}/D_{Li}) in the mixtures of glymes and Li[TFSA] is close to 1.0 as in the case of the experimental self-diffusion coefficients. The nearly identical self-diffusion coefficients of glyme and Li^+ indicate the formation of the stable $[Li(glyme)]^+$ complexes in the mixtures. The self-diffusion coefficients calculated for ions and glyme in the mixture of G4 and Li[TFSA] are substantially larger than

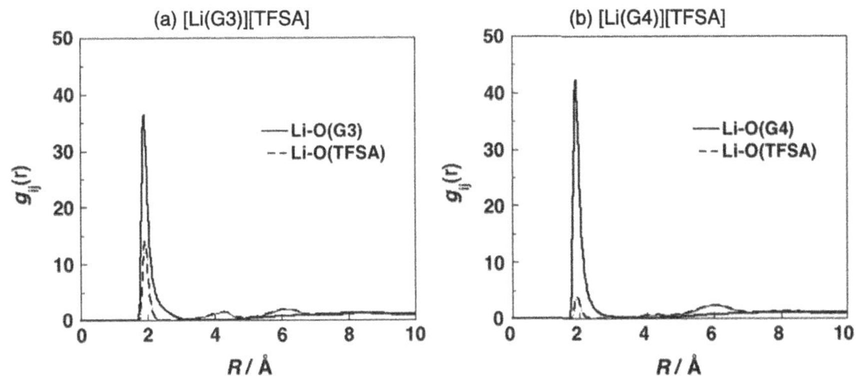

Fig. 3 Site–site intermolecular radial distribution functions in the mixture of G3 and G4 with Li[TFSA] at 303 K. Radial distribution function between the Li^+ and oxygen atoms of the glymes and that between the Li^+ and oxygen atoms of the $[TFSA]^-$ anions

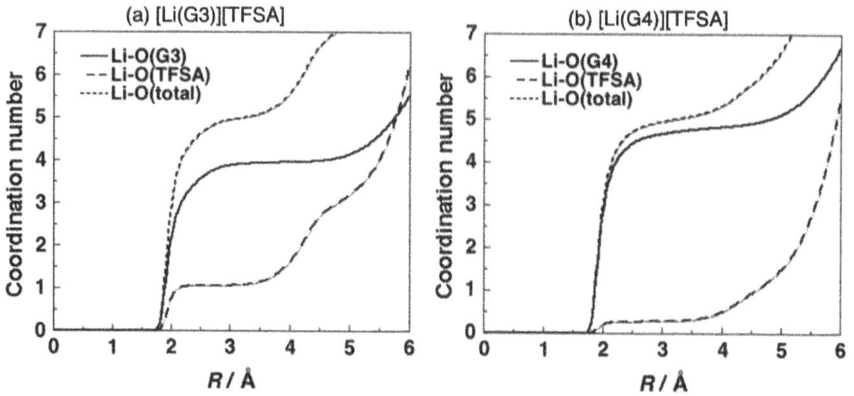

Fig. 4 Accumulated coordination number of Li^+ in the mixture of G3 and G4 with Li[TFSA] at 303 K

Table 4 Self-diffusion coefficient calculated for equimolar mixtures of glymes and Li[TFSA]

	D_{glyme} [a]	D_{Li} [a]	D_{TFSA} [a]
Calc			
[Li(G3)][TFSA]	0.43	0.43	0.24
[Li(G4)][TFSA]	0.78	0.78	0.73
exp[b]			
[Li(G3)][TFSA]	0.77	0.77	0.54
[Li(G4)][TFSA]	1.26	1.26	1.22

[a]Self-diffusion coefficient in 10^{-7} cm^2 s^{-1}. Calculated values at 403 K and experimental values at 303 K. [b]Reference [8]

those in the mixture of G3 and Li[TFSA] [6]. The self-diffusion coefficients calculated for the ions and glyme in the two mixtures accurately describe the experimental trend of the glyme chain length dependence of self-diffusion coefficients, although the calculated self-diffusion coefficients were smaller than the experimental values as in the cases of other ionic liquids. The magnitude of the interaction between cation and anion is one of the important factors controlling the diffusion of ions in ionic liquids [7]. The diffusion of ions becomes slow as the interaction between cation and anion becomes strong. The weaker interaction between $[Li(G4)]^+$ and $[TFSA]^-$ anion compared with that between $[Li(G3)]^+$ and $[TFSA]^-$ anion is apparently the cause of the faster diffusion of ions in the mixture of G4 and Li[TFSA] compared with the mixture of G3 and Li[TFSA].

5 Stability of Li[glyme]$^+$ Complex in Equimolar Mixtures of Glymes and Li[TFSA]

In equimolar mixture of Li[TFSA] and glymes, stable Li[glyme]$^+$ complex formation was evident in the structure of the mixture as well as self-diffusion coefficients. However, [Li(glyme)]$^+$ complexes should have a finite lifetime and eventually exchange the partner by a new glyme or Li moiety. From the interaction analysis in Table 1, the stabilization energies (ΔE_{form}) are significantly large values compared to thermal fluctuation. However, in the bulk liquids, the effective association free energy of Li$^+$ and glyme should be different from the stabilization energy in vacuum. We can evaluate the associate free energy or the potential of mean force, $W(R)$, as a function of distance, R, between centers of mass of Li$^+$ and glyme, using the radial distribution function, $g(R)$; $W(R) = -k_B T \ln g(R)$, where k_B is the Boltzmann constant and T is temperature. The potential mean forces are plotted for Li–G3 and Li–G4, respectively, in Fig. 5. In both cases, there is an energy barrier at around 5 Å, which separates bound and unbound states. Energy minimum is found in the bound state at around 0 and 1 Å for G4 and G3 cases, respectively. The lower energy minimum at the bound state for G4 indicates the higher stability as Li-G4 complex, which results in a higher diffusion of Li–G4 than of Li–G3. As evident from Fig. 4, Li$^+$ ion is not fully covered by the four oxygen atoms of G3, and Li$^+$ is coordinated by the oxygen atom of TFSA. In contrast, Li$^+$ is almost fully covered by the five oxygen atoms of G4, reducing the direct contact to TFSA. This explains the difference found in the stability of bound state of Li-glyme complex between G3 and G4. The effective energy barriers to overcome to remove the G3 and G4 from Li are 5 and 7 kcal/mol, respectively, which are much smaller than the stabilization energies

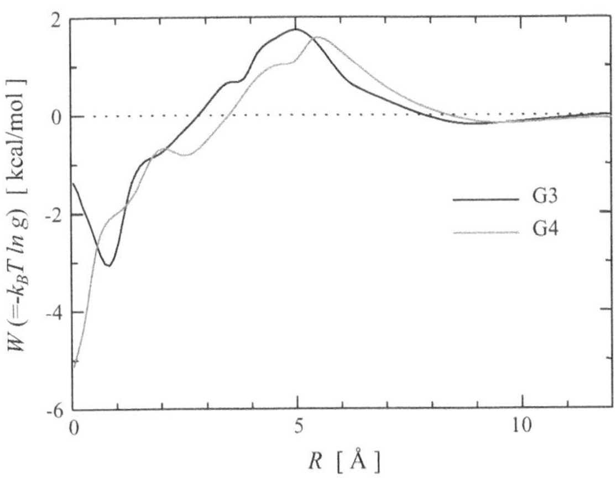

Fig. 5 Potential of mean force, W, of glyme around Li$^+$ in the equimolar mixture of glyme (G3 or G4) and Li[TFSA]. Reprinted with permission from Ref. [9]. Copyright 2018, American Institute of Physics

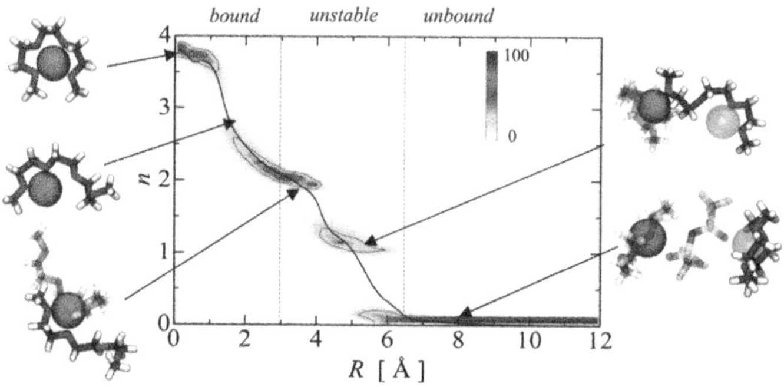

Fig. 6 The two-dimensional contour map of the probability density with respect to the coordination number of the oxygen atoms of G3 around Li^+, n, and the distance between the centers of mass of G3 and Li^+, R. The color bar shows probability in %, which are normalized at each R. The solid line indicates the average value of n at each R. Representative snapshots of the Li^+-G3 pairs are given along with the surrounding molecules (transparent). The color codes of atoms are Li: orange, C: grey, O: red, and H: white. Reprinted with permission from Ref. [9]. Copyright 2018, American Institute of Physics

(ΔE_{form}) in vacuum. This is simply because of the interaction with the surrounding molecules around the focused Li-glyme pairs.

Figure 6 shows the two-dimensional contour map of the probability distribution with respect to the coordination number of the G3 oxygen to the Li^+ and the distance between the centers of mass of G3 and the Li^+ ion. We used a threshold distance of 3.25 Å for the coordination between the glyme oxygen and Li^+ in this analysis. When the distance is very short ($R < 1$ Å), almost all oxygen atoms bind to Li^+ ion as shown in the snapshot, resulting in the coordination number $n \sim 4$. When the distance exceeds 3 Å, the coordination number is lower than 2, meaning the Li^+ ion associates with another glyme most likely. However, Fig. 5 suggests that as long as the distance R is less than 5 Å (where $n \sim 1$), the pair does not exceed the free energy barrier. In any cases, in this unstable region, Li^+ should bind to multiple glymes, which might drive an exchange the partner of the Li[glyme]$^+$ complex.

6 Ion Exchange Dynamics

A significant reduction of the association free energy of Li-glyme complex in the mixture of glyme and Li[TFSA], compared to the stabilization energy in vacuum, should be explained by the interaction with the surrounding molecules. As suggested from Fig. 6, the Li-glyme pair has another surrounding glyme to coordinate to, when the pair distance is getting longer.

Figure 7 shows snapshots of representative exchange event observed between two pairs of Li(G3) complexes. In the panel (a), initially detected two separated two pairs of Li(G3) are separated by TFSI (shown as thin transparent sticks). However, due to

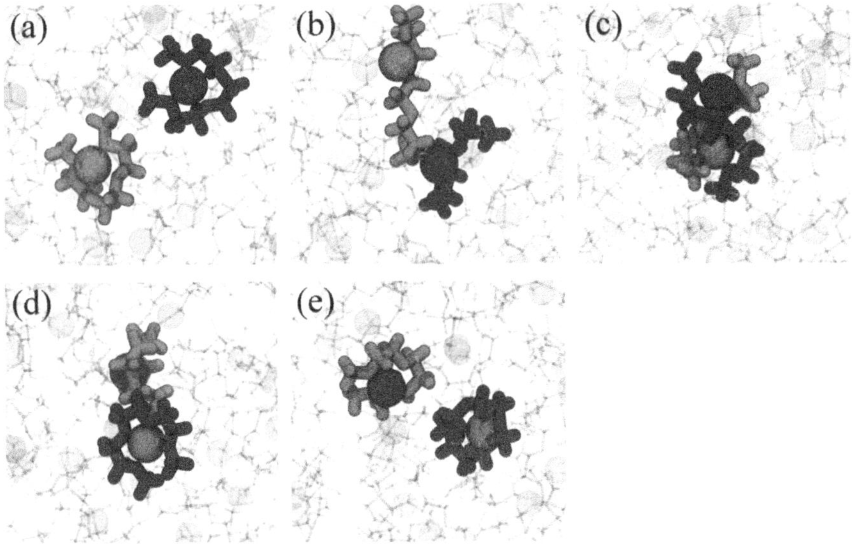

Fig. 7 Snapshots from a representative exchange event between two pairs of $[Li(G3)]^+$ complexes. Initial two pairs of Li(G3) complexes are colored by red and blue, respectively. The surrounding molecules are depicted in transparency. Reprinted with permission from Ref. [9]. Copyright 2018, American Institute of Physics

thermal fluctuation, G3 sometimes show an extended conformation, where some of the oxygen atoms coordinate to another Li^+ ion around, as shown in the panel (b). Although most of the 2:2 Li-G complex structure found in the panel (b) dissociate to regenerate the original Li(G3) complexes, it was occasionally stabilized in the symmetrically bound state shown in the panel (c). It is worth noting that a similar double helical structure was found in the crystal structure in [Li(G4)][AsF6]. Once the symmetric double helical structure attained, it was almost the equal probability to forward the exchange process [as shown in panel (d) and (e)] or revert to the original two Li(G3) complexes.

We measured the rate of exchanging dynamics of neighboring ion (or Li-glyme) pairs. We use the time-correlation function of the state function; $C(t) = \langle S(t)S(0)\rangle/\langle S^2\rangle$, where $S(t)$ is the state function at time t to be 1 for bound state and 0 for non-bound state. The correlation time of the function $C(t)$ gives a measure of lifetime of neighboring molecules in the system. The estimated lifetime, τ, of neighboring molecules is listed in Table 5. The lifetime of Li-glyme pair is much

Table 5 Lifetime of neighboring molecules in nanoseconds

	$\tau_{\text{Li-G}}$	$\tau_{\text{Li-anion}}$	$\tau_{\text{Li-Li}}$	$\tau_{\text{anion-anion}}$
[Li(G3)][TFSA]	72 ± 6	2.5 ± 0.1	3.5 ± 0.1	4.6 ± 0.3
[Li(G4)][TFSA]	87 ± 7	2.5 ± 0.1	3.2 ± 0.2	3.0 ± 0.2

longer than that of any ion pairs. The above-mentioned pair exchanging event of Li-glyme complex is rather a rare event, compared to ion exchange dynamics. However, exchange event of Li-glyme complex does take place in a longer time regime, which does not conflict with the fact that similar self-diffusion coefficients are obtained for Li and glyme, because they diffuse together for several tens of nanoseconds.

7 Summary

Computational methods such as quantum mechanical calculations and molecular dynamics simulations are becoming powerful methods for studying structures and transport properties of liquids and their mechanism. In this chapter, we summarized our recent studies on the mixtures of glymes and Li[TFSA] salt. Our computational analysis reveals detailed molecular level information on the liquid structures and dynamics of ions in the mixtures. We hope that our findings will contribute to accelerating the development of new batteries.

References

1. Tsuzuki, S., Shinoda, W., Seki, S., Umebayashi, Y., Yoshida, K., Dokko, K., Watanabe, M. (2013). Intermolecular interactions in Li^+-glymes and Li^+-glymes-TFSA-complexes: Relationship with physicochemical properties of [Li(glymes)][TFSA] ionic liquids. *ChemPhysChem, 14,* 1993–2001.
2. Stone, A. J., Alderton, M. (1985) Distributed multipole analysis—Methods and applications. *Molecular Physics, 56*, 1047–1064.
3. Stone, A. J. (1985). Distributed polarizabilities. *Molecular Physics, 56,* 1065–1082.
4. Stone, A. J. (2013). *The theory of intermolecular forces* (2nd ed.). Oxford University Press: Oxford.
5. Yoshida, K., Nakamura, M., Kazue, Y., Tachikawa, N., Tsuzuki, S., Seki, S., Dokko, K., Watanabe, M. (2011). Anomalous oxidative-stability enhancement and charge transport mechanism in glyme-lithium salt equimolar complexes. *Journal of the American Chemical Society, 133,* 13121–13129.
6. Tsuzuki, S., Shinoda, W., Matsugami, M., Umebayashi, Y., Ueno, K., Mandai, T., Seki, S., Dokko, K., Watanabe, M. (2015). Structures of $[Li(glyme)]^+$ complexes and their interactions with anions in equimolar mixtures of glymes and Li[TFSA]: Analysis by molecular dynamics simulations. *Physical Chemistry Chemical Physics, 17*, 126–129.
7. Tsuzuki, S. (2012). Factors controlling the diffusion of ions in ionic liquids. *ChemPhysChem, 13,* 1664–1670.
8. Zhang, C., Ueno, K., Yamazaki, A., Yoshida, K., Moon, H., Mandai, T., Umebayashi, Y., Dokko, K., Watanabe, M. (2014). Chelate effects in glyme/lithium bis(trifluoromethanesulfonyl) amide solvate ionic liquids. I Stability of solvate cations and correlation with electrolyte properties. *The Journal of Physical Chemistry B, 118,* 5144–5153.
9. Shinoda, W., Hatanaka, Y., Hirakawa, M., Okazaki, S., Tsuzuki, S., Ueno, K., Watanabe, M. (2018). Molecular dynamics study of thermodynamic stability and dynamics of $[Li(glyme)]^+$ complex in lithium-glyme solvate ionic liquids. *The Journal of Chemical Physics, 148,* 193809/1-8.

Lithium Metal Anode

Naoki Tachikawa, Nobuyuki Serizawa, and Yasushi Katayama

Abstract Deposition and dissolution of lithium metal have been investigated in an equimolar mixture of lithium bis(trifluoromethylsulfonyl)amide (LiTFSA) and glyme [triglyme (G3) or tetraglyme (G4)] solvate ionic liquid. The limiting current for deposition of lithium was not observed probably because of the high concentration of lithium species and a decrease in the local viscosity by the liberation of glyme. On the other hand, the dissolution of lithium was limited due to an increase in the local viscosity by the formation of $[Li(TFSA)_2]^-$. The formation of solid electrolyte interphase (SEI) was suggested to form by the cathodic decomposition of the solvate ionic liquids by electrochemical quartz crystal microbalance. Lithium phosphorous oxynitride (LiPON) thin film was found to act as the artificial SEI, which prevented the cathodic decomposition of the solvate ionic liquids and enabled the deposition and dissolution of lithium. The cycle performance of deposition and dissolution of lithium was found to be improved by coating a Cu substrate with vapor-grown carbon fiber.

Keywords Lithium metal anode · Solvate ionic liquid · Solid electrolyte interphase · Lithium phosphorous oxynitride · Vapor-grown carbon fiber

1 Lithium Metal Anode with Solvate Ionic Liquid

Anode materials having high specific capacity, such as Li and Si, are required in order to best utilize the capabilities of the sulfur cathode. Among the various anode materials available, a lithium metal anode is favorable for realizing a large energy density in rechargeable lithium-sulfur batteries because of its high theoretical specific capacity of 3861 mAh g^{-1}. At the lithium metal anode, the lithium ion in the electrolyte is electrochemically reduced to deposit lithium metal during the charge reaction, and the dissolution of the deposited lithium metal occurs during the discharge reaction. Whisker-like or dendritic deposition of lithium metal is known to occur, leading to low coulombic efficiency of the lithium metal anode due to isolation of the

N. Tachikawa (✉) · N. Serizawa · Y. Katayama
Keio University, Tokyo, Japan

K. Kanamura (ed.), *Next Generation Batteries*,
https://doi.org/10.1007/978-981-33-6668-8_28

electrical contact of the deposited lithium with the substrate. Moreover, whisker-like growth of lithium metal causes short circuits in the cells by penetration of the lithium metal through the separator.

The electrolyte plays a crucial role in the performance of lithium metal anodes. Lithium metal is thermodynamically unstable when used with almost all organic solvents and ionic liquids [1]. The reductive decomposition reaction of the electrolyte with lithium metal results in the formation of a kind of surface film on the electrode, the so-called solid electrolyte interphase (SEI) [2]. The SEI acts as both a lithium ion conductor and an electron insulator, enabling the deposition/dissolution of lithium metal through the SEI and protecting the electrolyte from further decomposition. Since the SEI is produced via the decomposition of the electrolyte, the thickness and chemical composition of the SEI tend to be heterogeneous. Lithium metal is preferentially deposited through thin spots in the heterogeneous SEI, leading to the formation of whisker-like or dendritic lithium metal due to local current concentration.

The selection of the proper solvents and lithium salts is necessary for use in rechargeable lithium-sulfur batteries. The compatibility of both the lithium metal anode and the sulfur cathode should be taken into account. Carbonate-based electrolytes have been used in practical lithium ion batteries. However, these electrolytes are incompatible with both the lithium anode and the sulfur cathode in the lithium-sulfur batteries. For example, in 1 M $LiPF_6$/ethylene carbonate (EC) + diethyl carbonate (DEC) (1:1 vol%), the morphology of the deposited lithium metal was dendritic and the coulombic efficiency of the lithium anode was less than 50% [3]. In the case of sulfur cathode, the carbonate-based electrolytes were decomposed by the nucleophilic attack of sulfides produced during the lithium-sulfur battery reaction [4]. Thus, ether-based electrolytes and ionic liquids are widely used in lithium-sulfur batteries.

Since Aurbach's report in 2009 [5], many researchers have employed ether-based electrolytes with $LiNO_3$ as additive for lithium-sulfur batteries. In the presence of $LiNO_3$, some favorable surface film on the lithium metal anode is formed, minimizing the redox shuttle of the sulfur species and enhancing the coulombic efficiency of both the sulfur cathode [5] and the lithium metal anode [6]. However, Watanabe group pointed out that the lithium-sulfur batteries with an equimolar mixture of triglyme (G3) and $LiNO_3$ (abbreviated as [Li(G3)]NO_3) showed poor coulombic efficiency owing to the cathodic decomposition reaction of NO_3^- on the carbon–sulfur composite cathode around 1.7 V versus Li|Li(I) [7]. Furthermore, the solubility toward lithium sulfides is much higher in [Li(G3)]NO_3 than in [Li(G3 or G4)]TFSA (G4 = tetraglyme, $TFSA^-$ = bis(trifluoromethylsulfonyl)amide) and [Li(G4)]BETA ($BETA^-$ = bis(perfluoroethylsulfonyl)amide). Therefore, [Li(G3)]NO_3 system does not fit the concept of lithium-sulfur batteries with sparingly soluble electrolytes. In studies using lithium metal anodes, electrolytes with LiFSA (FSA^- = bis(fluorosulfonyl)amide) have attracted much attention due to formation of a suitable SEI on the lithium metal anode. Qian [8] reported that smooth deposit of lithium metal was obtained, and a high coulombic efficiency of 98.4% for the lithium deposition/dissolution process was exhibited by highly concentrated electrolytes [4 M LiFSA in 1,2-dimethoxyethane (G1 or monoglyme)] compared with a normal concentration of electrolytes (1 M LiFSA in G1). The preferable SEI formed in concentrated LiFSA electrolytes is reported to be due to a low concentration of reactive uncoordinated solvent molecules, and the low resistance of SEI is due to the reduction reaction of FSA^- [8, 9]. Although there are no reports on sulfur cathodes in [Li(glyme)]FSA, Park, a member of Watanabe's group, pointed out that lithium-sulfur batteries using FSA^--based ionic liquid showed rapid capacity fading with cycling due to the irreversible side reaction between FSA^- and lithium sulfides [10]. The electrolytes with LiFSA are available only when using sulfur-encapsulated micropore carbon cathodes in order to prevent contact between FSA^- and lithium sulfides [11, 12].

One of the difficulties in achieving a large energy density in rechargeable lithium-sulfur batteries with sparingly soluble electrolytes is the limitation of available electrolytes. Ueno et al. proposed that [Li(G3 or G4)]TFSA and [Li(G4)]BETA solvate ionic liquids enabled stable battery performance in lithium-sulfur cells because of the significantly low solubility of lithium sulfides [7]. Watanabe's group reported that the coulombic efficiency for deposition/dissolution of lithium metal increased with increasing concentrations of LiTFSA in G4 [13]. The coulombic efficiency for the lithium anode in 2.8 M LiTFSA/G4 (i.e., [Li(G4)]TFSA solvate ionic liquid) was 91% at the 50th cycle, which is much higher than the value of 26% in 1 M LiTFSA/G4 (LiTFSA:G4 = 1:4 by mol). Highly concentrated electrolytes of [Li(glyme)]TFSA solvate ionic liquids have potential use for both sulfur cathodes and lithium metal anodes.

2 Fundamental Electrochemistry of the Lithium Metal Anode in Solvate Ionic Liquids

Recently, the use of highly concentrated electrolytes is emerging as one of the strategies for designing next-generation batteries owing to their unique physicochemical properties and interfacial chemistry [8, 14–17]. One of the key factors is few and/or negligible amounts of uncoordinated solvent molecules in highly concentrated electrolytes. In general, the deposition and dissolution reaction of lithium metal in low-concentrated electrolytes below 1 M can be written as.

$$\mathrm{Li^+} + \mathrm{e^-} \rightleftharpoons \mathrm{Li} \tag{1}$$

On the other hand, the activity of the solvent molecules should be considered in highly concentrated electrolytes. For example, in [Li(glyme)]TFSA solvate ionic liquids [18], the equilibrium reaction of Li(I)/Li should be modified as follows:

$$[\mathrm{Li(glyme)}]^+ + \mathrm{e^-} \rightleftharpoons \mathrm{Li} + \mathrm{glyme} \tag{2}$$

This means that the activities of $[Li(glyme)]^+$ and uncoordinated glyme should be taken into account when investigating lithium metal anodes using [Li(glyme)]TFSA solvate ionic liquids.

A three-electrode cell with [Li(G4)]TFSA was used to measure the fundamental deposition and dissolution reaction of lithium metal [19]. Lithium metal was used as the working, reference and counter electrode. Figure 1 shows the relationship between the applied potentials and stead-state current densities for the deposition and dissolution of lithium metal. When the working electrode was stepped to a negative potential, a negative current corresponding to the deposition reaction of lithium metal decayed with time, becoming a steady-state current after several hundred seconds. This current decay was caused by the growth of the diffusion layer. Steady-state currents for the cathodic reaction increased as the potential became more negative. On the other hand, the steady-state currents for the anodic dissolution reaction of lithium metal were independent of the applied potential between 0.5 and 1.0 V. The limiting current was 0.5 mA cm^{-2}. The limiting current during the anodic reaction was attributed to the diffusion limit of Li^+ and/or ligands that coordinate to Li^+. During the dissolution reaction of lithium metal under high overpotential conditions, the concentration of Li^+ increases as $[Li(TFSA)_2]^-$ [20] forms in the vicinity of the electrode because [Li(G4)]TFSA has an extremely small amount of uncoordinated G4 [21]. An increase in $[Li(TFSA)_2]^-$ leads to an increase in the viscosity of the electrolytes [22]. It is suggested that the change in viscosity at the electrode interface affects the Li^+ flux, resulting in the diffusion-limited current densities.

The details of the electrochemical deposition and dissolution process of lithium metal were investigated in situ using an impedance electrochemical quartz crystal microbalance (EQCM) [23]. EQCM measurements provide information on both

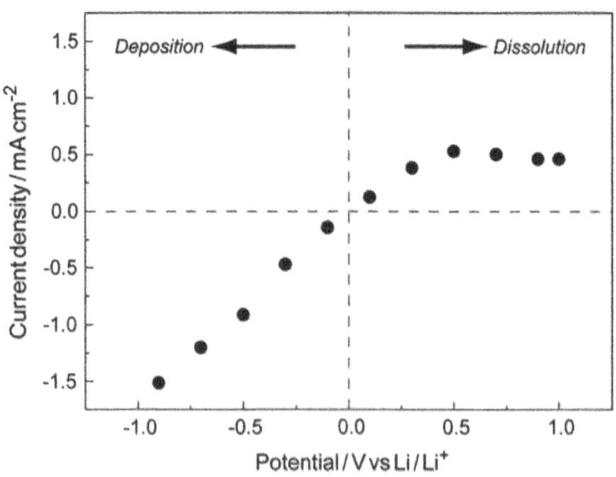

Fig. 1 Relationship between the applied potentials and steady-state current densities of Li deposition and dissolution in [Li(G4)]TFSA measured with a three-electrode cell at 30 °C. Reprinted with permission from *J. Electrochem. Soc.*, 159, A1005 (2012). Copyright 2012, The Electrochemical Society

the mass change of the electrode and the physical properties [viscosity (η_L) and density (ρ_L)] of the electrolyte near the electrode. The deposition and dissolution reaction of lithium metal in [Li(G3)]TFSA was successfully monitored by EQCM using a Ni-coated quartz crystal electrode at a scan rate of 50 mV min^{-1}, as shown in Fig. 2. The cathodic current attributable to the deposition of lithium metal was observed below 0 V, and the dissolution of lithium metal can be seen at the potential range of 0.0–0.5 V (Fig. 2a). The mass change (Δm) of the electrode was calculated using the Sauerbrey equation with correction for changes in η_L and ρ_L (Fig. 2b). The value of m increased during the cathodic reaction and decreased during the anodic reaction, corresponding to deposition and dissolution of lithium metal on the electrode, respectively. Importantly, Δm was higher than the value calculated from the amount of electric charge during cyclic voltammetry and did not revert to 0.0 μg

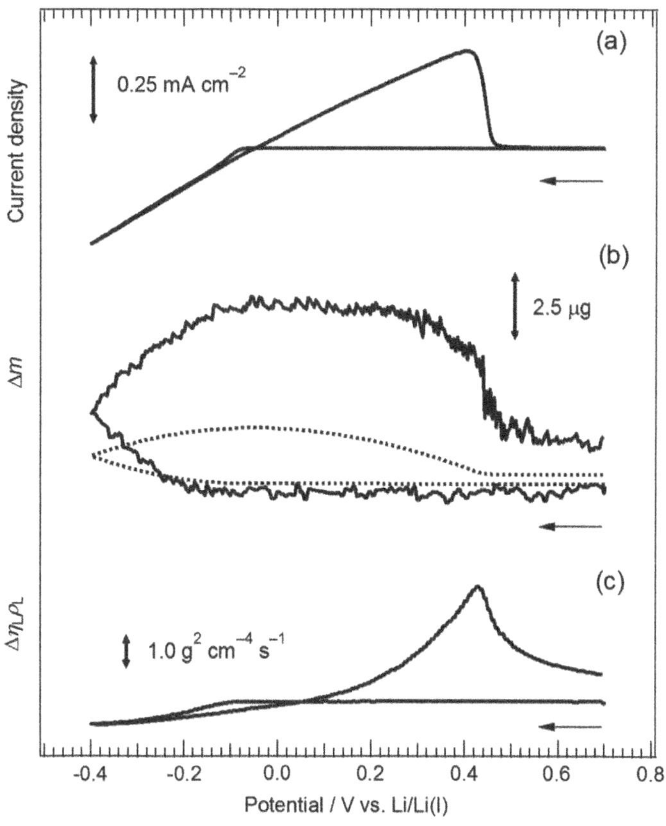

Fig. 2 EQCM measurement during cyclic voltammetry with a Ni-coated quartz crystal electrode in [Li(G3)]TFSA at 25 °C; **a** cyclic voltammogram, **b** mass change (Δm) estimated from the frequency change and the electric charge (dotted line), and **c** the change in $\eta_L\rho_L$. Scan rate: 50 mV min^{-1}. Reprinted with permission from *J. Electrochem. Soc.*, 160, A1529 (2013). Copyright 2013, The Electrochemical Society

(initial value of Δm) after cyclic voltammetry, suggesting that an SEI was formed on the electrode in [Li(G3)]TFSA. The change in the product of η_L and ρ_L ($\Delta\eta_L\rho_L$), which represents the change in the physical properties of the electrolyte near the electrode, started to decrease below 0.0 V along with the cathodic current in the cyclic voltammogram (Fig. 2c). The viscosity and density depended on the molar ratio of LiTFSA and G3. The viscosity and density decreased as the molar ratio of LiTFSA decreased [22–24]. Thus, the decrease in $\Delta\eta_L\rho_L$ can be explained by a decrease in LiTFSA at the electrode interface. In other words, $[Li(G3)]^+$ was reduced to lithium metal, generating uncoordinated G3 near the electrode. On the other hand, $\Delta\eta_L\rho_L$ dramatically increased during the anodic scan in the cyclic voltammetry, indicating that the concentration of Li^+ near the electrode increased during the dissolution reaction of lithium metal. An increase in Li^+ near the electrode may lead to formation of $[Li(TFSA)_2]^-$, as described above. These EQCM results support the possibility of the diffusion-limited current during the dissolution process of lithium metal.

Compared with low concentrated electrolytes, local changes in the physical properties were significant under electrochemical measurements, especially under high overpotential conditions. There is a possibility that the local changes in physical properties influence the rate performance of lithium-sulfur batteries in terms of the Li^+ diffusion process.

3 Lithium Phosphorus Oxynitride Modified Electrode in a Solvate Ionic Liquids

Designing lithium metal/electrolyte interfaces offers possibilities for better understanding of electrochemical reactions. It has been suggested that an SEI is formed in [Li(glyme)]TFSA by EQCM [23]. However, the SEI has not yet been characterized. An artificial SEI is useful for studying the nature of the SEI because the reductive decomposition of the electrolytes that forms the SEI can be avoided. To date, anode materials coated with lithium phosphorus oxynitride (LiPON), a Li^+ ion conductor, have been investigated as the artificial SEI [25, 26]. A Ni electrode coated with LiPON thin film was investigated in our research. The LiPON thin film was prepared by radio frequency magnetron sputtering on a Ni substrate. The thickness of the LiPON thin film was 0.9 ± 0.3 μm. Figure 3 shows the change in voltage under galvanostatic deposition and dissolution of the lithium metal on the Ni electrode with and without the LiPON thin film using [Li(G3)]TFSA [27]. In the case of the Ni electrode without the LiPON thin film, several minutes were necessary to reach a potential below 0 V. Reactions at the voltages higher than 0 V correspond to cathodic decomposition of the electrolytes. The amount of electric charge consumed during the decomposition reaction was 0.036 C cm^{-2}. In the case of the Ni electrode coated with LiPON thin film, however, the cell voltage quickly dropped below 0 V and remained around –0.1 V during galvanostatic deposition, suggesting that Li^+ can transfer through the LiPON thin film and that lithium metal can deposit without the

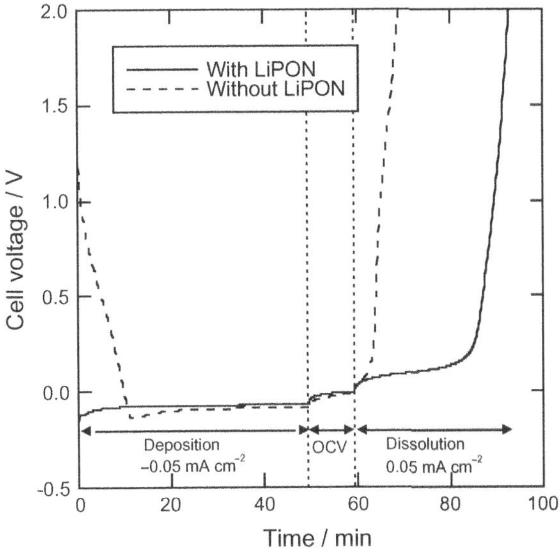

Fig. 3 Change of the cell voltage of a Li|[Li(G3)]TFSA|Ni with and without LiPON coin cell under galvanostatic deposition and deposition at 25 °C. Deposition: −0.05 mA cm^{-2} (0.148 C cm^{-2}). Dissolution: 0.05 mA cm^{-2} (Cut-off voltage: 2 V). Reprinted with permission from *Electrochemistry*, 83, 846 (2015). Copyright 2015, The Electrochemical Society of Japan

cathodic decomposition of the electrolyte. The LiPON thin film was found to function as an artificial SEI in [Li(G3)]TFSA. The dissolution reaction of the deposited lithium metal was also possible through the LiPON thin film. The cell using the Ni electrode with LiPON thin film showed a coulombic efficiency of 67%, which is higher than that without LiPON thin film (19%). This improvement could be related to the suppression of the irreversible decomposition of the electrolytes. However, cracks in the LiPON thin film were generated during the deposition and dissolution reaction, resulting in exposure of the reactive deposited lithium metal to the electrolyte. Artificial SEIs based on LiPON are useful for understanding the nature of SEIs, but are not suitable for practical application.

4 Design of the Interface Between the Lithium Metal Anode and Solvate Ionic Liquids

A serious problem with lithium anode is whisker-like and dendritic growth of lithium metal, associated with the volumetric expansion of the deposits. It has been reported that the introduction of a matrix structure for lithium deposition sites effectively improves coulombic efficiency in organic electrolytes [28–30]. Carbon fibers can provide a framework for the matrix. Vapor-grown carbon fiber (VGCF®-H) is a multi-wall carbon nanotube with a high aspect ratio. A carbon fiber composite composed of VGCF®-H and poly(vinylidene fluoride) was coated on the Cu substrate. This composite was used and investigated as the matrix for the deposition of lithium metal in [Li(G4)]TFSA [31]. The thickness of the VGCF®-H modified layer was approximately 10 μm. The VGCF®-H on the Cu substrate was tangled, forming a porous network structure. When the electric charge for deposition was 0.5 C cm^{-2}, the coulombic efficiency using a bare Cu|[Li(G4)]TFSA|Li cell increased within the first several cycles and became 90% at the 20 cycle (Fig. 4). However, fluctuation of the cell voltage and significant capacity fading were observed after 40 cycles, suggesting a short circuit of the cell caused by the deposited lithium metal. In contrast, reversible cycles without short circuiting was possible using the VGCF®-H modified Cu|[Li(G4)]TFSA|Li cell. A coulombic efficiency of 98% at the 100th cycle was achieved with the VGCF®-H modified Cu electrode. While whisker-like deposits were observed on the bare Cu substrate after 20 cycles, the porous structure of the VGCF®-H network was mainly observed using the cell operated with the VGCF®-H. Deposition of lithium metal with the VGCF®-H modified Cu electrode took place in the void based on the VGCF®-H network, preventing short circuiting. VGCF®-H is multi-wall carbon nanotube with a high aspect ratio. Therefore, the cylindrical basal plane of the VGCF®-H was mainly in contact with [Li(G4)]TFSA. In the case of the VGCF®-H modified Cu electrode, the lithium deposition sites are considered to be

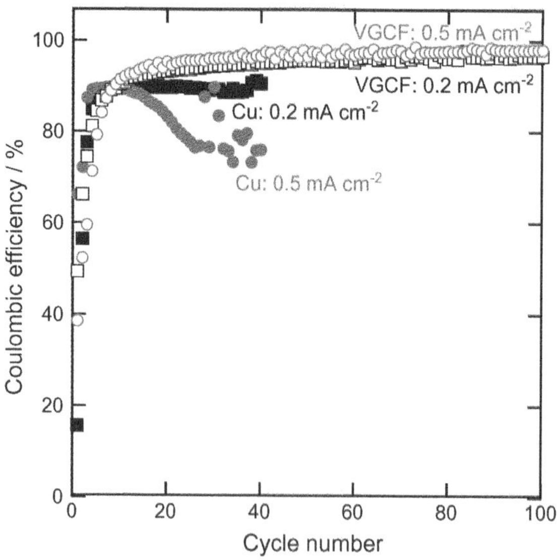

Fig. 4 Cycle dependence of the coulombic efficiency for dissolution of Li deposits on a VGCF®-H-modified Cu electrode in [Li(G4)]TFSA. Current density: ± 0.2 and 0.5 mA cm^{-2}. Electric charge: 0.5 C cm^{-2}. Cut-off voltage: 2 V. Reprinted with permission from *Electrochemistry*, 85, 667 (2017). Copyright 2017, The Electrochemical Society of Japan

on (i) the Cu substrate, (ii) the deposited lithium, and (iii) the cylindrical basal plane of the VGCF®-H. Greater reactivity of the Cu substrate and deposited lithium metal is expected, compared with the basal plane of the VGCF®-H [32]. Thus, deposition of lithium metal preferentially occurred on the Cu substrate and/or deposited lithium rather than the cylindrical basal plane of the VGCF®-H. The porous network structure of VGCF®-H may contribute to (i) mitigating the probability of short circuiting and (ii) improving the coulombic efficiency caused by support of electrical contact with dead lithium.

To achieve large specific capacity of lithium-sulfur batteries, the electric charge for deposition of lithium metal was 3.6 C cm^{-2}, corresponding to 1.0 mAh cm^{-2}. Figure 5 shows the charge/discharge test on the VGCF®-H-modified Cu electrode [31]. The deposited lithium after the first charge was observed inside of the void of the VGCF®-H matrix. The charge/discharge cycle achieved a coulombic efficiency of 90% at 15 cycles. In contrast to the small electric charge (0.5 C cm^{-2}) for deposition, the overpotential of the cell gradually increased with cycling. The change in the overpotential of the cell operated at a large electric charge may be due to local consumption of $[Li(G4)]^+$ during charging, and local generation of $[Li(TFSA)_2]^-$ during discharging near the electrode. Further development is needed to optimize the composition of the solvate ionic liquid electrolytes in order to control their physical properties in the vicinity of the electrode.

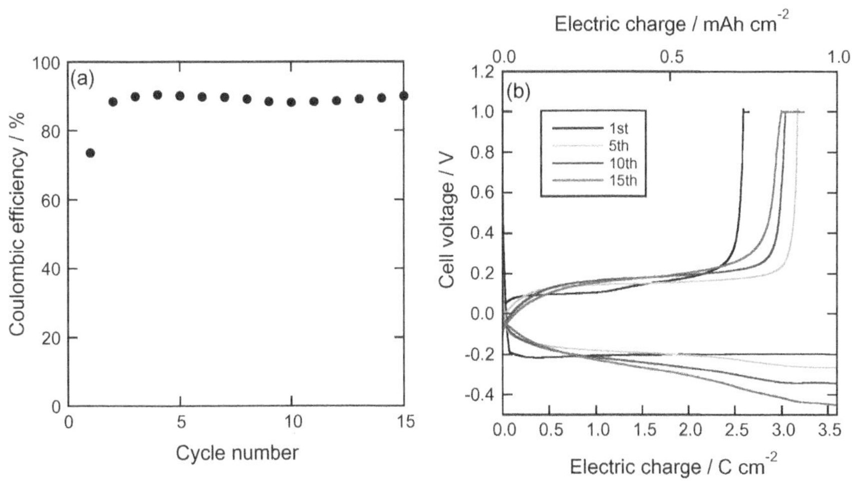

Fig. 5 **a** Cycle dependence of the coulombic efficiency for dissolution of Li deposits and **b** the charge/discharge curves on a VGCF®-H-modified Cu electrode in [Li(G4)]TFSA. Current density for charge: –0.5 mA cm^{-2}, electric charge: 3.6 C cm^{-2}. Current density for discharge: 0.5 mA cm^{-2} (cut-off voltage: 1 V), followed by applying a constant voltage of 1 V for 1 h. Reprinted with permission from *Electrochemistry*, 85, 667 (2017). Copyright 2017, The Electrochemical Society of Japan

References

1. Aurbach, D., Zinigrad, E., Cohen, Y., & Teller, H. (2002). A short review of failure mechanisms of lithium metal and lithiated graphite anodes in liquid electrolyte solutions. *Solid State Ionics, 148,* 405–416.
2. Peled, E. (1979). The electrochemical behavior of alkali and alkaline earth metals in nonaqueous battery systems-the solid electrolyte interphase model. *Journal of the Electrochemical Society, 126,* 2047–2051.
3. Aurbach, D., Zaban, A., Schechter, A., Ein-Eli, Y., Zinigrad, E., & Markovsky, B. (1995). The study of electrolyte solutions based on ethylene and diethyl carbonates for rechargeable Li batteries. *Journal of the Electrochemical Society, 142,* 2873–2882.
4. Gao, J., Lowe, M. A., Kiya, Y., & Abruña, H. D. (2011). Effects of liquid electrolytes on the charge-discharge performance of rechargeable lithium/sulfur batteries: Electrochemical and in-situ X-ray absoption spectroscopic studies. *Journal of Physical Chemistry C, 115,* 25132–25137.
5. Aurbach, D., Pollak, E., Elazari, R., Salitra, G., Kelley, C. S., & Affinito, J. (2009). On the surface chemical aspects of very high energy density, rechargeable Li-sulfur batteries. *Journal of the Electrochemical Society, 156,* A694–A702.
6. Li, W., Yao, H., Yan, K., Zheng, G., Liang, Z., Chiang, Y. M., & Cui, Y. (2015). The synergetic effect of lithium polysulfide and lithium nitrate to prevent lithium dendrite growth. *Nature Communications, 6,* 7436.
7. Ueno, K., Park, J. W., Yamazaki, A., Mandai, T., Tachikawa, N., Dokko, K., & Watanabe, M. (2013). Anionic effects on solvate ionic liquid electrolytes in rechargeable lithium-sulfur batteries. *Journal of Physical Chemistry C, 117,* 20509–20516.
8. Qian, J., Henderson, W. A., Xu, W., Bhattacharya, P., Engelhard, M., Borodin, O., & Zhang, J. G. (2015). High rate and stable cycling of lithium metal anode. *Nature Communications, 6,* 6362.
9. Sodeyama, K., Yamada, Y., Aikawa, K., Yamada, A., & Tateyama, Y. (2014). Sacrificial anion reduction mechanism for electrochemical stability improvement in highly concentrated Li-salt electrolyte. *Journal of Physical Chemistry C, 118,* 14091–14097.
10. Park, J. W., Ueno, K., Tachikawa, N., Dokko, K., & Watanabe, M. (2013). Ionic liquid electrolytes for lithium-sulfur batteries. *Journal of Physical Chemistry C, 117,* 20531–20541.
11. Salitra, G., Markevich, E., Rosenman, A., Talyosef, Y., Aurbach, D., & Garsuch, A. (2014). High-performance lithium-sulfur batteries based in ionic-liquid electrolytes with bis(fluorolsufonyl)imide anions and sulfur-encapsulated highly disordered activated carbon. *ChemElectroChem, 1,* 1492–1496.
12. Takahashi, T., Yamagata, M., & Ishikawa, M. (2015). A sulfur-microporous carbon composite positive electrode for lithium/sulfur and silicon/sulfur rechargeble batteries. *Progress in Natural Science, 25,* 612–621.
13. Tachikawa, N., Yoshida, K., Dokko, K., Watanabe, M. (2011). *Electrochemical properties of lithium metal anode in glyme-Li salt equimolar complexes.* Presented at the 52nd battery symposium in Japan, Tokyo: #3G21.
14. Suo, L., Hu, Y. S., Li, H., Armand, M., & Chen, L. (2013). A new class of solvate-in-salt electrolyte for high-energy rechargeable metallic lithium batteries. *Nature Communications, 4,* 1481.
15. Yamada, Y., Furukawa, K., Sodeyama, K., Kikuchi, K., Yaegashi, M., Tateyama, Y., & Yamada, A. (2014). Unusual stability of acetonitrile-based superconcentrated electrolytes for fast-charging lithium-ion batteries. *Journal of the American Chemical Society, 136,* 5039–5046.
16. Yamada, Y., & Yamada, A. (2017). Superconcentrated electrolytes to create new interfacial chemistry in non-aqueous and aqueous rechargeable batteries. *Chemistry Letters, 46,* 1056–1064.
17. Watanabe, M., Dokko, K., Ueno, K., & Thomas, M. L. (2018). From ionic liquids to solvate ionic liquids: Challenges and opportunities for next generation battery electrolytes. *Bulletin of the Chemical Society of Japan, 91,* 1660–1682.

18. Moon, H., Tatara, R., Mandai, T., Ueno, K., Yoshida, K., Tachikawa, N., et al. (2014). Mechanism of Li ion desolvation at the interface of graphite electrode and glyme-Li salt solvate ionic liquids. *Journal of Physical Chemistry C, 118,* 20246–20256.
19. Yoshida, K., Tsuchiya, M., Tachikawa, N., Dokko, K., & Watanabe, M. (2012). Correlation between battery performance and lithium ion diffusion in glyme-lithium bis(trifluoromethanesulfonyl)amide equimolar complexes. *Journal of the Electrochemical Society, 159,* A1005–A1012.
20. Umebayashi, Y., Mitsugi, T., Fukuda, S., Fujimori, T., Fujii, K., Kanzaki, R., et al. (2007). Lithium ion solvation in room-temperature ionic liquids involving bis(trifluoromethanesulfonyl)imide anion studied by Raman spectroscopy and DFT calculations. *The Journal of Physical Chemistry B, 111,* 13028–13032.
21. Ueno, K., Tatara, R., Tsuzuki, S., Saito, S., Doi, H., Yoshida, K., Mandai, T., Matsugami, M., Umebayashi, Y., Dokko, K., Watanabe, M. (2015). Li^+ solvation in glyme-Li salt solvate ionic liquids. *Physical Chemistry Chemical Physics 17,* 8248–8257
22. Hirayama, H., Tachikawa, N., Yoshii, K., Watanabe, M., & Katayama, Y. (2015). Ionic conductivity and viscosity of solvate ionic liquids composed of glymes and excess lithium bis(trifluoromethylsulfonyl)amide. *Electrochemistry, 83,* 824–827.
23. Serizawa, N., Seki, S., Takei, K., Miyashiro, H., Yoshida, K., Ueno, K., et al. (2013). EQCM measurement of deposition and dissolution of lithium in glyme-Li salt molten complex. *Journal of the Electrochemical Society, 160,* A1529–A1533.
24. Yoshida, K., Tsuchiya, M., Tachikawa, N., Dokko, K., & Watanabe, M. (2011). Change from glyme solutions to quasi-ionic liquids for binary mixtures consisting of lithium bis(trifluoromethanesulfonyl)amide and glymes. *Journal of Physical Chemistry C, 115,* 18384–18394.
25. Li, J., Dudney, N. J., Nanda, J., & Liang, C. (2014). Artificial solid electrolyte interphase to address the electrochemical degradation of silicon electrodes. *ACS Applied Materials & Interfaces, 6,* 10083–10088.
26. Furuya, R., Tachikawa, N., Yoshii, K., Katayama, Y., & Miura, T. (2015). Deposition and dissolution of lithium through lithium phosphorus oxynitride thin film in some ionic liquids. *Journal of the Electrochemical Society, 162,* H634–H637.
27. Tachikawa, N., Furuya, R., Yoshii, K., Watanabe, M., & Katayama, Y. (2015). Deposition and dissolution of lithium through lithium phosphorus oxynitride thin film in lithium bis(trifluoromethylsulfonyl)amide-glyme solvate ionic liquid. *Electrochemistry, 83,* 846–848.
28. Ji, X., Liu, D. Y., Prendiville, D. G., Zhang, Y., Liu, X., & Stucky, G. D. (2012). Spatially heterogeneous carbon-fiber papers as surface dendrite-free current collectors for lithium deposition. *Nano Today, 7,* 10–20.
29. Zhang, G., Lee, S. W., Liang, Z., Lee, H. W., Yan, K., Yao, H., et al. (2014). Interconnected hallow carbon nanospheres for stable lithium metal anodes. *Nature Nanotechnology, 9,* 618–623.
30. Kang, H. K., Woo, S. G., Kim, J. H., Lee, S. R., & Kim, Y. J. (2015). Conductive porous carbon film as a lithium metal storage medium. *Electrochimica Acta, 176,* 172–178.
31. Tachikawa, N., Kasai, R., Yoshii, K., Watanabe, M., & Katayama, Y. (2017). Electrochemical deposition and dissolution of lithium on a carbon fiber composite electrode in a solvate ionic liquid. *Electrochemistry, 85,* 667–670.
32. Kneten, K. R., & McCreery, R. L. (1992). Effects of redox system structure on electron-transfer kinetics at ordered graphite and glassy carbon electrode. *Analytical Chemistry, 64,* 2518–2524.

Silicon LeafPowder® Anode

Masakazu Haruta, Takayuki Doi, and Minoru Inaba

Abstract The anode properties of Si nano-flake powder (Si LeafPowder®) were investigated in an ethylene carbonate (EC)-based solution and in a solvate ionic liquid of an equimolar complex of lithium bis(trifluoromethanesulfonyl)imide (LiTFSI) and tetraglyme (G4) ([Li(G4)][TFSI]) for use as an anode in Si–Li_2S batteries. The nano-flake structure of 100 nm in thickness gave a high initial capacity of ca. 2600 mAh g^{-1} and good cyclability. The addition of film-forming additives such as fluoroethylene carbonate (FEC) and vinylene carbonate (VC) significantly improved the cyclability; however, swelling of the electrode was found to be a serious problem to be solved for practical use. A Li pre-doping method using a solution of Li-naphthalene complex successfully compensated the large irreversible capacity in the initial cycle. It was also found that a pre-film formed in an FEC electrolyte effectively suppresses the increase in overpotential on cycling and improves the cyclability in [Li(G4)][TFSI].

Keywords Si anode · Nano-flake · Pre-doping · Cyclability · Pre-film formation

The use of silicon as an anode material is very promising for realizing advanced secondary batteries with high energy densities because Li–Si alloy anode provides a high gravimetric and a volumetric capacity (3578 mAh g^{-1} and 2194 mAh cm^{-3}, respectively, based on $Li_{15}Si_4$) at sufficiently low discharge potentials (ca. 0.4 V vs. Li/Li^+ on average) [1–3]. However, Li–Si alloy anodes have encountered a serious problem, i.e., poor capacity retention, on charge/discharge (alloying/dealloying) cycles. The poor cyclability is attributable to crack formation and pulverization of Si due to large volume expansion and contraction upon charging and discharging, respectively. It is thus important to suppress the physical stress induced by the large volume changes for improving the cyclability of Si anodes. Following this concept, various kinds of nano-structured Si anodes such as nanopowders [4, 5], nanowires [6, 7], and thin films [8, 9] have been designed to overcome the poor cyclability. Moreover, the formation of a stable solid electrolyte interphase (SEI) on the anodes

M. Haruta · T. Doi · M. Inaba (✉)
Department of Molecular Chemistry and Biochemistry, Doshisha University, 1-3 Tatara-Miyakodani, Kyotanabe 610-0321, Kyoto, Japan
e-mail: minaba@mail.doshisha.ac.jp

K. Kanamura (ed.), *Next Generation Batteries*,
https://doi.org/10.1007/978-981-33-6668-8_29

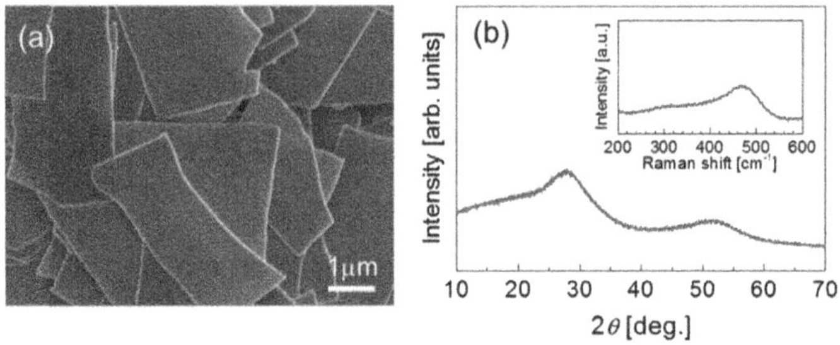

Fig. 1 **a** SEM image and **b** XRD pattern of Si nano-flake powders (Si LeafPowder®, Si-LP). The inset of (**b**) shows the Raman spectrum of Si-LP

is necessary to suppress the unfavorable irreversible reactions such as electrolyte decomposition during charge/discharge cycling. It is widely known that the cyclability of graphite and Si anodes can be improved by the addition of film-forming additives such as vinylene carbonate (VC) and fluoroethylene carbonate (FEC), which promote the formation of a stable SEI that functions as an effective passive layer [10–13].

To achieve a Si anode with a high capacity and good cyclability, we designed a Si powder with a nano-flake shape, which is called Si LeafPowder® (Si-LP, OIKE & Co., Ltd.) as shown in Fig. 1a [14–16]. The Si-LP is prepared using thin-film-based techniques. A Si film of 100 nm in thickness is deposited on a polymer-coated metal-base-sheet by electron-beam evaporation with a roll-to-roll process. The film deposition conditions are optimized to obtain an amorphous Si. Then the Si film is peeled from the polymer sheet and crushed into flake powder with a lateral size of 3–5 μm to obtain Si-LP with an amorphous structure, which is confirmed by X-ray diffraction (XRD) and Raman spectrum (Fig. 1b).

The charge and discharge characteristics of a Si-LP electrode (100 nm in Si-LP thickness, Si-LP:KetjenBlack:CMCNa = 83:6:11) in 1 M $LiPF_6$ dissolved in a mixture of ethylene carbonate (EC) and diethyl carbonate (DEC) without additives are shown in Fig. 2a. A high initial discharge capacity of ca. 2600 mAh g^{-1} was obtained. The potential changed monotonously on both charging and discharging without any clear plateaus, which indicates the absence of crystalline $Li_{15}Si_4$-phase formation [17, 18]. The variations of discharge capacity with cycle number up to 200 cycles are summarized in Fig. 2b. The cyclability of the Si-LP electrode was superior to conventional Si powder electrodes (not shown, less than 20 cycles) [14]; however, the discharge capacity decreased to half of the initial value at the 100th cycle in the absence of additives. By the addition of fluoroethylene carbonate (FEC, 10 wt.%) as a film-forming additive, the capacity retention of the Si-LP electrode has significantly improved, indicating suppression of the electrolyte decomposition by the FEC-derived SEI on the Si surface [19, 20]. Similar improvement was observed for another additive, vinylene carbonate (VC, 10 wt.%) [20].

Si-LPs of different thicknesses (50–400 nm) were prepared and their charge and discharge characteristics were tested to understand the effects of the thickness on the

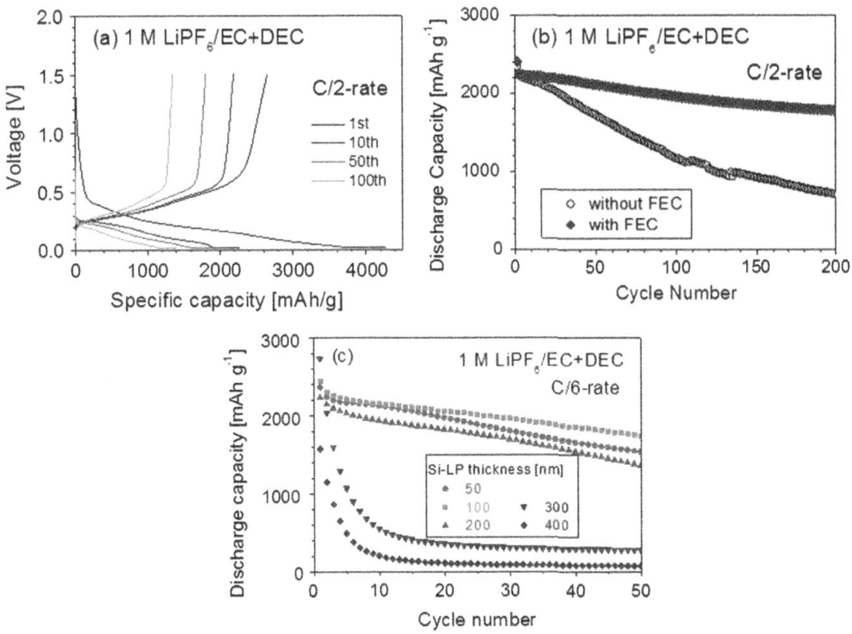

Fig. 2 **a** Charge and discharge characteristics of Si-LP (100 nm) anode at C/2 rate in 1 M $LiPF_6$/EC + DEC (1:1). **b** Variation of the discharge capacity with cycle number of Si-LP (100 nm) with and without 10 wt.% FEC at C/2 (From Ref. [20]. Reproduced with permission from Elsevier, Ltd.). **c** Cyclability of the Si-LPs of different thicknesses (50–400 nm) at C/6 in 1 M $LiPF_6$/EC + DEC (1:1) (without FEC). (From Ref. [15]. Reproduced with permission from Elsevier, Ltd.)

cyclability. The results are shown in Fig. 2c [15]. Si-LPs of 50–200 nm in thickness exhibited good cyclability, whereas the discharge capacities of the 300- and 400-nm-thick Si-LPs decreased rapidly in the initial 10 cycles. SEM observation of the electrodes after 50 cycles revealed that Si-LP particles of 300 and 400 nm in thickness were significantly fractured to be smaller (in lateral size). The fracture of the thicker Si-LPs is considered to be caused by a high physical stress in the Si-LP particles owing to a large difference in Li content between the surface and the inside during charging and discharging, which resulted in the observed poor capacity retention. On the other hand, cracking of the Si-LP particles of 50–200 nm in thickness was not observed after 50 cycles, although they were significantly deformed by repeated charging/discharging cycles as discussed later. Hereafter 100-nm-thick Si-LP is used as a standard because of its good cyclability, and the results based on the 100-nm-thick Si-LP are used for further discussion.

Figure 3 compares cross-sectional views of the Si-LP electrodes before and after 50 cycles in 1 M $LiPF_6$/EC + DEC (1:1) with and without 10 wt.% FEC. Although Si-LP particles were flat and stacked uniformly before cycling, they were significantly deformed and distorted after 50 cycles [20]. The thickness of the Si-LP composite electrode was 3.2 μm before cycling, but swelled to 20 and 14 μm after

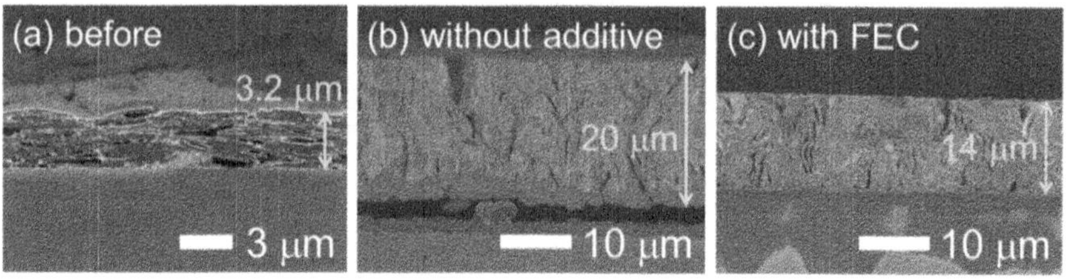

Fig. 3 Cross-sectional SEM images of the Si-LP electrodes (**a**) before and after 50 cycles in 1 M $LiPF_6$/EC + DEC (**b**) without and (**c**) with 10 wt.% FEC (From Ref. [20]. Reproduced with permission from Elsevier, Ltd.)

50 cycles in the absence and presence of FEC, respectively. This electrode swelling was attributable not only to the deformation of Si-LP by expansion and shrinkage upon cycling, but also the accumulation of decomposition products of the electrolyte between the Si-LP particles upon cycling, the latter of which is also a serious problem in practical use. The swelling was suppressed by FEC addition owing to the inhibition of electrolyte decomposition by the FEC-derived SEI. However, further suppression of the electrode swelling is necessary for practical cell applications. Preliminary study revealed that the electrode swelling with cycling can be suppressed by the optimization of the lateral size and the oxygen content of the Si-LP [unpublished data].

As shown in Fig. 2a, the Si-LP electrode had a large irreversible capacity (Q_{irr}, over 1000 mAh g^{-1}) in the first cycle, which is a common issue of Si anodes for practical cells because the large Q_{irr} wastes mobile lithium ions originally stored in cathode materials. The large Q_{irr} of Si anode is attributable to the reduction of the surface SiO_x layer, and the electrolyte decomposition and SEI formation in the first charging. Li pre-doping using lithium metal has been reported for compensating the large irreversible capacity [21]. We developed another Li pre-doping method using a solution of Li-naphthalene complex dissolved in tetrahydrofuran (THF) [22]. Here naphthalene is reduced by the Li metal to form a Li-naphthalene complex, which works as a reductant and a Li source for the Si anode as:

$$\mathrm{Li} + \mathrm{naphthalene} \rightarrow \mathrm{Li}^+ - [\mathrm{naphthalene}]^{-\bullet} (\mathrm{at\,Li\,metal})$$

$$x\mathrm{Li}^+ - [\mathrm{naphthalene}]^{-\bullet} + \mathrm{Si} \rightarrow \mathrm{Li_xSi} + x(\mathrm{naphthalene})(\mathrm{at\,Si\,anode}).$$

Charge and discharge curves and Q_{irr} values of the Si-LP electrodes in the first cycle after pre-doped for 1 h in the Li-naphthalene complex/THF solutions of different concentrations are shown in Fig. 4. The value of Q_{irr} decreased with the concentration of the pre-doping solution, meaning that lithium was alloyed with the Si-LP and the Li content increased with the concentration of Li-naphthalene complex. These results indicate that the degree of pre-doping can be controlled by the concentration of naphthalene (and the doping time). The initial Q_{irr} (1162 mAh g^{-1}) for

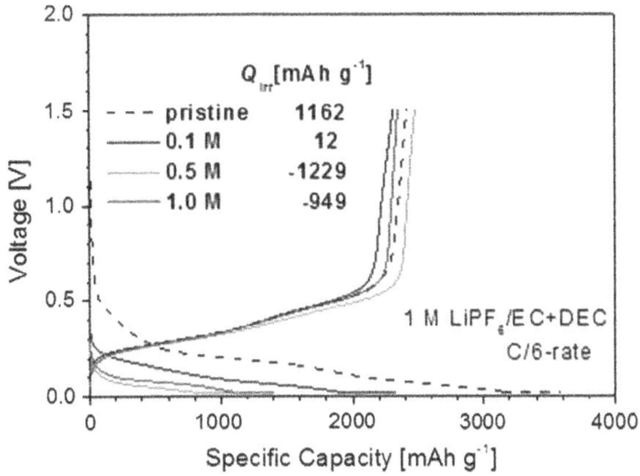

Fig. 4 Charge and discharge curves of the pristine and the pre-doped Si-LP electrodes at C/6 rate in 1 M $LiPF_6$/EC + DEC. Pre-doping was performed for 1 h using 0.1, 0.5 and 1.0 M Li-naphthalene complex/THF solutions before the charge/discharge test. Irreversible capacity (Q_{irr}) values are also shown, where a minus value means excessive pre-doping (From Ref. 22. Reproduced with permission from The Electrochemical Society of Japan.)

the pristine Si-LP electrode was successfully compensated by pre-doping for 1 h using the 0.1 M solution; Q_{irr} of the pre-doped Si-LP electrode was reduced to only 12 mAh g^{-1}. The Li pre-doping method using Li-naphthalene complex/THF solutions is a useful technique for compensating the large Q_{irr}, not only for the Si-LP but also for other types of Si anodes.

A combination of silicon anode and sulfur cathode, i.e., Si–Li_2S battery, will be a promising battery system with a high energy density in the near future, though Li pre-doping of either the cathode or the anode is necessary [22, 23]. It is widely known that sulfur cathodes have a serious drawback, i.e., dissolution of polysulfide anions (e.g. S_8^{2-}), resulting in poor cyclability in conventional electrolyte solutions. Dokko and Watanabe et al. have reported that the dissolution of polysulfide anions was successfully suppressed by using a solvate ionic liquid of an equimolar complex of lithium bis(trifluoromethanesulfonyl)imide (LiTFSI) and tetraglyme (G4) (hereafter called [Li(G4)][TFSI]) [24, 25]. In the ALCA SPRING project, Si-S batteries using the solvate ionic liquid as an electrolyte are being developed, and hence the compatibility of the Si-LP anode and the solvate ionic liquid as an electrolyte is important for realizing a high energy density and a long-term cycle life [26, 27].

Charge and discharge properties of Si-LP were measured in [Li(G4)][TFSI] diluted with a hydrofluoroether (HFE, 1,1,2,2-tetrafluoroethyl-2,2,3,3-tetrafluoropropyl ether) solvent to reduce the electrolyte viscosity ([Li(G4)][TFSI]/HFE = 1:4) [26]. The charge and discharge characteristics are shown in Fig. 5a. The Si-LP electrode exhibited a reversible capacity of ca. 2200 mAh g^{-1} at the first cycle; however, the discharge capacity decreased with cycling and dropped to ca. 700 mAh g^{-1} at the 100th cycle. Furthermore, the overpotential significantly increased upon cycling.

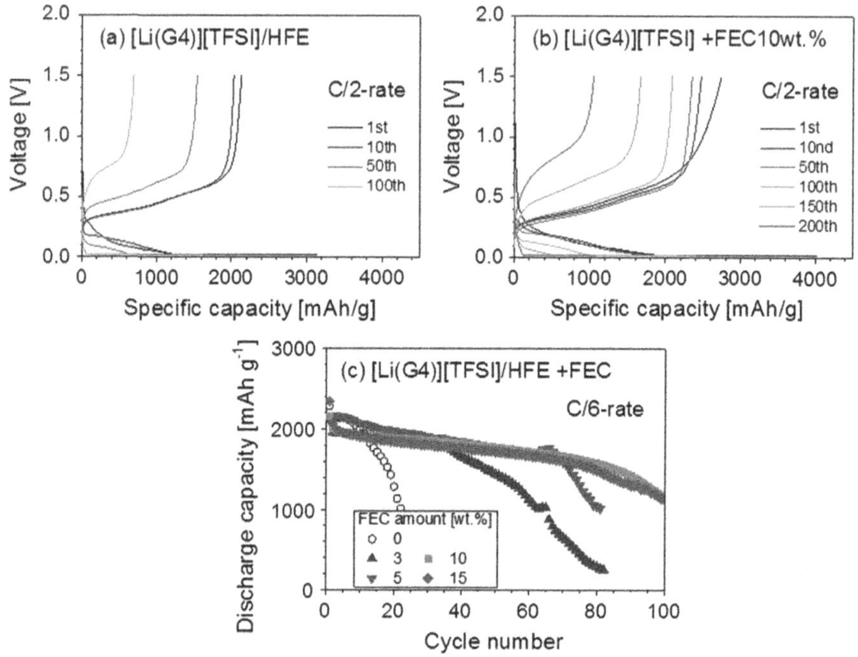

Fig. 5 Charge and discharge characteristics of the Si-LP electrodes at C/2 in [Li(G4)][TFSI]/HFE (**a**) without and **b** with 10 wt.% FEC (From Ref. 26. Reproduced with permission from The Electrochemical Society of Japan.) **c** Variations of the discharge capacity of the Si-LP electrodes at C/6 rate in [Li(G4)][TFSI]/HFE with FEC of different amounts (0–15 wt.%)

As mentioned earlier, the SEI-forming additives are effective for improving cyclability in the EC-based electrolyte, and hence FEC was added at 10 wt.% to the [Li(G4)][TFSI]/HFE electrolyte. As shown in Fig. 5b, the capacity retention improved and the increase in overpotential during cycling was suppressed by adding FEC up to the 100th cycle. However, the discharge capacity dropped and the overpotential abruptly increased after the 100th cycle. The discharge capacity decreased to ca. 1000 mAh g^{-1} at the 200th cycle. Figure 5c shows the cyclability of the Si-LP electrodes in [Li(G4)][TFSI]/HFE with FEC of different amounts. The cyclability strongly depended on the amount of FEC, indicating that the FEC was exhausted by the continuous decomposition of FEC upon cycling. The capacity retention of the Si-LP electrode improved with FEC amount; however, the improvement peaked at 10 wt.% and no further improvement was observed at 15 wt.%. These results indicate that a stable SEI layer was not formed from FEC in the [Li(G4)][TFSI]/HFE electrolyte, and thus, a long-term cyclability cannot be achieved solely by adding SEI-forming additive.

To obtain a stable pre-film (i.e. artificial SEI) on the Si-LP, the composite electrode was charged and discharged once in 1 M $LiPF_6$ dissolved in neat FEC [27]. After one cycle in 1 M $LiPF_6$/FEC, the electrode was rinsed with 1,2-dimethoxyethane (DME)

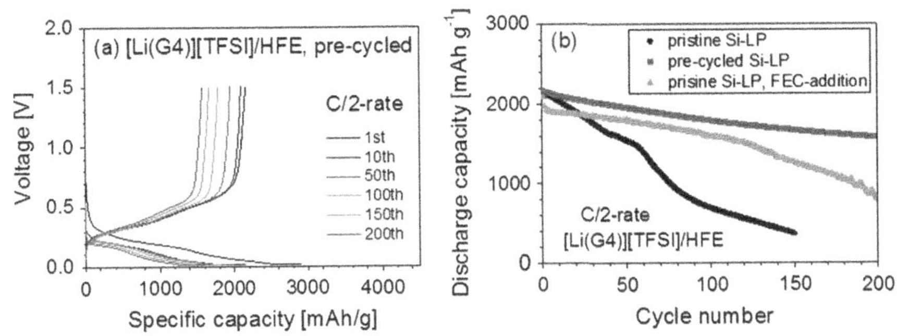

Fig. 6 **a** Charge and discharge characteristics of the pre-cycled Si-LP electrode at C/2 in [Li(G4)][TFSI]/HFE. **b** Variations of the discharge capacity of the pristine and pre-cycled Si-LP electrodes at C/2 rate in [Li(G4)][TFSI]/HFE. Data for the pristine Si-LP electrode in [Li(G4)][TFSI]/HFE + 10 wt.% FEC are also shown in (**b**) (From Ref. 27. Reproduced with permission from The Electrochemical Society, Inc.)

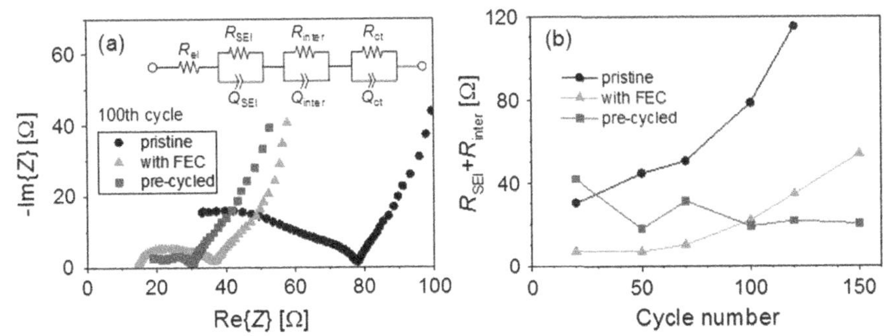

Fig. 7 **a** Nyquist plots in the 100th cycle for the pristine Si-LP electrode in [Li(G4)][TFSI]/HFE with and without FEC, and the pre-cycled Si-LP electrode in [Li(G4)][TFSI]/HFE. The impedance spectra were acquired at 0.1 V during charging (From Ref. 27. Reproduced with permission from The Electrochemical Society, Inc.) **b** Variations of $R_{SEI} + R_{inter}$ with cycle number

to remove residual solvent and salt. Then the charge and discharge characteristics of the pre-cycled Si-LP anode were measured in the [Li(G4)][TFSI]/HFE electrolyte that did not contain FEC. The results are shown in Fig. 6. The pre-cycled Si-LP anode exhibited an initial capacity of 2200 mAh g^{-1}, which was comparable to that of the pristine Si-LP anode. In contrast to the FEC-added system (Fig. 5b), the increase in the overpotential during cycling was suppressed up to the 200th cycle (Fig. 6a). The pre-cycled Si-LP demonstrated a superior cycle performance up to the 200th cycle as shown in Fig. 6b, even though FEC was not added in the electrolyte [27]. Even at the 200th cycle, the pre-cycled Si-LP retained a high capacity of ca. 1600 mAh g^{-1}.

Electrochemical impedance spectroscopy (EIS) was performed to understand the degradation mechanism of the Si-LP electrodes [27]. Nyquist plots of the impedance spectra obtained at 0.1 V during charging at the 100th cycle are shown in Fig. 7a.

The impedance spectra consisted of three semicircles, which were analyzed using an equivalent circuit made of the impedances of the SEI layer, the interparticle electronic contact, and the charge-transfer reaction as shown in the inset of Fig. 7a [20, 28]. The pristine Si-LP electrode in [Li(G4)][TFSI]/HFE exhibited large SEI and interparticle resistances (R_{SEI} and R_{inter}, respectively) in comparison with the pristine Si-LP in the presence of FEC and the pre-cycled Si-LP in the absence of FEC.

Variations of the sum of R_{SEI} and R_{inter}, derived from EIS spectra, with cycle number are summarized in Fig. 7b. The value of $R_{SEI} + R_{inter}$ of the pristine electrode significantly increased with cycle number in [Li(G4)][TFSI]/HFE. This indicates rapid growth of the SEI layer owing to reductive decomposition of the electrolyte during charge and discharge cycles. The addition of FEC to the [Li(G4)][TFSI]/HFE electrolyte significantly reduced R_{SEI} and R_{inter}, suggesting that electrolyte decomposition was inhibited by the SEI derived from FEC. However, these increased rapidly after 100 cycles, implying exhaustion of the FEC followed by acceleration of electrolyte decomposition on Si-LP. Though $R_{SEI} + R_{inter}$ for the pre-cycled electrode was higher than those for the other two electrodes at the 20th cycle, the increase in $R_{SEI} + R_{inter}$ during cycling was much smaller, and $R_{SEI} + R_{inter}$ remained low even at the 150th cycle. This indicated that the pre-film formed on the Si-LP in 1 M $LiPF_6$/FEC effectively inhibited the electrolyte decomposition during long-term cycling.

It is not clear why the pre-cycled Si-LP exhibits a superior cycle performance even in the [Li(G4)][TFSI]/HFE electrolyte without film-forming additives. Our preliminary analytical data revealed that the pre-film consisted of fine LiF grains (ca. 10 nm) distributed uniformly on the Si-LP [27]. It is hence considered that organo-fluorine compounds fixed on the LiF grains functioned as a stable SEI and provided long-term cyclability in the [Li(G4)][TFSI]/HFE electrolyte. Unfortunately, the pre-cycling process may not be suitable for commercial production of practical Si-S batteries. It is therefore important to develop a more practical method for producing effective pre-films (consisting of LiF and organo-fluorine compounds) as artificial SEIs are needed to realize Si-S batteries.

References

1. Boukamp, B. A., Lesh, G. C., & Huggins, R. A. (1981). All-solid lithium electrodes with mixed-conductor matrix. *Journal of the Electrochemical Society, 128,* 725–729.
2. Huggins, R. A. (1999). Lithium alloy negative electrodes. *Journal of Power Sources, 81–82,* 13–19.
3. Wu, H., & Cui, Y. (2012). Designing nanostructured Si anodes for high energy lithium ion batteries. *Nano Today, 7,* 414–429.
4. Li, H., Huang, X., Chen, L., Wu, Z., & Liang, Y. (1999). A high capacity nano Si composite anode material for lithium rechargeable batteries. *Electrochemical and Solid-State Letters, 2,* 547–549.
5. Kim, H., Seo, M., Park, M.-H., & Cho, J. (2010). A critical size of silicon nano-anodes for lithium rechargeable batteries. *Angewandte Chemie International Edition, 49,* 2146–2149.

6. Chan, C. K., Peng, H. L., Liu, G., McIlwrath, K., Zhang, X. F., Huggins, R. A., & Cui, Y. (2008). High-performance lithium battery anodes using silicon nanowires. *Nature Nanotechnology, 3,* 31–35.
7. Wu, H., Chan, G., Choi, J. W., Ryu, I., Yao, Y., McDowell, M. T., et al. (2012). Stable cycling of double-walled silicon nanotube battery anodes through solid–electrolyte interphase control. *Nature Nanotechnology, 7,* 310–315.
8. Beaulieu, L. Y., Ebrman, K. W., Turner, R. L., Krause, L. J., & Dahn, J. R. (2001). Colossal reversible volume changes in lithium alloys. *Electrochemical and Solid-State Letters, 4,* A137–A140.
9. Maranchi, J. P., Hepp, A. F., & Kumta, P. N. (2003). High capacity, reversible silicon thin-film anodes for lithium-ion batteries. *Electrochemical and Solid-State Letters, 6,* A198–A201.
10. Aurbach, D., Gamolsky, K., Markovsky, B., Gofer, Y., Schmidt, M., & Heider, U. (2002). On the use of vinylene carbonate (VC) as an additive to electrolyte solutions for Li-ion batteries. *Electrochimica Acta, 9,* 1423–1439.
11. Ota, H., Sakata, Y., Inoue, A., & Yamaguchi, S. (2004). Analysis of vinylene carbonate derived SEI layers on graphite anode. *Journal of the Electrochemical Society, 151,* A1659–A1669.
12. Choi, N.-S., Yew, K. H., Lee, K. Y., Sung, M., Kim, H., & Kim, S.-S. (2006). Effect of fluoroethylene carbonate additive on interfacial properties of silicon thin-film electrode. *Journal of Power Sources, 161,* 1254–1259.
13. Peled, E., & Menkin, S. (2017). Review—SEI: past, present and future. *Journal of the Electrochemical Society, 164,* A1703–A1719.
14. Saito, M., Nakai, K., Yamada, T., Takenaka, T., Hirota, M., Kamei, A., et al. (2011). Si thin platelets as high-capacity negative electrode for Li-ion batteries. *Journal of Power Sources, 196,* 6637–6643.
15. Saito, M., Yamada, T., Yodoya, C., Kamei, A., Hirota, M., Takenaka, T., et al. (2012). Influence of Li diffusion distance on the negative electrode properties of Si thin flakes for Li secondary batteries. *Solid State Ionics, 225,* 506–509.
16. Inaba, M., Haruta, M., Saito, M., & Doi, T. (2017). Silicon nano-flake powder as an anode for the next generation lithium-ion batteries: Current status and challenges. *Electrochemistry, 85,* 623–629.
17. Obrovac, M. N., & Christensen, L. (2004). Structural changes in silicon anodes during lithium insertion/extraction. *Electrochemical and Solid-State Letters, 7,* A93–A96.
18. Hatchard, T. D., & Dahn, J. R. (2004). In situ XRD and electrochemical study of the reaction of lithium with amorphous silicon. *Journal of the Electrochemical Society, 151,* A838–A842.
19. Haruta, M., Okubo, T., Masuo, Y., Yoshida, S., Tomita, A., Takenaka, T., et al. (2017). Temperature effects on SEI formation and cyclability of Si nanoflake powder anode in the presence of SEI-forming additives. *Electrochimica Acta, 224,* 186–193.
20. Haruta, M., Hioki, R., Moriyasu, T., Tomita, A., Takenaka, T., Doi, T., & Inaba, M. (2018). Morphology changes and long-term cycling durability of Si flake powder negative electrode for lithium-ion batteries. *Electrochimica Atca, 267,* 94–101.
21. Okubo, T., Saito, M., Yodoya, C., Kamei, A., Hirota, M., Takenaka, T., et al. (2014). Effects of Li pre-doping on charge/discharge properties of Si thin flakes as a negative electrode for Li-ion batteries. *Solid State Ionics, 262,* 39–42.
22. Yoshida, S., Masuo, Y., Shibata, D., Haruta, M., Doi, T., & Inaba, M. (2015). Li pre-doping of amorphous silicon electrode in Li-naphthalene complex solutions. *Electrochemistry, 83,* 843–845.
23. Li, Z., Zhang, S., Zhang, C., Ueno, K., Yasuda, T., Tatara, R., et al. (2015). One-pot pyrolysis of lithium sulfate and graphene nanoplatelet aggregates: in situ formed Li_2S/graphene composite for lithium–sulfur batteries. *Nanoscale, 7,* 14385–14392.
24. Yoshida, K., Nakamura, M., Kazue, Y., Tachikawa, N., Tsuzuki, S., Seki, S., et al. (2013). Oxidative-stability enhancement and charge transport mechanism in glyme-lithium salt equimolar complexes. *Journal of the American Chemical Society, 133,* 13121–13129.
25. Dokko, K., Tachikawa, N., Yamauchi, K., Tsuchiya, M., Yamazaki, A., Takashima, E., et al. (2013). Solvate ionic liquid electrolyte for Li–S batteries. *Journal of the Electrochemical Society, 160,* A1304–A1310.

26. Haruta, M., Masuo, Y., Moriyasu, T., Tomita, A., Sakakibara, C., Kamei, A., et al. (2015). Cycle performances of Si-flake-powder anodes in lithium salt-tetraglyme complex electrolytes. *Electrochemistry, 83,* 837–839.
27. Haruta, M., Moriyasu, T., Tomita, A., Takenaka, T., Doi, T., & Inaba, M. (2018). Pre-film formation and cycle performance of silicon-flake-powder negative electrode in a solvate ionic liquid for silicon-sulfur rechargeable batteries. *Journal of the Electrochemical Society, 165,* A1874–A1879.
28. Guo, J., Sun, A., Chen, X., Wang, C., & Manivannan, A. (2011). Cyclability study of silicon–carbon composite anodes for lithium-ion batteries using electrochemical impedance spectroscopy. *Electrochimica Acta, 56,* 3981–3987.

Electrochemically Deposited Si–O–C Anode

Seongki Ahn and Toshiyuki Momma

Abstract In the past couple of decades, carbon-based material such as graphite is widely used as an anode for lithium storage battery. However, with the development of larger power demand of industries such as electric vehicles and energy storage systems, to increase the charge/discharge capacity of batteries alternative anodes are required. In this chapter, we discuss the synthesis of Si–O–C composite by electrodeposition and its electrochemical behaviors, and current issues as an alternative anode for lithium–sulfur batteries. Also, the strategies to access the current problems and possible solutions are provided.

Keywords Anode materials · Silicon-based anode · Si–O–C · Electrodeposition

Silicon is a promising anode material because of its advantages such as eco-friendless, abundance in nature, and low working potential (~0.5 V vs. Li/Li^+). Especially, its higher theoretical capacity than commercial graphite anode is one of the great advantages of an anode when used as energy storage devices (silicon, $Li_{4.4}Si$: 4200 mAh g^{-1}; graphite, Li_6C: 372 mAh g^{-1}) [1–5]. However, silicon and silicon-based composites suffer from a large volume change during lithiation and de-lithiation (~400%), resulting in capacity fading caused by pulverization of silicon. In addition, the low electric conductivity of silicon is one of the major reason for the degraded performance of lithium-ion batteries (LIBs) and lithium–sulfur batteries (LSBs) [6–8]. To overcome these issues, Osaka and Momma groups have reported a novel silicon-base anode material including silicon, oxygen, and carbon, denoted as Si–O–C composite, synthesized by electrodeposition in the organic solvent [9–11]. As a silicon resource, $SiCl_4$ was used and the chemical reaction of decomposition of $SiCl_4$ in the organic solvent during electrodeposition is described as follows:

$$SiCl_4 + 4e^- \rightarrow Si + 4Cl^- \left(\text{ca. } 1.3\,\text{V vs. } Li/Li^+\right)$$

Figure 1b shows a curve of linear voltammogram of the solvent containing 0.5 mol dm^{-3} of $SiCl_4$ and 0.5 mol dm^{-3} of tetrabutylammonium perchlorate in propylene carbonate (PC) for electrodeposition with and without $SiCl_4$. A reductive

S. Ahn · T. Momma (✉)
Faculty of Science and Engineering, Waseda University, Okubo 3-4-1, Shinjuku, Tokyo, Japan
e-mail: momma@waseda.jp

K. Kanamura (ed.), *Next Generation Batteries*,
https://doi.org/10.1007/978-981-33-6668-8_30

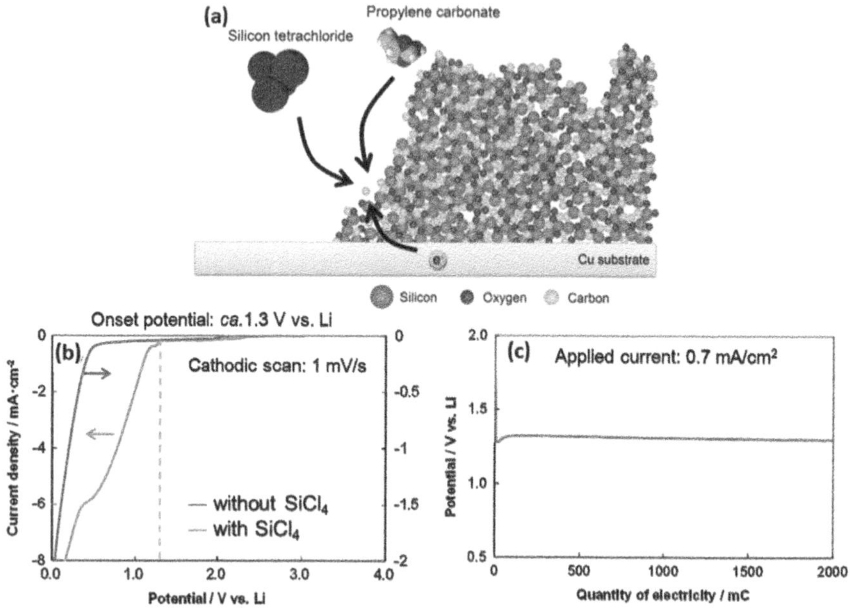

Fig. 1 **a** Schematic diagram of electrodeposition of the Si–O–C composite using $SiCl_4$ in the PC solvent, **b** Linear sweep voltammogram of electrodeposition of Si–O–C composite, and **c** Chronopotentiogram during the electrodeposition of Si–O–C composite with charge density of 2 C cm^{-2}. Reproduced from Ref. [10] by permission of The Royal Society of Chemistry

current was observed at ca. 1.3 V versus Li/Li^+ because of decomposition of $SiCl_4$, indicating that the decomposition of $SiCl_4$ and deposition of Si–O–C composite are carried out at ca. 1.3 V versus Li/Li^+. Based on this result, the electrodeposition was conducted with a current density of 0.7 mA cm^{-2} to maintain the potential value at ca. 1.3 V versus Li/Li^+ as shown in Fig. 1c. As a synthesizing method for electrode, electrodeposition techniques have unique features: (a) the active material can be directly deposited onto current collector without any additives like carbon conductivity material and binders which can decrease gravimetric energy and power density; (b) homogeneously deposited elements can be obtained, leading to structural stability compared to the physical manufacturing method of the electrode. Thus, electrodeposition techniques are widely used for various energy storage devices, not only LIBs but also other devices such as supercapacitors [12–15].

Figure 2a illustrates the cross-sectional images of the Si–O–C composite measured by field emission scanning electron microscopy with element mappings of silicon, oxygen, and carbon. It can be confirmed that the three elements are homogeneously deposited on the Cu substrate, not only on the surface but also entire of Si–O–C composite. Figure 2b shows a thickness change of Si–O–C composite depending on cycling numbers from 1 to 100 cycles. There is a large experimental

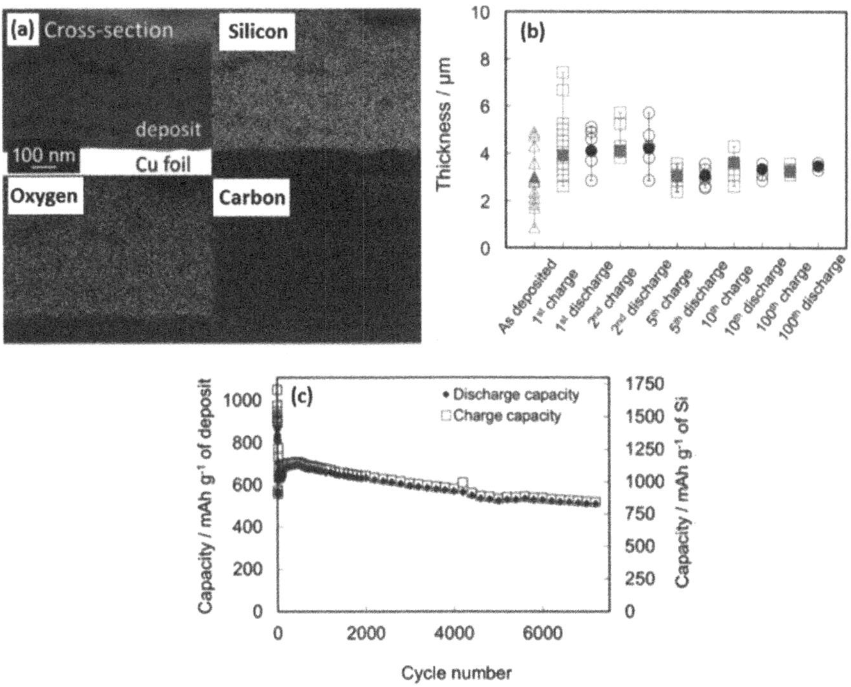

Fig. 2 **a** Cross-sectional STEM image of Si–O–C composite with element mapping, **b** Thickness change of Si–O–C composite for 100 cycles, and **c** Capacity cyclability for 7200 cycles with a current density of 250 μA cm^{-2}. **a** and **c**: Reproduced from Ref. [10] with permission of The Royal Society of Chemistry, **b**: Reprinted from Ref. [16], Copyright 2013, with permission from Elsevier

error of thickness change, however, the transformation tendency of thickness can be discussed. The average thickness of the as-deposited Si–O–C composite is ca. 3 μm. After first charge and discharge, the thickness increased up to ca. 3.9 and 4.1 μm, respectively. From first to second cycle, the non-uniform variation can be observed. This is attributed to solid-electrolyte interphase (SEI) formation on the surface of Si–O–C composite. After the 10th cycle, the thickness change of Si–O–C composite can be stable with a uniform variation. In addition, it can be confirmed that several lithium compounds such as Li_2O, Li_2CO_3, and inorganic phases formation were formed during charge and discharge process analyzed by scanning transmission electron microscopy, electron energy-loss spectroscopy, and X-ray photoelectron spectroscopy analysis [16]. The homogeneous dispersity of elements in the organic/inorganic compound can act as a buffer matrix to reduce internal stress during silicon volume change. Because of its advantages, the Si–O–C composite can show an outstanding cyclability for 7200 cycles with good discharge capacity of 842 mAh g^{-1} as shown in Fig. 2c. Despite this superb cycle performance of Si–O–C, there are some limits for utilizing for practical application as an anode for LSBs such as weak structural stability, low electric conductivity, and low productivity of active materials. To meet these requirements, many attempts have been studied and reported.

1 Improvement of Si–O–C Anode and Its Application for LSBs

As described above, the Si–O–C composite has homogeneous dispersity of not only silicon but also oxygen and carbon which act as buffer matrix to reduce internal stress during silicon volume change, resulting in a good discharge capacity for several thousand cycles. Despite this advantage of Si–O–C composite, it faces several challenges for its practical application as large-scale battery because of low electrical conductivity of silicon and weak structural stability. To tackle these issues, various attempts have been tried to improve the material and electrochemical characteristics of Si–O–C composite for the future battery system. As a facile way to increase the structural stability of Si–O–C composite, three-dimensional (3D) substrates have been reported. Figure 3 shows the various types of 3D substrate for electrodeposition of Si–O–C composite.

The 3D Ni-macro-nanocone substrate was synthesized by anodizing and electrodeposition techniques on commercial Cu substrate [17, 18]. The 3D Ni-micro-nanocone substrate has a micro-sized flower-like shape which has an average diameter of ca. 3 μm in Fig. 4a. In the inset image, it is revealed that the micro-sized

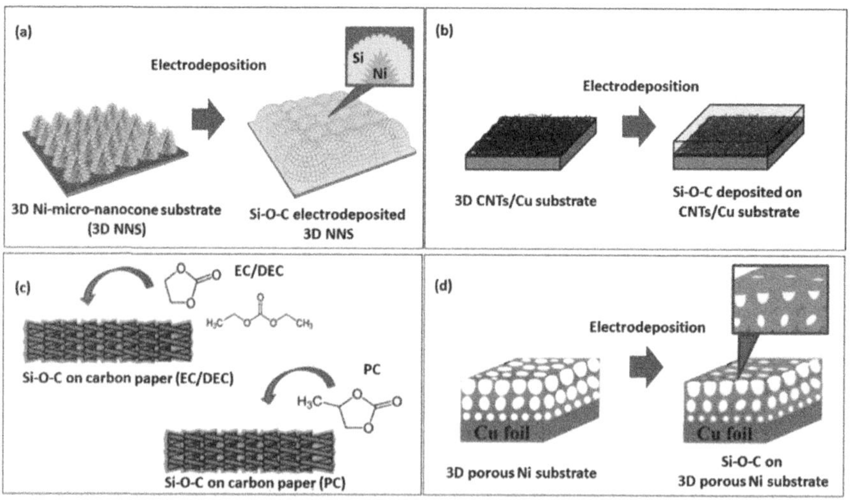

Fig. 3 Various types of 3D structured substrate for Si–O–C composite. **a** 3D Ni-micro-nanocone substrate, **b** 3D CNTs anchor layer, **c** Carbon paper, and 3D porous Ni substrate. **a**, **b**, and **d**: Reprinted from Refs. [17, 19, 26], Copyright 2013, 2016, 2014, with permission from Elsevier. **c**: Reprinted from Ref. [21] with permission of ECS

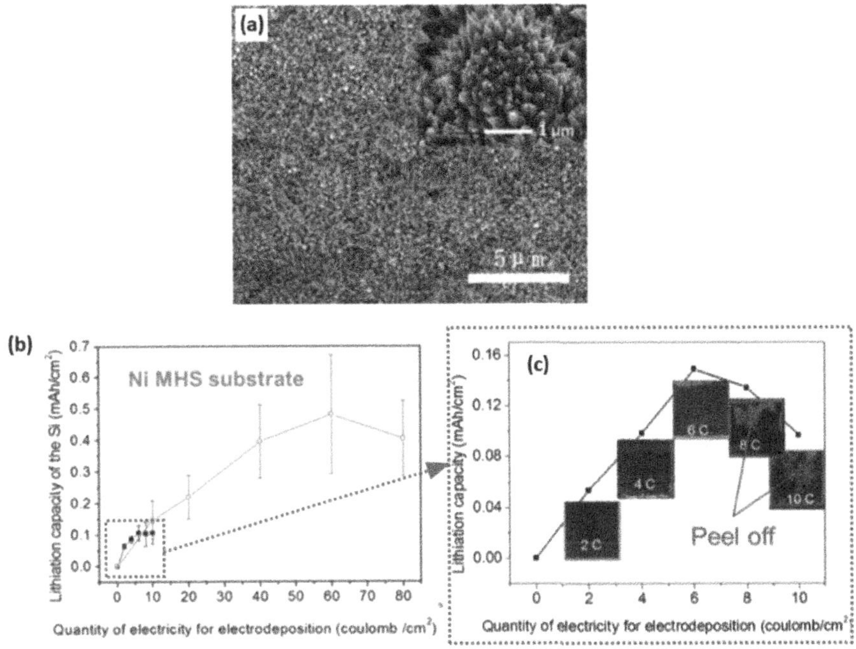

Fig. 4 **a** SEM image of 3D Ni-micro-nanocone substrate. Lithiation capacity of Si–O–C composite deposited on (**b**) 3D Ni-micro-nanocone substrate or **c** as-received Cu substrate. Reprinted from Ref. [17], Copyright 2013, with permission from Elsevier

flower-like shape cones have many nano-sized cones which have average height and diameter of ca. 300 and 200 nm, respectively, on their surface. These unique structural characteristics of 3D Ni-micro-nanocone substrate can be expected to increase adhesion strength between Si–O–C composite and substrate.

Figure 4b illustrates a comparison of discharge areal capacity of Si–O–C composite deposited on smooth surface of Cu substrate or 3D Ni-micro-nanocone substrate. It is clear that the Si–O–C composite deposited on 3D Ni-micro-nanocone substrate shows higher discharge areal capacity than Si–O–C composite fabricated on as-received Cu substrate. In addition, Fig. 4c shows that peeling off of Si–O–C from as-received Cu substrate started at a charge density of 6 C cm^{-2} because of weak adhesion strength, and thus leading to decreasing of discharge areal capacity.

In 2016, Ahn et al. reported carbon nanotubes (CNTs) anchor layer fabricated by high-voltage electrophoretic deposition (HVEPD) on Cu substrate (Fig. 3b) [19, 20]. For the HVEPD, $CoCl_2$ was used as a metal additive to impart positive charge on the surface of CNTs and holding the CNTs from the surface of the Cu substrate. The CNTs/Cu substrate offers a higher surface area than the as-received Cu substrate. This implies that the CNTs/Cu substrate has high chemical reaction sites during electrodeposition, resulting in increase of amounts of deposited silicon at same charge density (Si–O–C/Cu: 48 $\mu g\ cm^{-2}$, Si–O–C/CNTs/Cu: 56 $\mu g\ cm^{-2}$, at the charge density of

2 C cm^{-2}). In addition, the CNTs layer enhances the structural stability of Si–O–C composites after silicon volume change by holding each Si–O–C composites, as well as an adhesion strength between Si–O–C composite and Cu substrate. Thus, Si–O–C composite can be deposited on the substrate without any structural defects such as crack and peeling off of Si–O–C composite even at a high charge density of 15 C cm^{-2} compared to Si–O–C composite deposition on as-received Cu substrate. Figure 5a shows a surface change of Si–O–C composite depending on charge density at 2, 8, and 15 C cm^{-2}. As described above, Si–O–C/CNTs/Cu displays well-deposited Si–O–C composite without any structural collapses. On the other hand, in the case of Si–O–C/Cu, the small cracks occurred on its surface at charge density of 8 C cm^{-2}; afterward, peeling off of Si–O–C composite can be observed at 15 C cm^{-2}. Figure 5b demonstrates a comparison of amounts of deposited silicon at a charge density of 2, 4, 8, and 15 C cm^{-2} on CNTs anchor layer and Cu substrate. It is clear that the deposited amounts of silicon on the CNTs anchor layer increases with an increase in charge density of electrodeposition, whereas Si–O–C composite on as-received Cu substrate shows decreased amounts of deposited silicon after charge density of 8 C cm^{-2}. In addition, CNTs anchor layer can play the role as not only structural supporting materials but also as an electric path during the charge/discharge process.

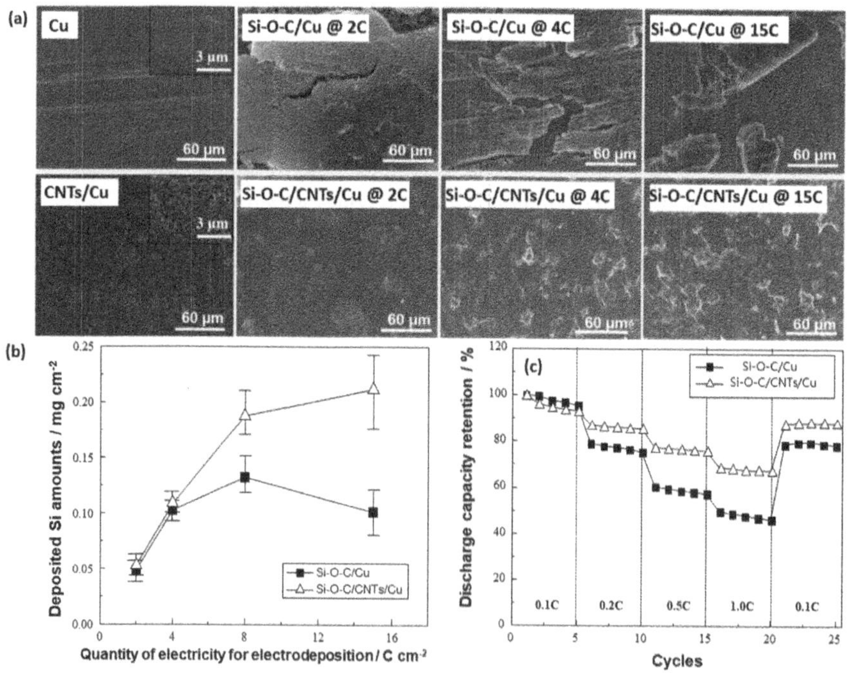

Fig. 5 **a** SEM images of Si–O–C composite deposited on as-received Cu and CNTs/Cu substrate, **b** Comparison of amounts of deposited silicon, and **c** C-rate performance at various C-rate condition from 0.1 to 1 C. Reprinted from Ref. [19], Copyright 2016 with permission from Elsevier

Figure 5c shows an improved rate performance of Si–O–C/CNTs/Cu because of the electric path.

As a different type of 3D structured substrate, carbon paper (CP) is used as a substrate for electrodeposition of Si–O–C composite (Fig. 3c) [21]. It is well known that the carbon materials have poor wettability with organic solvents [22, 23]. Thus, surface treatment using a sulfuric acid-hydrogen peroxide mixture solution was conducted before electrodeposition. The acid treatment can improve the surface wettability of CP with the solvent. However, acid treatment with carbon material impairs the carbon networks such as exfoliation of graphite layers [24]. In 2001, Jeong et al. have reported that the poor compatibility of graphite-based anode with PC electrolyte is because of Li^+ ions and PC molecules co-intercalated into graphite layers, thus resulting in exfoliation of graphite layer and degradation of cell performance (Fig. 6a) [25]. To investigate this issue, two kinds of solvent, PC and ethylene carbonate/diethyl carbonate (EC/DEC, 1:1 v/v) were used as an electrolyte for electrodeposition of Si–O–C composite. The Si–O–C composite deposited on

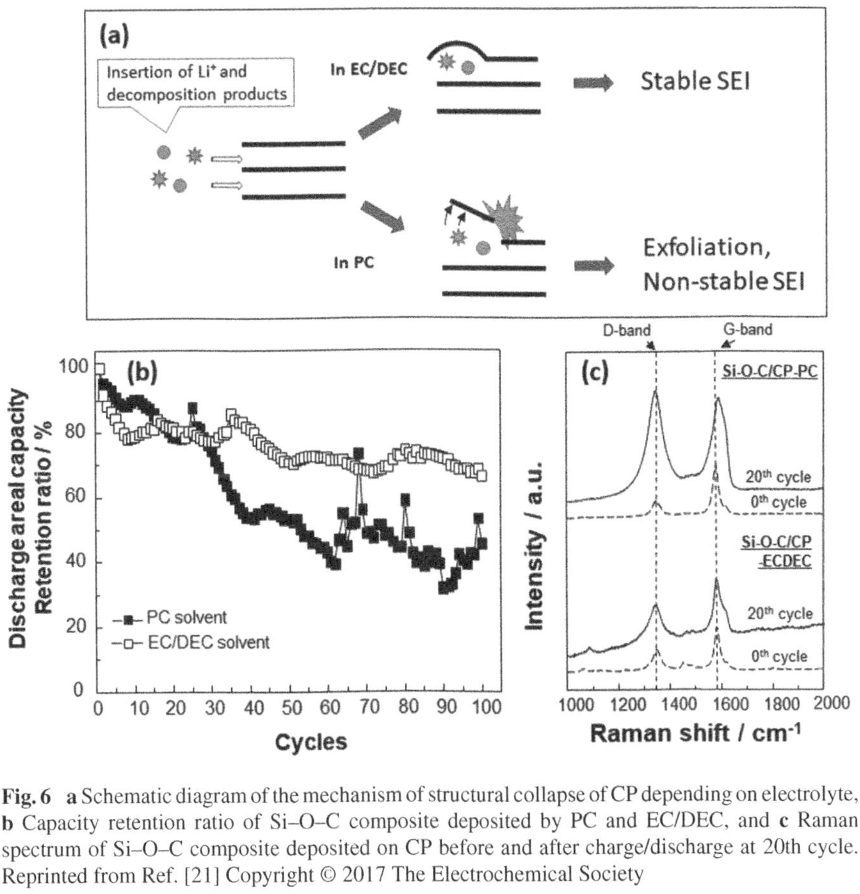

Fig. 6 **a** Schematic diagram of the mechanism of structural collapse of CP depending on electrolyte, **b** Capacity retention ratio of Si–O–C composite deposited by PC and EC/DEC, and **c** Raman spectrum of Si–O–C composite deposited on CP before and after charge/discharge at 20th cycle. Reprinted from Ref. [21] Copyright © 2017 The Electrochemical Society

CP by EC/DEC delivered a better cyclability than Si–O–C composite synthesized by PC electrolyte. The cyclability of Si–O–C composite synthesized by PC or EC/DEC electrolyte is shown in Fig. 6b. At initial cycles, Si–O–C composite (PC) delivered higher capacity retention than Si–O–C composite (EC/DEC). However, after 20 cycles, a sharp decline of discharge capacity retention can be observed. On the other hand, Si–O–C composite (EC/DEC) shows relatively stable retention ratio. Figure 6c illustrates a Raman spectrum of Si–O–C composite deposited by PC or EC/DEC before and after charge/discharge test. It is clear that the D/G band intensity of Si–O–C composite (PC) is higher than another sample. This implies that the carbon structural defects of graphite might occur in the charge/discharge process. Despite this shortcoming, through this attempt, an outstanding areal capacity could be obtained over 2.7 mA h cm^{-2}. These results propose a possibility to apply as an appropriate anode material for large-scale production of the battery cell.

Qian et al. have reported a 3D porous Ni substrate to accommodate a volume change of silicon during lithiation and de-lithiation in Fig. 3d [26]. The 3D porous Ni substrate was fabricated by two-step electrodeposition using $NiCl_2 \cdot 6H_2O$ as a Ni source. Figure 7a demonstrates a surface morphology characteristics of 3D porous Ni substrate and Si–O–C composite deposited on it. It is clear that pores are homogeneously distributed on the substrate. The average pore diameters are ca. 5–7 μm. As described above, silicon suffers a large volume change up to 400% and it is the main reason for pulverization of electrode and rapid capacity fading. However, as shown in the SEM images, the 3D porous Ni substrate and Si–O–C composite which have a micro-scale porous structure can be expected to accommodate a huge volume change

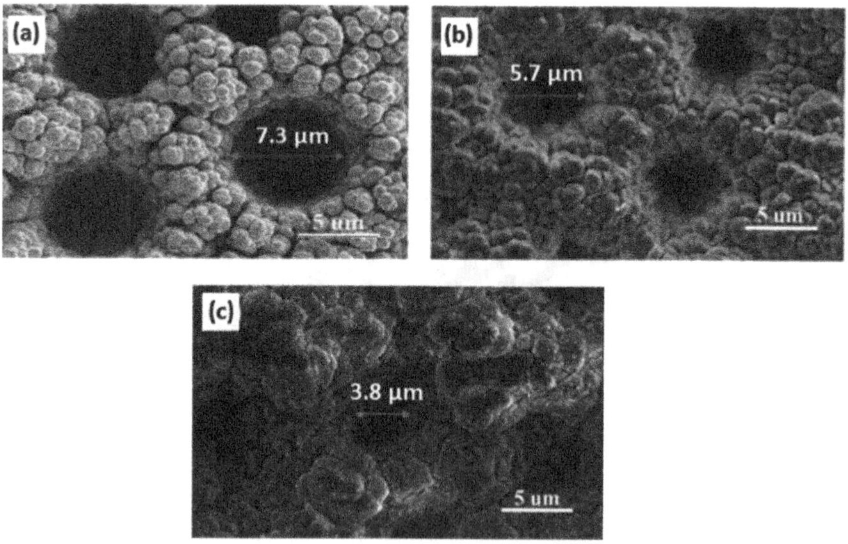

Fig. 7 SEM image of (**a**) 3D porous Ni substrate, **b** Si–O-C composite deposited on it, and **c** Si–O–C composite after charge/discharge cycling after 100 cycles. Reprinted from Ref. [26], Copyright 2014, with permission from Elsevier

of silicon, resulting in improvement of cell performance. Figure 7b, c illustrates the Si–O–C composite before and after charge/discharge for 100 cycles. Even after cycling, the porous structure can be maintained, implying that the porous structure can accommodate the volume change of silicon during charge and discharge cycling. Therefore, the Si–O–C composite deposited on 3D porous Ni substrate can show a good rate-performance tested from 0.5 to 10 C-rate for 55 cycles. This result indicates that the porous structure by 3D porous Ni substrate enhances electrochemical stability even at high C-rate condition.

The structural stability of Si–O–C composite and volume change of silicon could be improved and accommodated by using 3D structured substrates. However, another issue such as low electric conductivity should also be developed for the practical application of Si–O–C composite as an anode for LSBs. To meet this requirement, Jeong et al. have proposed a two-step potentiostatic techniques to fabricate a co-deposited nano-sized Cu particles and electrodeposition of Si–O–C composite to increase the electrical conductivity [27]. To fabricate the Si–O–C composite with nano-sized Cu particles, denoted as *n*-Cu/Si–O–C, first, the dissolution of Cu ions from the Cu substrate is conducted. Afterward, the electrodeposition of Si–O–C composite was carried out on the Cu substrate which has nano-sized Cu particles on their surface. Both processes were performed by anodic and cathodic reactions at ca. 3.6 and 1.3 V versus Li/Li^+, respectively (Fig. 8a). As aforementioned, the electrodeposition technique is a good method to form a composite which has homogeneous dispersity of elements. The *n*-Cu/Si–O–C composite has a uniform dispersity of copper, silicon, oxygen, and carbon in entire of the composite. The *n*-Cu/Si–O–C composite delivered a higher discharge capacity and retention ratio than Si–O–C composite (*n*-Cu/Si–O–C: 1390 mAh g^{-1} at 92nd cycle, Si–O-C: 1000 mAh g^{-1} at 125th cycle, Fig. 8c). This is because the enhanced electrical conductivity by using nano-sized Cu particles leads to active material utilization during electrodeposition.

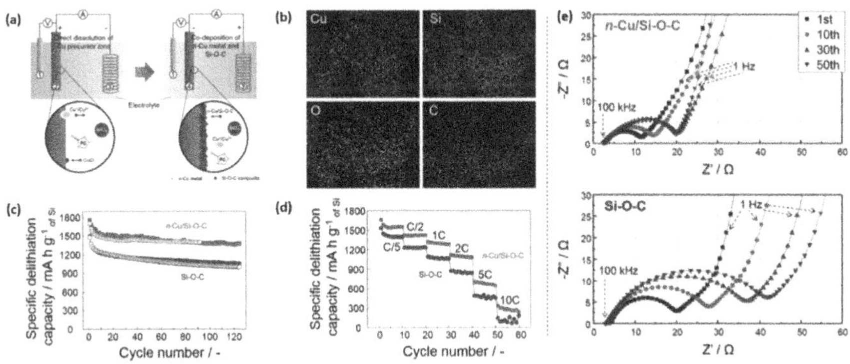

Fig. 8 **a** Schematic diagram of synthesizing of the n-Cu/Si–O–C composite, **b** Element mappings of the n-Cu/Si–O–C composite, **c** cyclability, **d** C-rate performance, and **e** Impedance analysis of the n-Cu/Si–O-C and Si–O–C composite. Reprinted from Ref. [27], Copyright 2016, with permission from Elsevier

Likewise, the *n*-Cu/Si–O–C composite shows a better C-rate performance than the Si–O–C composite tested at C-rate condition from 0.2 to 10 C (Fig. 8d). For further investigation of the effect of nano-sized Cu particles, electrochemical impedance spectroscopy was conducted using the *n*-Cu/Si–O–C and Si–O–C composite. Figure 8e shows the Nyquist plots of those samples. These are suppressed semi-circles from the high to mid-frequency regions which represent a charge transfer resistance (Rct). And the inclined line at the low-frequency regions demonstrates Li^+ diffusion in the solid active materials. From these results, it is clear that the Rct of the *n*-Cu/Si–O–C composite was remarkably smaller than the one of Si–O–C composite. This implies that the co-deposited nano-sized Cu particles can act as supporting material to enhance the electrical conductivity of the entire Si–O–C composite. As a result, electrochemical degradation of Si–O–C composite caused by the low electrical conductivity of silicon was improved by co-deposition of nano-sized Cu particles with Si–O–C composite.

For the LSB full-cell design, in 2014, Agostini et al. have reported LSB full-cell consisting of a carbon-coated lithium sulfide cathode (Li_2S) and the Si–O–C composite [28]. As mentioned above, the Si–O–C composite has advantages as an anode material such as outstanding cyclability for several thousand cycles, good retention ratio, and high discharge capacity than commercial graphite-based anodes. Moreover, the Si–O–C composite has low lithiation/de-lithiation potential, indicating the possibility to obtain the high overall voltage. Also, high capacity of Si–O–C is almost double that of the cathode. This feature implies easy handling for the cell balancing between cathode and anode for the utilization of full-cell design (Fig. 9a). In the case of Li_2S/Si–O–C LSBs, the full-cell is discharged state, therefore, for the full-cell test, activation charge process is required as in the following:

$$Li_2S + Si \leftrightarrow Li_{2-x} + Li_xSi$$

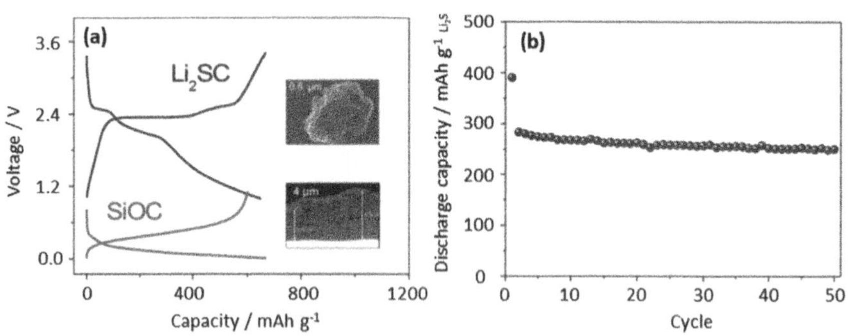

Fig. 9 **a** Charge and discharge curves of Li_2S and Si–O–C composite, and **b** cyclability of Li_2S/Si–O–C full-cell. Reprinted with permission from Ref. [28]. Copyright 2014 American Chemical Society

This reaction is reversed in the discharge process and the total charge and discharge cycling are as follows:

$$Li_2S + Si \underset{charge}{\overset{discharge}{\leftrightarrow}} Li_{2-x} + Li_xSi$$

Figure 9b demonstrates a cyclability of LSBs consisting of Li_2S and Si–O–C composite tested at a current density of 232 mA g_{Li2S}^{-1}. It is confirmed that the LSB cell has a stable capacity of 280 mAh g_{Li2S}^{-1} for 50 cycles. The energy density of LSBs is of 390 Wh kg^{-1}. This study proposes a possibility for LSB full-cell design using Si–O–C composite as an anode.

Despite the applicability of Si–O–C composite as an anode for LSB full-cell, utilization of Si–O–C composite in the glyme-based electrolyte which has attracted as a promising electrolyte to suppress the dissolution of polysulfide had not been reported before. In 2017, Seko et al. have reported the electrochemical performance of Si–O–C composite in the glyme-based ionic liquid electrolytes such as Li(G3)TFSI or Li(G4)TFSI (triglyme: G3 and tetraglyme: G4) with carbonate-based additive [29]. In the case of Si–O–C half-cell in Li(G3)TFSI or Li(G4)TFSI without additives, it shows a poor cyclability and capacity retention of ca. 70% at the 100th cycle. On the other hand, other samples which have additives demonstrate an improved electrochemical performance, including good capacity retention of 80–90% at the 100th cycle. The reason for this improved performance of Si–O–C composite is well-formed SEI on the surface of Si–O–C composite derived from additives. However, when the additive was used before activation of Si–O–C composite, the half-cell shows lower discharge capacity than Si–O–C composite which has no additive. This is attributed to blocking of the activation process, involving the formation of Li_xSi by reductive decomposition of additives. Thus, additives should be used after activation of Si–O–C composite for the good electrochemical performance.

2 Conclusions and Perspectives

In this chapter, we have discussed the background and current challenges of Si–O–C composite. As we described above, the Si–O–C composite has the possibility as an anode for LSBs because of its outstanding electrochemical characteristics such as long-term stability with good discharge capacity retention. This excellent performance is attributed to the homogeneous dispersity of silicon, oxygen, and carbon which can play a role as a buffer matrix to reduce the internal stress during volume change of silicon. However, electrodeposition for synthesizing of Si–O–C composite suffers low productivity of silicon by low adhesion strength between Si–O–C composite and substrate, low electric conductivity, and pulverization of Si–O–C composite by the volume change of silicon. To alleviate these issues, various attempts such as using 3D structural substrates, the addition of metal clusters and electrolyte additives have tried for the practical application of Si–O–C composite for the LSB full-cell. However, through these means, the structural and electrochemical weakness of Si–O–C composite could be potentially developed. But, in order to future progress the applicability of Si–O–C composite for the large-scale productivity of LSBs, a lot of studies are still required. Firstly, a more profound understanding of the synthesizing and alloy/de-alloy mechanism of the Si–O–C composite is required. This may be performed through in situ microscopy techniques such as transmission electron microscopy, scanning electron microscope, and confocal laser scanning microscopy. These analysis techniques are helpful to understand the morphology evolution of Si–O–C composite during charge and discharge cycling. Secondly, additive materials such as electrolyte additives should be utilized to optimize the full-cell performances. Finally, comprehensive studies for the structural design of Si–O–C composite are important to enhance the electrochemical performance of Si–O–C composite with high areal capacity. Overall, we predict rapid growth of more advanced and high-performance Si–O–C composite as an anode for LSBs tailored with advanced technologies to realize high-performance LSBs as a power source in the future. We wish this chapter would propose insight into the approaches needed to solve the aforementioned challenges of Si–O–C composite for the next generation of sustainable LSBs.

References

1. Xiao, J., Xu, W., Wang, D., Choi, D., Wang, W., Li, X., et al. (2010). *Journal of The Electrochemical Society, 157,* A1047–A1051.
2. Xu, C., Lindgren, F., Philippe, B., Gorgoi, M., Björefors, F., Edström, K., & Gustafsson, T. (2015). *Chemistry of Materials, 27,* 2591–2599.
3. Takamura, T., Uehara, M., Suzuki, J., Sekine, K., & Tamura, K. (2006). *Journal of Power Sources, 158,* 1401–1404.
4. Ge, M., Rong, J., Fang, X., & Zhou, C. (2012). *Nano Letters, 12,* 2318–2323.
5. Obrovac, M. N., & Christensen, L. (2004). *Electrochemical and Solid-State Letters, 7,* A93–A96.
6. Li, X., Gu, M., Hu, S., Kennard, R., Yan, P., Chen, X., et al. (2014). *Nature Communications, 5,* 4105.
7. Choi, S., Kwon, T.-W., Coskun, A., & Choi, J. W. (2017). *Science, 357,* 279–283.
8. Zang, J.-L., & Zhao, Y.-P. (2012). *Composites Part B: Engineering, 43,* 76–82.
9. Momma, T., Aoki, S., Nara, H., Yokoshima, T., & Osaka, T. (2011). *Electrochemistry Communications, 13,* 969–972.
10. Nara, H., Yokoshima, T., Momma, T., & Osaka, T. (2012). *Energy and Environmental Science, 5,* 6500–6505.
11. Osaka, T., Nara, H., Momma, T., & Yokoshima, T. (2014). *Journal of Materials Chemistry A, 2,* 883–896.
12. Chen, L., Tang, Y., Wang, K., Liu, C., & Luo, S. (2011). *Electrochemistry Communications, 13,* 133–137.
13. Wu, G., Li, N., Zhou, D.-R., Mitsuo, K., & Xu, B.-Q. (2004). *Journal of Solid State Chemistry, 177,* 3682–3692.
14. Miller, J. R., Outlaw, R. A., & Holloway, B. C. (2010). *Science, 329,* 1637–1639.

15. Wu, M.-S., Huang, C.-Y., & Lin, K.-H. (2009). *Journal of Power Sources, 186,* 557–564.
16. Nara, H., Yokoshima, T., Otaki, M., Momma, T., & Osaka, T. (2013). *Electrochimica Acta, 110,* 403–410.
17. Hang, T., Nara, H., Yokoshima, T., Momma, T., & Osaka, T. (2013). *Journal of Power Sources, 222,* 503–509.
18. Hang, T., Mukoyama, D., Nara, H., Yokoshima, T., Momma, T., Li, M., & Osaka, T. (2014). *Journal of Power Sources, 256,* 226–232.
19. Ahn, S., Jeong, M., Yokoshima, T., Nara, H., Momma, T., & Osaka, T. (2016). *Journal of Power Sources, 336,* 203–211.
20. Ahn, S., Nara, H., Yokoshima, T., Momma, T., & Osaka, T. (2019). *Materials Letters, 245,* 200–203.
21. Ahn, S., Jeong, M., Miyamoto, K., Yokoshima, T., Nara, H., Momma, T., & Osaka, T. (2017). *Journal of The Electrochemical Society, 164,* A355–A359.
22. Lim, S. C., Jang, J. H., Bae, D. J., Han, G. H., Lee, S., Yeo, I.-S., & Lee, Y. H. (2009). *Applied Physics Letters, 95,* 264103.
23. Fang, B., Wei, Y. Z., Maruyama, K., & Kumagai, M. (2005). *Journal of Applied Electrochemistry, 35,* 229–233.
24. Zhong, X., Wang, R., Liu, L., Kang, M., Wen, Y., Hou, F., et al. (2012). *Nanotechnology, 23,* 505712.
25. Jeong, S.-K., Inaba, M., Mogi, R., Iriyama, Y., Abe, T., & Ogumi, Z. (2001). *Langmuir, 17,* 8281–8286.
26. Qian, X., Hang, T., Nara, H., Yokoshima, T., Li, M., & Osaka, T. (2014). *Journal of Power Sources, 272,* 794–799.
27. Jeong, M., Ahn, S., Yokoshima, T., Nara, H., Momma, T., & Osaka, T. (2016). *Nano Energy, 28,* 51–62.
28. Agostini, M., Hassoun, J., Liu, J., Jeong, M., Nara, H., Momma, T., et al. (2014). *ACS Applied Materials & Interfaces, 6,* 10924–10928.
29. Seko, S., Nara, H., Jeong, M., Yokoshima, T., Momma, T., & Osaka, T. (2017). *Electrochimica Acta, 243,* 65–71.

S_8 Cathode

Kaoru Dokko

Abstract In this chapter, the porous composite cathodes for Li–S batteries are overviewed. The composite sulfur cathodes are typically composed of elemental sulfur S_8, carbon support, and a polymer binder. The effects of carbon support and polymer binder in the sulfur composite cathode on the charge/discharge performance of Li–S cell with solvate ionic liquid electrolytes are reviewed. The electrochemical conversion reaction of S_8 cathode in Li–S cell can be written as $S_8 + 16Li^+ + 16e^- \rightarrow 8Li_2S$. The volume of the active material expands ca. 80% during the discharge reaction; therefore, the pores of carbon support should accommodate the volume expansion of the active material. The porous structure of the carbon support gives a significant effect on the discharge capacity of the S_8 cathode. In the porous composite cathode, the binder interconnects the carbon particles to form an electron conduction network, and the surface of S_8 is partially covered with the polymer. Therefore, the compatibility of the polymer binder with the electrolyte has a significant effect on the charge/discharge performance of the S_8 cathode. The design of the chemical structure of polymer binder and the distribution of binder within the composite electrode are important for accomplishing both high capacity and long cycle life of a Li–S cell.

Keywords Porous electrode · Carbon support · Inverse opal · Polymer binder · Poly(vinyl alcohol)

1 Introduction

The elemental sulfur, cyclooctasulfur molecule S_8, can be electrochemically converted to Li_2S in aprotic electrolytes. S_8 and Li_2S are electronic insulators; as a result, electronically conductive supporting materials, typically porous carbons, are mixed with S_8 to prepare sulfur cathode for Li–S cells [1, 2]. The electrochemical conversion reaction of S_8 cathode in Li–S cell can be written as $S_8 + 16Li^+ + 16e^- \rightarrow 8Li_2S$, and the average electrode potential for the reaction is ~2.1 V versus Li/Li^+.

K. Dokko (✉)
Department of Chemistry and Life Science, Yokohama National University, 79-5 Tokiwadai, Hodogaya-ku, Yokohama 240-8501, Japan
e-mail: dokko-kaoru-js@ynu.ac.jp

K. Kanamura (ed.), *Next Generation Batteries*,
https://doi.org/10.1007/978-981-33-6668-8_31

However, the conversion reaction involves a multistep reduction process, forming reaction intermediates, i.e., lithium polysulfides (Li_2S_m, $m = 2$–8). As shown in Fig. 1, Li–S cell shows two voltage plateaus at ~2.4 and ~2.0 V during galvanostatic discharge [2]. The higher discharge plateau at ~2.4 V is assigned to the reduction of S_8 via a four-electron process: $S_8 + 4Li^+ + 4e^- \rightarrow 2Li_2S_4$. The lower plateau at ~2.0 V is assigned to the further reduction of longer lithium polysulfides Li_2S_m to yield shorter sulfides, such as Li_2S_2 and Li_2S. During the charging of the cell, reverse reactions are assumed to occur.

It is well known that the longer Li_2S_m ($m = 4$–8) can dissolve into conventional liquid electrolytes containing excess solvent [3]. Conventional liquid electrolytes for Li–S batteries are typically composed of ether solvents, such as 1,2-dimethoxyethane

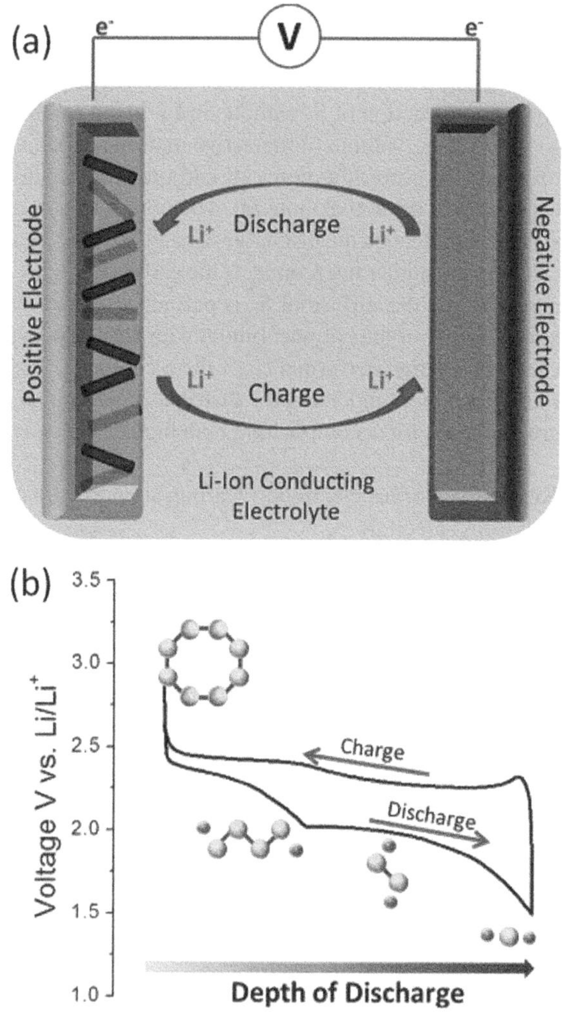

Fig. 1 **a** Schematic illustration of a typical Li–S cell. **b** A typical voltage vs. capacity plot for Li–S cell. Reprinted with permission from Ref. [2]. Copyright 2013 American Chemical Society

(DME) and 1,3-dioxolane (DOL), and Li salt. The dissolution of Li_2S_m causes a rapid capacity decay (i.e., short cycle life) and unfavorable side reactions in Li–S cell. The redox shuttle effect, where Li_2S_m acts as a redox shuttle through reduction at the lithium metal anode and re-oxidation at the sulfur cathode, brings about a low Coulombic efficiency of discharge/charge of the cell. Certain electrolytes, such as ionic liquids [4, 5] and solvate ionic liquids (SILs) [6, 7], exhibit poor solvating ability toward Li_2S_m, suppressing the dissolution of Li_2S_m. This is effective in achieving a high Coulombic efficiency of discharge/charge and a long charge/discharge cycle life of Li–S cells [7, 8].

In this chapter, the effects of materials, such as carbon supports and polymer binder, comprising S_8 composite cathode, on the charge/discharge performance of Li–S cell with SIL electrolytes are reviewed.

2 Carbon Supports for S_8

As mentioned above, carbon supports are commonly used to fabricate S_8 cathode. The active material, S_8, and carbons are typically in powder form, and the cathodes for Li–S cells are prepared by mixing S_8, carbon, and polymer binder. To increase the reactive surface area of the S_8 cathode, carbons with high specific surface areas such as activated carbons, mesoporous carbons, and Ketjen black have been used. To fabricate a composite material composed of S_8 and a porous carbon, a melt-diffusion method is frequently used [9]. The melting point of S_8 is relatively low at 115 °C. The impregnation of sulfur into the pores of carbon can be performed by the heat treatment of the mixture of S_8 powder and carbon powder at just above the melting point of sulfur [9]. During the heat treatment, the sulfur melt is absorbed into the pores of carbon by capillary forces. By cooling the mixture to room temperature, sulfur solidifies and shrinks in the pores, and solid sulfur comes into intimate contact with the pore walls.

The carbons with ordered porous structures can be synthesized using template methods [9–12]. Figure 2 shows the scanning electron microscopy (SEM) image of an inverse opal carbon (IOC) [10–12]. IOCs can be synthesized using a silica opal template composed of monodispersed colloidal silica particles. Organic materials such as poly(furfuryl alcohol) and organic ionic liquids can be impregnated into the voids between the silica particles, and the organic materials can be carbonized by heat treatment in an inert atmosphere. After carbonization, the silica particles are etched out using an HF solution, and an IOC with a three-dimensionally ordered porous structure can be obtained. By changing the particle size of the silica template, the pore size of the IOC can be controlled. By decreasing the particle size of the silica, the resulting pore diameter decreases, and the specific surface of the IOC increases.

Using IOCs with different pore diameters, S_8/carbon composite electrodes were prepared. Hereafter, the IOC with a pore diameter of x nm is denoted as IOC-x. For reference, S_8/acetylene black electrode was also prepared. Figure 3 shows discharge and charge curves of S_8/IOC electrodes [11]. The electrolyte was a solvate ionic

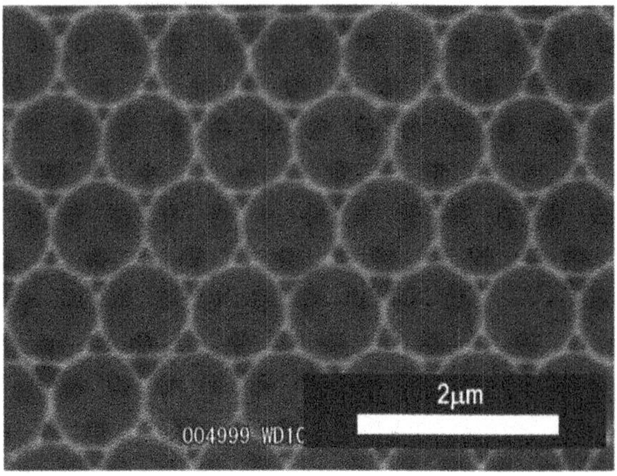

Fig. 2 SEM image of IOC-1000. Reproduced from Ref. [11] with the permission of The Royal Society of Chemistry

liquid [Li(tetraglyme)][N(SO_2CF_3)$_2$] (hereafter [Li(G4)][TFSA]), and the counter electrode was Li metal foil. The specific discharge capacity of the S_8 electrode significantly changes depending on the porous structures of the carbons. The specific surface areas of IOC-100, IOC-1000, and acetylene black were 1186, 582, and 65 m^2/g, respectively. The discharge capacity of S_8 electrode increased with the increasing of specific surface area of carbon support. The carbon with a higher specific surface area is expected to provide higher amounts for the triple-phase junctions of S_8/carbon/electrolyte. S_8 is an insulator, and the redox reaction of S_8 takes place within the vicinity of the triple-phase junctions. Therefore, homogeneous distribution of S_8 on the porous carbon support is crucial for increasing the utilization and reactivity of S_8 in the composite cathode. It is worth mentioning that both the specific surface area of the carbon support and the pore volume significantly influence the discharge capacity (utilization of the active material) of the S_8/carbon composite cathode. Figure 4 shows the discharge capacities of S_8 electrodes with carbon supports having different pore volumes [12]. As increasing the pore volume of carbon support, the discharge capacity increases. The volume of the active material expands ca. 80% during the discharge reaction of $S_8 \rightarrow 8Li_2S$ in the composite cathode; therefore, the pores of carbon support should accommodate the volume expansion of the active material. In the case of SIL electrolyte, the solubility of reaction intermediate Li_2S_m is very low, and most of the Li_2S_m remains in the solid-state during discharge and charge reactions. If the pores of the carbon supports are completely filled with active materials such as S_8 and Li_2S_m, the active materials cannot expand further. In addition, the clogging of the pores causes an increase in the resistance for the ion conduction of electrolyte among the pores. This would cause a low utilization of active materials in the composite cathode. Therefore, the design of the porous structure of the carbon support is important to increase the discharge capacity of the S_8/carbon composite cathode.

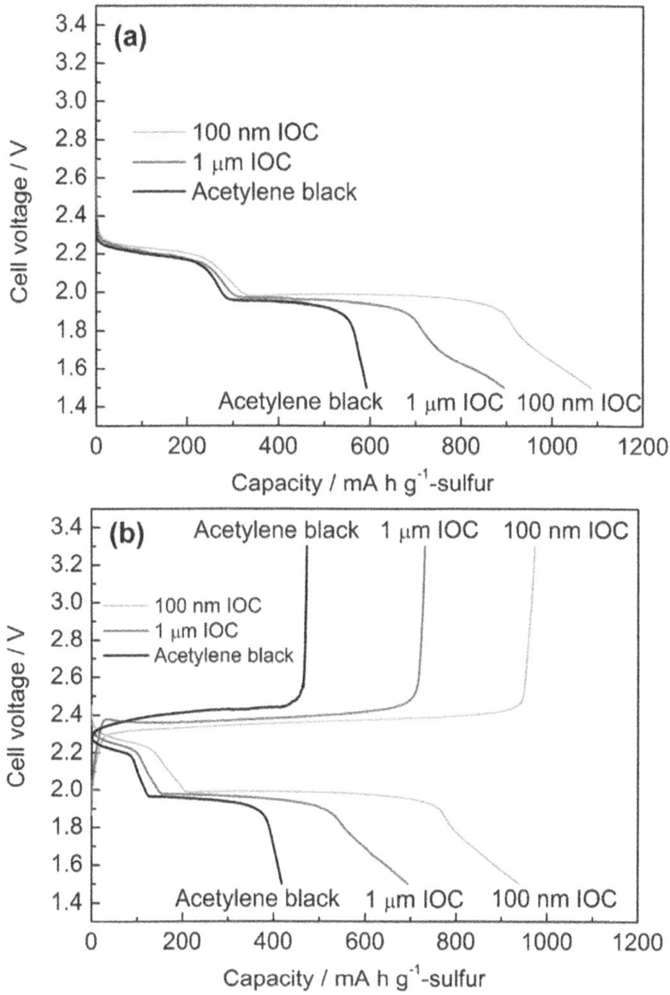

Fig. 3 Charge–discharge curves of Li–S cells at **a** 1st discharge and **b** 2nd charge/discharge. Current density; 139 mA g^{-1} sulfur. Reproduced from Ref. [11] with permission of The Royal Society of Chemistry

3 Effects of Polymer Binder in S_8/Carbon Composite Cathode

The S_8/carbon cathode is composed of S_8, carbon powder, and a polymer binder, and the composite cathode has a porous structure. The ion transport takes place in the pores during the electrochemical reaction of S_8. In the porous composite cathode, the binder interconnects the carbon particles to form an electron conduction network, and the surface of S_8 is partially covered with the polymer. Therefore, the compatibility of

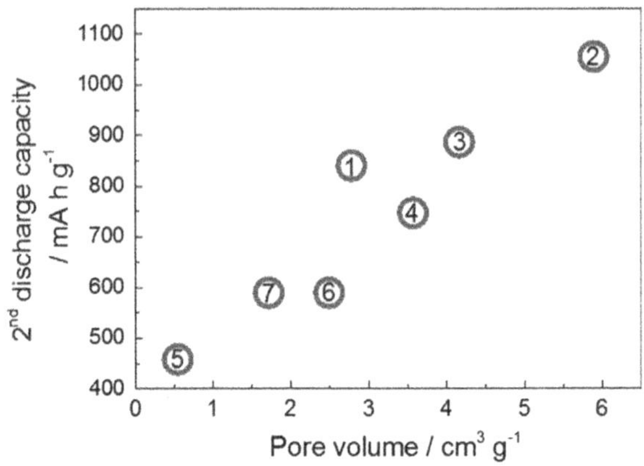

Fig. 4 Discharge capacities of S_8/carbon composite cathodes in a SIL electrolyte [Li(G4)][TFSA]. Carbon supports with different pore volumes were used to prepare the composite cathodes. The weight ratio of S_8 and carbon in the electrodes was controlled at 67:33. Reprinted with permission from Ref. [12]. Copyright 2016 American Chemical Society

the polymer binder with the electrolyte has a significant effect on the charge/discharge performance of Li–S cell. Here, we show the effects of interactions between electrolyte and polymer binders on the electrochemical reactions of S_8 cathode. We found that the discharge capacity and rate capability of S_8/carbon cathode change depending on the compatibility of the polymer binder with the electrolyte.

In this work, poly(vinyl alcohol) (PVA) with different degrees of saponification was used as a binder to fabricate the composite cathode [13]. As is well known, PVA is synthesized by the hydrolysis (saponification) of poly(vinyl acetate) (PVAc). Hereafter, we denote the PVA binder with degree of saponification $x\%$ as PVA-x. PVA-x with $x < 100$ contains vinyl acetate moieties within the polymer chain. By decreasing the degree of saponification, the fraction of acetate moieties increases. The degree of saponification can be controlled, and PVA-x binders with various x values are commercially available. Figure 5 shows the discharge capacities of Li–S cells with an SIL electrolyte [Li(G4)][TFSA], measured at various current densities [13]. As the degree of saponification of PVA binder decreases in S_8/carbon cathode, the charge and discharge capacities of the Li–S cell increase. We examined the compatibility of PVA-x binder with the [Li(G4)][TFSA]. PVA-100 is insoluble in the electrolyte and is not swollen because of the strong hydrogen bonds formed between hydroxyl groups of PVA-100 molecules. The S_8 covered with PVA-100 in the composite cathode becomes electrochemically inert because Li^+ ion cannot gain access to the active material, resulting in relatively low capacity, even at low current density.

We tested the compatibility of PVAc (PVA-0) with the electrolyte and found that PVAc dissolves into the electrolyte. PVA-x with $x < 100$ contains vinyl acetate moieties within the polymer chain. Therefore, the compatibility of PVA-x with [Li(G4)][TFSA] changes depending on the degree of saponification. PVA-x was

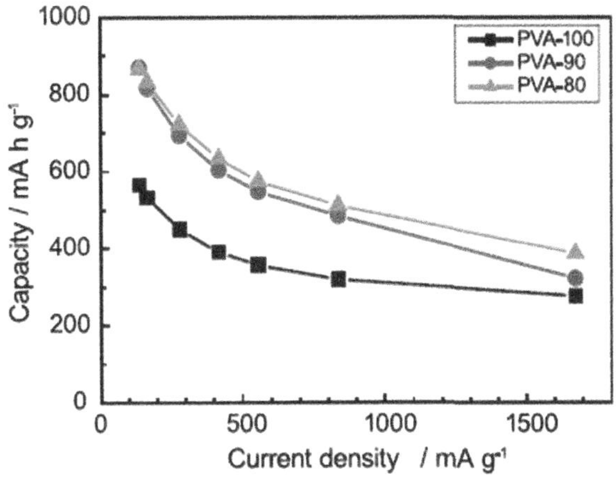

Fig. 5 Discharge capacities of Li–S cells with [Li(G4)][TFSA] electrolyte and PVA-*x* binders, measured at various current densities at 30 °C. The cells were charged up to 3.3 V at a current density of 139 mA g^{-1} prior to each discharge test. Reprinted with permission from Ref. [13]. Copyright 2016, with permission from Elsevier

mixed with an excess amount of [Li(G4)][TFSA] electrolyte, and the electrolyte uptake was estimated by the weight change of PVA-*x*. The electrolyte uptake of PVA-*x* increased with decreasing degree of saponification. Therefore, we concluded that [Li(G4)][TFSA] could penetrate into the acetate moiety of the PVA-*x* polymer chain, and the PVA-*x* was swollen with the electrolyte. The swelling of PVA-*x* binder can enhance ion transport in the porous S_8/carbon composite cathode during charge and discharge reactions, leading to the increase in the charge and discharge capacities of the Li–S cell. The rate capability of Li–S cell was also improved by the swelling of PVA-*x* binder.

It is worth mentioning that retaining the porous structure of the S_8/carbon composite cathode is also important for the Li–S cells. During the discharge reaction of the S_8/carbon cathode, the volume of active material increases ca. 80% because of the complete conversion of $S_8 \rightarrow 8Li_2S$. The volume change of active material is repeated during the charge/discharge cycles of Li–S cell, which may occasionally loosen the electrical contacts between carbon and the active material. The electrically isolated active material cannot contribute to the charge/discharge reaction, and the capacity of the cell is decreased. In addition, the repetition of the volume expansion and shrinkage of the active material may bring about the gradual collapse of the porous structure, if the binder does not maintain a tight binding between the carbon particles in the electrode. The disconnection between carbon particles deteriorates the electronic conduction path within the composite electrode, leading to the increase in resistance of S_8/carbon composite cathode. The stiff polymer binders are useful in retaining the electronic conduction path and porous structure of the composite electrode. However, an active material covered with a stiff polymer, which is incompatible with electrolyte, cannot contribute to the charge or discharge reaction because the electrolyte cannot approach the active material. Therefore, the design of the chemical structure of polymer binder and the distribution of binder within the composite electrode are important for accomplishing both high capacity and long cycle life.

4 Effect of Porosity of S_8/Carbon Composite Cathode

To achieve a higher gravimetric energy density in Li–S cell, the amount of electrolyte should be minimized. Actually, the density of S_8 is as low as 2.07 g/cm^3. The density of a representative SIL [Li(G4)][TFSA] is 1.40 g/cm^3 [14]. Therefore, as the volume fraction of the electrolyte in the cell increases, both volumetric and gravimetric energy densities of Li–S cell decrease. To minimize the amount of electrolyte in the cell, the porosity of the S_8/carbon composite electrode should be as small as possible. However, ionic conduction path for the electrochemical reaction in the composite cathode is needed for the electrochemical reactions of the active material. Figure 6 shows the discharge capacities of S_8/carbon composite electrodes with different porosities [15]. The porosity of the electrode significantly influences the discharge capacity of the electrode. The discharge capacity is dramatically decreased when the porosity is lower than 30%. As the porosity decreases, the Li^+ ion flux from the electrolyte in the porous cathode for the electrochemical reactions reduces. This causes a lower utilization of active material within the composite cathode, resulting in a low discharge capacity of the cell. Therefore, the design of the porous structure of S_8/carbon composite cathode and developing electrolytes, which enables a high Li^+ ion flux even if the cathode porosity is low, is crucial for realizing Li–S batteries with high energy density and long life.

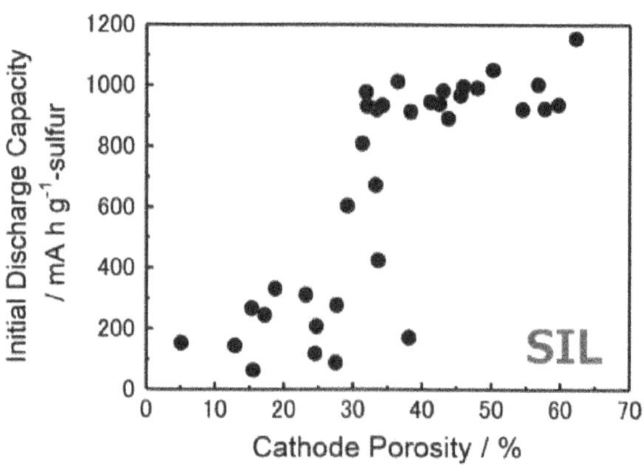

Fig. 6 Discharge capacities of Li–S cells using an SIL electrolyte [Li(G4)][TFSA] and S_8/Ketjen black composite cathodes with different porosities. Discharge/charge tests were performed at 30 °C under a current density of 100 μA cm^{-2}. Reproduced from Ref. [15] with permission of the Electrochemical Society of Japan

References

1. Manthiram, A., Fu, Y., Chung, S.-H., Zu, C., & Su, Y.-S. (2014). *Chemical Reviews, 114,* 11751–11787.
2. Evers, S., & Nazar, L. F. (2013). *Accounts of Chemical Reserach, 46,* 1135–1143.
3. Zhang, S., Ueno, K., Dokko, K., & Watanabe, M. (2015). *Advanced Energy Materials, 5,* 1500117.
4. Park, J.-W., Yamauchi, K., Takashima, E., Tachikawa, N., Ueno, K., Dokko, K., & Watanabe, M. (2013). *The Journal of Physical Chemistry C, 117,* 4431–4440.
5. Park, J.-W., Ueno, K., Tachikawa, N., Dokko, K., & Watanabe, M. (2013). *The Journal of Physical Chemistry C, 117,* 20531–20541.
6. Ueno, K., Park, J.-W., Yamazaki, A., Mandai, T., Tachikawa, N., Dokko, K., & Watanabe, M. (2013). *The Journal of Physical Chemistry C, 117,* 20509–20516.
7. Dokko, K., Tachikawa, N., Yamauchi, K., Tsuchiya, M., Yamazaki, A., Takashima, E., et al. (2013). *Journal of The Electrochemical Society, 160,* A1304–A1310.
8. Seki, S., Serizawa, N., Takei, K., Umebayashi, Y., Tsuzuki, S., & Watanabe, M. (2017). *Electrochemistry, 85,* 680–682.
9. Ji, X., Lee, K. T., & Nazar, L. F. (2009). *Nature Materials, 8,* 500–506.
10. Tabata, S., Isshiki, Y., & Watanabe, M. (2008). *Journal of The Electrochemical Society, 155,* K42–K49.
11. Tachikawa, N., Yamauchi, K., Takashima, E., Park, J.-W., Dokko, K., & Watanabe, M. (2011). *Chemical Communications, 47,* 8157–8159.
12. Zhang, S., Ikoma, A., Li, Z., Ueno, K., Ma, X., Dokko, K., & Watanabe, M. (2016). *ACS Applied Materials and Interfaces, 8,* 27803–27813.
13. Nakazawa, T., Ikoma, A., Kido, R., Ueno, K., Dokko, K., & Watanabe, M. (2016). *Journal of Power Sources, 307,* 746–752.
14. Yoshida, K., Tsuchiya, M., Tachikawa, N., Dokko, K., & Watanabe, M. (2011). *The Journal of Physical Chemistry C, 115,* 18384–18394.
15. Matsumae, Y., Obata, K., Ando, A., Yanagi, M., Kamei, Y., Ueno, K., et al. (2019). *Electrochemistry, 87,* 254–259.

S-Encapsulated Micropore Carbon Cathode

Masashi Ishikawa, Yoshifumi Egami, and Tomohiro Shimizu

Abstract This chapter describes sulfur (S)-encapsulated micropore carbon as a cathode material for lithium (Li) S batteries. Although S cathodes have an extremely large theoretical capacity of 1672 mAh g^{-1}, there have been critical issues preventing practical use: elution of polysulfide intermediates, irreversible reactions of them with an electrolyte, and so on. To solve these problems, extensive studies have been conducted for limiting the elution of Li polysulfides from S cathodes. For instance, various nanocarbon materials, polymers, and their precursors, etc., have been tested to confine S stably. This chapter focuses mainly on activated carbons especially microporous carbons, whose pores have a diameter less than 2 nm. Such microporous carbons can effectively confine S to prevent the intermediate elution and undesirable side reactions with electrolytes. In that case, charge and discharge curves exhibit a single plateau with a gentle slope, which are very similar to the curves observed in an all-solid-state Li–S battery. There are some critical factors to govern the cycle performance of S cathodes based on microporous carbons: pore distribution, pore volume, loadable S content, particle size, and fine structure of applied microporous carbons. The chapter also introduces a relatively simple and easy procedure to obtain microporous carbons from polysaccharides with heat treatment and alkali activation.

Keywords Lithium–sulfur battery · Sulfur positive · Activated carbon · Microporous carbon · Alkali activation

Studies to apply sulfur as a positive electrode material have been conducted by groups such as Brummer et al. [1] and Yamin et al. [2, 3] since the late 1970s. The discharge potential of sulfur is about 2.2 V versus Li^+/Li, which is low compared to that of $LiCoO_2$, etc., but it has an extremely large theoretical capacity of 1672 mAh g^{-1}.

M. Ishikawa (✉) · Y. Egami
Department of Chemistry and Materials Engineering, Faculty of Chemistry, Materials and Bioengineering, Kansai University, 3-3-35 Yamate-cho, Suita 564-8680, Japan
e-mail: masaishi@kansai-u.ac.jp

T. Shimizu
Department of Mechanical Engineering, Faculty of Engineering Science, Kansai University, 3-3-35 Yamate-cho, Suita 564-8680, Japan

K. Kanamura (ed.), *Next Generation Batteries*,
https://doi.org/10.1007/978-981-33-6668-8_32

As this value is much better than that of the positive currently used in lithium-ion batteries, it is believed that high-energy–density secondary batteries can be constructed. Furthermore, sulfur is abundantly present in the earth's crust, there is no concern of exhaustion, and it is available as a by-product of petroleum refining at a low price. Despite the advantages of using sulfur as a positive electrode material, it has so far been difficult to achieve practical use because the following issues remain: (1) elution of lithium polysulfide, Li_2S_x (x: 8–4), which is a reaction intermediate, into an electrolyte and the associated redox shuttle, (2) irreversible reactions between lithium polysulfide and some electrolytes (conventional electrolytes for lithium-ion batteries), (3) a large volume change of 180% during charge and discharge, (4) low electron conductivity of sulfur (5×10^{-30} S cm^{-1}). In particular, the problems resulting from the elution of lithium polysulfide prevent practical use. Many studies have, therefore, been conducted not to elute lithium polysulfide from the positive electrode.

1 Elution Suppression Approach by Positive Electrode Material Improvement

In order to use sulfur which is an insulator as a positive electrode active material, a method of combining sulfur with an electron conductive material has been proposed. Examples of electron conductive substance include carbon nanotubes, graphene, porous carbon, conductive polymers, polyacrylonitrile, metal oxides, etc. [4]. Among them, a method of combining porous carbon and sulfur is known to be effective. By supporting sulfur in the pores of porous carbon, the primary size of sulfur can be limited, and an electron conduction path can be secured. Furthermore, the capillary force can suppress the elution of lithium polysulfide from the inside of the pore, and the redox shuttle can be prevented to some extent.

Activated carbon, which is a typical example of porous carbon, has pores of various sizes. The pores are classified into micropores having a pore diameter of 2 nm or less, mesopores of 2–50 nm, and macropores of 50 nm or more. L. F. Nazar et al. reported a positive electrode combining sulfur with mesoporous carbon having a three-dimensional ordered structure with a pore diameter of about 2–5 nm. This material achieved a charge/discharge capacity exceeding 1000 mAh g^{-1} and improved cycle stability [5]. Generally, however, the positive electrode using mesoporous carbon would not completely suppress the elution of lithium polysulfide, and there is a problem that the charge/discharge coulombic efficiency is low.

On the other hand, researches using microporous carbon have also been conducted for the purpose of further suppressing the elution. X. P. Gao et al. successfully attained good cycle stability by combining sulfur with microporous carbon with a pore diameter below 1 nm [6, 7]. The positive electrode using such a composite of microporous carbon and sulfur does not exhibit a typical two-stage discharge curve shown in Fig. 1. Instead, it has a gentle plateau throughout the reaction (the profile

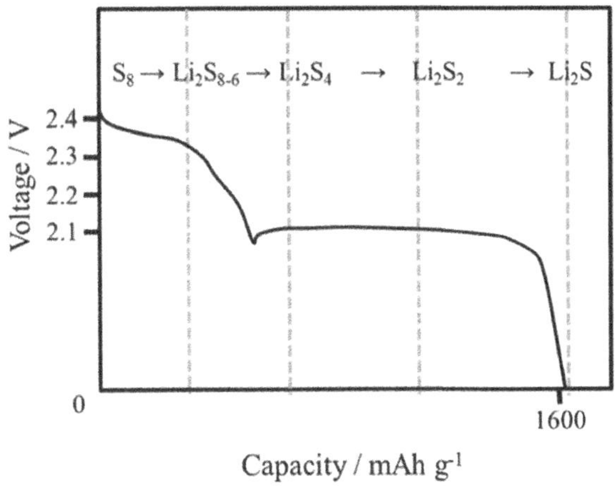

Fig. 1 Potential profile associated with lithiation process at sulfur positive

is shown later). In addition, since it can be charged and discharged without a large capacity fading over 500 cycles, it is considered that the elution of lithium polysulfide is substantially suppressed.

When a cyclic carbonate solvent is applied to a usual sulfur positive electrode, there is a problem of irreversible reactions in which eluted polysulfide anions nucleophilically attack carbonate-based solvents to form organic sulfur compounds such as thioether [8]. However, a specific sulfur positive electrode where sulfur is supported by microporous carbon has also been reported as an example exhibiting a superior cycle characteristic even in the presence of a carbonate-type solvent which has been considered unusable. The origin of the elution suppression effect of microporous carbon would be that it causes strong adsorption with lithium polysulfide [7], and that electrolyte solvents cannot penetrate the pore and thus the reduction (discharge) reaction proceeds between solid and solid phases [9]. It is also considered for the origin that the formation reaction of soluble Li_2S_x (x: 8–4) does not occur because sulfur is present as stable $Li_2S_{4\text{–}2}$ in the pore [10, 11]. As described above, when a composite of microporous carbon and sulfur is used as a positive electrode material, excellent cycle stability can be obtained, but there is a problem that the content of sulfur is modest because microporous carbon has a small pore volume. The lower weight ratio of sulfur in the composite leads to a decrease in energy density per electrode weight. In addition, sulfur has a lower density than current positive electrode materials, and therefore, there is also a problem that improvement in volumetric energy density becomes difficult if a large amount of porous carbon with a low density is used.

While studies are being conducted to physically suppress the elution of lithium polysulfide by combining porous carbon and sulfur, there are also other studies to suppress the elution by some chemical interaction. D. Wang et al. reported the improvement of cycling stability by suppressing lithium polysulfide elution by combining nitrogen-doped mesoporous carbon with sulfur [12]. Furthermore, Y. G. Guo et al. reported that the combination of boron-doped porous carbon and sulfur improved cycle stability [13]. From the viewpoint of the positive electrode material, it is necessary to completely suppress the elution of lithium polysulfide and increase the amount of the active material in the electrode in order to put the lithium-sulfur battery into practical use. However, no solution has yet been found that satisfies these two problems.

2 Activated Carbon

Activated carbon is a substance having pores surrounded by carbon atoms. This material is mainly used for purification of water and air, separation of mixed gas. Activated charcoal is characterized by having many pores that are the size of molecules. These pores are spaces that have no charge but have strong van der Waals forces from adjacent carbon atoms; hence strong adsorption force [14]. The activated carbon pores are classified as micropores with diameters of 2 nm or less, mesopores with 2–50 nm, and macropores with a diameter of 50 nm or more. It is called nanopore (<100 nm) including this pore classification.

Some activated carbons exhibit adsorption performance as they exist naturally, such as anthracite, and may be used as they are, but are basically produced by carbonizing and activating the raw materials to satisfy the required performance. Carbonization is a general term for carbon-enriched chemical reactions in which bonds are cleaved by heat treatment of organic substances and recombined into more stable bonds such as decomposition, polycondensation, and aromatic cyclization. However, activated carbon formed by carbonization develops only insufficient pores for the intended use in most cases, and hence an activation process is required. Activation is a method of developing pores by selectively gasifying carbon components or using chemicals. There are two main methods of activation: gas activation and chemical activation.

Gas activation is a technique that uses carbon dioxide or water vapor as an activator. These activators select carbon atoms from the porous carbon structure according to the endothermic stoichiometric formula, such as $C + CO_2 \rightarrow 2CO$ for carbon dioxide activation and $C + H_2O \rightarrow CO + H_2$ for steam activation [14]. On the other hand, zinc chloride, phosphoric acid, potassium hydroxide, potassium carbonate, etc., are used as activators for chemical activation. In particular, when powdered activated carbon is produced, chemical activation is suitable, and an impregnation operation is used to ensure contact between the precursor and the chemical. In the impregnation process, chemicals are dissolved in water to form an aqueous solution, which is effectively mixed with the precursor to effectively hydrate and expand the channels in the precursor structure, making the reactants closer to the interior of the particles [15]. In addition, activation using an alkali such as potassium hydroxide is called alkali activation. With potassium hydroxide, the dehydration reaction proceeds at 300–500 °C, and the activation reaction proceeds at 600–800 °C.

It is necessary to solve various problems in order to use sulfur as a positive electrode active material. Among these, suppression of lithium polysulfide elution is a top priority for improving battery characteristics. M. Ishikawa et al. are working on suppression of elution from the cathode material side, and as a previous study, applied commercially available porous carbon rich in microporosity to a sulfur cathode. Then, stabilization of cycle characteristics and charge/discharge efficiency in a lithium-sulfur battery was achieved [16–18]. Commonly used commercial microporous carbon has a small pore volume and can only fill sulfur up to about 30–50 wt.% [16]. Since the ideal sulfur loading in carbon in lithium-sulfur batteries is considered to be 60 wt.% or more, it is necessary to increase the pore volume. In the conventional activated carbon activation method, it is difficult to increase the pore volume while selectively forming micropores. Therefore, from here on, carbonization and alkali activation are performed using sucrose and glucose, which are known to selectively form micropores of 1 nm or less, as a carbon material precursor. The following process gives us a typical and useful information about a porous activated carbon supporting sulfur.

3 Preparation of Microporous Activated Carbon Suitable for Sulfur Cathode

Here, glucose, which is a monosaccharide, and sucrose, which is a disaccharide, are used as raw materials for activated carbon. Glucose is made from starch that is separated from corn. Sucrose is the main component of sugar and is one of the few natural chemicals consumed as it is, just like water and sodium chloride. These sugars, glucose, and sucrose are dissolved in 6 mol dm^{-3} sulfuric acid, respectively, to prepare 10 wt.% glucose solution and 10 wt.% sucrose solution. Transfer each solution to a round bottom flask and reflux at 120 °C for 10 h. Thereafter, the suspension is filtered, and the precursor is washed with pure water. The washed precursor is dried in air for about 14–15 h at 100 °C. Both are brown powders after drying. Next, put this powder in a horizontal tubular furnace, raise the temperature to 800 °C at a rate of 10 °C min^{-1} in an argon gas atmosphere at a flow rate of 50 mL min^{-1}, and hold it for 2 h to carbonize. Carbon powder obtained from glucose is designated glAC, and carbon powder obtained from sucrose is designated suAC.

4 Pore Characteristics of Microporous Activated Carbon

Figure 2a shows the nitrogen adsorption and desorption isotherms for glAC and suAC. The initial gradient of the isotherm where the relative pressure P/P_0 is 0–0.05 represents the pore size of the micropore, and the steeper gradient indicates that the pore has a smaller pore size [14]. The adsorption/desorption isotherms in Fig. 2a both have linearly increased adsorption in the range of P/P_0 between 0 and 0.05, so that pores with smaller pore diameters are formed among the micropores. In

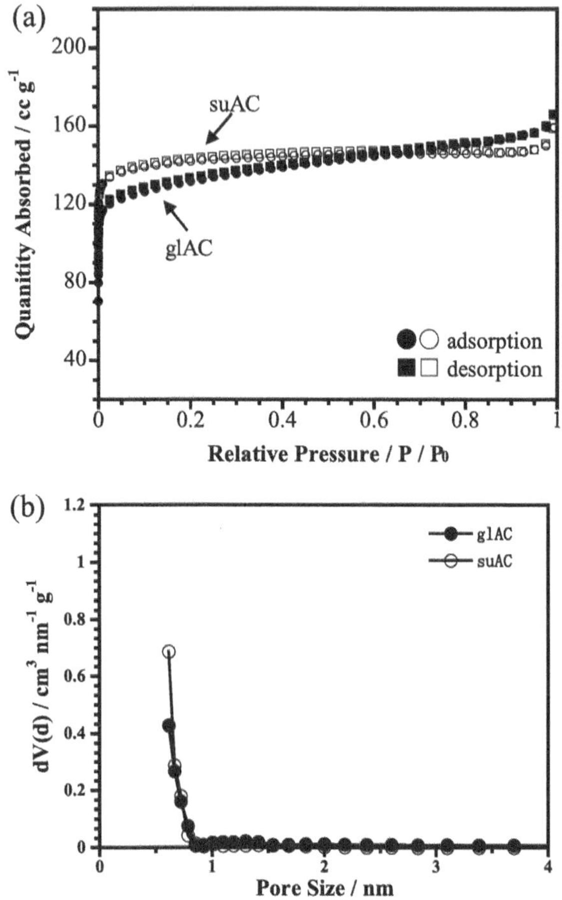

Fig. 2 N_2 adsorption/desorption isotherm for glAC and suAC **a**; pore size distribution calculated by using QSDFT method for glAC and suAC **b**

addition, since the amount of adsorption goes to the maximum without any inflection point, it shows the type I isotherm in the IUPAC classification for the six types of isotherm [19], suggesting a typical micropore carbon. Figure 2b shows the pore size distribution calculated by the QSDFT method. From the pore size distribution, it can be confirmed that many micropores are formed below 2 nm. The micropores formed in glAC and suAC are thought to have been formed by CO, CO_2, and H_2O gases produced during the pyrolysis of glucose/sucrose [4]. However, when carbonized without passing through the reflux process with sulfuric acid, such micropores are not formed because CO, CO_2, and H_2O gases are generated abruptly. It is thought that by dehydrating glucose/sucrose in advance by refluxing, gas is slowly generated during the carbonization process and micropores are formed. The BET specific surface areas of glAC and suAC are 515 and 556 $m^2\ g^{-1}$, respectively, and the pore volume is 0.229, 0.213 $cm^3\ g^{-1}$. Here, the theoretically possible sulfur loadings in glAC and suAC are 32.0 and 30.4%. The formula for calculating the theoretical sulfur loading is shown below as a calculation formula of theoretical sulfur loading per gram of carbon:

$$\text{Theoretical loading percentage of sulfur} = 100[V \times M(\text{carbon}) \times \rho(\text{sulfur})]/[\{V \times M(\text{carbon}) \times \rho(\text{sulfur})\} + M(\text{carbon})]$$
$$\left[V : \text{pore volume}\left(\text{cm}^3\ \text{g}^{-1}\right), M : \text{mass of carbon}(\text{g}), \rho : \text{density of sulfur}\left(\text{g cm}^{-3}\right)\right]$$

However, if sulfur is packed too much into the activated carbon pores, sulfur will come out from the carbon pores due to a volume change of about 1.8 times due to reduction to Li_2S during discharge. Therefore, this volume change should be taken into account when loading sulfur into the carbon pores, so that even if all of the supported sulfur is reduced to Li_2S, the volume is below the total pore volume of the activated carbon. Here, considering the volume change of 1.8 times, the sulfur loading rate when the sulfur (or Li_2S) volume does not exceed the pore volume is taken as the actual sulfur loading rate. The approximate sulfur loadings of glAC and suAC are about 20.6 and 19.5%, respectively. The formula for the actual sulfur loading is shown below as a calculation formula for the approximate sulfur loading per gram of carbon:

$$\text{Optimal loading percentage of sulfur} = 100[V \times M \times \rho(\text{lithium sulfide}) \times (\text{ratio of sulfur to lithium sulfide})] /[\{V \times M \times \rho(\text{lithium sulfide}) \times (\text{ratio of sulfur to lithium sulfide})\} + M]$$
$$\left[V : \text{pore volume}\left(\text{cm}^3\ \text{g}^{-1}\right), M : \text{mass of carbon}(\text{g}), \rho : \text{density of lithium sulfide}\left(\text{g cm}^{-3}\right)\right]$$

Comparing the ratio of micropore volume to the total pore volume of glAC and suAC, glAC is 84, suAC is 99.1%; suAC has higher microporosity. Therefore, glAC has a larger pore volume and can support more sulfur, but suppression of lithium polysulfide elution in an electrolyte is a priority issue, so suAC using sucrose as the raw material is more suitable for the sulfur positive electrode. However, the pore volume of glAC and suAC is smaller than that of previous studies [16], and further increase in pore size is necessary. Therefore, next, alkali activation is further performed on these materials.

5 Alkali Activation for Micropore Preparation for Activated Carbon

Activation is performed with mixture ratios of suAC and KOH as 1:1, 1:2, 1:3 (by weight). First, the required amount of KOH and pure water are mixed to form a saturated aqueous solution. After stirring this saturated aqueous solution and suAC for a short time using a rotating/revolving mixer, the mixed solution is transferred to a beaker and dried at 100 °C for several hours. When some moisture evaporates and becomes sticky, it is transferred to an alumina boat and pre-dried in an argon atmosphere at a gas flow rate of 50 ml L^{-1} at a temperature of 100 ºC for 2 h. After that, it is activated under the condition that the heating rate is 10 ºC min^{-1} and the activation temperature is 800 ºC for 1 h. After activation, it is washed with hydrochloric acid, filtered with suction, further neutralized with excess pure water, and dried overnight at 100 °C. Activated charcoals obtained by this preparation method are denoted as suAC1 (the same weight of KOH as carbon), suAC2 (twice the weight of KOH as much as carbon), suAC3 (3 times the weight of KOH as much as carbon).

Figure 3a shows the nitrogen adsorption and desorption isotherms of suAC1, suAC2, and suAC3. The initial gradient of the isotherm where the relative pressure P/P_0 is 0–0.05 represents the micropore [14]. All of the nitrogen adsorption and

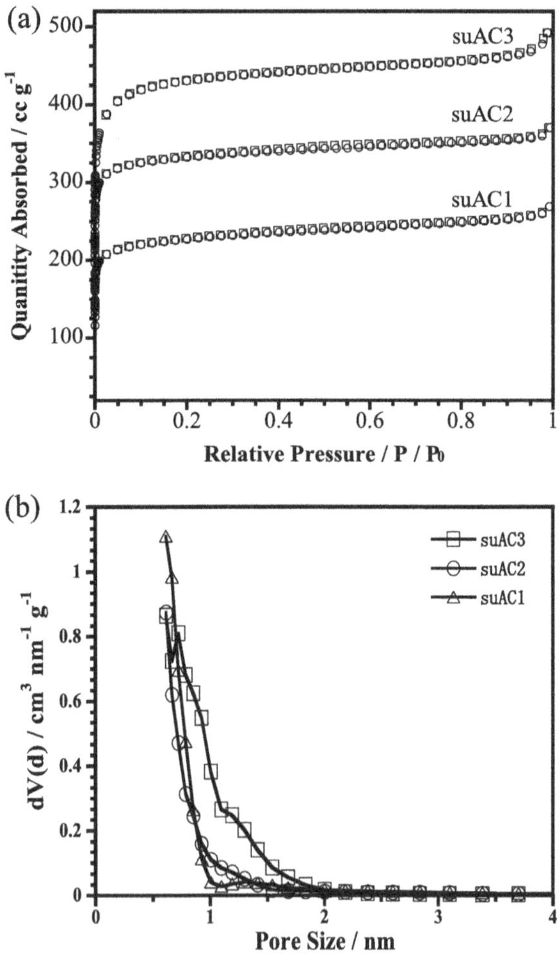

Fig. 3 N_2 adsorption/desorption isotherm for suAC1, suAC2, and suAC3 **a**; pore size distribution calculated by using QSDFT method for suAC1, suAC2, and suAC3 **b**

desorption isotherms in Fig. 3a are linearly increasing in the range of P/P_0 between 0 and 0.05, so that pores with smaller pore diameters are formed among the micropores. Figure 3b shows the pore size distribution of suAC1, suAC2, and suAC3 calculated by the QSDFT method. From the pore size distribution, it can be confirmed that many micropores are formed below 2 nm. The BET specific surface areas of suAC1, suAC2, and suAC3 are 850, 1280, and 1693 $m^2 g^{-1}$, respectively, and the pore volume is 0.366, 0.519, 0.682 $cm^3 g^{-1}$. Here, the theoretically possible sulfur loadings in suAC1, suAC2, and suAC3 are 42.9, 51.6, and 58.4%. The approximate sulfur loadings for suAC1, suAC2, and suAC3 are about 29.6, 37.0, and 43.6%, respectively. SuAC1, suAC2, and suAC3 using the alkali activator have higher BET surface area and pore volume than suAC. This is because the pores grow due to the activator KOH, and it can be confirmed that the pores develop more when the amount of the activator is larger.

Figure 4 shows secondary electron (SE) images of suAC, suAC1, suAC2, and suAC3 in the SEM. The shape of suAC shown in Fig. 4a is spherical, and many particles with a size of about 0.5–1.5 μm can be confirmed. It is considered that sucrose aggregated during refluxing with sulfuric acid, and a shape close to a spherical shape was formed in order to keep the interfacial energy stable. Figure 4b, c show the SE images of suAC1 and suAC2. Both shapes are spherical like suAC, and many particles of the same size can be confirmed. Furthermore, in the SE image of

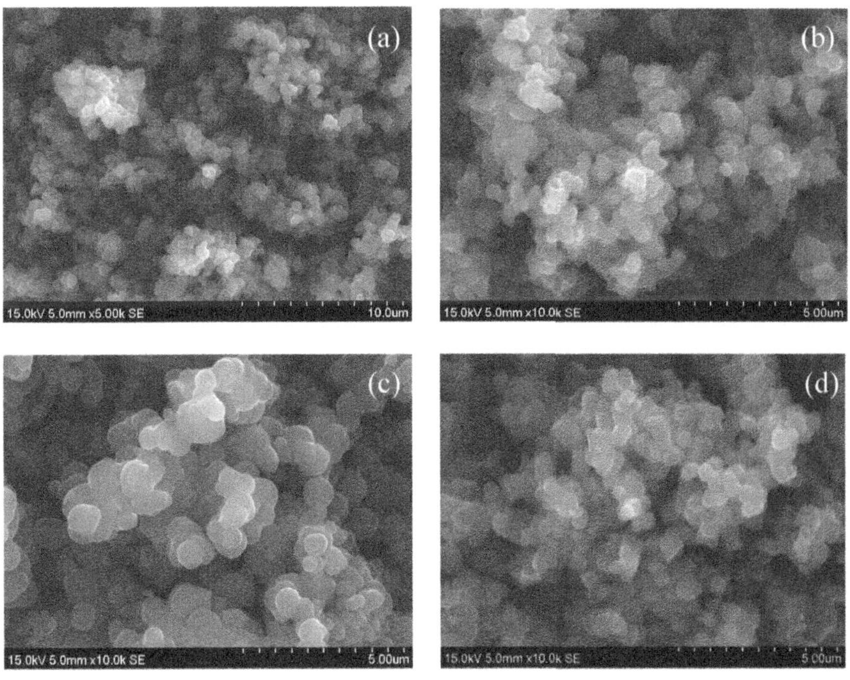

Fig. 4 SEM images of suAC **a**, suAC1 **b**, suAC2 **c**, suAC3, **d** particles

suAC3 is shown in Fig. 4d, the shape is spherical like suAC, suAC1 and suAC2, but a rough surface can be seen on the suAC3 carbon surface. This is probably because the graphene ends of carbon were eroded by more activators than suAC1 and suAC2. Furthermore, it is assumed that the carbon surface was destroyed during the desorption process of metallic potassium intercalated in the graphene layer. Among the prepared microporous activated carbons, suAC2 and suAC3 are considered suitable for sulfur loading carbon because their sulfur loading considering the volume expansion during charging and discharging exceeds 30%.

6 Microporous Carbon Confining Sulfur

To prepare a composite material of suAC2 or suAC3 and sulfur, suAC2 and sulfur are mixed in a weight ratio of suAC2: S = 67:33, while suAC3 is mixed with sulfur in a weight ratio of suAC3: S = 58:42. The heat treatment was performed at 155 °C for 5 h and then at 300 °C for 6 h. Sulfur is melted during the heating process at 155 °C and permeated into the microporous activated carbon pores by capillary action. In the 300 °C heating process, sulfur remaining on the surface of the microporous activated carbon is sublimated and removed. The composites are prepared in this way, and materials obtained by combining suAC2 and suAC3; sulfur are denoted as suAC2-S and suAC3-S, respectively.

Figure 5 shows the SE (secondary electron) and BSE (backscattered electron) images of suAC3-S. It was thought that the original spherical shape was crushed because microporous activated carbon and sulfur were kneaded in a mortar at the compounding stage, but the same spherical shape as before compounding can be confirmed from the SE image (Fig. 4c, d). Therefore, it is considered that there was no change in the shape of the microporous activated carbon in the composite process. The BSE image produces a contrast between light and dark depending on the atomic weight, and the brightness increases as the atomic weight increases. If sulfur is not

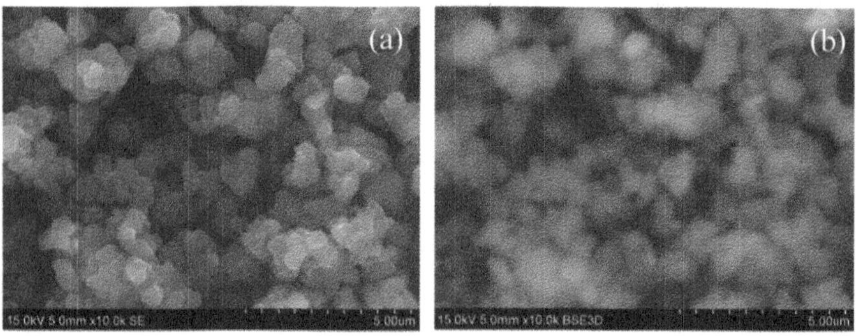

Fig. 5 SEM images of suAC3-S particles as secondary electron image **a**, backscattered electron image **b**

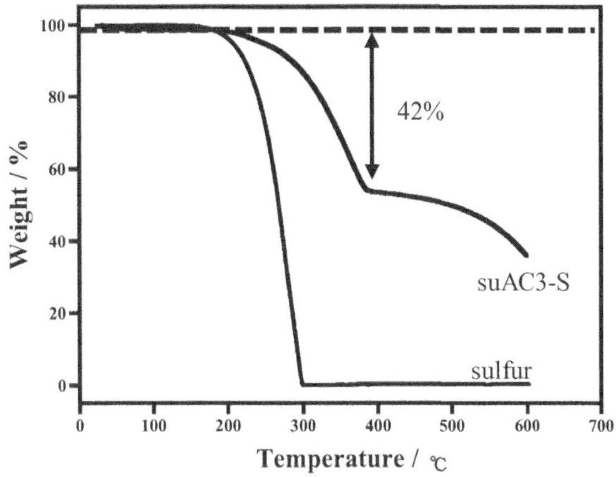

Fig. 6 TGA at temperature increase rate of 10 °C min^{-1} in Ar gas for sulfur and suAC3-S powder

supported in the microporous activated carbon pores, the sulfur remains mixed with the carbon. Therefore, a contrast between carbon and sulfur appears, and sulfur can be confirmed on the surface of the microporous activated carbon. However, since there is no change in contrast in the BSE image, sulfur melts and diffuses into the microporous activated carbon pores in the 155 °C heat treatment. Even if then residual sulfur remains on the surface of the microporous activated carbon, residual sulfur sublimates in the following 300 °C heat treatment. It is, therefore, considered that no sulfur remains except for the micropores. Figure 6 shows the thermogravimetric analysis results of elemental sulfur and suAC3-S. It can be seen that sulfur alone has a weight loss between 200 and 300 °C, indicating sublimated sulfur. On the other hand, suAC3-S gradually loses weight from 250 to 400 °C. This suggests that the sulfur supported in the microporous activated carbon pores is sublimated, and the thermal stability of sulfur is improved by supporting sulfur in the pores. The sulfur loading in suAC2-S and suAC3-S was found to be about 33 and 42%, respectively, from the weight loss ratio of sulfur. Since the approximate sulfur loadings for suAC2 and suAC3 are 37% and 44%, respectively, the sulfur loading measured by TGA can follow the volume expansion during discharge. The loss curve for the second stage of the composite is the weight loss due to pyrolysis of the carbon itself.

7 Charge/discharge Behavior of Microporous Carbon Confining Sulfur

Figure 7 shows the charge and discharge curves and cycle characteristics of suAC3-S in the solvated ionic liquid. The charge/discharge curve shown in Fig. 7a shows a one-step discharge plateau, indicating the unique behavior that occurs when sulfur is loaded into microporous carbon. The initial discharge capacity is about 2180 mAh g^{-1}, and the reversible capacity in the second cycle is about 1450 mAh g^{-1}.

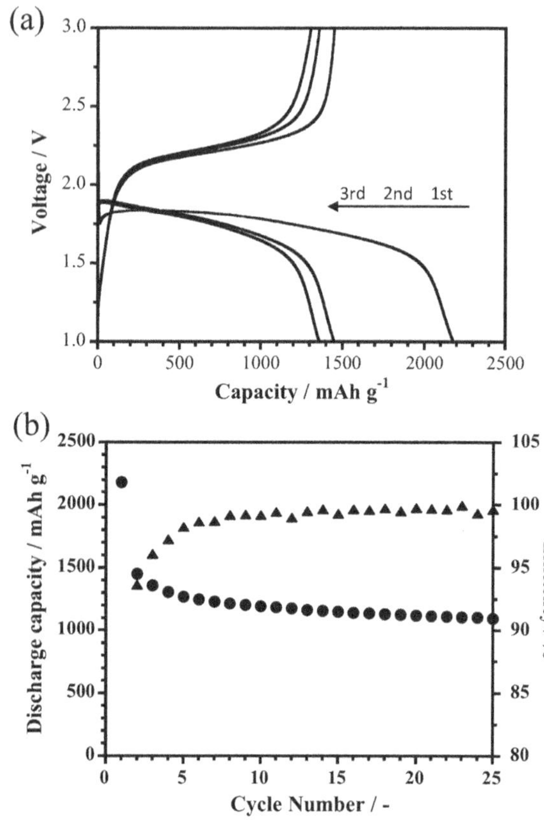

Fig. 7 Charge–discharge curves **a** and discharge capacity with coulombic efficiency (**b**) at 0.1C-rate for suAC3-S cathode in LiTFSI:G4:HFE (10:8:40 by mol) electrolyte with Li metal anode

As shown in Fig. 7b, the capacity retention from the second cycle is 75.1% after 25 cycles, indicating a stable cycle. Furthermore, the average charge/discharge efficiency of 2–25 cycles is 98.9%, which is a stable charge/discharge efficiency.

Figure 8 shows the charge and discharge curves and cycle characteristics of suAC3-S in a carbonate electrolyte. As can be seen from the charge/discharge curve shown in Fig. 8a, the second and subsequent cycles show a one-step discharge plateau, and this solvent system also exhibits the behavior peculiar to microporous carbon. In general, when sulfur is supported on porous carbon having a large pore diameter such as mesopores, the charge/discharge curve shows a two-stage discharge plateau. First, in the discharge plateau at 2.4–2.3 V versus Li^+/Li, S_8 molecules are cleaved by reduction to form Li_2S_8 and Li_2S_6 [20]. Furthermore, Li_2S_4 is formed in a slow reaction before reaching the discharge plateau near 2.1 V versus Li^+/Li, and is reduced to Li_2S_2 and Li_2S in the plateau of 2.1 V versus Li^+/Li [20]. However, as shown in Fig. 8a, the charge/discharge curve of suAC3-S is only a one-step discharge plateau, so it is considered that sulfur exists in the micropores as S_{4-2}. In this case, charging and discharging would proceed without the formation reaction of Li_2S_x (x: 8–4), which is easily dissolved in the electrolyte. Even if Li_2S_x (x: 8–4) is formed, suAC3

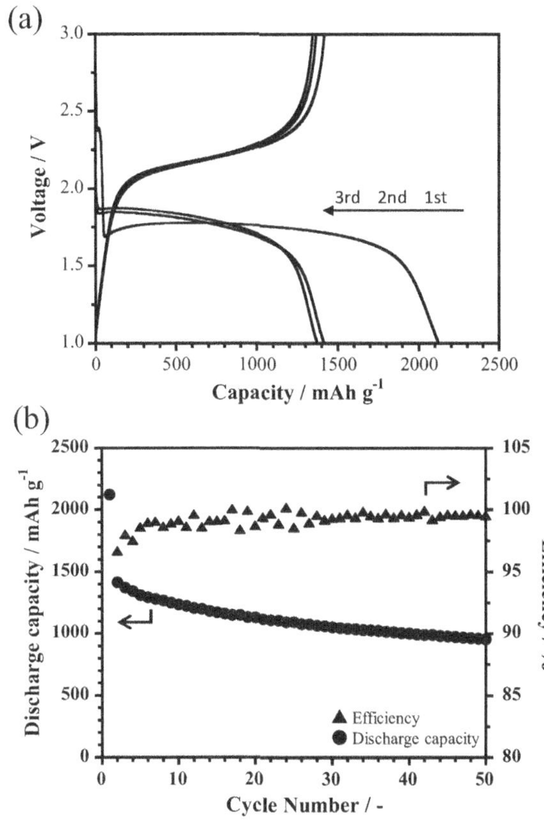

Fig. 8 Charge–discharge curves **a** and discharge capacity with coulombic efficiency (**b**) at 0.1C-rate for suAC3-S cathode in 1 mol dm^{-3} $LiPF_6$/EC:DMC (1:1 by vol.) electrolyte with Li metal anode

with micropores spatially restricts the penetration of the electrolyte into the pores, creating an environment that does not cause solvation. Such a situation may also prevent the elution of Li_2S_x (x: 8–4).

However, a two-stage discharge plateau can be confirmed in the first discharge. This may be caused by a reaction that shows elution because sulfur is supported in a small amount of mesopores present in suAC3 and the carbonate solvent and sulfur come into contact during discharge. Normally, sulfur and the carbonate solvent cause an undesirable side reaction and the cathode cannot be cycled. However, since there is little contact with sulfur and the formation of a passive film due to the side reaction is insignificant, the second and subsequent cycles may continue. The initial discharge capacity is about 2123 mAh g^{-1}, and the second cycle capacity is about 1414 mAh g^{-1}. It can be confirmed that the initial irreversible capacity is about 709 mAh g^{-1}. The reversible capacity after the second cycle is about 1414 mAh g^{-1} corresponding to about 84.5% of the theoretical capacity of sulfur (1672 mAh g^{-1}). As shown in Fig. 8b, the capacity retention rate from the second cycle is 67.2% after 50 cycles. The average charge/discharge efficiency of 2–50 cycles is 99.2%, and the charge/discharge efficiency is not deteriorated. The suAC3-S system shows a stable cycle, which suggests that the elution of lithium polysulfide can be suppressed.

8 Sensitive Micropore Factors Affecting Charge/discharge Performance

Figure 9a shows the charge/discharge curve of the 10th cycle of suAC2-S in the ether electrolyte. The charge/discharge curve here shows a one-stage discharge plateau, which is characteristic of microporous carbon. As can be seen from this result as well as the other results, the suAC2-S positive electrode shows a one-step plateau in all of the three types of electrolytes: carbonate electrolyte, solvated ionic liquid, and ether electrolyte, and suppresses the elution of lithium polysulfide. Figure 9b shows

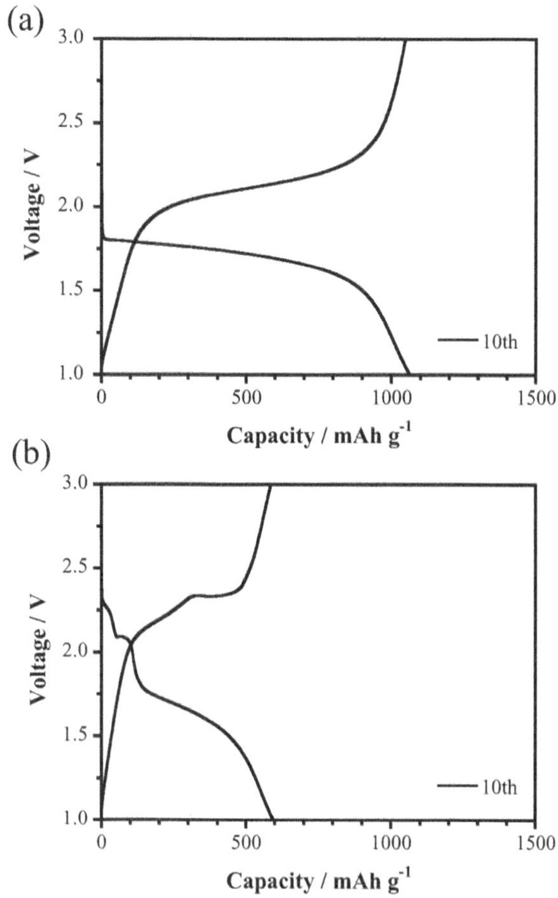

Fig. 9 Charge–discharge curves for suAC2-S **a** and suAC3-S, **b** cathodes at 0.1C-rate in 1 mol dm^{-3} LiTFSI / DME:DOL (1:1 by vol.) electrolyte with Li metal anode at 10th cycle

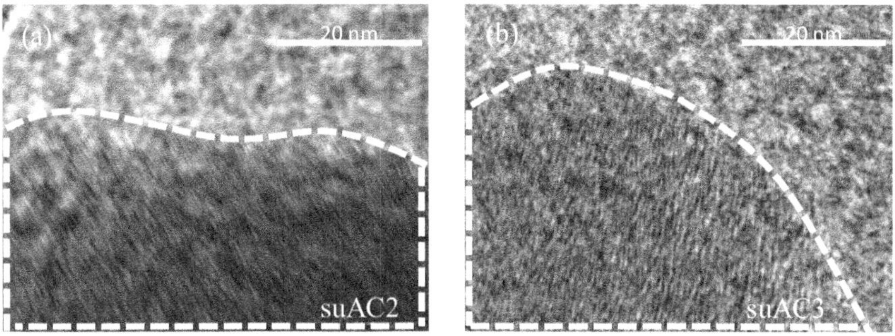

Fig. 10 TEM images of suAC2-S **a** and suAC3-S, **b** particles

the 10th cycle charge/discharge curve of suAC3-S in the ether electrolyte. Unlike the charge and discharge curves of suAC2-S, a multi-stage plateau can be confirmed. Both suAC2-S and suAC3-S have micropores of 2 nm or less, and the difference is not very large. Nevertheless, there is a considerable difference in behavior when ether solvents are used. To elucidate the origin of such sensitive differences, we have to get an insight into microporous carbon from a viewpoint of fine structure consideration.

Figure 10 shows the TEM images of suAC2 and suAC3. It can be confirmed that both the suAC2 and suAC3 carbons are partially crystalline. Furthermore, the surface spacing is about 0.5 nm for suAC2 and about 0.6 nm for suAC3, indicating that the surface spacing of suAC3 using more activators is wider. The observation of the local structure by TEM shows that the layered structure is partially developed. As the XRD pattern measurement essentially showed amorphous, the crystal pattern in the TEM image implies slit-type pores in suAC2 and suAC3.

As discussed so far, the plateau near 2.3 V seen in the early stage of discharge of a general sulfur cathode represents a reductive reaction from cyclic S_8 sulfur to a long-chain polysulfide that is soluble in the electrolyte solvent. The second-stage discharge plateau is associated with further reductive reactions from long-chain polysulfides to Li_2S_2 and Li_2S. Furthermore, unlike the general two-stage discharge plateau of the sulfur positive electrode, there are some that show a gentle one-stage charge/discharge plateau. What progresses in charge and discharge in this one-stage plateau is a behavior seen in a lithium-sulfur solid battery using a solid electrolyte, in other words, a behavior seen in a system where the solvent and sulfur do not come into direct contact. Therefore, when the charge/discharge behavior shows a one-step plateau even in a system using a liquid electrolyte, the lithium ion after desolvation is present in a solvent-free environment in the pores. This situation should provide a pseudo-solid-phase reaction.

In the behavior with ether electrolytes, elution is suppressed because the electrolyte solvent could not spatially enter the slit pores of suAC2-S with small pores, while suAC3-S causes a multi-step reaction because a free electrolyte solvent enters the pores. The suAC3-S positive electrode showed a two-stage plateau in the first cycle when the carbonate system was used, and a one-stage plateau in the second and subsequent cycles (Fig. 8a). Therefore, in the suAC3-S positive electrode using the carbonate system, there was a part where the electrolyte solution could invade during the first discharge, and the sulfur and the electrolyte solvent contacted each other to cause a side reaction. Fortunately, however, only a thin passive film was formed, and the subsequent cycles behaved in one step where the electrolyte and sulfur do not contact any more.

Among micropores, the relationship between pores and electrolyte is an important factor, and in order to suppress elution and obtain stable cycle characteristics, a structure that prevents sulfur in the activated carbon pores from contacting the electrolyte is desirable. As explained above with reference to actual examples, there is a very delicate structure factor for microporous activated carbon having micropores of 2 nm or less, and it is necessary to proceed with development in such consideration.

References

1. Rauh, R. D., Abraham, K. M., Pearson, G. F., Surprenant, J. K., & Brummer, S. B. (1979). A lithium/dissolved sulfur battery with organic electrolyte. *Journal of The Electrochemical Society, 126,* 523.
2. Yamin, H., Penciner, J., Gorenshtain, A., Elam, M., & Peled, E. (1985). The electrochemical behavior of polysulfide in tetrahydrofuran. *Journal of Power Sources, 14,* 129.
3. Yamin, H., & Peled, E. (1983). Electrochemistry of a nonaqueous lithium/sulfur cell. *Journal of Power Sources, 9,* (281).
4. Jung, D. S., Hwang, T. H., Lee, J. H., Koo, H. Y., Shakoor, R. A., Kahraman, R., et al. (2014). Hierarchical porous carbon by ultrasonic spray pyrolysis yields stable cycling in lithium−sulfur battery. *Nano Letters, 14,* 4418.
5. Ji, X., Lee, K. T., & Nazar, L. F. (2009). A highly ordered nanostructured carbon-sulphur cathode for lithium-sulphur batteries. *Nature Materials, 8,* 500.
6. Lai, C., Gao, X. P., Zhang, B., Yan, T. Y., & Zhou, Z. (2009). Synthesis and electrochemical performance of sulfur/highly porous carbon composite. *The Journal of Physical Chemistry C, 113,* 4712.
7. Zhang, B., Qin, X., Li, G. R., & Gao, X. P. (2010). Enhancement of long stability of sulfur cathode by encapsulating sulfur into micropores of carbon. *Energy and Environmental Science, 3,* 1531.
8. Yim, T., Park, M. S., Yu, J. S., Kim, K. J., Im, K. Y., Kim, J. H., et al. (2013). Effect of chemical reactivity of polysulfide toward carbonate-based electrolyte on the electrochemical performance of Li-S batteries. *Electrochimica Acta, 107,* 454.
9. Wang, D. W., Zhou, G., Li, F., Wu, K. H., Lu, G. Q., Cheng, H. M., & Gentle, I. R. (2012). A microporous–mesoporous carbon with graphitic structure for a high-rate stable sulfur cathode in carbonate solvent-based Li–S batteries. *Physical Chemistry Chemical Physics, 14,* 8703.
10. Xin, S., Gu, L., Zhao, N.-H., Yin, Y.-X., Zhou, L.-J., Guo, Y.-G., & Wan, L.-J. (2012). Smaller sulfur molecules promise better lithium−sulfur batteries. *Journal of The American Chemical Society, 134,* 18510.
11. Moon, S., Jung, Y. H., Jung, W. K., Jung, D. S., Choi, J. W., & Kim, D. K. (2013). "Encapsulated monoclinic sulfur for stable cycling of Li–S rechargeable batteries. *Advanced Materials, 25,* 6547.
12. Song, J., Xu, T., Gordin, M. L., Zhu, P., Lv, D., Jiang, Y.-B., et al. (2014). Nitrogen-doped mesoporous carbon promoted chemical adsorption of sulfur and fabrication of high-areal-capacity sulfur cathode with exceptional cycling stability for lithium-sulfur batteries. *Advanced Functional Materials, 24,* 1243.

13. Yang, C.-P., Yin, Y.-X., Ye, H., Jiang, K.-C., Zhang, J., & Guo, Y.-G. (2014). Insight into the effect of boron doping on sulfur/carbon cathode in lithium–sulfur batteries. *ACS Applied Materials and Interfaces, 6,* 8789.
14. Hayashi, M., & Kawashita, Y. (2011). *Handbook of activated carbon: Structure control and characterization*. Maruzen.
15. Molina-Sabio, M., Almansa, C., & Rodriguez-Reinoso, F. (2003). Phosphoric acid activated carbon discs for methane adsorption. *Carbon, 41,* 2113.
16. Takahashi, T., Yamagata, M., & Ishikawa, M. (2015). A sulfur–microporous carbon composite positive electrode for lithium/sulfur and silicon/sulfur rechargeable batteries. *Progress in Natural Science: Materials International, 25,* 612
17. Usuki, S., Uchida, S., Matsui, Y., Yamagata, M., Hinago, H., & Ishikawa, M. (2017). Preparation of micropore-rich high surface area activated carbon from N-doped carbon precursor and its application to positive electrode in lithium-sulfur battery. *Electrochemistry, 85,* 650.
18. Okabe, S., Uchida, S., Matsui, Y., Yamagata, M., & Ishikawa, M. (2017). Performance enhancement of rechargeable sulfur cathode utilizing microporous activated carbon composite. *Electrochemistry, 85,* 671.
19. Sing, K. S. W., Everett, D. H., Haul, R. A. W., Moscou, L., Pierotti, R. A., Rouquerol, J., & Siemieniewska, T. (1985). Reporting physisorption data for gas/solid systems with special reference to the determination of surface area and porosity. *Pure and Applied Chemistry, 57,* 603.
20. Mikhaylik, Y. V., & Akridge, J. R. (2004). Polysulfide shuttle study in the Li/S battery system. *Journal of The Electrochemical Society, 151,* A1969.

Li_2S Cathode

Toshikatsu Kojima and Nobuhiko Takeichi

Abstract The recent research studies of the Li_2S cathodes are reviewed. The cathode showed a high capacity of about 1000 mAh/g, and maintaining about 600 mAh/g after one hundred cycles thanks to nanocompositing, carbon coating, and new electrolyte components or additives. Expensive Li_2S production can be reduced by the heat-treatment of cheap carbon and Li_2SO_4. Precise investigations of the synthesis methods of the Li_2S-graphene composite are described.

Keywords Li_2S · Cathode · Graphene · Li_2SO_4

1 Short Review: Investigations of Li_2S Cathode

Sulfur is one of the most promising cathode materials for lithium-ion batteries, because it has a high capacity of 1672 mAh/g and is cheap, abundant, and benign to the environment. It needs, however, anode materials such as ones that contain lithium or lithium metal itself. There have been obstacles to the realization of the lithium anode because of dendrites and short-circuits. Although there have been many research studies involving sulfur cathode materials, only a few have involved the activity for a large-scale Li–S battery. On the other hand, the sulfur changed to Li_2S when it is discharged and captures two lithium ions. Li_2S itself also has a high capacity of 1166 mAh/g, and it has been recognized as one of the promising cathode materials. Cathodes containing Li_2S can provide the lithium ion, therefore, conventional anode materials such as graphite, SiO, and Si can be combined to make a battery. Recently, Li_2S-related cathodes [1] are gathering attention and not only sulfur but also the Li_2S cathode have entered the stage to be compared to commercial LIBs [2].

T. Kojima (✉) · N. Takeichi
Research Institute of Electrochemical Energy, National Institute of Advanced Industrial Science and Technology (AIST), 1-8-31 Midorigaoka, Ikeda 563-8577, Osaka, Japan
e-mail: kojima-t@aist.go.jp

N. Takeichi
e-mail: n.takeichi@aist.go.jp

K. Kanamura (ed.), *Next Generation Batteries*,
https://doi.org/10.1007/978-981-33-6668-8_33

Several Li_2S cathodes were studied in all-solid-state batteries [3–9]. Cathodes of Li_2S with acetylene black, Cu, FeS_2, or graphite made by milling have been examined but the capacities of Li_2S were not fully extracted. This may be due to the large volume change during charge/discharge producing disconnections between the particles of the active material and solid electrolyte.

As for the liquid electrolyte cell, Yang et al. reported that a high potential barrier (~1 V) exists at the beginning of the Li_2S charging. By applying a higher voltage cutoff, this barrier can be overcome and Li_2S becomes active, and after the first activation, subsequent charge processes need no activation [10, 11]. For the sake of the activation at a higher voltage, the electrolytes and liquids used for Li_2S cathodes are very different from ordinary LIBs. Not $LiPF_6$, but LiTFSI is mainly used for the Li_2S cathode. Common electrolyte liquid components, such as EC, DMC, and PC, were scarcely used, instead, ethers, such as 1,2-dimethoxyethane (DME), 1,3-dioxolane (DOL), 1,3-dioxane (DIOX), and tetra(ethylene glycol) dimethylether (TEGDME), have been used. Ionic liquid, such as N-methyl-N-butylpyrrolidinium bis(trifluoromethane sulfonyl)imide (PYR14TFSI), have also been used.

It is also important for Li_2S to have a small particle size and make good contact with the conductive additive in order to have an electric contact. One method is simple milling or slurry making with the Li_2S and conductive additives [10, 12–17].

Coating methods have often been used. Li_2S can be dissolved in ethanol and the solution used for Li_2S coating on carbon nanotubes, graphene, graphene oxide (GO), reduced graphene oxide (rGO), etc. [18–25].

Sulfur has also been well investigated for coating several conductive materials. Sulfur coating several materials such as mesoporous carbon, graphene, GO, etc., was reduced using n-butyllithium, LiH, lithium powder, or $LiEt_3BH$ to form fine Li_2S suitable for electrochemical reactions [26–32].

Organo-Li_2S composites have been made by heat treatment using PAN, PVP, and rGO with Li_2S or Li_2S_6 [33–36].

From the viewpoint of the realization of a Li_2S cathode, the expensive Li_2S production should be economically scaled-up. Cheap and scalable Li_2S-carbon composition syntheses have been developed by heat-treating the mixture of carbon and Li_2SO_4 [37–41]. These materials showed about 800 mAh/gLi_2S.

Recent investigations of the Li_2S cathodes are summarized in Table 1. The cathode showed a high capacity of about 1000 mAh/g, and maintaining about 600 mAh/g after one hundred cycles due to nanocompositing, carbon coating, and using new electrolyte components or additives. The performance and durability of the Li_2S cathodes are steadily improving, therefore, a cheap and massive production technology for Li_2S itself or Li_2S cathodes will be important. In the following subsection, we will present an investigation of the Li_2S cathode production.

Table 1 Reported Li_2S cathodes, preparation, methods, capacity, and cyclability

Author	Refs.	Cathode composition (preparation, method)	Electrolyte	Capacity @ 1st	Capacity @nth cycles
Takeuchi T	3	Li_2S-AB (milling and Spark-Plasma-Sintering, all-solid-state cell)	Li_3PO_4–Li_2S–SiS_2	920 mAh/gLi_2S	740 mAh/gLi_2S@10
Takeuchi T	4	Li_2S-AB (8:2weight ratio) (milling and Spark-Plasma-Sintering)	1MLiTFSI in TEGDME[a]	370 mAh/g	350 mAh/g@10
Takeuchi T	5	Li_2S-Fe_2S composite (milling and Spark-Plasma-Sintering)	1M$LiPF_6$(EC/DMC1:1)	740 mAh/g	500mAh/g@10
Takeuchi T	6	Li_2S-C composite (milling and Spark-Plasma-Sintering, Li_2S/Graphite all-solid-state full cell)	70Li_2S-30P_2S_5	740 mAh/gLi_2S	–
Hayashi A	7	Li_2S-Cu(3:1molar ratio), (milling, In anode, all-solid-state cell)	80Li_2S-20P_2S_5	490 mAh/g	360 mAh/g@20
Nagao M	8	Li_2S:AB:solid electrolyte (25:25:50 weight ratio) (milling, In anode, all-solid-state cell)	75Li_2S-25P_2S_5	620 mAh/gLi_2S	650 mAh/gLi_2S@10
Eom M	9	Li_2S-VGCF nanocomposite (EtOH soln. coating, In-Li anode, all-solid-state cell)	78Li_2S-22P_2S_5	470 mAh/gLi_2S	590 mAh/gLi_2S@700
Yang Y	10	Slurry of Li_2S:Super P:PVdF (40:45:15wt%) in NMP coated on carbon paper	1MLiTFSI in DME:DOL1:1 + 0.2MLi_2S_8	600 mAh/g	430 mAh/g@50

(continued)

Table 1 (continued)

Author	Refs.	Cathode composition (preparation, method)	Electrolyte	Capacity @ 1st	Capacity @nth cycles
Liu J	12	Li_2S-KB (8:2weight ratio) (milling, slurry cast with AB, PVP, and PEO)	$LiCF_3SO_4$,TEGDME,$LiNO_3$	740 mAh/gLi_2S	460 mAh/gLi_2S@50
Cai K	13	Li_2S:DenkaBlack:MWCNT (75:20:5 weight ratio) (milling, SBR binder)	1MLiTFSI in 1:1TEGDME:PYR14TFSI[b]	1144 mAh/gLi_2S	600 mAh/gLi_2S@50
Meini S	14	Slurry of Li_2S:carbon black:PVdF(59:33:8wt%) in NMP coated on Al disk.(decamethylferrocene as reaction mediator additive to electrolyte)	10%LiTFSI in DME:DOL1:1 + 2%$LiNO_3$	750 mAh/gLi_2S	500 mAh/gLi_2S@150
Xu R	15	Li_2S:carbon:PVdF (40:50:10wt%)	1MLiTFSI in DME:DOL1:1 + 1wt%$LiNO_3$	1300 mAh/gLi_2S	1000 mAh/gLi_2S @65
Hoanga VC	16	Slurry of Li_2S:SuperP:PVDF(60:30:10) coated on Al foil then graphene slurry was coated	1MLiTFSI in DME:DOL1:1 + $LiNO_3$	1050 mAh/gLi_2S	600 mAh/gLi_2S@50
Fu Y	17	Li_2S -CNT paper sandwiched electrode	1 M $LiCF_3SO_3$ in DME:DOL1:1 + 0.1M$LiNO_3$	838 mAh/g	680 mAh/g @100
Wu F	18	MWCNT-linked Li_2S(EtOH soln.deposition)	7MLiTFSI in 1:1DME: DIOX[c]	900 mAh/gLi_2S	540 mAh/gLi_2S@50
Wu F	19	Li_2S- graphene (ethanol soln. of Li_2S coating and CVD carbon coating, PVP binder)	1MLiTFSI in 1:1DME: DIOX + LiI	800 mAh/gLi_2S	750 mAh/gLi_2S@700

(continued)

Table 1 (continued)

Author	Refs.	Cathode composition (preparation, method)	Electrolyte	Capacity @ 1st	Capacity @nth cycles
Qiu Y	20	Li_2S-nitridated graphene (ethanol soln. of Li_2S_3 coating and calcination, PVP binder)	1MLiTFSI in 1:1DME: DIOX + $LiNO_3$	850 mAh/gLi_2S	500 mAh/gLi_2S@500
Chen L	21	N-doped carbon coated-Li_2S (nano Polystyrene template coated Li_2S in EtOH soln.then Pyrrole coating, heat-treated at 600 °C)	1MLiTFSI in DME:DOL1:1 + 1wt%$LiNO_3$	915 mAh/g	500 mAh/g@300
Wang C	22	Li_2S-reduced graphene oxide paper (EtOH soln. coating)	0.6MLiTFSI in DME:DOL1:1 + 0.4M$LiNO_3$	1102 mAh/gLi_2S	462 mAh/gLi_2S@200(5C)
Wang DH	23	Li_2S-PVP-3D rGO (EtOH soln. coating, carbonization using PVP) (rGO: reduced graphene oxide)	1MLiTFSI in DME:DOL1:1 + 1wt%$LiNO_3$	856 mAh/g	563 mAh/g@100
He J	24	Li_2S flexible 3D graphene foam (EtOH soln. coat)	1MLiTFSI in DME:DOL1:1 + 0.2M$LiNO_3$	942 mAh/g	775 mAh/g@100
Seh ZW	25	Li_2S-GO (Li_2S and GO (75: 25 wt%) mixed in ethyl acetate)	1MLiTFSI in DME:DOL1:1 + 1wt%$LiNO_3$	782 mAh/gLi_2S	690 mAh/gLi_2S @150
Yang Y	26	Li_2S-mesoporous carbon(S + mesoporous carbon, n-butyllithium reduction)	1MLiTFSI in DME:DOL1:1	580 mAh/g	350 mAh/g@20
Zhang K	27	Li_2S-thermally exfoliated graphene nanosheet (TG) (TG + S, $LiEt_3BH$ reduction)	1MLiTFSI in DME:DOL1:1 + 0.2M$LiNO_3$	1119 mAh/gLi_2S	200 mAh/gLi_2S@1500

(continued)

Table 1 (continued)

Author	Refs.	Cathode composition (preparation, method)	Electrolyte	Capacity @ 1st	Capacity @nth cycles
Nan C	28	Carbon coated Li_2S (S *submicron particle*, $LiEt_3BH$ reduction, CVD carbon coating)	1MLiTFSI in 2:1:1PYR14TFSI:DOL:DME + 1wt% $LiNO_3$	972 mAh/gLi_2S	417 mAh/g@400
Hwa Y	29	Carbon coated Li_2S-GO composite(S + GO, $LiEt_3BH$ reduction, CVD carbon coating)	1MLiTFSI in 2:1:1PYR14TFSI:DOL:DME	650 mAh/gLi_2S	200 mAh/gLi_2S@1500
Jiao Z	30	Core–shell $Li_2S@Li_3PS_4$ + graphene aerogel (S reduced by $LiEt_3BH$, $3Li_2S + P_2S_5 = 2Li_3PS_4$ in THF, dispersed in graphene aerogel)	1MLiTFSI in DME:DOL1:1 + 0.1 mol/L$LiNO_3$	934 mAh/g	486 mAh/g@100
Li X	31	Li_2S-heat-treated PAN (S + PAN + LiH, milling, heat-treated at 1000 °C)	1MLiTFSI in DME:DOL1:1 + 2wt%$LiNO_3$	971 mAh/g	570 mAh/g@200
Zheng S	32	Li_2S/ microporous carbon (Sulfur /microporous carbon reduced by Li powder)(Li_2S/MC-graphite full cell, 570 mAh/g@150cy)	1.0 M $LiPF_6$ in EC/DEC 1:1	770 mAh/g	670 mAh/g@900
Guo J	33	Li_2S-heat-treated PAN(Li_2S + S + PAN, heat-treated at 600 °C)	1MLiTFSI in 1:1PYR14TFSI: TEGDME	870 mAh/gLi_2S (10 mA/g)	510 mAh/g@20(200 mA/g)

(continued)

Table 1 (continued)

Author	Refs.	Cathode composition (preparation, method)	Electrolyte	Capacity @ 1st	Capacity @nth cycles
Wu F	34	Heat-treated PVP-coated CB-Li_2S (PVP and CB-Li_2S mixed in Pyrrole, heat-treated at 600 °C) (CB:carbon black)	LiTFSI in DME:DOL1:1	1560 mAh/gS	1200 mAh/gS@100
Chen L	35	Heat-treated PVP coated-Li_2S (PVP and Li_2S composite made in EtOH soln. heat-treated at 700 °C)	1MLiTFSI in DME:DOL1:1 + 1wt%$LiNO_3$	1020 mAh/g	665 mAh/g@200
Han K	36	Li_2S-rGO (rGO + Li2S + S, heat-treated at 700 °C) (rGO:reduced graphene oxide)	1MLiTFSI in DME:DOL1:1 + 0.2M$LiNO_3$ + 0.1MLi_2S	982 mAh/gLi_2S	315 mAh/gLi_2S@100
Yang Z	37	Li_2S-C (Li_2SO_4 + resorcinol + HCHO composite, heat-treatment at 900 °C)	1MLiTFSI in tetraglyme + $Li_2S_{3.5}$	320 mAh/g	300 mAh/g@40
Wang DH	38	Li_2S-carbon paper (Li_2SO_4 reduction using carbon paper, CVD *carbon coating)*	1MLiTFSI in DME:DOL1:1 + 1wt%$LiNO_3$	820 mAh/g	430 mAh/g@100
Liu J	39	Li_2S-ketjen black (Li_2S:KB = 8:2 milling, Li_2S-KB:AB:PVP:PEO = 75:10:10:5)	$LiCF_3SO_3$: TEGDME = 1:4mole ratio, + 1wt%$LiNO_3$	740 mAh/g	460 mAh/g@50
Li Z	40	Li_2S-graphene (Li_2SO_4 coated graphene, Li_2SO_4 reduction using graphene at 781 °C)	LiTFSI: tetraglyme(1:1) + HFE[d]	693 mAh/gLi_2S	508 mAh/gLi_2S@40

(continued)

Table 1 (continued)

Author	Refs.	Cathode composition (preparation, method)	Electrolyte	Capacity @ 1st	Capacity @nth cycles
Li Z	41	Li_2S-graphene (Li_2SO_4 coated graphene, Li_2SO_4 reduction using graphene at 781 °C, milling, Li_2S/graphite full cell)	LiTFSI: tetraglyme(1:0.8) + HFE	809 mAh/gLi_2S	450 mAh/gLi_2S@100

[a]TEGDME:tetra(ethylene glycol) dimethylether, [b]PYR14TFSI: N-methyl-N-butylpyrrolidinium bis(trifluoromethane sulfonyl)imide, [c]DIOX: 1,3-dioxane, [d]HFE: 1,1,2,2-tetrafluoroethyl 2,2,3,3-tetrafluoropropyl ether

2 Li_2S Synthesis

Pure Li_2S is synthesized by the reaction of sulfur and lithium in liquid ammonia. Pure chemicals are very expensive and the price is getting higher and higher because Li_2S is utilized as one of the base materials for solid electrolytes with high conductivity, such as Li_6PS_5Cl or $Li_{10}GeP_2S_{12}$. Lower cost synthesis methods have been investigated. They utilize hydrogen sulfide, which is produced during petroleum refining, with lithium salts, i.e., LiOH, Li_2O, and Li_2CO_3. These synthesis methods can be realized only in a chemical plant in which deadly poisonous hydrogen sulfide can be utilized. In order to obtain the Li_2S cathode material, the following reaction (1) is suitable for a small scale and is safe in a laboratory setup [37–41], in which relatively cheap Li_2SO_4 was reduced by carbon to Li_2S at 780–900 °C.

$$Li_2SO_4 + 2C = Li_2S + 2CO_2 \quad (1)$$

From the point of view of an electrochemical reaction, non-conductive Li_2S as well as sulfur needs a conductive additive or be composited with conductive carbonaceous materials not only for conduction, but also prevent dissolving the Li_2S_x into the electrolyte. Cathodes of the Li_2S composite have been investigated.

The sulfur cathode is much easier to discharge or charge than the Li_2S one, therefore, there are a greater number of reports about the sulfur electrode properties and performance than that of Li_2S. On the other hand, Li_2S is very hygroscopic thus it should be handled only in a glove box, it is difficult to be dispersed in the electrode and needs a high voltage activation in the first charge [10, 11]. Li (Zhe) et al. succeeded in obtaining good performance of the Li_2S covered graphene composite by reaction (1) at 783 °C under flowing Ar [40, 41]. They first made a Li_2SO_4 covered graphene composite by recrystallization of Li_2SO_4 on a graphene surface from a Li_2SO_4 aqueous solution by dropwise adding ethanol. In the composite, Li_2S was well dispersed on the graphene and this provided a high reversible capacity of about 800 mAh/g and good cyclability, therefore, the composite is one of the best Li_2S cathode materials.

On an experimental scale, the electrochemical cell utilizes an amount of about 2 mg of cathode material (total capacity, 1.6 mAh). In the Advanced Low Carbon Technology Research and Development Program (ALCA), materials should be evaluated from the point of view of an actual battery. The materials are made to be a slurry-coated electrode for a 1 Ah coffee-bag cell with about a 1.25 g cathode active material, combined with a real anode such as graphite [6, 32, 41], SiO, or Si. The batteries also undergo not only charging/discharging but also safety tests. For such battery production and evaluation, an amount of at least 100 g of active material is needed. At this time, the material and work require an experimental scale-up production of 1 g by 100 times to 100 g. We have encountered many problems with the scale-up synthesis such as byproduct impurities, choice, and preparation of scale-up reactor vessels, which will be mentioned later. For industrial production, maintaining the performance of the material, scale-up, abbreviation of procedures, decreasing energy and materials, and total cost reduction should be considered. Some new, refined, and changed methods have been done. Although only a 100 g production is the first step, the discussions about our scale-up activity will provide knowledge for an industrial scale-up. In this article, we report how to make the Li_2S-graphene composite.

2.1 *The Original Experimental Method of the Li_2S-graphene Composite*

The method by Li et al. [40, 41] regarding the original experimental method of the Li_2S-graphene composite is depicted in Fig. 1. In this method, ethanol was dropwise added in order to gradually decrease the solubility of the lithium sulfate in an aqueous solution then, $Li_2SO_4 \cdot H_2O$ was precipitated and covered the surface of the dispersed graphene to produce a precursor of the $Li_2SO_4 \cdot H_2O$-graphene composite. The precursor was heat treated at 783 °C under flowing Ar, then the SO_4^{2-} ion was reduced by carbon of the graphene to produce CO_2 and Li_2S remained on the surface of the residual graphene.

The advantages of the original method are that the thin Li_2S covered on the large area provided by the graphene which also acts as the current collector. However, the production of the method is only 1.3 g. There were several obstacles to the scale-up production method in order to supply the material for coffee-bag battery testing.

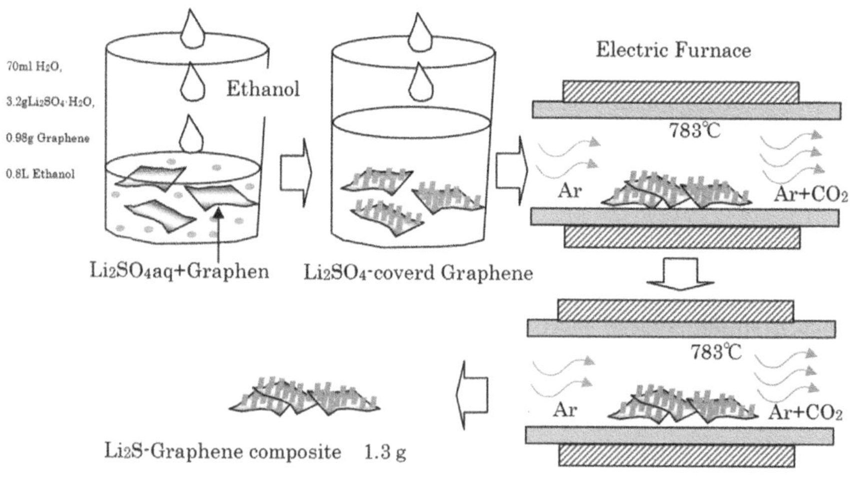

Fig. 1 Original synthesis scheme of the Li_2S-graphene composite [40, 41]

2.2 *The Obstacles for the Scale-Up Synthesis*

We followed the original method [41] for making the precursor of the $Li_2SO_4 \cdot H_2O$-graphene composite because the method can evenly coat the $Li_2SO_4 \cdot H_2O$ on the graphene and this directly affects the performance of the obtained composite. In order to scale-up, the concentration of lithium sulfate and the amount of graphene in the precipitation process was ten times that of the original method (Fig. 1). Graphene of 9.6 g was dispersed in a 140 ml solution of water containing 32 g of $Li_2SO_4 \cdot H_2O$, then 2 L of ethanol was dropwise added to obtain the $Li_2SO_4 \cdot H_2O$-graphene composite. After drying the composite, the 32 g had a relatively large volume, which required a larger electric furnace.

The precursor was calcined at 783 °C under flowing Ar, then the Li_2S-graphene composite was obtained. An XRD of the obtained product is shown in Fig. 2. It can be seen that a large part of the product changed to Li_2S, however, some impurities such as remaining raw material, Li_2SO_4 (PDF 01-073-1250), and the byproducts of Li_2O (PDF 00-012-0254) and Li_2CO_3 (PDF 00-022-1141). In the original method,

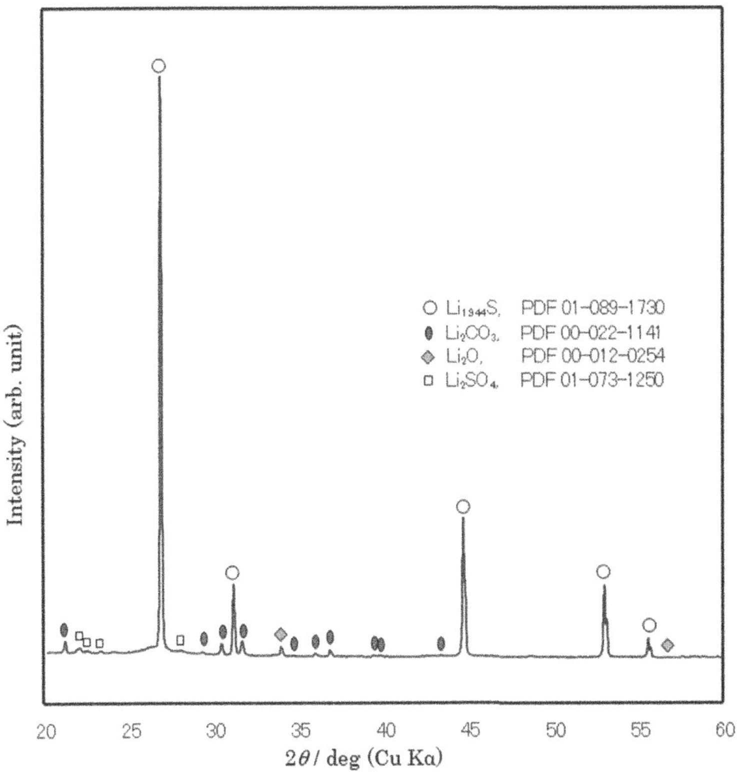

Fig. 2 XRD of the synthesized Li_2S and identified the impurity phases

only Li_2O was the byproduct. These impurities suggested that the scale-up precursor should be needed some cares.

In order to reduce the rest of the lithium sulfate, milling of the precursor for 1 h at 500 rpm using five 20mm ϕ zirconia balls and a zirconia pot (Frich, PULVERISETTE 6) was found to be effective. This process produced good contact and dispersion between the graphene and lithium sulfate.

The mechanisms of the Li_2O and Li_2CO_3 production should be the following reactions (2) and (3).

$$Li_2S + CO_2 = Li_2O + CO + S \tag{2}$$

$$Li_2O + CO_2 = Li_2CO_3 \tag{3}$$

The carbon dioxide, which is produced by the reduction reaction of lithium sulfate by carbon, can oxidize the Li_2S to sulfur and lithium oxide, and react with the Li_2O to form lithium carbonate. The sulfur can be seen as a deposit in the outlet gas pipe, where the gas is cooled below about 150 °C. Such a deposition of sulfur would block the gas from the reactor vessel, which should be paid special caution such as using an outlet pipe with a larger inner diameter, cleaning the outlet pipe, and monitoring the pressure in the reactor vessel.

The origin of the byproducts of Li_2O and Li_2CO_3 should be CO_2, therefore, it was considered effective to reduce these byproducts by removing the CO_2 as quickly as possible. It is effective to use the reaction under vacuum conditions.

In some cases under flowing Ar, a ball-shaped lump of 3-mm diameter can be seen, thought to be a mixture of Li_2S–Li_2SO_4–Li_2CO_3, that melted at the reaction temperature. We tried to use vacuum conditions instead of the Ar flow conditions.

In order to remove the composite of CO_2, a heat-treatment chamber under vacuum was employed as shown in Fig. 3. There happened to be another problem for the heat treatment under the vacuum conditions. The CO_2 gas accumulated in the powder and the powder of Li_2SO_4 -graphene was blown out by the CO_2 gas to the vacuum pump. Making pellets of Li_2SO_4-graphene could prevent the blow-out of the powder. Pelletizing the precursor also effectively contacts the Li_2SO_4 and graphene. There is little space for the accumulation of the evolved CO_2 gas, and instead, small cracks help the evolved CO_2 gas to escape.

The 10-times scale-up synthesis procedures are summarized in Fig. 4. The new procedures are milling of the Li_2SO_4-graphene composite, followed by pelletizing, and heat treatment under vacuum conditions. The synthesis procedures provided 12.6 g of the Li_2S-graphene composite, which contained relatively reduced Li_2O as shown in Fig. 5a.

It was reported that [41] in order to enhance the performance, they milled 0.4 g of the Li_2SO_4-graphene composite using a Fritsch PULVERISETTE 7, 192 sets at 600 rpm for 10 and 5 min rest using 120 balls of 4 mm ϕ. It took about 48 h for the 0.4 g milling. It was difficult to mill a large amount of the composite for such a long time, so we only milled the composite using the Fritsch Pulverisette 6, 20 sets at 500 rpm

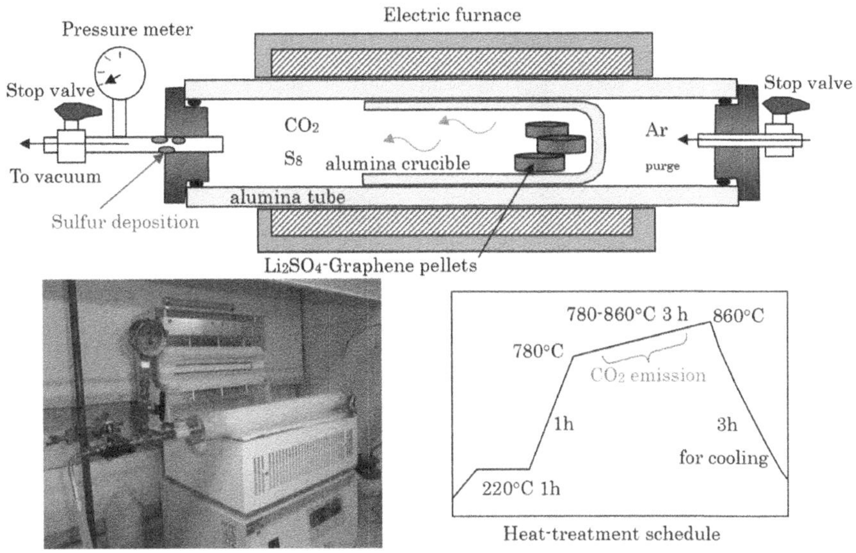

Fig. 3 Reaction chamber, alignment, and heat-treatment schedule for the Li_2S-graphene synthesis

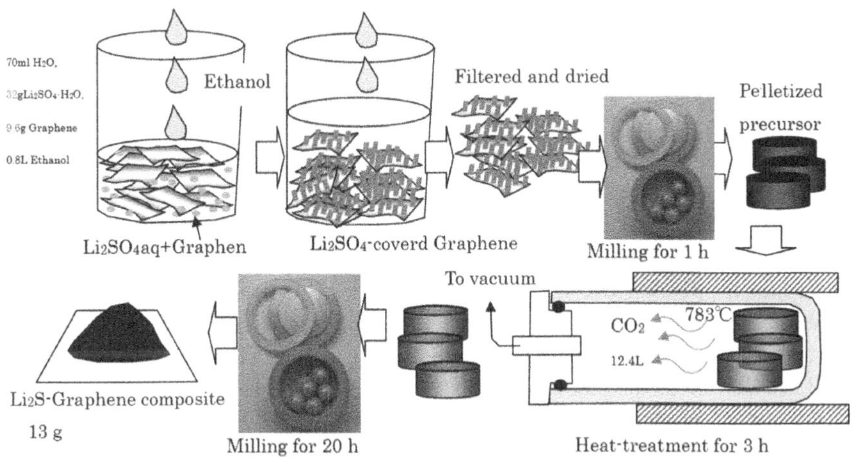

Fig. 4 Scale-up synthesis scheme

for 60 and 5 min rest using 5 balls of 20 mm ϕ. After this treatment, the XRD profile of the Li_2S was broadened as shown in Fig. 5b. The performance of the composite was evaluated by Prof. Watanabe's research group [40, 41] and their results showed about an 80% capacity of the original composite [41]. The decrease in capacity was thought to originate from the short milling treatment. We ran the processes 12 times and succeeded in producing 100 g of the Li_2S-graphene composite.

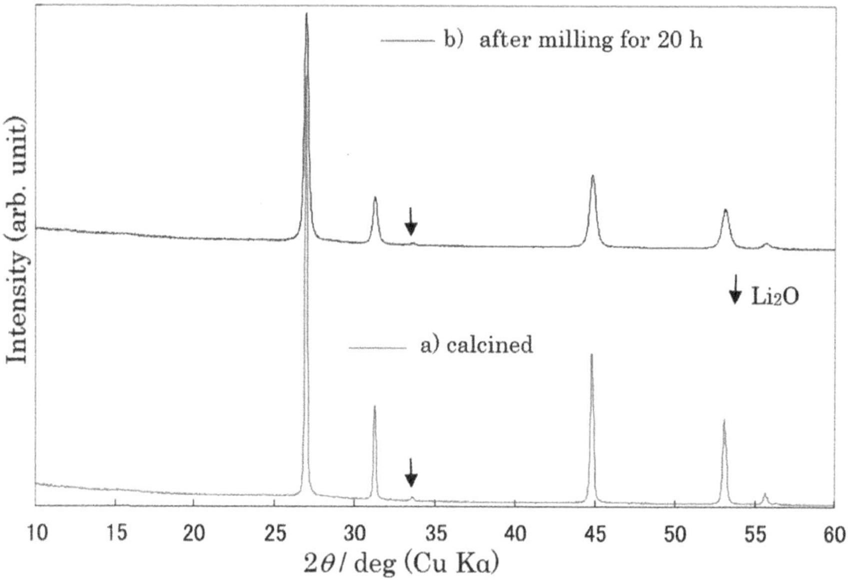

Fig. 5 The effect of milling: XRD of Li_2S-graphene composite **a** calcined under vacuum condition and **b** obtained after milling for 20 h

2.3 *Investigation for a More Convenient Method*

The described precursor production of Li_2SO_4-graphene consists of many processes such as dispersion and stirring of the graphene, dropwise ethanol addition, precipitation, filtration, drying, milling, pelletizing, which takes a total of 3 days and 2 L of expensive ethanol to make 12.5 g of the Li_2S-graphene composite. Therefore, in future industrial production, it will be favorable to only mix the Li_2SO_4 and graphene.

We also tried to develop a mixing precursor production using a larger milling pot of 500 ml and four balls of 40 mm ϕ set in the Fritsch PULVERISETTE 6 at 350 rpm for 15 h. Ten times the amount of the precursor [41], 32 g $Li_2SO_4{\cdot}H2O$, and 9.8 g of a graphene mixture was obtained in one batch and one night process. We also obtained almost the same Li_2S-Graphene composite with little byproduct as shown in Fig. 5 by the vacuum heat treatment in Fig. 4.

It took one day for the heat-treatment process because the bulky precursor needed a relatively large electric furnace, which took 2 h for heat up, 3 h for the reaction, and 4 h for cool down. Another problem was air exposure of the obtained Li_2S-Graphene composite during its transfer to a glove box. We also found a convenient method for the heat treatment and preventing air exposure as shown in Fig. 6. The temperature of the gold crucible in a Potable Vacuum Reaction Vessel (PVRV) can reach 900 °C in ten minutes. The reaction from Li_2SO_4 to Li_2S can be monitored using a pressure meter, which indicates the pressure of evolved CO_2 during the reaction when the outlet valve is closed. It was found that the heat treatment of 1 g of precursor takes

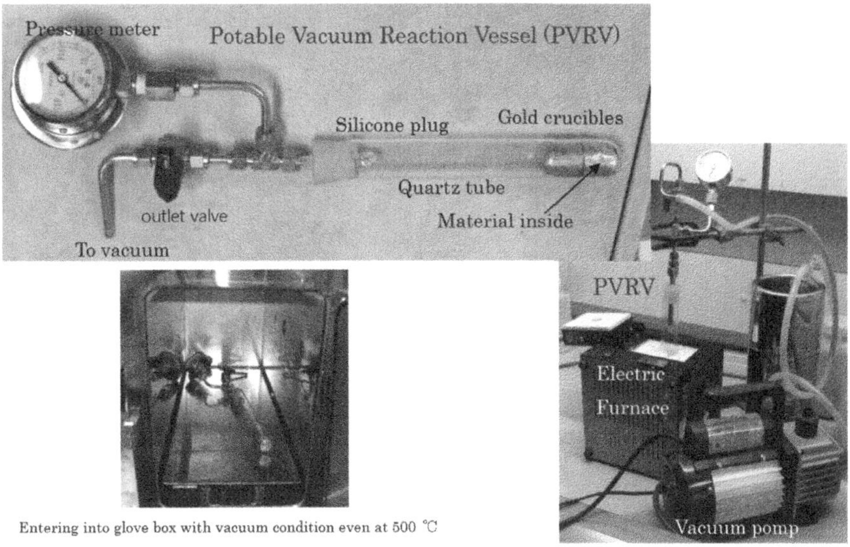

Fig. 6 Pictures of reaction chamber and alignment for the Li_2S-graphene synthesis

only 15 min. The PVRV can be transferred into a glove box even at 500 °C with the product keeping inside and under vacuum, therefore, the product is free from air exposure.

The heat-treatment time was decreased from 9 h to within 0.5 h. The quick heat treatment also reduced the byproduct, such as Li_2O, as shown in Fig. 7.

Although the PVRV can be scaled-up, the heat treatment of 8 g of precursor took 1 h to complete the reaction. It should be noted that more precursor takes more time because the gas evolving reaction requires heat to occur.

3 Summary and Future Prospect

The original method of the 1.3 g Li_2S-Graphene composite production was partly changed to a 10 times scale-up to 12.5 g in one batch, milling and pelletizing of the Li_2SO_4-graphene precursor, and heat treatment under vacuum conditions. Based on this scale-up method and refinement, we succeeded in obtaining 100 g of the Li_2S-Graphene composite necessary for making the coffee-bag batteries. We also developed convenient methods such as the precursor milling production and portable vacuum reaction vessel. This method will be developed for kg and ton unit production and also used for new material developments.

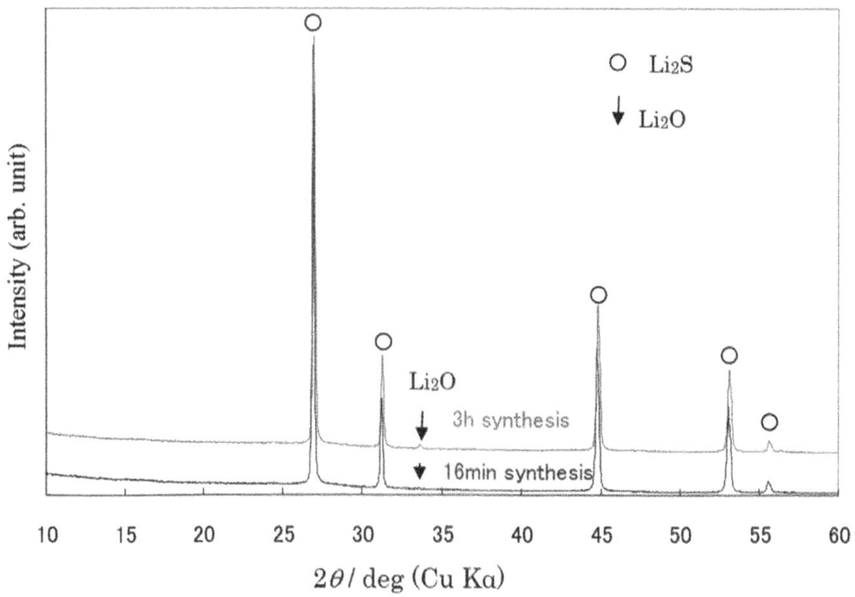

Fig. 7 XRD of Li_2S-graphene produced by a 3-h synthesis and 16-min synthesis

References

1. Son, Y., Lee, J. S., Son, Y., Jang, J. H., & Cho, J. (2015). Recent advances in lithium sulfide cathode materials and their use in lithium sulfur batteries. *Advanced Energy Materials, 5,* 1500110.
2. Berg, E. J., & Trabesinger, S. (2018). Viability of polysulfide-retaining barriers in Li–S battery. *Journal of the Electrochemical Society, 165,* A5001–A5005.
3. Takeuchi, T., Kageyama, H., Nakanishi, K., Tabuchi, M., Sakaebe, H., Ohta, T., et al. (2010). All-solid-state lithium secondary battery with Li_2S–C composite positive electrode prepared by spark-plasma-sintering process. *Journal of the Electrochemical Society, 157,* A1196–A1201.
4. Takeuchi, T., Sakaebe, H., Kageyama, H., Senoh, H., Sakai, T., & Tatsumi, K. (2010). Preparation of electrochemically active lithium sulfide–carbon composites using spark-plasma-sintering process. *Journal of Power Sources, 195,* 2928–2934.
5. Takeuchi, T., Kageyama, H., Nakanishi, K., Inada, Y., Katayama, M., Ohta, T., et al. (2012). Improvement of cycle capability of FeS_2 positive electrode by forming composites with Li_2S for ambient temperature lithium batteries. *Journal of the Electrochemical Society, 159,* A75–A84.
6. Takeuchi, T., Kageyama, H., Nakanishi, K., Ohta, T., Sakuda, A., Sakaebe, H., et al. (2014). Rapid preparation of Li_2S-P_2S_5 solid electrolyte and its application for graphite/Li_2S all-solid-state lithium secondary battery. *ECS Electrochemistry Letters, 3,* A31–A55.
7. Hayashi, A., Ohtsubo, R., Ohtomo, T., Mizuno, F., & Tatsumisago, M. (2008). All-solid-state rechargeable lithium batteries with Li_2S as a positive electrode material. *Journal of Power Sources, 183,* 422–426.
8. Nagao, M., Hayashi, A., Tatsumisago, M., Ichinose, T., Ozaki, T., Togawa, Y., & Mor, S. (2015). Li_2S nanocomposites underlying high-capacity and cycling stability in all-solid-state lithiumesulfur batteries. *Journal of Power Sources, 274,* 471–476.

9. Eom, M., Son, S., Park, C., Noh, S., Nichols, W. T., & Shin, D. (2017). High performance all-solid-state lithium-sulfur battery using a Li_2S-VGCF nanocomposite. *Electrochimica Acta, 230,* 279–284.
10. Yang, Y., Zheng, G., Misra, S., Nelson, J., Toney, M. F., & Cui, Y. (2012). High-capacity micrometer-sized Li_2S particles as cathode materials for advanced rechargeable lithium-ion batteries. *Journal of the American Chemical Society, 134,* 15387–15394.
11. Vizintin, A., Chabanne, L., Tchernychova, E., Arcon, I., Stievano, L., Aquilanti, G., et al. (2017). The mechanism of Li_2S activation in lithium-sulfur batteries: Can we avoid the polysulfide formation? *Journal of Power Sources, 344,* 208–217.
12. Liu, J., Nara, H., Yokoshima, T., Momma, T., & Osaka, T. (2015). Li_2S cathode modified with polyvinylpyrrolidone and mechanical milling with carbon. *Journal of Power Sources, 273,* 1136–1141.
13. Cai, K., Song, M. K., Cairns, E. J., & Zhang, Y. (2012). Nanostructured Li_2S–C composites as cathode material for high-energy lithium/sulfur batteries. *Nano Letters, 12,* 6474–6479.
14. Meini, S., Elazari, R., Rosenman, A., Garsuch, A., & Aurbach, D. (2014). The use of redox mediators for enhancing utilization of Li_2S cathodes for advanced Li–S battery systems. *The Journal of Physical Chemistry Letters, 5,* 915–918.
15. Xu, R., Zhang, X., Yu, C., Ren, Y., Li, J., & Belharouak, I. (2014). Paving the way for using Li_2S batteries. *ChemSusChem, 7,* 2457–2460.
16. Hoanga, V. C., Do, V. D., Naha, I. W., Lee, C., Cho, W. I., & Oh, I. H. (2016). Facile coating of graphene interlayer onto Li_2S as a high electrochemical performance cathode for lithium sulfur battery. *Electrochimica Acta, 210,* 1–6.
17. Fu, Y., Su, Y. S., & Manthiram, A. (2014). Li_2S-carbon sandwiched electrodes with superior performance for lithium-sulfur batteries. *Advanced Energy Materials, 4,* 1300655–1300659.
18. Wu, F., Magasinski, A., & Yushin, G. (2014). Nanoporous Li_2S and MWCNT-linked Li_2S powder cathodes for lithium-sulfur and lithium-ion battery chemistries. *Journal of Materials Chemistry, A2,* 6064–6070.
19. Wu, F., Lee, J. T., Zhao, E., Zhang, B., & Yushin, G. (2016). Graphene–Li_2S–carbon nanocomposite for lithium–sulfur batteries. *ACS Nano, 10,* 1333–1340.
20. Qiu, Y., Rong, G., Yang, J., Li, G., Ma, S., Wang, X., et al. (2015). Highly nitridated graphene-Li_2S cathodes with stable modulated cycles. *Advanced Energy Materials, 5,* 1501369.
21. Chen, L., Liu, Y., Dietz-Rago, N., & Shaw, L. L. (2015). Bottom-up, hard template and scalable approaches toward designing nanostructured Li_2S for high performance lithium sulfur batteries. *Nanoscale, 7,* 18071–18080.
22. Wang, C., Wang, X., Yang, Y., Kushima, A., Chen, J., Huang, Y., & Li, J. (2015). Slurryless Li_2S/reduced graphene oxide cathode paper for high-performance lithium sulfur battery. *Nano Letters, 15,* 1796–1802.
23. Wang, D. H., Xie, D., Yang, T., Zhong, Y., Wang, X. L., Xia, X. H., et al. (2016). Li_2S@C composite incorporated into 3D reduced graphene oxide as a cathode material for lithium-sulfur batteries. *Journal of Power Sources, 313,* 233–239.
24. He, J., Chen, Y., Lv, W., Wen, K., Li, P., Qi, F., et al. (2016). Highly-flexible 3D Li_2S/graphene cathode for high-performance lithium sulfur batteries. *Journal of Power Sources, 327,* 474–480.
25. Seh, Z. W., Wang, H., Liu, N., Zheng, G., Li, W., Yao, H., & Cui, Y. (2014). High-capacity Li_2S-graphene oxide composite cathodes with stable cycling performance. *Chemical Science, 5,* 1396–1400.
26. Yang, Y., McDowell, M. T., Jackson, A., Cha, J. J., Hong, S. S., & Cui, Y. (2010). New nanostructured li_2s/silicon rechargeable battery with high specific energy. *Nano Letters, 10,* 1486–1491.
27. Zhang, K., Wang, L., Hu, Z., Cheng, F., & Chen, J. (2014). Ultrasmall Li_2S nanoparticles anchored in graphene nanosheets for high-energy lithium-ion batteries. *Scientific Reports, 4,* 6467.
28. Nan, C., Lin, Z., Liao, H., Song, M. K., Li, Y., & Cairns, E. J. (2014). Durable carbon-coated Li_2S core-shell spheres for high performance lithium/sulfur cells. *Journal of the American Chemical Society, 136,* 4659–4663.

29. Hwa, Y., Zhao, Y. J., & Cairns, E. J. (2015). Lithium sulfide (Li_2S)/graphene oxide nanospheres with conformal carbon coating as a high-rate, long-life cathode for Li/S cells. *Nano Letters, 15,* 3479–3486.
30. Jiao, Z., Chen, L., Si, J., Xu, C., Jiang, Y., Zhu, Y., et al. (2017). Core-shell $Li_2S@Li_3PS_4$ nanoparticles incorporated into graphene aerogel for lithium-sulfur batteries with low potential barrier and overpotential. *Journal of Power Sources, 251,* 331–337.
31. Li, X., Gao, M., Du, W., Ni, B., Wu, Y., Liu, Y., et al. (2017). A mechanochemical synthesis of submicron-sized Li_2S and a mesoporous Li_2S/C hybrid for high performance lithium/sulfur battery cathodes. *Journal of Materials Chemistry A, 5,* 6471–6482.
32. Zheng, S., Chen, Y., Xu, Y., Yi, F., Zhu, Y., Liu, Y., et al. (2013). In situ formed lithium sulfide/microporous carbon cathodes for lithium-ion batteries. *ACS nano, 7,* 10995–11003.
33. Guo, J., Yang, Z., Yu, Y., Abruña, H. D., & Archer, L. A. (2013). Lithium–sulfur battery cathode enabled by lithium–nitrile interaction. *Journal of the American Chemical Society, 135,* 763–767.
34. Wu, F., Kim, H., Magasinski, A., Lee, J. T., Lin, H. T., & Yushin, G. (2014). Harnessing steric separation of freshly nucleated Li_2S nanoparticles for bottom-up assembly of high-performance cathodes for lithium-sulfur and lithium-ion batteries. *Advanced Energy Materials, 4,* 1400196.
35. Chen, L., Liu, Y., Zhang, F., Liu, C., & Shaw, L. L. (2015). PVP-assisted synthesis of uniform carbon coated Li_2S/CB for high-performance lithium-sulfur batteries. *ACS Applied Materials and Interfaces, 7,* 25748–25756.
36. Han, K., Shen, J., Hayner, C. M., Ye, H., Kung, M. C., & Kung, H. H. (2014). Li_2S-reduced graphene oxide nanocomposites as cathode material for lithium sulfur batteries. *Journal of Power Sources, 251,* 331–337.
37. Yang, Z., Guo, J., Das, S. K., Yu, Y., Zhou, Z., Abruña, H. D., & Archer, L. A. (2013). In situ synthesis of lithium sulfide–carbon composites as cathode materials for rechargeable lithium batteries. *Journal of Materials Chemistry A, 1,* 1433–1440.
38. Wang, D. H., Xie, D., Yang, T., Zhong, Y., Wang, X. L., Xia, X. H., Gu, C. D., & Tu, J. P. (2016). Conversion from Li_2SO_4 to $Li_2S@C$ on carbon paper matrix: A novel integrated cathode for lithium-sulfur batteries. *Journal of Power Sources,* 475–480.
39. Liu, J., Nara, H., Yokoshima, T., Momma, T., & Osaka, T. (2015). Micro-scale Li_2S–C composite preparation from Li_2SO_4 for cathode of lithium ion battery. *Electrochimica Acta, 183,* 70–77.
40. Li, Z., Zhang, S., ZhangC, UenoK., YasudaT, TataraR., Dokko, K., & Watanabe, M. (2015). One-pot pyrolysis of lithium sulfate and grapheme nanoplatelet aggregates: In situ formed Li_2S/graphene composite for lithium–sulfur batteries. *Nanoscale, 7,* 14385–14392.
41. Li, Z., Zhang, S., Terada, S., Ma, X., Ikeda, K., Kamei, Y., et al. (2016). Promising cell configuration for next-generation energy storage: Li_2S/graphite battery enabled by a solvate ionic liquid electrolyte. *ACS Applied Materials and Interfaces, 8,* 16053–16062.

Lithium–Sulfur Batteries

Shiro Seki, Hiromitsu Takaba, Yuki Ishino, and Keitaro Takahashi

Abstract There is a need to develop new electrolytes for lithium–sulfur (Li–S) batteries. From the viewpoint of battery performance, control of interfacial stability between the sulfur electrode and electrolyte is an important issue for achieving a long cycle-life. Stable charge–discharge operation of the prepared Li–S cell consisting of a Li negative electrode with $[Li(triglyme)]^+$ $[bis(trifluoromethanesulfonyl)amide]^-$ (TFSA) as the electrolyte was achieved over 800 cycles. This result indicates that the dissolution of Li_2S_x in the electrolyte, a major cause of degradation in Li–S batteries, was effectively suppressed in this system. To prevent degradation (dissolution) of Li_2S_x, we investigated the effects of changing the composition ratio of the lithium salt (LiTFSA) and the low-molecular-weight ether (CH_3–O–$(C_2H_4O)_3$–CH_3) in the system. A highly electrochemically stable Li salt glyme solvated ionic liquid electrolyte was realized, with long-lived, robust Li solvation complexes by increasing the Li salt composition ratio. In Li–S batteries based on the electrolyte, the positive electrode had low degradation (i.e., low solubility of Li_2S_x in the electrolyte), which led to a longer cycle-life for the Li–S batteries.

Keywords Li–S battery · Cycle performance · Electrolyte composition

1 Introduction

Li-ion batteries (LIBs) have high energy densities, and potential for scale-up enabling applications in consumer electric devices and larger systems, such as electric vehicles and stationary use [1, 2]. Continuously increasing demand for better performance (in particular, energy density) and cost-effectiveness motivates the development of batteries with high-performance at a lower cost. To improve the charge–discharge capacity and energy density of battery systems, Li (metal)-molecular sulfur (elemental S) batteries have been investigated for next-generation secondary-battery systems [3–5]. Elemental S (S_8) forms compounds with Li during

S. Seki (✉) · H. Takaba · Y. Ishino · K. Takahashi
Department of Environmental Chemistry and Chemical Engineering, School of Advanced Engineering, Kogakuin University, 2665-1 Nakano-machi, Hachioji 192-0015, Tokyo, Japan
e-mail: shiro-seki@cc.kogakuin.ac.jp

K. Kanamura (ed.), *Next Generation Batteries*,
https://doi.org/10.1007/978-981-33-6668-8_34

charge–discharge process and is abundant, inexpensive, and nontoxic. Therefore, S_8 is suitable for use as a positive active material for Li secondary batteries. The overall reaction between S_8 and Li^+ is

$$S_8 + 16Li^+ + 16e^- \rightarrow 8Li_2S, \quad (1)$$

where elemental S_8, which is positive electrode (PE) material, forms compounds with Li during multistep charge–discharge operations. The theoretical capacity of S_8 is estimated to be 1672 mAh g^{-1}, which is almost 10 times as great as that of other transition metal compound-based electrode materials, such as Li_xCoO_2 ($0.5 < x < 1$; 137 mAh g^{-1}), $Li_xMn_2O_4$ ($0 < x < 1$; 148 mAh g^{-1}). However, S_8-PE has the following intrinsic drawbacks: a large volumetric expansion during Li insertion, low conductivities of S_8 and Li_2S_x ($2 < x < 8$, lithium polysulfide intermediate), slow electrode kinetics between S_8 and Li_2S_x, and dissolution of Li_2S_x from the S_8-PE into an electrolyte [6]. The dissolved Li_2S_x diffuses to the negative electrode (NE) side and is oxidized by the redox shuttle mechanism, which leads to rapid capacity degradation and low Coulombic efficiency of the cell [7, 8]. Highly ordered, nanostructured S_8/carbon composites, surface coatings, and additives have been investigated in attempts to suppress the dissolution of Li_2S_x and enhance the conductivity of the cell [9–14].

Solvate ionic liquids (SILs), which consist of a low-molecular-weight ether (glyme) and a Li salt at a high concentration, have been reported to be effective for Li–S batteries [15, 16]. These electrolytes behave like ionic liquids (ILs) but are classed as a new type of electrolyte species [17, 18]. SILs have higher thermal and electrochemical stabilities than pure glyme, because the glyme molecule is strongly coordinated to Li^+ to form the stable $[Li(glyme)]^+$ complex cation. The dissolution of Li_2S_x is suppressed in these systems and stable charge–discharge of the S_8/carbon composite PE is achieved because of the low Lewis acidity of $[Li(glyme)]^+$ and low Lewis basicity of $[TFSA]^-$. Here, the insertion and extraction of Li ions and S-PE in SILs are investigated through the use of cyclic voltammetry (CV) under various potential sweep conditions. Moreover, strategies for improving the battery performance of the S-PE are examined in relation to the capacity retention and Coulombic efficiency.

2 Cycle Lifetime of Li–S Batteries [19]

Figure 1 shows the dependence of the CV results for the prepared S_8/KB composite PE in SIL ($[Li(G3)_1TFSA]$) (G3: triglyme, TFSA: bis(trifluoromethanesulfonyl)amide) on the potential sweep rate (a) and potential range (b). At least three reductive current peaks appear in the CV (rate: 10 mV s^{-1}); these correspond to gradual Li insertion into the S_8-PE. The reactions in the high (2.4–2.0 V) and low (~2.0 V) voltage regions are attributed to the reduction

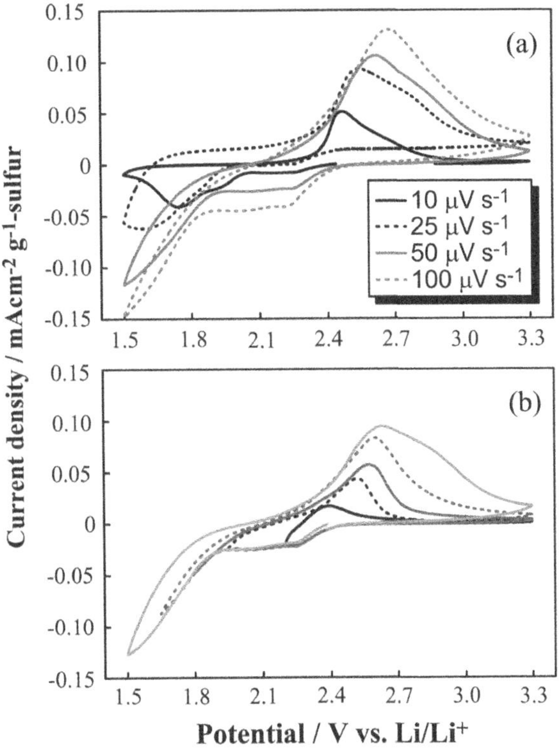

Fig. 1 Cyclic voltammograms of a sulfur/carbon composite electrode in $[Li(G3)_1TFSA]$ under various potential sweep rates (**a**) and various lower voltage limits (rate: 50 $\mu V\ s^{-1}$) (**b**) at 303 K. Reprinted from Electrochemistry 85, Seki S, Serizawa N, Takei K, Umebayashi Y, Tsuzuki S, Watanabe M, Long-cycle-life lithium-sulfur batteries with lithium solvate ionic liquids, p. 681, Copyright (2017) with permission The Electrochemical Society of Japan

reaction of S_8 to Li_2S_x, and Li_2S_x to Li_2S, respectively. In contrast, only one broad oxidative current peak with a shoulder, which corresponds to Li extraction from the lithiated S_8-PE, was observed in the reverse scan, even at a scan rate of 10 mV s^{-1}. The oxidative peak broadened during the sweep toward lower voltage. Although the depth of the Li doping was shallower for higher potential sweep rates, the oxidative peak broadened with increasing sweep rates. This result indicates that the electrode kinetics of the oxidative Li extraction process is slower than those of reductive Li insertion. The electrochemical properties of active materials were previously investigated using composite electrodes [20] with several types of specific surface area by inverse opal carbon and these studies have shown that the charge–discharge behavior in the Li–S system strongly depends on the composition and structure of the composite electrodes. The current study further shows that the electrode kinetics of Li extraction are slower than those of Li insertion on the composite electrodes. It is possible to discuss the Li insertion and extraction mechanism in these electrolytes in more detail with the use of single-particle electrochemical measurement methods [21].

Figure 2a shows the charge–discharge curves of the $[S_8\text{-PE}|[Li(G3)_1TFSA]|Li$ metal NE] cell at a current of 92.9 mA g^{-1}-sulfur ($C/18$). The initial discharge and charge cycles are defined as the first and second cycles, respectively. A utilization of

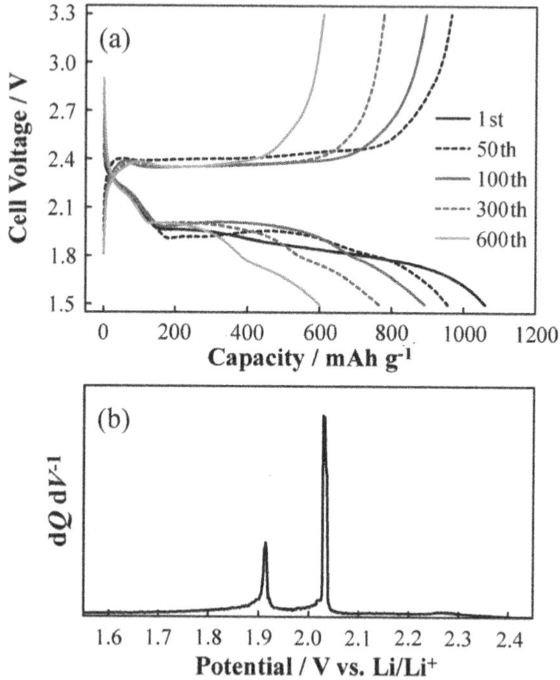

Fig. 2 Charge–discharge profiles with *C*/18 (**a**) and differential discharge curve with *C*/100 (**b**) of the [S_8 positive electrode|[Li(G3)$_1$TFSA]|Li metal negative electrode] cell at 303 K Reprinted from Electrochemistry 85, Seki S, Serizawa N, Takei K, Umebayashi Y, Tsuzuki S, Watanabe M, Long-cycle-life lithium-sulfur batteries with lithium solvate ionic liquids, p. 681, Copyright (2017) with permission The Electrochemical Society of Japan

greater than 50% indicates partial reduction to Li_2S in this system (initial discharge capacity: 1060 mAh g^{-1}). The discharge curve contained two voltage regions at 2.3 and 2.0 V, and the differential curve of the discharge at *C*/100 is shown in Fig. 2b. The three *dQ*/*dV* peaks (1.91, 2.03, 2.26 V) suggest that Li insertion took place over at least three steps, which is consistent with the results shown in Fig. 1. In contrast, the charging profile did not have an obvious voltage plateau, which lowered the energy efficiency.

The cycle performance for charge–discharge capacity and Coulombic efficiency of the cell is shown in Fig. 3. The Coulombic efficiency was defined as the ratio of the discharge (Li doping into S_8) capacity to the charging (Li de-doping of Li_2S_x) capacity. The charge–discharge operation was stable at 800 cycles and the calculated Coulombic efficiency of the cell was more than 97.5% in all cycles. The discharge capacities were 821, 702, and 459 mAh g^{-1} after 200, 400, and 800 cycles, which correspond to 78%, 66%, and 43% of the initial discharge capacity, respectively. Although there were no radical changes in the design of our composite S_8-PE, the stable charge–discharge operation indicates that the dissolution of Li_2S_x into the electrolyte was suppressed in [Li(G3)$_1$TFSA], and that the redox shuttle mechanism

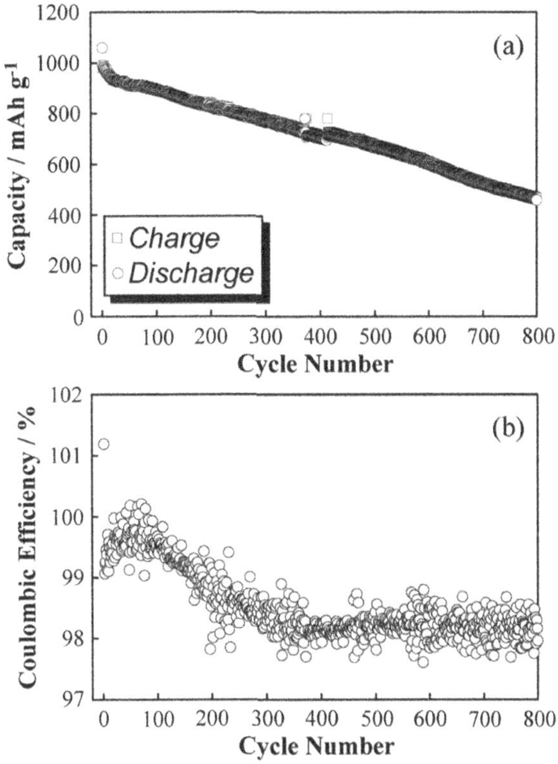

Fig. 3 Battery performances of the [S_8 positive electrode|[$Li(G3)_1$TFSA]|Li metal negative electrode] cell (**a**). Relationship between cycle number and charge and discharge capacity, (**b**). Coumlombic efficiency Reprinted from Electrochemistry 85, Seki S, Serizawa N, Takei K, Umebayashi Y, Tsuzuki S, Watanabe M, Long-cycle-life lithium-sulfur batteries with lithium solvate ionic liquids, p. 682, Copyright (2017) with permission The Electrochemical Society of Japan

did not occur. In this system, the low Lewis acidity and basicity of probably $[Li(G3)]^+$ and $TFSA^-$ also effectively suppress the dissolution of Li_2S_x. Optimization of the electrolyte composition and cell fabrication processes (e.g., the PE composition, the roll-press conditions, and current collector) will be investigated in future work. These modifications are expected to further improve the performance of Li–S batteries with SILs.

3 Electrolyte Design of High-Performance Li–S Batteries [22]

Generally, SILs are 1:1 molar complexes of a multi-oxygen-containing glyme and a Li salt under static conditions. In charge–discharge processes at electrode interfaces, the complex with a 1:1 molar ratio is destroyed and free glyme is temporally formed. Free glyme readily reacts with Li_2S_x or causes its dissolution into the electrolyte. We propose that the use of a non-equimolar (i.e., high Li salt concentration) SILs can compensate for local composition changes and suppress free glyme formation [23],

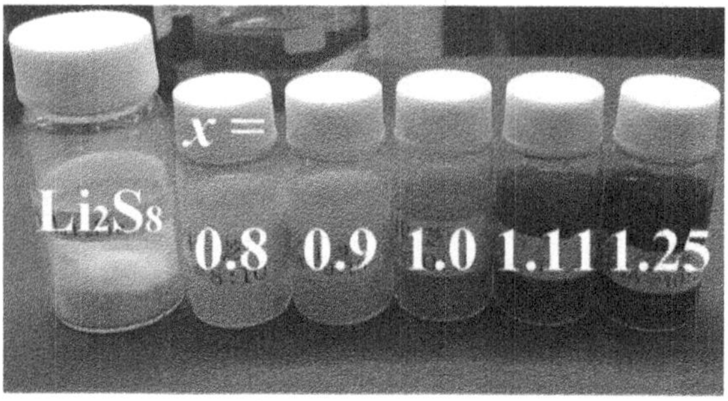

Fig. 4 Photographs of the Li_2S_8 powder and five electrolytes, i.e., [Li(G3)$_x$TFSA] ($x = 0.8$, 0.9, 1.0, 1.11, and 1.25) saturated with Li_2S_8 at room temperature Reprinted from Energy Technology, Ishino Y, Takahashi K, Murata W, Umebayashi Y, Tsuzuki S, Watanabe M, Kamaya M, Seki S, Effect of electrolyte composition on performance and stability of lithium–sulfur batteries, 1,900,197–2, Copyright (2019) with permission WILEY-VCH

although high-concentration electrolytes have been reported to be effective for high-performance Li–S batteries. We also investigated the solubility of Li_2S_x ($x = 8$) into SILs of various compositions and the correlation between Li–S battery performances and SIL composition.

Figure 4 shows photographs of Li_2S_8 powder and five electrolytes, i.e., [Li(G3)$_x$TFSA] ($x = 0.8$, 0.9, 1.0, 1.11, and 1.25), saturated with Li_2S_8. The color of the electrolyte solution gradually changed as the LiTFSA concentration increased and Li_2S_x saturated. Generally, Li_2S_8 is a major reaction intermediate at the PE during charge–discharge operations and easily dissolves into the electrolyte solution. We controlled the Li_2S_x solubility by changing the ratio of the solvent (G3) to that of the dissolved salt (LiTFSA). We have also reported suppression of the formation of free glyme by high LiTFSA at high concentrations, i.e., a G3:LiTFSA molar ratio greater than 1:1 [23]. The solubility of Li_2S_x in the formed complex cation [Li(G3)$^+$] is extremely low but is high in the free solvent, G3 that is transiently generated. These observations provide direct evidence of this mechanism.

Figure 5a shows photographs [Li(G3)$_x$TFSA] ($x = 0.8$, 0.9, 1.0, 1.11, and 1.25) solutions of Li_2S_8; the solutions were diluted four times with 1,1,2,2-tetrafluoroethyl 2,2,3,3-tetrafluoropropyl ether (HFE). Addition of HFE did not change the order of color tones in Fig. 5. Hence, HFE does not change the solvated structure of the electrolyte and only has a simple dilution effect without any notable interactions [24]. In practical applications in batteries, a combination of high-mobility reactive ionic species (in this study, Li) and an electrolyte with high electrochemical stability is important, and the addition of HFE might be used to improve the performance of Li–S cells. Figure 5b shows the UV–vis spectra of [Li(G3)$_x$TFSA] ($x = 0.8$, 0.9, 1.0, 1.11, and 1.25) solutions of Li_2S_8, which were diluted four times with HFE. At $x = 1.11$ and 1.25, i.e., systems with a large amount of free glyme [25], the UV absorbance maximum values were above the measurement limit, i.e., 400 nm. As the LiTFSA concentration increased, the peak decreased to be approximately 260 nm (attributed Li_2S_8) [26].

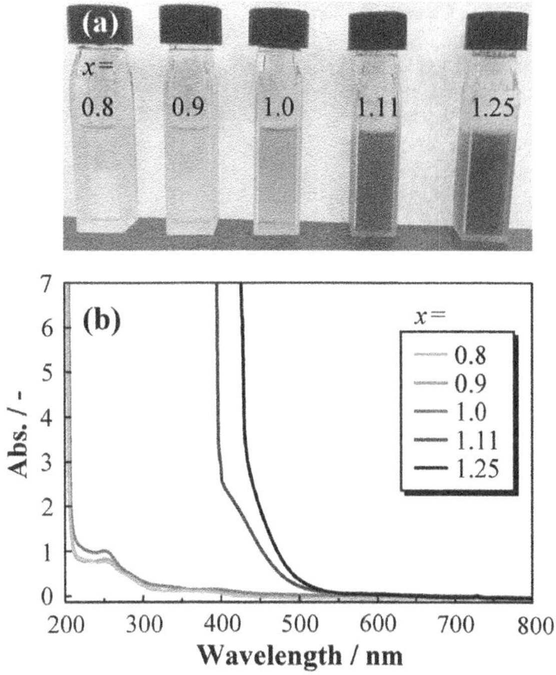

Fig. 5 (**a**) Photographs and (**b**) UV–vis spectra of Li_2S_8 dissolved $[Li(G3)_xTFSA]$ (x = 0.8, 0.9, 1.0, 1.11, and 1.25) and diluted four times with HFE at room temperature Reprinted from Energy Technology, Ishino Y, Takahashi K, Murata W, Umebayashi Y, Tsuzuki S, Watanabe M, Kamaya M, Seki S, Effect of electrolyte composition on performance and stability of lithium–sulfur batteries, 1,900,197–2, Copyright (2019) with permission WILEY-VCH

Figure 6a shows the charge/discharge curves of the $[S_8\text{-PE}|[Li(G3)_xTFSA]|Li$ metal NE] cell. The temperature was set to be 318 K to eliminate viscosity effects and accelerate cycling life tests. In this study, the initial discharge and charge cycles were defined as the first and second cycles, respectively. The initial discharge capacity of each Li–S cell ($n = 3$) was in the range of 1100–1300 mAh g^{-1}, which corresponds to 65–80% of the theoretical capacity of elemental S8. The discharge curve contained two voltage plateaus at 2.2 and 2.0 V. Figure 6 also shows the discharge capacities (b), calculated relative capacities (c), and Coulombic efficiencies (d) of the $[S_8\text{-PE}|[Li(G3)_xTFSA]|Li$ metal NE] cells during cycling. The cycling performances of the prepared cells clearly improved as the LiTFSA concentration increased. These results reflect the solubility of Li_2S_x, which is one of the main factors in the degradation of Li–S batteries. The retained discharge capacities after 50 cycles were 43.0% ($[Li(G3)_{1.25}TFSA]$), 45.4% ($[Li(G3)_1TFSA]$), and 64.7% ($[Li(G3)_{0.8}TFSA]$). However, the measurement temperature was relatively high, which affects the solubility of Li_2S_x. A similar improvement in Coulombic efficiency was observed as the LiTFSA concentration increased. The cell with $[Li(G3)_{0.8}TFSA]$ retained a Coulombic efficiency greater than 98% after more than 50 cycles; this is a high-value for a Li–S battery operated at high temperature. These results confirm that excess LiTFSA (carrier ionic species) in G3 (glyme-based solvent) led to the formation of rigid complex cations and these ionic species compensated for charge/discharge

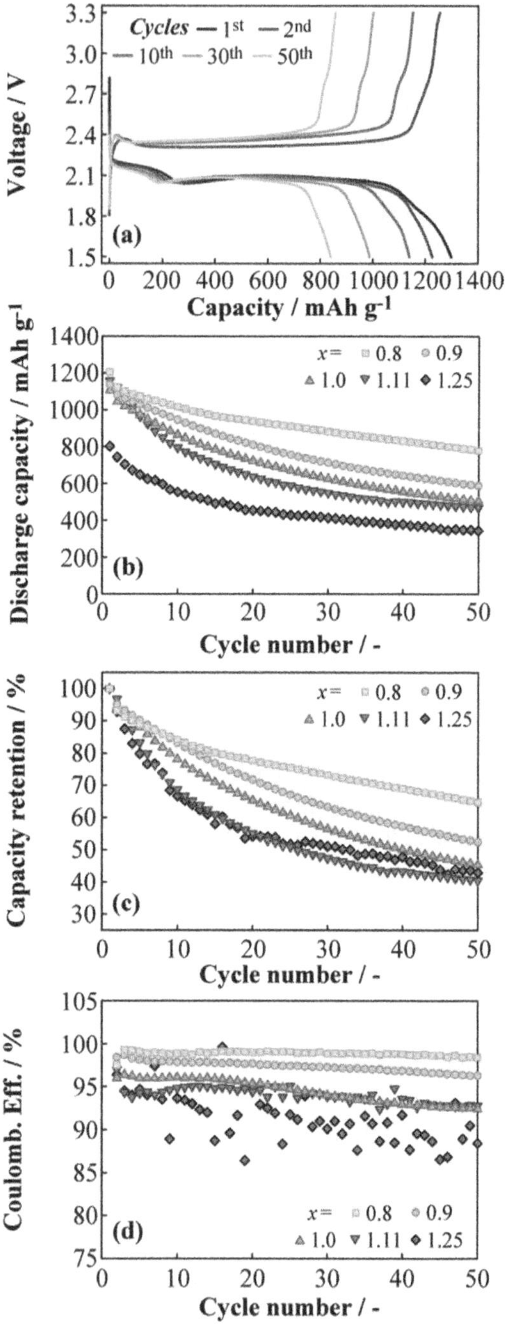

Fig. 6 (**a**) Charge–discharge profiles of [S_8 positive electrode|[Li(G3)$_1$TFSA]|Li metal negative electrode] cells. (**b**) Cycle number dependences of discharge capacities per positive electrodes. (**c**) Calculated relative capacities. (**d**) Coulombic efficiencies at 318 K (voltage range: 1.5–3.3 V, current: *C*/12). Each plot was obtained from the average of all cells in (**b–d**) Reprinted from Energy Technology, Ishino Y, Takahashi K, Murata W, Umebayashi Y, Tsuzuki S, Watanabe M, Kamaya M, Seki S, Effect of electrolyte composition on performance and stability of lithium–sulfur batteries, 1,900,197–3, Copyright (2019) with permission WILEY-VCH

of S_8 and Li_2S. This behavior improves the electrolyte/PE (S_8) interface in the charge/discharge process and leads to longer cycling lifetimes of Li–S batteries.

4 Summary

Multistep Li insertion into S_8-PEs was observed in $[Li(G3)_xTFSA]$, which is consistent with the discharge profile of the Li–S cells. Stable cell charge–discharge operation with no major modifications of the S_8/carbon composite suggests that the properties of the electrolyte play an important role in suppressing the dissolution of Li_2S_x and the redox shuttle mechanism. As a result, a favorable charge–discharge operation of 800 cycles was achieved. We have also demonstrated that the SIL concept can be used in the development of Li–S cells. Moreover, the electrolyte composition (x values, $[Li(G3)_xTFSA]$) can control the chemical stability of reactive intermediates at the Li_2S_x electrode and provide high-performance Li–S cells for next-generation battery systems.

References

1. Larcher, D., & Tarascon, J.-M. (2015). Towards greener and more sustainable batteries for electrical energy storage. *Nature Chemistry, 7,* 19–29.
2. Nitta, N., Wu, F., Lee, J. T., & Yushin, G. (2015). Li-ion battery materials: present and future. *Materials Today, 18,* 252–264.
3. B. Scrosati, J. Hassoun, & Y.-K. Sun. (2011). Lithium-ion batteries. A look into the future. *Energy and Environmental Science, 4,* 3287–3295
4. Bruce, P. G., Freunberger, S. A., Hardwick, L. J., & Tarascon, J.-M. (2012). Li–O_2 and Li–S batteries with high energy storage. *Nature Materials, 11,* 19–29.
5. Ji, X., & Nazar, L. F. (2010). Advances in Li–S batteries. *Journal of Materials Chemistry, 10,* 9821–9826.
6. Rauh, R. D., Shuker, F. S., Marston, J. M., & Brummer, S. B. (1977). Formation of lithium polysulfides in aprotic media. *Journal of Inorganic and Nuclear Chemistry, 39,* 1761–1766.
7. Shim, J., Striebel, K. A., & Cairns, E. J. (2002). The lithium/sulfur rechargeable cell. *Journal of the Electrochemical Society, 149,* A1321–A1325.
8. Mikhaylik, Y. V., & Akridge, J. R. (2004). Polysulfide shuttle study in the Li/S battery system. *Journal of the Electrochemical Society, 151,* A1969–A1976.
9. Ji, X., Lee, K. T., & Nazar, L. F. (2009). A highly ordered nanostructured carbon–sulphur cathode for lithium–sulphur batteries. *Nature Materials, 8,* 500–506.
10. Takeuchi, T., Sakaebe, H., Kageyama, H., Senoh, H., Sakai, T., & Tastumi, K. (2010). Preparation of electrochemically active lithium sulfide–carbon composites using spark-plasma-sintering process. *Journal of Power Sources, 195,* 2928–2934.
11. Cao, Y. L., Li, X. L., Aksay, I. A., Lemmon, J., Nie, Z. M., Yang, Z. G., & Liu, J. (2011). Sandwich-type functionalized graphene sheet-sulfur nanocomposite for rechargeable lithium batteries. *Physical Chemistry Chemical Physics, 13,* 7660–7665.
12. Zheng, G., Ynag, Y., Cha, J. J., Hong, S. S., & Cui, Y. (2011). Hollow carbon nanofiber-encapsulated sulfur cathodes for high specific capacity rechargeable lithium batteries. *Nano Letters, 11,* 4462–4467.

13. Choi, J.-W., Cheruvally, G., Kim, D.-S., Ahn, J.-H., Kim, K.-W., & Ahn, H.-J. (2008). Rechargeable lithium/sulfur battery with liquid electrolytes containing toluene as additive. *Journal of Power Sources, 183,* 441–445.
14. Aurbach, D., Pollak, E., Elazari, R., Salitra, G., Kelley, C. S., & Affinito, J. (2009). On the surface chemical aspects of very high energy density, rechargeable Li–sulfur batteries. *Journal of the Electrochemical Society, 156,* A694–A702.
15. Yoshida, K., Nakamura, M., Kazue, Y., Tachikawa, N., Tsuzuki, S., Seki, S., et al. (2011). Oxidative-stability enhancement and charge transport mechanism in glyme–lithium salt equimolar complexes. *Journal of the American Chemical Society, 133,* 13121–13129.
16. Dokko, K., Tachikawa, N., Yamauchi, K., Takashima, E., Park, J., Ueno, K., et al. (2013). Solvate ionic liquid electrolyte for Li–S batteries. *Journal of the Electrochemical Society, 160,* A1304–A1310.
17. Henderson, W. A., McKenna, F., Khan, M. A., Brooks, N. R., Young, V. G., & Frech, R. (2005). Glyme−lithium bis(trifluoromethanesulfonyl)imide and glyme−lithium bis(perfluoroethanesulfonyl)imide phase behavior and solvate structures. *Chemistry of Materials, 17,* 2284–2289.
18. Angell, C. A., Ansari, Y., & Zhao, Z. (2011). Ionic liquids: past, present and future. *Faraday Discussions, 154,* 9–27.
19. Seki, S., Serizawa, N., Takei, K., Umebayashi, Y., Tsuzuki, S., & Watanabe, M. (2017). Long-cycle-life lithium-sulfur batteries with lithium solvate ionic liquids. *Electrochemistry, 85,* 680–682.
20. N. Tachikawa, K. Yamauchi, E. Takashima. J-W. Park, K. Dokko, & M. Watanabe. (2011). Reversibility of electrochemical reactions of sulfur supported on inverse opal carbon in glyme–Li salt molten complex electrolytes. *Chemical Communications, 47,* 8157–8159
21. Dokko, K., Horikoshi, S., Itoh, T., Nishizawa, M., Mohamedi, M., & Uchida, I. (2000). Microvoltammetry for cathode materials at elevated temperatures: Electrochemical stability of single particles. *Journal of Power Sources, 90,* 109–115.
22. Y. Ishino, K. Takahashi, W. Murata, Y. Umebayashi, S. Tsuzuki, M. Watanabe, M. Kamaya, & S. Seki. (2019). Effect of electrolyte composition on performance and stability of lithium–sulfur batteries. *Energy Technology, 1900197.*
23. Seki, S., Serizawa, N., Takei, K., Tsuzuki, S., Umebayashi, Y., Katayama, Y., et al. (2016). Effects of non-equimolar lithium salt glyme solvate ionic liquid on the control of interfacial degradation in lithium secondary batteries. *RSC Advances, 6,* 33043–33047.
24. Saito, S., Watanabe, H., Ueno, K., Mandai, T., Seki, S., Tsuzuki, S., et al. (2016). Li^+ local structure in hydrofluoroether diluted Li-glyme solvate ionic liquid. *The Journal of Physical Chemistry B, 120,* 3378–3387.
25. Ueno, K., Tatara, R., Tsuzuki, S., Saito, S., Doi, H., Yoshida, K., et al. (2015). Li^+ solvation in glyme–Li salt solvate ionic liquids. *Physical Chemistry Chemical Physics, 17,* 8248–8257.
26. Kawase, A., Shirai, S., Yamoto, Y., Arakawa, R., & Takata, T. (2014). Electrochemical reactions of lithium–sulfur batteries: an analytical study using the organic conversion technique. *Physical Chemistry Chemical Physics, 16,* 9344–9350.

Li–S Battery Using Li_2S Cathode

Masayoshi Watanabe

Abstract A major drawback of Li–S cells is the dendritic growth of Li metal and the formation of dead Li that occurs in cells during the charging/discharging processes, which causes the loss of electrochemically active Li, the depletion of the electrolyte, and the internal short circuit of the cell, and therefore prevents the safe use and long cycling-life of Li–S batteries. Using Li_2S (theoretical capacity of 1166 mAh/g) and Li-free materials as cathode and anode materials, respectively, is a strategy that could overcome this concern. The use of Li-free anode materials such as graphite, Sn, and Si, prevent the formation of Li dendrites and thus improved cell safety. In this chapter, the fabrication of graphite|solvate ionic liquid|Li_2S and Si|solvate ionic liquid|Li_2S cells and their performance are presented.

Keywords Lithium sulfide (Li_2S) cathode · Graphite anode · Si anode · Solvate ionic liquid · Sparingly solvating electrolyte

1 Introduction

The rapidly increasing demand for stationary energy storage and electric vehicles requires the development of "beyond Li-ion batteries" featuring high energy densities that would exceed those of state-of-the-art Li-ion batteries [1]. Li–S batteries, which present very high theoretical capacities and energy densities, are considered to be some of the most promising candidates for this. Based on the complete conversion of S_8 into Li_2S ($S_8 + 16Li^+ + 16e^- \rightarrow 8Li_2S$), the sulfur cathode could deliver the theoretical capacity of 1672 mAh/g [2–6]. One significant disadvantage of the sulfur cathode is the dissolution of lithium polysulfides (LPSs), which form during the discharge reactions of the sulfur cathode (see Chap. 25 and Chap. 31) [7–9]. This results in the low discharge/charge cycle stability owing to the loss of active materials from the cathode and low Coulombic efficiency of the discharge/charge process owing to the "redox-shuttle effect". As part of the ALCA-SPRING project,

M. Watanabe (✉)
Institute of Advanced Sciences, Yokohama National University, Yokohama, Japan
e-mail: mwatanab@ynu.ac.jp

K. Kanamura (ed.), *Next Generation Batteries*,
https://doi.org/10.1007/978-981-33-6668-8_35

we focused on ionic liquids (ILs) and solvate IL electrolytes, which could suppress the dissolution of LPSs owing to very low coordinating properties of the cations and anions [10, 11], as LPSs are ionic materials and for the dissolution coordination of solvents (ions) to LPSs is essential. A 1:1 complex of tetraglyme (G4) and Li[TFSA] (where $[TFSA]^-$ is $[N(SO_2CF_3)_2]^-$) is a representative solvate IL, comprising the $[Li(G4)]^+$ complex cation and $[TFSA]^-$ anion [12]. [Li(G4)][TFSA] mixed with 1,1,2,2-tetrafluoroethyl 2,2,3,3-tetrafluoropropyl ether, a hydrofluoroether (HFE), which has sparing solubility for LPSs and relatively high ionic conductivity, could enable the stable and long-term operation of Li–S cells [11], as discussed somewhere in this chapter.

Another major drawback of Li–S cells is the dendritic growth of Li metal and the formation of dead Li that occurs in cells during the charging/discharging processes particularly at high current densities (see Chap. 28). This causes the loss of electrochemically active Li, the depletion of the electrolyte, and the internal short circuit of the cell, and therefore prevents the safe use and long cycling-life of Li–S batteries. Using Li_2S (theoretical capacity of 1166 mAh/g) and Li-free materials as cathode and anode materials, respectively, is a strategy that could overcome this concern [13–16]. As indicated in Chapter "Li_2S Cathode" (Chap.33), a Li_2S/graphene nanocomposite electrode, which exhibited a high discharge capacity of ~700 mAh/g and good cycle stability, was prepared in this study [17]. The use of Li-free anode materials such as graphite, Sn, and Si, prevent the formation of Li dendrites and thus improved cell safety. However, the electrochemical intercalation/deintercalation of Li^+ ions into/from graphite is highly electrolyte-selective, and cannot usually take place in ether-based electrolytes, which are commonly used for Li–S batteries, because the cointercalation of the solvent induces exfoliation of graphite [18]. Most of the successful Li^+ ion intercalation/deintercalation into/from graphite electrodes was performed using carbonate electrolytes, which have been typically utilized for Li-ion batteries [19]. A small amount of ethylene carbonate (EC) was reductively decomposed at the graphite electrode during the initial charging process, and the decomposition products formed a solid electrolyte interphase (SEI) layer on the surface of graphite. The SEI acted as passivation layer, suppressing the subsequent reductive decomposition of EC as well as the cointercalation of solvent molecules into graphite. Carbonate electrolytes used for Li-ion batteries, however, are chemically incompatible with most cathodes used for Li–S batteries, because carbonates can adversely react with the sulfide anion of LPSs via nucleophilic attacks to the carbonate molecules [20]. Consequently, previous attempts to fabricate graphite-based Li–S batteries (graphite|electrolyte|Li_2S) required the adoption of special cathodes, typically S-encapsulated micro-pore carbon cathodes [21], which were compatible with the carbonate-based electrolytes, as discussed in Chap. 32. Another alternative method is based on a complicated full cell configuration consisting of a two-phase electrolyte system [graphite|electrolyte 1 (carbonate)||electrolyte 2 (ether)|Li_2S], wherein the carbonate electrolyte (which is compatible with the graphite anode) and the ether electrolyte (which is compatible with the Li_2S cathode) have to be separated by a ceramic solid electrolyte layer [22]. Therefore, the development of an electrolyte suitable for the stable operation of graphite anode-based Li–S full cells is an intriguing challenge. In this study, solvate ILs were demonstrated to act as single-phase electrolytes, which were compatible with the graphite anode, where the electrochemical Li^+ ions intercalation/deintercalation would proceed reversibly [23, 24]. The performance of graphite|solvate IL|Li_2S cells is described in the first part of this section, where the reversible intercalation/deintercalation as well as the suppression of the dissolution of LPSs were simultaneously enabled by the use of the solvate IL.

In addition to graphite, Si as Li-alloying metal is a more promising anode material candidate owing to its high capacity of 3580 mAh/g [25, 26]. However, the large change in the volume of Si during lithiation (charging) and delithiation (discharging) resulted in mechanical cracking or pulverization and poor electrochemical reversibility [25, 27]. As indicated in Chap. 29, amorphous Si nanoflake powder (Si LeafPowder®, denoted as Si-LP) with a thickness of 50–200 nm has been utilized in this project [28, 29]. The short Li-diffusion-length facilitated the uniform alloying and de-alloying of Li on thin Si-LP layers. This resulted in the relaxation of the stress caused by the change in volume of Si, which consequently reduced the mechanical degradation of Si-LP and rendered good charge/discharge cycle stability to the cell. The battery performance of Si-LP/Li_2S cell using a solvate IL electrolyte is described later in this chapter.

2 Graphite|Electrolyte|Li_2S Cells

In this project, we prepared Li_2S/graphene composites [17] via one-pot pyrolysis of lithium sulfate (Li_2SO_4) in the presence of excess graphene nanoplatelet aggregates (GNAs). Herein, the GNAs served as the carbon source to chemically reduce Li_2SO_4 and generate Li_2S at 781 °C ($Li_2SO_4 + 2C \rightarrow 2CO_2\uparrow + Li_2S$), while the residual GNAs formed thin graphene layers, via in situ etching, which could immobilize Li_2S and yield Li_2S/graphene composites exhibiting intimate contact between the highly conductive graphene and active material [17]. Consequently, the resulting Li_2S/graphene composites were suitable as cathode materials in graphite|electrolyte|Li_2S cells [30].

The Li_2S/graphene composite, carbon black, and polyvinylpyrrolidone (PVP, mol. wt. = 1200 kDa) were dispersed into *N*-methyl-2-pyrrolidinone (NMP), and the mixture was stirred for 48 h. The mass ratio of Li_2S, carbon (including graphene and carbon black), and PVP was 60:30:10. This slurry was absorbed using carbon fiber paper (CFP), and dried in vacuum for 16 h at 80 °C. The Li_2S loading of the CFP-supported Li_2S electrode sheet was ~ 2.2 mg/cm^2 [30]. To obtain graphite anodes, artificial graphite (90 wt%), acetylene black (3 wt%), and poly(vinylidene fluoride) (7 wt%) were mixed with NMP to form a slurry, which was coated onto Cu foil current collectors and then dried in an oven for 12 h at 80 °C. Prior to use, the obtained graphite electrodes were compressed to enhance their electrical conductivity, and further dried in vacuum for 24 h at 80 °C. The graphite loading on the electrode was 4.5 mg/cm^2 [30].

A graphite|electrolyte|Li_2S cell was fabricated [30], using the Li_2S composite, 1 M $[Li(G4)_1][TFSA]$/HFE, and graphite as cathode, electrolyte, and anode, respectively, as presented in Fig. 1a. Figure 1b illustrates the cyclic voltammetry (CV) curves for the first three cycles, which clearly demonstrate the stable and repeatable oxidation/reduction current. The initial anodic scan corresponded to the so-called initial activation process of Li_2S. During the subsequent anodic scans, only one broad peak was observed at approximately 2.4 V, which could be assigned to the typical oxidation process of Li_2S. By contrast, two peaks at 1.9 and 2.1 V, which were attributed to the stepwise reduction of elemental S, were observed during all cathodic scans. Figure 1c depicts the charge/discharge curves of the Li_2S/graphite full cell where 1 M $[Li(G4)_1][TFSA]$/HFE was used as electrolyte. The two discharge plateaus and the charge plateau observed for this full cell were in good agreement with the redox peaks observed in the CV curves (Fig. 1b). The full cell delivered the initial discharge capacity of 628 mAh/g at 1/12 C and maintained the capacity of 463 mAh/g and Coulombic efficiency of 97% after 20 cycles. The successful

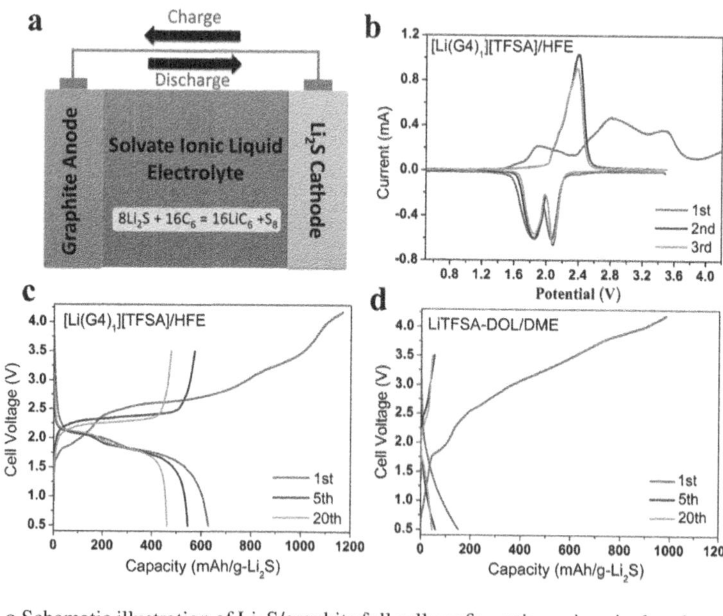

Fig. 1 **a** Schematic illustration of Li_2S/graphite full cell configuration using single solvate IL electrolyte. **b** Typical cyclic voltammetry curves of graphite/Li_2S full cell using graphite, Li_2S/graphene, and 1 M $[Li(G4)_1][TFSA]$/HFE as anode, cathode, and electrolyte, respectively, at a scan rate of 0.04 mV/s. The potential was scanned first from the open-circuit voltage (OCV) to 4.2 V and then between 3.5 and 0.5 V. Galvanostatic charge/discharge curves of Li_2S/graphite full cells using **c** 1 M $[Li(G4)_1][TFSA]$/HFE and **d** 1 M Li[TFSA]/DOL/DME as electrolytes. The full cells were initially charged from the OCV to 4.2 V at 1/48 C, and then cycled in the voltage range of 0.5–3.5 V at 1/12 C. Here, G4 is tetraglyme, [TFSA] is $[N(SO_2CF_3)_2]^-$, HFE is hydrofluoroether, DOL is 1,3-dioxolane, and DME is 1,2-dimethoxyethane. Reprinted with permission from Ref. [30]. Copyright 2016, American Chemical Society

charge/discharge process was attributed to the stable and reversible Li^+ ions intercalation/deintercalation into/from graphite electrodes and sparing solubility of LPSs in 1 M $[Li(G4)_1][TFSA]$/HFE (Fig. 2) [24]. The solvate IL could greatly reduce the shuttle effect, in addition to being compatible with the graphite anode. The most commonly used 1 M Li[TFSA]/DOL/DME (where DOL is 1,3-dioxolane and DME is 1,2-dimethoxyethane) electrolyte could not be recharged at all under the same conditions (Fig. 1d). Although the high capacity of 990 mAh/g was obtained during the first charge, the subsequent charging/discharging processes delivered quite low discharge capacities. The decrease in capacity was mainly ascribed to the cointercalation of the ether solvent into graphite, and the subsequent exfoliation of graphite.

Figure 3 illustrates the charge/discharge curves of the graphite|electrolyte|Li half cells. The charge and discharge capacities of the graphite electrodes increase as the G4/Li[TFSA] molar ratio decreased. For example, the 1 M $[Li(G4)_{0.6}][TFSA]$/HFE and 1 M $[Li(G4)_{0.8}][TFSA]$/HFE electrolytes could deliver charge capacities of 349 and 319 mAh/g at 1/20 C during the third cycle, respectively, which were higher than the capacity delivered by 1 M $[Li(G4)_1][TFSA]$/HFE (231 mAh/g). The increased capacities in solvate ILs containing excess Li[TFSA] could be attributed to the significantly reduced activity (concentration) of free G4 in these electrolytes [23, 24]. Cointercalation could be suppressed by decreasing the activity of free glyme in the electrolyte. Instead, the desolvation of Li^+ ions could occur at the graphite/electrolyte interface, and Li^+ ions could be reversibly intercalated into graphite. For cells using 1 M $[Li(G4)_x][TFSA]$/HFE ($x \leq 1$) as electrolytes, the desolvation of Li^+ ions took place at the graphite anode during charging, while the solvation of Li^+ ions occurred at the Li cathode.

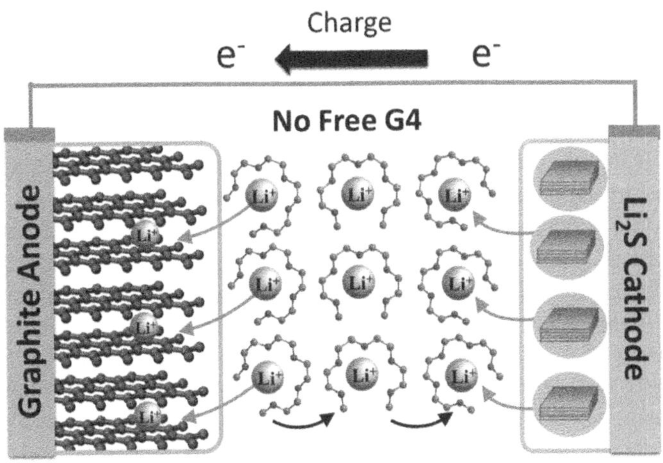

Fig. 2 Schematic illustration of charging process of graphite/Li_2S full cell using solvate IL electrolytes (1 M $[Li(G4)_x][TFSA]$/HFE for $x = 0.6$, 0.8, and 1, where G4 is tetraglyme, [TFSA] is $[N(SO_2CF_3)_2]^-$, and HFE is hydrofluoroether). Reprinted with permission from Ref. [30]. Copyright 2016, American Chemical Society

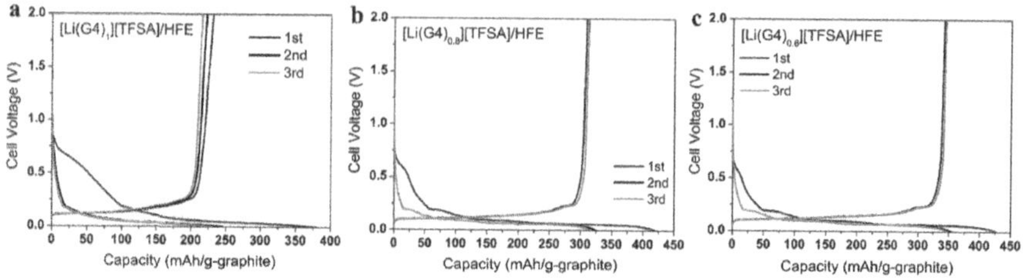

Fig. 3 Galvanostatic charge/discharge curves of Li/graphite half cells using 1 M $[Li(G4)_x][TFSA]$/HFE as electrolytes during the first three cycles at the current rate of 1/20 C: **a** $x = 1$, **b** $x = 0.8$, and **c** $x = 0.6$. Here, G4 is tetraglyme, [TFSA] is $[N(SO_2CF_3)_2]^-$, and HFE is hydrofluoroether. Reprinted with permission from Ref. [30]. Copyright 2016, American Chemical Society

Graphite/Li_2S full cells using 1 M $[Li(G4)_{0.8}][TFSA]$/HFE and 1 M $[Li(G4)_{0.6}][TFSA]$/HFE as electrolytes were also assembled, and their charge/discharge profiles are displayed in Figs. 4a, b [30]. Initial discharge capacities of 809 and 709 mAh/g, and 20th-cycle discharge capacities of 586 and 644 mAh/g were obtained, respectively, which were higher than those obtained when 1 M $[Li(G4)_1][TFSA]$/HFE was used as an electrolyte. As depicted in Fig. 3, this appeared to be caused by the enhanced reversibility of the Li^+ ions intercalation/deintercalation process into/from the graphite electrodes and the increase in capacity associated with the decrease in G4/Li[TFSA] molar ratios. The $[Li(G4)_1][TFSA]$/HFE and $[Li(G4)_{0.8}][TFSA]$/HFE electrolytes could maintain discharge capacities of 343 and 410 mAh/g at 1/12 C after 100 cycles (Fig. 4c), respectively [30]. By contrast, for the $[Li(G4)_{0.6}][TFSA]$/HFE electrolyte, the smaller discharge capacity of 231 mAh/g and very strong polarization, were obtained during the 100th cycle, and were associated with the low ionic conductivity of the electrolyte. All Li_2S/graphite full cells demonstrated high Coulombic efficiencies over 100 cycles (Fig. 4d), owing to the reduced dissolution of polysulfides in solvate ILs [30]. The rate capabilities of Li_2S/graphite full cells using different electrolytes at variable current rates were also investigated (Fig. 4e) [30]. The corresponding results reflected the fact that the electrochemical performance of the full cells was determined by both the ionic conductivity of the electrolyte and Li^+ ions intercalation/deintercalation properties into/from the graphite anode. The discharge capacities of the cell using $[Li(G4)_{0.8}][TFSA]$/HFE as electrolyte at the different rates of 1/12, 1/6, and 1/3 C were in the ranges of 786–650, 581–552, and 406–403 mAh/g, respectively, which were significantly higher than those of the cells using $[Li(G4)_1][TFSA]$/HFE and $[Li(G4)_{0.6}][TFSA]$/HFE as electrolytes. The best rate capability of the cell using $[Li(G4)_{0.8}][TFSA]$/HFE as electrolyte at low current rates could be attributed to its balanced Li^+ ions intercalation properties (Fig. 3) and ionic conductivity. However, upon further increasing the current rate to 1 and 2 C, the higher ionic conductivity of 1 M $[Li(G4)_1][TFSA]$/HFE than that of 1 M $[Li(G4)_{0.8}][TFSA]$/HFE appeared to dominate the rate performance. Very recently, further development of this cell in terms of the increase in Li_2S loading in

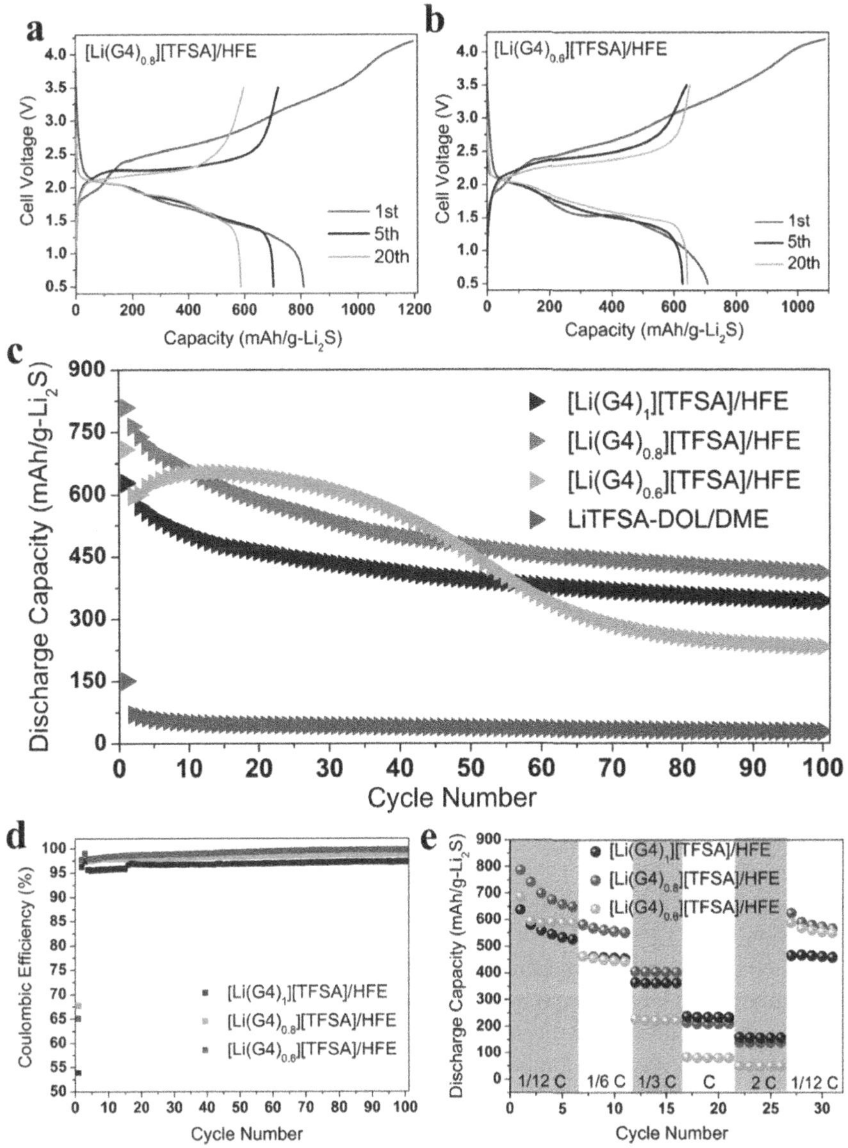

Fig. 4 Galvanostatic charge/discharge curves of graphite/Li_2S full cells using **a** 1 M $[Li(G4)_{0.8}][TFSA]/HFE$ and **b** 1 M $[Li(G4)_{0.6}][TFSA]/HFE$ as electrolytes. **c** Discharge capacity change and **d** Coulombic efficiency of graphite/Li_2S full cells during cycling. The full cells were initially charged at 1/48 C, and then cycled at 1/12 C. **e** Rate capability of Li_2S/graphite full cells at various rates from 1/12 to 2 C. Here, G4 is tetraglyme, [TFSA] is $[N(SO_2CF_3)_2]^-$, and HFE is hydrofluoroether. Reprinted with permission from Ref. [30]. Copyright 2016, American Chemical Society

the cathode and the decrease in electrolyte amounts was made by using other carbon precursors for Li_2S/C composites than GNAs [31].

3 Si|Electrolyte|Li_2S Cells

We synthesized 100-nm-thick Si-LP using the electron beam vapor deposition method (Chap. 29) [28, 29]. A homogenous slurry was formed by mixing Si-LP (83.3 wt%), Ketjenblack (5.6 wt%), and carboxymethyl cellulose sodium salt (11.1 wt%) in water, and was then pasted onto Cu foil. After vacuum drying overnight at 80 °C, the composite electrode sheet featuring the Si loading of 0.9 mg/cm^2 was obtained. The same Li_2S/graphene composite electrode that was used for the graphite|electrolyte|Li_2S cell [30] was used to assemble Si|[Li(G4)][TFSA]/4HFE|Li_2S cells.

The charge–discharge profiles of an Si|[Li(G4)][TFSA]/4HFE|Li_2S cell are illustrated in Fig. 5a [32]. The large overpotential during the initial charging process originated from the activation process of Li_2S, as discussed above. The initial charge capacity of 1188 mAh/g, which was close to the theoretical value of Li_2S (1166 mAh/g), suggested that Li_2S was sufficiently activated. However, the following discharge capacity (first) was decreased to 702 mAh/g, which was probably due to: (i) the volumetric shrinkage of Li_2S during the first charging process leading to the loss of electrical contact between the active material and conductive carbon and (ii) the irreversible consumption of Li at the Si anode side during the initial cycle. However, during the following cycles, a very flat charge plateau at 2.1 V and two well-defined discharge plateaus at 1.8 and 1.4 V were observed. After 40 cycles, the cell delivered the discharge capacity of 403 mAh/g and the average Coulombic efficiency of 96% (Fig. 5b) [32]. The relatively high Coulombic efficiency indicated again that the shuttle effect was effectively suppressed in the cell owing to the very low solubility of LPSs in [Li(G4)][TFSA]/4HFE [11, 15]. The decrease in capacity of the cell during cycling appeared to be mainly attributed to the degradation of the Si electrode. The large change in volume of Si during the charge/discharge process and the side reaction of the electrolyte at the surface of the Si electrode could have been the origin of the deterioration, as discussed in Chap. 29.

To evaluate the rate capability of the Si/Li_2S full cell, the charge and discharge rates were changed from 1/12 to 2 C (Fig. 5c) [32]. The discharge capacity of the cell decreased as the current density increased, however, the cell still delivered the capacity of 292 mAh/g after 200 cycles at 1 C (Fig. 5c), which implied that the cell presented good cycle stability at high current densities. The rate capability and cycle performance of the cell could be attributed to the short Li-diffusion-length in Si-LP, intimate contact between Li_2S and graphene, and high ionic conductivity of the [Li(G4)][TFSA]/HFE electrolyte. The short Li diffusion length of Si-LP contributed to the charge/discharge of the anode at high current densities. The intimate contact between Li_2S and graphene lowered the resistance of the composite cathode and enabled the redox reaction of the insulating Li_2S. The high ionic conductivity of the

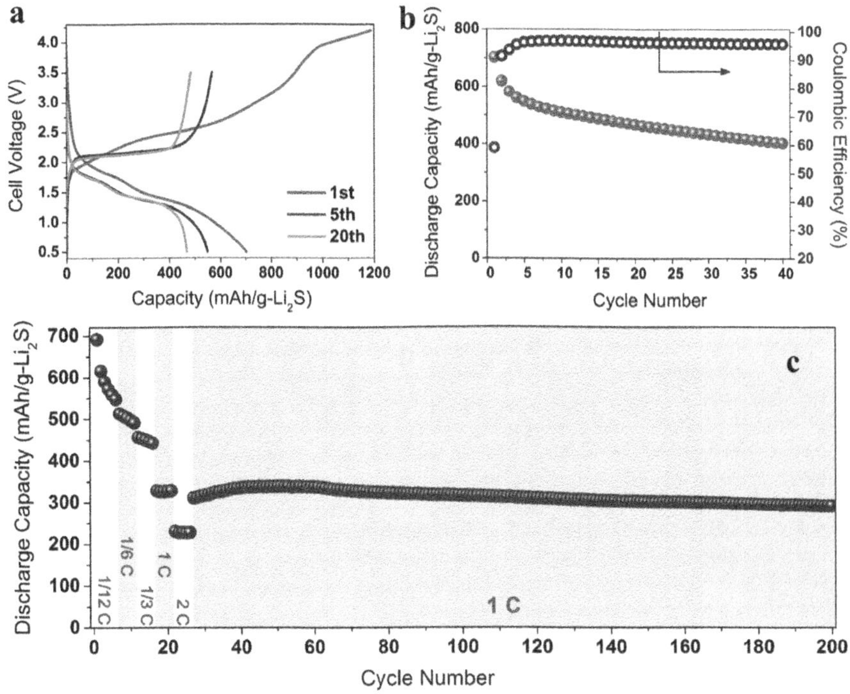

Fig. 5 **a** Charge/discharge curves of Si|[Li(G4)$_1$][TFSA]/4HFE|Li_2S cell. The cell was initially charged to 4.2 V at 1/48 C and then discharged and charged between 0.5 and 3.5 V at 1/12 C rate in subsequent cycles. **b** Discharge capacity of Si/Li_2S cell during cycling at 1/12 C. **c** Rate capability of Si/Li_2S cell. Here, G4 is tetraglyme, [TFSA] is $[N(SO_2CF_3)_2]^-$, and HFE is hydrofluoroether. Reprinted with permission from Ref. [32]. Copyright 2016, Electrochemical Society of Japan

electrolyte reduced the internal resistance of the cell and allowed to provide the large Li^+ ions flux required for the electrode reactions.

After 200 charge/discharge cycles, the Si/Li_2S cell was disassembled to observe the surfaces of the electrodes (Fig. 6) [32]. The morphology of the Li_2S/graphene composite particles and highly porous structure of the carbon fiber network of the CFP was well preserved for the cycled Li_2S cathode. On the contrary, some agglomerations of Si flakes and coverage of flakes by the decomposition products of the electrolyte were observed for the cycled Si anode, however, no obvious cracks were observed on Si-LP.

4 Future Challenges and Prospects

Graphite|electrolyte|Li_2S cells were constructed by combining Li_2S/graphene composite, solvate IL, and graphite as cathode, electrolyte, and anode, respectively,

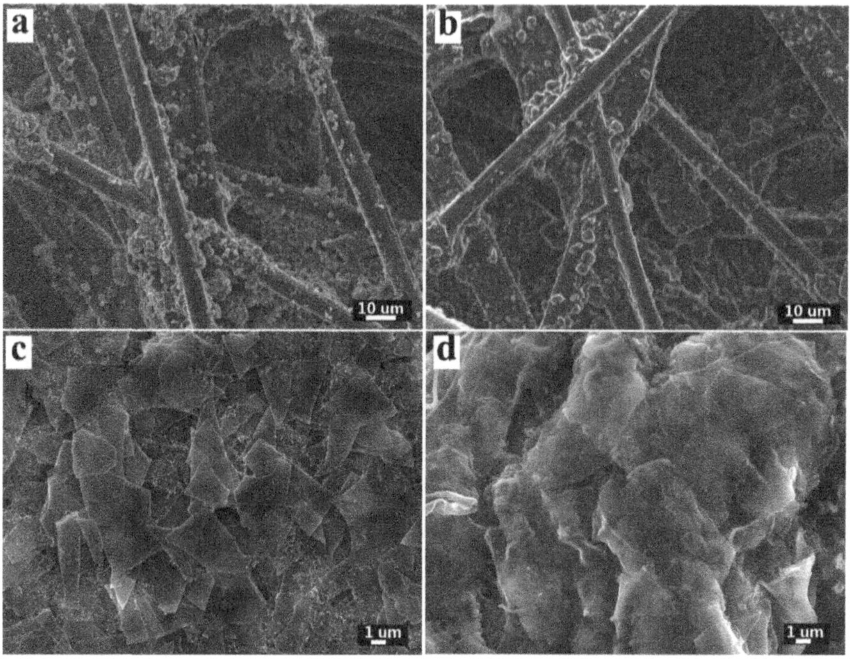

Fig. 6 Field-emission scanning electron microscopy images of **a** fresh Li_2S cathode, **b** Li_2S cathode after 200 cycles, **c** fresh Si anode, and **d** Si anode after 200 cycles. Reprinted with permission from Ref. [32]. Copyright 2016, Electrochemical Society of Japan

using simple cell configurations, where solvate IL, acting as single-phase electrolyte, could enable the effective Li^+ ions intercalation/deintercalation into/from graphite electrodes and could simultaneously inhibit the severe shuttle effect of LPSs. The ionic conductivity and Li^+ ions intercalation/deintercalation properties of the electrolytes were determined to greatly affect the battery performance. Moreover, $[Li(G4)_{0.8}][TFSA]$/HFE, which possessed both balanced conductivity and intercalation/deintercalation capability, exhibited the highest 1st- and 20th-cycles discharge capacities of 809 and 586 mAh/g, respectively, in conjunction with the high Coulombic efficiency of up to 99%. However, it may be difficult that the energy density of this type of cell exceeds that of state-of-art Li-ion batteries.

To increase the possible energy density of Li_2S batteries, an Si|electrolyte|Li_2S cell was fabricated using Si-LP, a solvate IL, and Li_2S/graphene as anode, electrolyte, and cathode, respectively. The cell exhibited the discharge capacity of 292 mAh/g at 1 C even after 200 cycles. The stable operation of the cell was enabled by the intimate contact between graphene and Li_2S in the cathode, low Li-diffusion-length two-dimensional Si-LP at the anode, and solvate IL, whereas the swelling of the anode caused by the formation of the thick SEI layer owing to the decomposition of the electrolyte was still challenging.

References

1. Bruce, P. G., Freunberger, S. A., Hardwick, L. J., & Tarascon, J. M. (2012). Li-O_2 and Li-S batteries with high energy storage. *Nature Materials, 11,* 19–29.
2. Ji, X., Lee, K. T., & Nazar, L. F. (2009). A highly ordered nanostructured carbon-sulphur cathode for lithium-sulphur batteries. *Nature Materials, 8,* 500–506.
3. Manthiram, A., Fu, Y., & Su, Y. S. (2013). Challenges and prospects of lithium-sulfur batteries. *Accounts of Chemical Research, 46,* 1125–1134.
4. Song, M. K., Cairns, E. J., & Zhang, Y. (2013). Lithium/sulfur batteries with high specific energy: Old challenges and new opportunities. *Nanoscale, 5,* 2186–2204.
5. Zhang, S. S. (2013). Liquid electrolyte lithium/sulfur battery: Fundamental chemistry, problems, and solutions. *Journal of Power Sources, 231,* 153–162.
6. Urbonaite, S., Poux, T., & Novák, P. (2015). Progress towards commercially viable Li-S battery cells. *Advanced Energy Materials, 5,* 1500118.
7. Nan, C., Lin, Z., Liao, H., Song, M. K., Li, Y., & Cairns, E. J. (2014). Durable carbon-coated Li_2S core-shell spheres for high performance lithium/sulfur cells. *Journal of The American Chemical Society, 136,* 4659–4663.
8. Wu, F., Lee, J. T., Fan, F., Nitta, N., Kim, H., Zhu, T., & Yushin, G. (2015). A hierarchical particle-shell architecture for long-term cycle stability of Li_2S cathodes. *Advanced Materials, 27,* 5579–5586.
9. Zhang, S., Ueno, K., Dokko, K., & Watanabe, M. (2015). Recent advance in electrolytes for lithium-sulfur batteries. *Advanced Energy Materials, 5,* 1500117.
10. Park, J. W., Ueno, K., Tachikawa, N., Dokko, K., & Watanabe, M. (2013). Ionic liquid electrolytes for lithium-sulfur batteries. *The Journal of Physical Chemistry C, 117,* 20531–20541.
11. Dokko, K., Tachikawa, N., Yamauchi, K., Tsuchiya, M., Yamazaki, A., Takashima, E., et al. (2013). Solvate ionic liquid electrolyte for Li-S batteries. *Journal of the Electrochemical Society, 160,* A1304–A1310.
12. Watanabe, M., Dokko, K., Ueno, K., & Thomas, M. L. (2018). From ionic liquids to solvate ionic liquids: Challenges and opportunities for next generation battery electrolytes. *Bulletin of the Chemical Society of Japan, 91,* 1660–1682.
13. Manthiram, A., Fu, Y., Chung, S. H., Zu, C., & Su, Y. S. (2014). Rechargeable lithium-sulfur batteries. *Chemical Reviews, 114,* 11751–11787.
14. Son, Y., Lee, J. S., Son, Y., Jang, J. H., & Cho, J. (2015). Recent advances in lithium sulfide cathode materials and their use in lithium sulfur batteries. *Advanced Energy Materials, 5,* 1500110.
15. Yang, Y., McDowell, M. T., Jackson, A., Cha, J. J., Hong, S. S., & Cui, Y. (2010). New nanostructured Li_2S/silicon rechargeable battery with high specific energy. *Nano Letters, 10,* 1486–1491.
16. Zhang, K., Wang, L., Hu, Z., Cheng, F., & Chen, J. (2014). Ultrasmall Li_2S nanoparticles anchored in graphene nanosheets for high-energy lithium-ion batteries. *Scientific Reports, 4,* 6467.
17. Li, Z., Zhang, S., Zhang, C., Ueno, K., Yasuda, T., Tatara, R., et al. (2015). One-pot pyrolysis of lithium sulfate and graphene nanoplatelet aggregates: In situ formed Li_2S/graphene composite for lithium-sulfur batteries. *Nanoscale, 7,* 14385–14392.
18. Lv, D., Yan, P., Shao, Y., Li, Q., Ferrara, S., Pan, H., et al. (2015). High performance Li-Ion sulfur batteries enabled by intercalation chemistry. *Chemical Communications, 51,* 13454–13457.
19. Aurbach, D., Talyosef, Y., Markovsky, B., Markevich, E., Zinigrad, E., Asraf, L., et al. (2004). Design of electrolyte solutions for li and li-ion batteries: A review. *Electrochimica Acta, 50,* 247–254.
20. Gao, J., Lowe, M. A., Kiya, Y., & Abruña, H. D. (2011). Effects of liquid electrolytes on the charge-discharge performance of rechargeable lithium/sulfur batteries: Electrochemical and in-situ x-ray absorption spectroscopic studies. *The Journal of Physical Chemistry C, 115,* 25132–25137.

21. Zheng, S., Chen, Y., Xu, Y., Yi, F., Zhu, Y., Liu, Y., et al. (2013). In situ formed lithium sulfide/microporous carbon cathodes for lithium-ion batteries. *ACS Nano, 7,* 10995–11003.
22. Wang, L., Wang, Y., & Xia, Y. (2015). A high performance lithium-ion sulfur battery based on a Li_2S cathode using a dual-phase electrolyte. *Energy and Environmental Science, 8,* 1551–1558.
23. Moon, H., Tatara, R., Mandai, T., Ueno, K., Yoshida, K., Tachikawa, N., et al. (2014). Mechanism of Li ion desolvation at the interface of graphite electrode and Glyme–Li salt solvate ionic liquids. *The Journal of Physical Chemistry C, 118,* 20246–20256.
24. Moon, H., Mandai, T., Tatara, R., Ueno, K., Yamazaki, A., Yoshida, K., et al. (2015). Solvent activity in electrolyte solution controls electrochemical reactions in Li-Ion and Li-Sulfur batteries. *The Journal of Physical Chemistry C, 119,* 3957–3970.
25. Park, C. M., Kim, J. H., Kim, H., & Sohn, H. J. (2010). Li-alloy based anode materials for Li secondary batteries. *Chemical Society Reviews, 39,* 3115–3141.
26. Hatchard, T. D., & Dahn, J. R. (2004). In situ XRD and electrochemical study of the reaction of lithium with amorphous silicon. *Journal of the Electrochemical Society, 151,* A838–A842.
27. Wu, H., Zheng, G., Liu, N., Carney, T. J., Yang, Y., & Cui, Y. (2012). Engineering empty space between Si nanoparticles for lithium-ion battery anodes. *Nano Letters, 12,* 904–909.
28. M. Haruta, Y. Masuo, T. Moriyasu, A. Tomita, C. Sakakibara, A. Kamei, M. Hirota, T. Takenaka, T. Doi, & M. Inaba. (2015). Cycle performances of Si-flake-powder anodes in lithium salt-tetraglyme complex electrolytes. *Electrochemistry, 83,* 837–839.
29. Saito, M., Yamada, T., Yodoya, C., Kamei, A., Hirota, M., Takenaka, T., et al. (2012). Influence of Li diffusion distance on the negative electrode properties of Si thin flakes for Li secondary batteries. *Solid State Ionics, 225,* 506–509.
30. Li, Z., Zhang, S., Terada, S., Ma, X., Ikeda, K., Kamei, Y., et al. (2016). Promising cell configuration for next-generation energy storage: Li_2S/graphite battery enabled by a solvate ionic liquid electrolyte. *ACS Applied Materials and Interfaces, 8,* 16053–16062.
31. Seita, T., Matsumae, Y., Liu, J., Tatara, R., Ueno, K., Dokko, K., & Watanabe, M. (2020). Graphite-lithium sulfide battery with a single-phase sparingly solvating electrolyte. *ACS Energy Letters, 5,* 1–7.
32. Z. Li, Y. Kamei, M. Haruta, T. Takenaka, A. Tomita, T. Doi, S. Zhang, K. Dokko K, M. Inaba, & M. Watanabe. (2016). Si/Li_2S battery with solvate ionic liquid electrolyte. *Electrochemistry, 84,* 887–890.

Scale-up Efforts

Tetsuya Osaka, Tokihiko Yokoshima, Hiroki Nara, Hitoshi Mikuriya, and Toshiyuki Momma

Abstract Both the high sulfur loading cathode and less electrolyte are the most important parameters to realize lithium-sulfur batteries with high energy density. A high sulfur loading cathode was achieved by using 3D structured aluminum foam to act as a support structure and current collector. Moreover, the energy density close to 200 Wh kg^{-1} was successfully obtained by decreasing the electrolyte in the pouch cell. 5 Ah cell was developed by using the six sheets of the cathode in the cell. Recently, a multi-layer pouch-type cell with a high capacity of 5 Ah and a high energy density of 300 Wh kg^{-1} was successfully developed with technically reoptimized preparation conditions. These results indicate the possibility of the development of a lithium-sulfur battery with discharge capacity and high energy density for commercial use.

Keywords Sulfur cathode · Aluminum foam current collector · Areal capacity · Electrolyte/sulfur (E/S) ratio · Theoretical vacancy filling rate

1 Background

Sulfur is a promising cathode material for the lithium-based battery because of its advantages such as eco-friendliness, abundance in nature, and a high theoretical capacity of 1675 mAh g^{-1}. The sulfur cathode possesses extremely high capacities with the ability to overcome the shortcomings of its low operating potential, thereby increasing its energy density. The sulfur cathode possesses many advantages for practical use: lithium-sulfur batteries have a theoretical energy density (reported to be around 2500 Wh kg^{-1} and around 2800 Wh L^{-1}) [1, 2] much higher than that of conventional graphite-$LiCoO_2$ batteries (387 Wh kg^{-1}), and sulfur is both abundant in nature and of low toxicity [2]. However, sulfur cathodes are not without their own

T. Osaka (✉) · T. Yokoshima · H. Nara · H. Mikuriya
Research Organization for Nano and Life Innovation, Waseda University, Tokyo, Japan
e-mail: osakatets@waseda.jp

T. Momma
Faculty of Science and Engineering, Waseda University, Tokyo, Japan

K. Kanamura (ed.), *Next Generation Batteries*,
https://doi.org/10.1007/978-981-33-6668-8_36

intrinsic problems such as low ionic and electronic conductivity, large volumetric expansion upon lithiation from S_8 to Li_2S (around 180%), the dissolution of so-called polysulfide intermediates (Li_2S_x where x represents a value between 4 and 8) [3], and low sulfur loading in the cathode [4]. Electrolytes, such as glyme-based ionic liquids, have also been investigated to suppress the dissolution of polysulfides [5]. This electrolyte system has received attention owing to its action as a sparingly solvating electrolyte for polysulfides. One issue that must be overcome before their practical use can be fully realized is that of low sulfur loading in the cathode and a subsequently low sulfur loading ratio in the entire cathode [4]. Hagen et al. determined that the areal sulfur loading of the majority of the investigated sulfur cathodes was less than 2.0 mg cm^{-2} [6]. Gao and Abruña demonstrated that a low sulfur loading ratio in the cathode of a lithium-sulfur battery (less than around 70%) resulted in the battery possessing a lower energy density than a conventional lithium-ion battery [7]. In an effort to remediate this effect, Zhou et al. have investigated a promising method allowing for the achievement of high sulfur loading using a graphene foam as a current collector, and have summarized the previously reported relationship between the sulfur loading and areal capacity [4]. The group was able to achieve a high sulfur loading of 10.1 mg cm^{-2} with a high areal capacity of 13.4 mAh cm^{-2}. However, the mass production of graphene foam is an industry that has not yet been established for practical use. The relationship between the sulfur loading and the areal capacity of the recently developed sulfur cathodes is shown in Fig. 1 [4, 8].

Fig. 1 The sulfur loading and the areal capacity of the cathodes [4, 8]

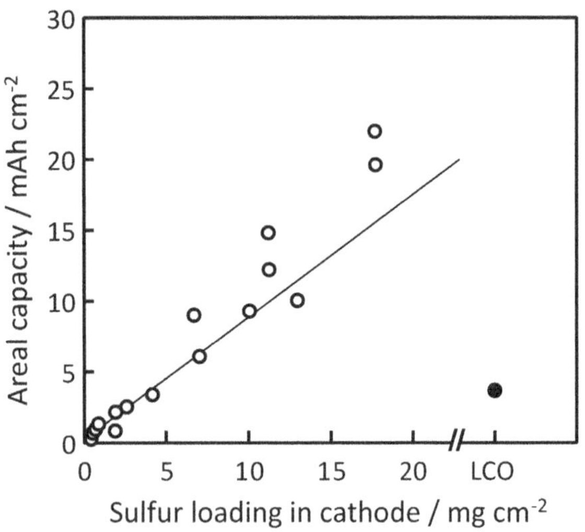

2 Preparation of a High Sulfur Loading Cathode

An attempt was made by the Osaka group to verify the possibility of achieving a lithium-sulfur battery with an energy density of 200 Wh kg^{-1} by using a high sulfur loading cathode [8]. In order to achieve a high sulfur loading cathode, 3D structured aluminum foam which is commercially available was selected to act as a support structure and current collector for this system. Celmet™ (Sumitomo Electric Industries, Ltd.) with 1.0 mm thick and 550 μm pores was chosen as the 3D structured metal foams because of a superior high porosity. Figure 2 shows charge-discharge curves (a) and charge-discharge cycle properties (b) of a single-layer pouch-type cell

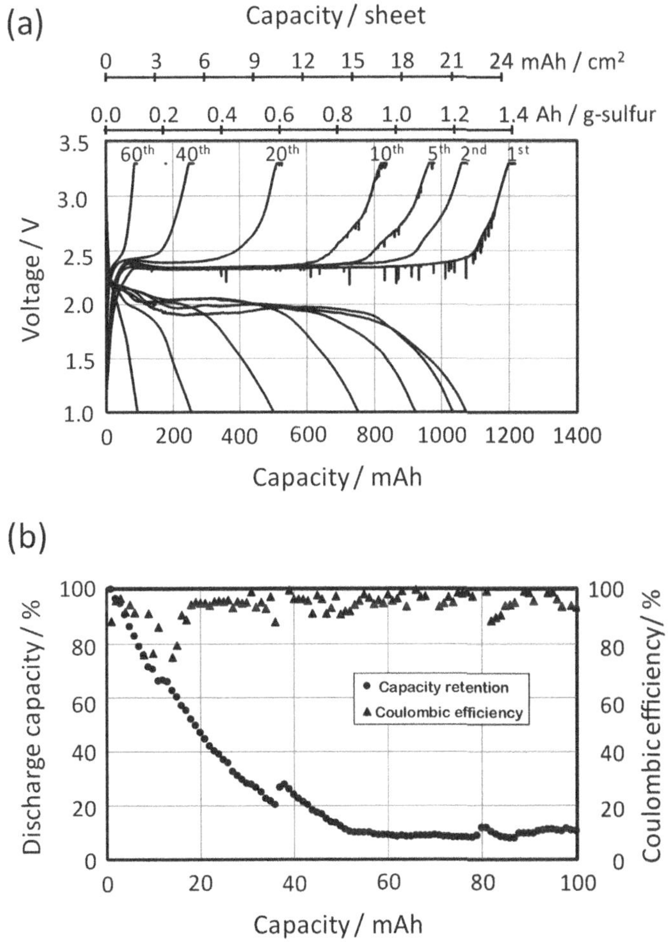

Fig. 2 Charge-discharge curves (**a**) and charge-discharge cycle properties (**b**) of lithium-sulfur batteries possessing high sulfur loading on Al foam. The number of the cathode was one layer

possessing high sulfur loading on the Al foam. In this study, sulfur/Ketjenblack (KB) composite was used as active materials in a 4–6 weight ratio. Carboxymethyl cellulose (CMC) and styrene-butadiene rubber (SBR) (CMC + SBR, at 2:1 weight ratio) were used for the binder. The full cell was assembled with the configuration of an S/KB on Al foam cathode, a Li metal on Cu foil anode, and a glyme-based electrolyte as a solvate ionic liquid. The dimension of the electrodes was 7.0 cm × 7.0 cm. The glyme-based electrolyte is composed of ethylene glycol dimethyl ether (G1), triethylene glycol diethyl ether (G3), lithium bis (trifluoromethanesulfonyl) imide (LiTFSI), and 1,1,2,2–tetrafluoroethyl 2,2,3,3–tetrafluoropropyl ether (HFE) ([Li(G1)$_2$][TFSI], [Li(G3)][TFSI], and HFE at 1:1:8 volume ratio). The amount of electrolyte used was as much volume as 260% of the theoretical vacancy of the cathode and the separator. The charge-discharge property of the sulfur cathodes was evaluated with a cut-off voltage between 1.0 and 3.3 V. This pouch cell shows over 1 Ah at the discharge rate of 0.01 C-rate, only containing one cathode sheet. Therefore, a high areal capacity of 21.9 mAh cm^{-2} at the discharge rate of 0.01 C-rate was demonstrated. A high sulfur loading of 17.5 mg cm^{-2} on Al form was realized and 1241 mAh (g-sulfur)$^{-1}$ was obtained. This high sulfur loading was accomplished using a filling technique that they have developed involving the deposit of a sulfur carbon composite slurry into a 3D structured metal foam current collector, the production of which has been industrially established. This result using the industrial technique will be expected to realize a 5 Ah cell using 5 or more sheets of the cathode.

3 Preparation of 5 Ah Cell

Figure 3 shows charge-discharge curves (a) and charge-discharge cycle properties (b) of a multi-layer pouch-type cell possessing high sulfur loading on Al foam. The number of cathodes was six. The other parameters were the same as those explained in Sect. 1. This pouch cell shows 5837 mAh at the discharge rate of 0.01 C-rate. The capacity of one sheet of the cathode was about 970 mAh. Therefore, a high areal capacity of 19.9 mAh cm^{-2} at the discharge rate of 0.01 C-rate was demonstrated. A high sulfur loading of 15.3 mg cm^{-2} on Al form was realized and 1350 mAh (g-sulfur)$^{-1}$ was obtained. However, the energy density of the stack (electrodes, separators, and electrolyte) in the cell was 103 Wh kg^{-1}, because of a large amount of electrolyte in the cell. In this investigation, the amount of electrolyte used was as much volume as 260% of the theoretical vacancy filling rate of the cathode and the separator for obtaining cycle performance better. The electrolyte/sulfur (E/S) ratio (μL mg^{-1}) was about 13.5. The theoretical energy density of the stack with minimum requirement parameter (theoretical vacancy filling rate = 100%, minimum amount of Li anode, thinner current collector, etc.) was about 200 Wh kg^{-1}. Therefore, to improve energy density, decreasing the electrolyte in the cell was carried out.

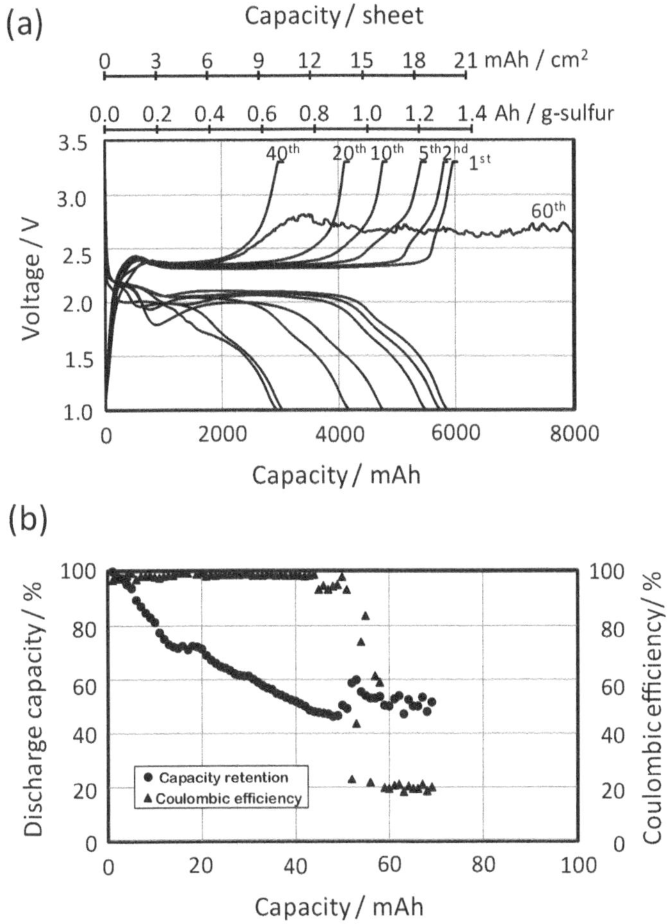

Fig. 3 Charge-discharge curves (**a**) and charge-discharge cycle properties (**b**) of lithium-sulfur batteries possessing high sulfur loading on Al foam. The number of cathodes was six

4 Preparation of Cells with 200 Wh/kg or Higher

Figure 4 shows charge-discharge curves of charge-discharge cycle properties of a single-layer pouch-type cell (a) and a multi-layer pouch-type cell (b) possessing high sulfur loading on Al foam. The dimensions of electrodes of the single-layer cell and multi-layer cell were 3.5 cm × 7.0 cm and 7.0 cm × 7.0 cm, respectively. The number of cathodes in the multi-layer cell was six. The theoretical vacancy filling rates of the single-layer cell and multi-layer cell were 103% and 106%, respectively, and the E/S ratios of the stack of the cells were 5.1 and 5.4, respectively. The discharge capacities of the cells were 491 mAh and 5398 mAh, respectively, and energy densities 183 Wh kg^{-1} and 180 Wh kg^{-1}, respectively, were obtained successfully. On the

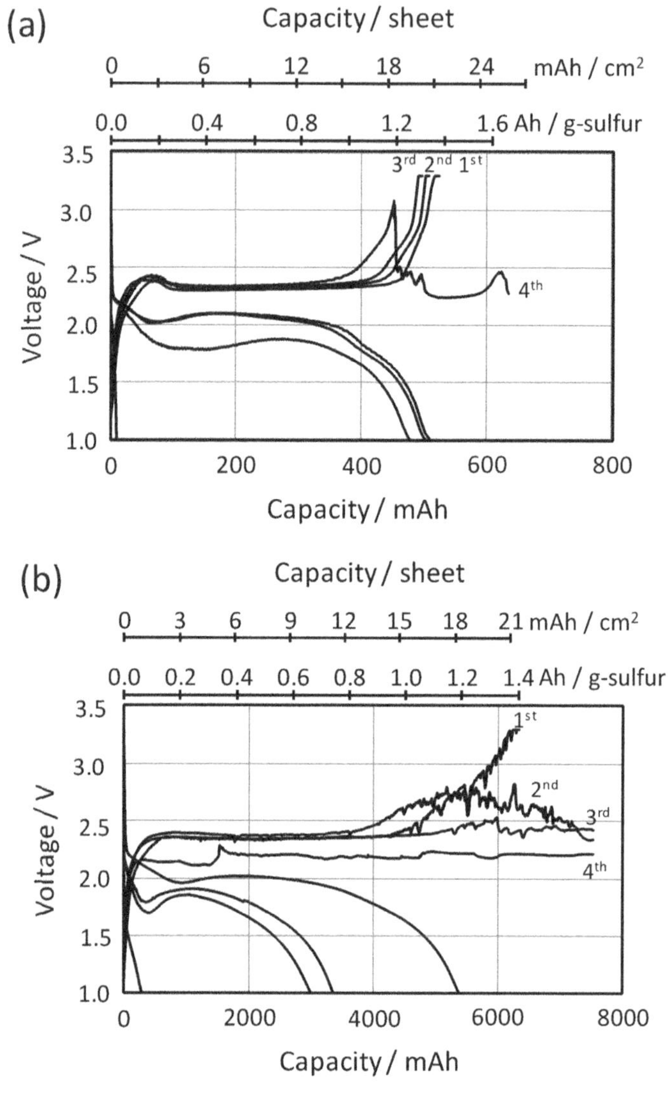

Fig. 4 Charge-discharge curves and charge-discharge cycle properties of a single-layer pouch-type cell (**a**) and a multi-layer pouch-type cell (**b**) possessing high sulfur loading on Al foam. The number of cathodes in **b** was six

other hand, the cycle performances dramatically degraded, achieving only three and one cycle of battery operation, respectively. These results indicated that a theoretical vacancy filling rate of about 100% was too low to show better cycle performance. Recently, we have technically reoptimized the composition and the preparation condition of the sulfur cathode. And a multi-layer pouch-type cell (six cathode layers, E/S ratio of 2.3, and 15.4 mg cm^{-2} of sulfur loading) with a high capacity of 5200 mAh and a high energy density of 301 Wh kg^{-1} (at the first cycle and discharge rate of 0.01 C-rate) was successfully achieved [9].

Energy density close to 200 Wh kg^{-1} was successfully obtained by decreasing the electrolyte in the pouch cell, where the high sulfur loading cathode and less electrolyte are the most important parameters to realize lithium-sulfur batteries with high energy density. Recently, a multi-layer pouch-type cell with a high capacity of 5 Ah and a high energy density of 300 Wh kg^{-1} was successfully developed with technically reoptimized preparation conditions, especially the high energy-density cathode with both high sulfur loading and less electrolyte was successfully developed. However, still some issues remain to develop lithium-sulfur batteries for commercial use. Here, some techniques for solving problems have already been reported, for example, Li^+ permeable and polysulfide nonpermeable organic film in a lithium-sulfur battery was developed [10], and a lithium-sulfur battery using sulfur cathode on Al foam using this high-function film at a high discharge rate of over 2 C-rate was demonstrated with an optimized condition for high-power output [11]. These results indicate the possibility of the development of lithium-sulfur batteries with discharge capacity and high energy density for commercial use.

References

1. Ji, X., Lee, K. T., & Nazar, L. F. (2009). A highly ordered nanostructured carbon–sulphur cathode for lithium–sulphur batteries. *Nature Materials, 8,* 500–506. https://doi.org/10.1038/nmat2460.
2. Hassoun, J., & Scrosati, B. (2010). A high-performance polymer tin sulfur lithium ion battery. *Angewandte Chemie International Edition, 49,* 2371–2374. https://doi.org/10.1002/anie.200907324.
3. Wang, K. X., Li, X. H., & Chen, J. S. (2015). Surface and interface engineering of electrode materials for lithium-ion batteries. *Advanced Materials, 27,* 527–545. https://doi.org/10.1002/adma.201402962.
4. Zhou, G., Li, L., Ma, C., Wang, S., Shi, Y., Koratkar, N., et al. (2015). A graphene foam electrode with high sulfur loading for flexible and high energy Li-S batteries. *Nano Energy, 11,* 356–365. https://doi.org/10.1016/j.nanoen.2014.11.025.
5. Dokko, K., Tachikawa, N., Yamauchi, K., Tsuchiya, M., Yamazaki, A., Takashima, E., et al. (2013). Solvate ionic liquid electrolyte for Li–S batteries. *Journal of the Electrochemical Society, 160,* A1304–A1310. https://doi.org/10.1149/2.111308jes.
6. Hagen, M., Dörfler, S., Fanz, P., Berger, T., Speck, R., Tübke, J., et al. (2013). Development and costs calculation of lithium-sulfur cells with high sulfur load and binder free electrodes. *Journal of Power Sources, 224,* 260–268. https://doi.org/10.1016/j.jpowsour.2012.10.004.
7. Gao, J., & Abruña, H. D. (2014). Key parameters governing the energy density of rechargeable Li/S batteries. *The Journal of Physical Chemistry Letters, 5,* 882–885. https://doi.org/10.1021/jz5001819.
8. Nara, H., Yokoshima, T., Mikuriya, H., Tsuda, S., Momma, T., & Osaka, T. (2017). The potential for the creation of a high areal capacity lithium-sulfur battery using a metal foam current collector. *Journal of the Electrochemical Society, 164,* A5026–A5030. https://doi.org/10.1149/2.0381701jes.
9. Mikuriya, H., Shiosaki, A., Ahn, S., Nara, H., Nakamura, N., & Osaka, T. (2020). 300 Wh/kg class high-energy Li-S laminated batteries using high-sulfur loading cathode on 3D mesh structured aluminum foam. In *Pacific RIM Meeting (PRiME2020), A02-0436.*

10. Nakamura, N., Yokoshima, T., Nara, H., Momma, T., & Osaka, T. (2015). Suppression of polysulfide dissolution by polypyrrole modification of sulfur-based cathodes in lithium secondary batteries. *Journal of Power Sources, 274,* 1263–1266. https://doi.org/10.1016/j.jpowsour.2014.10.192.
11. Nakamura, N., Yokoshima, T., Nara, H., Mikuriya, H., Ahn, S., Shiosaki, A., et al. (2021). High-rate and high sulfur-loaded lithium-sulfur batteries with a polypyrrole-coated sulfur cathode on a 3D aluminum foam current collector. *Materials Letters,285, 129115.* https://doi.org/10.1016/j.matlet.2020.129115.

Lithium-Air Battery

Lithium–Air Battery System

Yoshimi Kubo

Abstract The lithium–air battery is "ultimate battery" with the highest theoretical energy density, using oxygen in the air as the cathode active material, and lithium metal as the anode active material, and are expected to be used in various applications such as electric vehicles, household batteries, and drones. However, many technical challenges remain, such as increased charging voltage and low cycle performance, which are mostly related to parasitic reactions within the cell. Further, in order to actually realize a high energy density, it is necessary to stack a large number of lithium–air cells as in the case of conventional lithium-ion batteries, which requires ensuring air passage. Here, the basic aspects of lithium–air batteries are described, as well as the newly developed "air-breathing" passive stack where air is breathed in and out through porous current collectors sandwiched between cathodes.

Keywords Lithium–air battery · Battery reaction · Cell stack · Air breathing · Passive stack

1 Introduction

The lithium–air battery (LAB) is the "ultimate rechargeable battery" with the highest theoretical energy density, using oxygen in the air as the cathode active material, and lithium metal as the anode active material. Lithium is an ideal anode active material because it is the lightest metal and has the most negative electrode potential. Since oxygen is taken in from the air rather than retaining inside the battery; the combination of these two elements constitutes the lightest battery. The battery capacity for the active material inside the battery (Li metal) is 3860 mAh/g (1170 mAh/g for Li_2O_2 after discharge), and the energy density calculated by multiplying the discharge voltage of ~2.7 V exceeds 10,000 Wh/kg (3100 Wh/kg for the discharged state). In the actual battery configuration, this value is smaller because the weight of the other materials is also considered. However, even if it is as low as 1/10,

Y. Kubo (✉)
National Institute for Materials Science, 1-1 Namiki, Tsukuba 305-0044, Japan
e-mail: KUBO.Yoshimi@nims.go.jp

K. Kanamura (ed.), *Next Generation Batteries*,
https://doi.org/10.1007/978-981-33-6668-8_37

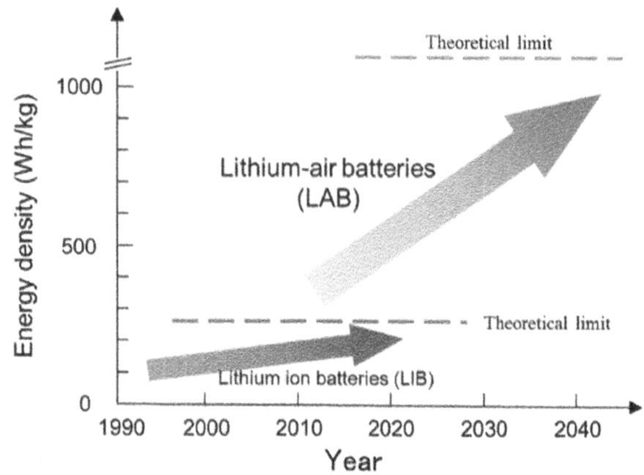

Fig. 1 Roadmap for energy density of lithium–air batteries

it becomes 1000 Wh/kg, which exceeds the theoretical limit (~250 Wh/kg) of the current lithium-ion battery (LIB) by several times. Lithium–air batteries are overwhelmingly lightweight with a large energy density. Figure 1 shows a development roadmap for energy density.

The demands on battery capacity (run time) are endless. It is well known that the LIB, which was put into practical use in 1991, contributed greatly to the popularization of portable telephones and notebook PCs due to its small size and lightness, which greatly exceeded those of conventional batteries. However, the demands on battery capacity remain high. Currently, LIBs comprise half the contents of smartphones, but they are insufficient for watching videos. In the future, there is no doubt that even more lightweight and larger capacity batteries will be necessary for eyeglass terminals, which will become important in the IoT society. The drone, which has recently been a hot topic, has a flight time of about 15 min, but there is a high demand for continuous flight exceeding 1 h.

Although the market for large LIBs for electric vehicles has rapidly increased since 2010, serious problems remain in terms of cruising distance and cost. To ensure a cruising range equivalent to that of a gasoline-powered vehicle, it is necessary to install about 100 kWh of energy. However, the weight of the LIB is calculated to be ~700 kg at present and ~500 kg even in the future. Conversely, lithium–air batteries can reduce the battery weight to 200 kg or less. LIBs also have resource problems. Particularly, it has been suggested that rare metals such as Ni and Co will become scarcer as the production volume increases. Conversely, since lithium–air batteries contain no Ni or Co, they pose no resource problem, and recycling is easy. Thus, considerable cost reductions are also possible.

Non-aqueous lithium–air batteries were first proposed by Abraham and Jiang in 1996 [1]. Since then, the instability of the air electrode reaction has posed a major problem [2, 3]. However, in around 2011, reasonably stable discharge and charge became possible with an ether-based electrolyte [4–9], and research became active worldwide. Consequently, the cell characteristics have been greatly improved, and the understanding of the cell reactions has deepened. However, the cycling performance remains below 100 cycles, which is a major obstacle to practical applications. The main problems are the increase in the charge voltage and the formation of Li dendrites, which will be discussed in detail in later sections.

2 Structure and Principle of the Lithium–Air Battery

Figure 2 shows a schematic of a lithium–air battery. Its structure is similar to that of a LIB, involving simple stacking of positive and negative electrodes with a separator in between, filled with electrolyte. However, there are small holes for taking in air (oxygen) on the positive electrode (cathode). Porous carbon is typically used as the current collector for the cathode. The electrolyte is contained in the pores of the porous cathode and separator. In the discharge reaction, the Li metal anode dissolves as Li^+ ions that react with oxygen in the cathode to produce solid lithium peroxide (Li_2O_2) on the carbon. Charging is the reverse reaction, in which Li_2O_2 on the cathode decomposes to release oxygen, and Li metal is plated on the anode. Reactions at both electrodes as well as the overall reaction are as follows (discharge to the right and charge to the left).

$$\text{Anode: Li} \leftrightarrow \text{Li}^+ + \text{e}^- \tag{1}$$

$$\text{Cathode: O}_2 + 2\text{Li}^+ + 2\text{e}^- \leftrightarrow \text{Li}_2\text{O}_2 \tag{2}$$

$$\text{Overall: 2Li} + \text{O}_2 \leftrightarrow \text{Li}_2\text{O}_2, \quad E = 2.96\,\text{V} \tag{3}$$

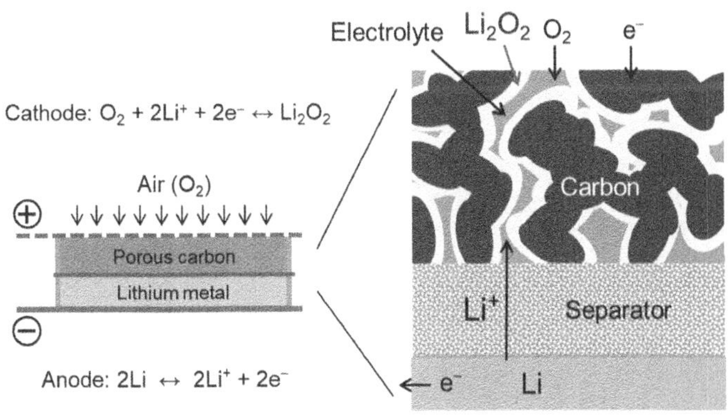

Fig. 2 Schematics of a lithium–air battery

As shown in reaction (3), the lithium–air battery extracts electrical energy from the free energy change of Li oxidation, and the theoretical voltage is 2.96 V. Interestingly, the reaction product is peroxide Li_2O_2 rather than oxide Li_2O. Reaction (2) shows the formation of peroxide ions (O_2^{2-}) by the two-electron reduction of oxygen, which incompletely dissociates the O–O bond of O_2. This is in sharp contrast to the fuel cell reaction, in which the O–O bond is completely dissociated by the four-electron reduction of oxygen to form water ($O_2 + 4H^+ + 4e^- \rightarrow 2H_2O$). However, in the absence of a platinum catalyst, the reaction is usually limited to two-electron reduction ($O_2 + 2H^+ + 2e^- \rightarrow H_2O_2$) to produce hydrogen peroxide. This suggests the existence of a reaction barrier between the two-electron and the four-electron reduction of oxygen, which requires a platinum catalyst to overcome it. However, the lithium–air battery based on the two-electron reduction requires no expensive noble metal catalysts.

Figure 3 shows the crystal structure of lithium peroxide, Li_2O_2 [10], which is a typical ionic crystal composed of Li^+ and O_2^{2-}, but a strong covalent O–O bond remains inside the peroxide ion (O_2^{2-}) with a slightly elongated O–O bond length. The charging reaction is the oxidative decomposition of Li_2O_2. Electrons are extracted from O_2^{2-} to produce O_2 gas, and Li^+ is eluted into the electrolyte. The discharge–charge reactions of lithium–air batteries are such two-electron reactions based on peroxides, which proceed without breaking O–O bonds. Although Li_2O_2 is theoretically an insulator, calculations predict that hopping conduction of hole carriers may occur if Li deficiency exists [11], and that Li deficiency may be stable on certain crystal surfaces [12]. Since Li_2O_2 deposited by discharge consists of nanocrystals of about 10 nm, it might exhibit some hole carrier conductivity due to Li deficiency at grain boundaries.

The discharge product, Li_2O_2, is produced on the cathode surface. Therefore, to increase the current density, the cathode should be a porous material with a large specific surface area. Additionally, it should have high porosity to store sufficient Li_2O_2. Nanocarbon materials such as carbon black and carbon nanotubes are promising conductive materials capable of forming such a super porous body having a large surface area and high porosity [13].

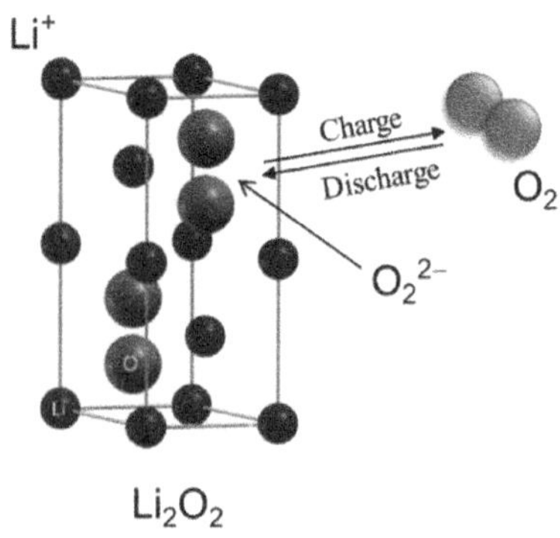

Fig. 3 Crystal structure (hexagonal) of lithium peroxide (Li_2O_2) and O_2 migration accompanying discharge and charge

Research on lithium–air batteries was initially performed using standard carbonate solvents for the LIBs, such as PC, EC, and DEC, but around 2010, it was found that these solvents were decomposed [14]. This is because the superoxide anion produced during the first step of the oxygen reduction reaction at the cathode ($O_2 + e^- \rightarrow O_2^-$) can break the C–O bond by nucleophilic attack on the carbon adjacent to the carbonate group of the solvent [15–17]. New solvents that could withstand the strong electron-donating properties of O_2^- were then sought, and a series of glycol diether solvents called glymes, $CH_3O(C_2H_4O)_nCH_3$, were found to be reasonably stable [4–9]. Among these, tetraglyme ($n = 4$), which has a high boiling point of 275 °C and a low vapor pressure, appears to be a suitable solvent for air batteries. Lithium salts, such as $LiCF_3SO_3$ and $Li(CF_3SO_2)_2N$ (LiTFSI), were used as solutes. The glyme solvents are also comparatively stable with Li metal. Research on the lithium–air battery has been greatly advanced in recent years by using these electrolytes.

3 High Energy Density Cell Stack

Although lithium–air batteries theoretically have a very high energy density, to achieve a high energy density in practical applications, it is necessary to stack a large number of lithium–air cells with a high density, as in conventional LIBs. However, for a lithium–air battery, the problem is that air must be supplied to the cathode while achieving high-density stacking. In a fuel cell system, gases are supplied to the electrodes through flow channels formed on a bipolar plate. Although this method can be applied to the lithium–air system, it decreases the energy density and increases the cost. Therefore, we have developed a "passive parallel stack," as shown in Fig. 4, as a novel stacking technology that can achieve both a high energy density and an air supply at a low cost [18].

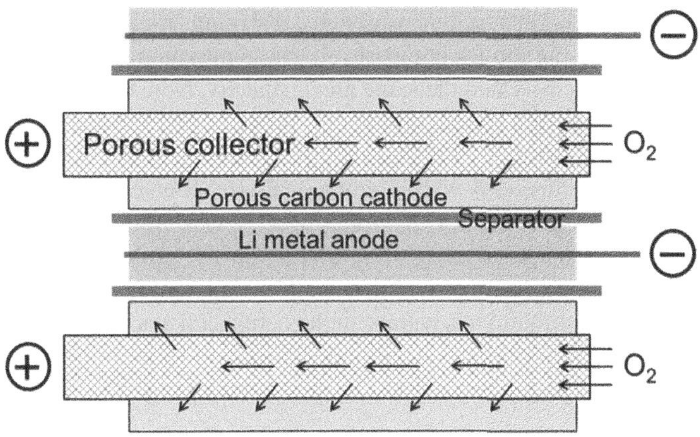

Fig. 4 Schematic of the air-breathing cell stack for lithium–air batteries

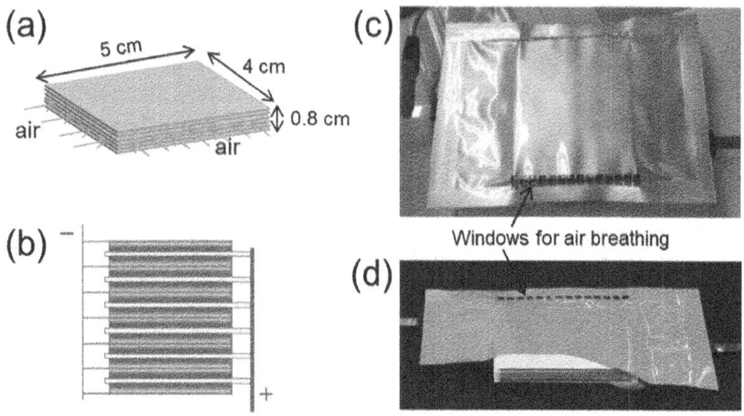

Fig. 5 Schematics of the general view (**a**) and stacking configuration (**b**) of the air-breathing ten-cell stack of the lithium–air battery. Photographs of the actual ten-cell stack (**c**) and its cutting model (**d**)

This stack configuration is the same as that in a LIB, in which electrodes are formed on both sides of the current collector, and the positive and negative electrode plates are laminated in parallel. The major difference is that the cathode current collector is very porous. Air is taken up from the periphery of the stack through the pores of the cathode current collector. Note that air (oxygen) is naturally breathed in by discharging and breathed out by charging with no additional mechanical devices, such as pumps. In such a passive air-breathing system, the driving force for the airflow is generated by the electrochemical reactions that consume/produce oxygen at the cathode. This stack structure is mass-producible, and the cost can be reduced.

Figure 5 shows a schematic and photographs of the developed stack. Ten cells with an electrode area of 4×5 cm are stacked with a thickness of 8 mm (total cell area is 200 cm^2), and the weight of the stack is about 16 g. Figure 6a shows the cycling performance of the ten-cell stack measured in a dry room. A parallel result for a coin cell is also shown in Fig. 6b for comparison. These measurements were performed at the same current density and the same areal capacity. Note that both results show very similar discharge and charge profiles for up to 16 or 17 cycles, although their cell areas differ by a factor of 100. These results clearly demonstrate that the passive air-breathing system functions well in a practical size. If a stack of this size works well, it should be easy to scale up. Furthermore, it was confirmed that this stack achieved a large discharge capacity of ~18 mAh/cm^2 by prolonged discharge. The calculated energy density of the stack reached 600 Wh/kg, demonstrating the high energy density of the lithium–air battery in the actual cell stack.

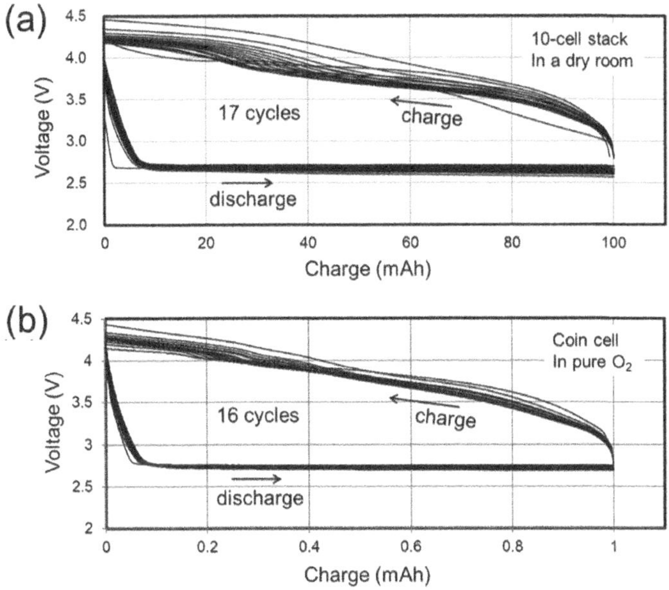

Fig. 6 **a** Cycling performance of the ten-cell stack (total cell area 200 cm^2) measured in a dry room at a constant current density of 0.05 mA/cm^2 with a discharge capacity of 0.5 mAh/cm^2. **b** A parallel result for a coin cell (cell area 2 cm^2) measured in pure O_2 is also shown for comparison; this cell was measured at the same current density and with the same discharge capacity

References

1. Abraham, K. M., & Jiang, Z. (1996). A polymer electrolyte-based rechargeable lithium/oxygen battery. *Journal of the Electrochemical Society, 143,* 1–5.
2. Girishkumar, G., McCloskey, B. D., Luntz, A. C., Swanson, S., & Wilcke, W. (2010). Lithium–air battery: Promise and challenges. *The Journal of Physical Chemistry Letters, 1,* 2193–2203.
3. Bruce, P. G., Freunberger, S. A., Hardwick, L. J., & Tarascon, J. M. (2012). Li–O_2 and Li–S batteries with high energy storage. *Nature Materials, 11,* 19–29.
4. Laoire, C. O., Mukerjee, S., Plichta, E. J., Hendrickson, M. A., & Abraham, K. M. (2011). Rechargeable lithium/TEGDME-$LiPF_6$ / O_2 battery. *Journal of the Electrochemical Society, 158,* A302–A308.
5. Hassoun, J., Croce, F., Armand, M., & Scrosati, B. (2011). Investigation of the O_2 electrochemistry in a polymer electrolyte solid-state cell. *Angewandte Chemie International Edition, 50,* 2999–3002.
6. McCloskey, B. D., Bethune, D. S., Shelby, R. M., Girishkumar, G., & Luntz, A. C. (2011). Solvents' critical role in nonaqueous lithium–oxygen battery electrochemistry. *The Journal of Physical Chemistry Letters, 2,* 1161–1166.
7. Freunberger, S. A., Chen, Y., Drewett, N. E., Hardwick, L. J., Bardé, F., & Bruce, P. G. (2011). The lithium–oxygen battery with ether-based electrolytes. *Angewandte Chemie International Edition, 50,* 8609–8613.
8. Lu, Y. C., Kwabi, D. G., Yao, K. P. C., Harding, J. R., Zhou, J., Zuin, L., & Shao-Horn, Y. (2011). The discharge rate capability of rechargeable Li–O_2 batteries. *Energy & Environmental Science, 4,* 2999–3007.

9. Jung, H. G., Hassoun, J., Park, J. B., Sun, Y. K., & Scrosati, B. (2012). An improved high-performance lithium-air battery. *Nature Chemistry, 4,* 579–585.
10. Cota, L. G., & de la Mora, P. (2005). On the structure of lithium peroxide, Li_2O_2. *Acta Crystallographica Section B, B61,* 133–136.
11. Ong, S. P., Mo, Y., & Ceder, G. (2012). Low hole polaron migration barrier in lithium peroxide. *Physical Review B, 85,* 081105(R).
12. Radin, M. D., Rodriguez, J. F., Tian, F., & Siegel, D. J. (2012). Lithium peroxide surfaces are metallic, while lithium oxide surfaces are not. *Journal of the American Chemical Society, 134,* 1093–1103.
13. Nomura, A., Ito, K., & Kubo, Y. (2017). CNT sheet air electrode for the development of ultra-high cell capacity in lithium-air batteries. *Scientific Reports, 7,* 45596.
14. Mizuno, F., Nakanishi, S., Kotani, Y., Yokoishi, S., & Iba, H. (2010). Rechargeable Li-air batteries with carbonate-based liquid electrolytes. *Electrochemistry, 78,* 403–405.
15. Freunberger, S. A., Chen, Y., Peng, Z., Griffin, J. M., Hardwick, L. J., Bardé, F., et al. (2011). Reactions in the rechargeable lithium–O_2 battery with alkyl carbonate electrolytes. *Journal of the American Chemical Society, 133,* 8040–8047.
16. Bryantsev, V. S., & Blanco, M. (2011). Computational study of the mechanisms of superoxide-induced decomposition of organic carbonate-based electrolytes. *The Journal of Physical Chemistry Letters, 2,* 379–383.
17. Okamoto, Y., & Kubo, Y. (2013). Ab initio calculations for decomposition mechanism of $CH_3O(CH_2CH_2O)NCH_3$ ($N = 1$–4) by the attack of O_2^- anion. *Journal of Physical Chemistry C, 117,* 15940–15946.
18. Kubo, Y., & Ito, K. (2014). Multicell stack of nonaqueous lithium–air batteries. *ECS Transactions, 62,* 129–135.

Air Cathode

Shin Mukai and Shinichiroh Iwamura

Abstract The air cathode of a lithium-air battery requires a large void space within it to efficiently store Li_2O_2 which is formed during discharging. It also requires a large surface area within it, along with efficient conductive paths which interconnect the surfaces, to smoothly provide/withdraw electrons during discharging/charging. Therefore, presently carbon paper coated with porous carbon powder is typically used. Carbon blacks are generally used as the porous carbon, as the flexibility of their porous structure usually leads to high capacity. However, this flexibility also leads to the instability of the cathode. Another interesting category of porous carbon is carbon gels, mesoporous carbon with a highly tunable structure. Carbon gels can be synthesized to have a large pore volume with a sufficient surface area, and also have a rigidity which ensures structural stability. Another interesting air cathode is a sheet made from fibrous carbon with a small diameter. Voids which are naturally formed within the sheet can be used to store Li_2O_2, and if the diameter of the fiber is small enough, a sufficient surface area can be provided by the carbon. As fibers naturally form a conductive network, an additional current collector is not necessary, so the weight of the cathode can also be significantly reduced.

Keywords Air cathode · Carbon black · Carbon gel · Carbon nanotube · Carbon nanofiber

1 Structure of the Air Cathode of LABs

When a lithium-air battery (LAB) is discharged, Li ions react with oxygen supplied from air and solid Li_2O_2 is deposited within the air cathode, and when a discharged battery is subsequently charged, the deposited Li_2O_2 decomposes, releasing Li ions

S. Mukai (✉) · S. Iwamura
Faculty of Engineering, Hokkaido University, N13W8 Kita-ku, Sapporo, Hokkaido 060-8628, Japan
e-mail: smukai@eng.hokudai.ac.jp

S. Iwamura
e-mail: iwamura@eng.hokudai.ac.jp

K. Kanamura (ed.), *Next Generation Batteries*,
https://doi.org/10.1007/978-981-33-6668-8_38

to the electrolyte and oxygen to the atmosphere. Therefore, the capacity of this battery system highly depends on the Li_2O_2 storage ability of the cathode.

Considering the principles of LAB systems, the air cathode should possess the following features to ensure efficient and stable operation:

- A 3-dimensional network of conductive paths within the cathode to enable rapid distribution and collection of electrons;
- A large void volume within the cathode to enable the storage of a large amount of Li_2O_2;
- A large surface area within the cathode to enable the uniform deposition of Li_2O_2.

Carbon materials are usually used to construct the air cathode, as carbon is an electrical conductor which is extremely light. Among the various forms of carbon, porous carbon is thought to be an ideal base material for the cathode, as they can provide both a large void volume and a large surface area. However, such types of carbon are usually synthesized in the form of particles, making it difficult to establish a reliable 3-dimensional network of conductive paths between them. Therefore, they are usually combined with a conductive material which can easily provide such a network. Metal foams can be used for this purpose [1], but as the density of metals is generally high, sheets formed by carbon fibers (a material also referred to as "carbon paper" or "carbon cloth") are presently widely used, due to their lightweight [2, 3]. Resins such as polyvinylidene difluoride (PVDF) are also introduced into the air cathode to ensure the binding of the carbon particles to the fibers forming the 3-dimensional network of conductive paths.

One thing that makes the designing of an air cathode difficult is the fact that the Li_2O_2 formed during discharging is basically an electrical insulator. This leads to the limitation in the thickness that this material can be deposited during discharging. Therefore, when selecting porous carbon for the cathode, not only the surface area and pore volume of the material, but also the size of the pores developed within them is important. Introducing large pores into carbon may lead to a significant increase in pore volume, but the Li_2O_2 capacity of carbon with only large pores is likely to be small, as it becomes much more difficult to completely fill pores with Li_2O_2 when its size is increased. Introducing small pores into carbon tends to lead to a significant increase in surface area, but as the contribution of small pores to the total pore volume of the material is relatively small, using small pores for Li_2O_2 storage is also not efficient. The inlets of small pores also tend to be easily blocked when binders are used to prepare the air cathode, making it difficult to use the the entire volume of them.

2 Nanocarbon Materials for the Air Cathode of LABs

A wide variety of porous carbon is available, so there are many candidates for the base material to be used to construct the air cathode of LABs. One popular category of carbon is carbon blacks. Carbon blacks are materials which are obtained through the incomplete combustion of heavy hydrocarbons, such as pitches from petroleum and coal, or through the thermal decomposition of hydrocarbons such as acetylene [4]. They are typically categorized into groups according to the process and/or raw material adopted for production. Typical groups are furnace blacks, channel blacks, thermal blacks and acetylene blacks.

The basic building block of carbon blacks are carbon nanoparticles having diameters roughly in the range of a few nanometers to a few hundred nanometers. These particles are frequently referred to as “primary particles.” These primary particles are fused by covalent bonds, and aggregates called “structures” are formed. The degree of aggregation depends on several factors, such as the method and raw material adopted for production. Typically, a few to a few hundred primary particles are included in a single structure. Such structures attract each other by van der Waals forces, and agglomerates having a size in the range of a few micrometers to a few hundred micrometers are formed.

Carbon blacks can be practically regarded as porous materials, as voids are naturally formed between their structures. The size of such voids depends on the size of the structures, so the size of the voids falls in the nanometer range. Moreover, as a large part of the primary particles which form the structures are exposed to the atmosphere, carbon blacks possess a fairly high surface area. Carbon blacks are mostly used to reinforce rubber used in tires, but as highly conductive carbon blacks with high purity are readily available now, it has become popular to use them as a conductive-aid in batteries.

Among highly conductive carbon blacks, “Ketjen Black” (KB) is an extremely popular product used for conductive aiding. KB is produced by activating (partially oxidizing) carbon black [5]. Due to this treatment, a large ratio of the inner part of the primary particles forming the structures of this material is removed. As the outer skin of the primary particles mostly contributes to electron transport, this material can show the same level of conductive enhancement at a lower amount of addition, when compared to other conductive carbon blacks.

Due to the unique features of conductive carbon blacks, especially KB, this material has been widely used as the base material of LAB air cathodes [6, 7]. As the agglomerates of this material are quite flexible, they tend to expand when they store Li_2O_2 during discharging. This leads to high capacity, but as expanded agglomerates tend to contract upon charging, such high capacities are a trade-off with structural stability.

Another unique category of porous carbon which is thought to be suitable for LAB air cathodes is carbon gels [8, 9]. Carbon gels are practically aggregates of carbon nanoparticles in which all of the nanoparticles are covalently bound together [10, 11]. The difference between carbon gels and structures of carbon blacks is that the size of the aggregates is much larger. Monolithic blocks of carbon gels with dimensions up to tens of centimeters can be synthesized.

Carbon gels are obtained by first synthesizing a hydrogel from resorcinol and formaldehyde through a sol-gel transition using a base as the catalyst, and subsequently drying and carbonizing the obtained hydrogel [10, 11]. A unique feature of this material is that it is composed of nanoparticles which size can be controlled by adjusting the concentration of the catalyst used for synthesizing the hydrogel precursor. As the size of the pores which are naturally formed between the nanoparticles depends on the size of the nanoparticles, the size of the pores of this material can be controlled by adjusting the size of the nanoparticles. The size of the nanoparticles forming carbon gels is typically a few to a few hundred nanometers, so the size of the pores falls into the mesopore size range which is 2–50 nm. Therefore, this carbon is widely known as a mesoporous carbon.

The similarity between the structure of carbon gels and carbon blacks suggests that this material can also be used as the base material for LAB air cathodes. Moreover, the rigidity in the structure will also allow us to use it to investigate how the pore structure of porous carbon affects its Li_2O_2 capacity.

The authors prepared 4 carbon gels with different pore structures by varying the catalyst concentration of their starting solution [9]. The concentration of the catalyst, sodium carbonate, is typically represented by the molar ratio between resorcinol and the catalyst (*R*/*C*), so hereafter, the samples will be labeled by this ratio.

Air cathodes were synthesized by coating these samples on carbon paper using PVDF. Figure 1 shows the nitrogen isotherms of the air cathodes synthesized using the 4 samples, and pore size distributions which were derived from the isotherms

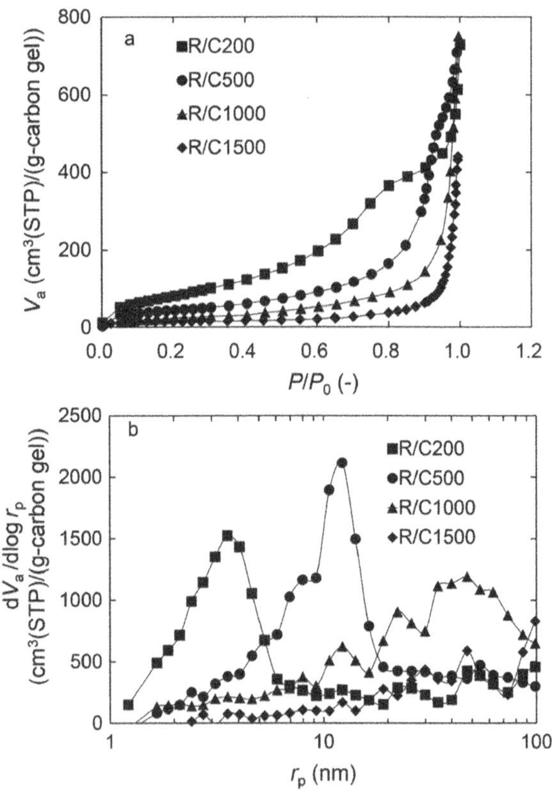

Fig. 1 N_2 adsorption isotherms (**a**) and pore size distributions (**b**) of cathodes prepared using carbon gels synthesized using different *R*/*C* values. Republished with permission of IOP Publishing, Ltd., from Sakai et al. [9]; permission conveyed through Copyright Clearance Center, Inc.

using the Dollimore-Heal method. As the adsorption capacity of carbon paper is negligibly small, the amount of nitrogen adsorbed is shown based on the weight of the carbon gel. Due to their micropores, carbon gels typically show a moderate increase in the amount of nitrogen adsorption in the low relative pressure range, but such increases could not be observed in the isotherms of the cathode samples. This indicates that the micropores were filled with or blocked by the binder and/or solvent used to synthesize the cathodes. All cathodes showed significant nitrogen uptakes at relative pressures over 0.6 indicating that they have developed meso-macropores within them. From the pore size distributions, it can be confirmed that the size of the meso-macropores within the cathodes increases with the increase in the *R*/*C* value of the carbon gel used to construct the cathode. Therefore, a series of cathodes with different pore sizes was successfully obtained.

Cells were constructed using the synthesized cathodes and lithium foil as the anode. A glass filter was used as the separator, and a solution of CF_3SO_3Li in tetraethylene glycol dimethyl ether (TEGDME) was used as the electrolyte (TEGDME: $CF_3SO_3Li = 4{:}1$, molar ratio). The cells were discharged and charged in the range of 2.0–4.4 V at a current density of 50 mA/(g-carbon gel) after a 10 h rest in an oxygen atmosphere. Figure 2 shows the discharge-charge curves of the 4 cathodes. The voltage during discharging was about 2.7 V, and when the discharged cells were subsequently charged, the voltage quickly increased to about 3.8 V, and then gradually increased up to the cutoff voltage of 4.4 V. The capacity of the cathodes first increased with the increase in the *R*/*C* value of the carbon gel used to synthesize the cathode, and then after the *R*/*C* value exceeded 500, the capacity started to decrease. Such results indicate that the specific capacity is governed not only by pore size, but also factors such as pore volume.

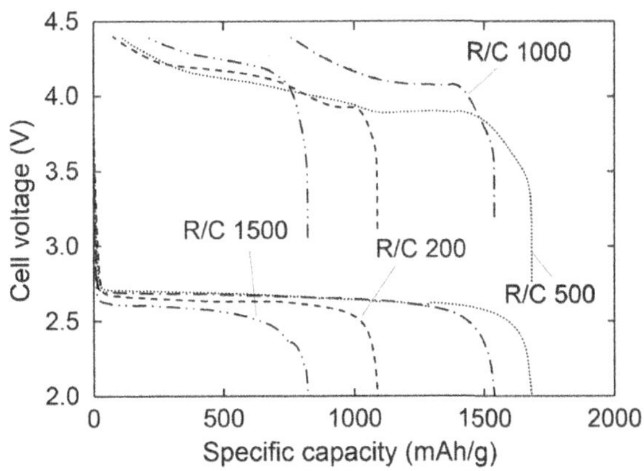

Fig. 2 Discharge-charge curves of cathodes prepared using carbon gels synthesized at different *R*/*C* values measured at 50 mA/g in 1 atm of O_2. The curves shown are those of the cycle in which the cathodes showed the largest discharge capacity. Republished with permission of IOP Publishing, Ltd., from Sakai et al. [9]; permission conveyed through Copyright Clearance Center, Inc.

In order to clarify how the pores are filled with Li_2O_2 during discharging, the cathodes were discharged to different depths, and the pore structure of the resulting cathodes was examined. Figure 3 shows the nitrogen adsorption isotherms and corresponding pore size distributions of *R*/*C*200, 500 and 1000 cathodes which were

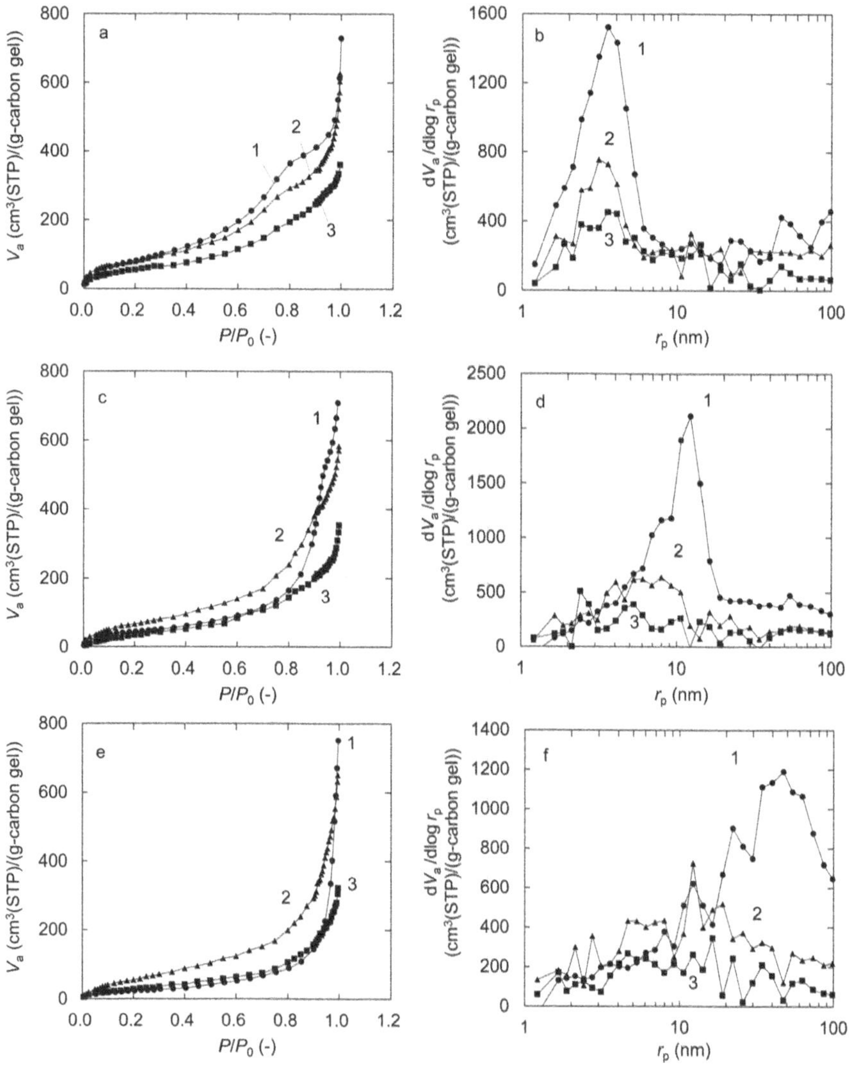

Fig. 3 N_2 uptakes and pore size distributions of cathodes prepared using carbon gels synthesized using different *R*/*C* values. **a** and **b** *R*/*C*200, **c** and **d** *R*/*C*500, and **e** and **f** *R*/*C*1000. Curves labeled 1 are of samples prior to discharging. Curves labeled 2 are of samples after discharging for 10 h. Curves labeled 3 are of samples after discharging to the predetermined end voltage (2.0 V). Republished with permission of IOP Publishing, Ltd., from Sakai et al. [9]; permission conveyed through Copyright Clearance Center, Inc.

discharged to different depths. The cathodes were either discharged for 10 h, or until the discharge voltage reached 2.0 V. In the *R/C*200 cathode, nitrogen uptake in the relative pressure range 0.20–0.96 significantly decreased after 10 h of discharging. Further discharging to 2.0 V led to further uptake decreases in the same pressure range, indicating that the mesopores of carbon gels in the cathode are filled during discharging. A decrease in the volume of larger pores was also observed, indicating that the voids which are naturally formed between carbon-gel particles can also contribute to Li_2O_2 storage. Interestingly, nitrogen uptake increased in the relative pressure range of about 0.2–0.9 in the *R/C*500 and 1000 cathodes discharged for 10 h. As the position of the peak in the pore size distributions also shifted to the left, it is assumed that larger pores in carbon gels get narrower due to the deposition of Li_2O_2. Eventually, nitrogen uptake in this pressure range decreases and becomes similar to that of a cathode prior to discharging.

The average thickness of the solid deposited in the cathodes (d_{Li2O2}) was estimated from the BET surface area of the cathode (S_{BET}), discharge capacity (Q) and the density of the deposit, assuming that only Li_2O_2 was deposited. The results are summarized in Table 1. It can be noticed that d_{Li2O2} increases with the increase in the *R/C* value and then levels off at around 5–6 nm. This thickness coincides with the calculated d_{Li2O2} value of a Li_2O_2 layer deposited on a flat surface of a glassy carbon at a current density of 1 μA cm^{-2} (4–5 nm) [12]. Therefore, it is assumed that the pores of cathodes synthesized using carbon gels of *R/C*500–1500 are sufficiently used for Li_2O_2 deposition. On the other hand, these results suggest that most of the pores of cathodes synthesized using *R/C*200 carbon gels are too small to be effectively utilized.

Next, the volume of the Li_2O_2 deposited in the cathodes was compared with the volumes of mesopores, meso-macropores and all of the pores in the cathode. The results are summarized in Fig. 4. The ratio between the volume of deposited Li_2O_2 and the volume of mesopores of the cathodes, $V_{Li2O2}/V_{mesopore}$, exceeds unity in some cases. Therefore, it can be concluded that macropores can also contribute to Li_2O_2 storage. The ratios between the volume of deposited Li_2O_2 and the volume of meso-macropores of the cathodes ($V_{Li2O2}/V_{meso\text{-}macro\ pore}$) are quite similar. A similar trend can be seen in the ratio between the volume of deposited Li_2O_2 and the total volume of the pores in the cathodes ($V_{Li2O2}/V_{total\ pore}$), which is quite natural as the volume of micropores in all of the cathodes is quite limited. Similar and fairly high

Table 1 Specific properties of carbon-gel cathodes

R/C	S_{BET} (m^2/g)	Q (mAh/g)	d_{Li2O2} (nm)
200	292	1088	1.5
500	157	1683	4.0
1000	102	1540	5.6
1500	55	821	5.6

Fig. 4 $V_{Li2O2}/V_{mesopore}$, $V_{Li2O2}/V_{meso\text{-}macro\ pore}$ and $V_{Li2O2}/V_{total\ pore}$ values of cathodes prepared using carbon gels synthesized using different *R*/*C* values. Republished with permission of IOP Publishing, Ltd., from Sakai et al. [9]; permission conveyed through Copyright Clearance Center, Inc.

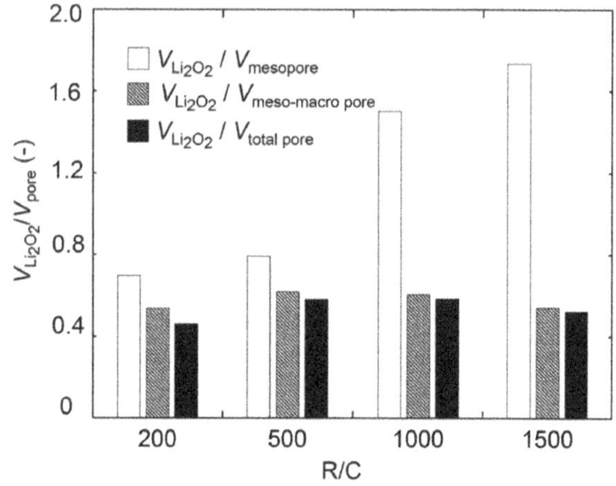

$V_{Li2O2}/V_{meso\text{-}macro\ pore}$ values indicate that increasing the meso-macropore volume of the cathode is effective to increase discharge capacity.

The $V_{Li2O2}/V_{total\ pore}$ values of various types of carbon materials are summarized in Table 2. The values for flexible materials such as carbon blacks, carbon nanotubes and graphene nanosheets tend to be higher than unity, indicating that the materials expand as they store Li_2O_2. The values for activated carbon are typically lower than 0.2, suggesting that microporous materials are not suitable for the air cathode for LABs. The values of carbon gels tend to be higher, and by carefully tuning their pore structure, over half of their total pore volume can be used for Li_2O_2 storage. Therefore, when carbon with a rigid structure is preferred, carbon gels are thought to be a suitable candidate for LAB air cathodes.

3 A New Approach for Air Cathode Designing

As mentioned above, particulate carbon is usually combined with carbon paper when used as the base material for LAB air cathodes. Although the weight of such a paper typically exceeds that of the particulate carbon combined with it, it can hardly contribute to Li_2O_2 storage, as the surface area that it can provide is extremely limited. Therefore, a new approach for air cathode designing which allows a more efficient use of the cathode material is desired.

In LAB air cathodes, a high surface area is preferred, but it has been shown that carbon with only a moderate surface BET area, about 100–200 m^2/g, can be effectively used as the base material for LAB air cathodes [20, 21]. Carbon paper with a similar or higher BET surface area can be obtained by reducing the diameter of the fibrous carbon used to synthesize it. It is expected that such carbon papers can be used as the air cathode without the aid of additional porous carbon.

Table 2 Pore volumes and capacities of lithium-air batteries of various carbon materials

Sample name	V_{total} (cm^3/g)	V_{meso} (cm^3/g)	i	Capacity (mAh/g)	V_{Li2O2} (cm^3/g)	V_{Li2O2}/V_{total} (–)	V_{Li2O2}/V_{meso} (–)	Material	References
ACPF800-15	1.04	0.84	70 mA/g	910	0.34	0.33	0.40	PF carbon gel	[13]
ACPF800-30	1.25	0.98	70 mA/g	1247	0.46	0.37	0.47	PF carbon gel	[13]
ACPF800-60	1.60	1.30	70 mA/g	1299	0.48	0.30	0.37	PF carbon gel	[13]
ACPF800-90	1.85	1.54	70 mA/g	1461	0.54	0.29	0.35	PF carbon gel	[13]
ACPF800-120	2.01	1.87	70 mA/g	1852	0.69	0.34	0.37	PF carbon gel	[13]
ACRF002-1123K	1.32	0.87	20 mA/g	630	0.23	0.17	0.26	RF carbon gel	[8]
ACRF002-1223K	1.12	0.75	20 mA/g	180	0.07	0.06	0.09	RF carbon gel	[8]
ACRF002-1273K	1.43	1.01	20 mA/g	740	0.28	0.20	0.28	RF carbon gel	[8]
ACRF003-1073K	1.2	0.95	20 mA/g	528	0.2	0.17	0.21	RF carbon gel	[8]
ACRF003-1123K	1.34	1.02	20 mA/g	880	0.33	0.25	0.32	RF carbon gel	[8]
ACRF003-1323K	2.20	1.73	20 mA/g	1290	0.48	0.22	0.28	RF carbon gel	[8]

(continued)

Table 2 (continued)

Sample name	V_{total} (cm^3/g)	V_{meso} (cm^3/g)	i	Capacity (mAh/g)	V_{Li2O2} (cm^3/g)	V_{Li2O2}/V_{total} (–)	V_{Li2O2}/V_{meso} (–)	Material	References
YP-50F	0.55		20 mA/g	264	0.1	0.18		Activated carbon	[14]
CEP21S	0.85		20 mA/g	124	0.05	0.06		Activated carbon	[14]
Carbon aerogel	2.53		20 mA/g	4155	1.55	0.61		Carbon aerogel	[14]
CRF1500-40	1.02	0.85		13,500	5.02	4.92	5.91	RF carbon gel	[15]
Carbon A	2.01	2.01	0.1 mA/cm^2	2220	0.83	0.41	0.41	Conductive carbon	[16]
Carbon B	1.92	1.34	0.1 mA/cm^2	360	0.13	0.07	0.10	Conductive carbon	[16]
Carbon C	0.82	0.22	0.1 mA/cm^2	22	0.01	0.01	0.05	Activated carbon	[16]
Carbon D	0.46	0.102	0.1 mA/cm^2	17	0.01	0.02	0.10	Activated carbon	[16]
GNSs		1.17	75 mA/g	8706	3.24		2.77	Graphene nanosheet	[17]
BP-2000		1.11	75 mA/g	1909	0.71		0.64	Carbon black	[17]
Vulcan XC-72		0.27	75 mA/g	1054	0.39		1.44	Carbon black	[17]

(continued)

Table 2 (continued)

Sample name	V_{total} (cm^3/g)	V_{meso} (cm^3/g)	i	Capacity (mAh/g)	V_{Li2O2} (cm^3/g)	V_{Li2O2}/V_{total} (–)	V_{Li2O2}/V_{meso} (–)	Material	References
Super C	0.22		210 mA/g	2600	0.97	4.41		Carbon black	[18]
CNTs	1.33		210 mA/g	5000	1.86	1.40		Carbon nanotube	[18]
GNRs	1.46		210 mA/g	5000	1.86	1.27		Graphene nanosheet	[18]
PGE1	1.9		200 mA/g	20,310	7.56	3.98		Graphene nanosheet	[19]
PGE2	1.5		200 mA/g	29,375	10.9	7.27		Graphene nanosheet	[19]
PGE3	1.15		200 mA/g	22,458	8.36	7.27		Graphene nanosheet	[19]

Republished with permission of IOP Publishing, Ltd., from Sakai et al. [9]; permission conveyed through Copyright Clearance Center, Inc.

As will be shown in the next chapter, air cathodes which show excellent performances can be obtained by using carbon paper synthesized from carbon nanotubes (CNTs) [22]. CNTs are nonporous, but as their diameter is only a few nanometers, voids which can be considered as meso-macropores are naturally formed between them. CNT papers are flexible materials, so they tend to expand like carbon blacks when they store Li_2O_2. However, as the 3-dimensional network that the CNTs form is thought to be much firmer than those that other types of flexible carbon such as carbon blacks and graphene nanosheets form, higher stability during discharging-charging cycles is expected.

Carbon paper synthesized from CNTs is an attractive material for LAB air cathodes, but the high production cost of CNTs is an obstacle for such materials to be widely used. One way to significantly reduce the production cost of such a paper is to use carbon nanofibers (CNFs) having diameters in the range of about ten nanometers to about a hundred nanometers instead of CNTs. Such a paper is still thought to be applicable to LAB air cathodes, as it has a fairly high BET surface area, over 100 m^2/g, and as voids with similar dimensions as those formed between CNTs are also formed within them.

Carbon paper can be easily obtained by first dispersing fibrous carbon into a suitable solvent and then vacuum-filtering the resulting suspension. However, it is usually difficult to obtain stable carbon paper from CNFs. In order to obtain a stable paper, the fibrous carbon used to synthesize it must have a fairly high aspect ratio. Both CNTs and CNFs are typically synthesized using an integrated process of catalysis and thermal chemical vapor deposition (CVD), which is often referred to as a catalytic CVD (CCVD) process [23]. They elongate using the catalysis of metal nanoparticles, and they thicken through thermal CVD. Whether CNTs or CNFs are obtained depends on the degree that the thermal CVD process is allowed to proceed during synthesis. This means that the lengths of both types of carbon are quite similar, but the aspect ratio of CNFs becomes one or two orders lower than that of CNTs, as the diameter of a CNF is one or two orders higher than that of a CNT. Figure 5a shows a photograph of a commercially available CNF after dispersion and filtration. It can be noticed that a self-standing paper cannot be obtained.

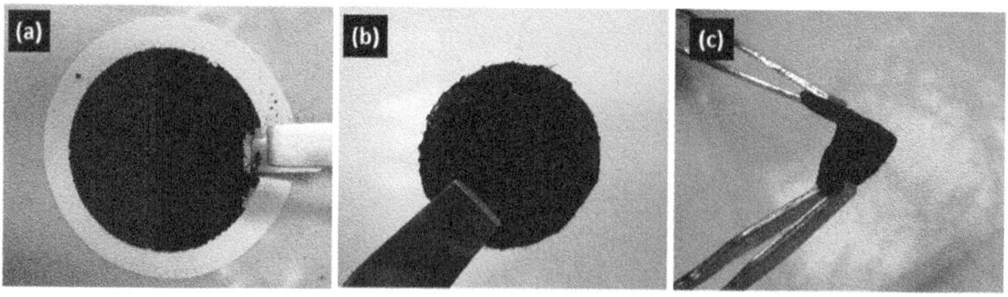

Fig. 5 Photographs of **a** commercially available CNF after filtration, **b** LPI-CNF paper, and **c** folded LPI-CNF paper. From ACS Omega 2018, 3, 691–697, https://pubs.acs.org/doi/10.1021/acsomega.7b01822. Further permissions related to this figure should be directed to the ACS

The authors have developed a method which allows the synthesis of CNFs with an extremely high aspect ratio, the Liquid Pulse Injection (LPI) technique [24]. Metal nanoparticles which show extremely high catalytic activities can be easily generated through this modified CCVD technique, so CNFs with aspect ratios 1–2 orders higher than those obtained through conventional synthesis methods can be easily obtained. Figure 5b shows a photograph of carbon paper synthesized from LPI-CNFs having an average diameter of about 20 nm. Self-standing carbon paper with high flexibility, as shown in Fig. 5c, can be easily obtained from such CNFs. Through nitrogen adsorption experiments, it was confirmed that this LPI-CNF paper has a significant volume of voids which correspond to meso-macropores.

To check the performance of this LPI-CNF paper as a LAB cathode, electrochemical tests were conducted at a current density of 50 mA/g. For comparison, similar tests were conducted using carbon paper synthesized using carbon fibers with a diameter of about 10 μm (CP), and a conventional cathode synthesized by coating KB on this carbon paper using a PVDF binder (KB-CP). Details of the tested cathodes are summarized in Table 3. Figure 6 shows the discharge curves of the 3 cathodes. Due to its low surface area, the discharge capacity of CP was extremely low, but the CNF carbon paper showed a high discharge capacity of about 920 mAh/(g-cathode) and that of KB-CP was about 420 mAh/(g-cathode). In the KB-CP cathode, KB mostly contributes to the discharge capacity so if the capacity is compared on the basis of the weight of active material, a high discharge capacity of about 6000 mAh/(g-KB)

Table 3 Weights and areal densities of electrodes and active materials

Sample	Weight of electrode (mg)	Weight of the loaded active material (mg)	Areal density of active materials (mg/cm^2)	Areal density of electrode (mg/cm^2)
LPI-CNF paper	12		6.0	6.0
CP	9.9			4.9
CP-KB	10	1.1	0.55	5.0

From ACS Omega 2018, 3, 691–697, https://pubs.acs.org/doi/10.1021/acsomega.7b01822. Further permissions related to this table should be directed to the ACS

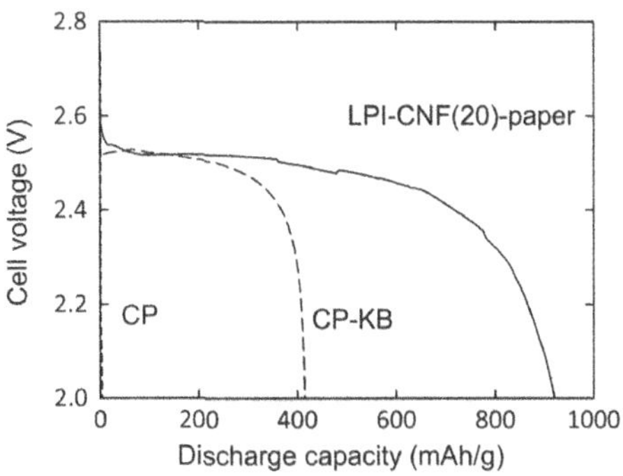

Fig. 6 Discharge curves of LPI-CNF paper, CP and CP-KB at a current density of 50 mA/(g-cathode). The cutoff voltage was set to 2.0 V. From ACS Omega 2018, 3, 691–697, https://pubs.acs.org/doi/10.1021/acsomega.7b01822. Further permissions related to this figure should be directed to the ACS

has been achieved. However, KB requires carbon paper to be used as the cathode material, so it is difficult to further increase the discharge capacity based on cathode weight, as long as carbon paper synthesized using thick carbon fibers is used to construct the cathode. It should be noted that the areal capacity of the CNF carbon paper was about 5.5 mAh/cm^2, which is comparable to 6.5 mAh/cm^2 (measured at 0.4 mA/cm^2), the value of a paper made of 12 mg of single-walled CNTs [22].

As the production cost of not only CNFs but also CNTs is decreasing year by year, papers made from CNTs and CNFs are thought to become a powerful candidate for LAB air cathodes which may dominate the field in the near future.

References

1. Beattie, S. D., Manolescu, D. M., & Blair, S. L. (2009). High-capacity lithium-air cathodes. *Journal of the Electrochemical Society, 156,* A44–A47.
2. Jung, H.-G., Hassoun, J., Park, J.-B., Sun, Y.-K., & Scrosati, B. (2012). An improved high-performance lithium-air battery. *Nature Chemistry, 4,* 579–585.
3. Christensen, J., Albertus, P., Sanchez-Carrera, R. S., Lohmann, T., Kozinsky, B., Liedtke, R., et al. (2012). A critical review of Li/air batteries. *Journal of the Electrochemical Society, 159,* R1–R30.
4. Donnet, J.-B., & Voet, A. (1976). *Carbon black physics, chemistry, and elastomer reinforcement*. New York: Marcel Dekker, Inc.
5. Hisashi, H. (2000). *Nihon Gomu Kyokaishi (in Japanese), 73,* 362–370.
6. Zhang, J. G., Wang, D. Y., Xu, W., Xiao, J., & Williford, R. E. (2010). Ambient operation of Li/air batteries. *Journal of Power Sources, 195,* 4332–4337.
7. Xiao, J., Wang, D. H., Xu, W., Wang, D. Y., Williford, R. E., Liu, J., & Zhang, J. G. (2010). Optimization of air electrode for Li/air batteries. *Journal of the Electrochemical Society, 157,* A487–A492.
8. Mirzaeian, M., & Hall, P. J. (2009). Preparation of controlled porosity carbon aerogels for energy storage in rechargeable lithium oxygen batteries. *Electrochimica Acta, 54,* 7444-7451.
9. Sakai, K., Iwamura, S., & Mukai, S. R. (2017). Influence of the porous structure of the cathode on the discharge capacity of lithium-air batteries. *Journal of the Electrochemical Society, 164,* A3075–A3080.
10. Al-Muhtaseb, S. A., & Ritter, J. A. (2003). Preparation and properties of resorcinol-formaldehyde organic and carbon gels. *Advanced Materials, 15,* 101-114.
11. Tamon, H., Ishizaka, H., Mikami, M., & Okazaki, M. (1997). Porous structure of organic and carbon aerogels synthesized by sol-gel polycondensation of resorcinol with formaldehyde. *Carbon, 35,* 791-796.
12. Viswanathan, V., Thygesen, K. S., Hummelshoj, J. S., Norskov, J. K., Girishkumar, G., McCloskey, B. D., & Luntz, A. C. (2011). Electrical conductivity in Li_2O_2 and its role in determining capacity limitations in non-aqueous Li-O_2 batteries. *The Journal of Chemical Physics, 135,* No. 214704.
13. Shitta-Bey, G. O., Mirzaeian, M., & Halla, P. J. (2012). The electrochemical performance of Phenol-Formaldehyde based activated carbon electrodes for lithium/oxygen Batteries. *Journal of the Electrochemical Society, 159,* A315-A320.
14. Ma, S. B., Lee, D. J., Roev, V., Im, D., & Doo, S. G. (2013). Effect of porosity on electrochemical properties of carbon materials as cathode for lithium-oxygen battery. *Journal of Power Sources, 244,* 494-498.
15. Wang, F., Xu, Y. H., Luo, Z. K., Pang, Y., Wu, Q. X., Liang, C. S., et al. (2014). A dual pore carbon aerogel based air cathode for a highly rechargeable lithium-air battery. *Journal of Power Sources, 272,* 1061-1071.

16. Kuboki, T., Okuyama, T., Ohsaki, T., & Takami, N. (2005). Lithium-air batteries using hydrophobic room temperature Ionic liquid electrolyte. *Journal of Power Sources, 146,* 766-769.
17. Li, Y., Wang, J., Li, X., Geng, D., Li, R. Y., & Sun, X. L. (2011). Superior energy capacity of graphene nanosheets for a nonaqueous lithium-oxygen battery. *Chemical Communications, 47,* 9438-9440.
18. Yin, J., Carlin, J. M., Kim, J., Li, Z., Park, J. H., Patel, B., et al. (2015). Synergy between metal oxide nanofibers and graphene nanoribbons for rechargeable lithium-oxygen battery cathodes. *Advanced Energy Materials, 5,* No. 1401412.
19. Sun, B., Huang, X., Chen, S., Munroe, P., & Wang, G. (2014). Porous graphene nanoarchitectures: An efficient catalyst for low charge-overpotential, long life, and high capacity lithium-oxygen batteries. *Nano Letters, 14,* 3145-3152.
20. Read, J. (2002). Characterization of the lithium/oxygen organic electrolyte battery. *Journal of the Electrochemical Society, 149,* A1190–A1195.
21. Yang, X. H., He, P., & Xia, Y. Y. (2009). Preparation of mesocellular carbon foam and Its application for lithium/oxygen battery. *Electrochemistry Communications, 11,* 1127–1130.
22. Nomura, A., Ito, K., & Kubo, Y. (2017). CNT sheet air electrode for the development of ultra-high cell capacity in lithium-air batteries. *Scientific Reports, 7,* No. 45596.
23. Endo, M. (2012). Carbon Nanotube Research: Past and future. *Japanese Journal of Applied Physics, 51,* No. 040001.
24. Mukai, S. R., Rikima, Y., Furukawa, R., & Ogino, I. (2013). Analysis of the growth behavior of carbon nanofibers synthesized using the liquid pulse injection technique. *Industrial and Engineering Chemistry Research, 52,* 15281–15286.

CNT Electrode

Akihiro Nomura

Abstract Lithium-Air Battery (LAB) is expected to develop lightweight and high capacity energy storages. The battery performance is much dominated by the air-breathing cathode, or air electrode, in which the oxygen reduction/evolution reactions occur. Among several nanocarbons proposed for air electrode materials, a carbon nanotube (CNT) sheet as air electrode provides the LAB cells ultra-high cell capacity as large as ~30 mAh/cm^2, which is ~15 times higher than that of current lithium-ion battery technology. The cell capacity can be even more extended, developing almost unlimited cell capacities by exploiting the flexible nature of the CNT sheet. In addition, optimizing the pore size distribution of the CNT sheet electrode brings ever-high rate capability to the LAB cells, and this also improves the cycle performance of the LAB cells. Flexible but tough CNT bundle network of the CNT sheet provides a new design of air electrode to pave the way for developing the LAB with ultra-high energy density.

Keywords Lithium-air battery · Air electrode · Carbon nanotube

The properties of Lithium-Air Batteries (LABs) are much dominated by the air-breathing cathode or the air electrode in which the fundamental electrochemical reaction of the LAB ($2Li^+ + O_2 + 2e^- \leftrightarrow Li_2O_2$) occurs. LAB cathodes, as with that of common batteries, require electrochemically active large surface areas to proceed with the battery reaction. They also require open pores to secure oxygen supply to the active surface. In LAB, solid discharge products of lithium oxides, typically lithium peroxide, Li_2O_2, are deposited on the surface during discharge. The open pores provide space for accumulated discharge products. Therefore, both high surface area and high porosity, such as sponge-like structure, are desirable for the cathode materials to ensure high battery performance. The cathode itself has no active materials but offers a scaffold for the LAB reaction, hence, it significantly influences the battery performance. The architecture of the LAB cathode is distinguished from other battery electrodes by its microscopic open pore structures.

A. Nomura (✉)
National Institute for Materials Science, 1-1 Namiki, Tsukuba 305-0044, Japan
e-mail: NOMURA.Akihiro@nims.go.jp

K. Kanamura (ed.), *Next Generation Batteries*,
https://doi.org/10.1007/978-981-33-6668-8_39

Several nanocarbons are practically available for constructing the microstructure of LAB cathodes. Carbon blacks [1, 2], graphene [3, 4], carbon aerogels [5, 6], carbon fibers [7, 8], and carbon nanotubes (CNTs) [9–11] have been proposed as LAB cathode materials. Some of these nanocarbons exhibit very high specific capacities of about 10,000 mAh/g_{carbon} due to their large surface areas (about 1,000 m^2/g). However, their lean carbon loadings, often below 1 mg/cm^2, limit their resulting cell capacities per electrode area to 10 mAh/cm^2 or lesser. The cell capacity per electrode area (mAh/cm^2) is essential in developing cells with high energy densities, thus, even larger areal capacities are needed to assemble practical LAB cells. Also, to withstand the repeated deposition/decomposition of discharge products, the microstructure of cathodes need to exhibit high resilience, however, such durability has not yet been satisfactorily developed nor evaluated in practical discharge/charge conditions.

CNTs, among other proposed nanocarbons, possess a unique character stemming from their nanofibrous structure. Powdery carbon materials such as Ketjen black® and Super-P® require rigid solidification to be utilized as LAB cathode, however, such firm solid electrodes often have a low tolerance for mechanical stresses triggered by the discharge solid deposition especially in deeply discharged states. In contrast, CNTs produces paper-like flexible sheets due to the tubular nature of their constituents. These sheets of CNTs can readily be used as LAB cathodes. The fibrous network of CNTs is also beneficial for retaining the conductive network of the electrode during the repeated deposition/decomposition of the discharge solids. In this chapter, recent developments in CNT sheets for LAB cathodes are reviewed.

1 Discharge Capacity of CNT Sheet Cathode

CNT sheets are typically prepared by the filtration of CNT dispersed slurry. Various microstructures can be obtained depending on the characteristics of the pristine CNT and the sheet fabrication process. Table 1 summarizes the characteristics of three CNT sheets discussed in this chapter. Figure 1a shows the SEM image of a CNT sheet (CNT_A) prepared from a single-walled CNT (SWCNT) produced by the eDIPS method. The method has superiority in quantity production of SWCNT with high crystallinity (*G*/*D* ratio, carbon crystallinity indicator based on the *G*-band (1590 cm^{-1}) to *D*-band (1340 cm^{-1}) intensity ratio as measured by Raman spectroscopy, is >100) and precise tube diameter control [12]. SWCNT with a 2 nm tube diameter is used for CNT_A. The prepared sheet is a flexible paper-like non-woven

Table 1 Characteristics of CNT sheets

	Raw CNTs	*G*/*D*	BET (m^2 g^{-1})	Density (g cm^{-3})
CNT_A	eDIPS, single walled	>100	300	0.45
CNT_B	Super-growth, single walled	~5	800	0.16
CNT_C	Super-growth, single walled	~5	600	0.08

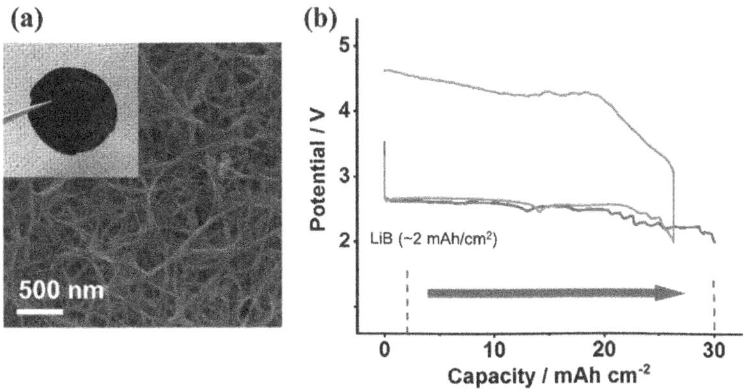

Fig. 1 **a** SEM image of the CNT sheet (CNT_A); a superimposed photographic image. **b** The discharge-charge profiles of LAB cell with CNT_A as the cathode at a low current density of 0.05 mA/cm^2 (from [11])

textile of SWCNT bundles aggregated to 50–200 nm, which could be directly used as LAB cathodes. The sheet density of the CNT_A was estimated to be 0.45 g/cm^3 by its sheet thickness and weight ratio. The sheet can have any CNT loadings in proportion to its thickness, but usually less than 10 mg/cm^2 or 200 μm in practice. The higher carbon loading of the CNT sheet, unlike in powdery carbons, is beneficial in obtaining higher areal capacity owing to its total effective surface area and pore volume. Figure 1b shows the discharge-charge curves of the LAB cells with CNT_A (CNT loading of 2.5 mg/cm^2) as the cathode. The cells initially displayed a discharge voltage of 2.6 V, and while charging, gradually increased voltage up to 4.5 V, which is the typical profile found in LABs. One distinctive feature of these discharge-charge curves is that a large discharge capacity of about 30 mAh/cm^2, which is approximately 15 times higher than the areal cell capacity of common Lithium-ion Battery (LiB, ~2 mAh/cm^2), is provided by the CNT sheet cathode [11]. The capacity is even higher than that of conventional LAB cells (about 10 mAh/cm^2). This enormously high cell capacity encourages the development of practical LAB with energy densities far exceeding those of existing batteries.

The flexibility and high strength of the CNT bundle network are believed to be responsible for the high cell capacity developed by the CNT sheet. While discharging, the CNT sheet cathode tends to swell due to the deposition of the discharge product. This enhances the continuous breathing of oxygen by the cathode. It also enhances the accumulation of the discharge products without clogging the open pore between the bundles. Figure 2a shows the photo image of the CNT sheet cathode after a 5 mAh/cm^2 discharge. It shows multiple bumps extruding into the holes perforated on the cathode case for oxygen intake. The cross-sectional SEM images of the sheet, as shown in Fig. 2b, reveal that the part of the CNT sheet that faced the open perforated holes [b, (i)] experienced an increase in thickness, but the part harnessed by the cathode case [b, (ii)] experienced no increase in thickness. The swelling ratio of the initial thickness at the protruding parts during the discharge-charge cycle is shown in

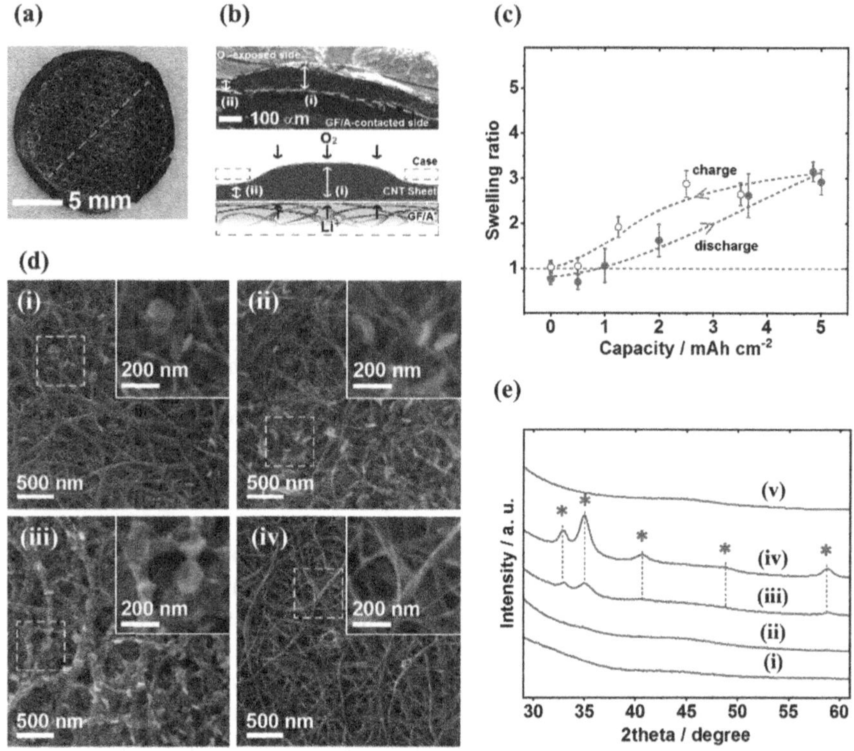

Fig. 2 **a** Photographic image of a typical CNT sheet (CNT_A) after the discharge experiment. The sheet was cut with precision scissors along the dotted line for cross-sectional SEM observation. **b** Typical cross-sectional SEM image of the CNT sheet (CNT_A) after the discharge experiment (up) and its schematic illustration (low). The orange dotted line highlights the edge of the cross-section. **c** Swelling ratios of CNT_A after the discharge. The ratios of CNT_A after charging from a 5 mAh/cm^2 discharged state were plotted as open symbols. **d** SEM images of CNT_A after discharges of 1 mAh/cm^2 (i), 2 mAh/cm^2 (ii), 5 mAh/cm^2 (iii), and after being fully charged following a 5 mAh/cm^2 discharge (iv). The dotted yellow squares show the magnified areas superimposed in each image. **e** XRD spectra of pristine CNT_A (i) after discharges of 1 mAh/cm^2 (ii), 2 mAh/cm^2 (iii), 5 mAh/cm^2 (iv), and after being fully charged following a 5 mAh/cm^2 discharge (v). The * symbols denote Li_2O_2 crystal reflections of 100, 101, 102, 004, and 110 at 2theta ≈ 32.9, 35.0, 40.6, 48.7, and 58.6, respectively (from [11])

Fig. 2c. As shown in Fig. 2, the sheet swelled at approximately threefold thickness at 5 mAh/cm^2 discharge but returned to its initial thickness after a full charge.

The swelling/shrinkage of the CNT sheet cathode is triggered by lithium peroxide (Li_2O_2) deposition/decomposition while discharging or charging. Figure 2d shows the magnified SEM image of the CNT sheet after (i) 1 mAh/cm^2, (ii) 2 mAh/cm^2, (iii) 5 mAh/cm^2 discharge, and (iv) full charge following the 5 mAh/cm^2 discharge. These images illustrate the toroidal particles of 50–150 nm diameter evolved on the CNT bundles in a way that the number of the particles increases while discharging and then completely decomposes after a full charge. The toroid shaped particles are characteristic morphology for lithium peroxide (Li_2O_2) as the discharge product often observed in a wide range of LAB cathodes reported so far. X-ray diffraction (XRD) analysis of the CNT sheets, shown in Fig. 2e, also infers that the particle deposits are assigned to Li_2O_2 and no other side reaction products are produced.

The bloating nature of CNT sheets can be exploited to further increase the discharge capacity of LAB cells. This can be achieved by preparing multiple layered cathodes of CNT sheets and gas diffusion layers (GDLs). GDL is a highly porous (80% porosity) piled layer of conductive carbon fibers of about 10 μm in diameter that is often used for a membrane electrode assembly (MEA) in fuel cells. Figure 3a shows the discharge profiles of LAB cells with the layered CNT/GDL cathode. The layered cathodes are prepared in three different ways and their respective structures are illustrated in the profile ((i), (ii), and (iii)). Figure 3b shows the plot of the discharge capacities against the corresponding total CNT loadings. The CNT loading of (i), a piece of CNT_A sandwiched between two GDL cathodes of 190 μm thickness, was varied from 2 to 6 mg/cm^2 by regulating the sheet thickness. At higher CNT loading, no further increase in the discharge capacity was observed, as shown in

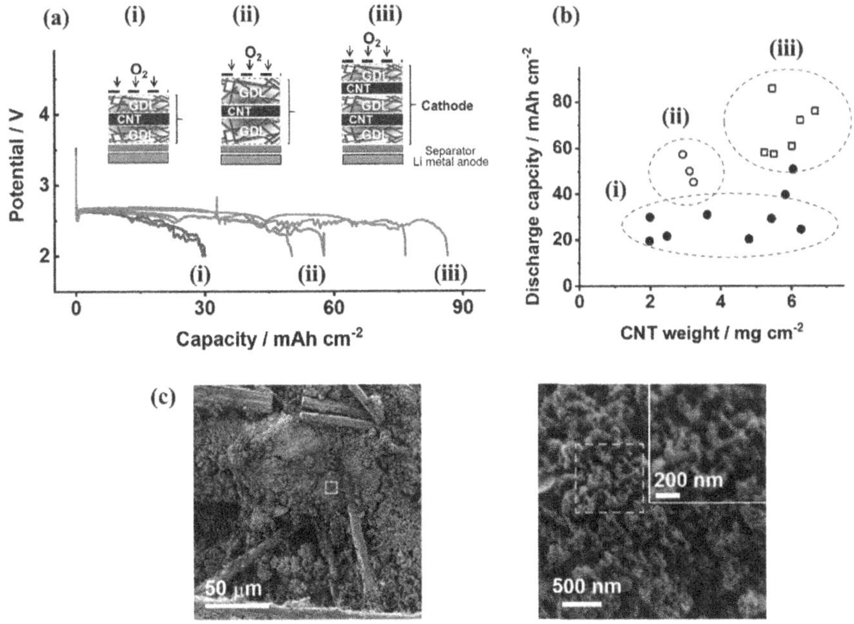

Fig. 3 **a** Discharge profiles of the LAB cells with layered CNT/GDL cathodes of (i, red line), (ii, purple line), and (iii, orange line). The layered cathodes are composed of CNT sheet (CNT_A) sandwiched between two GDLs of 190 μm thick (i), or 370 μm thick GDLs (ii). The layered cathode is also composed of alternating layers of two CNT sheets and three GDLs of 190 μm thick (iii). **b** Discharge capacities of LAB cells with layered CNT/GDL cathodes ((i) closed circle, (ii) open circle, and (iii) open square) plotted against their CNT loadings per electrode area. **c** SEM images of the GDL after 20 mAh/cm^2 discharge (low (left) and high (right) magnifications). The yellow dotted squares show the magnified areas (from [11])

Fig. 3b. This could be attributed to the limitation of oxygen diffusion in the direction of the sheet thickness [13]. However, inserting thicker GDLs (ii)—a piece of CNT_A sandwiched between two 370 μm thick GDLs, or increasing the number of pieces (iii)—two pieces of CNT_A sandwiched alternately between three 190 μm thick GDLs significantly enhanced the discharge capacities. The capacity of the electrode area increased with increasing CNT weight by 10,000–15,000 mAh/g_{CNT}, achieving almost unlimited capacity (provisionally up to 82 mAh/cm^2 attained).

Since GDL itself contributes a negligible amount of discharge, the discharge capacity is mostly developed by the LAB reaction in the CNT sheet. However, GDL supports the sheet in developing high cell capacity by increasing the overall porosity of the layered cathode. The integrated porous structure feeds oxygen into the CNT with the enhanced facility and provides volumetric space for the CNT swelling to accommodate the discharge products. The GDLs filled with bloated CNT layers are often observed when the layered cathodes are taken out from the cells after discharge. As depicted in the SEM image (Fig. 3c), the CNT bundles penetrated the gap in between the carbon fiber in the GDL, thereby, allowing the deposition of a massive amount of discharge products. The image also shows the GDL part after discharge, showing a densely filled granular deposit between the gaps of the carbon fibers of the GDL. The magnified image reveals particles of about 100 nm in size captured by the thin fibers of the CNT. XRD measurement confirmed that the deposited particles are Li_2O_2 as an exclusive product.

2 Rate and Cycle Properties of LAB Cells with CNT Sheet Cathode

The multiple layered cathode in Fig. 3a inspires the porosity modulation of the CNT sheet for achieving batteries of good properties with a lightweight cathode. Among several CNT production methods, the Super-Growth method is a good method for the mass production of SWCNT, as with the e-DIPS method. The Super-Growth method yields extremely long SWCNT (>100 μm) with ultra-high purity [14], but is a little bit inferior in crystallinity compared with those produced by the e-DIPS. The SWCNT is obtained in a vertically aligned state from the planar substrate. Utilizing this oriented collective flocculate results in CNT sheets with a highly porous structure. Increasing the porosity and pore size results in improved rate and cycle performance, as well as improved cathode capacity.

Two other types of CNT sheets, CNT_B and CNT_C, are prepared from Super-Growth SWCNTs that have increased sheet porosity from CNT_A. The characteristics of the sheets are also tabulated in Table 1. The vertically aligned Super-Growth CNT powder forms wide CNT bundles (~10 μm) resulting in large open pores in the sheet. The sheet density of CNT_B is 0.16 g/cm^3, which is one-third of that of CNT_A made up of 50–200 nm bundles. The sheet density can be further decreased by varying the widened CNT bundles, providing the CNT_C with a sheet density of

0.08 g/cm^3. This extremely low sheet density implies a very high porosity of >90% in CNT_C, however, it still maintains the adequate ease of handling as an electrode due to its super long CNT. Although the lower crystallinity of the Super-Growth SWCNT results in a lower *G/D* ratio (~5) for CNT_B and CNT_C, there is no detrimental effect on the sheet electrode conductivity.

Figure 4a, b show the discharge profiles of LAB cells with the CNT_B and CNT_C cathodes, respectively, recorded at various discharge rates. The cells were assembled with a simple cathode configuration comprised of a stack of single thin GDL (110 μm thick) at the oxygen exposed side and a CNT sheet (~2 mg/cm^2 CNT loading) at the separator side. Although both cells are prepared in the same manner, they result in distinct rate properties. At low discharge rates, below 1 mA/cm^2, both cathodes exhibit comparable discharge capacities of approximately 15 mAh/cm^2 or 7000 mAh/g_{CNT}. However, at higher discharge rates, CNT_C cathodes exhibit higher capacity than the CNT_B cathodes. A plot of the discharge capacities against the discharge rates is shown in Fig. 4c. In general, as the discharge rate increases, the

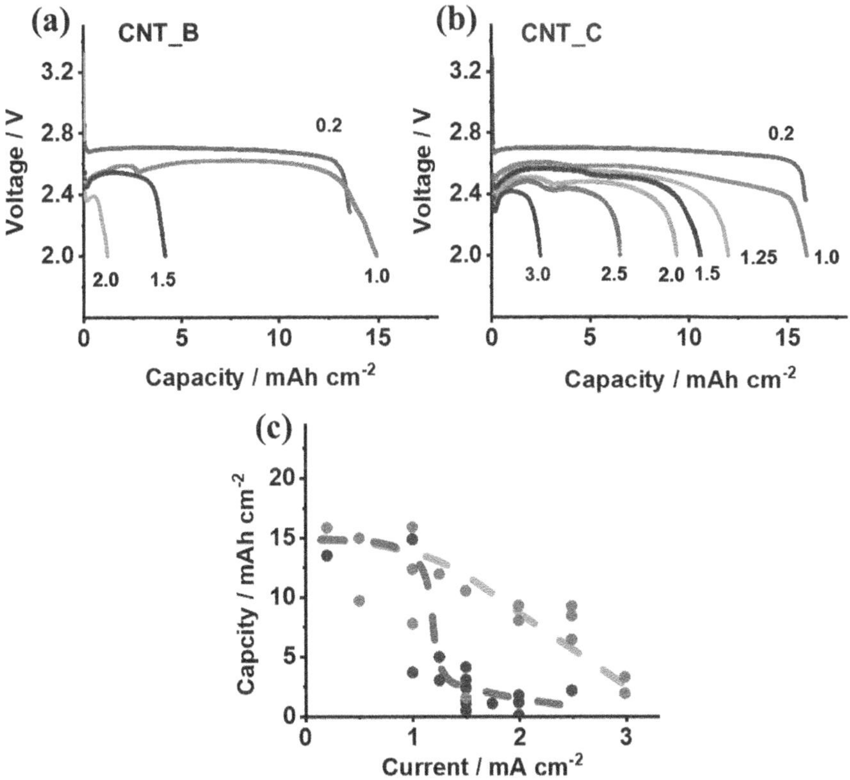

Fig. 4 **a, b** Discharge profiles of LAB cells with CNT_B (**a**) and CNT_C (**b**) cathodes. The numbers in the figures denote the applied discharge rate in a unit of mA/cm^2. **c** The discharge capacities of LAB cells with CNT_B (blue circle) and CNT_C (red circle) cathodes (Color figure online)

battery capacity decreases because it becomes more difficult for battery reactants to catch up with the reaction. The capacity of both LAB cells also decreases with increasing rates. However, at rates of over 2 mA/cm^2, the cells with CNT_C cathode still maintain high capacities, whereas those with CNT_B cathode rapidly decay resulting in very low capacities. This infers that CNT_C cathodes are better than the CNT_B cathode in fabricating practical LAB cells. The discharge capacities obtained by CNT_C, specifically of approximately 10 mAh/cm^2 at a high rate of 2 mA/cm^2, is obviously higher than ever considered for LAB cells.

The pore structures of the CNT sheets are investigated to reveal the rate performance of the resulting LAB cells. The pore size distribution of CNT sheets depicted by a mercury porosimeter is shown in Fig. 5a. While CNT_A shows little mercury intrusion to exhibit a negligible volume of pores in the sheet, low-density CNT sheets, CNT_B, and CNT_C allow some mercury intrusions, indicating their lower sheet density or increased porosity by forming macropores with the Super-Growth CNT. CNT_B has 0.1–10 μm pores whereas CNT_C contains a much larger volumetric amount of pores in the size of 1–100 μm. The expanded pores of CNT_C are considered to have enhanced its rate performance.

The pore size distribution is also evaluated by nitrogen adsorption measurement, which is effective for analyzing micropores in the nanometer scale. Figure 5b presents the pore size distribution of the CNT sheets, obtained by the nitrogen adsorption measurement in the BJH method. The distribution reveals 10–100 nm pores in CNT_A which corresponds to the gaps between the CNT bundles shown in the SEM image (Fig. 1a). The total volume of a pore is, however, low resulting in the relatively high sheet density of CNT_A. The micropores of CNT_B are almost collapsed, leaving tiny micropores below 10 nm. Hence, most of the CNT_B pores fall into the 0.1–10 μm pores estimated by mercury intrusion. In contrast, CNT_C contains a larger number of pores in the range of 10–100 nm, in addition to the wide-open

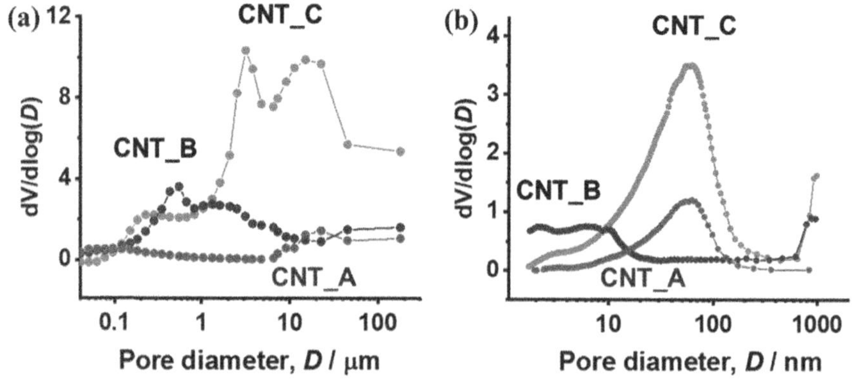

Fig. 5 **a**, **b** Pore size distribution of CNT_A (green), CNT_B (blue), and CNT_C (red), measured by mercury intrusion (**a**) and nitrogen adsorption (**b**) (Color figure online)

pores of 1–100 μm by mercury intrusion. The BET surface area of CNT_B is somewhat higher than that of CNT_C since smaller pores critically affect the surface area. However, CNT_C which develops the high discharge rate is distinguished by its two distinctive pores in nanometer and micrometer scales.

There are reports that pores in a mesoscale (typically 2–50 nm) crucially determine the capacity of the LAB cathode [7, 8, 15]. The surface of the pores in the mesoscale or wider is dominantly effective to inhale and diffuse oxygen to the active surface to proceed with the LAB reaction. Pores below the mesoscale contribute little to the battery reaction due to the short supply of oxygen. Although these were obtained from powdery carbon LAB cathodes at a low capacity region, it is valid for CNT sheet cathodes at a high capacity region. CNT_B and CNT_C have comparable total surface areas, hence, both cathodes exhibit a similar amount of discharge capacity at low discharge rates at which oxygen is adequately supplied to follow the reaction regardless of their pore sizes. However, at higher rates, it becomes more difficult for CNT_B to transport oxygen to its micropore surface and hence to abruptly decrease its cathode capacity. On the other hand, the two distinctive pores in the range of 10–100 nm and 1–100 μm of CNT_C efficiently adsorb oxygen gas and supply the dissolved oxygen to the reactive carbon surface to maintain its intrinsic capacity, even at a higher discharge rate. This implies that CNT bundles comprising the sheet cathodes can be appropriately designed to develop practical cell capacity with high power density.

The expanded pore distribution is also beneficial for the discharge-charge cycle operation. The discharge-charge curves of LAB cells with the CNT_A and CNT_C cathodes are shown in Fig. 6a, b, respectively. The profiles were recorded at a rate of 0.4 mAh/cm^2 and a cycle capacity of 4 mAh/cm^2. To suppress charge overpotential in the cycle runs, lithium salts of LiBr/$LiNO_3$, which serve as redox mediators (RMs) [16, 17], were added to the electrolyte. The CNT_C cell steadily maintained a discharge voltage of 2.6 V and effectively suppressed the charge voltage, running 14 cycles before reaching the cutoff voltage of 2.0 V. In contrast, the cell with the CNT_A cathode showed higher charge voltage and reached a cutoff voltage of 2.0 V

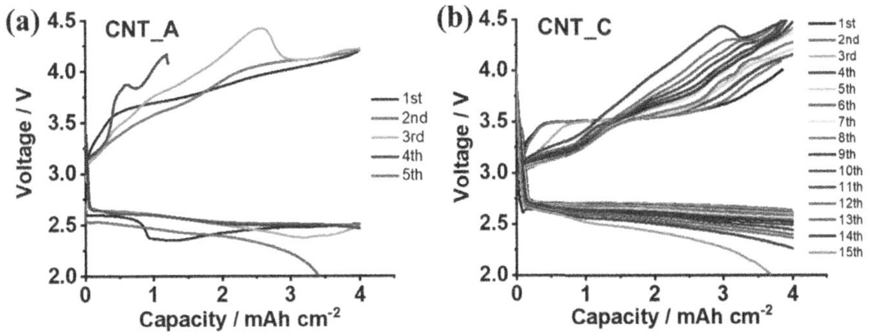

Fig. 6 **a**, **b** Discharge-charge profiles of LAB cells with CNT_A (**a**) and CNT_C (**b**) as cathodes. The profiles are recorded at a rate of 0.4 mAh/cm^2 and a cycling capacity of 4 mAh/cm^2

at the 5th discharge. These results demonstrate that the very porous CNT_C significantly improves the cycle performance of LAB cells. CNT sheets can be designed with various pore structures to further enhance the rate and cycle performances. This is the promising potential of the CNT sheet as a LAB cathode.

3 Future Works of CNT Electrode for LABs

In this chapter, the characteristics of CNT electrodes are reviewed, introducing the superior properties of the CNT sheet cathode for developing ultra-high cathode capacities, high discharge rate performance, and improved cycle ability in LAB cells. These results indicate the CNT sheets are promising candidates for cathode materials. As the highly porous structure of CNT sheets is the critical aspect for developing superior cathode performance, they should be properly fabricated in the sheet with acceptable weight and thickness. In addition, the oxidative stability of carbon, durability to physical stress triggered by repeated deposition/decomposition of discharge product, and compatibility to pressure constraint in cell assembly need to be assessed to obtain durable sheet cathodes. The evolution of the LAB materials including the CNT sheet cathodes would result in practical secondary LABs.

References

1. Jung, H. G., Hassoun, J., Park, J. B., Sun, Y. K., & Scrosati, B. (2012). An improved high-performance lithium-air battery. *Nature Chemistry, 4*(7), 579–585.
2. Kubo, Y., & Ito, K. (2014). Multicell stack of nonaqueous lithium-air batteries. In J. W. Fergus (Ed.), *17th International Meeting on Lithium Batteries, Vol. 62. ECS Transactions, Vol. 1* (pp. 129–135). Pennington: Electrochemical Soc Inc.
3. Wang, Z. L., Xu, D., Xu, J. J., Zhang, L. L., & Zhang, X. B. (2012). Graphene oxide gel-derived, free-standing, hierarchically porous carbon for high-capacity and high-rate rechargeable Li-O_2 batteries. *Advanced Functional Materials, 22*(17), 3699–3705.
4. Xin, X., Ito, K., & Kubo, Y. (2016). Graphene/activated carbon composite material for oxygen electrodes in lithium–oxygen rechargeable batteries. *Carbon, 99*, 167–173.
5. Mirzaeian, M., & Hall, P. J. (2009). Preparation of controlled porosity carbon aerogels for energy storage in rechargeable lithium oxygen batteries. *Electrochimica Acta, 54*(28), 7444–7451.
6. Sakai, K., Iwamura, S., & Mukai, S. R. (2017). Influence of the porous structure of the cathode on the discharge capacity of lithium-air batteries. *Journal of the Electrochemical Society, 164*(13), A3075–A3080.
7. Mitchell, R. R., Gallant, B. M., Thompson, C. V., & Shao-Horn, Y. (2011). All-carbon-nanofiber electrodes for high-energy rechargeable Li-O_2 batteries. *Energy & Environmental Science, 4*(8), 2952–2958.
8. Sakai, K., Iwamura, S., Sumida, R., Ogino, I., & Mukai, S. R. (2018). Carbon paper with a high surface area prepared from carbon nanofibers obtained through the liquid pulse injection technique. *ACS Omega, 3*(1), 691–697.

9. Mitchell, R. R., Gallant, B. M., Shao-Horn, Y., & Thompson, C. V. (2013). Mechanisms of morphological evolution of Li_2O_2 particles during electrochemical growth. *The Journal of Physical Chemistry Letters, 4*(7), 1060–1064.
10. Nasybulin, E. N., Xu, W., Mehdi, B. L., Thomsen, E., Engelhard, M. H., Masse, R. C., et al. (2014). Formation of interfacial layer and long-term cyclability of Li-O_2 batteries. *ACS Applied Materials & Interfaces, 6*(16), 14141–14151.
11. Nomura, A., Ito, K., & Kubo, Y. (2017). CNT sheet air electrode for the development of ultra-high cell capacity in lithium-air batteries. *Scientific Reports, 7,* 45596.
12. Saito, T., Ohshima, S., Okazaki, T., Ohmori, S., Yumura, M., & Iijima, S. (2008). Selective diameter control of single-walled carbon nanotubes in the gas-phase synthesis. *Journal of Nanoscience and Nanotechnology, 8*(11), 6153–6157.
13. Dutta, A., Ito, K., & Kubo, Y. (2019). Establishing the criteria and strategies to achieve high power during discharge of a Li-air battery. *Journal of Materials Chemistry A, 7*(40), 23199–23207.
14. Hata, K., Futaba, D. N., Mizuno, K., Namai, T., Yumura, M., & Iijima, S. (2004). Water-assisted highly efficient synthesis of impurity-free single-walled carbon nanotubes. *Science, 306*(5700), 1362–1364.
15. Tran, C., Yang, X. Q., & Qu, D. Y. (2010). Investigation of the gas-diffusion-electrode used as lithium/air cathode in non-aqueous electrolyte and the importance of carbon material porosity. *Journal of Power Sources, 195*(7), 2057–2063.
16. Xin, X., Ito, K., & Kubo, Y. (2017). Highly efficient Br^-/NO_3^- dual-anion electrolyte for suppressing charging instabilities of Li-O_2 batteries. *ACS Applied Materials & Interfaces, 9*(31), 25976–25984.
17. Xin, X., Ito, K., Dutta, A., & Kubo, Y. (2018). Dendrite-free epitaxial growth of lithium metal during charging in Li-O_2 batteries. *Angewandte Chemie International Edition, 57*(40), 13206–13210.

Electrolytes and General Properties of Glyme-Based Electrolytes for Rechargeable Li–Air Batteries

Morihiro Saito

Abstract In this chapter, general properties and recent studies of glyme-based electrolytes for non-aqueous lithium–air batteries (LABs) will be introduced and discussed together with and through our research data by the ALCA SPRING project. Especially for understanding ion transport in the glyme-based electrolytes, a pulsed gradient spin echo (PGSE) NMR method was applied to measure self-diffusion coefficients, D, of Li^+ ion and anions and tetraglyme (G4) solvent individually. Apparent degree of dissociation, α_{app}, of the Li salts, transference number, t_{Li+}, and diffusion radius ratios of Li^+, r_{Li+}/r_{G4}, and anions, r_{Li+}/r_{G4}, against G4 solvent were estimated by using the D values for the electrolytes. Namely, the analysis methodology of ion transport by a combination with the PGSE-NMR method and the conventional measurements of ionic conductivity, σ, and viscosity, η, will be also introduced and explained as well as the general properties of the electrolytes. Moreover, a proposal of examples for new LAB electrolytes will be presented, which is based on the results of the above analysis. In addition, the positive effect of O_2 gas supplied from an air electrode will be also suggested to suppress the reduction decomposition of the electrolytes on the surface of the Li metal electrode for LABs.

Keywords Glyme-based electrolytes · Ion transport analysis · PGSE-NMR · Self-diffusion coefficients · Mobility and number of carrier ions · Effect of O_2 gas

1 Introduction

1.1 Electrolytes for Li–Air Batteries

As described in the preceding chapters, non-aqueous rechargeable lithium (Li)–air (O_2) batteries (LABs) have attracted much attention as potential high energy storage devices for electric vehicles (EVs) as well as for stationary systems. These batteries

M. Saito (✉)
Department of Materials and Life Science, Seikei University, 3-3-1 Kichijojikitamachi, Musashino-shi, Tokyo 180-8633, Japan
e-mail: mosaito@st.seikei.ac.jp

K. Kanamura (ed.), *Next Generation Batteries*,
https://doi.org/10.1007/978-981-33-6668-8_40

are desirable because they can exhibit high theoretical specific energy densities up to 3505 Wh kg^{-1}, which is approximately nine times the maximum value reported for conventional Li-ion batteries (LIB) (387 Wh kg^{-1}) [1]. The first LAB system was reported by Abraham and co-workers and was based on a gel polymer electrolyte containing propylene carbonate (PC) as the solvent [2]. Recently, there have been many attempts to improve the performance of such cells. Organic carbonate solvents such as PC, ethylene carbonate (EC), and diethyl carbonate (DEC) were initially used in LABs because of their low volatility, compatibility with the Li metal negative electrode (NE), and high oxidation stability above 4.5 V versus Li/Li^+. However, these compounds were found to be readily decomposed by superoxide ($O_2^{\cdot-}$) radicals formed during the discharge process [3–6], producing byproducts such as $C_3H_6(OCO_2Li)_2$, Li_2CO_3, HCO_2Li, CH_3CO_2Li, CO_2, and H_2O in conjunction with the formation of Li_2O_2 [7]. Ether-based electrolytes incorporating 1,2 dimethoxyethane (DME or G1), diglyme (G2), triglyme (G3), or tetraglyme (G4) as the solvent have also been widely applied in non-aqueous LABs [8]. These ethers exhibit high oxygen solubility and relatively low dielectric constants, which result in lower reactivity toward $O_2^{\cdot-}$ radicals compared with carbonate-based electrolytes [9]. Various studies have been done and have confirmed the production of Li_2O_2 in these systems following discharge (Table 1). Unfortunately, G1 and G2 are not suitable for practical applications because of their high vapor pressures at room temperature, and so G4-based electrolytes containing Li salts such as $LiSO_3CF_3$ (LiOTf) and $LiN(SO_2CF_3)_2$ (LiTFSI) are commonly utilized in LAB research. Even so, recent studies have demonstrated that glyme-based electrolytes also have stability problems due to the other decomposition mechanisms such as electrochemical oxidation during charging and chemical decomposition on the Li_2O_2 surfaces [10, 11]. As a result, the round-trip efficiency of O_2 for LABs is still low (at approximately

Table 1 Previous investigations of glyme-based electrolytes for LABs

Electrolyte composition	Experimental details	References
1 M LiTFSI/G1, 1 M LiTFSI/G4	Air electrode and Li metal NE; assessment of air electrode catalysts and LAB durability	[13–15, 22, 29, 33, 46]
1 M LiTFSI/G2 with RM	Use of LiI, LiBr, TFF, TEMPO, MPT, and others as RMs	[38–43]
1 M LiOTf/G4 with $LiNO_3$, 1 M $LiNO_3$/G4	Suppression of Li dendrite growth and electrolyte decomposition	[44, 45]
1 M $LiNO_3$/G4 with RM	$LiNO_3$ salt electrolytes with RMs; suppression of shuttle effect and improvement of RM effect	[47, 48]
4 M LiFSI/G1, $[Li(G3)_1]$TFSI	Concentrated Li salt electrolytes; suppression of Li dendrite growth and improved durability against high over-potential during charging	[49–51]

60%), and so glyme-based electrolytes are not optimal for use in LABs [1, 12]. However, glyme-based electrolytes are still one of the better candidates for LAB systems because the majority of alternative electrolytes tend to decompose under severe operating conditions. If the overpotential associated with the electrochemical oxidation of Li_2O_2 can be lowered, it may be possible to employ glyme-based electrolytes in LABs. On this basis, there has been a significant quantity of research work devoted to developing suitable air electrode catalysts that exhibit high catalytic activity not only for the oxygen reduction reaction (ORR) but also for the oxygen evolution reaction (OER) during discharging and charging, respectively [13, 14]. As an example, nanoparticles of noble metals (Pt, Au) [15–21] and metal oxides (Co_3O_4, RuO_2, IrO_2) [22–28] loaded on nano-carbon materials (such as Ketjen black, carbon nanotubes, and graphene), all of which have previously been investigated for use in polymer electrolyte fuel cells (PEFCs), have been examined. Perovskite nanoparticles [29–32] and Mn oxides [33–37] have also been assessed as catalysts with the aim of enhancing cell performance.

During the operation of a LAB, solid-state catalysts are typically covered with a coating of Li_2O_2 during the discharge stage. Therefore, the development of redox mediators (RMs) to oxidatively decompose Li_2O_2 at the electrolyte side of the electrolyte/electrode interface has been examined [38–43]. Both inorganic (LiI [38, 39], LiBr [40]) and organic (including tetrathiafulvalene (TTF) [41], 2,2,6,6-tetramethylpiperidinyloxyl (TEMPO) [42], and N-methylphenothiazine (MPT) [43]) RMs have been intensively researched as approaches to reducing the large charging overpotential in LABs, and the details of RM systems intended for LABs are discussed in the following section.

With the aim of suppressing the growth of Li dendrites at the Li metal negative electrode (NE), $LiNO_3$ salt has been used to glyme-based electrolytes so as to stabilize the NE surface via the formation of a protective Li_2O layer [44, 45]. Our own group has also demonstrated the importance of the oxidation of the NE surface using other Li salt electrolytes in Li | Li symmetric cells under an O_2 atmosphere [46]. In this prior work, a protective Li_2O layer was successfully formed on the NE surface and was found not only to effectively suppress Li dendrite growth but also to reduce electrolyte decomposition. The synergy effect of $LiNO_3$/G4-based electrolytes with RMs was also demonstrated to exhibit both the suppression of shuttle effect, i.e. self-discharge by RMs, and the improvement of the RM effect [47, 48].

Watanabe et al. developed glyme-based electrolytes containing high concentrations of Li salts [49, 50], termed solvate ionic liquids, for LAB and Li-S battery systems. These electrolytes displayed interesting properties, such as relatively low Li polysulfide solubility and high electrochemical stability, as a result of the strong interactions between Li^+ ions and the G3 and G4 solvents. Both battery systems containing the solvate ionic liquids also demonstrated stable cell performance. In further work related to concentrated Li salt electrolytes, Qian et al. reported that DME-based electrolytes having high concentrations of $LiN(SO_2F_3)_2$ (LiFSI) showed highly efficient Li dissolution and deposition at the Li metal NE without Li dendrite growth [51]. These studies thus suggest new possibilities for glyme-based electrolytes. Therefore, further research concerning glyme-based electrolytes will be

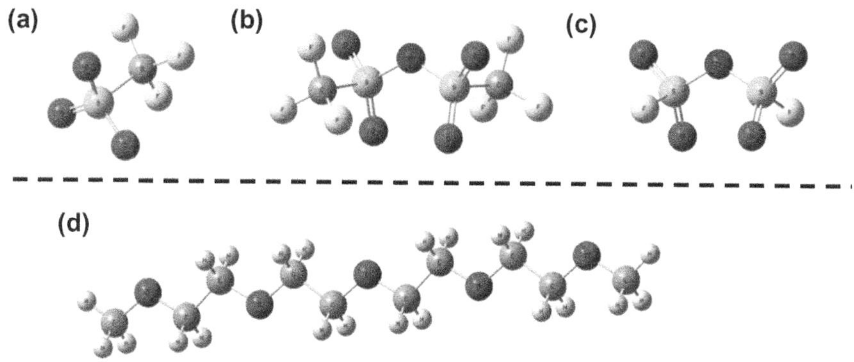

Fig. 1 Chemical structures of the anions and G4 solvent used in this study, as optimized by DFT calculation using the B3LYP/6-311+G** basis set. **a** $CF_3SO_3^-$ (OTf^-), **b** $N(SO_2CF_3)_2^-$ ($TFSI^-$), **c** $N(SO_2F)_2^-$ (FSI^-), and **d** tetraglyme (G4). Red: O, yellow: S, light blue: F, gray: C, dark blue: N, and white: H

vital to realizing the rapid ion transport required for next-generation battery systems, including non-aqueous LABs.

In this chapter, the general properties of glyme-based electrolytes incorporating the Li salts LiOTf, LiTFSI, and LiFSI will be introduced (Fig. 1), focusing on LAB systems. The effects of the Li salt concentration will also be examined, using the common salt LiTFSI as an example, and two approaches to enhancing the ionic conductivity of glyme-based electrolytes will be introduced [52]. In addition, the effects of introducing gaseous O_2 to a LAB cell on Li dissolution and deposition behavior at the Li metal NE will be discussed [46].

2 Requirements for LAB Electrolytes

As discussed in Sect. 1, the electrolyte plays an important role in the LAB, and the cell performance will be significantly changed by variations in the electrolyte properties. In general, a battery electrolyte should exhibit a number of specific properties. These include high ionic conductivity over a wide temperature range, rapid transport of the relevant carrier ions (such as Li^+ ions in LABs), a wide electrochemical window (so as to show suitable durability to oxidation at the air electrode and reduction at the Li metal NE), and low flammability and explosive potential. In the case of LAB systems, low vapor pressure to inhibit drying of the battery and high chemical stability together with minimal reactivity with $O_2^{\cdot -}$ radicals are also required. In addition, all Li metal batteries, including LABs, must effectively suppress Li dendrite growth during discharge/charge cycling. Therefore, an optimal electrolyte composition that promotes the formation of a suitable solid-electrolyte interphase (SEI) film is also a necessity.

In Sect. 3, the characteristics of ion transport in glyme-based electrolytes are mainly discussed, together with the means of improving the performance of non-aqueous rechargeable LABs.

3 General Properties of Glyme-Based Electrolytes

3.1 Ionic Conductivity and Viscosity

The ionic conductivity, σ, viscosity, η, and density, ρ, of battery electrolytes containing various Li salts and solvents are typically used to characterize these materials. The effects of the type and concentration of Li salts on these variables have also been examined, as functions of temperature, with the aim of determining the optimal electrolyte properties for battery applications. Figures 2 and 3 plot the σ and η values for various glyme-based electrolytes as functions of reciprocal temperature. The data in Fig. 2 demonstrate that σ is significantly affected by both the Li salt and the salt concentration. For example, the LiFSI electrolyte exhibited the highest σ values over the entire temperature range, while the LiTFSI electrolyte had its highest σ at a concentration of 1.0 M. The value of σ can be determined from the equation:

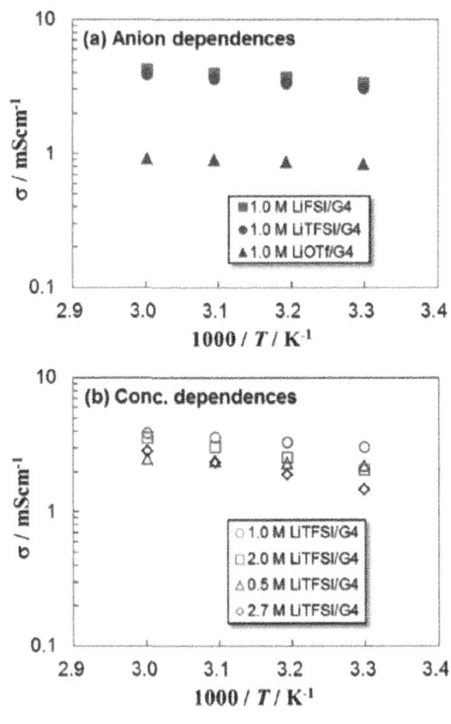

Fig. 2 Ionic conductivity, σ, of glyme-based electrolytes as functions of reciprocal temperature with variations in **a** anion and **b** Li salt concentration

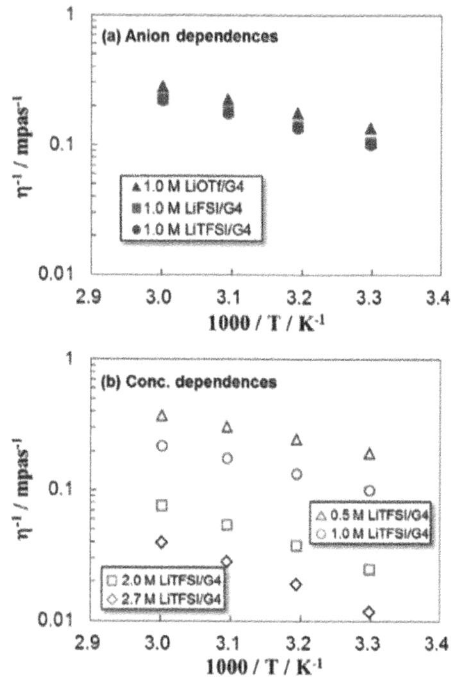

Fig. 3 Inverse viscosity, η^{-1}, of glyme-based electrolytes as functions of reciprocal temperature with variations in **a** anion and **b** Li salt concentration

$$\sigma = \sum q_i \times \mu_i \times n_i, \tag{1}$$

where q_i, μ_i, and n_i are the charge, mobility, and number per unit volume of each carrier ion, *i*, respectively. In a Li battery, q_i essentially has a value of 1 for both Li^+ ions and anions, and so the ionic conductivity depends on the mobility and number of carrier ions. The μ_i values of ions and solvent are basically proportional to the inverse number of electrolyte viscosity η^{-1}. However, as shown in Fig. 3, the η^{-1} values do not show the same trends as the σ data, indicating that the value of n_i is an important aspect of ion transport in glyme-based electrolytes. Therefore, it is helpful to discuss the contributions of both μ_i and n_i to σ as a means of understanding the solution structure and the ionic conduction mechanism in these electrolytes.

3.2 Correlation Between the Self-diffusion Coefficient and Ionic Conductivity

The determination of the self-diffusion coefficients, *D*, of ions in various solvents using pulsed gradient spin echo (PGSE) NMR is a helpful means of evaluating the individual mobility, μ_i, of each charge carrier. This is possible because the value of *D* is proportional to μ_i according to the relationship:

$$D = \mu RT/zF. \tag{2}$$

Therefore, the contribution of μ to σ can be assessed based on D values, and Fig. 4 plots the D values for various glyme-based electrolyte components as functions of reciprocal temperature. It is evident that G4 has the highest D value, followed by the anions and finally Li^+ ions. The magnitude of D is also dependent on the type and concentration of Li salt employed. The contribution of μ_i to the ionic conductivity of the electrolyte was ascertained by examining the correlations between σ and $(D_{Li+} + D_{anion})$ for various electrolytes, as shown in Fig. 5. Although a proportional correlation was confirmed at a concentration of 1.0 M for all electrolytes, this relationship did not hold at higher Li salt concentrations and the plots were curved when employing LiTFSI. This result suggests that increases in η have an effect on σ, possibly by decreasing the contribution of the μ_i values of the Li^+ ions and anions.

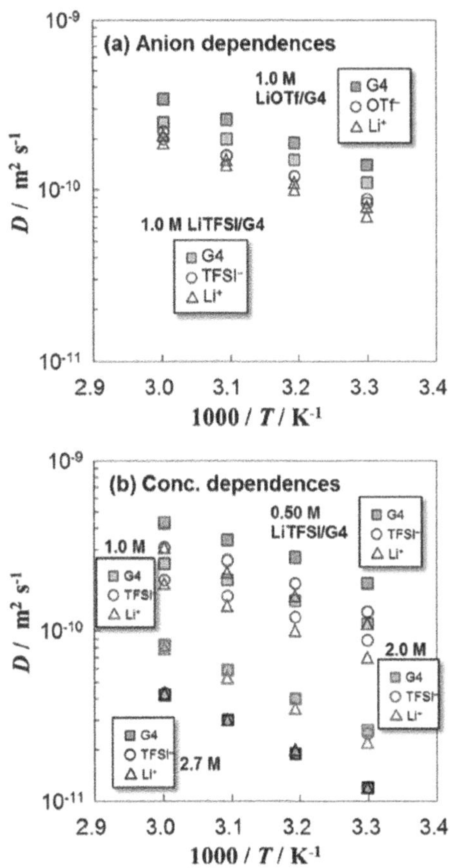

Fig. 4 Self-diffusion coefficients, D, of ions and G4 in glyme-based electrolytes as functions of reciprocal temperature with variations in **a** anion and **b** Li salt concentration

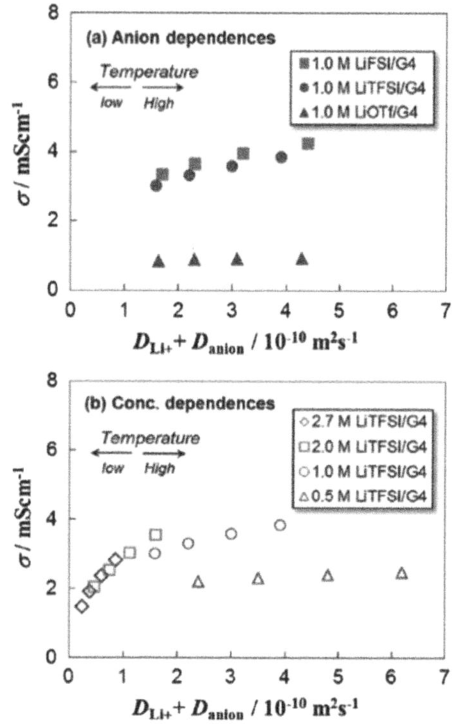

Fig. 5 Correlations between σ and $D_+ + D_-$ values in glyme-based electrolytes with variations in **a** anion and **b** Li salt concentration

3.3 Evaluations of Li Salt Dissociation by Walden Plots and Apparent Dissociation Degree

The contribution of n_i to σ can be discussed based on Walden plots of η against molar ionic conductivity, Λ_{imp}, as well as plots of the apparent degree of dissociation, α_{app}, of the Li salts. These parameters can be calculated from the equations [52–55]:

$$\Lambda_{\mathrm{NMR}} = \mathrm{Ne}^2(D_+ + D_-)/kT \tag{3}$$

and

$$\alpha_{\mathrm{app}} = \Lambda_{\mathrm{imp}}/\Lambda_{\mathrm{NMR}}, \tag{4}$$

where Λ_{NMR} is the molar ionic conductivity calculated from the D values for both Li^+ ions (D_+) and anions (D_-). The α_{app} values can be estimated from the ratios of Λ_{imp} to Λ_{NMR} because Λ_{NMR} includes the effects of all diffusing species, and thus does not distinguish between charged (isolated) and paired ions, while Λ_{imp} is solely associated with the effects of the former. Figures 6 and 7 present the Walden plots and the α_{app} values for the various glyme-based electrolytes assessed in this

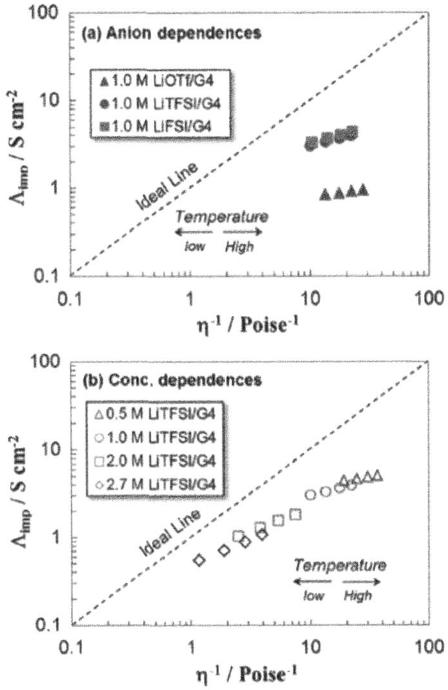

Fig. 6 Walden plots for the glyme-based electrolytes with variations in **a** anion and **b** Li salt concentration

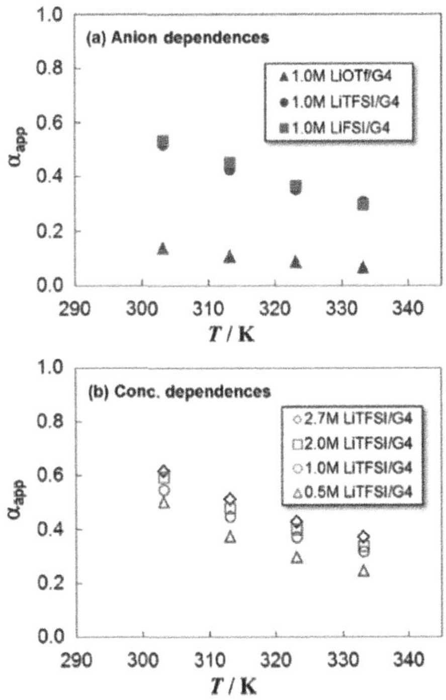

Fig. 7 Apparent Li salt dissociation degrees, α_{app}, estimated from self-diffusion coefficients, D, in glyme-based electrolytes with variations in **a** anion and **b** Li salt concentration

work. Generally, Walden plots will exhibit a larger downward deviation from the ideal line if the Li salt is less dissociated. Therefore, the LiTFSI and LiFSI both had a greater degree of dissociation compared with LiOTf, and the same trend is evident in the α_{app} values (Fig. 7). These data confirm that the contribution of n_i to σ was larger for LiTFSI and LiFSI. In fact, the σ values associated with the LiOTf electrolyte were significantly lower than those for the LiTFSI and LiFSI electrolytes, demonstrating a relatively large contribution of n_i compared with μ_i for these glyme-based electrolytes. The data also show that the degree of dissociation of the Li salts decreased as the temperature increased for all the electrolytes. This result indicates that the extent of Li salt dissociation was reduced and the contribution of μ_i for each ion was enhanced as the temperature was increased.

3.4 Transference Number and Diffusion Radius of Li^+ Ions

In addition to considering the contributions of μ_i and n_i to σ, the transference number, t_{Li+}, and the ratios of the diffusion radius of Li^+ and anion to that of G4, r_{Li+}/r_{G4} and r_{anion}/r_{G4}, were estimated from the D values for Li^+ , the anions and G4. The t_{Li+} values were calculated using the equation:

$$t_{Li+} = D_{Li+}/(D_{Li+} + D_{anion}). \tag{5}$$

The r_{ion}/r_{G4} ratios were determined from the Stokes–Einstein equation:

$$D = kT/c\pi\eta r_{ion}, \tag{6}$$

where k is Boltzmann's constant, T is the temperature (K), η is the viscosity of the electrolyte (Pa s^{-1}), r_{ion} is the Stokes radius for the solvated ion (m), and c is a constant that ranges between 4 and 6 for slip and stick boundary conditions, respectively [56]. The values of c and η were assumed to be the same for the ions and for G4 and so r_{ion}/r_{G4} could be simply defined as [55]

$$r_{ion}/r_{G4} = D_{G4}/D_{ion}. \tag{7}$$

This ratio provides important information regarding the solution structure, assuming that the Li^+ ions are surrounded by coordinated solvent molecules. Table 2 summarizes the t_{Li+} and r_{ion}/r_{G4} values obtained for the glyme-based electrolytes. All the t_{Li+} values are between 0.43 and 0.50 at 30 °C, and thus are relatively high compared with those reported for the organic carbonate electrolytes used in LIBs [53]. Interestingly, all the r_{ion}/r_{G4} values were found to be larger than those determined from the Van der Waals radii of the respective ions. This result implies that either the Li^+ ions strongly interacted with the G4 molecules to form solvated $Li^+(G4)_x$ species and/or the counter anions contributed to the formation of ion pairs. As a result, the Li^+ ions diffused more slowly than the anions in the electrolytes. In addition, the

Table 2 Transference numbers, t_{Li+}, and diffusion radius ratios, r_{Li+}/r_{G4} and r_{anion}/r_{G4}, in the glyme-based electrolytes at 30 °C

	LiOTf (1.0 M)	LiTFSI (1.0 M)	LiFSI (1.0 M)	LiTFSI (0.5 M)	LiTFSI (2.0 M)	LiTFSI (2.7 M)
t_{Li+}	0.49	0.44	0.43	0.46	0.47	0.50
r_{Li+}/r_{G4}	1.8	1.6	1.6	1.7	1.2	1.0
r_{anion}/r_{G4}	1.7	1.3	1.2	1.5	1.0	1.0

r_{Li+}/r_{G4} and r_{anion}/r_{G4} values for the LiOTf electrolyte were both larger than those determined for the other Li salt electrolytes, suggesting the formation of a greater proportion of ion pairs and larger solvated ion structures. Finally, the r_{ion}/r_{G4} values were found to decrease with increases in the Li salt concentration, indicating that the degree of dissociation increased along with the concentration.

4 A Proposal for New LAB Electrolytes

The above results indicate that the σ values of glyme-based electrolytes are determined by both μ_i and n_i. In particular, increasing the degree of Li salt dissociation is quite important because the dielectric constants of the glyme solvents are lower than those of the carbonate-type solvents used in LIBs. In addition, the α_{app} value was determined to increase along with the Li salt concentration. On the basis of these results, we attempted to design a new electrolyte system, using G1. G1 was selected partly because its dielectric constant (7.2) is similar to that of G4 (7.9) while its η value (0.455 mPa s) is lower (4.05 mPa s). Here, we introduce two representative models for glyme-based electrolytes, focusing on improving μ_i and increasing n_i.

With the intent of improving μ_i, a portion of G4 in a 1.0 M LiFSI/G4 electrolyte was replaced with G1 (having a lower η). As a result of this change, the η value for the electrolyte decreased and σ was improved, albeit with a slight decrease in α_{app} (Fig. 8a–c). In this case, the effect of μ_i exceeded that of n_i. The data also showed that a 3.5 M LiFSI/G1 electrolyte exhibited the highest σ in spite of a decrease in D (Fig. 8d, e). The value of α_{app} was found to increase as the Li salt concentration was increased from 1.0 to 3.5 M (Fig. 8). Thus, variations in n_i had the greatest effect on σ. From these results, it is apparent that changing both μ_i and n_i can produce new glyme-based electrolytes. In addition, a combination of conventional analytical methods and the PGSE-NMR technique has been shown to represent a useful approach to electrolyte design.

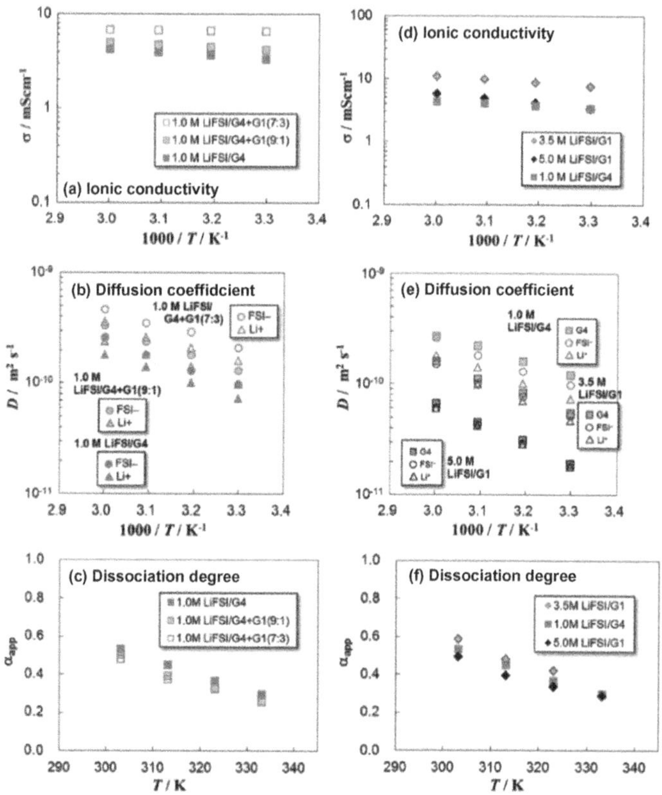

Fig. 8 **a** and **d** Ionic conductivities, σ, **b** and **e** self-diffusion coefficients, D, and **c** and **f** apparent extents of Li salt dissociation, α_{app}, for glyme-based electrolytes as functions of reciprocal temperature

5 Effects of O_2 on Li Dissolution/Deposition at the Li Metal NE

Sections 1–4 examined ionic conduction in glyme-based electrolytes intended for non-aqueous LAB systems, including for the other next-generation batteries. However, the stable operation of the Li metal NE is also a key aspect of the reliable operation of these devices. Specifically, the development of next-generation batteries, including Li-sulfur and all-solid-state batteries, will require the suppression of Li dendrite growth. Consequently, the formulation of new electrolytes based on concentrated solutions [51] or the incorporation of Cs^+ [57] and quaternary ammonium cations has been studied. In addition, as discussed in Sect. 1, $LiNO_3$ salt-used electrolytes are quite effective at suppressing the growth of Li dendrites due to the formation of a Li_2O layer on the surface of the Li metal NE [47, 48]. Our own research group has also demonstrated that gaseous O_2 introduced at the air electrode migrates into the electrolyte and reacts with the surface of the NE to generate a good protective layer consisting of Li_2O, even when using other Li salts [46].

Figure 9 shows the results of Li dissolution/deposition tests using a Li | Li symmetric cell under either Ar or O_2 atmospheres [46]. In the case of a 1.0 M LiOTf/G4 electrolyte under Ar, electrolyte decomposition clearly occurred at the surface of the Li metal electrode, with essentially no Li dissolution or deposition. However, upon introducing O_2, the electrolyte decomposition was suppressed and the dissolution and deposition processes were highly stable. In contrast, a 1.0 M LiTFSI/G4 electrolyte did not generate a large current due to electrolyte decomposition under either Ar or O_2. The data show that the overvoltage was somewhat reduced following the addition of O_2. It was also apparent that the charge transfer resistance, R_{ct}, was reduced by introducing O_2 to the Li | Li symmetric cells containing either LiOTf or LiTFSI (Fig. 10). This result demonstrates that the Li metal NE reaction can proceed more smoothly and in a more stable manner following the formation of a Li_2O layer. In the case of a $LiNO_3$/G4 electrolyte, the protective Li_2O layer was formed more efficiently and both Li dendrite growth and electrolyte decomposition were greatly suppressed [48, 58]. By effectively using a protective Li_2O layer, both safety and cell performance can therefore be improved so as to allow the practical applications of non-aqueous LAB systems.

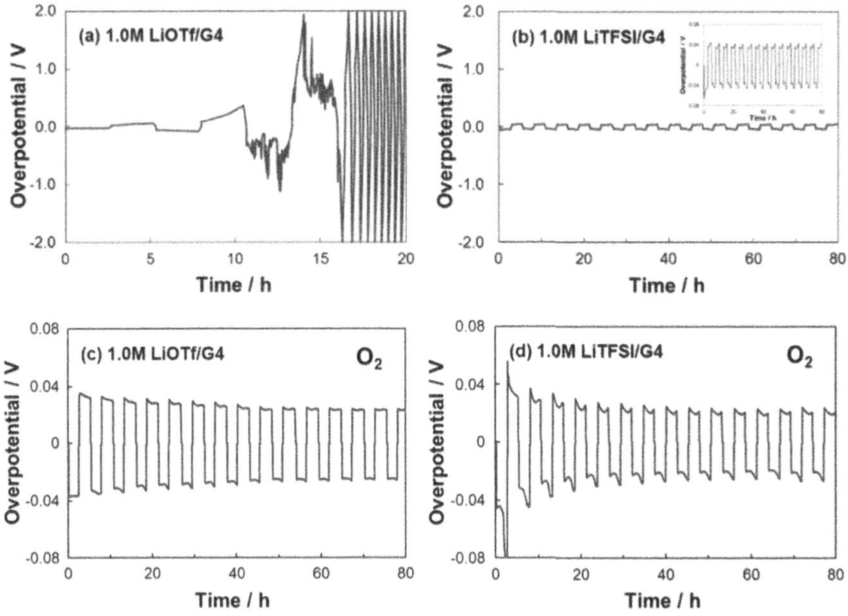

Fig. 9 Li | Li polarization curves obtained from Li | Li symmetric cells under **a**, **b** Ar and **c**, **d** O_2 at 30 °C

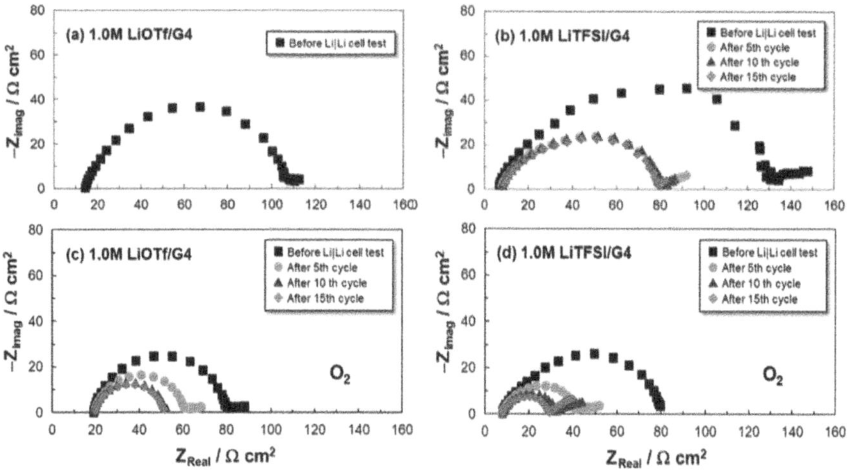

Fig. 10 AC impedance plots obtained from Li | Li symmetric cells in **a**, **b** Ar and **c**, **d** O_2 at 30 °C

6 Conclusions

In this chapter, the general properties of glyme-based electrolytes were discussed and the requirements for practical applications were examined. Compared with conventional organic carbonate electrolytes, glyme-based materials offer several advantages, including relatively low reactivity with the Li metal NE and with $O_2^{\cdot -}$ radicals and, especially in the case of G4-based electrolytes, a low vapor pressure for open-type batteries such as LABs. Although there has been significant research concerning these materials with regard to their use in next-generation batteries, their relatively high viscosities and low dielectric constants tend to inhibit rapid ion transport as well as the Li salt dissociation required to generate carrier ions. In addition, the growth of Li dendrites must be slowed and the durability to attack by $O_2^{\cdot -}$ radicals must be improved prior to the commercialization of LAB systems. This chapter primarily focused on the former aspect of LAB improvement and suggested concepts for the optimization of glyme-based electrolytes. The use of PGSE-NMR was shown to be an effective approach to understanding the relationship between ion transport and solution structure in such electrolytes. Continued progress in the development of these materials is expected, with the eventual aim of fabricating non-aqueous, rechargeable LABs.

Acknowledgements This work was supported under the Project for Next Generation Batteries in Advanced Low Carbon Technology Research and Development (ALCA-SPRING JPMJAL1301) by JST, Japan.

References

1. Bruce, P. G., Freunberger, S. A., Hardwick, L. J., & Tarascon, J. M. (2012). Li–O_2 and Li–S batteries with high energy storage. *Nature Materials, 11,* 19–29.
2. Abraham, K. M., & Jiang, Z. A. (1996). A polymer electrolyte-based rechargeable lithium/oxygen battery. *Journal of the Electrochemical Society, 143*(1), 1–5.
3. Mizuno, F., Nakanishi, S., Kotani, Y., Yokoishi, S., & Iba, H. (2010). *Rechargeable Li-Air Batteries with Carbonate-Based Liquid Electrolytes. Electrochemistry, 78*(5), 403–405.
4. McCloskey, B. D., Bethune, D. S., Shelby, R. M., Grishkumar, G., & Luntz, A. C. S. (2011). Solvents' critical role in nonaqueous lithium–oxygen battery electrochemistry. *The Journal of Physical Chemistry Letters, 2*(10), 1161–1166.
5. Xu, W., Xu, K., Viswanathan, V. V., Towne, S. A., Hardy, J. S., Xiao, J., et al. (2011). Reaction mechanisms for the limited reversibility of Li–O_2 chemistry in organic carbonate electrolytes. *Journal of Power Sources, 196*(22), 9631–9639.
6. Bryantsev, V. S., Giordani, V., Walker, W., Blanco, M., Zecevic, S., & Sasaki, K., et al. (2011). Predicting solvent stability in aprotic electrolyte Li–Air batteries: nucleophilic substitution by the superoxide anion radical ($O_2^{\bullet -}$). *Journal of Physical Chemistry A, 115*(44), 12399–12409.
7. Freunberger, S. A. (2011). Reactions in the rechargeable lithium–O_2 battery with alkyl carbonate electrolytes. *Journal of the American Chemical Society, 133*(20), 8040–8047.
8. Aurbach, D., Daroux, M., Faguy, P., & Yeager, E. (1991). Journal of electroanalytical chemistry and interfacial electrochemistry. *Journal of Electroanalytical Chemistry, 297*(1), 225–244.
9. Gunasekara, I., Mukerjee, S., Plichta, E. J., Hendrickson, M. A., & Abraham, K. M. (2015). A Study of the influence of lithium salt anions on oxygen reduction reactions in Li-Air batteries. *Journal of the Electrochemical Society, 162*(6), A1055–A1066.
10. Freunberger, S. A., Chen, Y., Drewett, N. E., Hardwick, L. J., Barde, F., & Bruce, P. G. (2011). The lithium–oxygen battery with ether-based electrolytes. *Angewandte Chemie International Edition, 50*(37), 8609–8613.
11. McCloskey, B. D., Bethune, D. S., Schelby, R. M., Grishkumar, G., & Luntz, A. C. S., (2011). Solvents' critical role in nonaqueous lithium-oxygen battery electrochemistry. *The Journal of Physical Chemistry Letters, 2*(10), 1161–1166.
12. Xu, W., Xiao, J., Wang, D., Zhang, J., & Zhang, J.-G. (2010). Crown ethers in nonaqueous electrolytes for lithium/air batteries. *Electrochemical and Solid-State Letters, 13*(4), A48–A51.
13. McCloskey, B. D., Scheffler, R., Speidel, A., Bethune, D. S., Shelby, R. M., & Luntz, A. C. (2011). On the efficacy of electrocatalysis in nonaqueous Li–O_2 batteries. *Journal of the American Chemical Society, 133*(45), 18038–18041.
14. McCloskey, B. D., Speidel, A., Scheffler, R., Miller, D. C., Viswanathan, V., Hummelshøj, J. S., Nørskov, J. K., & Luntz, A. C. (2012). Twin problems of interfacial carbonate formation in nonaqueous Li−O_2 batteries. *The Journal of Physical Chemistry Letters, 3,* 997–1001.
15. Yeager, E. (1984). Electrocatalysts for O_2 reduction. *Electrochimica Acta, 29*(11), 1527–1537.
16. Lu, Y.-C., Gasteiger, H. A., Parent, M. C., Chiloyan, V., & Shao-Horn, Y. (2010). The influence of catalysts on discharge and charge voltages of rechargeable li-oxygen batteries. *Electrochemical and Solid-State Letters, 13*(6), A69–A72.
17. Lu, Y.-C., Xu, Z., Gasteiger, H. A., Chen, S., Hamad-Schifferli, K., & Shao-Horn, Y. (2010). Platinum-gold nanoparticles: a highly active bifunctional electrocatalyst for rechargeable lithium-air batteries. *Journal of the American Chemical Society, 132*(35), 12170–12171.
18. Wang, W., Wang, R. F., Ji, S., Feng, H., Wang, H., & Lei, Z. (2010). Pt overgrowth on carbon supported PdFe seeds in the preparation of core–shell electrocatalysts for the oxygen reduction reaction. *Journal of Power Sources, 195*(11), 3498–3503.
19. Wang, H., Wang, R., Li, H., Wang, Q., Kang, J., & Lei, Z. (2011). Facile synthesis of carbon-supported pseudo-core@shell PdCu@Pt nanoparticles for direct methanol fuel cells. *International Journal of Hydrogen Energy, 36*(1), 839–848.
20. Wang, J., Wu, H., Gao, D., Miao, S., Wang, G., & Bao, X. (2015). High-density iron nanoparticles encapsulated within nitrogen-doped carbon nanoshell as efficient oxygen electrocatalyst for zinc–air battery. *Nano Energy, 13,* 387–396.

21. Sevim, M., Francia, C., Amici, J., Vankova, S., Şener, T., & Metin, Ö. (2016). Bimetallic MPt (M: Co, Cu, Ni) alloy nanoparticles assembled on reduced graphene oxide as high performance cathode catalysts for rechargable lithium-oxygen batteries. *Journal of Alloys and Compounds, 683,* 231–240.
22. Débart, A., Bao, J., Armstrong, G., & Bruce, P. G. (2007). An O_2 cathode for rechargeable lithium batteries: The effect of a catalyst. *Journal of Power Sources, 174,* 1177–1182.
23. Débart, A., Paterson, A. J., Bao, J., & Bruce, P. G. (2008). α-MnO_2 Nanowires: a catalyst for the O_2 electrode in rechargeable lithium batteries. *Angewandte Chemie International Edition, 47,* 4521–4524.
24. Wang, L., Zhao, X., Lu, Y., Xu, M., Zhang, D., & Ruoff, R. S., et al. (2011). $CoMn_2O_4$ spinel nanoparticles grown on graphene as bifunctional catalyst for lithium-air batteries. *Journal of the Electrochemical Society, 158*(12), A1379–A1382.
25. Xu, J., Gao, P., & Zhao, T. S. (2012). Non-precious Co_3O_4 nano-rod electrocatalyst for oxygen reduction reaction in anion-exchange membrane fuel cells. *Energy & Environmental Science, 5,* 5333–5339.
26. Suntivich, J., Perry, E. E., Gasteiger, H. A., & Shao-Horn, Y. (2013). The influence of the cation on the oxygen reduction and evolution activities of oxide surfaces in alkaline electrolyte. *Electrocatalysis, 4,* 49–55.
27. Grande, L., Paillard, E., Hassoun, J., Park, J. B., Lee, Y. J., & Sun, Y.-K., et al. (2015). The lithium/air battery: still an emerging system or a practical reality? *Advanced Materials, 27,* 784–800.
28. Jung, C. Y., Zhao, T. S., Zeng, L., & Tan, P. (2016). Vertically aligned carbon nanotube-ruthenium dioxide core-shell cathode for non-aqueous lithium-oxygen batteries. *Journal of Power Sources, 331,* 82–90.
29. Fu, Z., Lin, X., Huang, T., & Yu, A. (2012). Nano-sized $La_{0.8}Sr_{0.2}MnO_3$ as oxygen reduction catalyst in nonaqueous Li/O_2 batteries. *Journal of Solid State Electrochemistry, 16,* 1447–1452.
30. Garcia, E. M., Tarôco, H. A., Matencio, T., Domingues, R. Z., & dos Santos, J. A. F. (2012). Electrochemical study of $La_{0.6}Sr_{0.4}Co_{0.8}Fe_{0.2}O_3$ during oxygen evolution reaction. *International Journal of Hydrogen Energy, 37,* 6400–6406.
31. Cheng, J., Zhang, M., Jiang, Y., Zou, L., Gong, Y., Chi, B., & Pu, J. (2016). Perovskite $La_{0.6}Sr_{0.4}Co_{0.2}Fe_{0.8}O_3$ as an effective electrocatalyst for non-aqueous lithium air batteries. *Electrochimica Acta, 191,* 106–115.
32. Francia, C., Amici, J., Tasarkuyu, E., Coşkun, A., Gül, Ö. F., & Şener, T. (2016). What do we need for the lithium-air batteries: a promoter or a catalyst? *International Journal of Hydrogen Energy, 41,* 20583–20591.
33. Cui, Z. H., & Guo, X. X. (2014). Manganese monoxide nanoparticles adhered to mesoporous nitrogen-doped carbons for nonaqueous lithium–oxygen batteries. *Journal of Power Sources, 267,* 20–25.
34. Cheng, H., & Scott, K. (2010). Carbon-supported manganese oxide nanocatalysts for rechargeable lithium-air batteries. *Journal of Power Sources, 195,* 1370–1375.
35. Debart, A., Paterson, A. J., Bao, J., & Bruce, P. G. (2008). α-MnO_2 Nanowires: A Catalyst for the O_2 electrode in rechargeable lithium batteries.*Angewandte Chemie, 120,* 4597–4600.
36. Kalubarme, R. S., Cho, M.-S., Yun, K.-S., Kim, T.-S., & Park, C.-J. (2011). Catalytic characteristics of MnO_2 nanostructures for the O_2 reduction process. *Nanotechnology, 22,* 395402–395407.
37. Ida, S., Thapa, A. K., Hidaka, Y., Okamoto, Y., Matsuka, M., Hagiwara, H., & Ishihara, T. (2012). Manganese oxide with a card-house-like structure reassembled from nanosheets for rechargeable Li-air battery. *Journal of Power Sources, 203,* 159–164.
38. Kim, D. S., & Park, Y. J. (2014). Effect of multi-catalysts on rechargeable Li–air batteries. *Journal of Alloys and Compounds, 591,* 164–169.
39. Kwak, W.-J., Hirshberg, D., Sharon, D., Shin, H.-J., Afri, M., & Park, J.-B., et al. (2015). Understanding the behavior of Li–oxygen cells containing liI. *Journal of Materials Chemistry A, 3*(16), 8855–8864.

40. Kwak, W.-J., Hirshberg, D., Sharon, D., Afri, M., Frimer, A. A., & Jung, H. G., et al. (2016). Li–O_2 cells with LiBr as an electrolyte and a redox mediator. *Energy & Environmental Science, 9*(7), 2334–2345.
41. Chen, Y., Freunberger, S. A., Peng, Z., Fontaine, O., & Bruce, P. G. (2013). Charging a Li-O_2 battery using a redox mediator. *Nature Chemistry, 5*(6), 489–494.
42. Bergner, B. J., Schürmann, A., Peppler, K., Garsuch, A., & Janek, J. (2014). TEMPO: a mobile catalyst for rechargeable Li-O_2 batteries. *Journal of the American Chemical Society, 136*(42), 15054–15064.
43. Feng, N., Mu, X., Zhang, X., He, P., & Zhou, H. (2017). Intensive study on the catalytical behavior of N-Methylphenothiazine as a soluble mediator to oxidize the Li_2O_2 cathode of the Li–O_2 battery. *ACS Applied Materials & Interfaces, 9*(4), 3733–3739.
44. Uddin, J., Bryantsev, V. S., Walker, V., Chase, G. V., & Addison, D. (2013). Lithium nitrate as regenerable SEI stabilizing agent for rechargeable Li/O_2 batteries. *The Journal of Physical Chemistry Letters, 4*(21), 3760–3765.
45. Carbone, L., Gobet, M., Peng, J., Devany, M., Scrosati, B., Greenbaum, S., & Hassoun, J. (2015). Polyethylene glycol dimethyl ether (PEGDME)-based electrolyte for lithium metal battery. *Journal of Power Sources, 299,* 460–464.
46. Saito, M., Fujinami, T., Yamada, S., Ishikawa, T., Otsuka, H., Ito, K., & Kubo, Y. (2017). Effects of li salt anions and O_2 gas on li dissolution/deposition behavior at li metal negative electrode for non-aqueous li-air batteries. *Journal of the Electrochemical Society, 164*(12), A2872–A2880.
47. Xin, X., Ito, K., & Kubo, Y. (2017). Highly efficient Br^-/NO_3^- dual-anion electrolyte for suppressing charging instabilities of Li–O_2 batteries. *ACS Applied Materials & Interfaces, 9*(31), 25976–25984.
48. Xin, X., Ito, K., Dutta, A., & Kubo, Y. (2018). Dendrite-free epitaxial growth of lithium-metal during charging in Li-O_2 batteries. *Angewandte Chemie International Edition, 57*(40), 13206.
49. Moon, H., Mandai, T., Tatara, R., Ueno, K., Yamazaki, A., & Yoshida, K., et al. (2015). Solvent Activity in Electrolyte Solutions Controls Electrochemical Reactions in Li-Ion and Li-Sulfur Batteries. *Journal of Physical Chemistry C, 119*(8), 3957-3970.
50. Kwon, H.-M., Thomas, M. L., Tatara, R., Oda, Y., Kobayashi, Y., & Nakanishi, A., et al. (2017). Stability of glyme solvate ionic liquid as an electrolyte for rechargeable Li−O_2 batteries.*ACS Applied Materials & Interfaces, 9*(7), 6014–6021
51. Qian, J., Henderson, W. A., Xu, W., Bhattacharya, P., Engelhard, M., Borodin, O., & Zhang, J. G. (2015). High rate and stable cycling of lithium metal anode. *Nature Communications, 6*(1), 6362.
52. Saito, M., Yamada, S., Ishikawa, T., Otsuka, H., Ito, K., & Kubo, Y. (2017). Factors influencing fast ion transport in glymebased electrolytes for rechargeable lithium–air batteries. *RSC Advances, 7,* 49031–49040.
53. Hayamizu, K., Aihara, Y., Arai, S., & Martinez, C. G. (1999). Pulse-gradient spin-echo ^{1}H, ^{7}Li, and ^{19}F NMR diffusion and ionic conductivity measurements of 14 organic electrolytes containing $LiN(SO_2CF_3)_2$. *The Journal of Physical Chemistry B, 103*(3), 519–524.
54. Aihara, Y., Sugimoto, K., Price, W. S., & Hayamizu, K. (2000). Ionic conduction and self-diffusion near infinitesimal concentration in lithium salt-organic solvent electrolytes. *The Journal of Chemical Physics, 113*(5), 1981–1991.
55. Hayamizu, K., Akiba, E., Banno, T., & Aihara, Y. (2002). ^{1}H, ^{1}H, ^{7}Li, ^{7}Li, and ^{19}F nuclear magnetic resonance and ionic conductivity studies for liquid electrolytes composed of glymes and polyetheneglycol dimethyl ethers of $CH_3O(CH_2CH_2O)_nCH_3CH_3O(CH_2CH_2O)_nCH_3$ (n=3–50) doped with $LiN(SO_2CF3)_2$. *The Journal of Chemical Physics, 117*(12), 5929–5939.
56. Evey, H., Bishop, A. G., Forsyth, M., & MacFarlane, D. R. (2000). Ion diffusion in molten salt mixtures. *Electrochimica Acta, 45*(8–9), 1279–1284.
57. Lee, C. K., & Park, Y. J. (2016). CsI as multifunctional redox mediator for enhanced li–air batteries. *ACS Applied Materials & Interfaces, 8*(13), 8561–8567.
58. Saito. M., Fujinami, T., Somiya, M., Hayashi, Y., Koyama, K., Otsuka, H., Ito, K., Kubo, Y., & Horiba, T. (2021). Comparison of lithium salt effect on negative electrodes and lithium–air cell performance. *Journal of The Electrochemical Society, 168,* 010520.

Electrolytes with Redox Mediators

Yoshimi Kubo

Abstract Although lithium–air batteries are attractive because of their huge energy density, further improvements in cycle performance are required for practical use. The most serious problems are the increase in charging voltage at the cathode and the formation of Li dendrites at the anode during the charging process. Soluble-type catalysts or redox mediators (RMs) are known to be effective in lowering the charge voltage to the redox potential, although they cannot inhibit Li dendrite formation. We have recently shown that LiBr–$LiNO_3$ mixed anion electrolytes can simultaneously alleviate both of these problems. LiBr effectively functions as an RM that lowers the charge voltage to ~3.5 V and suppresses side reactions. $LiNO_3$ acts synergistically with LiBr to form a thin, dense, and uniform SEI film on the surface of Li metal. As a result, Li dendrite was effectively suppressed and even epitaxial growth of Li metal was observed.

Keywords Lithium air battery · Electrolyte · Redox mediator · LiBr · $LiNO_3$ · Dendrite

1 Problems of Battery Reactions

Figure 1a shows typical discharge–charge cycling properties of a lithium–air battery. The discharge occurs at ~2.7 V, which is slightly lower than the theoretical voltage of 2.96 V, and proceeds at an almost constant voltage. However, the charging voltage gradually increases from ~3 V to over 4 V. In situ observation of this process by synchrotron radiation X-ray diffraction showed that the diffraction intensity of the Li_2O_2 crystalline phase increased and decreased almost in proportion to the discharge and charge capacities, and no diffraction peaks of other by-products (such as LiOH or Li_2CO_3) were observed [1]. Additionally, FIB-SEM observation of the cathode revealed that the fine pores of the porous carbon were filled with a precipitate on discharging, which decomposed and disappeared on subsequent charging. These

Y. Kubo (✉)
National Institute for Materials Science, 1-1 Namiki, Tsukuba 305-0044, Japan
e-mail: KUBO.Yoshimi@nims.go.jp

K. Kanamura (ed.), *Next Generation Batteries*,
https://doi.org/10.1007/978-981-33-6668-8_41

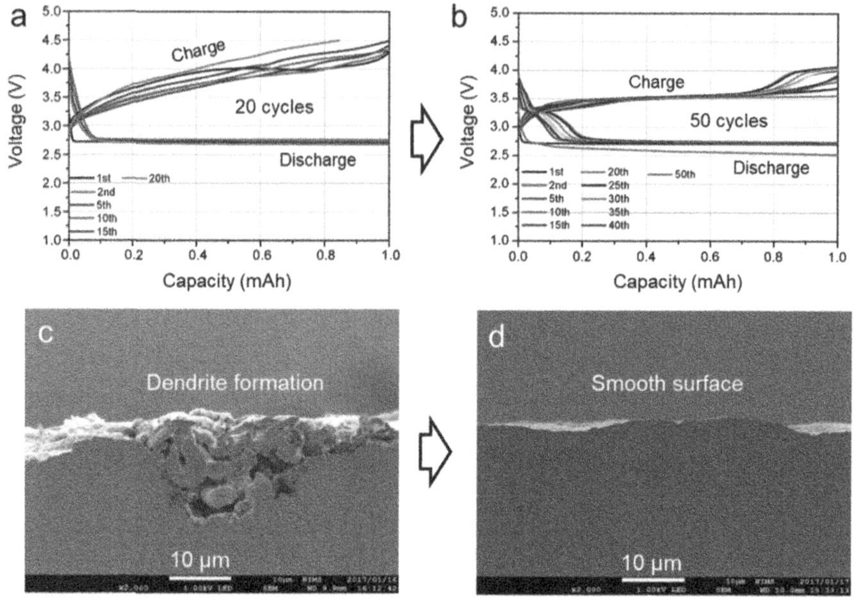

Fig. 1 Discharge–charge cycling performance of Li–O_2 batteries (**a**, **b**), and cross-sectional SEM images of Li metal anode after 20 cycles (**c**, **d**). Cells were cycled in **a**, **c** 1 M $LiCF_3SO_3$ and **b**, **d** 0.05 M LiBr–1 M $LiNO_3$ electrolytes with tetraglyme solvent. Current density was 0.1 mA/cm^2, and areal capacity was 0.5 mAh/cm^2

results clearly showed that the discharge and charge reactions of the lithium–air battery were mainly due to the formation and decomposition of solid Li_2O_2. Particularly, the discharge reaction appeared almost ideal with a small overpotential. The ratio of the discharge capacity to oxygen consumption (e^-/O_2) was reasonably consistent with the two-electron reaction (2.0) [2], and chemical analysis showed almost no deterioration of the electrolyte after discharge.

However, many problems remain with respect to the charging process. One is that the charge voltage increases above 4 V, and side reactions such as decomposition of the electrolyte become significant. Oxygen generation is less than the theoretical value, and CO_2 is generated at the end of charging [3–6]. On the anode side, the well-known lithium dendrite growth occurs (Fig. 1c). These two issues, namely the increased charge voltage (overvoltage) and generation of lithium dendrites, are the main challenges associated with current lithium–air batteries and the cause of cycle degradation.

The reason why the charge voltage increases as charging progress is only partially understood. Since Li_2O_2 can react with electrolyte or carbon substrates, it has been suggested that the reaction products such as lithium carbonate, lithium formate, and lithium acetate may be generated at the interfaces to form a barrier [7]. The presence of these by-products was confirmed by chemical analysis and Fourier transform infrared spectroscopy (FTIR) and was also supported by the release of CO_2 at the end of charging (>4 V). These side reactions are unnoticeable up to around 3.5 V but tend to increase gradually above this value [8]. Another possible cause of side reactions is the generation of singlet oxygen, $^1O_2$ [9–11]. This is known to be extremely reactive, and its formation energy is ~1 eV higher than that of the normal triplet oxygen, $^3O_2$. If $^1O_2$ is produced during the charge reaction, the equilibrium potential will increase by ~0.5 V (because of the two-electron reaction) to ~3.5 V versus Li/Li^+. Therefore, the generation of singlet oxygen that induces side reactions will become significant from around 3.5 V. In any case, to suppress the side reactions, it is essential to maintain the charge voltage below 3.5 V. Lowering the charge voltage is also important in enhancing the round trip energy efficiency.

2 Effect of Redox Mediators

It is noted that adding a redox mediator (RM) to the electrolyte is effective in reducing the charge voltage (overvoltage). A RM is a molecule or ion that has a redox potential, E_{redox}, that is slightly higher than the equilibrium potential of Li_2O_2 [12–14]. When the charge potential exceeds E_{redox}, the RM is first electrochemically oxidized to RM^+, which then chemically oxidizes Li_2O_2 ($2RM^+ + Li_2O_2 \rightarrow 2RM + 2Li^+ + O_2$). Thus, the cathode potential is fixed near E_{redox}. During this process, RM^+ returns to the original RM, which can then be reused as a homogeneous catalyst. Organic molecules such as TTF (E_{redox} ~ 3.6 V) and TEMPO (~3.74 V) have been studied as RMs, but because of their issues with cycle stability, more chemically stable LiI (2.9–3.55 V) and LiBr (3.5–3.9 V) were recently studied [15].

We recently found that a combined electrolyte of LiBr and $LiNO_3$ could effectively suppress both the charging voltage and Li dendrite formation [16]. Figure 1b shows the cycling properties using this electrolyte (0.05 M LiBr–1 M $LiNO_3$/tetraglyme). Compared with the conventional electrolyte solution (1 M $LiCF_3SO_3$/tetraglyme) shown in Fig. 1a, the charge voltage was greatly reduced to around 3.5 V and the cycle number was extended to 50. Since the charge voltage was consistent with the redox potential of Br^-/Br_3^- (E_{redox} ~ 3.5 V), LiBr was considered to function as a RM. Figure 2 shows the results of differential electrochemical mass spectrometry (DEMS) of the released gas during charging. For the conventional electrolyte shown in Fig. 2a, the amount of oxygen generated was only about 60% of the theoretical value ($e^-/O_2 = 2.0$), and a large amount of CO_2 was released at the end of charging, which appeared to be due to the decomposition of by-products.

However, for the new electrolyte shown in Fig. 2b, the amount of oxygen generated increased to over 80% of the theoretical value, and hardly any CO_2 was released. Chemical analysis of the electrolyte after the cycle test also confirmed that the side reaction was reduced by more than one order of magnitude. These results clearly show that the RM effect of LiBr drastically reduces the charge voltage and suppresses the side reactions. However, the fact that the amount of oxygen generated remained lower than the theoretical value suggests that the side reactions are incompletely

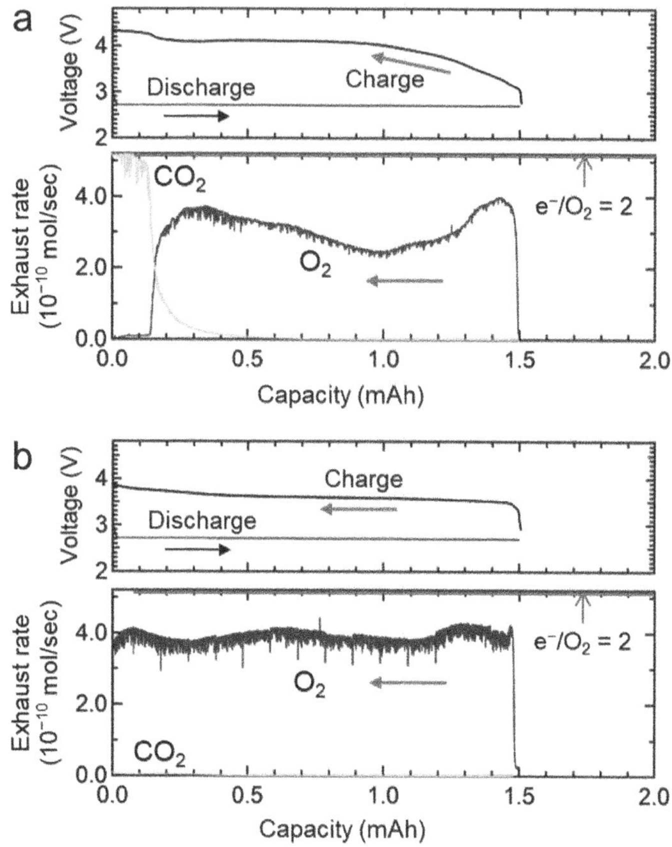

Fig. 2 Discharge–charge voltage curves (upper) and the corresponding DEMS results of O_2 and CO_2 evolution during charging in **a** 1 M $LiCF_3SO_3$ and **b** 0.05 M LiBr–1 M $LiNO_3$ electrolytes with tetraglyme solvent at a current of 0.1 mA (cathode area of 2 cm^2). Red lines ($e^-/O_2 = 2$) indicate the theoretical O_2 evolution rate according to the two-electron reaction

suppressed, and the detailed elucidation of this reaction remains a challenge for the future.

3 Mixed Anion Electrolyte

The LiBr–$LiNO_3$ electrolyte was found to be effective for both the cathode and Li metal anode. As shown in Fig. 1d, the surface of the Li metal anode remained smooth with no generation of dendrites even after 20 cycles. This effect was remarkable only in the mixed anion system combining LiBr and $LiNO_3$ [16]. $LiNO_3$ is known to oxidize and passivate the Li surface ($2Li + NO_3^- \rightarrow Li_2O + NO_2^-$) and is expected to prevent a "shuttle effect" in which RM^+ is discharged on the anode side [17, 18]. However, in this case, $LiNO_3$ should exhibit a synergistic effect with LiBr, because it promotes both passivation and dendrite suppression. The important point is that a uniform Li_2O-like film as thin as 0.1 μm formed on the Li metal surface. This oxide-based thin and uniform film functioned as an excellent SEI (solid electrolyte interphase), resulting in uniform Li plating and suppression of dendrite growth.

Figure 3 shows a reaction model of this LiBr–$LiNO_3$ mixed anion electrolyte. As described above, Br^- functions as the RM of the charging reaction at the cathode. On the other hand, NO_3^- oxidizes the Li metal surface and changes to nitrite (NO_2^-). This reaction ends when NO_3^- is depleted in conventional closed systems. However, since oxygen is always supplied in air batteries, NO_2^- can immediately react with dissolved oxygen and return to NO_3^- [18, 19]. That is, the cycle of $NO_3^- \rightarrow NO_2^- \rightarrow NO_3^-$ is repeated in an oxygen atmosphere; thus, the Li surface is always exposed to a strongly oxidizing environment. This situation is considered to play an important role in the formation of dense Li_2O-like films (SEI) that can be quickly repaired even if broken. However, this alone gradually increases the SEI film thickness. In fact, the SEI film becomes thicker during the cycle test in the $LiNO_3$-only electrolyte, as shown in Fig. 4 [16].

Thus, what role does LiBr play? One possibility is that LiBr accelerates the decomposition of the SEI film during discharge. It is well-known that the halide ions break and corrode passivation films of, for example, stainless steel. Although the mechanism is only partially understood, it has been suggested that halide ions penetrate the oxide and replace oxygen due to the high affinity between halogens and metals. If so, the surface film of Li will be eroded by Br^- during discharge, as with electropolishing. This means that the Li anode surface is in a state where the formation and decomposition of oxide films compete in a strongly oxidizing environment. Consequently, a thin, dense, and uniform surface film (SEI) is formed. One of the causes

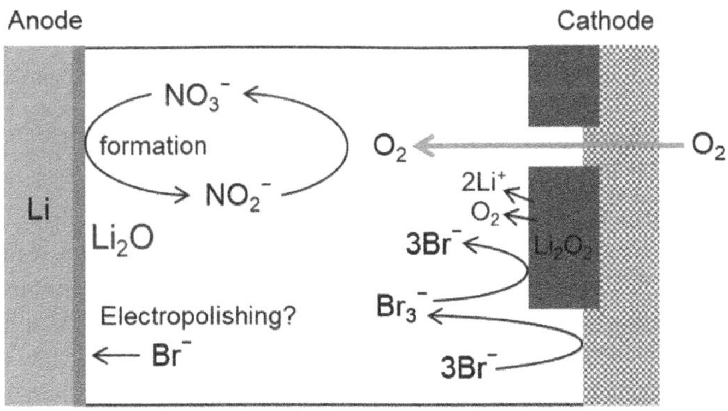

Fig. 3 Possible reactions in the LiBr–$LiNO_3$ electrolyte of lithium–air batteries. Br^- acts as a redox mediator at the cathode during charging. NO_3^- continues to oxidize the Li metal surface in the presence of oxygen. Br^- may induce the decomposition of the Li_2O layer during discharge

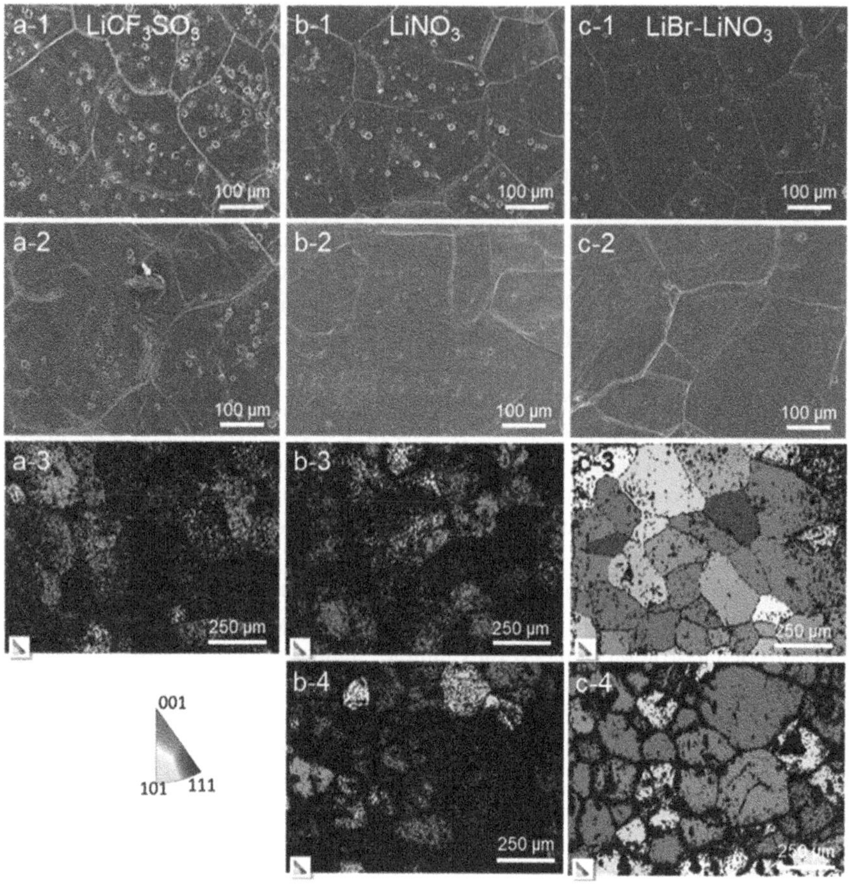

Fig. 4 SEM images and EBSD mapping of the Li foils in different electrolytes with tetraglyme solvent after discharge (1 and 3) and after recharge (2 and 4). Parts 1 and 2 are SEM images. Parts 3 and 4 are EBSD mapping images (inverse pole figure maps from the *Z* direction). **a** Tested in 1 M $LiCF_3SO_3$; **b** tested in 1 M $LiNO_3$; **c** tested in 0.05 M LiBr–1 M $LiNO_3$. The Li-foil anodes (area 2 cm^2) were discharged/recharged at a current of 0.1 mA with a fixed capacity of 4 mAh, corresponding to a Li stripping/plating thickness of 10 μm

of Li dendrites appears to be the roughness of the surface film. If this is thin, dense, and uniform, with a uniform ionic conductivity, uniform Li deposition may occur.

As an interesting phenomenon related to this, we recently found that the lithium metal anode was "epitaxially grown" [19]. Figure 5a–c shows top-view and cross-sectional SEM images of as-received, discharged (stripped), and charged (plated) lithium metal anodes in the LiBr–$LiNO_3$ electrolyte. The stripping/plating thickness of lithium metal was set to 10 μm. The surface of a commercial Li foil (Fig. 5a) was covered with a thick uneven film over 0.5 μm, while the surface film after 10 μm stripping was as thin as 0.1 μm and uniform as shown in Fig. 5b. Grain

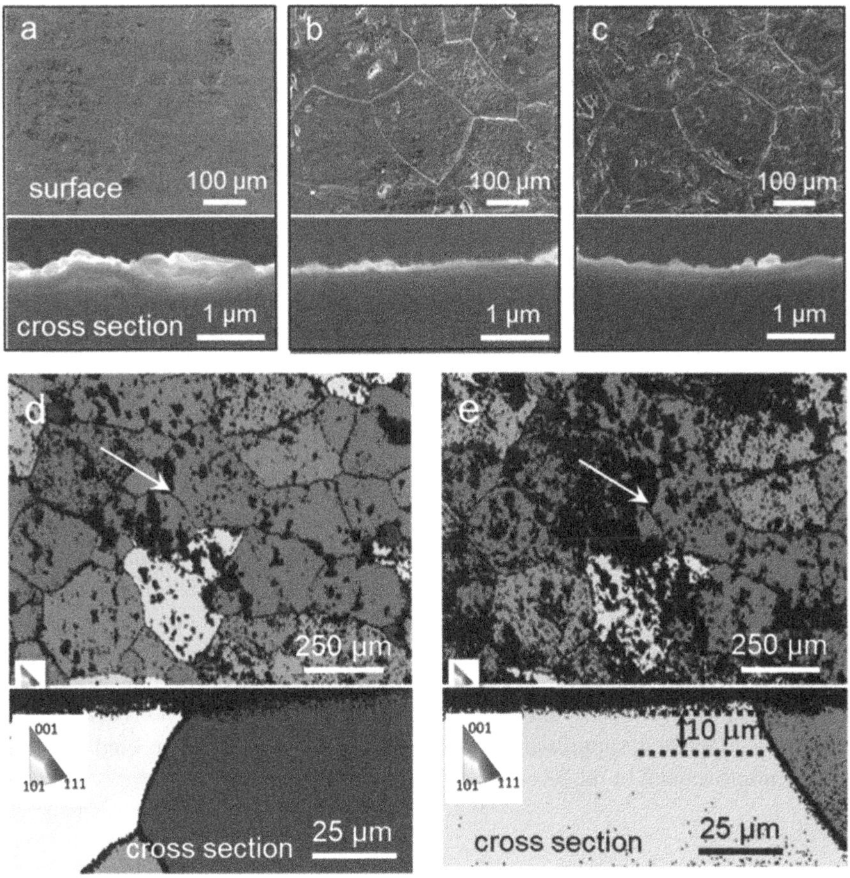

Fig. 5 Top-view and cross-sectional SEM images of Li foils: **a** as-received, **b** after stripping, and **c** after plating. Top-view and cross-sectional EBSD mapping images of Li foils: **d** after stripping, and **e** after plating. The white arrows in **d** and **e** indicate the same point on the Li metal surface. The Li stripping/plating thickness was set to 10 μm with a current density of 0.2 mA/cm^2 in the symmetrical Li | Li sprit-type cells with 0.05 M LiBr–1 M $LiNO_3$/tetraglyme electrolyte under O_2 atmosphere

boundaries of Li metal are clearly observed in the surface SEM image. This sample was subsequently plated to 10 μm, and the same spot on the surface was observed as shown in Fig. 5c. It was found that the grain boundary pattern was exactly the same despite the 10 μm Li plating. This is surprising considering that electrodeposited textures are usually very fine. It is also noted that the surface film remained thin and uniform. These results strongly suggest that Li^+ ions were epitaxially deposited at the Li/SEI interface through the SEI film. EBSD (electron backscatter diffraction) measurements were also performed to accurately determine the crystal orientation of each grain. Figure 5d, e shows EBSD mapping of the surface and cross-section of the lithium metal anode after discharging and charging. The crystal orientation at each point is indicated by a color code. (Black indicates that orientation analysis was impossible due to grain boundaries or a thick surface coating. The fact that orientation analysis was possible indicates that the surface film was very thin.) Comparison of the surfaces after discharging (Fig. 5d) and after charging (Fig. 5e) shows that the crystal grain shape and crystal orientation are perfectly consistent. This is clear evidence that the lithium metal anode was epitaxially grown during charging. There is no discontinuity even in the cross-sectional EBSD in the region of 10 μm depth from the surface. Such epitaxial growth of the lithium metal anode was observed only when a thin and uniform SEI film was formed by the LiBr–$LiNO_3$ electrolyte under an oxygen atmosphere. This is an important guideline for future dendrite suppression.

4 Conclusion

Although lithium–air batteries are attractive because of their huge energy density, further improvement of their cycle performance is necessary for practical use. The most serious problems are the increase in the charge voltage in the cathode and the accompanying side reactions, and the generation of Li dendrites in the anode. It was recently found that LiBr–$LiNO_3$ mixed anion electrolytes can simultaneously alleviate both of these problems. LiBr functions effectively as a RM in reducing the charge voltage to ~3.5 V and suppressing side reactions. The thin, dense, and uniform SEI film formed by the synergistic effect of LiBr and $LiNO_3$ is considered to be very effective in suppressing Li dendrites. These results are expected to lead to further improvement of the electrolyte for lithium–air batteries.

References

1. Song, C., Ito, K., Sakata, O., & Kubo, Y. (2018). *Operando* structural study of non-aqueous Li–air batteries using synchrotron-based X-ray diffraction. *RSC Advances, 8*, 26293–26299.
2. McCloskey, B. D., Scheffler, R., Speidel, A., Girishkumar, G., & Luntz, A. C. (2012). On the mechanism of nonaqueous Li–O_2 electrochemistry on C and its kinetic overpotentials: Some implications for Li–air batteries. *Journal of Physical Chemistry C, 116*, 23897–23905.
3. McCloskey, B. D., Bethune, D. S., Shelby, R. M., Mori, T., Scheffler, R., Speidel, A., et al. (2012). Limitations in rechargeability of Li-O_2 batteries and possible origins. *The Journal of Physical Chemistry Letters, 3*, 3043–3047.
4. McCloskey, B. D., Valery, A., Luntz, A. C., Gowda, S. R., Wallraff, G. M., Garcia, J. M., et al. (2013). Combining accurate O_2 and Li_2O_2 assays to separate discharge and charge stability limitations in nonaqueous Li–O_2 batteries. *The Journal of Physical Chemistry Letters, 4*, 2989–2993.
5. Tsiouvaras, N., Meini, S., Buchberger, I., & Gasteiger, H. A. (2013). A novel On-line mass spectrometer design for the study of multiple charging cycles of a Li-O_2 battery. *Journal of the Electrochemical Society, 160*, A471–A477.
6. Beyer, H., Meini, S., Tsiouvaras, N., Piana, M., & Gasteiger, H. A. (2013). Thermal and electrochemical decomposition of lithium peroxide in non-catalyzed carbon cathodes for Li–air batteries. *Physical Chemistry Chemical Physics: PCCP, 15*, 11025–11037.

7. McCloskey, B. D., Speidel, A., Scheffler, R., Miller, D. C., Viswanathan, V., Hummelshøj, J. S., et al. (2012). Twin problems of interfacial carbonate formation in nonaqueous Li–O_2 batteries. *The Journal of Physical Chemistry Letters, 3*, 997–1001.
8. Ottakam Thotiyl, M. M., Freunberger, S. A., Peng, Z., & Bruce, P. G. (2012). The carbon electrode in nonaqueous Li–O_2 cells. *Journal of the American Chemical Society, 135*, 494–500.
9. Hassoun, J., Croce, F., Armand, M., & Scrosati, B. (2011). Investigation of the O_2 electrochemistry in a polymer electrolyte solid-state cell. *Angewandte Chemie International Edition, 50*, 2999–3002.
10. Wandt, J., Jakes, P., Granwehr, J., Gasteiger, H. A., & Eichel, R. A. (2016). Singlet oxygen formation during the charging process of an aprotic lithium–oxygen battery. *Angewandte Chemie International Edition, 55*, 6892–6895.
11. Mahne, N., Schafzahl, B., Leypold, C., Leypold, M., Grumm, S., Leitgeb, A., et al. (2017). Singlet oxygen generation as a major cause for parasitic reactions during cycling of aprotic lithium–oxygen batteries. *Nature Energy, 2*, 17036.
12. Chen, Y., Freunberger, S. A., Peng, Z., Fontaine, O., & Bruce, P. G. (2013). Charging a Li–O_2 battery using a redox mediator. *Nature Chemistry, 5*, 489–494.
13. Park, J. B., Lee, S. H., Jung, H. G., Aurbach, D., & Sun, Y. K. (2018). Redox mediators for Li–O_2 batteries: Status and perspectives. *Advanced Materials, 30*, 1704162.
14. Kwak, W. J., Kim, H., Jung, H. G., Aurbach, D., & Sun, Y. K. (2018). Review—A comparative evaluation of redox mediators for Li-O_2 batteries: A critical review. *Journal of the Electrochemical Society, 165*, A2274–A2293.
15. Liang, Z., & Lu, Y.-C. (2016). Critical role of redox mediator in suppressing charging instabilities of lithium–oxygen batteries. *Journal of the American Chemical Society, 138*, 7574–7583.
16. Xin, X., Ito, K., & Kubo, Y. (2017). Highly efficient Br^-/NO_3^- dual-anion electrolyte for suppressing charging instabilities of Li$-O_2$ batteries. *ACS Applied Materials & Interfaces, 9*, 25976–25984.
17. Aurbach, D., Pollak, E., Elazari, R., Salitra, G., Kelley, C. S., & Affinito, J. (2009). On the surface chemical aspects of very high energy density, rechargeable Li–sulfur batteries. *Journal of the Electrochemical Society, 156*, A694–A702.
18. Uddin, J., Bryantsev, V. S., Giordani, V., Walker, W., Chase, G. V., & Addison, D. (2013). Lithium nitrate as regenerable SEI stabilizing agent for rechargeable Li/O_2 batteries. *The Journal of Physical Chemistry Letters, 4*, 3760–3765.
19. Xin, X., Ito, K., Dutta, A., & Kubo, Y. (2018). Dendrite-free epitaxial growth of lithium metal during charging in Li–O_2 batteries. *Angewandte Chemie International Edition, 57*, 13206–13210.

Mg Rechargeable Battery

Novel Mg Rechargeable Battery Cathodes: Chevrel to Spinel

Tetsu Ichitsubo and Shunsuke Yagi

Abstract In this chapter, it is shown that spinel oxides such as $MgCo_2O_4$ work as cathode materials for Mg rechargeable batteries with a high redox potential about 2–3 V versus Mg^{2+}/Mg on the basis of the similarity between spinel and rocksalt structures (Okamoto et al., Adv. Sci., 1500072, 2015, [1]). The Mg insertion into spinel lattices occurs via "insertion and push-out" process to form a rocksalt phase in the spinel mother phase. For example, by utilizing the valence change from Co(III) to Co(II) in $MgCo_2O_4$, Mg insertion occurs at a considerably high potential of about 2.9 V versus Mg^{2+}/Mg, and similarly, it occurs at around 2.3 V versus Mg^{2+}/Mg with the valence change from Mn(III) to Mn(II) in $MgMn_2O_4$. In addition, Mg^{2+} ions originally in $MgMn_2O_4$ and $MgCr_2O_4$ can be extracted to some extent because of the robust host structure. The "insertion and push-out" process proposed here provides a new design of cathode materials for Mg rechargeable batteries, and various approaches are introduced to develop cathode materials based on this mechanism in the subsequent chapters.

Keywords Chevrel · Spinel · Rocksalt · High voltage

1 Introduction

It is well known that Chevrel compounds such as Mo_6S_8 show a good cyclability as a cathode for Mg rechargeable batteries [2, 3]; however, the redox potential of the Chevrel compounds is low about 1.0–1.2 versus Mg^{2+}/Mg, resulting in a low energy density less than 150 mWh g^{-1} (currently 370 mWh g^{-1} in Li-ion

T. Ichitsubo (✉)
Institute for Materials Research, Tohoku University, 2-1-1 Katahira, Aoba-ku, Sendai 980-8577, Japan
e-mail: tichi@imr.tohoku.ac.jp

S. Yagi
Institute of Industrial Science, The University of Tokyo, 4-6-1 Komaba, Meguro-ku, Tokyo 153-8505, Japan
e-mail: syagi@iis.u-tokyo.ac.jp

K. Kanamura (ed.), *Next Generation Batteries*,
https://doi.org/10.1007/978-981-33-6668-8_42

battery electrode energy density of $LiCoO_2$ vs. graphite). Despite that several candidates have been reported for the Mg rechargeable battery cathode materials [4–7], there are few cathode materials that can work at ambient temperature except for Chevrel compounds. It has been also reported that the redox potential of the Chevrel compounds is maintained at a high level of about 3 V versus Mg^{2+}/Mg without using Grignard reagent-based tetrahydrofuran (THF) electrolytes containing Cl^- ions [8, 9]; however, the cyclability of the Chevrel compounds becomes very low. Therefore, unless cathode materials with a higher potential, a capacity, and a cyclability that can accommodate Mg ions are sought out, Mg^{2+} rechargeable batteries would not be comparable to Li-ion batteries specifically in terms of the energy density. Thus, in order to change the energy storage paradigm, we have to seek new cathode materials for Mg rechargeable batteries.

2 Spinel Oxides as Cathode Candidates for Mg Rechargeable Batteries

Mg spinel oxides are expected as candidates for cathode materials of Mg rechargeable batteries. As shown in Fig. 1, the lattice sites in the spinel structure are generally denoted as 8*a*, 16*d* (cation sites), and 32*e* (oxygen sites) in the Wyckoff position in the space group No. 227 ($Fd\overline{3}m$), while those in the rocksalt structure are denoted as 16*c*, 16*d* (cation sites) and 32*e* (oxygen sites) when it is assigned to the same space group. In other words, a spinel structure can be regarded as a rocksalt structure whose 16*c* sites are vacant and instead the 8*a* sites are occupied by cations. Therefore, it is expected that Mg cations can be inserted onto 16*c* vacant sites in the spinel lattice [10–12]. In fact, in our recent study, we have demonstrated the insertion of Mg cations into some spinel oxides such as $MgCo_2O_4$, $MgMn_2O_4$, $MgFe_2O_4$, $MgCr_2O_4$, and Co_3O_4 at around 3 V versus Mg^{2+}/Mg at 150 °C via "insertion and push-out" mechanism shown in Fig. 1 using CsTFSA-based ionic liquids [13, 14]. In this chapter, we discuss the feasibility of Mg insertion/extraction into/from the spinel oxides.

3 Redox Behavior of Spinel Oxides

As illustrated in Fig. 2a (upper), all the electrochemical tests were conducted at 150 °C using a beaker cell with CsTFSA-based ionic liquids reported by Hagiwara et al. [14]. Mg^{2+} ions can be inserted into the Chevrel compounds in an $Mg(TFSA)_2$/CsTFSA binary ionic liquid because the electrolytic dissociation of $Mg(TFSA)_2$ is favorable although the electrodeposition of Mg hardly occurs in this electrolyte [8].

Using the experimental result that hcp Mg metal can be electrodeposited in (Mg/Li/Cs)-TFSA ternary ionic liquids [14, 15], the redox potential of Mg^{2+}/Mg redox couple was deduced by reducing the Li composition from the ternary ionic

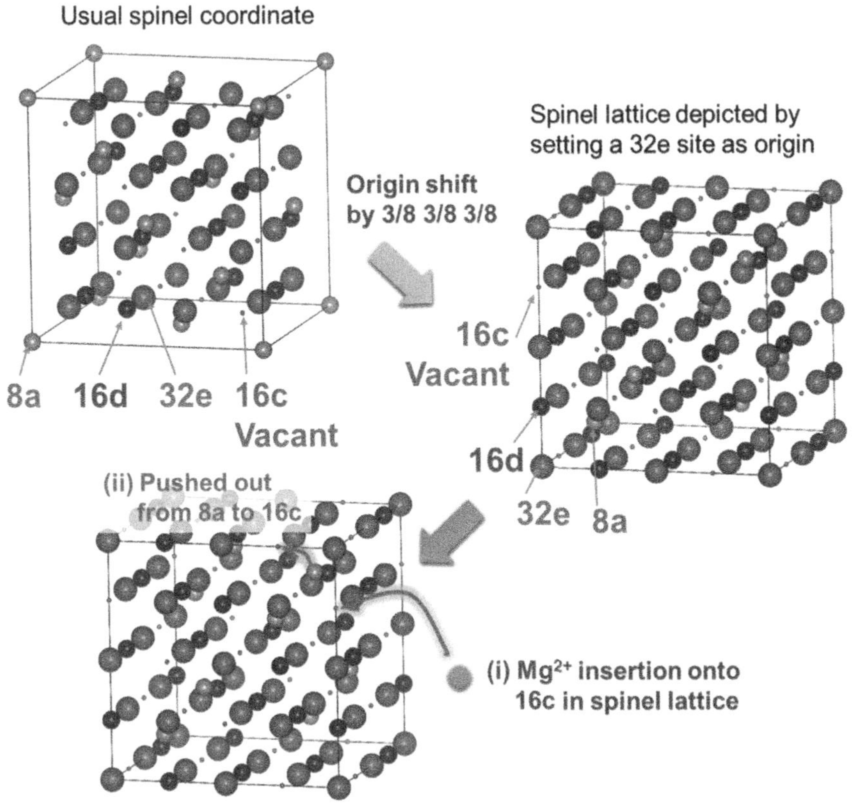

Fig. 1 Insertion process of Mg^{2+} ions into spinel structure. (Upper left) usual spinel coordinate. (middle right) setting 32*e* site (Wyckoff position) for oxygen in the space group No. 227 as origin. (Lower left) cations originally located at 8*a* site moves to an adjacent 16*c* site due to the repulsion between the cations with the insertion of an Mg^{2+} ion into a 16*c* site

liquids as "0.5 V versus Li^+/Li in RE $\approx$ 0 V versus Mg^{2+}/Mg" as shown in Fig. 2a (lower). Figure 2b shows the cyclic voltammogram obtained for $MgCo_2O_4$. One would consider the conventional reaction, i.e., the extraction of Mg^{2+} ions from the host material, $MgCo_2O_4 \Leftrightarrow Mg_{1-x}Co_2O_4 + x(Mg^{2+} + 2e^-)$. Such a reaction may occur above 4.4 V versus Li^+/Li in RE, but this cation extraction from $MgCo_2O_4$ appears to be difficult in terms of the structural stability and anodic limit of the potential window of the binary ionic liquid. Apart from this usual extraction, it is clearly seen that direct Mg insertion can occur into the host $MgCo_2O_4$ without a pre-charge process, and then the cation extraction is observed during a charging process; the equilibrium redox potential is about 2.9 V versus Mg^{2+}/Mg (3.4 V vs. Li^+/Li in RE), which is also in agreement with the ab initio calculation [1]. Inductively coupled plasma (ICP) analysis after the electrochemical test also supports the insertion of Mg^{2+} ions for the Mg-Li rocking-chair type dual-salt battery [16]. Figures 2c–f show cyclic voltammograms measured in the (Mg10/Cs90)-TFSA binary ionic liquid at

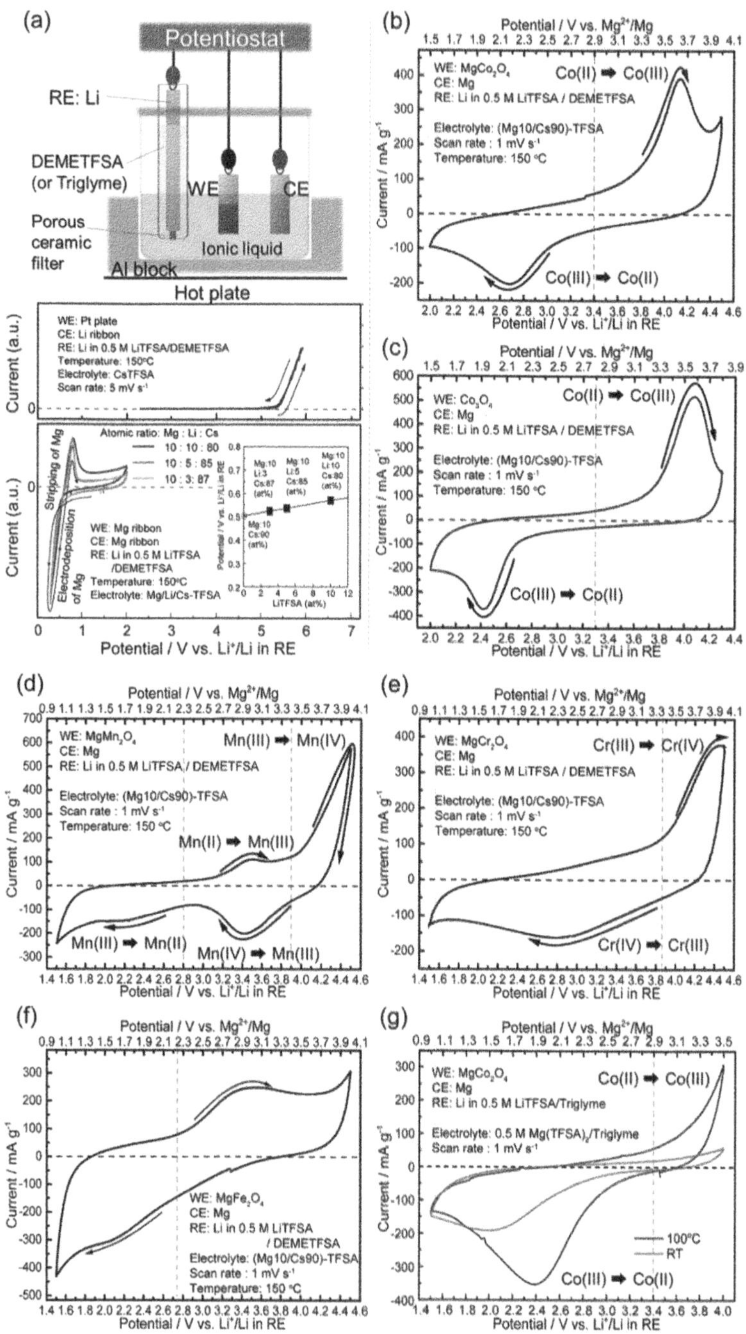
(a)
Potentiostat
RE: Li
DEMETFSA
(or Triglyme)
WE
CE
Porous ceramic filter
Ionic liquid
Al block
Hot plate
WE: Pt plate
CE: Li ribbon
RE: Li in 0.5 M LiTFSA/DEMETFSA
Temperature: 150°C
Electrolyte: CsTFSA
Scan rate: 5 mV s-1
Current (a.u.)
Atomic ratio: Mg : Li : Cs
10 : 10 : 80
10 : 5 : 85
10 : 3 : 87
Stripping of Mg
Electrodeposition of Mg
WE: Mg ribbon
CE: Mg ribbon
RE: Li in 0.5 M LiTFSA /DEMETFSA
Temperature: 150°C
Electrolyte: Mg/Li/Cs-TFSA
LiTFSA (at%)
Potential / V vs. Li+/Li in RE
(b)
Potential / V vs. Mg2+/Mg
WE: MgCo2O4
CE: Mg
RE: Li in 0.5 M LiTFSA / DEMETFSA
Electrolyte: (Mg10/Cs90)-TFSA
Scan rate : 1 mV s-1
Temperature: 150 °C
Co(II) ➡ Co(III)
Co(III) ➡ Co(II)
Current / mA g-1
Potential / V vs. Li+/Li in RE
(c)
Potential / V vs. Mg2+/Mg
WE: Co3O4
CE: Mg
RE: Li in 0.5 M LiTFSA / DEMETFSA
Electrolyte: (Mg10/Cs90)-TFSA
Scan rate : 1 mV s-1
Temperature: 150 °C
Co(II) ➡ Co(III)
Co(III) ➡ Co(II)
Current / mA g-1
Potential / V vs. Li+/Li in RE
(d)
Potential / V vs. Mg2+/Mg
WE: MgMn2O4
CE: Mg
RE: Li in 0.5 M LiTFSA / DEMETFSA
Electrolyte: (Mg10/Cs90)-TFSA
Scan rate : 1 mV s-1
Temperature: 150 °C
Mn(III) ➡ Mn(IV)
Mn(II) ➡ Mn(III)
Mn(III) ➡ Mn(II)
Mn(IV) ➡ Mn(III)
Current / mA g-1
Potential / V vs. Li+/Li in RE
(e)
Potential / V vs. Mg2+/Mg
WE: MgCr2O4
CE: Mg
RE: Li in 0.5 M LiTFSA / DEMETFSA
Electrolyte: (Mg10/Cs90)-TFSA
Scan rate : 1 mV s-1
Temperature: 150 °C
Cr(III) ➡ Cr(IV)
Cr(IV) ➡ Cr(III)
Current / mA g-1
Potential / V vs. Li+/Li in RE
(f)
Potential / V vs. Mg2+/Mg
WE: MgFe2O4
CE: Mg
RE: Li in 0.5 M LiTFSA / DEMETFSA
Electrolyte: (Mg10/Cs90)-TFSA
Scan rate : 1 mV s-1
Temperature: 150 °C
Current / mA g-1
Potential / V vs. Li+/Li in RE
(g)
Potential / V vs. Mg2+/Mg
WE: MgCo2O4
CE: Mg
RE: Li in 0.5 M LiTFSA/Triglyme
Electrolyte: 0.5 M Mg(TFSA)2/Triglyme
Scan rate : 1 mV s-1
Co(II) ➡ Co(III)
Co(III) ➡ Co(II)
100°C
RT
Current / mA g-1
Potential / V vs. Li+/Li in RE

◂**Fig. 2** **a** Three-electrode beaker cell setup (upper) using (Mg10/Cs90)-TFSA ionic liquid, the cathode active material (WE), a Mg ribbon (CE), and Li metal immersed in a 0.5 M LiTFSA/DEMETFSA electrolyte in a glass tube separated with a porous ceramic filter (RE). Cyclic voltammogram (lower) measured at 150 °C in the mixture of $Mg(TFSA)_2$, LiTFSA, and CsTFSA salts with various concentrations. The potential window of the CsTFSA ionic liquid (middle). Cyclic voltammograms measured at a scan rate of 1 mV s^{-1} at 150 °C for **b** $MgCo_2O_4$, **c** Co_3O_4, **d** $MgMn_2O_4$, **e** $MgCr_2O_4$, **f** $MgFe_2O_4$, and **g** $MgCo_2O_4$ in a 0.5 M $Mg(TFSA)_2$/triglyme electrolyte at room temperature and 100 °C

150 °C for other spinel oxides, such as Co_3O_4, $MgMn_2O_4$, $MgCr_2O_4$, and $MgFe_2O_4$. The cyclic voltammogram for Co_3O_4 (Fig. 2c) is very similar to that of $MgCo_2O_4$, where the extraction of Co^{2+} ions is not observed during the anodic scan from the open circuit potential (OCP), whereas Mg^{2+} ions cations were inserted into the spinel Co_3O_4, which was confirmed by EDS elemental analysis (data not presented here). In contrast, the extraction of Mg^{2+} ions from $MgMn_2O_4$ is observed in Fig. 2d during the first anodic scan from the OCP, and two redox-peak couples corresponding to the insertion/extraction of Mg^{2+} ions are observed at around 3.4 and 2.3 V versus Mg^{2+}/Mg, corresponding to the valence changes of Mn(IV) to Mn(III) and Mn(III) to Mn(II), respectively. A similar behavior was observed in Fig. 2e for $MgCr_2O_4$, i.e., the valance change from Cr(IV) to Cr(III), whereas the valence change of Cr(III) to Cr(II) was hardly observed. We also confirmed that $MgFe_2O_4$ can be used as a potential cathode material, but clear redox peaks were not observed in Fig. 2f. The insertion/extraction potential is about 2.7 V versus Li^+/Li in RE, which is lower than the redox potential (about 3.4 V vs. Li^+/Li) of Fe cations in the olivine $LiFePO_4$. The temperature was set at 150 °C for the experimental results described above. Figure 2g shows the cyclic voltammograms measured for $MgCo_2O_4$ in a triglyme electrolyte containing 0.5 M $Mg(TFSA)_2$ at RT and 100 °C [17]. The insertion of Mg^{2+} ions into the spinel $MgCo_2O_4$ was observed in each case (60 mAh g^{-1} at RT and 105 mAh g^{-1} at 100 °C) below 3.4 V versus Li^+/Li in RE. After the cathodic sweep to 1.5 V versus Li^+/Li in RE, the anodic current corresponding to the extraction of Mg^{2+} ions was markedly observed above 3.4 V versus Li^+/Li in RE.

For the comprehension of the cation insertion mechanism, structural analyses were conducted for $MgCo_2O_4$, Co_3O_4, $MgMn_2O_4$, and $MgFe_2O_4$. As shown in XRD profiles in Fig. 3a (left), spinel and rocksalt phases were recognized after the insertion of Mg^{2+} ions by the discharge of about 120 mAh g^{-1}. Further insertion of Mg^{2+} ions with about 210 mAh g^{-1} results in a rocksalt single phase. The fact that the spinel phase disappears even at such an incomplete discharge amount less than the theoretical value (260 mAh g^{-1}) indicates that the rocksalt phase includes a certain amount of cation vacancies, i.e., a solid-solution phase with off-stoichiometry. The structure completely reverts to the spinel structure by the charge after the discharge of 120 mAh g^{-1}. As shown in the corresponding XANES spectra around the Co K-edge in Fig. 3a (right), a part of Co(III) cations in the spinel phase is reduced to Co(II) after the discharge of about 120 mAh g^{-1} and oxidized to Co(III) after charge. The Mg insertion/extraction behavior into/from Co_3O_4 in Fig. 3b is very similar to that

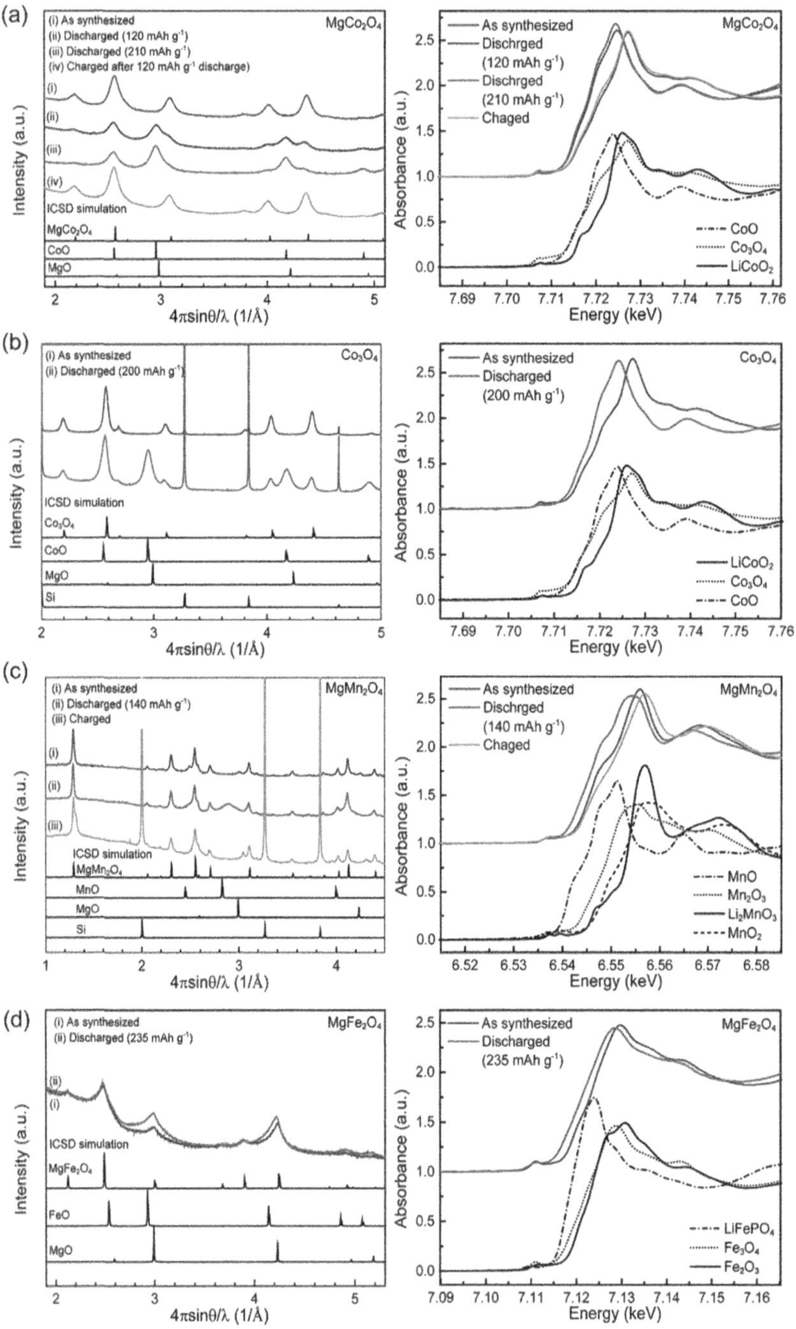

Fig. 3 XRD profiles (left) and XANES spectra (right) measured for **a** $MgCo_2O_4$, **b** Co_3O_4, **c** $MgMn_2O_4$, and **d** $MgFe_2O_4$

of $MgCo_2O_4$. In contrast, we observe both the insertion of Mg^{2+} ions into a pristine $MgMn_2O_4$ and the extraction of Mg^{2+} ions from the pristine $MgMn_2O_4$ as can be seen in Fig. 3c. In the XRD profile, broad peaks corresponding to the rocksalt phase are observed, whereas several new peaks are seen after the extraction. In the XANES spectra (Mn K-edge), the white line of as-synthesized $MgMn_2O_4$ shifts to a lower energy after discharge and shifts to a higher energy after charge. As seen in the XRD profile in Fig. 3d, the peak positions for $MgFe_2O_4$, FeO, and MgO are significantly close. After the insertion of Mg^{2+} ions, most of the peaks move to lower angles possibly because of FeO, and the white line (Fe K-edge) shifts to lower angles. The XANES profile for $MgFe_2O_4$ after the insertion of Mg^{2+} ions is similar to those for Fe_2O_3 and Fe_3O_4 but not to the Fe K-edge profile for the olivine $LiFePO_4$ structure.

In our recent research, we have successfully improved the cyclability by mixing Fe and Mn cations in $Mg(Mn_{1-x}Fe_x)_2O_4$ spinel oxides [18]. The substitution by Fe cations leads to a phase transition from tetragonal $MgMn_2O_4$ to cubic $Mg(Mn_{1-x}Fe_x)_2O_4$ ($x = 0.2, 0.4, 0.6, 0.8, 1$), and furthermore, decreases the catalytic activity for the oxidative decomposition of the electrolyte, which is both favorable for the improvement of the cyclability. As shown in Fig. 4, the anodic current due to the oxidative decomposition of the electrolyte was observed at potentials more positive than 2.9 V versus Mg^{2+}/Mg. The current decreases with increasing Fe content in $Mg(Mn_{1-x}Fe_x)_2O_4$. Thus, Fe cations are catalytically less active than Mn in the oxidative decomposition of the electrolyte. Using the low catalytic activity of the

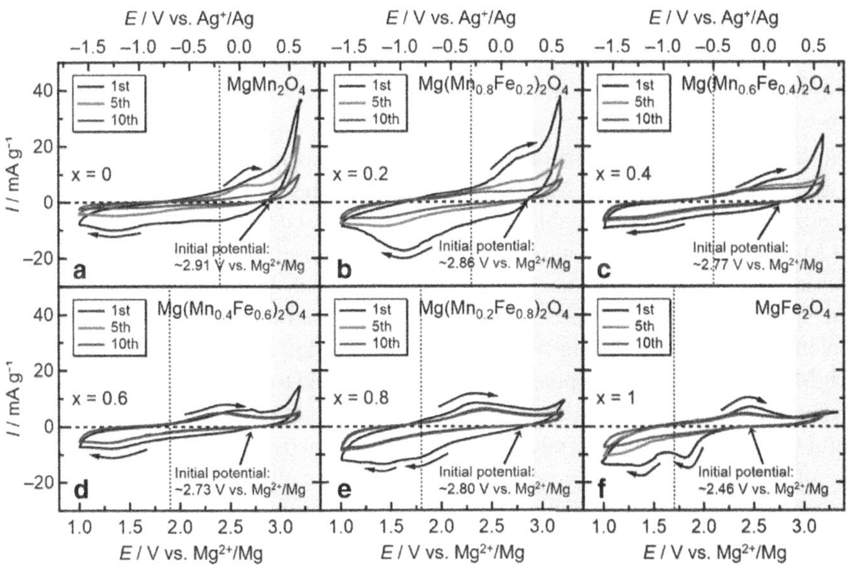

Fig. 4 Cyclic voltammograms of $Mg(Mn_{1-x}Fe_x)_2O_4$ synthesized at 500 °C for 24 h: **a** $x = 0$, **b** 0.2, **c** 0.4, **d** 0.6, **e** 0.8, **f** 1 in 0.3 M $[Mg(G4)][TFSA]_2/P_{13}TFSA$ measured at 100 °C. CE: Mg ribbon, RE: Ag wire in G3 containing 0.01 M $AgNO_3$ and 0.10 M $Mg(TFSA)_2$. Scan rate was 25 μV s^{-1}

Fe ions for the oxidative electrolyte decomposition, the cyclability of Mg rechargeable batteries could be significantly improved by suppressing the anodic electrolyte decomposition. Although suitable electrolytes for Mg rechargeable batteries with a sufficiently wide potential window remain unavailable, this approach is effective to improve battery performance independent of electrolyte stability.

By Rietveld analysis in our previous report with RIETAN-FP software [1, 19], it was found that Mg^{2+} ions are inserted into 16*c* sites in the spinel structure, and the cations originally located at the 8*a* sites are pushed out to the 16*c* sites, resulting in the formation of a rocksalt structure. This "insertion and push-out" process, is close to the Li-insertion mechanism in spinel oxide materials, which is expressed as follows [10, 11]:

$$\underset{\text{Spinel}}{MgCo_2O_4} + x\left(Mg^{2+} + 2e^-\right) \leftrightarrow \underset{\text{Spinel}}{(1-x)MgCo_2O_4} + \underset{\text{Rocksalt}}{xMg_2Co_2O_4}$$

Slight structural change or atomic rearrangement must be accompanied by the insertion. Therefore, the crystal lattice undergoes the spinel-to-rocksalt transition with the atomic-level two-phase equilibrium around the Mg-inserted 16*c* sites, leading to coherent structural change as illustrated in Fig. 1. This partial structural change would be favorable to maintain the lattice structure. In the subsequent chapters, various approaches are introduced to develop cathode materials based on this "insertion and push-out" mechanism.

4 Conclusions

We have clarified the insertion and extraction behaviors of Mg^{2+} ions for spinel oxides, such as $MgCo_2O_4$, $MgMn_2O_4$, $MgFe_2O_4$, $MgCr_2O_4$, and Co_3O_4 as cathode candidates for Mg rechargeable batteries. The proposed mechanism of the insertion of Mg^{2+} ions into spinel oxide lattices was termed "insertion and push-out" process. An atomic-level coherent phase transition proceeds with dual-phase reaction between the spinel and rocksalt phases. Mg^{2+} ions are inserted into spinel oxides at a high potential (about 2.9 V vs. Mg^{2+}/Mg for $MgCo_2O_4$). Mg^{2+} ions originally contained in $MgMn_2O_4$ and $MgCr_2O_4$ spinels can also be extracted to some extent at about 3.4 V versus Mg^{2+}/Mg. By changing the kind of transition cations, the redox potential and structure can be modified in addition to the catalytic activity for oxidative electrolyte decomposition. Specifically, Fe ions has a suppressive effect on oxidative electrolyte decomposition, leading to the improvement of cyclability.

The "insertion and push-out" mechanism proposed here provides a new strategy for designing future cathode active materials for Mg rechargeable batteries. Furthermore, some of the spinel oxides such as $MgCo_2O_4$ allows the insertion of both the Mg and Li ions, which enables us to design a new type of rechargeable battery, "rocking-chair type Mg-Li dual-salt battery" [16].

References

1. Okamoto, S., Ichitsubo, T., Kawaguchi, T., Kumagai, Y., Oba, F., Yagi, S., et al. (2015). *Advanced Science,* 1500072.
2. Aurbach, D., Lu, Z., Schechter, A., Gofer, Y., Gizbar, H., Turgeman, R., et al. (2000). *Nature, 407,* 724–727.
3. Cheng, Y., Parent, L. R., Shao, Y., Wang, C., Sprenkle, V. L., Li, G., & Liu, J. (2014). *Chemistry of Materials, 26,* 4904.
4. Novák, P., & Desilvestro, J. (1993). *Journal of the Electrochemical Society, 140,* 140.
5. Gregory, T. D., Hoffman, R. J., & Winterton, R. (1990). *Journal of the Electrochemical Society, 137,* 775.
6. Ichitsubo, T., Adachi, T., Yagi, S., & Doi, T. (2011). *Journal of Materials Chemistry, 21,* 11764.
7. Yagi, S., Fukuda, M., Ichitsubo, T., Nitta, K., Mizumaki, M., & Matsubara, E. (2015). *Journal of the Electrochemical Society, 162*(12), A2356–A2361.
8. Ichitsubo, T., Yagi, S., Nakamura, R., Ichikawa, Y., Okamoto, S., Sugimura, K., et al. (2014). *Journal of Materials Chemistry A, 2*(36), 14858–14866.
9. Wan, L. F., Perdue, B. R., Apblett, C. A., & Prendergast, D. (2015). *Chemistry of Materials, 27*(17), 5932–5940.
10. Thackeray, M. M., David, W. I. F., & Goodenough, J. B. (1982). *Materials Research Bulletin, 17,* 785.
11. Ohzuku, T., Ueda, A., & Yamamoto, N. (1995). *Journal of the Electrochemical Society, 142,* 1431.
12. Yagi, S., Morinaga, T., Togo, M., Tsuda, H., Shio, S., & Nakahira, A. (2016). *Materials Transactions, 57*(1), 42–45.
13. Hagiwara, R., Tamaki, K., Kubota, K., Goto, T., & Nohira, T. (2008). *Journal of Chemical and Engineering Data, 53,* 355.
14. Gao, B., Nohira, T., Hagiwara, R., & Wang, Z. (2014). Molten salts chemistry and technology. In M. Gaune-Escard & G. M. Haarberg (Eds.) (Chap. 5.4). Hoboken, NJ: Wiley.
15. Oishi, M., Ichitsubo, T., Okamoto, S., Toyoda, S., Matsubara, E., Nohira, T., & Hagiwara, R. (2014). *Journal of the Electrochemical Society, 161,* A943.
16. Ichitsubo, T., Okamoto, S., Kawaguchi, T., Kumagai, Y., Oba, F., Yagi, S., et al. (2015). *Journal of Materials Chemistry A, 3,* 10188.
17. Fukutsuka, T., Asaka, K., Inoo, A., Yasui, R., Miyazaki, K., Abe, T., et al. (2014). *Chemistry Letters, 43,* 1788.
18. Han, J., Yagi, S., & Ichitsubo, T. (2019). *Journal of Power Sources, 435,* 226822.
19. Izumi, F., & Momma, K. (2007). *Solid State Phenomena, 130,* 15.

New Cathode Materials with Spinel and Layered Structures

Yasushi Idemoto, Naoya Ishida, and Naoto Kitamura

Abstract From previous researches on a cathode material of the Mg rechargeable battery, spinel-type oxides can be regarded as promising candidates for the cathode material. $MgCo_2O_4$ with a disordered spinel structure has drawn much attention as a typical cathode recently, but the cathode property is not sufficiently high for practical use. To overcome this problem, we have tried to improve its cathode property by two kinds of approaches. One is a partial substitution of another transition metal for Co in $MgCo_2O_4$, and a part of results on Mn-substituted materials, $MgCo_{2-x}Mn_xO_4$, is described in Sect. 1. In addition, we have focused on vanadium-based materials and recently succeeded in finding a new family of the cathode material, i.e., $Mg_{1+x}V_{2-x}O_4$ with a spinel structure. As an example of the vanadium-based materials, a cathode property and a crystal structure of $Mg_{1.5}V_{1.5}O_4$ are introduced briefly in Sect. 2. To gain a deeper understanding of the atomic structures of the spinel-type materials, we have performed neutron and synchrotron X-ray total scattering measurements. In Sect. 3, we present an analytical result of an atomic structure of $MgCo_2O_4$ nanoparticle. A layered material can be considered to be promising as a cathode material of the Mg rechargeable battery, as in the case of a lithium ion battery. Recently, we have applied delithiated layered material, i.e., $Li_{0.13}Mn_{0.54}Ni_{0.13}Co_{0.13}O_{2-\delta}$ to the Mg rechargeable battery and then demonstrated that the samples showed 273 mA h g^{-1} at a first discharge process. The result of cathode property is described in detail in Sect. 4.

Keywords Cathode material · Spinel structure · Layered structure · Electronic structure · Local structure

Y. Idemoto (✉) · N. Ishida · N. Kitamura
Tokyo University of Science, Noda, Chiba, Japan
e-mail: idemoto@rs.tus.ac.jp

K. Kanamura (ed.), *Next Generation Batteries*,
https://doi.org/10.1007/978-981-33-6668-8_43

1 Cathode Property and Crystal Structure of $MgCo_{2-x}Mn_xO_4$

Spinel-type $MgCo_{2-x}Mn_xO_4$ (x = 0.1, 0.2, 0.4) was synthesized as a new cathode material for use in magnesium secondary batteries [1], using an inverse co-precipitation method. The obtained product was determined to have a spinel structure (space group $Fd\overline{3}m$) based on powder X-ray diffraction data. A Rietveld analysis of synchrotron X-ray diffraction data showed that Mg, Co, and Mn in this material were distributed in a disordered manner, meaning that cation mixing had occurred. Synchrotron X-ray diffraction and neutron diffraction showed that the presence of Co atom in the Mg diffusion pathways in the $MgCo_{2-x}Mn_xO_4$ (x = 0.1 to 0.4) samples was reduced compared to that in $MgCo_2O_4$. This occurred because the proportion of Mg atoms in these same pathways was increased as a result of the substitution of Mn atoms. The results of a MEM (maximum entropy method) analysis (Fig. 1) demonstrated that the covalent nature of the M(8*a*)-O bond decreased with increasing Mn substitution, such that a greater number of Mg atoms would be expected to be eliminated during the charging process.

Charge-discharge testing using $MgCo_{2-x}Mn_xO_4$/AZ31 cells with Ag reference electrodes demonstrated that the cycling performance of x = 0.4 sample was improved after the second cycle during charge and discharge tests. The valence state of the Co atoms in these materials became lower with increasing substitution of tetravalent Mn atoms based on XANES and XAFS data. The improved cycling performance of this material is ascribed to the replacement of a portion of the Co atoms with Mn. The stability of the crystal structure was investigated based on first-principles calculations and the results showed that a model in which Mn occupied only the 16*d* sites was the most stable. The ordered Mg/Co/Mn structure of this material would be expected to facilitate the diffusion of Mg^{2+} ions throughout the cathode material in a magnesium secondary battery.

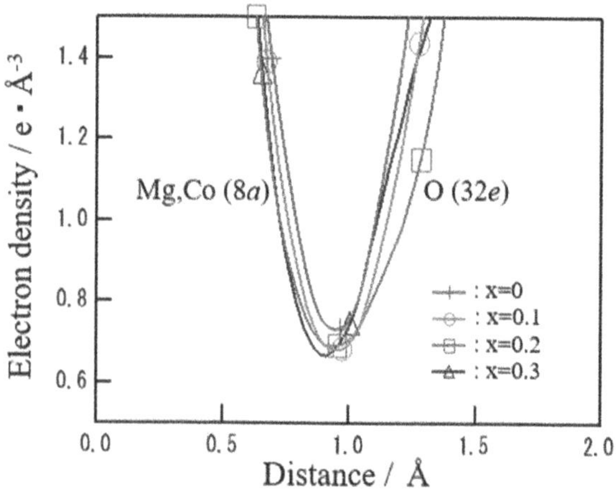

Fig. 1 Line profiles of electron densities between 8*a* and 32*e*, in $MgCo_{2-x}Mn_xO_4$ (x = 0, 0.1, 0.2, 0.4)

2 Cathode Property and Crystal Structure of $Mg_{1.5}V_{1.5}O_4$

Ni substitution was performed into a spinel-type compound containing V, and characterization and evaluation of the battery characteristics were performed [2]. $Mg(Mg_{1.5}V_{1.4}Ni_{0.1})O_4$ was synthesized by a solid-phase method under high vacuum. The product was assigned to a spinel structure with space group $Fd\bar{3}m$ based on the powder X-ray diffraction spectrum. Synthesized materials of uniform composition were confirmed from elemental mapping by STEM-EDS.

Rietveld analysis on synchrotron X-ray diffraction data suggested that MgO, which is a starting material, remains in the samples. The proportion is around 25% in $Mg(Mg_{0.5}V_{1.5})O_4$, and the proportion of MgO increased with the amount of Ni substitution. In terms of battery characteristics, charging and discharging was able to be performed repeatedly on all of the samples, and it was confirmed that reversible charging and discharging could be performed on Ni substituted samples.

The electronic structure was analyzed by MEM using the result of Rietveld analysis, which showed that Mg could be easily inserted into the $Mg(Mg_{1.5}V_{1.4}Ni_{0.1})O_4$ and that the host structure was stable due to the high covalency of M(16*d*)-O. MO_6 octahedral strain was alleviated by increasing the amount of substitution of Ni. Furthermore, in the $Mg(Mg_{0.5}V_{1.4}Ni_{0.1})O_4$ sample, Mg desorbs easily from the 8a site, which may stabilize the structure of the 16d-32e site, and this sample exhibited the best cycle characteristics and 19 cycles of charging and discharging were performed (Fig. 2). The Coulomb efficiency was roughly 1, and the capacity retention rate increased with the number of cycles. At this time, the discharge capacity of the 19th cycle was 66 mAh/g, and the energy density was 142 Wh/kg. Since the XANES spectra of the powder samples and electrodes after discharge showed that redox of

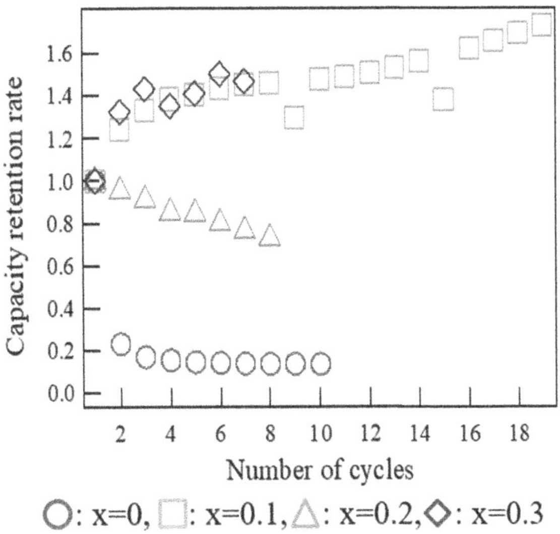

Fig. 2 Cycle efficiency, capacity retention ratio of $Mg(Mg_{0.5}V_{1.5-x}Ni_x)O_4$, ○: x = 0, □: x = 0.1, △: x = 0.2, ◇: x = 0.3

the V occurred with charging, this showed that oxidation of V contributed to desorption of Mg in the first charging. The above indicates that the $Mg(Mg_{0.5}V_{1.4}Ni_{0.1})O_4$ fabricated in this work is a candidate for a magnesium rechargeable battery cathode material that has good battery characteristics.

3 Local Structure of $MgCo_2O_4$ Nanoparticle

As mentioned above, $MgCo_2O_4$ with a disordered spinel structure, where Mg occupy not only a tetrahedral site (8*a* site), but also an octahedral site (16*d* site), has been applied for the Mg rechargeable battery. Since the discharge and charge process accompany Mg^{2+} insertion and deinsertion, respectively, an atomic structure of the nanoparticle must play an important role in the electrochemical behavior. Generally, information on an atomic structure of a crystal is obtained by a structure analysis using Bragg peaks in a diffraction profile. In the case of a nanoparticle, however, a detailed atomic structure cannot be revealed by this strategy, due to broad Bragg peaks. Therefore, we have performed neutron and synchrotron X-ray total scattering measurements (diffraction measurements with a wide range of a scattering vector, Q) for $MgCo_2O_4$-based nanoparticles [3]. By normalizing the measured data, Faber-Ziman structure factors, $S(Q)$, were obtained, and then the structure factors were analyzed on the basis of the Monte Carlo method, which is the so-called reverse Monte Carlo (RMC) modeling [4, 5]. In the modeling, a distribution of Mg and Co, which occupy the same crystallographic sites in the spinel structure, was determined by swapping these atoms in a simulation box, and the atomic positions were optimized by moving the atoms randomly.

Figure 3 shows the modeling result, in which $S_{box}(Q)$ means structure factors convolved by taking a box size (1512 atoms) into account, and Fig. 3c depicts an atomic configuration snapshot obtained by the RMC modeling. As can be seen in Fig. 3, all the experimental data could be reproduced by using the obtained atomic configuration snapshot. To extract structural features from the modeling result, a condensed view was constructed from the snapshot (Fig. 4). By comparing the condensed view with the average atomic position refined by the Rietveld analysis, it was indicated that cations at tetrahedral sites (8*a* site) tend to spread toward vacant octahedral sites (16*c* site) in the spinel structure. Since cations at the 8*a* site are pushed out to the 16*c* site at a discharge process as mentioned above, such cation distribution around the tetrahedral sites has a positive influence on the cathode property of $MgCo_2O_4$ nanoparticle.

To gain quantitative information on the atomic configuration, we calculated octahedral distortions (bond-angle variances, σ^2) of MgO_6 and CoO_6 in the spinel structure from the snapshot. Although these values cannot be distinguished by the Rietveld analysis, the RMC modeling using total scattering data demonstrated that the distortion of MgO_6 ($\sigma^2 = 157.7$ deg^2) was larger than that of CoO_6 ($\sigma^2 = 115.2$ deg^2). Since the larger distortion made the structure unstable, a positional exchange (a cation mixing) of Mg and Co should be suppressed for stable Mg insertion/deinsertion at

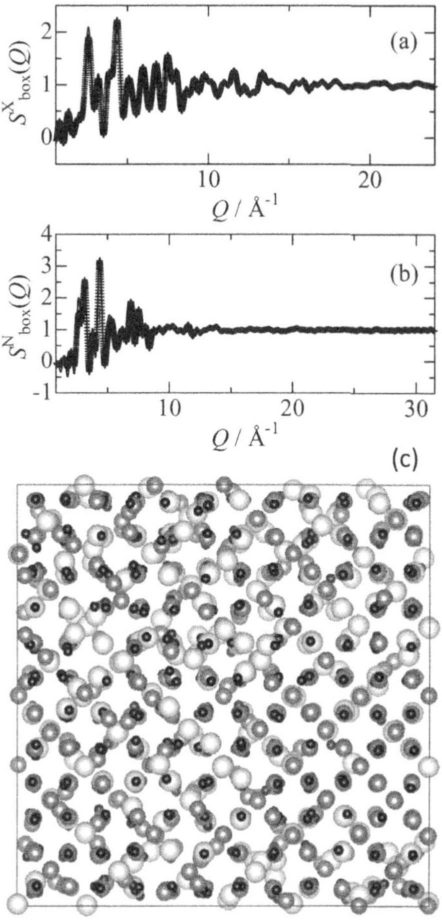

Fig. 3 **a** X-ray, and **b** neutron structure factors of $MgCo_2O_4$ nanoparticle. Plus marks and solid lines represent the experimental data and RMC model, respectively. **c** Atomic configuration snapshot of $MgCo_2O_4$. Light and dark gray spheres represent Mg and Co, respectively. Bragg spheres represent O Reproduced from Ref. [3] with permission from the Royal Society of Chemistry

least in the case of $MgCo_2O_4$-based nanoparticles. Indeed, Mn-substituted samples with a lower cation mixing amount exhibited better cathode property [1], as described in Sect. 1. We also evaluated volumes around the vacant site (16*c* site) in addition to MgO_6 and CoO_6 volumes, and then found that volumes of CoO_6 were smaller than those around the vacant sites. This result implies that the diffusion of Co from the tetrahedral site to the vacant site during a discharging process is quite difficult due to a large volume mismatch in comparison with Mg. From this viewpoint, the cation mixing should be suppressed to achieve better electrochemical property at least in the case of $MgCo_2O_4$-based nanoparticles.

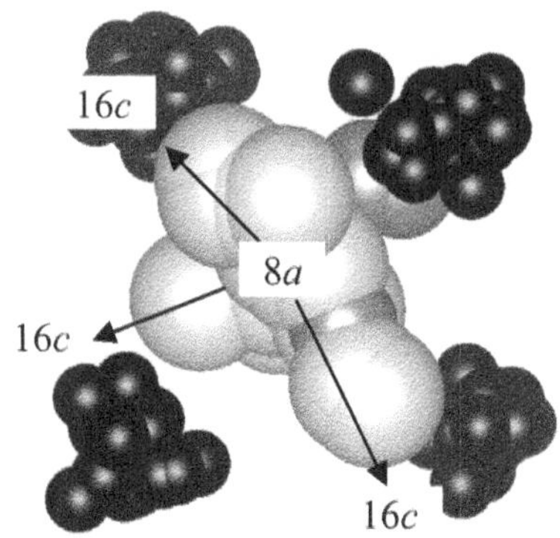

Fig. 4 Atom distribution around a tetrahedral site (8*a* site) in the spinel structure. Light and dark gray spheres represent Mg and Co, respectively. Bragg spheres represent O

4 Cathode Property and Crystal Structure of $Li_{0.13}Mn_{0.54}Ni_{0.13}Co_{0.13}O_{2-\delta}$

It was reported that LIB cathode material Li_2FeSiO_4, post its electrochemical delithiation, could be discharged versus Mg metal anode to synthesize $MgFeSiO_4$ delivering over 300 mAh g^{-1} for Mg insertion reaction [6]. Thus, by releasing Li from the positive electrode materials for the LIBs, there is a possibility that the delithiated materials can be used as a cathode active material for rechargeable Mg batteries [6, 7]. Therefore, we also focused on lithium rich cathode material $Li_{1.2}Mn_{0.54}Ni_{0.13}Co_{0.13}O_2$, which is known to deliver high discharge capacity exceeding 250 mAh g^{-1} in the Li battery configuration. We have reported that crystal and electronic structures and the thermodynamic stabilities were studied for the chemically or electrochemically delithiated $Li_{1.2-x}Mn_{0.54}Ni_{0.13}Co_{0.13}O_{2-\delta}$ [8]. Although there is no knowledge about the insertion and desorption of Mg for the delithiated one, the high capacity such as shown in the LIB is expected for rechargeable Mg battery. The chemical oxidation was performed to obtain the $Li_{1.2-x}Mn_{0.54}Ni_{0.13}Co_{0.13}O_{2-\delta}$ in a non-aqueous environment. The NO_2-BF_4 was added as an oxidizing agent so that the concentration was prepared to 0.2 M in acetonitrile and the reaction time was 24 h. After the NO_2BF_4 was dissolved, the $Li_{1.2}Mn_{0.54}Ni_{0.13}Co_{0.13}O_2$ of 0.5 g was added and the mixture was stirred for 24 h at room temperature in an argon filled glove box. The products were filtered and washed with acetonitrile to remove the remaining salt before they were dried in a vacuum atmosphere. The objectives were to evaluate the electrode property of delithiated $Li_{1.2-x}Mn_{0.54}Ni_{0.13}Co_{0.13}O_{2-\delta}$ in the Mg battery configuration, to analyze the crystal and electronic structure changes before and after Mg insertion, and to reveal the charge/discharge mechanism from the viewpoint of the valence change of transition metals [9].

The metal compositions for pristine $Li_{1.2}Mn_{0.54}Ni_{0.13}Co_{0.13}O_2$ and delithiated $Li_{1.2-x}Mn_{0.54}Ni_{0.13}Co_{0.13}O_{2-\delta}$ determined by ICP-AES were

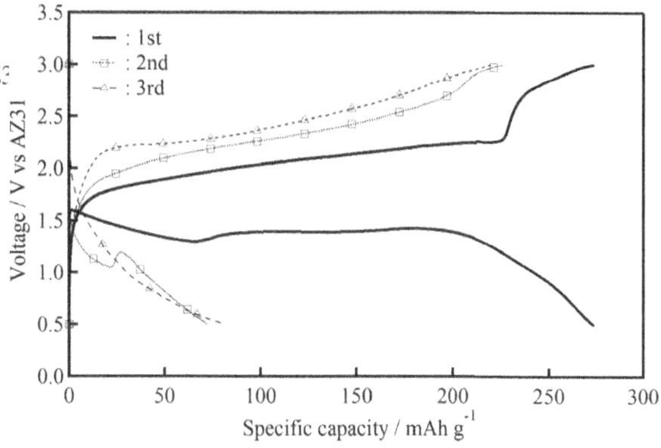

Fig. 5 Discharge and charge curves of $Li_{0.13}Mn_{0.54}Ni_{0.13}Co_{0.13}O_{2-\delta}$/AZ31 cell using 0.5 M $Mg(TFSI)_2/CH_3CN$ measured at 60 °C

$Li_{1.199(6)}Mn_{0.531(4)}Ni_{0.137(1)}Co_{0.132(1)}O_2$ and $Li_{0.125(7)}Mn_{0.532(5)}Ni_{0.137(1)}Co_{0.131(1)}O_{2-\delta}$, respectively. The metal composition of the pristine was calculated based on the total two metal compositions, whereas the lithium composition of chemically delithiated one was calculated based on the same transition metal compositions as pristine.

Figure 5 shows the discharge and charge curves of $Li_{0.13}Mn_{0.54}Ni_{0.13}Co_{0.13}O_{2-\delta}$/AZ31 cell using 0.5 M $Mg(TFSA)_2/CH_3CN$ measured at 60 °C. The 1.3 V plateau versus Mg alloy at first discharge and high capacity of 273 mAh g^{-1} were exhibited. The intercalated Mg was estimated to be about 0.5 atoms per formula unit which corresponded to +1.0 charge. However, the discharge capacities after the second cycle significantly decreased because of the overpotential and narrow potential window of the electrolyte. It was inferred that the higher charge capacities than discharge one after the second cycle were due to the side reaction in the electrolyte.

The synchrotron XRDs were performed for pristine, chemically delithiated, and first discharged materials. The pristine was analyzed by Rietveld method using *C2/m* space group, whereas the delithiated $Li_{0.13}Mn_{0.54}Ni_{0.13}Co_{0.13}O_{2-\delta}$ and the $Mg_{0.5}Li_{0.13}Mn_{0.54}Ni_{0.13}Co_{0.13}O_{2-\delta}$ electrode after first discharge were used with *C2/m* and *R*-3*m* space groups. The fitted patterns and refined crystal structure parameters were analyzed within the R_{wp} of 6.43–7.54%. The volumetric expansion from 0.09853(2) nm^3 to 0.10127(5) nm^3 after first discharge was attributed to the increase of the ionic radius derived from the reduction of transition metals accompanied by Mg insertion. Based on the refined structural parameters, the Mg was inserted more in the *R*-3 *m* structure after the first discharge whereas the inserted Mg in the *C2/m* structure was found to be occupied at the transition metal layers of 4 *g* and 2*b* sites. It is considered that the insertion of Mg into the transition metal layer is a major cause of capacity deterioration.

XAFS measurements were carried out to investigate the valence change between before discharge, i.e., the chemically delithiated material, and after discharge, i.e., Mg inserted $Mg_{0.5}Li_{0.13}Mn_{0.54}Ni_{0.13}Co_{0.13}O_{2-\delta}$. The XANES spectra of Mn, Co, and Ni are shown in Fig. 6. The standard Mn-, Ni- Co-oxides were used for the rough estimation of valence state in transition metals. Comparing the sample with chemically

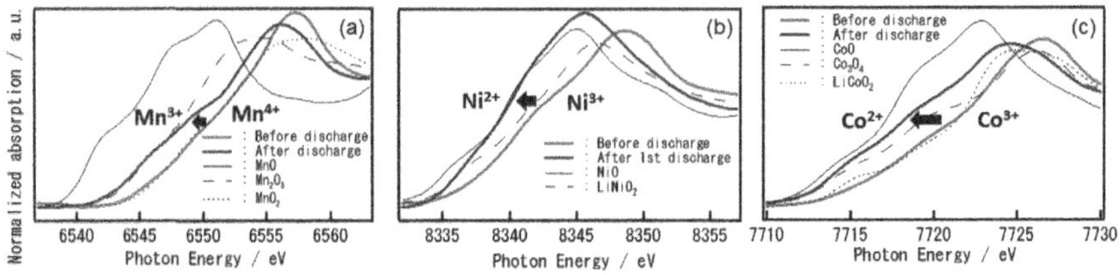

Fig. 6 XANES spectra at **a** Mn *K*-edge, **b** Ni *K*-edge and **c** Co *K*-edge before discharge, the $Li_{0.13}Mn_{0.54}Ni_{0.13}Co_{0.13}O_{2-\delta}$, and after discharge, the Mg_y $Li_{0.13}Mn_{0.54}Ni_{0.13}Co_{0.13}O_{2-\delta}$

delithiated one and the electrode after initial discharge, the entire spectrum shifted to lower energy side at all edges, Mn is from tetravalent to trivalent, Co is from trivalent to nearly divalent, Ni revealed that the valence was lowered to divalent. These results were in agreement with the reduction of transition metals accompanied by 0.5 pfu Mg insertion. It was established that the Mg was electrochemically inserted as shown in the decreasing valence of the transition metals with initial discharge and the expanded volume derived from the increase of ionic radius of reduced Mn, Ni, and Co. Therefore, further study for the delithiated layered materials will promote the development of high capacity and rechargeable cathode materials in Mg battery.

References

1. Idemoto, Y., Mizutani, Y., Ishibashi, C., Ishida, N., & Kitamura, N. (2019). *Electrochemistry, 87,* 220–228.
2. Idemoto, Y., Kawakami, N., Ishida, N., & Kitamura, N. (2019). *Electrochemistry, 87,* 281–288.
3. Kitamura, N., Tanabe, Y., Ishida, N., & Idemoto, Y. (2019). *Chemical Communications, 55,* 2517–2520.
4. McGreevy, R. L., & Pusztai, L. (1988). *Molecular Simulation, 1,* 359–367.
5. Tucker, M. G., Keen, D. A., Dove, M. T., Goodwin, A. L., & Hui, Q. (2007). *Journal of Physics: Condensed Matter, 19,* 335218.
6. Orikasa, Y., Masese, T., Koyama, Y., Mori, T., Hattori, M., Yamamoto, K., et al. (2014). *Sciences Report, 4,* 5622.
7. Huang, Z.-D., Masese, T., Orikasa, Y., Morib, T., & Yamamoto, K. (2015). *RSC Advances, 5,* 8598–8603.
8. Ishida, N., Tamura, N., Kitamura, N., & Idemoto, Y. (2016). *Journal of Power Sources, 319,* 255–261.
9. Ishida, N., Nishigami, R., Kitamura, N., & Idemoto, Y. (2017). *Chemistry Letters, 46,* 1508–1511.

A Facile Wet-Process for Preparing Mg–Mn Spinel Nanoparticles as Cathodes for Rechargeable Mg-Ion Batteries

Hiroaki Kobayashi

Abstract Reducing the particle size of cathode materials is effective to improve the rate capability of Mg-ion batteries. In this chapter, we report a facile wet-process of an alcohol reduction process for nanoparticle preparation. Ultrasmall cubic Mg–Mn spinel oxide nanoparticles with a particle size of approximately 5 nm were successfully synthesized in an ethanol solution within 30 min at room temperature. Though the particles were aggregated to form large secondary particles, the aggregation could be suppressed by covering the particles with graphene. The composite exhibited a specific capacity of 230 mAh g^{-1}, and well cycled more than 100 times without any large capacity loss even at a moderated current density.

Keywords Alcohol reduction process · Nanoparticle · Mg–Mn spinel · Cathode · Mg-ion battery

Spinel oxides have a high redox potential and a relatively low diffusivity among various cathode materials for Mg-ion batteries [1]. Especially, $MgMn_2O_4$ spinels can proceed both Mn^{3+}/Mn^{4+} redox and Mn^{3+}/Mn^{2+} redox to exhibit a high theoretical capacity of 540 mAh g^{-1} in the common electrochemical window of electrolytes (<3.5 V versus Mg/Mg^{2+}) [2]. Recently, $MgMn_2O_4$ with particle sizes of 10–120 nm were synthesized by sol-gel method, and their cathode performances were investigated (Table 1) [3–9]. Some reports exhibited high specific capacities, but so far, there are no reports where high capacity could be obtained at high rates with an anhydrous electrolyte.

Downsizing the particle size is one of the most effective ways to improve the specific capacity and the rate capability for suppressing the slow Mg^{2+} diffusion in solids [10]. Therefore, alternative synthetic methods that enable nanoparticles below 10 nm are required for achieving a breakthrough in such cathodes. Here we introduce a facile wet-process of alcohol reduction process for synthesizing ultrasmall Mg–Mn spinel oxides and their cathode performances in Mg-ion battery.

H. Kobayashi (✉)
Institute of Multidisciplinary Research for Advanced Materials, Tohoku University, 2-1-1 Katahira, Aoba-ku, Sendai 980-8577, Japan
e-mail: h.kobayashi@tohoku.ac.jp

K. Kanamura (ed.), *Next Generation Batteries*,
https://doi.org/10.1007/978-981-33-6668-8_44

Table 1 Comparison of the cathode performances for $MgMn_2O_4$ spinels prepared by sol-gel method

Calcination condition	Particle size/nm	Electrolyte	Discharge capacity/mAh g^{-1}	Current density/mA g^{-1}	References
400 °C, 10 h	17	aq. 3 M $Mg(NO_3)_2$	150	60 μA	[3]
400 °C, 10 h	17	0.5 M $Mg(ClO_4)_2$/EC-DEC	120	10 μA	[3]
400 °C, 10 h	11	3 M H_2O + 0.5 M $Mg(TFSA)_2$ + 0.5 M DPGDME/AN	220 186	27 135	[4]
400 °C, 10 h	11	0.5 M $Mg(TFSA)_2$ + 0.5 M DPGDME/AN	220 80	27 135	[4]
550 °C, 10 h	31	3 M H_2O + 0.5 M $Mg(TFSA)_2$ + 0.5 M DPGDME/AN	70 50	27 135	[4]
550 °C, 10 h	31	0.5 M $Mg(TFSA)_2$ + 0.5 M DPGDME/AN	70 50	27 135	[4]
650 °C, 12 h	40	0.5 M $Mg(ClO_4)_2$/AN	100	4	[5]
550 °C, 2 h	10	aq. 1 M $MgSO_4$	262	100	[6]
550 °C, 10 h	30	aq. 0.5 M $MgCl_2$	212	50	[7]
800 °C, 5 h	Not indicated	aq. 1 M $Mg(NO_3)_2$	263	30	[8]
700 °C, 24 h	120	$Mg(TFSA)_2$/EC-PC	225	1	[9]

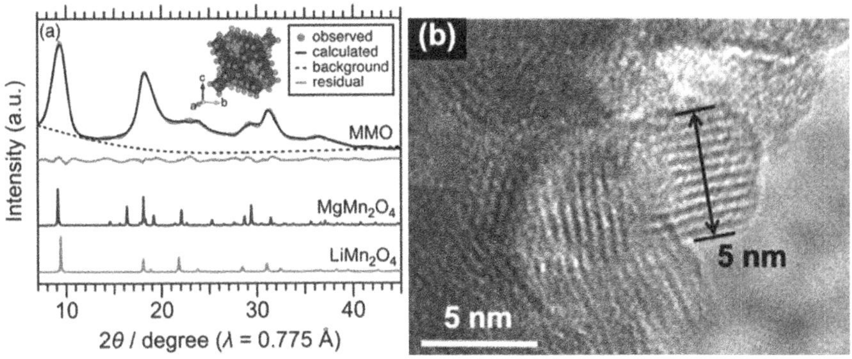

Fig. 1 **a** Synchrotron XRD patterns, **b** a TEM image Reproduced from Ref. [14] by permission of The Royal Society of Chemistry

Alcohol reduction process is commonly applied for synthesizing metal nanoparticles from metal salt precursors, and this process is able to apply to oxide nanoparticles preparations from high-valent metal ions. Recently our group applied it for various Mn oxides and showed high catalytic activity for aerobic oxidation reactions [11–13]. In the same way, Mg–Mn spinel oxide (MMO) was synthesized by a reduction of a precursor $n\text{-}Bu_4NMnO_4$ in an anhydrous $MgCl_2$ solution in ethanol within 30 min at room temperature [14]. According to XRD patterns and a TEM image as shown in Fig. 1a, b, the obtained MMO particles have a very small primary particle size of approximately 5 nm which is much smaller than that of previously reported $MgMn_2O_4$ nanoparticles (>10 nm) synthesized by sol-gel method [3]. These ultra-small particles are likely to have been obtained using this wet-process by suppressing the dissolution-recrystallization at ambient temperature and in an anhydrous solvent, while the sol-gel method contains a calcination step at which crystal growth should proceed.

Although the XRD pattern of MMO is likely attributable to the spinel structure, it was similar to that of cubic $LiMn_2O_4$ spinel rather than that of tetragonal $MgMn_2O_4$ spinel. Mn *K*-edge XANES spectra (Fig. 2a) and the pseudo-radial distribution functions (*p*-RDF) obtained by a Fourier transform (FT) of k^2-weighted EXAFS spectra (Fig. 2b) also exhibit similar behaviors, suggesting MMO has a cubic structure due to the suppression of Jahn–Teller distortion around MnO_6 octahedra. This suppression derives from the nonstoichiometric formula of MMO. ICP elemental analyses reveal that Mg/Mn (mol/mol) in MMO was 0.40, indicating that reduction from Mn^{7+} to Mn^{3+} does not proceed completely and nonstoichiometric cubic spinel $Mg_{0.80}Mn_2O_4$ is obtained via alcohol reduction process. The XRD pattern in Fig. 1a was successfully fitted to $Mg_{0.81(3)}Mn_2O_4$ with the *Fd*–3*m* space group, though the error of refined parameters should become large with such broad XRD patterns. The estimated structure is closely similar to the meta-stable cubic-$MgMn_2O_4$ phase or $Mg_yMn_2O_4$ solid solution phase between λ-MnO_2 and $MgMn_2O_4$ [15, 16]. Since the reduction reaction of MnO^{4-} ion proceeds rapidly to form nanoparticles without heating-up, once

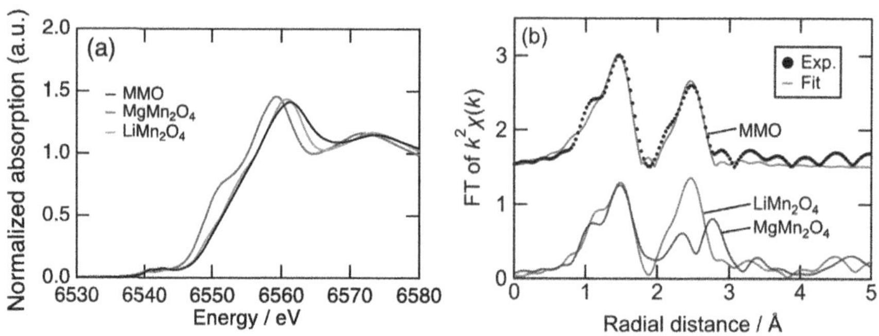

Fig. 2 **a** Mn *K*-edge XANES spectra and **b** pseudo-radial distribution functions (*p*-RDF) obtained from the Fourier transform (FT) of k^2-weighted Mn *K*-edge EXAFS spectrum, where the *k*-range of the FT was 3 ~ 14 $Å^{-1}$ with a Hanning window of 1 $Å^{-1}$ Reproduced from Ref. [14] by permission of The Royal Society of Chemistry

a meta-stable phase is formed, the phase transition to a more stable phase hardly proceeds. The cubic phase is obtained probably due to the rapid nucleation of MMO particles at room temperature.

Figure 3a shows voltage curves of MMO electrodes with a 0.5 M $Mg(ClO_4)_2/CH_3CN$ electrolyte at a current density of 10 mA g^{-1} at 25 °C using a three-electrode cell with an activated carbon counter electrode and an Ag/Ag^+ reference electrode (+2.6 V versus Mg/Mg^{2+}). No plateau was observed during discharge/charge and the discharge capacity was 60 mAh g^{-1}, which was much smaller than that of an ideal one electron reaction per Mn in MMO (280 mAh g^{-1}). According to a SEM image of MMO particles, MMO particles are strongly aggregated to form submicron secondary particles. These large aggregates form in the drying process and inhibit both electron conduction and Mg^{2+} ion diffusion, causing a low utilization ratio of active materials. To suppress the aggregation, graphene was used as an aggregation inhibitor and an electron conducting additive. The composite of MMO with graphene (MMO-G) was easily prepared by adding graphene to the reaction solution. The MMO-G electrode exhibited gentle slopes and a reversible capacity of more than 200 mAh g^{-1}. Since the contribution of the electric double-layer capacitor (EDLC) of the graphene is negligible, the enhancement of the specific capacity derives from the increase in the amount of the redox-active MMO particles. The obtained discharge capacity of 230 mAh g^{-1} corresponds to 0.83 electrons transferred per Mn, indicating that the average valence state of Mn changed from +3.20 to +2.37. Figure 3b shows Mn *K*-edge XANES spectra of the MMO-G electrodes. The Mn *K*-edge was shifted to lower energies during discharge and almost recovered after charge, indicating the reversible redox reaction of Mn. The Mn *K*-edge position at the full discharge was approximately in the middle between those of MnO and Mn_2O_3, supporting the valence state of Mn between +3 and +2.

Even at a moderate current density of 190 mA g^{-1} at 25 °C, the MMO-G electrode exhibited a reversible discharge/charge with little change in the voltage curves

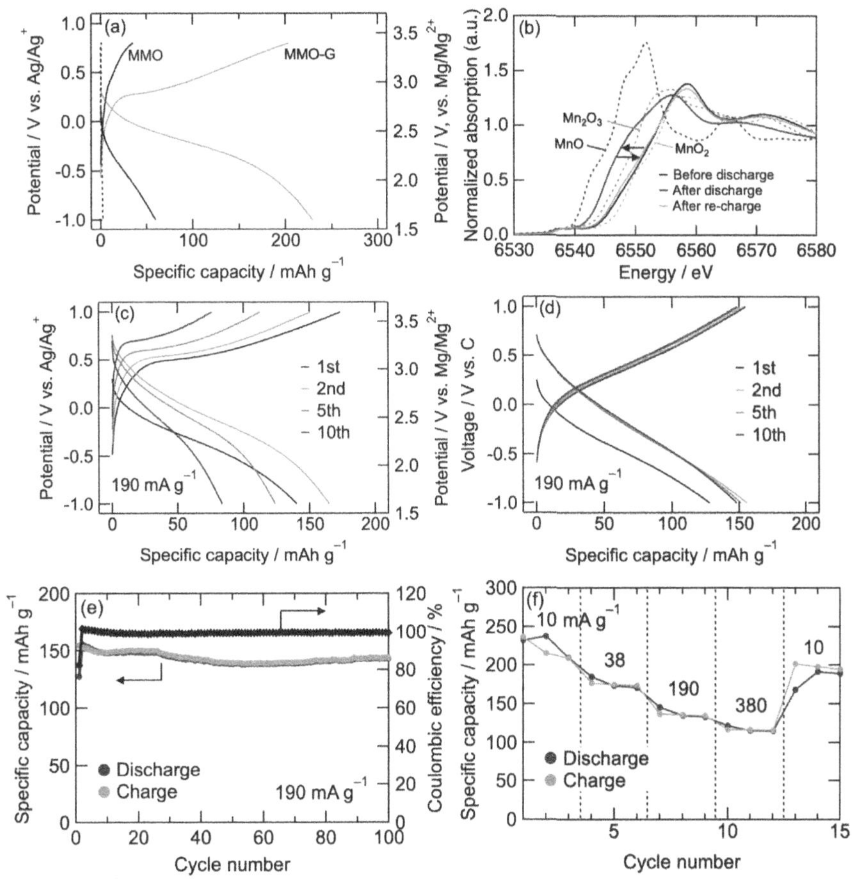

Fig. 3 **a** Voltage curves of MMO, MMO-G, and graphene (dotted lines). **b** Mn *K*-edge XANES spectra of MMO-G cathodes. **c** Voltage curves of MMO-G cathode at 190 mA g^{-1} using a 3-electrode cell or **d** a coin-type cell. **e** Cyclability and **f** rate capability tests of MMO-G cathode using a coin-type cell Reproduced from Ref. [14] by permission of The Royal Society of Chemistry

(Fig. 3c). However, the reversible capacity gradually decreased with the number of cycles, probably due to the dissolution of Mn to the electrolyte or the exfoliation of the electrode from the current collector during discharge/charge. To reduce the amount of electrolyte and add a confining pressure on the electrode for suppressing the dissolution and exfoliation, a cycle test with a 2032 coin-type cell was investigated. Though the voltage curves with the coin cell slightly reflected the EDLC behavior of the activated carbon counter electrode (0.0 V versus Ag/Ag^{+}), the cell was well cycled more than 100 times without any large capacity loss (Fig. 3d, e). Furthermore, as shown in Fig. 3f, the cell exhibited a relatively high rate capability even at 380 mA g^{-1} (corresponding to a 1.4 C-rate) compared with $MgMn_2O_4$ prepared by the sol-gel method (Table 1), indicating that the Mg^{2+} intercalation/deintercalation proceeded rapidly due to small diffusion paths in the ultrasmall MMO particles covered by graphene prepared by the alcohol reduction method. Both downsizing the primary particles of active materials and suppressing the aggregation of particles are important for further development of cathodes in Mg-ion batteries.

References

1. Mao, M., Gao, T., Hou, S., & Wang, C. (2018). *Chemical Society Reviews, 47,* 8804–8841.
2. Okamoto, S., Ichitsubo, T., Kawaguchi, T., Kumagai, Y., Oba, F., Yagi, S., et al. (2015). *Advanced Science, 2,* 1500072.
3. Cabello, M., Alcántara, R., Nacimiento, F., Ortiz, G., Lavela, P., & Tirado, J. L. (2015). *CrystEngComm, 17,* 8728–8735.
4. Yin, J., Brady, A. B., Takeuchi, E. S., Marschilok, A. C., & Takeuchi, K. J. (2017). *Chemical Communications, 53,* 3665–3668.
5. Truong, Q. D., Devaraju, M. K., Tran, P. D., Gambe, Y., Nayuki, K., Sasaki, Y., et al. (2017). *Chemistry of Materials, 29,* 6245–6251.
6. Tao, S., Huang, W., Liu, Y., Chen, S., Qian, B., & Song, L. (2018). *Journal of Materials Chemistry A, 6,* 8210–8214.
7. Liu, G., Chi, Q., Zhang, Y., Chen, Q., Zhang, C., Zhu, K., et al. (2018). *Chemical Communications, 54,* 9474–9477.
8. Banu, A., Sakunthala, A., Thamilselvan, M., Kumar, P. S., Suresh, K., & Ashwini, S. (2019). *Ceramics International, 45,* 13072–13085.
9. Zainol, N. H., Hambali, D., Osman, Z., Kamarulzaman, N., & Rusdi, R. (2019). *Ionics, 25,* 133–139.
10. Ling, C., Zhang, R., & Mizuno, F. (2016). *ACS Applied Materials & Interfaces, 8,* 4508–4515.
11. Miyamoto, Y., Kuroda, Y., Uematsu, T., Oshikawa, H., Shibata, N., Ikuhara, Y., et al. (2015). *Scientific Reports, 5,* 15011.
12. Miyamoto, Y., Kuroda, Y., Uematsu, T., Oshikawa, H., Shibata, N., Ikuhara, Y., et al. (2016). *ChemNanoMat, 2,* 297–306.
13. Nakai, S., Uematsu, T., Ogasawara, Y., Suzuki, K., Yamaguchi, K., & Mizuno, N. (2018). *ChemCatChem, 10,* 1096–1106.
14. Kobayashi, H., Yamaguchi, K., & Honma, I. (2019). *RSC Advances, 9,* 36434–36439.
15. Feng, Z., Chen, X., Qiao, L., Lipson, A. L., Fister, T. T., Zeng, L., et al. (2015). *ACS Applied Materials & Interfaces, 7,* 28438–28443.
16. Chen, W., Zhan, X., Luo, B., Ou, Z., Shih, P.-C., Yao, L., et al. (2019). *Nano Letters, 19,* 4712–4720.

Synthesis of Structured Spinel Oxide Positive Electrodes to Improve Electrochemical Performance

Hiroaki Imai

Abstract In this chapter, a novel structural design concept of positive electrodes is shown for improving the electrochemical performance of magnesium rechargeable batteries. Here, two types of structured spinel oxides, such as spinel $MgCo_2O_4$ mesocrystals consisting of iso-oriented nanometer-sized subunits and hierarchical porous frameworks of $MgMn_2O_4$ with interconnected bimodal pores, are proposed as potential cathode materials. The capacity of the active materials improves with a decrease in the overpotential because the reactivity and the electron and ion conductivities are highly enhanced in the designed architectures of the cathode materials.

Keywords Mesocrystal · Nanocrystal · Sol-gel

1 Introduction

Spinel oxides, such as $MgCo_2O_4$, were proposed as prospective candidates for a positive-electrode material for magnesium rechargeable batteries (MRBs) because these oxides have high theoretical capacity and high redox potential [1]. An improved insertion process is vital to enhance the electrochemical properties of MRBs, because the mobility of the divalent ion is very low in solid states [2, 3]. Nanosizing primary building blocks was reported to improve the electrochemical properties of active materials [4–6]. Unfortunately, the degradation of their working capacity and stability is common due to random aggregation and segregation of the active materials in the electrodes. The hierarchical structure design of active materials is essential to bring out their true competence. The present section proposes two types of structured spinel oxides for an MRB cathode material: spinel $MgCo_2O_4$ mesocrystals, consisting of iso-oriented nanometer-sized subunits [7], and hierarchical porous frameworks of

H. Imai (✉)
Keio University, Yokohama, Japan
e-mail: hiroaki@applc.keio.ac.jp

K. Kanamura (ed.), *Next Generation Batteries*,
https://doi.org/10.1007/978-981-33-6668-8_45

$MgMn_2O_4$ with interconnected bimodal pores [8]. These specific structures are effective in bringing out the capability of the active material. The design concept of positive electrodes would be applicable for improving the electrochemical performance of a variety of metal-anode secondary batteries.

2 Synthesis and Characterization of Structured Spinel Oxides

Mesocrystals consisting of iso-oriented nanometer-sized subunits were shown to be effective for improvement of the electrochemical performance of electrode materials [9]. Spinel $MgCo_2O_4$ mesocrystals (MCO-Ms) (Fig. 1b, c) were successfully obtained from calcite-type $CoCO_3$ grown in cobalt-ion-containing agar microgels ~50 μm in size [7]. Deformed rhombohedral $CoCO_3$ crystals comprised of nanograins smaller than ~5 nm (Fig. 1a) were produced in the microgels after a reaction with NH_4HCO_3. The desired $MgCo_2O_4$ product was then obtained following reaction of the $CoCO_3$ precursor with $Mg(NO_3)_2$ and subsequent heating at 350 °C. As shown in the scanning electron microscope (SEM) and transmission electron microscope (TEM) images (Fig. 1b, c), rhombohedral porous architectures 200–300 nm in size and consisting of nanometer-sized units of ~10 nm in diameter were obtained by the reaction. The short arcs observed in the selected area electron diffraction (SAED) pattern (Fig. 1d) indicate that the product is a spinel $MgCo_2O_4$ mesocrystal consisting of the nanoscale units likely being arranged along the same crystallographic direction. The nitrogen adsorption isotherms revealed a high specific surface area (110–125 $m^2\ g^{-1}$) and an average pore diameter of 7–8 nm for MCO-Ms. Random nanoparticles of spinel $MgCo_2O_4$ were prepared as a reference using the inverse coprecipitation method (MCO-ICs) as shown in Fig. 1e. The specific surface areas of MCO-ICs calcined at 500 °C for 24 h and at 350 °C for 12 h were 52 and 207 $m^2\ g^{-1}$, respectively.

A propylene oxide-driven sol-gel method has been applied for the fabrication of hierarchically structured spinel oxides [10–12]. Hierarchically porous $MgMn_2O_4$ was prepared using a propylene oxide-driven sol-gel method for an MRB cathode

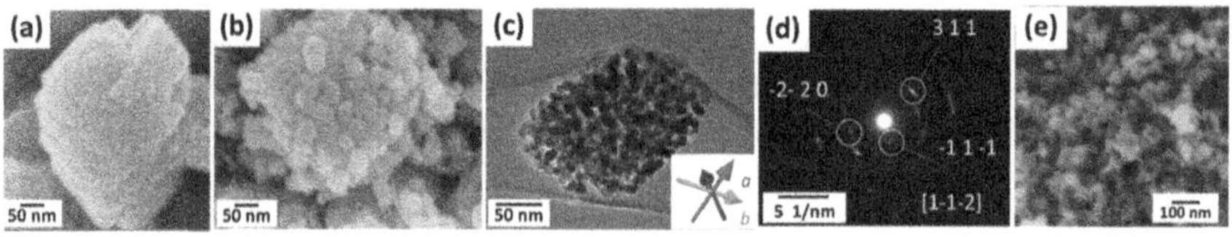

Fig. 1 SEM (**a**, **b**, **e**) and TEM (**c**) images and the SAED pattern (**d**) of a calcite-type $CoCO_3$ precursor (**a**), spinel $MgCo_2O_4$ mesocrystals (MCO-Ms) (**b**–**d**) obtained from the precursor, and $MgCo_2O_4$ nanoparticles prepared using the inverse coprecipitation method (MCO-ICs) (**e**) Reproduced from Ref. [7] with permission by Elsevier

Fig. 2 SEM (i) and TEM (ii) images and schematic illustrations (iii) of MMO-SGs (**a**) and MMO-ICs (**b**). Insets in (**a**-ii) and (**b**-ii) are FFT patterns derived from the lattice fringes from red-square regions in TEM images Reproduced from Ref. [8] with permission by Elsevier

material. The porous frameworks comprised of ~10 nm amorphous nanoparticles were produced as the xerogel precursors of spinel oxides through the freeze-drying of organic-metal complex wet gels after the exchange of pore liquid with cyclohexane. Figure 2a shows sol-gel-derived spinel-type $MgMn_2O_4$ powders (MMO-SGs) that were obtained by calcining the xerogel precursors at 300 °C for 5 h in air. The porous frameworks of MMO-SGs were comprised of nanoparticles ~7 nm in diameter (Fig. 2a–i). The formation of the spinel phase was confirmed by a fast Fourier transform (FFT) pattern derived from the crystal lattice fringes from the red-square region of the TEM image (Fig. 2a–ii). Thus, spinel oxide nanoparticles were prepared through relatively low-temperature calcination. The bimodal pores in involved hierarchical frameworks were found to be distributed in the micrometer (1–10 μm) and nanometer (20–100 nm) regions, as illustrated in Fig. 2a–iii. The micrometer and nanometer pores are derived from interconnecting voids in the xerogel network and interparticle spaces in the frame, respectively. The specific surface area was estimated to be ~150 m^2/g from the nitrogen adsorption isotherms. Figure 2b shows spinel-type $MgMn_2O_4$ powders that were synthesized using the inverse coprecipitation method (MMO-ICs) [1] as a reference. The powder of MMO-ICs prepared by calcination at 600 °C was comprised of micrometer-scale aggregates of spinel oxide particles ~30 nm in diameter (Fig. 2b–i, iii). The FFT pattern obtained from the red-square region of the TEM image indicates that the particles are a single crystal of the spinel phase (Fig. 2b–ii). The specific surface area was estimated to be ~20 $m^2\ g^{-1}$.

3 Electrochemical Properties of Structured Spinel Oxides

The electrochemical performance of the structured spinel oxides, such as mesocrystals and hierarchical porous frame works, was evaluated using a three-electrode beaker cell in an argon-filled glove box. Galvanostatic discharge/charge measurements were performed at a current density of 10.4 mA g^{-1} with a battery testing system at 100 °C. The working electrode composed of 60 wt% spinel oxides, 30 wt% acetylene black, and 10 wt% polytetrafluoroethylene binder was pressed onto an aluminum mesh. A magnesium alloy containing 3 wt% aluminum and 1 wt% zinc (AZ31) and a magnesium ribbon was used for a counter electrode for $MgCo_2O_4$ and $MgMn_2O_4$, respectively. Ionic liquid-based electrolytes containing $Mg[TFSA]_2$ and triglyme (G3) or tetraglyme (G4) [13, 14] were used for the measurements.

Figure 3 shows the first discharge curves of MCO-Ms with MCO-ICs having various specific surface areas. Here, the properties of active materials were evaluated from the first discharge curves to avoid the effects of the decomposition reaction. Basically, the MCO-Ms obtained by calcination at 350 °C for 2 and 4 h performed better than MCO-ICs with heating at 350 and 500 °C. The initial discharge potential of MCO-ICs increased with increasing specific surface area, but their discharge capacity clearly decreased. The improvement in the properties cannot be ascribed simply to a high specific surface area for the electrochemical reaction. Generally, magnesium ion diffusion is known to be very slow, and the reduction of the spinel oxide is strongly inhibited as the reduction proceeds. The enhanced performance of the mesocrystals (MCO-Ms) can be attributed to a combination of their high specific surface area and their single-crystalline porous framework that exhibits higher reactivity.

Figure 4 shows the discharge/charge operation conducted for hierarchically porous $MgMn_2O_4$ (MMO-SG) and a reference (MMO-IC). The charging process was

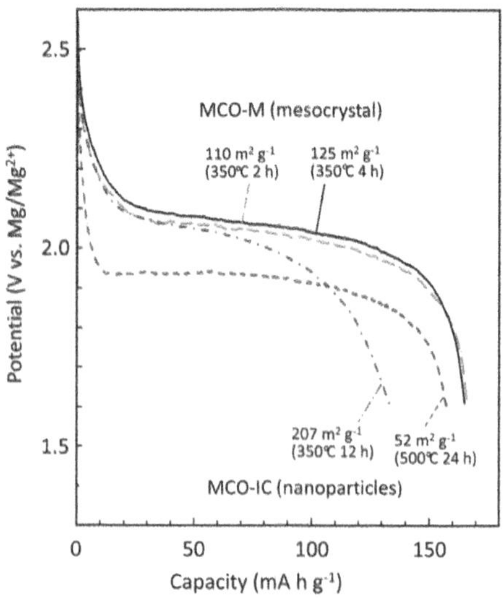

Fig. 3 The first discharge curves of MCO-Ms calcined at 350 °C for 2 and 4 h and MCO-ICs calcined at 350 °C for 12 h and 500 °C for 24 h Reproduced from Ref. [7] with permission by Elsevier

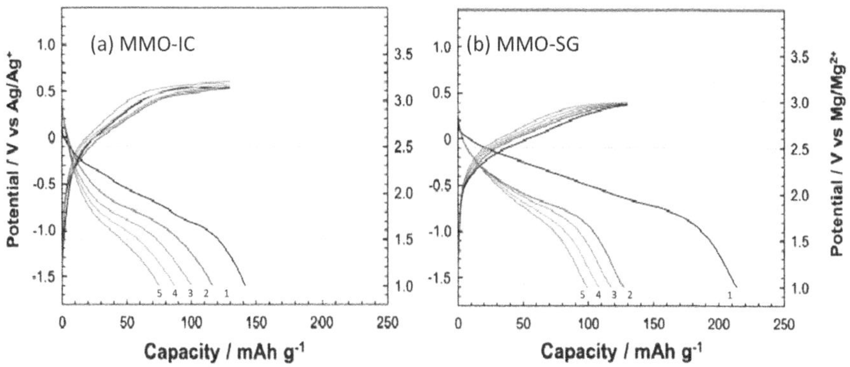

Fig. 4 Discharge/charge curves of MMO-IC (**a**) and MMO-SG (**b**) Reproduced from ref. [8] with permission by Elsevier

conducted within one-half of the theoretical capacity (130 mAh g^{-1}) to prevent decomposition of the electrolytes. A plateau around the reduction potential of about 2 V (versus Mg/Mg^{2+}) was observed in the discharge curve for both MMO-IC and MMO-SG. The discharge and charge potentials for MMO-SG were higher and lower than those of MMO-IC, respectively. Thus, the overpotential for the electrochemical reactions was lowered by the porous framework. The total capacity of MMO-SG was 214 mAh g^{-1} at the first discharge operation and gradually decreased to 1.87 V and 99 mAh g^{-1} after the fifth cycle. However, the capacity of MMO-IC decreased from 142 mAh g^{-1} to 75 mAh g^{-1} during the cycles. The ion diffusion is assisted by the submicron pores of the hierarchical frameworks with the penetration of an electrolyte. The improved conduction path that is constructed by the penetration of carbon nanoparticles into the bimodal pores of the spinel oxide framework is deduced to prevent degradation of the capacity during the cycle.

4 Conclusions

Two types of structured spinel oxides, mesocrystals and porous frameworks with a bimodal pore structure, were proposed as potential cathode materials for magnesium rechargeable batteries. The superiority of the electrochemical performance of structured spinel oxides was demonstrated with the galvanometric discharge/charge operation. The capacity of the active materials was found to improve with a decrease in the overpotential by the designed architectures because of the enhancement of the reactivity and the electron and ion conductivities in the cathode materials. The structural design of the electrode materials would be important in developing metal-anode rechargeable battery systems.

References

1. Okamoto, S., Ichitsubo, T., Kawaguchi, T., Kumagai, Y., Oba, F., Yagi, S., et al. (2015). *Advanced Science, 2,* 1500072.
2. Yoo, H. D., Shterenberg, I., Gofer, Y., Gershinsky, G., Pour, N., & Aurbach, D. (2013). *Energy & Environmental Science, 6,* 2265.
3. Shterenberg, I., Salama, M., Gofer, Y., Levi, E., & Aurbach, D. (2014). *MRS Bulletin, 39,* 453.
4. Yin, J., Brady, A. B., Takeuchi, E. S., Marschilok, A. C., & Takeuchi, K. J. (2017). *Chemical Communications, 53,* 3665.
5. Truong, Q. D., Devaraju, M. K., & Honma, I. (2017). *Journal of Power Sources, 361,* 195.
6. Kim, J.-S., Chang, W.-S., Kim, R.-H., Kim, D.-Y., Han, D.-W., Lee, K.-H., et al. (2015). *Journal of Power Sources, 273,* 210.
7. Kotani, Y., Ise, R., Ishii, K., Mandai, T., Oaki, Y., Yagi, S., et al. (2018). *Journal of Alloys and Compounds, 739,* 793.
8. Ishii, K., Doi, S., Ise, R., Mandai, T., Oaki, Y., Yagi, S., et al. (2020). *Journal of Alloys and Compounds, 816,* 152556.
9. Dang, F., Hoshino, T., Oaki, Y., Hosono, E., Zhou, H., & Imai, H. (2013). *Nanoscale, 5,* 2352.
10. Gash, A. E., Tillotson, T. M., Satcher, J. H., Poco, J. F., Hrubesh, L. W., & Simpson, R. L. (2001). *Chemistry of Materials, 13,* 999.
11. Cui, H., Zayat, M., & Levy, D. (2005). *Journal of Sol-Gel Science and Technology, 35,* 175.
12. Zhang, M., Guo, S., Zheng, L., Zhang, G., Hao, Z., Kang, L., et al. (2013). *Electrochimica Acta, 87,* 546.
13. Terada, S., Mandai, T., Suzuki, S., Tsuzuki, S., Watanabe, K., Kamei, Y., et al. (2016). *Journal of Physical Chemistry C, 120,* 1353.
14. Mandai, T., Tatesaka, K., Soh, K., Choudhary, A., Tateyama, Y., Ise, R., et al. (2019). *Physical Chemistry Chemical Physics: PCCP, 21,* 12100.

High-Temperature Conductivity Measurements of Magnesium-Ion-Conducting Solid Oxide Using Mg Metal Electrodes

Koichi Kajihara, Hayato Nagano, Takaoki Tsujita, Hirokazu Munakata, and Kiyoshi Kanamura

Abstract The conductivity of polycrystalline pellets of Mg^{2+}-ion-conducting solid oxide $Mg_{0.5-x}(Zr_{1-x}Nb_x)_2(PO_4)_3$ ($x = 0.15$) with monoclinic β-$Fe_2(SO_4)_3$-type structure was measured with non-blocking Mg metal electrodes at 350 °C in vacuum using a high-temperature cell, which avoided the exposure of Mg electrodes to air before and during the measurements. At 350 °C dc and ac conductivities measured with Mg electrodes were consistent with ac conductivity measured with Pt electrodes. These observations evidenced that this compound is a pure Mg^{2+}-ion conductor and stable in contact with Mg.

Keywords $Mg_{0.35}(Zr_{0.85}Nb_{0.15})_2(PO_4)_3$ · High-temperature dc and ac conductivity measurements with non-blocking Mg metal electrodes in vacuum · Pure Mg^{2+}-ion conduction

1 Introduction

The metals and alloys of Mg are typical non-blocking electrodes for Mg^{2+}-ion-conducting solids. They have been used for the conductivity measurements of several Mg^{2+}-ion-conducting solids up to 150 °C [1, 2]. At high temperatures, however, they are easily oxidized to insulating MgO. In addition, Mg metal and alloy electrodes have high reducing power, and their oxidation may be accompanied by the reductive decomposition of Mg^{2+}-ion-conducting solids under the electrodes. Because of such difficulties, the migration of Mg^{2+} ions in Mg^{2+}-ion-conducting solids has typically been evaluated by cross-sectional elemental analysis [3–6] and Tubandt electrolysis [3, 4, 6]. Thus, the high-temperature conductivity measurements of Mg^{2+}-ion-conducting solids using Mg electrodes remained a challenge.

K. Kajihara (✉) · H. Nagano · T. Tsujita · H. Munakata · K. Kanamura
Department of Applied Chemistry for Environment, Graduate School of Urban Environmental Sciences, Tokyo Metropolitan University, 1-1 Minami-Osawa, Hachioji, Tokyo 192-0397, Japan
e-mail: kkaji@tmu.ac.jp

K. Kanamura (ed.), *Next Generation Batteries*,
https://doi.org/10.1007/978-981-33-6668-8_46

Zirconium phosphate $Mg_{0.5}Zr_2(PO_4)_3$ [7, 8] and its derivatives [2–4, 9] with monoclinic β-$Fe_2(SO_4)_3$-type structure have been studied most often as Mg^{2+}-ion-conducting solid oxides. These compounds consist of the alternate bridging of octahedral ZrO_6 and tetrahedral PO_4 units at the two-coordinated apical O atoms. This local bonding rule of the octahedral and tetrahedral units is the same as that of the Na super ionic conductor (NASICON)-type structure [10], whereas the topology of the resultant polyhedral framework is different between these compounds.

Partial substitution of Nb into Zr in $Mg_{0.5}Zr_2(PO_4)_3$ significantly increases the conductivity and the highest conductivity was observed at $x = 0.15$ for $Mg_{0.5-x}(Zr_{1-x}Nb_x)_2(PO_4)_3$ [3, 4].

2 Synthesis and Conductivity Measurements of $Mg_{0.5-x}(Zr_{1-x}Nb_x)_2(PO_4)_3$ ($x = 0.15$) [11]

Simple solid-state reaction of MgO, ZrO_2, Nb_2O_5, and $NH_4H_2PO_4$ powders and subsequent sintering at 1350 °C yielded the polycrystalline pellets of $Mg_{0.35}(Zr_{0.85}Nb_{0.15})_2(PO_4)_3$. The sum of weight fractions of impurity phases (ZrP_2O_7 and $Zr_2P_2O_9$) was less than 2 wt% and relative density was ~0.78.

To perform conductivity measurements with Mg electrodes a homemade high-temperature vacuum sample cell evacuated to ~10^{-4} Pa was developed. This sample cell is small enough to be carried in a globe box equipped with a vacuum deposition chamber, and enables to set a sample with Mg electrodes while avoiding its exposure to air.

Figure 1 shows Nyquist plot measured with Mg electrodes at 350 °C. The capacitance of the large semicircle at the left (high-frequency) side calculated from the relation $C = (2\pi fR)^{-1}$ was ~3.4×10^{-11} F and such a small C value is typical of the bulk component. The size and shape of the semicircle for the bulk component were similar to that measured with blocking Pt electrodes. However, a linear tail in the low-frequency region was absent, demonstrating the non-blocking nature of Mg electrodes. In the low-frequency region, a small semicircle was observed. The slow response (~0.5 Hz) and large capacitance (~1.7×10^{-5} F) suggest that this component is attributed to the charge transfer process associated with the dissolution and precipitation of Mg at the electrode-oxide interface. The series resistance of the two semicircles was ~2.6×10^5 Ω and the total ac conductivity, σ_{ac}, was ~1.1×10^{-6} S cm^{-1}.

Inset of Fig. 1 shows dc polarization profile recorded after the measurement of the Nyquist plot shown in Fig. 1. A constant current of 2 μA was applied. The polarity of current was altered every 30 s, whereas the polarity change did not induce voltage spikes. Voltage reached to a steady-state value of ~±0.52 V in ~15 s, and its time constant was consistent with the frequency associated with the small semicircle (~0.5 Hz) observed in the Nyquist plot (Fig. 1). The dc conductivity, σ_{dc}, calculated from the steady-state voltage was ~1.1×10^{-6} S cm^{-1} and was equal to σ_{ac}. These

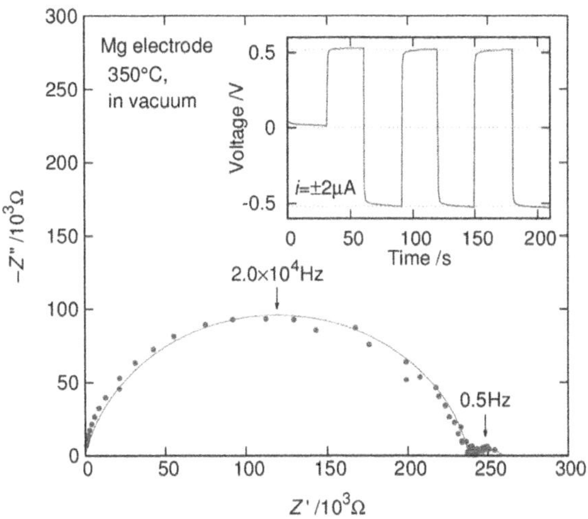

Fig. 1 Nyquist plot and dc polarization profile (inset) of a polycrystalline $Mg_{0.35}(Zr_{0.85}Nb_{0.15})_2(PO_4)_3$ pellet measured using Mg electrodes at 350 °C in vacuum. Solid semicircle was drawn by fitting to the Cole-Cole equation [12] Reprinted with permission from Ref. [11]. Copyright 2017 The Electrochemical Society

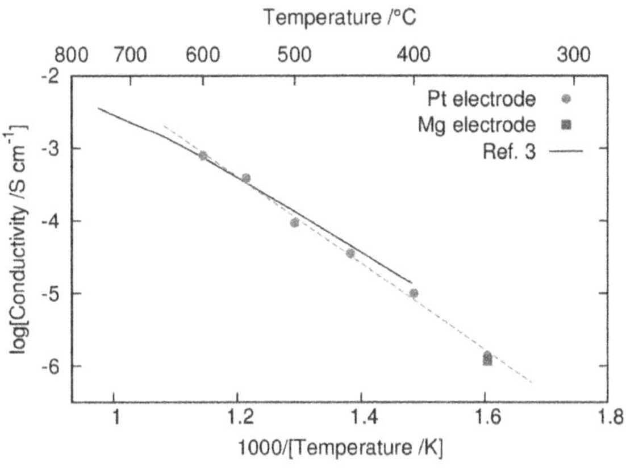

Fig. 2 Arrhenius plot of ac conductivities of a polycrystalline $Mg_{0.35}(Zr_{0.85}Nb_{0.15})_2(PO_4)_3$ pellet. Dashed line was drawn by fitting of the data measured with Pt electrodes. Solid line shows data taken from Ref. [3] Reprinted with permission from Ref. [11]. Copyright 2017 The Electrochemical Society

observations indicate that the transport number of Mg^{2+} ions is ~1 and confirms that $Mg_{0.35}(Zr_{0.85}Nb_{0.15})_2(PO_4)_3$ is a pure Mg^{2+}-ion conductor stable in contact with Mg.

Figure 2 shows Arrhenius plot of σ_{ac}. The activation energy of σ_{ac} measured with Pt electrodes was ~1.18 eV. These results were consistent with earlier studies on $Mg_{0.35}(Zr_{0.85}Nb_{0.15})_2(PO_4)_3$ polycrystalline ceramics [3]. The σ_{ac} and σ_{dc} values measured with Mg electrodes at 350 °C (~1.1 × 10^{-6} S cm^{-1}) were slightly smaller than the value measured with Pt electrodes at 350°C (~1.4 × 10^{-6} S cm^{-1}), implying the partial thermal oxidation of Mg electrodes with residual oxygen species during measurements.

Acknowledgement This work was supported by ALCA-SPRING project of Japan Science and Technology Agency (JST).

References

1. Higashi, S., Miwa, K., Aoki, M., & Takechi, K. (2014). *Chemical Communications, 50,* 1320.
2. Anuar, N. K., Adnan, S. B. R. S., & Mohamed, N. S. (2014). *Ceramics International, 40,* 13719.
3. Imanaka, N., Okazaki, Y., & Adachi, G. (2000). *Electrochemical and Solid-State Letters, 3,* 327.
4. Imanaka, N., Okazaki, Y., & Adachi, G. (2001). *Ionics, 7,* 440.
5. Omote, A., Yotsuhashi, S., Zenitani, Y., & Yamada, Y. (2011). *Journal of the American Ceramic Society, 94,* 2285.
6. Tamura, S., Yamane, M., Hoshino, Y., & Imanaka, N. (2016). *Journal of Solid State Chemistry, 235,* 7.
7. Ikeda, S., Takahashi, M., Ishikawa, J., & Ito, K. (1987). *Solid State Ionics, 23,* 125.
8. Nomura, K., Ikeda, S., Ito, K., & Einaga, H. (1992). *Bulletin of the Chemical Society of Japan, 65,* 3221.
9. Adamu, M., & Kale, G. M. (2016). *The Journal of Physical Chemistry C, 120,* 17909.
10. Goodenough, J. B., Hong, H. Y.-P., & Kafalas, J. A. (1976). *Materials Research Bulletin, 11,* 203.
11. Kajihara, K., Nagano, H., Tsujita, T., Munakata, H., & Kanamura, K. (2017). *Journal of the Electrochemical Society, 164,* A2183.
12. Cole, K. S., & Cole, R. H. (1941). *The Journal of Physical Chemistry, 9,* 341.

Magnesium Metal and Intermetallic Anodes

Masaki Matsui

Abstract Magnesium metal is an ideal negative electrode material for rechargeable batteries because of its high volumetric capacity and low equilibrium potential. The non-dendritic growth during the deposition process of the magnesium metal is also a significant advantage for battery applications. In this chapter, we focus on the electrochemical properties of the magnesium metal and intermetallic anodes. We briefly review the electrolyte solutions to understand the compatibilities of the magnesium metal and intermetallic anodes. Then we discuss the passivation layer of the magnesium metal anode based on our previous work using in situ FTIR and XPS. We also review the advantage of the intermetallic compounds as alternate negative electrode materials.

Keywords Magnesium metal anode · Intermetallic anode · Passivation layer

1 Introduction

Magnesium metal is an ideal negative electrode material for lithium-ion batteries, because of its high volumetric capacity (3833 mAh cm^{-3}), low equilibrium potential (−2.3 V versus SHE), and non-dendritic growth during the deposition process. The electrodeposition of the magnesium metal in high-temperature molten salts is studied from the early nineteenth century and still remains a standard production process of magnesium metal, on the other hand, the electrodeposition of magnesium metal for rechargeable battery applications is studied since the late twentieth century [1]. Early work by Gregory et al. was based on organomagnesium compounds such as Grignard reagent [2]. Aurbach et al. designed various electrolyte solutions, so-called "magnesium organohaloaluminates", which show excellent reversibility and relatively wide electrochemical window [3]. After those works, various new classes of electrolyte solutions were developed over the last decade. In addition, a series of analytical or

M. Matsui (✉)
Department of Chemical Science and Engineering, Kobe University, 1-1 Rokkodai-Cho, Nada-Ku, Kobe 657-8501, Japan
e-mail: matsui@godzilla.kobe-u.ac.jp

K. Kanamura (ed.), *Next Generation Batteries*,
https://doi.org/10.1007/978-981-33-6668-8_47

theoretical studies were also carried out to understand the electrochemical deposition/dissolution process of magnesium metal. In the present chapter, we review the recent progress of the magnesium metal anode and magnesium-based intermetallic compounds. We start with a brief introduction of the electrolyte solutions, because the choice of the electrolyte solution is very limited compared with lithium or sodium system. In this part, we categorize the electrolyte solutions based on the compatibility with magnesium metal. Secondly, we focus on the passivation layer, which determines the compatibility. Finally, we introduce the intermetallic anodes which show excellent compatibility with wide ranges of electrolyte solutions.

2 Overview of the Electrolyte Solutions

In this section, we briefly review the electrolyte solutions for the electrodeposition of magnesium. As described in the introduction, the electrodeposition of the magnesium metal in the high-temperature molten salt is a common process for the production of the magnesium metal. Typically, molten halide electrolytes such as $MgCl_2$-KCl binary or $MgCl_2$-KCl-NaCl ternary system are used as the electrolysis bath at the operation temperature 700–750 °C. The electrochemically active species in the molten halide electrolytes are complex anions such as $MgCl_6^{4-}$, $MgCl_4^{2-}$, and $MgCl_3^{-}$.

The electrodeposition of the magnesium metal at room temperature is originally studied using electrolyte solutions containing organomagnesium compounds such as Grignard reagent because the magnesium metal is not deposited in the electrolysis bath containing simple magnesium salts such as magnesium perchlorate. Jolibois firstly reported the room temperature electrodeposition of the magnesium metal using a solution of diethylmagnesium (Et_2Mg) and magnesium iodide (MgI_2) in diethyl ether (Et_2O) [4]. Electrolysis of a Grignard reagent was reported by Gaddum and French [5]. They reported that benzylmagnesium chloride shows better coulombic efficiency than phenylmagnesium bromide, and proposed a possible deposition mechanism. Twenty-five years later, Connor et al. successfully deposited magnesium metal using organomagnesium-free electrolysis baths [6]. There, the magnesium is electrodeposited from in situ formed $Mg(BH_4)_2$ solution from the mixture of $MgBr_2$ and $LiBH_4$ solution. In this work, the coulombic efficiency of the deposition process was significantly improved to 90% compared with the previous works.

Gregory et al. firstly performed a comprehensive study on rechargeable magnesium batteries in 1990 [2]. Their electrolyte solutions, Grignard reagents (MeMgCl and EtMgCl) with a small amount of aluminum chloride in THF, showed excellent coulombic efficiency of 100% as shown in Table 1. Interestingly, they claimed that dendritic growth of the magnesium occurs in some of the electrolyte solutions. Ten years after, Aurbach et al. finally succeeded in the development of a prototype of the rechargeable magnesium battery, using Bu_2Mg-$EtAlCl_2$ complex salt (Dichloro Complex: DCC) in THF as the electrolyte solution [7]. The electrolyte solution showed improved anodic stability >2.0 V versus Mg, while the

Table 1 Magnesium plating from organomagnesium chloride/$AlCl_3$/THF baths at 1.25 mA cm^{-2} Reproduced with permission from ref Gregory et al. [2]. Copyright 1990, The Electrochemical Society

R group	[RMgCl] (mol L^{-1})	[$AlCl_3$] (mol L^{-1})	Current efficiency, %		Mg deposition quality
			Anodic	Cathodic	
Methyl	1.5	0.10	100	101	Dendritic
Ethyl	0.8	0.10	98.9	100	Excellent, small grain
Ethyl	1.5	0.10	99.5	100	Excellent, large grain
Ethyl	2.0	0.20	90.1	–	Very dendritic
Butyl	1.0	0.1	104	98.4	Excellent, small grain

normal Grignard reagent such as BuMgCl solution is only stable below 1.5 V versus Mg. The electrolyte solution coupled with the Mo_3S_4 Chevrel phase allowed stable charge/discharge cycles >500. The excellent cycling performance suggests no dendritic growth of the magnesium metal anodes. They also made a continuous effort for the improvement of the magnesium organohaloaluminate electrolytes and discovered all phenyl complex (APC) electrolytes which shows high anodic stability of >3.0 V versus Mg [8]. Kim et al. developed a non-nucleophilic electrolyte [$Mg_2(\mu\text{-}Cl)_3\cdot 6THF$][$HMDSAlCl_3$] for Mg-S battery [9]. In this paper, they discussed the formation process of the electrochemically active [$Mg_2(\mu\text{-}Cl)_3\cdot 6THF$]$^+$ cation. Also, the anodic stability of the [$Mg_2(\mu\text{-}Cl)_3\cdot 6THF$][$HMDSAlCl_3$] electrolyte solution is improved via the removal of [$HMDS_2Mg$] species during the recrystallization process. Even with the improvement of the anodic stability of the electrolyte solutions, the electrolyte solutions containing halides show corrosive properties to various metals at high electrode potential in their following paper. Therefore, the practical voltage window of the electrolyte solutions containing halide still remains <2.5 V [10].

Magnesium bis(trifluoromethansulfonyl amide) ($Mg(TFSA)_2$) dissolved in glyme is reported as a halide-free electrolyte solution, by several research groups [11–13]. Figure 1 shows cyclic voltammograms of 0.5 mol L^{-1} $Mg(TFSA)_2$ in butyl methyl triglyme solution ($Mg(TFSA)_2$ solution) taken by the author's group. The CV shows very high overpotential and poor coulombic efficiencies during the deposition/dissolution process. The poor reversibility is due to the decomposition of the $TFSA^-$ anion during the deposition process. The reversibility can be improved by the addition of $MgCl_2$ [14]. In this case, the dimer [$Mg_2(\mu\text{-}Cl)_3$]$^+$ cation analogs are probably the electrochemically active species.

Major progress in the halide-free electrolyte was made by Tutusaus et al. in 2015 [15]. They developed a novel electrolyte salt magnesium monocarborane (MMC) and confirmed excellent reversibility without corrosion. A fluorinated alkoxyborate electrolyte $Mg[B(HFIP)_4]_2$ developed by Zhao-Karger et al. also shows stable cycling performance with high oxidation stability [16].

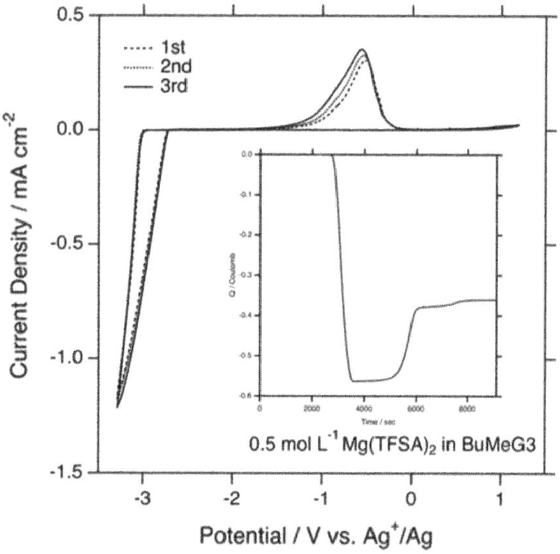

Fig. 1 A cyclic voltammogram of 0.5 mol L^{-1} $Mg(TFSA)_2$ in BuMeG3 with a sweep rate of 1.0 mV sec^{-1} using Pt electrode taken by the authors' group

We think the overall progress of the electrolyte solutions for the magnesium metal anode can be summarized in three categories as shown in Table 2. The GEN 1 electrolytes are "organomagnesium compounds", just for the electrodeposition. The GEN 2 electrolytes are "controlled complex compounds", for the improved reversibility and voltage window, and the GEN 3 electrolytes are "novel anions" for a halide-free system including $Mg(TFSA)_2$-glyme solutions. One important point to remember is that all the electrolytes are dissolved in "ether-based solvents", which is due to the reduction stability of the solvent molecules. Since the cathodic decomposition of the electrolyte solution passivates the surface of the magnesium metal anode, all the chemical species in the electrolyte solution need to be thermodynamically

Table 2 Summary of the peak assignments of the in situ FTIR spectra for the 0.25 M $EtMgCl$-$2Et_2AlCl$ in THF solution during the 1st cathodic scan

cm^{-1}	
	Positive Peaks
1062	C–O–C asymmetric stretching vibration of free THF
1025	C–O–C asymmetric stretching vibration of solvated THF
899	C–H wagging vibration of THF
854	C–H wagging vibration of THF
	Negative Peaks
1074	C–O–C asymmetric stretching vibration of adsorbed THF
921	C–H bending vibration of adsorbed THF

stable at the equilibrium potential of the magnesium metal. Practically the ether-based solvents are the only choice to avoid passivation layer formation, because the passivation layer is the most critical problem for the compatibility of magnesium metal anode and going to be discussed in the following sections.

3 Surface Morphologies of the Electrodeposited Magnesium

"No dendritic growth" is the most attractive aspect for the research of "Rechargeable Magnesium Batteries". Even though the cycling performance of the prototype Mg/DCC electrolyte/Mo_3S_4 cell by Aurbach et al. already proves the magnesium metal anode is a practical battery anode [7], there still exists the risk: "all the metals have some chance to form the dendrite during the crystal growth process". Therefore, we need to have a discussion concerning the surface morphology of the electrodeposited magnesium metal.

Though the early work of Gregory et al. shows a beautiful SEM image of the electrodeposited magnesium with hexagonal shaped crystals, they also reported the dendritic growth of magnesium metal from two electrolyte solutions: 1.5 mol L^{-1} MeMgCl with 0.1 mol L^{-1} $AlCl_3$ in THF and 2.0 mol L^{-1} EtMgCl with 0.2 mol L^{-1} $AlCl_3$ in THF as shown in Table 1 [2]. There is no discussion concerning the dendritic growth of magnesium metal in the paper, the results, however, indicate that the electrolyte composition has some influence on the crystal growth process.

Aurbach et al. reported the surface morphology change during the electrodeposition process of magnesium metal using in situ STM [17]. The deposited magnesium metal formed "pyramidal shaped" crystals, which also reflects the hcp structure of the magnesium metal. Similar surface morphologies of the electrodeposited magnesium are reported by various research groups. Matsui carried out a comparative study of the surface morphology of the magnesium metal and lithium metal [18]. The electrodeposited magnesium shows smooth surface morphologies without any dendritic growth, while the lithium films deposited under the same current density have whisker-shaped surface morphologies as shown in Fig. 2. Further crystallographic study of the electrodeposited magnesium metal provides helpful information concerning the crystal growth process. The electrodeposited magnesium shows that the preferred orientation of the electrodeposited magnesium changes from (001) plane to (100) plane, increasing the deposition current. In case the crystal growth rate is enough low, the crystals tend to minimize the surface energy. Hence, the deposited magnesium metal at a low current density (0.5 or 1.0 mA cm^{-2}) shows a preferred orientation of (001) plane, which is the closed packed layer of the hcp structure. On the other hand, the crystal growth rate along with c-axis is slower than that along a/b-axis, therefore, the (001) preferred orientation cannot be maintained at a high current density (2.0 mA cm^{-2}). The fast electrodeposition of magnesium consequently initiates the preferred orientation of (100) plane to maximize the crystal growth rate. The

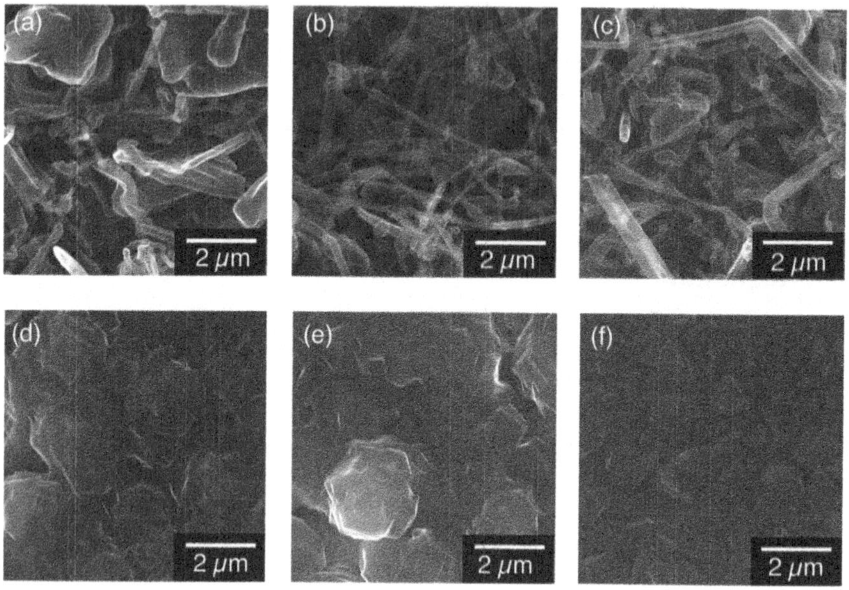

Fig. 2 SEM images of electrodeposited **a–c** lithium and **d–f** magnesium metal at various deposition current: **a**, **d** 0.5 mA cm^{-2}, **b**, **e** 1.0 mA cm^{-2}, and **c**, **f** 2.0 mA cm^{-2} Reprinted with permission from ref. [18]. Copyright 2010, Elsevier

change of the preferred orientation indicates that the faster electrodeposition may initiate the dendritic growth of magnesium metal.

In this paper, the possible reason for the non-dendritic growth of magnesium metal is also discussed. He found that the overpotential of the magnesium deposition/dissolution process significantly increases at a low concentration of the electrolyte solution as shown in Fig. 3a–c. Since the locally focused current is avoided by increasing the overpotential by consumption of Mg^{2+} ions, the random nucleation process consequently takes place during the initial stage of the electrodeposition process as shown in Fig. 3d.

Another reason for the non-dendritic magnesium metal deposition is reported by Ling et al. by DFT modeling [19]. Figure 4a is a schematic representation of the high dimensional phases (2D plate, 3D bulk particle) and low dimensional phases (1D, whisker/dendrite). They calculated the change of free energies with different morphologies: the high dimensional phases and the low dimensional phases for magnesium metal and lithium metal. Since the low dimensional phases have more surface area in the cluster, the energy change is represented by removing one atom from the bulk phase to the surface as shown in Fig. 4b. The DFT calculation results show that magnesium metal intrinsically prefers high dimensional growth compared with lithium metal, due to the strong Mg-Mg bond.

The surface morphology of the magnesium metal is also dependent upon the electrolyte solution. The electrolyte containing the dimer complex typically shows

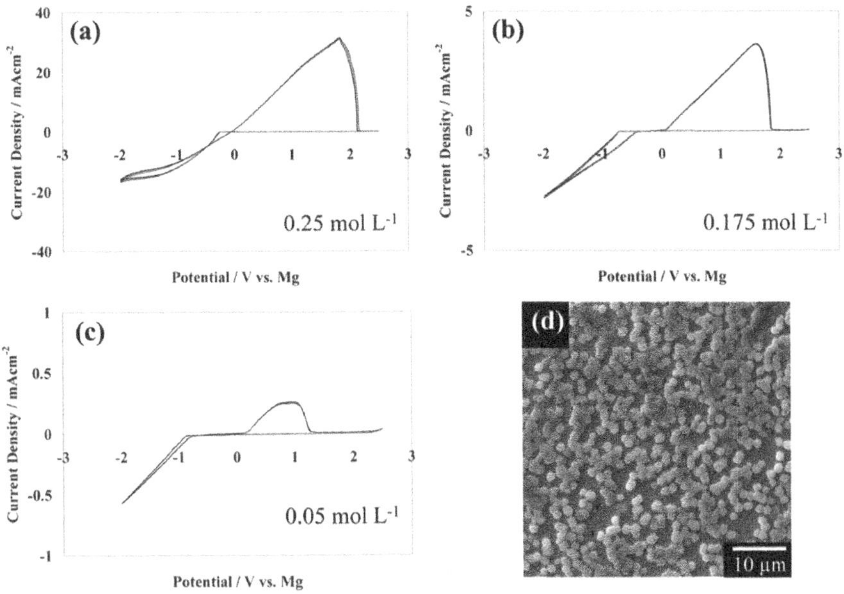

Fig. 3 Cyclic voltammograms of DCC type solution with different Mg^{2+} concentration **a–c** and **d** SEM image of the early stage of Mg deposition process Reprinted with permission from ref. [18]. Copyright 2010, Elsevier

pyramidal shaped grains, while the magnesium metal deposited in the $Mg(TFSA)_2$ in glyme solution has aggregated hemispherical particles as shown in Fig. 5 [20]. The pyramidal shaped grain reflects the well crystalline hcp structure of the magnesium metal, while the hemispherical particles are formed by the aggregation of nano-sized grains due to the simultaneous passivation of the surface of the grains during the electrodeposition process. On the other hand, Shterenberg et al. reported highly crystalline magnesium deposition from 0.25 mol L^{-1} of $Mg(TFSA)_2$ and 0.5 mol L^{-1} of $MgCl_2$ in DME solution [14]. It suggests that the high coulombic efficiency of >90% is necessary for the smooth crystal growth process of magnesium metal. They also confirmed an edgy morphology of the magnesium grains deposited at 5.0 mA cm^{-2}. The trend of the morphology change is consistent with the studies with the organohaloaluminate electrolytes.

The magnesium metal deposited in a state-of-art electrolyte MMC solution shows even more rough and edgy surface morphology [15]. An SEM image of the magnesium metal deposited at a high current density of 5.0 mA cm^{-2} from the MMC electrolyte solution is shown in Fig. 6. The increased surface area of the plate-shaped magnesium grains suggests the deposition rate is almost approaching the limit of the growth rate. Once the current density exceeds the growth rate of the metal, the dendritic growth is supposed to occur. Therefore, advanced electrolyte solutions may reinitiate the dendritic growth of magnesium metal.

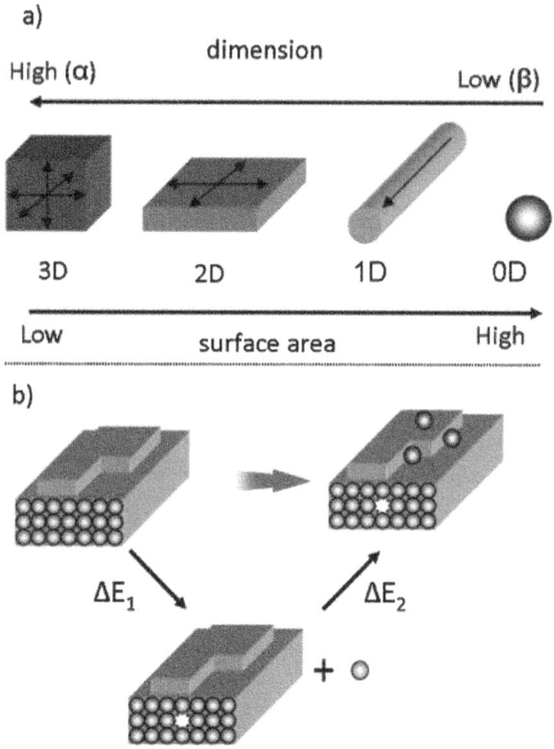

Fig. 4 **a** Schematic of high dimensional phases (phase α) and low dimensional phases (phase β). **b** Move of one atom from the bulk phase to the surface. The red, blue, and green circles show the atom in the bulk phase Reprinted with permission from ref. [19] 2012. Copyright 2012, Elsevier

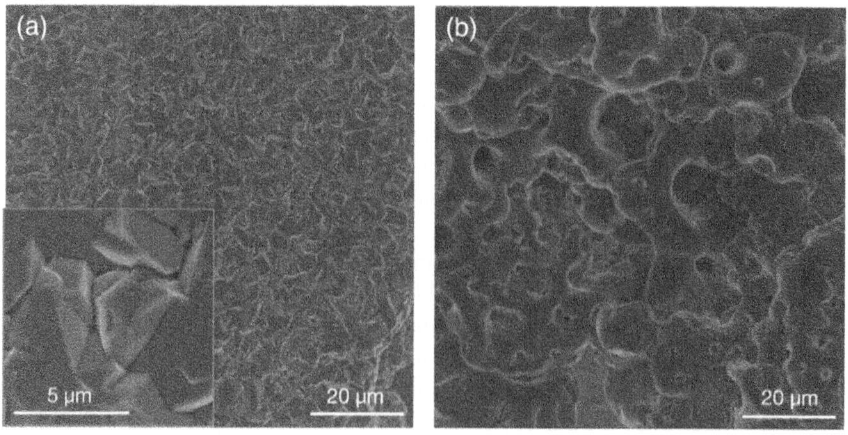

Fig. 5 SEM images of the electrodeposited magnesium from DCC electrolyte (**a**) and $Mg(TFSA)_2$-based electrolyte (**b**) taken by author's group

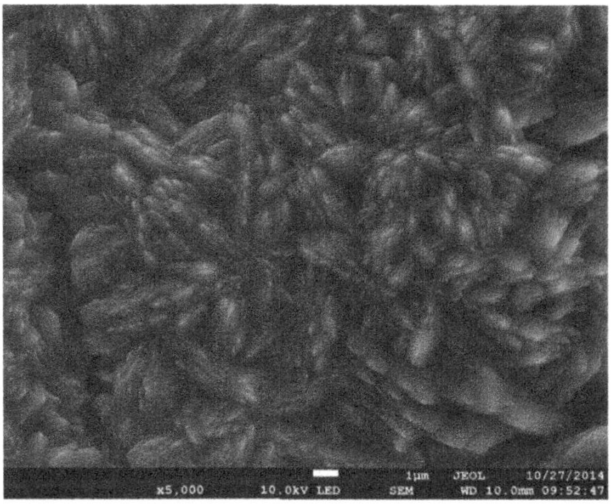

Fig. 6 SEM image of electrodeposited magnesium from 0.75 mol L^{-1} MMC/G4 solution at 5.0 mA cm^{-2} Reprinted with permission of the corresponding author from supporting information of ref. [15] 2015

The locally focused current is the trigger of the dendritic growth of the metal deposition. In the case of the magnesium metal, unexpected passivation layer also causes the locally focused current. Ding et al. reported the short circuit of the magnesium/$Mg(TFSA)_2$-glyme/magnesium symmetric cell via dendritic growth of magnesium metal [21]. The EDS mapping indicated the formation of the passivation layer limits the active surface area of the deposited magnesium metal resulting in the initiation of the focused current in the cell. The result clearly suggests that the passivation-free surface is crucial to avoid the focused current and dendritic growth during the deposition process. Even with the passivation-free surface, focused current initiates the dendritic growth of the magnesium. The recent report by Davidson et al. proved that the strong electric field at the corner of the electrode becomes a preferred site of the focused current resulting in the formation of a dendrite of magnesium metal shown in Fig. 7 [22].

Fig. 7 Magnesium dendrite deposited from 0.5 mol L^{-1} MeMgCl in THF solution Reprinted with permission from ref. [22] 2018. Copyright 2018, American Chemical Society

In this section, we reviewed the surface morphologies of electrodeposited magnesium metal as a potential problem for battery applications. Even though all the metals have some possibility to form the dendrite, magnesium metal shows relatively uniform deposition even at high current density compared to other metals. Furthermore, the sluggish mobility of Mg^{2+} ions in the solid phase limits the current density of the magnesium metal anode. Therefore, the dendrite formation is still not a critical problem compared with the passivation issue discussed in the following section.

4 Passivation Layer and Possible SEI Layer

The passivation layer is a severe issue when using magnesium metal as a battery anode. Even though Peled claims that all the alkaline metal and alkaline earth metal form SEI layers [23], he reports that the passivation layer is formed, rather than an SEI layer, at the surface of the Mg anode in the Mg-MnO_2 dry cell. The poor Mg^{2+} mobility in the passivation layer limits the electrochemical activity of the magnesium metal. Lu et al. conducted electrochemical and spectroscopic studies of the magnesium metal in various common electrolyte solutions such as $Mg(ClO_4)_2$ in ACN or PC. The Mg dissolution can take place only via a breakdown of the passivation layer during the anodic polarization [24]. In addition, the passivation layer may be restored during the cathodic scan, the electrodeposition of magnesium may not take place again. They also carried out EQCM measurements to confirm no surface layer formation during the electrodeposition/dissolution process of magnesium in DCC electrolyte solution [25]. These works suggest that the magnesium anode has to have an SEI-free interphase.

Our group reported a spectroscopic study for the characterization of the surface layer in an organohaloaluminate-based electrolyte solution (All Ethyl Complex: AEC solution) and a $Mg(TFSA)_2$ in butyl methyl triglyme solution ($Mg(TFSA)_2$ solution) using in situ FTIR and XPS [20].

We developed a diamond ATR-based in situ FTIR cell using a diamond ATR disc (DuraSamplIR, Smiths Detection) as shown in Fig. 8. We deposit a Pt thin film electrode on the diamond window using DC sputtering coater. The in situ FTIR measurement was carried out by an internal reflection geometry. Total internal reflection of the infrared beam occurs at the surface of the diamond crystal with the appearance of the evanescent wave. The evanescent wave typically has 0.5–1.0 μm penetration depth into the electrolyte solution to irradiates the molecules at the vicinity of the Pt thin film working electrode. Each FTIR spectrum was taken by the single beam mode with 512 of accumulation time. We calculate the subtractive normalized interfacial FTIR (SNIFTIR) spectrum with the following equation:

$$\Delta R = (R_{n+1} - R_n)/R_n \quad (1)$$

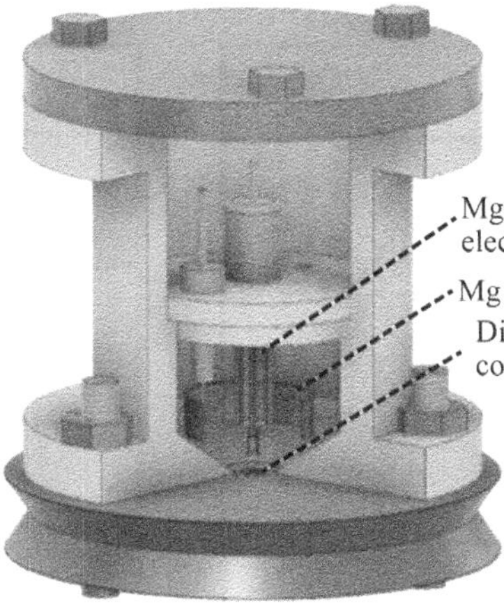

Fig. 8 Schematic representative of diamond ATR-based in situ FTIR cell developed by author's group from ref. [20] 2017 Copyright 2017, The Electrochemical Society

where R_n is a reference reflective spectrum and R_{n+1} is a reflective spectrum at a target electrode potential. The calculated SNIFTIR spectrum has positive and negative peaks; the positive peaks are corresponding to the disappeared species at the vicinity of the electrode during the two measurements. The negative peaks are corresponding to the newly formed species such as the surface layer formed at the surface of the electrode. The obtained SNIFTIR spectra have adsorption and desorption of the chemical species at the vicinity of the electrode.

Figure 9 shows a series of in situ FTIR spectra of the AEC solution during CV measurements. The SNIFTIR spectra observed the structural changes of the surface layer at >0.5 V versus Mg quasi-reference electrode. Several positive and negative peaks observed around 1078–860 cm^{-1} are assigned to THF molecules. The summary of the peak assignment is shown in Table 2. We also observe the peaks corresponding to the structural change of the interphase at higher electrode potential than that of the cathodic scan. Even though the spectra appear at different electrode potential, we think the structural change of the interphase is reversible, because the typical SNIFTIR spectra during the anodic scan have flipped the shape of the spectra during the cathodic scan. Furthermore, the SNIFTIR spectra in the 2nd cycle show similar reversible changes as shown in Fig. 9b; therefore, we think the SEI layer is not formed at the interphase in the AEC solution.

On the other hand, the $Mg(TFSA)_2$ solution forms a passivation layer during the electrodeposition process of magnesium metal. Figure 10 shows SNIFTIR spectra of $Mg(TFSA)_2$ solution during a CV measurement. Several positive and negative peaks are observed at 0 V versus Mg quasi-reference electrode and more negative potential. The positive peaks at 1346, 1328, 1041, 919 cm^{-1} are assigned to SO_2. The peaks at 1182, 759 cm^{-1} are assigned to CF_3 in $TFSA^-$ anion, respectively. Those peaks remain in the SNIFTIR spectra until the electrode potential reached the lower potential limit. Table 3 shows a summary of the peak assignments for the $Mg(TFSA)_2$ solution. The following SNIFTIR spectra taken at the 1st anodic scan, however, show almost no clear peaks up to the open circuit potential. The

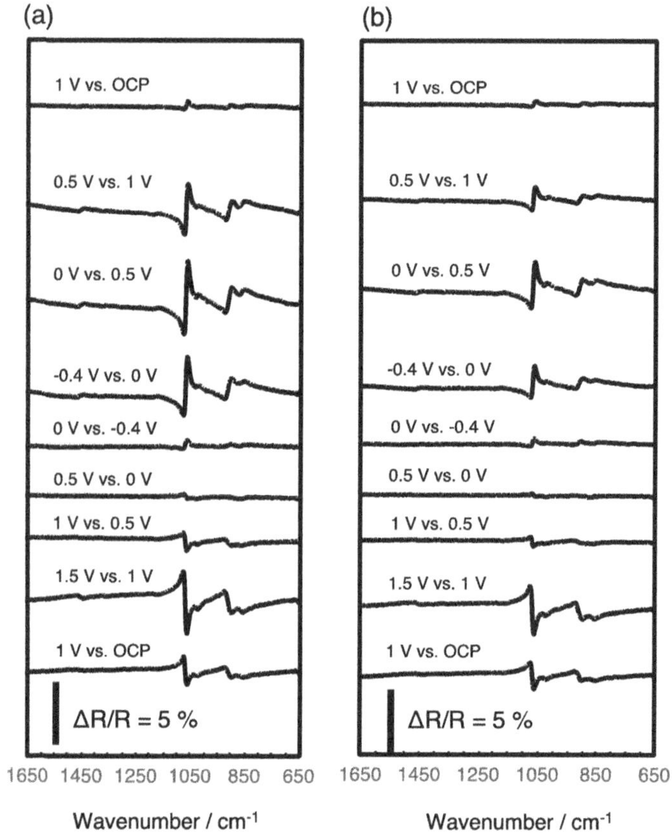

Fig. 9 In situ FTIR spectra of DCC type solution during the CV measurement: **a** 1st cycle, **b** 2nd cycle

flat SNIFTIR spectra suggest the formation of a stable surface layer during the 1st cathodic scan. Furthermore, the SNIFTIR spectra for the 2nd CV measurement shows even more flat spectra both cathodic and anodic sweep as shown in Fig. 10b. We think the surface layer on the Pt thin film electrode during the 1st cathodic scan is not like the surface layer so-called "solid electrolyte interphase (SEI) layer" but a "passivation layer," because the Pt thin film electrode seems to be mostly inactive in the following electrochemical process. Also, most of the peaks observed in the SNIFTIR spectra during the 1st cathodic scan are corresponding to the $TFSA^-$ anion, thus we think the decomposed species passivates the Pt electrode.

We also performed XPS analyses to characterize the passivation layer. To obtain depth profiles of the surface layer on magnesium metal, an argon ion sputtering was subsequently carried out after each measurement. The XPS spectra of a magnesium foil immersed in the AEC solution and the $Mg(TFSA)_2$ solution are shown in Fig. 11. A broad peak is observed in the Cl 2p spectra at 199 eV, suggesting that a magnesium

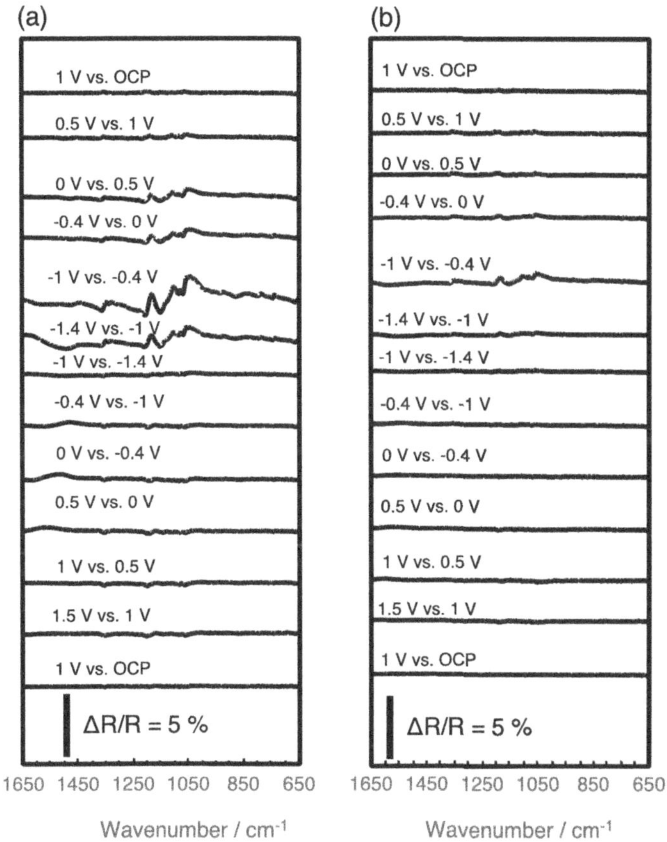

Fig. 10 In situ FTIR spectra of $Mg(TFSA)_2$-based solution during the CV measurement: **a** 1st cycle, b 2nd cycle

dimer complex formed in the EAC electrolyte adsorbed at the surface of the magnesium as shown in Fig. 11a. Since the amount of the remaining Cl species is minimal, we think that the XPS spectra do not show clear evidence of the decomposition of the EAC electrolyte at the surface of the magnesium metal. Figure 11b shows the F1s XPS spectra of a magnesium foil immersed in the $Mg(TFSA)_2$ solution. The F 1s XPS spectra have a peak corresponding to the $TFSA^-$ anion residue at 689.5 eV. During the argon sputtering process, the peak at 689.5 eV gradually disappeared, and another peak assigned to MgF_2 appeared at 686.2 eV. The formation of MgF_2 clearly shows the reduction of the $TFSA^-$ anion at the surface of the magnesium metal. The decomposition of the electrolyte also hinders the crystal growth of the magnesium metal during the deposition process; as a result, the deposited magnesium has poor crystallinity. Furthermore, the reduction also leads to the poor coulombic efficiency of the electrodeposition process. Therefore, we need to suppress the decomposition

Table 3 Summary of the peak assignments of the in situ FTIR spectra for the 0.5 M $Mg(TFSA)_2$ in BuMeG3 solution during the 1st cathodic scan

cm^{-1}	
	Positive Peaks
1346	O=S=O bending vibration of the TFSA anion
1328	O=S=O bending vibration of the TFSA anion
1182	Antisymmetric CF_3 stretching vibration of the TFSA anion
1100	C–O–C antisymmetric stretching vibration of the BuMeG3
1051	S–N–S stretching vibration of the TFSA anion
1041	O=S=O symmetric stretching vibration of the TFSA anion
979	C–O–C symmetric stretching vibration of the BuMeG3
919	O=S=O symmetric stretching vibration of the TFSA anion
782	C–S stretching vibration of the TFSA anion
759	CF_3 bending vibration of the TFSA anion
736	CH_2 rocking vibration of the BuMeG3
	Negative Peaks
1361	O=S=O bending vibration of the decomposed TFSA anion
1209	Antisymmetric CF_3 stretching vibration of the decomposed TFSA anion
1149	C–O–C antisymmetric stretching vibration of the adsorbed/decomposed BuMeG3
1083	S–N–S stretching vibration of the decomposed TFSA anion
1066	O=S=O symmetric stretching vibration of the decomposed TFSA anion

of the anion to utilize the reversible deposition/dissolution of magnesium metal as a battery anode.

A theoretical study by Rajput et al. predicts that the anodic and cathodic stability of anions [26]. In this work, the $TFSA^-$ anion shows excellent cathodic/anodic stability compared with other anions. However, once the $TFSA^-$ anions and Mg^{2+} cations form contact ion pairs (CIP): $[Mg^{2+}TFSA^-]^+$, the dissociation energy of the C-S bond in the $TFSA^-$ anion significantly drops via the formation of partially reduced CIP: $[Mg^+TFSA^-]$. The results show that the CIP is easily reduced during the electrodeposition process. The work proposes that the solvation structure needs to be considered to design electrolyte solutions with high cathodic stability against magnesium metal.

The surface passivation of magnesium metal anode is also affected by the small amount of water in the electrolyte solution. Connell et al. conducted a series of studies concerning the passivation layer formation [27]. They report that even the trace levels of H_2O ($\leq$3 ppm) can affect the kinetics of the magnesium deposition/dissolution process in the $Mg(TFSA)_2$-glyme system. Moreover, the passivation layer is simultaneously formed on the electrodeposited magnesium metal. They also found that the formation of the $Mg–Cl^+$(ad.) or $MgCl_2$ layer suppresses the passivation of the magnesium metal by H_2O molecules.

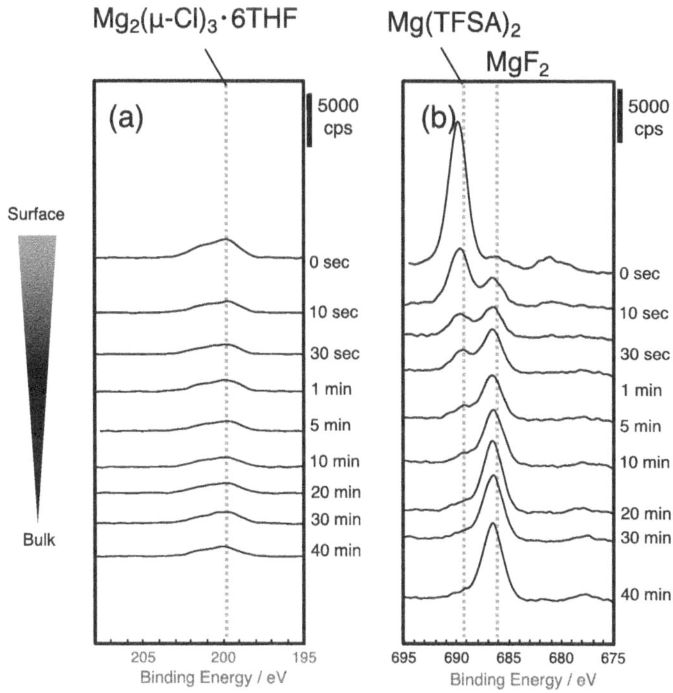

Fig. 11 XPS spectra for magnesium metal anodes. **a** Cl2p XPS spectra of a magnesium metal immersed in DCC type solution, **b** F1s XPS spectra for a magnesium metal immersed in the $Mg(TFSA)_2$ solution Reprinted with permission from ref. [20] 2017. Copyright 2017, The Electrochemical Society

Cathodic decomposition of solvent molecules is another trigger for the passivation process. Lowe and Siegel investigated possible reaction pathways of the reduction process of DME molecules at the surface of magnesium metal using first-principles calculations [28]. They proved highly exothermic and rapid decomposition of DME molecules at the surface of the magnesium metal with the evolution of ethylene gas. On the other hand, the surface of the ionic MgO and $MgCl_2$ has a limited impact on solvent decomposition. It was claimed there is some possibility that the $MgCl_2$ is integrated into the SEI layer.

Again, the formation of the passivation layer at the surface of magnesium metal is a critical issue as an overall direction for the development of the electrode/electrolyte interphase is focusing on "passivation-free" system. The naked surface of the magnesium metal is, however, highly reactive against the electrolyte solutions. Therefore, further materials design for the controlled formation of the interphase may be crucial for the development of the rechargeable magnesium batteries.

5 Intermetallic Anodes

The problems of the passivation layer of magnesium metal anodes have initiated another research direction toward "Mg-ion" batteries using alternate anode active materials. As already discussed in the previous section, the magnesium metal easily forms the passivation layer in most of the conventional electrolyte solutions, hence the intermetallic compounds become as alternative anode active materials for the Mg-ion batteries. Arthur et al. firstly reported that the electrodeposited $Bi_{1-x}Sb_x$ alloys show reversible magnesiation/demagnesiation process in half-cell tests [29]. The alloys, except pure Sb, showed excellent cyclability at 1C rate. Furthermore, the Mg_3Bi_2 intermetallic anode shows reversible magnesiation/demagnesiation in a conventional electrolyte solution as shown in Fig. 12. The electrolyte solution: 1.0 mol L^{-1} $Mg(TFSA)_2$ in acetonitrile solution, does not show reversible deposition/dissolution of the magnesium metal. Of course, the Mg_3Bi_2 anode sacrifices the energy density of the battery, due to its high equilibrium potential of 0.28 V versus Mg, and its low specific capacity of 385 mAh g^{-1}/Bi. The Mg_3Bi_2, however, still exhibits a high volumetric capacity of 1906 mAh cm^{-3}/Mg_3Bi_2 due to the high

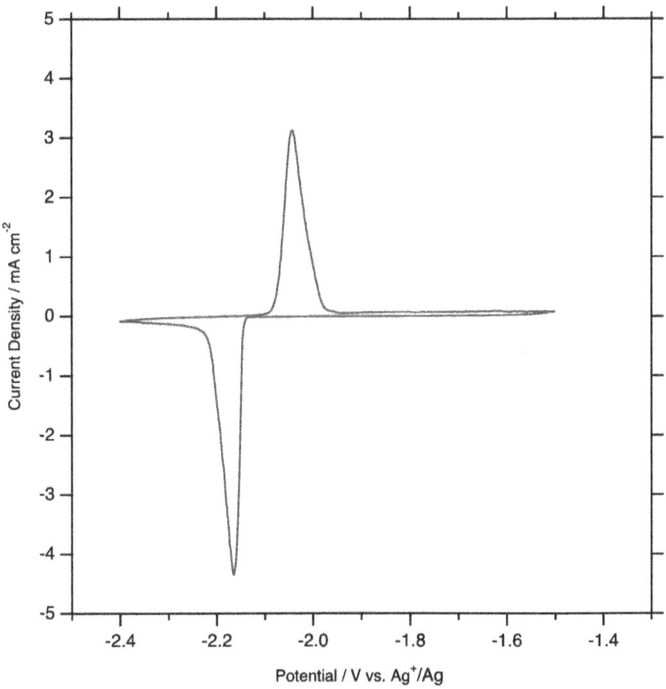

Fig. 12 A typical cyclic voltammogram of the Mg_3Bi_2 electrode in 1.0 mol L^{-1} $Mg(TFSA)_2$ in acetonitrile solution taken by author's group

atomic weight of bismuth. The volumetric capacity is comparable with lithium metal (2000 mAh g^{-1}).

As usual, the volume expansion/shrinkage of the alloy/intermetallic anodes is a critical issue for the reversibility of the batteries, because of the pulverization of the anode particles. The Mg_3Bi_2 shows volume expansion of 196% accompanied by the magnesiation process; thus a particle size control improves the capacity retention during the charge/discharge cycle. Shao et al. synthesized Bi nanotubes and confirmed the improvement in rate capability and capacity retention [30]. The TEM image after the discharging process shows that nano-sized domains ≤10 nm are formed during the initial magnesiation process, without losing the electrical contact. The reduced diffusion length contributes to the better rate capability of the electrode. The reaction mechanism of the Mg_3Bi_2 intermetallic anode during the electrochemical magnesiation/demagnesiation process was studied by Murgia et al. [31]. A typical two-phase reaction, which is normally predicted from the binary phase diagram of the Bi–Mg system, was confirmed by GITT analyses and in operando XRD as shown in Fig. 13. Recent DFT calculation studies by Jung and Han indicate a low migration barrier of the Mg^{2+} ions in the Mg_3Bi_2 solid phase [32]. Between two neighboring tetrahedral sites, the migration barrier of the Mg^{2+} ion is 0.32 eV, while it is 0.80 eV between two octahedral sites. Lee et al. reported that the migration barrier of the

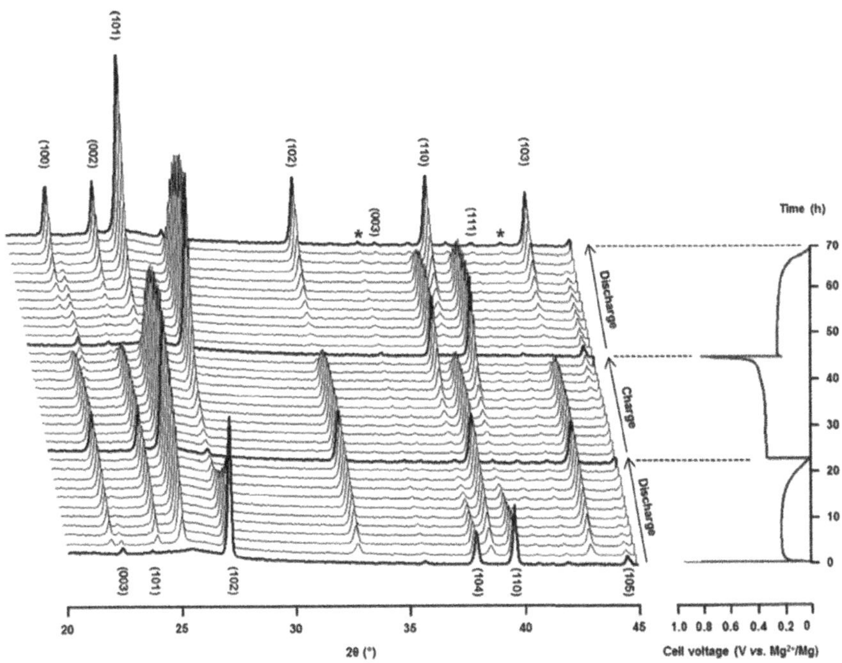

Fig. 13 In operando XRD patterns during the first magnesiation/demagnesiation of a Bi/Mg half-cell. The clear biphasic reaction is directly confirmed Reprinted with permission from ref Murgia et al. 2015 [31, 37]. Copyright 2015, The Royal Society of Chemistry

Mg^{2+} ions between the 1*a* octahedral site and the 2*d* tetrahedral site is also comparably low: 0.34 eV [33]. It shows the anisotropic diffusion pathway along (100) plane through the tetrahedral 2*d* sites of the Mg_3Bi_2 is preferably formed.

The biggest advantage of the intermetallic anode is compatibility against conventional electrolyte solutions, which typically form a passivation layer on magnesium metal. We fabricated Mg_3Bi_2 thin film electrodes to investigate the correlation between the electrochemical properties and the surface layer [34]. Figure 14 shows cyclic voltammograms of the Mg_3Bi_2 thin film electrodes in various electrolyte solutions: 1.0 mol L^{-1} $Mg(TFSA)_2$ in AN, 0.5 mol L^{-1} $Mg(TFSA)_2$ in BuMeG3 and 0.5 mol L^{-1} $Mg(TFSA)_2$ in dimethoxyethane: G1 (Kishida). The Mg_3Bi_2 thin film shows reversible redox reaction in all three electrolyte solutions.

The excellent reversibility of the Mg_3Bi_2 thin films in the three electrolyte solutions indicates that the surface of the Mg_3Bi_2 electrodes is not passivated in these electrolyte solutions. Then we performed XPS analyses of the Mg_3Bi_2 and the magnesium metal thin film immersed in the electrolyte solution to confirm the absence of the passivation layer. Despite our expectations, both of the thin films showed the formation of MgF_2. The F1s XPS spectra shown in Fig. 15 have peaks at 688.8 eV, corresponding to the residue of the TFSA anion at the surface of the electrodes. The TFSA residue gradually decreased and mostly disappeared after 10 s of the argon sputtering. Another peak corresponding to the MgF_2 simultaneously appeared at 685.7 eV during the further sputtering process. The formation of the MgF_2 layer is a direct evidence of the cathodic decomposition of the TFSA anion in the electrolyte solution. Even though the relatively weak peak intensity of the MgF_2 on the Mg_3Bi_2 suggests the formation of a thinner or less dense surface layer, the formation of MgF_2 proves that the equilibrium potential of the Mg_3Bi_2 (0.28 V versus Mg^{2+}/Mg) is still low enough to reduce the TFSA anion.

The Mg_3Bi_2 thin film electrode shows 100 mV of overpotential in the BuMeG3 solution, even with the formation of the MgF_2 layer. Moreover, the similarity of the surface layer on the Mg_3Bi_2 and magnesium metal thin films suggests that the MgF_2

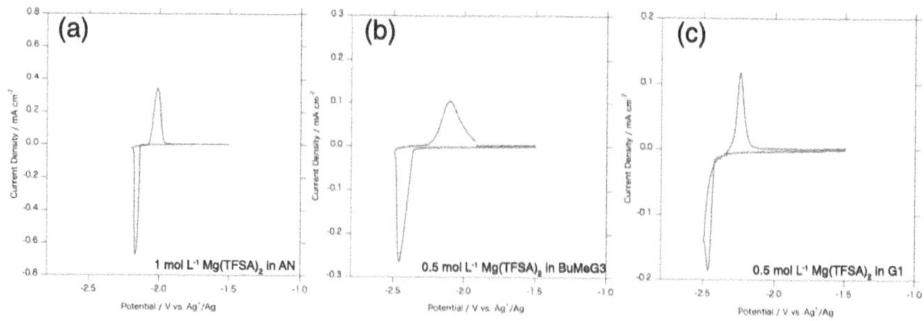

Fig. 14 Cyclic voltammograms of the Mg_3Bi_2 thin film electrodes in **a** 1 mol L^{-1} $Mg(TFSA)_2$ in AN solution, **b** 0.5 mol L^{-1} $Mg(TFSA)_2$ in BuMeG3 solution, and **c** 0.5 mol L^{-1} $Mg(TFSA)_2$ in G1 solution. Reference electrode is a silver wire immersed in 0.01 mol L^{-1} $AgNO_3$ AN solution for (**a**) and 0.01 mol L^{-1} AgTfO solution in BuMeG3/G1 solutions for (**b**) and (**c**). Counter electrode is Pt foil for (**a**) and Mg foil for (**b**) and (**c**). The sweep rate of the voltammetry is 0.1 mV sec^{-1}

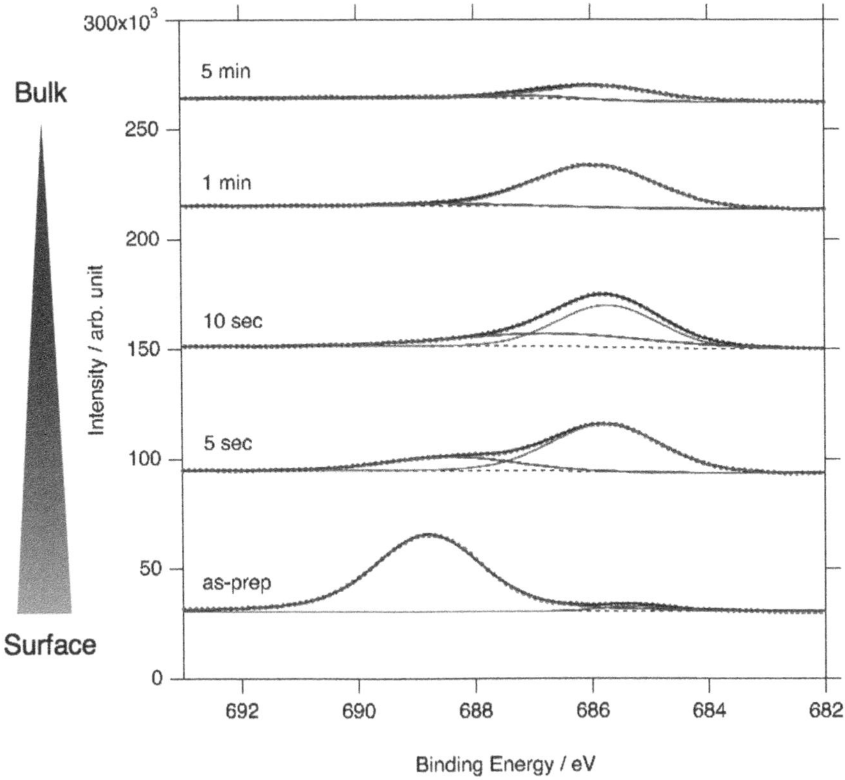

Fig. 15 F1s XPS spectra for a Mg_3Bi_2 thin film electrode immersed in 0.5 mol L^{-1} $Mg(TFSA)_2$ in BuMeG3 solution for 24 h. Reprinted with permission from ref. [34] 2019. Used under CC BY

does not completely passivate the surface of both electrodes. Therefore, we think that the origin of the high overpotential of the magnesium deposition/dissolution process in the glyme-based electrolyte solution is not the resistance of the surface layer, but the activation energy of the actual electrode reaction process, which consists of several reaction steps such as adsorption/desorption of Mg^{2+} ions, desolvation/solvation, electron-transfer, surface and bulk diffusion of Mg atoms, and so forth. Among these reaction steps, we speculate the electron-transfer process significantly contributes to the overpotential, because the biggest difference of the electrode reactions between the Mg_3Bi_2 and the magnesium metal, is the oxidation state of the Mg^{2+} ions. As presented in the Mg2p XPS spectra in Fig. 15a, the Mg^{2+} ions in the electrolyte solution are not reduced to the Mg atoms in the case of Mg_3Bi_2, on the other hand, the Mg^{2+} ions are reduced to the magnesium metal during the electrodeposition process. Further, analytical studies are necessary to be conducted to determine the origin of the overpotential.

We also performed a comparative study of the Mg_3Bi_2 and Mg_3Sb_2 synthesized by the solid-state process. Despite the same crystal structure of Mg_3Sb_2 and Mg_3Bi_2,

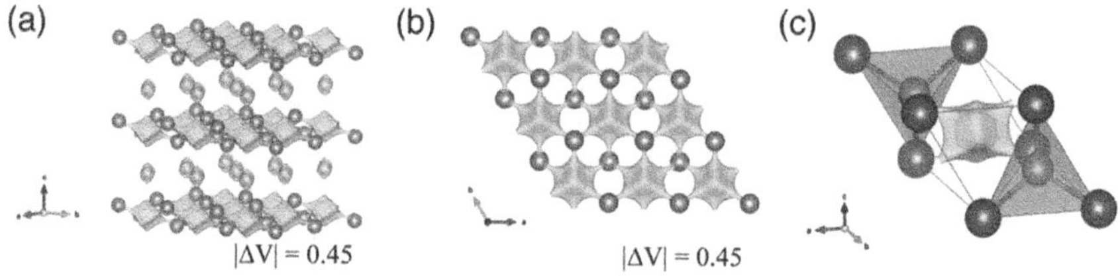

Fig. 16 BVS mapping of Mg_3Bi_2 intermetallic anode visualizing the migration pathways of Mg^{2+} ions Reprinted with permission from ref. [34] 2019. Used under CC BY

the electrochemical activity of the Mg_3Sb_2 was almost negligible in an acetonitrile-based electrolyte solution. The electrochemical inactivity of the Mg_3Sb_2 is not owing to the passivation layer, because both surfaces of the Mg_3Bi_2 and Mg_3Sb_2 are not essentially passivated even in an ambient atmosphere. Instead, the BVS mapping of the Mg_3Bi_2 and Mg_3Sb_2 indicates that relatively fast Mg^{2+} diffusion in the Mg_3Bi_2 stimulates the electrochemical activity, as shown in Fig. 16. Also, the migration pathways of the Mg^{2+} ions in the Mg_3Bi_2 are in good agreement with the DFT calculation results.

Other intermetallic anodes have also been investigated by various groups. The Mg_2Sn reported by Singh et al. has a theoretical capacity of 900 mAhg^{-1}/Sn during the magnesiation process [35]. The difficulty of the Mg_2Sn intermetallic anode is the reported sluggish rate capability and reversibility. The theoretical value of the magnesiation capacity was obtained only at 0.002 C and only 35% of the magnesium could be extracted. This is probably due to the slow kinetics of the Mg^{2+} in the inverse-fluorite structure of Mg_2Sn. On the other hand, the Mg_2Pb, which has the same inverse-fluorite structure as Mg_2Sn, shows excellent reversibility delivering ≈90% of the theoretical value [36]. The Pb/Mg half-cell exhibited a reversible capacity of approx. 450 mAh g^{-1}/Pb at 60 °C at a current density of C/50. Since the Pb is also a heavy element similar to Bi, the volumetric capacity of the Mg_2Pb is high, theoretically 2300 mAh L^{-1}. The low equilibrium potential of 121 mV versus Mg metal is also an advantage in terms of energy density. Mg-In binary alloy shows a reversible capacity of 450 mAh g^{-1} at the charge/discharge rate of C/100, reported by Murgia et al. [37]. They confirmed the expected two-phase reaction of the Mg-In binary system using in operando XRD. The Mg-In anode showed severe capacity fading at the high charge/discharge rate.

The Mg_3Bi_2, as the first investigated alternative intermetallic anode, still shows the most preferable electrochemical properties among all the intermetallic compounds. Series of studies suggest that the mobility of the Mg^{2+} in the solid phase is the key property of the intermetallic anodes, rather than the passivation layer discussed in the magnesium metal anode.

6 Summary

In this chapter, we discussed the electrochemical properties of magnesium metal and intermetallic anodes. In the early stage of the research, since few numbers of the organomagnesium compounds were available for the electrolyte solutions, the deposition mechanism of magnesium metal was only based upon the speculations. Also, it was not easy to claim the advantage of the rechargeable magnesium batteries as "beyond lithium-ion" system. During the last decade, several new classes of electrolytes improved the electrochemical properties of magnesium anode. A series of contributions from advanced spectroscopy and theoretical studies in the last decade gave deep insights into the reaction mechanism of magnesium-based anodes. Electrolyte solutions containing dimer complexes: $[Mg_2(\mu\text{-}Cl)_3]^+$ show excellent reversibility, while the practical electrochemical windows are limited by the corrosion problem. Despite the wide electrochemical window series of the glyme coordinates: $[\text{Mg-Glymes}]^{2+}$ in $Mg(TFSA)_2$-based system, CIP formation sacrifices the coulombic efficiency of the deposition/dissolution process. Even though state-of-the-art electrolyte solutions solve these problems, still, ether-based molecules are the matter of choice for the solvent. Intermetallic anode expands the choice of electrolyte solutions, with a sacrifice of energy density. We think those contributions enable us to propose a possible rechargeable magnesium battery system in near future.

References

1. Kipouros, G. J., & Sadoway, D. R. (1987). The chemistry and electrochemistry of magnesium production. In G. Mamantov, C. B. Mamantov, J. Braunstein (Eds.), *Advances in molten salt chemistry* (Vol. 6, pp 127–209). Amsterdam: Elsevier.
2. Gregory, T. D., Hoffman, R. J., & Winterton, R. C. (1990). Nonaqueous electrochemistry of magnesium: Applications to energy storage. *Journal of the Electrochemical Society, 137*(3), 775–780. https://doi.org/10.1149/1.2086553.
3. Aurbach, D., Weissman, I., Gofer, Y., & Levi, E. (2003). Nonaqueous magnesium electrochemistry and its application in secondary batteries. *Chemical Record, 3*(1), 61–73. https://doi.org/10.1002/tcr.10051.
4. Jolibois, M. P. (1912). Fomula of the organomagnesium derivative and magnesium hydride. *Comptes Rendus, 155,* 353–355.
5. Gaddum, L. W., & French, H. E. (1927). The electrolysis of grignard solutions1. *Journal of the American Chemical Society, 49*(5), 1295–1299. https://doi.org/10.1021/ja01404a020.
6. Connor, J. H., Reid, W. E., & Wood, G. B. (1957). Electrodeposition of metals from organic solutions: V. electrodeposition of magnesium and magnesium alloys. *Journal of The Electrochemical Society 104*(1), 38–41.
7. Aurbach, D., Lu, Z., Schechter, A., Gofer, Y., Gizbar, H., Turgeman, R., et al. (2000). Prototype systems for rechargeable magnesium batteries. *Nature, 407*(6805), 724–727.
8. Aurbach, D., Suresh, G. S., Levi, E., Mitelman, A., Mizrahi, O., Chusid, O., et al. (2007). Progress in rechargeable magnesium battery technology. *Advanced Materials, 19*(23), 4260–4267. https://doi.org/10.1002/adma.200701495.
9. Kim, H. S., Arthur, T. S., Allred, G. D., Zajicek, J., Newman, J. G., Rodnyansky, A. E., et al. (2011). Structure and compatibility of a magnesium electrolyte with a sulphur cathode. *Nature Communications, 2,* 427. https://doi.org/10.1038/ncomms1435.

10. Muldoon, J., Bucur, C. B., Oliver, A. G., Sugimoto, T., Matsui, M., Kim, H. S., et al. (2012). Electrolyte roadblocks to a magnesium rechargeable battery. *Energy & Environmental Science, 5*(3), 5941–5950. https://doi.org/10.1039/c2ee03029b.
11. Ha, S.-Y., Lee, Y.-W., Woo, S. W., Koo, B., Kim, J.-S., Cho, J., et al. (2014). Magnesium(II) Bis(trifluoromethane sulfonyl) imide-based electrolytes with wide electrochemical windows for rechargeable magnesium batteries. *ACS Applied Materials & Interfaces, 6*(6), 4063–4073. https://doi.org/10.1021/am405619v.
12. Fukutsuka, T., Asaka, K., Inoo, A., Yasui, R., Miyazaki, K., Abe, T., et al. (2014). New magnesium-ion conductive electrolyte solution based on triglyme for reversible magnesium metal deposition and dissolution at ambient temperature. *Chemistry Letters, 43*(11), 1788–1790. https://doi.org/10.1246/cl.140704.
13. Orikasa, Y., Masese, T., Koyama, Y., Mori, T., Hattori, M., Yamamoto, K., et al. (2014). High energy density rechargeable magnesium battery using earth-abundant and non-toxic elements. *Scientific Reports, 4,* 5622. https://doi.org/10.1038/srep05622.
14. Shterenberg, I., Salama, M., Yoo, H. D., Gofer, Y., Park, J.-B., Sun, Y.-K., et al. (2015). Evaluation of (CF3SO2)2N–(TFSI) based electrolyte solutions for Mg batteries. *Journal of the Electrochemical Society, 162*(13), A7118–A7128. https://doi.org/10.1149/2.0161513jes.
15. Tutusaus, O., Mohtadi, R., Arthur, T. S., Mizuno, F., Nelson, E. G., & Sevryugina, Y. V. (2015). An efficient halogen-free electrolyte for use in rechargeable magnesium batteries. *Angewandte Chemie International Edition, 54*(27), 7900–7904. https://doi.org/10.1002/anie.201412202.
16. Zhao-Karger, Z., Gil Bardaji, M. E., Fuhr, O., & Fichtner, M. (2017). A new class of non-corrosive, highly efficient electrolytes for rechargeable magnesium batteries. *Journal of Materials Chemistry A, 5*(22), 10815–10820. https://doi.org/10.1039/C7TA02237A.
17. Aurbach, D., Cohen, Y., & Moshkovich, M. (2001). The study of reversible magnesium deposition by in situ scanning tunneling microscopy. *Electrochemical and Solid-State Letters, 4*(8), A113–A116.
18. Matsui, M. (2011). Study on electrochemically deposited Mg metal. *Journal of Power Sources, 196*(16), 7048–7055. https://doi.org/10.1016/j.jpowsour.2010.11.141.
19. Ling, C., Banerjee, D., & Matsui, M. (2012). Study of the electrochemical deposition of Mg in the atomic level: Why it prefers the non-dendritic morphology. *Electrochimica Acta, 76,* 270–274. https://doi.org/10.1016/j.electacta.2012.05.001.
20. Kuwata, H., Matsui, M., & Imanishi, N. (2017). Passivation layer formation of magnesium metal negative electrodes for rechargeable magnesium batteries. *Journal of the Electrochemical Society, 164*(13), A3229–A3236. https://doi.org/10.1149/2.1191713jes.
21. Ding, M. S., Diemant, T., Behm, R. J., Passerini, S., & Giffin, G. A. (2018). Dendrite growth in Mg metal cells containing Mg(TFSI)2/Glyme electrolytes. *Journal of the Electrochemical Society, 165*(10), A1983–A1990.
22. Davidson, R., Verma, A., Santos, D., Hao, F., Fincher, C., Xiang, S., et al. (2019). Formation of magnesium dendrites during electrodeposition. *ACS Energy Letters, 4*(2), 375–376. https://doi.org/10.1021/acsenergylett.8b02470.
23. Peled, E. (1979). The electrochemical behavior of alkali and alkaline earth metals in nonaqueous battery systems—The solid electrolyte interphase model. *Journal of the Electrochemical Society, 126*(12), 2047–2051.
24. Lu, Z., Schechter, A., Moshkovich, M., & Aurbach, D. (1999). On the electrochemical behavior of magnesium electrodes in polar aprotic electrolyte solutions. *Journal of Electroanalytical Chemistry, 466*(2), 203–217. https://doi.org/10.1016/S0022-0728(99)00146-1.
25. Aurbach, D., Schechter, A., Moshkovich, M., & Cohen, Y. (2001). On the mechanisms of reversible magnesium deposition processes. *Journal of the Electrochemical Society, 148*(9), A1004–A1014.
26. Rajput, N. N., Qu, X., Sa, N., Burrell, A. K., & Persson, K. A. (2015). The coupling between stability and ion pair formation in magnesium electrolytes from first-principles quantum mechanics and classical molecular dynamics. *Journal of the American Chemical Society, 137*(9), 3411–3420. https://doi.org/10.1021/jacs.5b01004.

27. Connell, J. G., Genorio, B., Lopes, P. P., Strmcnik, D., Stamenkovic, V. R., & Markovic, N. M. (2016). Tuning the reversibility of Mg anodes via controlled surface passivation by H_2O/Cl^- in organic electrolytes. *Chemistry of Materials, 28*(22), 8268–8277. https://doi.org/10.1021/acs.chemmater.6b03227.
28. Lowe, J. S., & Siegel, D. J. (2018). Reaction pathways for solvent decomposition on magnesium anodes. *The Journal of Physical Chemistry C, 122*(20), 10714–10724. https://doi.org/10.1021/acs.jpcc.8b01752.
29. Arthur, T. S., Singh, N., & Matsui, M. (2012). Electrodeposited Bi, Sb and Bi1-xSbx alloys as anodes for Mg-ion batteries. *Electrochemistry Communications, 16*(1), 103–106. https://doi.org/10.1016/j.elecom.2011.12.010.
30. Shao, Y., Gu, M., Li, X., Nie, Z., Zuo, P., Li, G., et al. (2014). Highly reversible Mg insertion in nanostructured Bi for Mg ion batteries. *Nano Letters, 14*(1), 255–260. https://doi.org/10.1021/nl403874y.
31. Murgia, F., Stievano, L., Monconduit, L., & Berthelot, R. (2015). Insight into the electrochemical behavior of micrometric Bi and Mg_3Bi_2 as high performance negative electrodes for Mg batteries. *Journal of Materials Chemistry A, 3*(32), 16478–16485. https://doi.org/10.1039/C5TA04077A.
32. Jung, S. C., & Han, Y.-K. (2018). Fast magnesium ion transport in the Bi/Mg_3Bi_2 two-phase electrode. *The Journal of Physical Chemistry C, 122*(31), 17643–17649. https://doi.org/10.1021/acs.jpcc.8b02840.
33. Lee, J., Monserrat, B., Seymour, I. D., Liu, Z., Dutton, S. E., & Grey, C. P. (2018). An ab initio investigation on the electronic structure, defect energetics, and magnesium kinetics in Mg_3Bi_2. *Journal of Materials Chemistry A, 6*(35), 16983–16991. https://doi.org/10.1039/C7TA11181A.
34. Matsui, M., Kuwata, H., Mori, D., Imanishi, N., & Mizuhata, M. (2019). Destabilized passivation layer on magnesium-based intermetallics as potential anode active materials for magnesium ion batteries. *Frontiers in Chemistry 7*(7). https://doi.org/10.3389/fchem.2019.00007.
35. Singh, N., Arthur, T. S., Ling, C., Matsui, M., & Mizuno, F. (2013). A high energy-density tin anode for rechargeable magnesium-ion batteries. *Chemical Communications, 49*(2), 149–151. https://doi.org/10.1039/c2cc34673g.
36. Periyapperuma, K., Tran, T. T., Purcell, M. I., & Obrovac, M. N. (2015). The reversible magnesiation of Pb. *Electrochimica Acta, 165,* 162–165. https://doi.org/10.1016/j.electacta.2015.03.006.
37. Murgia, F., Weldekidan, E. T., Stievano, L., Monconduit, L., & Berthelot, R. (2015). First investigation of indium-based electrode in Mg battery. *Electrochemistry Communications, 60,* 56–59. https://doi.org/10.1016/j.elecom.2015.08.007.

Magnesium Batteries: Electrolyte

Minato Egashira

Abstract Electrolyte for magnesium secondary batteries must meet various requirements. In particular, the promotion of reversible magnesium deposition–dissolution, the negative electrode reaction, is the most important property of the electrolyte. The magnesium electrode process is not reversible and sometimes exhibits large overpotential, at the dissolution (discharge) reaction in conventional electrolyte systems, mainly due to the existence of surface passivation layer. Some magnesium halides soluble in non-aqueous solvents, such as Grignard reagent and its mixture with Lewis acid, promote reversible magnesium deposition without overpotential via multinuclear magnesium halide complex (μ-complex), although their insufficient anodic stability and corrosive nature inhibit their use in practical battery systems. Recent progress is on the understanding of solution structure of non-aqueous magnesium salt electrolyte. As a result, a solvent-separated ion pair (SSIP) between magnesium ion and glyme solvent molecules is more favorable compared with a contact ion pair. Magnesium salts with some boron-based anions effectively provide SSIP with weakly coordinating glyme solvents, and thus promote reversible magnesium electrode process without overpotential.

Keywords Reversible magnesium deposition · Magnesium halide · Solvent-separated ion pair

Development of practical magnesium battery electrolyte has some difficulties. Comparison of magnesium electrolyte with lithium battery electrolyte is effective to understand some difficulties associated with offering a suitable electrolyte for magnesium batteries. Presently, primary and secondary lithium batteries use non-aqueous electrolyte of various kinds with a lithium salt dissolved in an organic solvent. Some lithium salts having large and charge-delocalized anions such as $LiPF_6$ are easily dissolved in various organic solvents such as alkyl carbonates. One can readily find a combination of a lithium salt and a solvent such that properties of a cell can be fabricated. The deposition and dissolution of lithium occurs reversibly in various

M. Egashira (✉)
Nihon University, Fujisawa, Japan
e-mail: egashira.minato@nihon-u.ac.jp

K. Kanamura (ed.), *Next Generation Batteries*,
https://doi.org/10.1007/978-981-33-6668-8_48

non-aqueous electrolytes even though the equilibrium potential of lithium deposition–dissolution is more negative than the stability limit of most solvents. The initial decomposition product of electrolyte precipitates on the electrode surface and acts as a protective film against further electrolyte decomposition, with promotion of lithium electrode processes by the lithium-ion mobility in the film. One must keep in mind that such a mechanism is available only for lithium with its monovalent nature.

For magnesium, the situation is entirely different. A divalent magnesium ion delivers stronger Coulombic attraction against an anion. Therefore, a magnesium salt has lower solubility in conventional aprotic solvents than the corresponding lithium salt. In many cases, poor reversibility of magnesium deposition–dissolution processes in the electrolyte occurs even if one is lucky enough to find a combination of an aprotic solvent and a magnesium salt with sufficient solubility. In most conventional magnesium salt–aprotic solvent electrolytes, the electrochemical deposition–dissolution of magnesium exhibits overpotential as large as 2 V [1, 2]. The surface of magnesium metal, similar to lithium metal, is known to be covered by the passivation layer containing oxidation products of magnesium, MgO, $Mg(OH)_2$, $MgCO_3$, and so on [3, 4]. The oxidation of the surface is unavoidable because of the low standard potential of magnesium. Unlike lithium, the transport of magnesium ion in the passivation layer is strongly constrained. Consequently, the passivation layer is an insulator not only for electrons but also for magnesium ions. Therefore, the surface passivation layer inhibits electrochemical deposition and dissolution of magnesium.

Even under such severe circumstances, several series of electrolyte solutions reportedly exhibit reversible magnesium deposition–dissolution with comparably low overpotential. An early finding is an ethereal solution of alkylmagnesium halide complex, which is known as a Grignard reagent and which is familiar in the area of organic synthesis, for the conversion of carboxy group to alcohol with an additional alkyl group. A Grignard reagent has generally been synthesized via reaction of magnesium metal and a corresponding alkyl halide in ether solution. The reaction occurs chemically without an external circuit. However, one can readily imagine that magnesium metal can be deposited electrochemically in such a solution. At the early stage, electrochemical magnesium deposition was confirmed in a Grignard reagent electrolyte [5]. Although Gregory et al. demonstrated clearly that a Grignard reagent is unstable toward positive electrode with higher potential than 2 V [2], Liebenow claimed that Grignard reagent is still a candidate for electrolyte because of the reversible magnesium deposition without overpotential using a proper substrate [6]. Aurbach et al. intensively investigated electrolytes used for magnesium batteries with modified electrolyte of alkylmagnesium halide complex. To enhance the electrochemical stability of Grignard reagents, the addition of a Lewis acid such as aluminum chloride $AlCl_3$ or aluminum alkylchlorides appears to be effective [7–9]. Mixed electrolytes of Grignard reagents RMgCl/THF with Lewis acids $AlR_{2n}Cl_{3-n}$ have been recognized as aluminum chloride complex (ACC) electrolytes. The ACC series materials not only have a more stable nature; they also exhibit more reversible magnesium deposition–dissolution. Aurbach et al. displayed that a cell having Mg/Chevrel phase Mo_6S_8 with ACC electrolyte works with reversible capacity of ca. 80 mAh g^{-1} by

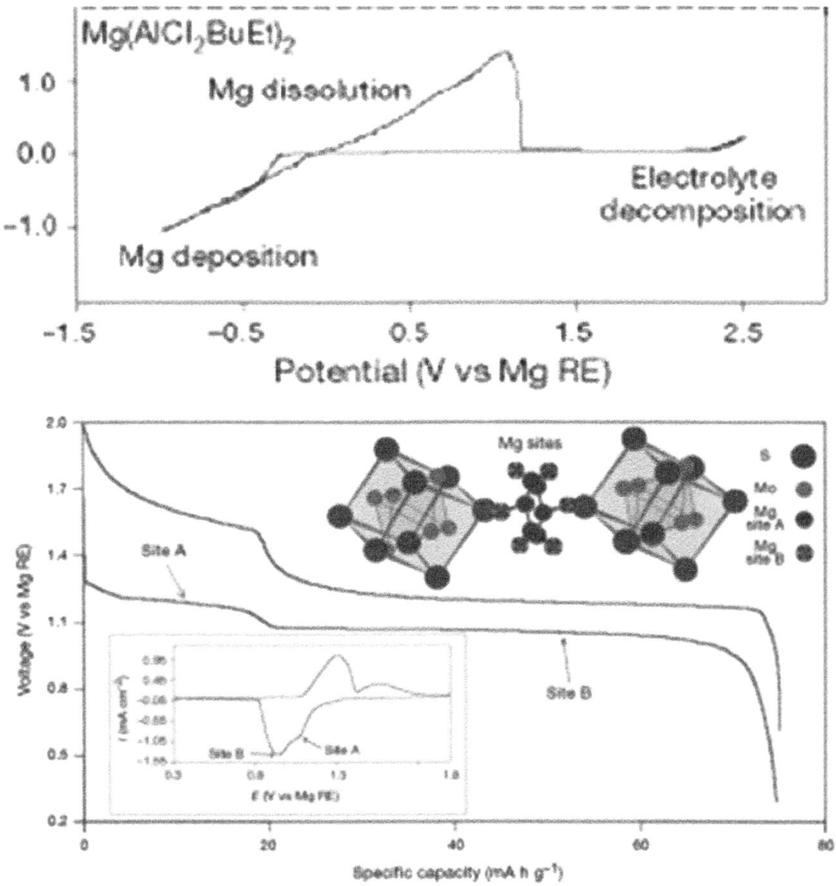

Fig. 1 (Upper) Cyclic voltammogram for magnesium deposition–dissolution in the ACC electrolyte; (lower) Charge–discharge profile of Chevrel phase Mo_6S_8 in the ACC electrolyte (from Ref. [7])

the mass of Mo_6S_8 and cell voltage of 2.4 V [7]. The typical cyclic voltammogram for magnesium deposition–dissolution and the charge–discharge profile of Mo_6S_8 in ACC electrolyte is shown in Fig. 1.

Grignard reagents are known to dissociate by the following Schlenk equilibrium.

$$2RMgCl \rightleftarrows R_2Mg + MgCl_2 \tag{1}$$

Together with Schlenk-type equilibrium (Eq. 1), the following reactions have been expected to occur.

$$2RMgCl \rightleftarrows R_2MgCl^- + MgCl^+ \tag{2}$$

$$MgCl^{+} + MgCl_2 \rightleftarrows Mg_2Cl_3^{+} \quad (3)$$

In case a Lewis acid $AlCl_3$ coexists, the following transmetallic reaction is also expected.

$$RMgCl + AlCl_3 \rightleftarrows MgCl^{+} + RAlCl_3^{-} \quad (4)$$

As a result of these reactions being forwarded to the right side, a multi-nuclear Mg–Cl complex ion $[Mg_2Cl_3(solvent)_6]^+$, denoted as μ-complex, can be formed. For most of the effective electrolytes against reversible magnesium deposition, the μ-complex has been observed in recrystallized salts. Furthermore, the μ-complex cation has been detected in some solutions [8, 10–13]. Therefore, this μ-complex has been assumed as an active species for the magnesium deposition. According to this widely accepted hypothesis, significant content of halide ion is the requirement for smooth deposition–dissolution of magnesium. Magnesium chloride $MgCl_2$ itself has a low solubility to aprotic solvent. Therefore, such active species can work only in a few electrolyte solutions. Grignard reagents can be assumed as concentrated aprotic solutions of magnesium halide complex salts. By the addition of Lewis acids, the Schlenk equilibrium is changed. Then the formation of $MgCl^+$, followed by the subsequent formation of multi-nuclear magnesium halide salt, is accelerated.

$$MgCl_2 + AlCl_3 \rightleftarrows MgCl^{+} + AlCl_4^{-} \quad (5)$$

Magnesium chloride complexes having organic ligands of several kinds, such as alkoxide and hexamethyldisilyl (HMDS), similarly promote smooth magnesium deposition–dissolution by providing moderate concentration [14–16]. Attractive all-inorganic magnesium halide solutions have been proposed recently, such as $MgCl_2$–$AlCl_3$ [17, 18], $MgCl_2$-Lewis acidic halides [18, 19], magnesium bis(trifluoromethane sulfonyl)amide ($Mg(TFSA)_2$) mixed with a chloride $MgCl_2$, $AlCl_3$, or $Al(C_2H_5)Cl_2$ in ethers such as THF and DME [20, 21]. These all-inorganic salt solutions are expected to be more chemically and electrochemically stable than organic-magnesium halide complex salts. However, some of these electrolytes require several tens of cycles to provide high Coulombic efficiency, perhaps for the rearrangement of magnesium–aluminum chloride complex.

Detailed magnesium deposition–dissolution mechanisms in these halide-containing aprotic electrolytes have been investigated using spectroscopic methods and using computational calculation [8, 11, 13, 22–26]. Multi-nuclear complex ions including μ-complex can be formed in a concentrated solution of Mg^{2+} and Cl^-. Some positively charged cations including the μ-complex can adsorb on the surface of a negatively charged electrode. By virtue of the charge-transfer reaction between a magnesium-containing cation and electrode, zero-valent magnesium can be isolated from the cluster ion and the ion can change into a neutral cluster. The resultant neutral cluster can participate in other equilibria of cluster ion formation. The deposited magnesium can be dissolved after the oxidation together with the inclusion of this magnesium species in a neutral Mg–Cl cluster and its conversion into monovalent cluster ions. This kind of Mg–Cl equilibrium can realize successive charge transfer of divalent magnesium. In addition, the hardness of magnesium cation as a Lewis acid is expected to promote the formation of such cluster ions, which might enable the reversible deposition–dissolution of magnesium without surface passivation.

Although the magnesium deposition–dissolution is possible with rather small overpotential in such chloride-containing ethereal electrolyte, the reaction is rather slow for battery applications [2, 11]. Such slow rate is observable in the cyclic voltammogram, as the steeper slope of magnesium deposition–dissolution couple compared to that of the lithium analogue. Matsui demonstrated that such a slow rate originates from the low rate of the growth of the deposit, in relation to the morphology of the deposit [11]. In addition, for practical application of chloride-containing electrolytes, the reactions on the side of a positive electrode of chloride ion become large barriers. Not only the intrinsic anodic reaction of chloride ion, but also the corrosion of a current collector and a package material has been reported as a difficulty preventing such electrolytes from practical usage as a high-voltage battery electrolyte [27, 28].

Alternative choices must be done as the combination from bulky magnesium salts with low dielectric solvents. The first report in 2014 concerned the electrolyte solution of $Mg(TFSA)_2$ with tetraglyme solvent [29, 30]. Glymes have weak Coulombic attraction with magnesium ion, a strong Lewis acid, while retaining strong coordination by multi-dentate coordination nature. The unique coordination facilitates the release of solvent from a solvated magnesium ion. The charge-delocalized TFSA anion and multi-dentate coordination nature of glymes help to improve the solubility in glyme solvent to the hyperconcentrated region, similarly to the well-known hyperconcentrated lithium salt solution proposed by Watanabe et al. as "solvated lithium ionic liquid" [31, 32]. Although the intrinsic stability of glymes toward oxidation is insufficient for practical use, the anodic stability is also expected to be improved to around 3 V from magnesium negative electrode by a hyperconcentrated solution. The typical cyclic voltammogram for magnesium deposition–dissolution at ambient temperature in $Mg(TFSA)_2$/triglyme electrolyte is shown in Fig. 2. As this voltammogram shows, magnesium deposition and dissolution can be promoted in such electrolyte.

The magnesium deposition–dissolution properties differ markedly according to the kind of glyme. Among glyme family members, monoglyme and diglyme have been proposed as better choices to promote rather reversible and low overpotential [33]. In the solution of glyme solvent, magnesium and TFSA ions form various ion pairs depending on the salt concentration and the kind of glyme solvent. The configuration of such ion pairs has been investigated using Raman spectroscopy and using computational simulation [34, 35]. Such ion-solvent clusters are classified into contact ion pairs (CIP), solvent-separated ion pairs (SSIP), and aggregates (AGG). Among the ion pairs, SSIP is regarded as favorable for reversible magnesium deposition. For lithium batteries, de-solvation from lithium ion is a rate-determining step in lithium ion intercalation. For magnesium batteries, the solvation must also be important on the overall electrode processes. SSIP might be beneficial in terms of

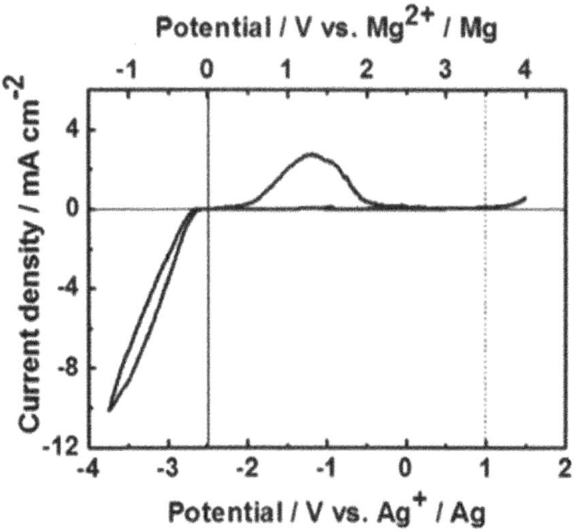

Fig. 2 Cyclic voltammogram for magnesium deposition–dissolution in $Mg(TFSA)_2$/triglyme electrolyte Reprinted from Ref. [30] by The Authors licensed under CC BY 4.0. (https://www.nature.com/articles/srep05622)

the de-solvation energy. Therefore, the overpotential depends strongly on the kind of glyme solvent and on the salt concentration.

However, the overpotential between deposition and dissolution is considerable. The oxidation current is far smaller than the reduction current, suggesting that the negative electrode process in this electrolyte is not reversible because of the by-reaction together with the magnesium deposition. The reductive decomposition of TFSA anion has been discussed as a by-reaction with magnesium deposition [36, 37]. Such insufficient stability of TFSA anion is unexpected from research experiences of lithium-ion batteries, for which LiTFSA shows good compatibility toward negative electrodes because of the existence of the surface protective film SEI on negative electrodes of lithium case and because of the stronger electrophilicity of magnesium. This is an example to display the difficulty of material selection of the magnesium battery electrolyte.

An alternative electrolyte system that is anticipated for practical magnesium batteries is an ethereal solution of boron hydride, organoborate, and boron-containing anion–magnesium salt. Gregory et al. attempted to dissolve several magnesium organoborates in THF. They found that magnesium dibutyldiphenylborate ($Mg(B(Bu_2Ph_2))_2$) has applicable solubility [2]. The electrolytic solution of $Mg(B(Bu_2Ph_2))_2$ in THF or mixed solvent THF/dimethoxyethane (DME) showed reversible magnesium deposition with approx. 0.3 V overpotential and compatibility in a Mg/Co_2O_3 cell. However, the anodic stability of this borate salt remains insufficient for 3 V batteries. A proposal was made in 2012 as the addition of magnesium borohydride $Mg(BH_4)_2$ in a $Mg(TFSA)_2$/glyme electrolyte system [38]. Although the addition of $Mg(BH_4)_2$ improves reversibility of the deposition–dissolution of magnesium with decreasing overpotential, $Mg(BH_4)_2$ entails several difficulties such as poor solubility in glyme solvents and high reactivity toward positive electrode and other cell components. Therefore, magnesium salts with boron-centered anions having a similar effect to that of $Mg(BH_4)_2$ have been prepared and examined for magnesium battery electrolytes. Among such attempts, a series of magnesium salts having carborane, cage-shape carbon–boron clusters shows interesting electrolyte behavior with markedly low magnesium deposition overpotential [39–41]. Recently a magnesium salt with tetrakis(hexafluoroisopropyloxy)borate anion has been proposed to provide almost ideal electrolyte properties, a stable potential region from magnesium metal to 4.5 V, and high conductivity [42]. These boron-based anions have weak coordination ability toward magnesium cation. Consequently, they are expected to provide SSIP. Boron-based salt family is anticipated for use as a mainstream of 3 V class magnesium battery electrolyte. An unavoidable difficulty related to their practical application is to find low-cost and abundant preparation methods of somewhat complicated structure of anion.

As described above, an electrolyte candidate must satisfy many requirements for use in 3 V class batteries. Consequently, the summary above becomes complicated against the author's intention. To index the achievements of the respective electrolyte candidates clearly, Song et al. have proposed a radar chart organized with six parameters: compatibility with magnesium negative electrode, anodic stability, stability toward current collector corrosion, compatibility toward higher voltage positive electrodes such as metal oxide, compatibility toward lower voltage positive electrodes such as Chevrel phase, and compatibility toward high-capacity conversion electrodes such as sulfur [43]. Such a radar chart is convenient for gaining a visual understanding of the achievement of each electrolyte. However, there must be some priority among these parameters. Electrolytes that are not compatible with reversible magnesium deposition cannot be applied to practical battery systems. For that reason, they must be omitted from electrolyte candidates. Inspired by the proposal of Song et al., properties of pentagonal radar charts are described here for a couple of electrolyte candidates that provide reversible magnesium deposition with low over-potential with five parameters, compatibility with Chevrel electrode, compatibility with high voltage electrode (high potential durability), stability toward corrosion of current collector, chemical and thermal stability, and material cost/facility of preparation with three digits. The expression of the pentagonal radar chart for Grignard reagent CH_3MgCl/THF and a carborane salt electrolyte is shown in Fig. 3.

The team ALCA-SPRING has attempted evaluation of the magnesium battery electrolyte for confirmation of the "standard" electrolyte, which is useful for the evaluation of magnesium battery materials. The strategy at the initial stage is the modification of candidates of two kinds: Grignard reagent RMgCl/THF and a magnesium salt–ether system, which were assumable sets in 2013. Starting from Grignard reagent, attempts have been made to improve its chemical stability by modification of the ligand structure. This attempt has been made, for example, by mixing designed ionic liquids in a Grignard reagent [44]. Magnesium chloride complexes having imide or alkoxy ligand are known to have higher stability than alkyl analogues because of the highly negative charge on the oxygen atom connected to magnesium [16], which is

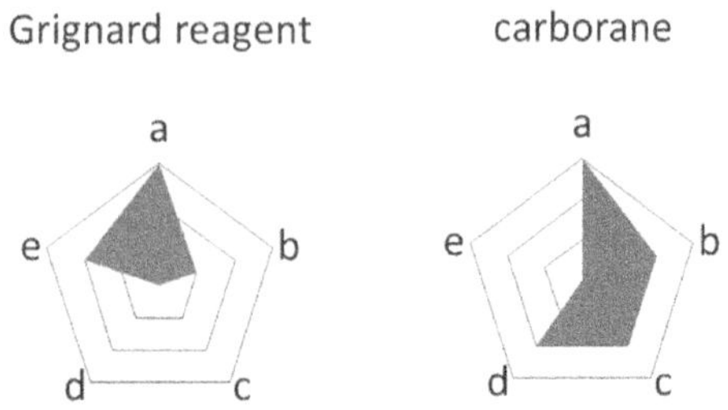

a. compatibility with Chevrel electrode
b. compatibility with high voltage electrode/ high voltage durability
c. stability toward corrosion of current collector
d. c hemical/ thermal stability
e. material cost/ facility of preparation

Fig. 3 Example pentagonal radar charts for magnesium battery electrolyte evaluation

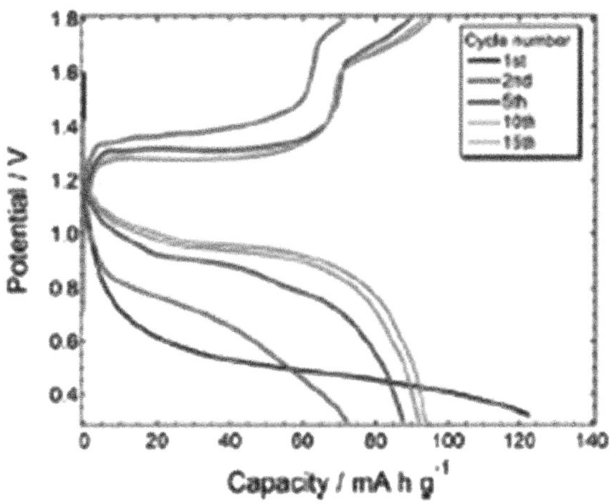

Fig. 4 Charge–discharge profile of Chevrel phase Mo_6S_8 in the ROMgCl-$Mg(TFSA)_2$/triglyme mixed electrolyte (from Ref. [45])

stable as anion and which therefore converts to magnesium halide species by equilibrium. Instead, alkoxy-magnesium chloride complex suffers from low solubility into glyme solvent. Alkoxy ligand has been modified for increased solubility. A novel electrolyte solution has been proposed by containing such modified magnesium chloride salts as additive to $Mg(TFSA)_2$/triglyme system [45]. Electrolyte solutions of this kind provide reversible magnesium deposition without overpotential, sufficient chemical stability, and compatibility toward Chevrel phase Mo_6S_8 to the reversible capacity of 90 mAh g^{-1} with a voltage plateau at 1.0 V at 373 K, as shown in Fig. 4. This electrolyte delivers reversible magnesium deposition with small overpotential. Moreover, it has sufficient chemical stability. It is therefore assumed as a standard

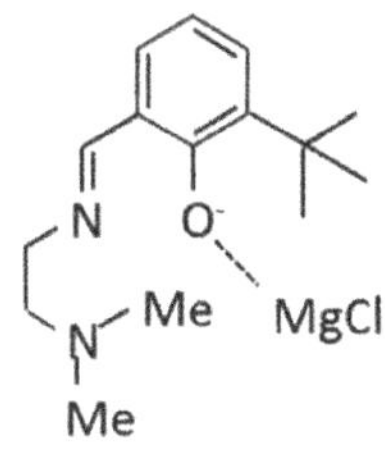

Fig. 5 Structure of phenoxyimine ligand L1

electrolyte for the development of magnesium battery materials, even though its low solubility limits the operation temperature to temperatures higher than 350 K.

The alcoxy-magnesium chloride complex solubility is rather low at ambient temperatures, but it has chemical stability. Further modification of ligand has been made with the strategy of a multi-dentate configuration. As a novel multi-dentate ligand, a functionalized phenoxyimine shown in Fig. 5 has been proposed. In this molecular structure, heteroatoms (O, N) bridged by rigid phenolic backbone act as a coordination site. Such multi-dentate coordination improves ambient-temperature solubility of the complex [46]. The phenoxyimine-MgCl complex salt, hereinafter denoted as L1MgCl, is prepared easily by mixing CH_3MgCl and corresponding iminophenol in THF solution. XRD study revealed that aggregate cations and anions with coordinated solvent molecule are formed in the crystal of the L1MgCl. Particularly, the aggregate cation, consisting of multiple magnesium ion, chloride ion, and THF molecule somewhat, resembles μ-complex $[Mg_2Cl_3(THF)_3]^+$. Such multinuclear cations are expected to form also in concentrated solution. The L1MgCl salt is stable against atmospheric oxygen and moisture, both at ambient temperature and at around 330 K. It is soluble in triglyme solvent up to 1 mol dm^{-3}. The electrolyte properties of the L1MgCl were evaluated with a mixed salt solution of the L1MgCl with $Mg(TFSA)_2$ in triglyme. Through cyclic voltammetry of magnesium deposition–dissolution in the L1MgCl-$Mg(TFSA)_2$/triglyme mixed salt solution, the redox current is observable with small overpotential, although the Coulombic efficiency is still slightly lower than original Grignard reagent electrolyte. The efficiency or other properties can be modified by changing the functional group of phenoxyimine ligand. Novel complex salts of this kind are useful not only for the development of magnesium chloride electrolyte systems but also compatible with practical low voltage batteries operating at ambient temperature. They are useful also for understanding the influence of coordination degree to magnesium ion on the electrolyte properties of such electrolyte systems.

Corrosion difficulties and low anodic stability are intrinsic to electrolytes containing chloride ion. Therefore, the development of chloride-free electrolyte is crucially important for high voltage practical magnesium battery systems. Optimization of the coordination status of magnesium ion is important for reversible magnesium deposition and for other electrolytic properties. For control of the coordination state of magnesium ion with ether oxygen, crown ethers are interesting host molecules because the guest ion size and the number of coordinating oxygen atoms can be prescribed. The addition of various crown ethers in $Mg(TFSA)_2$ solutions in various solvents has been attempted. Under certain conditions, the addition of crown ether changes magnesium deposition behavior drastically [47]. For example, although THF solvent electrolyte dissolving 0.5 mol dm^{-3} $Mg(TFSA)_2$ would not exhibit magnesium deposition–dissolution at ambient temperature, this solution, with the addition of 0.5 mol dm^{-3} of 18-crown-6 ether, provides reversible redox current with approx. 1 V overpotential. It is particularly interesting that the addition of 15-crown-5, having an adequate size of internal space for magnesium ion, is ineffective for such improvement. The addition effect of crown ether is dependent to some degree on the kind of solvent. Clarification by behavior of magnesium ions coordinated by various crown ethers in solution and at magnesium surface has been attempted using spectrometric and computational methods. Crown ether molecules tend to coordinate with magnesium ion superior to solvent molecules, and forming complex cations. Some oxygen atoms participate in the coordination with retaining flexibility for the crown ether molecule. The flexibility of coordination of magnesium ion with 18-crown-6 helps de-solvation from magnesium on the electrode surface. It also accelerates smooth deposition. The magnesium ion–18-crown-6 ether complex can work also in ionic liquid solvent having N-methyl-N-propylpiperidinium (PP13) cation and TFSA anion, although the Coulombic efficiency is not satisfactory. More or less, the magnesium ion–18-crown-6 ether complex is assumed to be an electrolyte component for reversible magnesium deposition–dissolution.

Solvent species are primarily responsible for the coordination environment of magnesium ion. Although glymes are used exclusively as solvents in the present stage, the electrochemical stability, solubility of effective magnesium salts, and high-temperature durability are insufficient. Sulfolane and dialkyl sulfone have been identified and proposed as novel co-solvent species. An electrolytic solution using mixed solvent of diglyme and ethylmethyl sulfone provides reversible magnesium deposition–dissolution with low overpotential, and stability even at 373 K. Such a novel solvent candidate might help the optimum design of electrolyte [48, 49].

The pentagonal radar charts for the electrolyte based on the alkoxide magnesium chloride complex, the phenoxyimine magnesium chloride complex, the 18-crown-6 additive, and ethylmethyl sulfone co-solvent are shown, respectively, in Fig. 6. One can readily understand that each electrolyte exhibits some improvement from its former version shown in Fig. 3. The combination of these components is also applicable to enhance these benefits.

As with other phases of material science, the computational science for battery electrolytes has progressed dramatically along with the evolution of computer processor speed. The formation of magnesium chloride–solvent complex in chloride-containing electrolytes, or the formation of coordinated ion pairs in magnesium salt-solvent system, is expected to follow a complicated equilibrium. It is difficult to clarify a rigorous solution structure even using spectroscopic methods. To visualize what occurs in the ACC electrolyte, Canepa et al. conducted classical molecular dynamic simulation for the bulk ACC electrolyte. Based on the results, they explained the NMR as well as electrochemical results by the equilibrium magnesium–aluminum–chloride–THF complex structure [25, 26]. Even though spectroscopic and electrochemical experimental studies are still necessary, computational

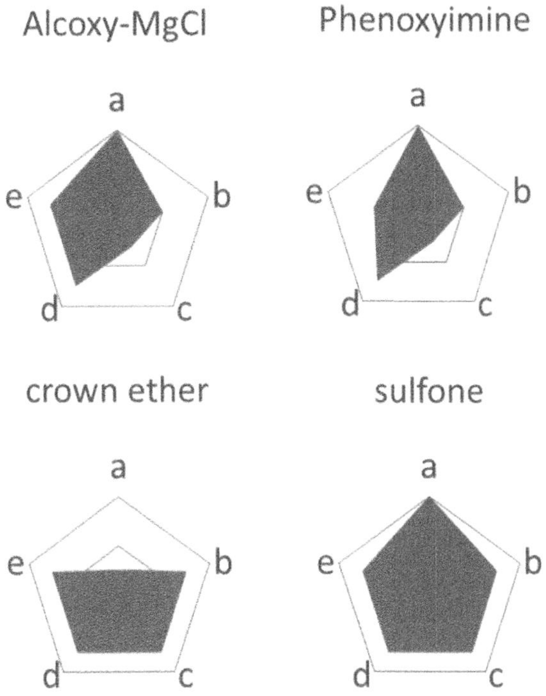

Fig. 6 Radar charts for magnesium battery electrolyte evaluation proposed in ALCA-SPRING. Parameters are the same as those shown in Fig. 3

methods have become crucially important tools for forecasting and understanding the behavior of magnesium battery electrolytes. In this project, the bulk structure of magnesium ion electrolytes and the electrode/electrolyte interface reaction have been clarified using molecular dynamic calculation [50].

Even the solidification of electrolyte is an important issue for magnesium batteries, although it is also difficult to assure the mobility of magnesium ion in a solid. Present studies appear to emphasize solid polymer and gel polymer electrolyte. Gel electrolytes containing a liquid fraction providing reversible magnesium deposition are expected to be compatible to quasi-all-solid state magnesium batteries. The kind of such liquid fraction is limited. Therefore, the variety of matrix/filler materials must be lined up. Such polymer backbones as a ring-cleavage polymerization product from THF, and "rotaxanes", a series of polymers with a unique topological character, serve flexible coordination circumstance to magnesium and the ability to hold the liquid fraction [51]. The environment of electrolyte in a cell, particularly beneath electrodes at a charged state, is severe for most organic polymers. In addition, some magnesium complex salts are reactive themselves. Regarded from this perspective, inactive inorganic matrices are attractive. This research project proposes silica nano-fibers prepared using an electrospun method for a unique inorganic matrix of quasi-solid electrolyte [52]. The simultaneous research development of liquid and gel electrolyte systems might serve to liberate cell design for advanced magnesium secondary batteries.

During the most recent couple of years, many reports have described the use of salts and solvent materials, spectroscopic characterization of bulk electrolyte and electrolyte–electrode interfaces, and computer simulations of electrode reactions. Therefore, a great variety of material selection has been developed. The reaction mechanisms have been understood rapidly during these years, even though many difficulties remain for their use in commercial batteries. In future magnesium secondary batteries, electrolyte materials are perhaps those which are not described in this chapter.

References

1. Brown, O. R., & McIntyre, R. (1986). The magnesium and magnesium amalgam electrodes in aprotic organic solvents. *A Kinetic Study, Electrochimica Acta, 30*, 627–633.
2. Gregory, T. D., Hoffman, R. J., & Winterton, R. C. (1990). Nonaqueous electrochemistry of magnesium application to energy storage. *Journal of the Electrochemical Society, 137,* 775.
3. Dubois, J.-E., Molle, G., Tourillon, G., & Bauer, P. (1979). *Tetrahedron Letters, 20,* 5069.
4. Lu, Z., Schechter, A., Moshkovich, M., & Aurbach, D. (1999). *Journal of Electroanalytical Chemistry, 466,* 203.
5. Gaddum, L. W., & French, H. E. (1927). The electrolysis of Grignard solutions. *Journal of the American Chemical Society, 49,* 1295.
6. Liebenow, C. (1997). Reversibility of electrochemical magnesium deposition from Grignard solutions. *Journal of Appled Electrochemistry, 27,* 221.
7. Aurbach, D., Lu, Z., Schechter, A., Gofer, Y., Glzbar, H., Turgeman, R., et al. (2000). Prototype systems for rechargeable magnesium batteries. *Nature, 407,* 724.
8. Aurbach, D., Turgeman, R., Chusid, O., & Gofer, Y. (2001). Spectroelectrochemical studies of magnesium deposition by in situ FTIR spectroscopy. *Electrochemistry Communications, 3,* 252.
9. Aurbach, D., Schechter, A., Moshkovich, M., & Cohen, Y. (2001). On the mechanisms of reversible magnesium deposition processes. *Journal of the Electrochemical Society, 148,* A1004.
10. Matsui, M. (2011). Study on electrochemically deposited Mg metal. *Journal of Power Sources, 196,* 7048.
11. Pour, N., Gofer, Y., Major, D. T., & Aurbach, D. (2011). Structural analysis of electrolyte solutions for rechargeable Mg batteries by stereoscopic means and DFT calculations. *Journal of the American Chemical Society, 133,* 6270.
12. Guo, Y. S., Zhang, F., Yang, J., Wang, F. F., NuLi, Y. N., & Hirano, S. I. (2012). *Energy and Environmental Science, 5,* 9100.
13. Arthur, T. S., Glans, P.-A., Matsui, M., Zhang, R., Ma, B., & Guo, J. (2012). Mg deposition observed by in situ electrochemical Mg K-edge X-ray absorption spectroscopy. *Electrochemistry Communications, 24,* 43.
14. Liebenow, C., Yang, Z., & Lobitz, P. (2000). The electrodeposition of magnesium using solutions of organomagnesium halides, amidomagnesium halides and magnesium organoborates. *Electrochemistry Communications, 2,* 641.
15. Kim, H. S., Arthur, T. S., Allred, G. D., Zajisek, J., Newman, J. G., Rodnyansky, A. E., et al. (2011). Structure and compatibility of a magnesium electrolyte with a sulphur cathode. *Nature Communications, 2,* 427.
16. Liao, C., Guo, B., Jiang, D., Custelcean, R., Mahurin, S. M., Sun, X.-G., et al. (2014). Highly soluble alkoxide magnesium salts for rechargeable magnesium batteries. *Journal of Materials Chemistry A, 2,* 581.

17. Liu, T., Shao, Y., Li, G., Gu, M., Hu, J., Xu, S., et al. (2014). A facile approach using $MgCl_2$ to formulate high performance Mg2+ electrolytes for rechargeable Mg batteries. *Journal of Materials Chemistry A, 2,* 3430.
18. Barile, C. J., Nuzzo, R. G., & Gewirth, A. A. (2015). Exploring salt and solvent effects in chloride-based electrolytes for magnesium electrodeposition and dissolution. *Journal of Physical Chemistry C, 119,* 13524.
19. Ha, J. H., Adams B., Cho, J.-H., Duffort, V., Kim, J. H., Chung, K. Y., Cho, B. W., Nazar, L. F., & Oh, S. H. (2016). A conditioning-free magnesium chloride complex electrolyte for rechargeable magnesium batteries. *Journal of Materials Chemistry A, 4,* 7160.
20. Shterenberg, I., Salama, M., Yao, H. D., Gofer, Y., Park, J.-B., Sun, Y.-K., et al. (2015). Evaluation of $(CF_3SO_2)_2N$-(TFSI) based electrolyte solutions for Mg batteries. *Journal of the Electrochemical Society, 162,* A7118.
21. Cheng, Y., Stolley, R. M., Han, K. S., Shao, Y., Arey, B. W., Washton, N. M., et al. (2015). Highly active electrolytes for rechargeable Mg batteries based on a $[Mg_2(\mu\text{-}Cl)_2]^{2+}$ cation complex in dimethoyethane. *Physical Chemistry Chemical Physics: PCCP, 17,* 13307.
22. Gofer, Y., Turgeman, R., Cohen, H., & Aurbach, D. (2003). XPS investigation of surface chemistry of magnesium electrodes in contact with organic solutions of organochloroaluminate complex salts. *Langmuir, 19,* 2344.
23. Nakayama, Y., Kudo, Y., Oki, H., Yamamoto, K., Kitajima, Y., & Noda, K. (2008). Complex structures and electrochemical properties of magnesium electrolytes. *Journal of the Electrochemical Society, 155,* A754.
24. Benmayza, A., Ramanathan, M., Arthur, T. S., Matsui, M., Mizuno, F., Guo, J., et al. (2013). Effect of electronic properties of a magnesium organohaloaluminate electrolyte on magnesium deposition. *Journal of Physical Chemistry C, 117,* 26881.
25. Canepa, P., Gautam, G. S., Malik, R., Jayaraman, S., Rong, Z., Zavadil, K. R., et al. (2015). Understanding the initial stages of reversible Mg deposition and stripping in inorganic nonaqueous electrolytes. *Chemistry of Materials, 27,* 3317.
26. Canepa, P., Jayaraman, S., Cheng, L., Rajput, N. N., Richards, W. D., Gautam, G. S., et al. (2015). Elucidating the structure of the magnesium aluminum chloride complex electrolyte for magnesium-ion batteries. *Energy and Environmental Science, 8,* 3718.
27. Yagi, S., Tanaka, A., Ichitsubo, T., & Matsubara, E. (2012). Electrochemical stability of metal electrodes for rechargeable magnesium deposition/dissolution in Tetrahydrofuran dissolving Ethylmagnesium Chloride. *ECS Electrochemical Letters, 1,* D11.
28. Muldoon, J., Bucur, C. B., Oliver, A. G., Zajicek, J., Allred, G. D., & Boggess, W. C. (2013). Corrosion of magnesium electrolytes: Chlorides—The culprit. *Energy and Environmental Science, 6,* 482.
29. Fukutsuka, T., Asaka, K., Inoo, A., Yasui, R., Miyazaki, K., Abe, T., et al. (2014). New magnesium-ion conductive electrolyte solution based on triglyme for reversible magnesium metal deposition and dissolution at ambient temperature. *Chemistry Letters, 43,* 1788.
30. Orikasa, Y., Masese, T., Koyama, Y., Mori, T., Hattori, M., Yamamoto, K., et al. (2014). High energy density rechargeable magnesium battery using earth-abundant and non-toxic elements. *Scientific Reports, 4,* 5622.
31. Yoshida, K., Tsuchiya, M., Tachikawa, N., Dokko, K., & Watanabe, M. (2011). Change from Glyme solutions to quasi-ionic liquids for binary mixtures consisting of lithium bis (Trifluoromethanesulfonyl) Amide and Glymes. *Journal of Physical Chemistry C, 115,* 18384.
32. Ueno, K., Yoshida, K., Tsuchiya, M., Tachikawa, N., Dokko, K., & Watanabe, M. (2012). Glyme–Lithium salt equimolar molten mixtures: Concentrated solutions or solvate ionic liquids? *Journal of Physical Chemistry B, 116,* 11323.
33. Ha, S.-Y., Lee, Y.-W., Woo, S. W., Koo, B., Kim, J.-S., Cho, J., et al. (2014). Magnesium(II) Bis(trifluoromethane sulfonyl) imide-based electrolytes with wide electrochemical windows for rechargeable magnesium batteries. *ACS Applied Materials and Interfaces, 6,* 4063.
34. Rajput, N. N., Qu, X., Sa, N., Burrell, A. K., & Persson, K. A. (2015). The coupling between stability and ion pair formation in magnesium electrolytes from first-principles quantum mechanics and classical molecular dynamics. *Journal of the American Chemical Society, 137,* 3411.

35. Kimura, T., Fujii, K., Sato, Y., Morita, M., & Yoshimoto, N. (2015). Solvation of magnesium ion in Triglyme-based electrolyte solutions. *Journal of Physical Chemistry C, 119,* 18911.
36. Yu, Y., Baskin, A., Valero-Vidal, C., Hahn, N. T., Liu, Q., Zavadil, K. R., et al. (2017). Instability at the electrode/electrolyte interface induced by hard cation chelation and nucleophilic attack. *Chemistry of Materials, 29,* 8504.
37. Kuwata, H., Matsui, M., & Imanishi, N. (2017). Passivation layer formation of magnesium metal electrodes for rechargeable magnesium batteries. *Journal of the Electrochemical Society, 164*(13), A3229.
38. Mohtadi, R., Matsui, M., Arthur, T. S., & Hwang, S.-J. (2012). Magnesium borohydride: From hydrogen storage to magnesium battery. *Angewandte Chemie International Edition, 51,* 9780.
39. Carter, T. J., Mohtadi, R., Arthur, T. S., Mizuno, F., Zhang, R., Shirai, S., et al. (2014). Boron Clusters as highly stable magnesium-battery electrolytes. *Angewandte Chemie, 53,* 3173.
40. Tutusaus, O., Mohtadi, R., Arthur, T. S., Mizuno, F., Nelson, E. G., & Sevryugina, Y. V. (2015) An efficient halogen-free electrolyte for use in rechargeable magnesium batteries. *Angewandte Chemie International Edition*, 54, 7900.
41. McArthur, S. G., Jay, R., Geng, L. X., Guo, J. C., & Lavallo, V. (2017). Below the 12-vertex: 10-vertex carborane anions as non-corrosive, halide free, electrolytes for rechargeable Mg batteries. *Chemical Communications, 53,* 4453.
42. Zhang, Z. H., Cui, Z. L., Qiao, L. X., Guan, J., Xu, H. M., Wang, X. G., et al. (2017). Novel design concepts of efficient Mg-ion electrolytes toward high-performance magnesium-selenium and magnesium-sulfur batteries. *Advanced Energy Materials, 7*(11), 1602055.
43. Song, J., Sahadeo, E., Noked, M., & Lee, S. B. (2016). Mapping the challenges of magnesium battery. *The Journal of Physical Chemistry Letters, 7,* 1736.
44. Kakibe, T., Yoshimoto, N., Egashira, M., & Morita, M. (2010). Optimization of cation structure of imidazolium-based ionic liquids as ionic solvents for rechargeable magnesium batteries. *Electrochemistry Communications, 12*(11), 1630.
45. Mandai, T., Akita, Y., Yagi, S., Egashira, M., Munakata, H., & Kanamura, K. (2017). A key concept of utilization of both non-Grignard magnesium chloride and imide salts for rechargeable Mg battery electrolytes. *Journal of Materials Chemistry A, 5,* 3152.
46. Egashira, M., Hiratsuka, K., Matsubara, K., Akita, Y., Munakata, H., & Kanamura, K. (2018). Multi-dentate phenoxyimine magnesium chloride complex for magnesium battery electrolyte. *Materials Today Energy, 9,* 279.
47. Sagane, F., Ogi, K., Konno, A., Egashira, M., & Kanamura, K. (2016). The effect of the cyclic ether additives to the ethereal electrolyte solutions for Mg secondary battery. *Electrochemistry, 84*(2), 76.
48. Senoh, H., Sakaebe, H., Sano, H., Yao, M., Kuratani, K., Takeichi, N., et al. (2014). *Journal of the Electrochemical Society, 161*(9), A1315.
49. JP Patent 2016-038845.
50. Hattori, M., Yamamoto, K., Matsui, M., Nakanishi, K., Mandai, T., Choudhary, A., et al. (2018). Role of coordination structure of magnesium ions on charge and discharge behavior of magnesium alloy electrode. *Journal of Physical Chemistry C, 122*(44), 25204.
51. Ab Aziz, A., Yoshimoto, N., Yamabuki, K., & Tominaga, Y. (2018). Ion-conductive, thermal and electrochemical properties of Poly(ethylenecarbonate)-Mg electrolytes with Glyme solution. *Chemistry Letters, 47*, 1258.
52. Yuuki, T., Konosu, Y., Ashizawa, M., Iwahashi, T., Ouchi, Y., Tominaga, Y., et al. (2017). Ionic liquid-based electrolytes containing surface-functionalized inorganic nanofibers for quasisolid lithium battery. *ACS Omega, 2*(3), 835.

Al and Zn Rechargeable Batteries

Aluminum and Zinc Metal Anode Batteries

Tetsuya Tsuda

Abstract Aluminum and zinc metal are expected to be anode active materials with high volumetric capacity for future secondary batteries. Theoretical volumetric capacities for Al and Zn are 5854 and 8046 mAh cm^{-3}, respectively. These metal anodes are usually handled in the Lewis acid-base type ionic liquid (IL) electrolytes with Al or Zn salts, e.g., $AlCl_3$–1-ethyl-3-methylimidazolium chloride ([C_2mim]Cl) and $ZnCl_2$–[C_2mim]Cl. However, the appropriate metal salt composition in the ILs is required to yield the metal anode reactions related to the charge-discharge process. These IL systems are good choices for conducting the studies on the Al and Zn metal anode battery because of adequate carrier anion density and very high coulomb efficiency for the anode reaction. Interestingly, their anode reactions are controlled by the anionic species, not cationic species, Al^{3+} and Zn^{2+}. Meanwhile, cathode active materials leave much to be improved, but some promising ones have recently been reported. The most important thing is that the full cells fabricated using the findings behave like commonly used secondary batteries, suggesting that these battery systems are worthy of careful study and possess tremendous potential to be one of the next-generation batteries.

Keywords Aluminum · Zinc · Metal anode · Ionic liquid · Battery

1 Introduction

Secondary battery is one of the most important energy devices for supporting our modern life. We unknowingly gain many benefits from the device, and its application fields are still expanding. Much will be expected of the secondary battery more than ever. Considering this, we have to create different types of secondary batteries, such as high capacity type, high power type, and inexpensive low environmental impact type that can take the place of conventional energy storage devices, in order to adapt to a wide variety of applications. As described in early chapters, the metal anodes

T. Tsuda (✉)
Department of Applied Chemistry, Graduate School of Engineering, Osaka University, 2-1 Yamada-Oka, Suita, Osaka 565-0871, Japan
e-mail: ttsuda@chem.eng.osaka-u.ac.jp

K. Kanamura (ed.), *Next Generation Batteries*,
https://doi.org/10.1007/978-981-33-6668-8_49

will contribute greatly to their development. For example, in terms of high capacity anodes, Li (2066 mAh cm^{-3}, 3861 mAh g^{-1}), Na (1128 mAh cm^{-3}, 1166 mAh g^{-1}), Mg (3833 mAh cm^{-3}, 2205 mAh g^{-1}), Zn (5854 mAh cm^{-3}, 820 mAh g^{-1}), and Al (8046 mAh cm^{-3}, 2980 mAh g^{-1}) are prospective candidates as shown in Fig. 1. If the resource-rich metals are required, Al (Abundance in crust: 80700 ~ 82300 ppm), Ca (36300 ~ 50000 ppm), Mg (20800 ~ 29000 ppm), and Na (23000 ~ 28300 ppm) are the most likely choices [1]. These metals will be widely used to produce the resource depletion-free secondary battery, which can adjust to the prediction that future secondary battery market will grow dramatically. Among them, Al has strong advantages in the theoretical volumetric capacity and the rich abundance. Now, a great number of research groups around the world are actively involved in the development of Al metal anode secondary battery. It is not an exaggeration to say that the article reported by Lin et al. triggered the Al battery research boom [2]. Note that the Al metal anode battery has great potential for becoming a future high volumetric capacity secondary battery exceeding Li-ion one. The same thing can be said with the Zn metal anode battery due to the favorable volumetric capacity of Zn metal, although there are insufficient researches on the Zn battery. In practical primary battery, Zn metal anode is very popular and it seems to be capable of adapting to future mass production of the Zn metal secondary battery. In this chapter, recent research and development on the Al and Zn metal anode-based secondary batteries, especially focusing on electrolytes and cathode active materials, is described in outline.

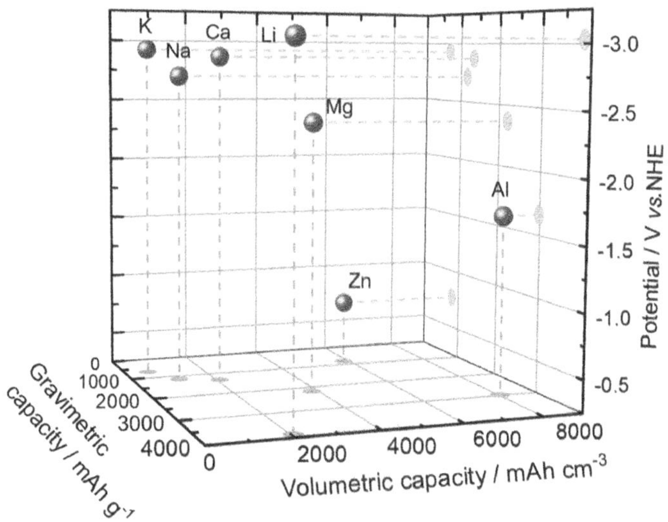

Fig. 1 Relationship among theoretical gravimetric capacity (mAh g^{-1}), theoretical volumetric capacity (mAh cm^{-3}), and standard electrode potential (versus NHE) for the possible metal anode

2 Electrolytes for Al and Zn Metal Anode Secondary Batteries

Li-ion battery normally uses the well-designed electrolyte that can proceed with the expected electrode reaction without difficulty. In the case of Al and Zn metal anode secondary battery, there is no such designed electrolyte. The nonaqueous electrolytes developed for Al and Zn electroplating are often employed as secondary battery electrolytes. That is, we have to know the electroplating electrolytes for putting them to full use. Electrodeposition potential of these metals is quite negative against the NHE. If the high coulomb efficiency on the metal deposition/stripping, i.e., the metal anode reaction in the secondary battery, is expected, nonaqueous electrolyte must be selected to obtain a favorable result. In addition, the electrolyte should have several following features; it can be used around room temperature; it does not contain volatile organic compounds (VOCs); it has flame retardant resistance, non-volatile nature, and high ionic conductivity. In fact, there are no many choices of the electrolyte to meet the conditions described above. In most cases, it is limited to ionic liquid (IL) and other similar non-volatile electrolytes, e.g., deep eutectic solvent (DES) (Fig. 2) [3]. As to the electrolyte for Al metal anode secondary battery, $AlCl_3$–1-ethyl-3-methylimidazolium chloride ([C_2mim]Cl) IL is commonly used, since there are sufficient thermodynamic data on this IL system in which the electroactive species for Al electroplating reach up to 3.3 mol L^{-1} [4–6]. Of course, physicochemical properties of the $AlCl_3$–[C_2mim]Cl vary with the composition ratios, because the ionic species and their concentrations also depend on the

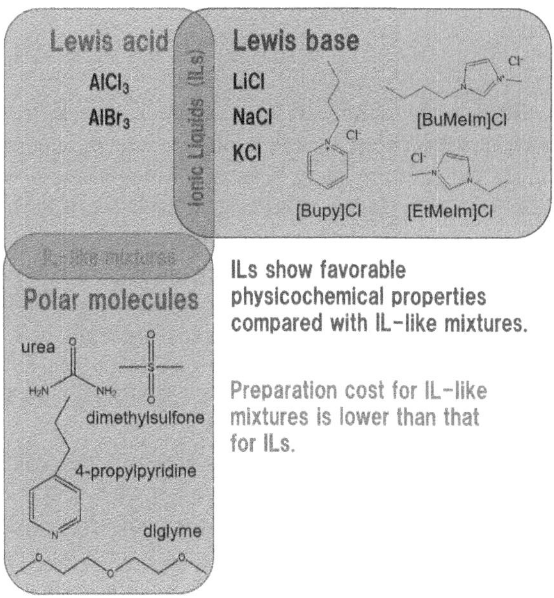

Fig. 2 Examples of commonly used polar molecules, Lewis acids, and Lewis bases for synthesizing ionic liquids and other similar liquid electrolytes

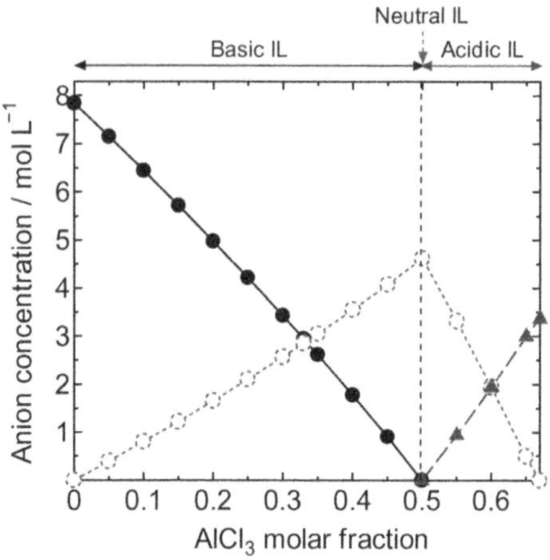

Fig. 3 Concentrations for anionic species (●: Cl^-, ○: $[AlCl_4]^-$, ▲: $[Al_2Cl_7]^-$) in the $AlCl_3$–$[C_2mim]Cl$ IL at different $AlCl_3$ molar fractions

ratios. Figure 3 exhibits anionic species and their concentrations at the different $AlCl_3$ molar fraction. When the $AlCl_3$ molar fraction (N_{AlCl_3}) is $0 < N_{AlCl_3} < 0.50$ (Lewis basic condition), Cl^- and $[AlCl_4]^-$ coexist in the Lewis basic IL. At $N_{AlCl_3} = 0.50$ (Lewis neutral condition) and $0.50 < N_{AlCl_3} \leq 0.67$ (Lewis acidic condition), respectively, only $[AlCl_4]^-$ and both $[AlCl_4]^-$ and $[Al_2Cl_7]^-$ exist in the ILs. In these three composition regions, only the Lewis acidic IL, which contains the $[Al_2Cl_7]^-$, can deposit Al metal electrochemically, suggesting that this is the only case where the Lewis acidic IL can be used as the electrolytes for electroplating and secondary battery. And then, Al electrodeposition/stripping reaction is as follows.

$$4[Al_2Cl_7]^- + 3e^- \rightleftharpoons Al + 7[AlCl_4]^- \quad (1)$$

In the Lewis acidic $AlCl_3$–$[C_2mim]Cl$, this electrode reaction reversibly proceeds with very low overpotential. It indicates that the electrolyte is very useful for Al metal anode. But, unfortunately, because limiting potential of the electrochemical oxidation reaction shown in Eq. (2) is ca. +2.50 V (versus Al(III)/Al), maximum theoretical voltage of the Al metal anode battery with this IL electrolyte is about 2.50 V. Even if the $[C_2mim]Cl$ is changed to other organic salts with electrochemically-stable cation, the anionic species for the electrode reactions at limiting potentials are exactly the same. Therefore, the battery voltage is little affected by the organic salt species.

$$4[AlCl_4]^- \rightleftharpoons 2[Al_2Cl_7]^- + Cl_2 + 2e^- \quad (2)$$

Recent years, IL-like mixtures, that are often called DESs and solvate ILs, have been developed. Some are composed of $AlCl_3$ and polar molecule (base). The preparation process is essentially the same as that for chloroaluminate IL, but in this case, polar organic molecule is employed as a substitute for organic salt. The typical polar molecules are urea ($(NH_2)_2CO$) [7, 8], dimethylsulfone ($DMSO_2$) [9, 10], diglyme

[11], 1,3-dimethyl-2-imidazolidinone [12], 4-propylpyridine [13], and so on. When these polar organic molecules are mixed with $AlCl_3$, two types of Al ionic species are produced by the following equilibrium reactions.

$$2AlCl_3 + n\mathrm{Base} \rightleftharpoons [AlCl_2 \cdot n\mathrm{Base}]^+ + [AlCl_4]^- \quad (3)$$

$$4AlCl_3 + 3\mathrm{Base} \rightleftharpoons \left[Al(\mathrm{Base})_3\right]^{3+} + 3[AlCl_4]^- \quad (4)$$

The generated Al complex cation is available for Al electrodeposition. In this situation, the ionic species for the Al deposition process are different from $[Al_2Cl_7]^-$ in Eq. (1). We have to consider that it has an impact on the morphology and current efficiency of the Al electrodeposit. Besides, in the composition region with free polar organic molecule, we should know that the free molecules may affect the physicochemical properties of the electrolyte and the electrode reactions. Adding excess $AlCl_3$ to the solution prepared by the reactions (Eq. (3) or (4)) generates $[Al_2Cl_7]^-$, like in chloroaluminate IL (Eq. (5)), and then, Al electrodeposition/stripping proceeds by the reaction given in Eq. (1).

$$[AlCl_4]^- + AlCl_3 \rightleftharpoons [Al_2Cl_7]^- \quad (5)$$

Production cost for the IL-like mixtures is less than one-fifth of that for chloroaluminate IL. This is clearly a major draw for us. Their drawback is insufficient physicochemical properties. Further development of the mixture system that has similar physicochemical properties comparable to the chloroaluminate IL is highly expected.

Zn metal can be electrochemically obtained using similar approach, since $ZnCl_2$ can also work as Lewis acid, like $AlCl_3$. Most electrolyte for the Zn electroplating/stripping is $ZnCl_2$–$[C_2mim]Cl$ [14, 15]. Other similar ones are ternary $AlCl_3$–$ZnCl_2$–$[C_2mim]Cl$ IL [16] and $ZnBr_2$–1-allyl-3-methylimidazolium bromide ([Allylmim]Br) [17]. As in the case of $AlCl_3$–$[C_2mim]Cl$ IL electrolyte, anionic Zn species in the halozincate ILs change with the Zn halide composition, and it leads to the variation in physicochemical properties and electrochemical behavior. As one of the examples, the electrochemical windows in different $ZnCl_2$–$[C_2mim]Cl$ ILs at 363 K are shown in Fig. 4. Needless to say, the different behavior is caused by the variation in anionic species. However, it is not easy to estimate the concentration of anionic species as a function of Zn halide composition, because various kinds of Zn complex anions can exist in the IL [15]. The handling temperature for electrochemical studies is usually higher than that for the chloroaluminate IL due to their viscous nature. Under such conditions, the electrode behavior is close to aluminum's one and the dendrite-free Zn metal can be obtained. Although small in number, there are several possible non-Lewis acid-base type IL electrolytes for Zn metal anode battery, too. These include non-halorozincate ILs with Zn salt (usually less than 0.1 M) [18–20] and with both Zn salt and water [21–23]. Compared to the electrolyte for Al metal anode battery, practical choices are not many currently.

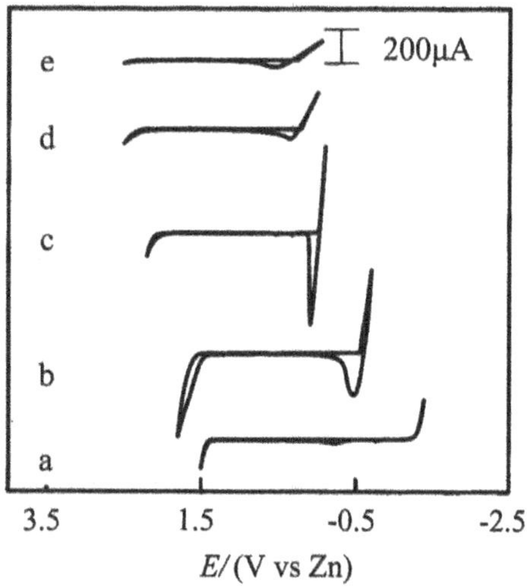

Fig. 4 Cyclic voltammograms recorded at a GC electrode in various $ZnCl_2$–[C_2mim]Cl IL electrolytes at 393 K. The molar ratio of $ZnCl_2$ to [C_2mim]Cl are **a** 1:3, **b** 1:2, **c** 1:1, **d** 2:1, and **e** 3:1. The scan rate was 50 mV s^{-1} Reprinted from ref. [15] Copyright 2002, with permission from Elsevier

3 Electrochemistry of Al and Zn Metal Anode for Secondary Batteries

As for the haloaluminate IL, a huge number of thermodynamic data are known, and there is the accumulation of handling techniques for the IL through a long history of Al electroplating in the IL [3, 24, 25]. Utilizing the information, Al metal anode can be handled with relative ease. However, we have to consider the chemical stability of the electrolyte component to the Al metal deposited from the electrolyte, since Al metal is a reactive material, not as much as Li and Mg. Finding the appropriate combination of Al metal anode and electrolyte is of considerable importance to apply the Al anode to the secondary battery system with high coulomb efficiency close to 100%. 1,3-dialkylimidazolium halides, which are commonly used for preparing the IL electrolytes as organic salts, meet the criterion. The point we should notice is that some haloaluminate ILs and mixtures are not suitable for the battery electrolyte because of the reason described above.

Figure 5 exhibits an example of chronopotentiogram concerning Al metal deposition/stripping process. While IR drop of ca. 0.1 V resulting from the use of the beaker cell is observed, electrode behavior itself is good and the coulomb efficiency is nearly 100%. In the usual case, applied current density for electroplating, i.e., charge process, required in battery system is less than one digit from that in electrolysis process. The charge-discharge process for Al metal anode would have no adverse impact on battery performance. But, under a limited amount of electrolyte usage in the coin-type and laminate-type cells, the variation in Lewis acidity at the interface between the chloroaluminate IL electrolyte and the electrode often make a major impact on the electrolyte properties. In extreme cases, the electrolyte becomes solid, as the phase diagram for $AlCl_3$–[C_2mim]Cl suggests [24]. This is directly

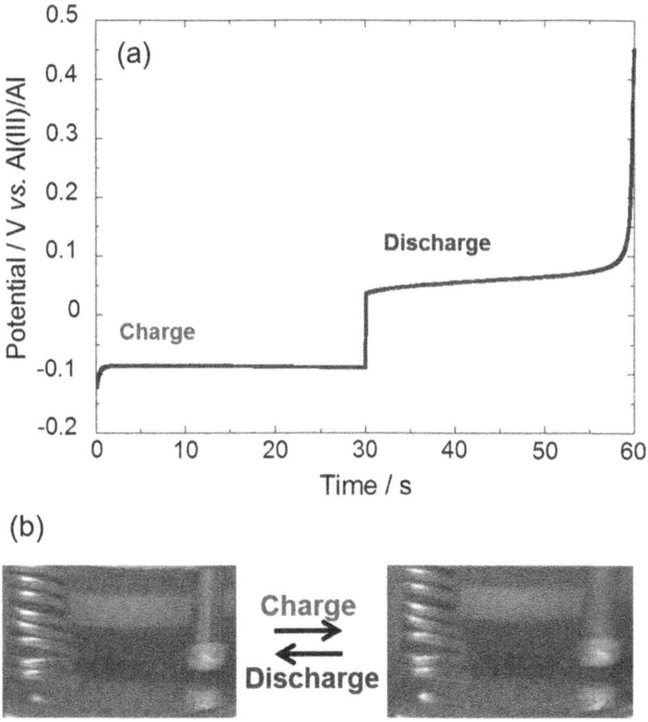

Fig. 5 **a** Chronopotentiogram concerning Al metal deposition/stripping process recorded at a Cu electrode in 60.0–40.0 mol% $AlCl_3$–[C_2mim]Cl IL electrolyte. The applied current density was ± 4 mA cm^{-2}. **b** Pictures of the Cu electrode during the Al metal deposition/stripping process

linked to the dendritic Al deposition due to the non-uniform current flow and should be avoided.

The same holds Zn metal deposition in the ILs consisting of Zn halide and organic salt, since similar Lewis acid-base type IL systems are exploited. Although it is well-known that Zn dendrite is easily obtained during the electrodeposition in aqueous solution, Zn deposits without dendrite are produced in the IL electrolytes [14, 15, 17]. Figure 6 shows SEM images of Zn electrodeposited in $ZnBr_2$–[Allylmim]Br IL at different temperatures [17]. Other electrodeposition conditions are exactly the same. Each crystal deposited on the substrate becomes larger by raising the IL temperature. The important point is that no dendrite appears. It suggests that the Zn halide-based ILs are favorable electrolyte for the Zn metal anode secondary battery.

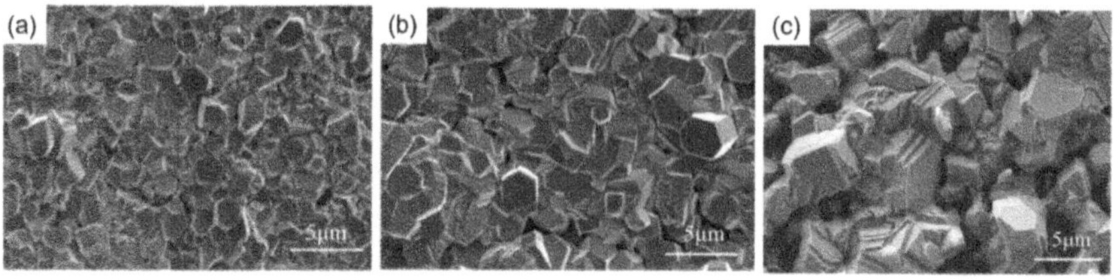

Fig. 6 SEM images of Zn electrodeposited on Pt substrates in $ZnBr_2$–[Allylmim]Br IL electrolyte (1.0 mol kg^{-1} $ZnBr_2$) at different temperatures: **a** 343 K, **b** 363 K, and **c** 383 K. The applied current density was −1 mA cm^{-2}. The deposition time was 2 h Reprinted from ref. [17] Copyright 2019, with permission from Elsevier

4 Cathodes for Al and Zn Metal Anode Secondary Batteries

Lewis acid-base type IL is normally used for the electrolytes in the Al metal anode secondary battery, because it shows better physicochemical properties than that for the IL-like mixtures. In such a case, both anode and cathode reactions are controlled by anions. This is essentially different from the case of Li-ion battery that uses Li^+ cation. We have to design the cathode reaction for the Al metal anode battery in a different approach from conventional cation-based battery systems. Pioneering researches on Al metal anode secondary battery were carried out by Gifford and Palmisano [26] and Takami and Koura [27] from 1980s to 1990s. Several cathode active materials were proposed, but the Al metal anode secondary battery was less noticed probably because of not enough battery performance compared to other battery systems. Recently, the situation has changed dramatically by the article published in Nature [2]. In the article, Lin et al. reported a new graphite cathode that has a self-standing monolith structure. This cathode shows a discharge capacity of ca. 60 mAh g^{-1} and a superior cyclability over 7000 cycles. Surprisingly the discharge capacity nearly unchanged up to the charge-discharge rate of 5000 mA g^{-1}. This remarkable finding would be attributed to the fact that serious damage of the cathode active material by volume expansion derived from the electrochemical anion intercalation/deintercalation during the charge-discharge process (Eq. (6)) is alleviated to some extent by the unique self-standing structure.

$$nC + [\text{Al anion}]^- \rightleftharpoons C_n[\text{Al anion}] + e^- \quad (\text{Al anion}: [AlCl_4]^- \text{ and } [Al_2Cl_7]^-) \tag{6}$$

We experimentally confirmed that morphology of graphite material is a key to design favorable graphite cathodes. As shown in Fig. 7, the cyclic voltammograms recorded at a graphite rod cathode show the redox waves originating in the anion intercalation/deintercalation reaction, but the cathode suffered a serious damage after several charge-discharge cycles and the damage was enhanced when the cut-off potential was over 2.15 V (Fig. 8). After the cycling test, black powder precipitated at the bottom of flask. This disadvantage can be conquered by the use of activated

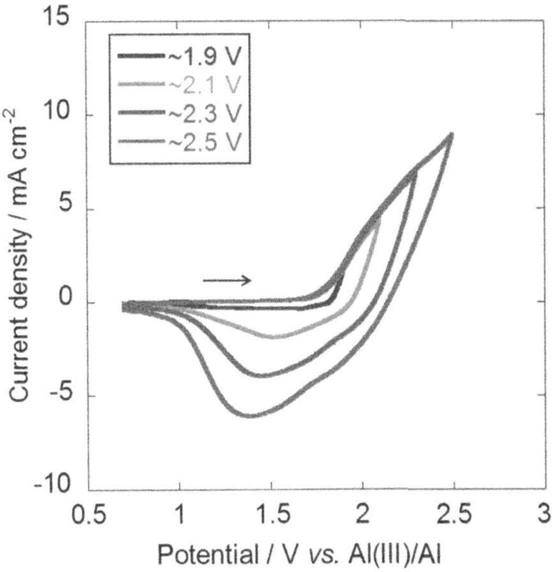

Fig. 7 Cyclic voltammograms recorded at different scan ranges in 60.0–40.0 mol% $AlCl_3$–[C_2mim]Cl IL electrolyte. The working electrode was a graphite rod. The scan rate was 10 mV s^{-1}. The temperature was 298 K

Fig. 8 SEM images of the graphite rod electrode before and after charge-discharge test in 60.0–40.0 mol% $AlCl_3$–[C_2mim]Cl IL electrolyte. The applied current density was ±0.4 mA cm^{-2}. The cut-off potential was 2.15 V (versus Al(III)/Al). (inset) Pictures of the graphite rod electrode at centimeter scale

carbon fiber cloth, but the electrode behavior was very similar to that for the electric double layer capacitor and it was hard to obtain a constant voltage during the charge-discharge process [28].

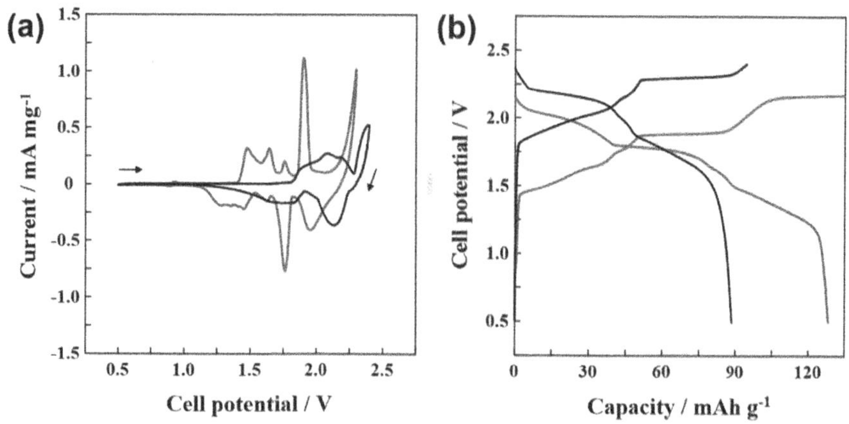

Fig. 9 **a** Cyclic voltammograms and **b** galvanostatic charge-discharge curves obtained from Al/Grafoil® cells with (—) 61.0–26.0–13.0 mol% $AlCl_3$–NaCl–KCl at 393 K and (—) 60.0–40.0 mol% $AlCl_3$–[C_2mim]Cl at 298 K. The scan rate and current density for each measurement was 0.5 mV s^{-1} and 100 mA g^{-1}, respectively Reproduced from Ref. [29] by permission of The Royal Society of Chemistry

As described above, the graphite rod is unfavorable as a cathode active material. Changing the rigid structure to moderately loosen one gives favorable cathode characteristics to the graphite electrode. An example is exfoliated graphite. Grafoil® is one of the commonly used exfoliated graphite materials. Cyclic voltammogram recorded at the Al/Grafoil® cells with 60.0–40.0 mol% $AlCl_3$–[C_2mim]Cl is shown in Fig. 9a [29]. The redox waves are attributed to the chloroaluminate anion intercalation and deintercalation to the Grafoil® electrode [30]. The charge-discharge curve recorded at the same cell is shown in Fig. 9b. At the current density of 100 mA g^{-1}, several plateaus appear and suggest the formation of graphite intercalation compounds with different stages [31]. The discharge capacity is nearly 90 mAh g^{-1} despite the high bulk density of the Grafoil® (ca. 1.12 g cm^{-3}). In order to reveal the temperature and electrolyte dependence of the electrode behavior, the same electrochemical analyses were conducted under different temperature and electrolyte conditions (Fig. 9a, b). Here, 61.0–26.0–13.0 mol% $AlCl_3$–NaCl–KCl electrolyte was used for the comparison. In the cyclic voltammogram, the pronounced peak separation and obvious current peaks related to the anion intercalation-deintercalation process are recognized. These peaks seem to move to negative potential. It means that the electrode reactions are more favorable in the chloroaluminate inorganic IL at 393 K than in the lower temperature $AlCl_3$–[C_2mim]Cl system. As expected from this result, larger discharge capacity was obtained at a discharge rate of 100 mA g^{-1}. The capacity, ca. 130 mAh g^{-1}, was approximately 1.5 times higher than that obtained in the $AlCl_3$–[C_2mim]Cl at 298 K. It is interesting to note that the enhancement of the reversible capacity and reduced onset voltage for intercalation was not observed for Grafoil® examined in $AlCl_3$–[C_2mim]Cl at elevated temperature conditions (over 353–393 K).

It strongly suggests that the intercalation ability is enhanced in the inorganic $AlCl_3$–NaCl–KCl electrolyte without a thermal effect. As a result, the Al/Grafoil® cell with the inorganic electrolyte delivered capacities of 124, 97, 90, and 63 mAh g^{-1} at rates of 1000, 2000, 4000, and 8000 mA g^{-1}, respectively. This high-rate performance is far superior to that measured in the organic IL electrolyte. Thus, we concluded that the inorganic electrolyte has a greater ability to maximize graphitic cathode properties and that the anion intercalation is not the only deciding factor for kinetic properties at elevated temperatures.

Gifford and Palmisano employed 100 mesh graphite powder as a cathode active material and prepared the graphite powder composite cathode using polysulfone binder [26]. The cathode shows the charge-discharge behavior, but the cathode performance, e.g., rate capability and discharge capacity, pales significantly in comparison to the self-standing graphite material proposed by Lin et al. By changing the graphite powder to commercially available flexible graphene nanoplatelets, the cathode performance is greatly enhanced [32, 33]. Further improvement is recognized by the use of conductive additives (acetylene black (AB), ketjen black (KB), and vapor grown carbon fiber (VGCF), KB and VGCF mixture (1:1 (ratio by weight))) [32]. Figure 10 exhibits chronopotentiograms recorded at the graphene nanoplatelet-polysulfone binder composite electrodes without and with the conductive additives.

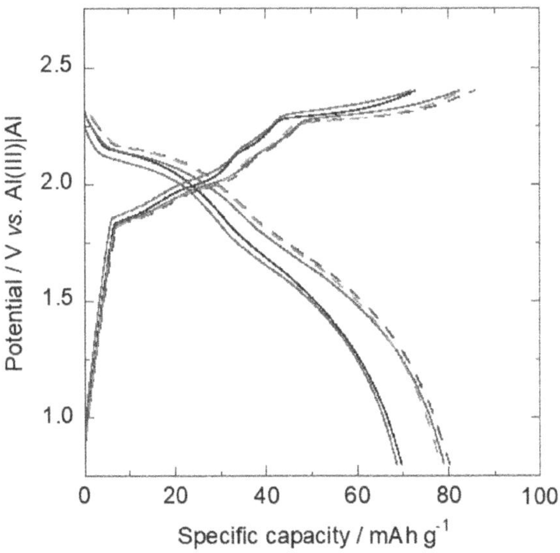

Fig. 10 Chronopotentiograms recorded at various graphene nanoplatelet composite electrodes in a 60.0–40.0 mol% $AlCl_3$–[C_2mim]Cl. The electrodes were (—) the graphene nanoplatelet composite one without a conductive additive and with (—) AB, (---) KB, (—) VGCF, and (---) VGCF/KB. The temperature was room temperature. The cut-off potentials were 2.40 and 0.80 V versus Al(III)/Al. The applied current density was 2000 mA g^{-1} Reproduced from Ref. [32] by permission of The Electrochemical Society of Japan

Charge-discharge behaviors of the composite electrodes without the conductive additive and with AB are very similar, indicating that AB does not work well as a conductive additive for the composite electrode. The other three composite electrodes with KB, VGCF, and KB/VGCF mixture show 10 mAh g^{-1} higher discharge capacity. Their discharge capacities are ca. 80 mAh g^{-1} at a current density of 2000 mA g^{-1}. VGCF-added cathode showed a better rate capability and capacity retention rate, but coulombic efficiency was insufficient. The disadvantage was mitigated by mixing VGCF and KB. Under such conditions, the discharge capacity was ca. 55 mAh g^{-1} at 10000 mA g^{-1} and the retention rate exceeded 65% to the value obtained at 1000 mA g^{-1}. Figure 11 shows the discharge behavior of the composite electrode

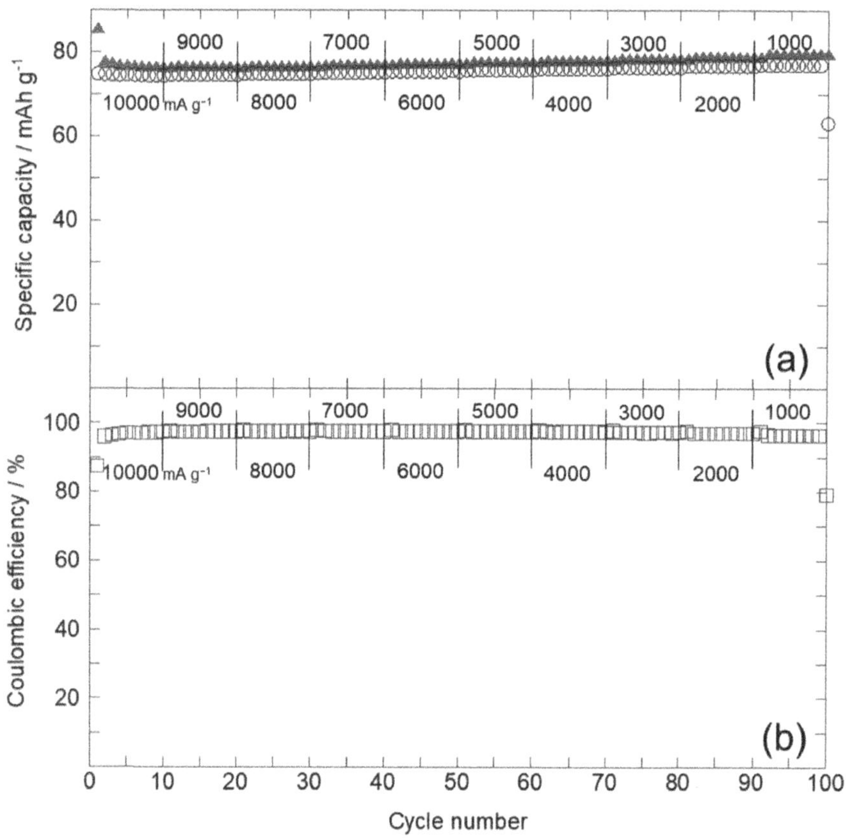

Fig. 11 **a** Charge-discharge capacity and **b** coulombic efficiency estimated from charge-discharge tests conducted at a graphene nanoplatelet composite cathode with a VGCF/KB mixture conductive additive in a 60.0–40.0 mol% $AlCl_3$–[C_2mim]Cl. The charge current density was 2000 mA g^{-1} and the discharge current densities were 1000–10000 mA g^{-1}. The specific capacity for (▲) charge and (○) discharge. The cut-off potentials were 2.40 and 0.80 V versus Al(III)/Al Reproduced from Ref. [32] by permission of The Electrochemical Society of Japan

with the VGCF/KB mixture after the charging process at 2000 mA g^{-1}. The capacities show almost the constant value, ca. 75 ~ 80 mAh g^{-1}, with high coulombic efficiencies at the discharge currents from 1000 to 10000 mA g^{-1}. In the same way as Li-ion battery, the role of conductive additives should not be forgotten.

Current cathode active materials for Al metal anode battery with Lewis acidic $AlCl_3$–[C_2mim]Cl IL electrolyte are roughly divided into four systems: graphite [2, 26, 29, 31–36], polymer [37, 38], conversion reaction [39–41], and sulfur [42–44]. At this time, most research on the cathode active material seems to be the graphite-based one, since the cathode reaction is very simple and stable. But the theoretical charge-discharge capacity is inferior to that for other similar intercalation/deintercalation systems, e.g., using Li^+ and K^+, because of anion's bulky structure. Typical discharge capacity of the graphite-based cathode materials for the Al secondary battery is about 60 ~ 80 mAh g^{-1}. It can be improved by designing the structure and some show the discharge capacity over 100 mAh g^{-1} [45–47]. Other three active materials also have a great potential. I leave out their detailed comments because several good reviews containing the information on the materials have recently been reported elsewhere. If you need further information on the cathode active materials, I strongly recommend you read the recent review articles [3, 48–50].

Because of a limited number of choices concerning the nonaqueous electrolytes for Zn metal anode secondary battery, it is quite unusual for the researches to be conducted on cathode active materials for the Zn battery. As far as I know, there is no report on the cathode for Zn secondary battery with the Lewis acid-base type IL electrolyte. For the Zn secondary battery with non-halozincate IL electrolytes, several cathode active materials are reported. Those are prussian blue analogue [22, 23], graphite [20], and MnO_2 [51]. Endres et al. reported that $FeFe(CN)_6$ delivers a reversible charge-discharge capacity and the discharge capacity is 120 mAh g^{-1} with coulomb efficiency of 99% at 10 mA g^{-1} [22]. This Zn battery system is faced with a big issue on rate capability, but the finding provides useful information to the scientists and engineers in the area of battery technology.

5 Concluding Remarks

Aluminum and zinc metal have already been used in various modern technologies, and these metals show very high theoretical volumetric capacity exceeding Li and Mg. Given the advantages, expectations for the secondary batteries with Al and Zn metal anode would increase more and more. In this context, the current status and issues of Al and Zn metal anode secondary batteries were described in this chapter. A decade ago, we could not imagine the drastic advances in researches on such battery systems. Now their full cells work like commercially available secondary batteries. Especially, Al metal anode secondary battery becomes recognized as the battery with unique characteristics that are difficult to find in conventional batteries (Fig. 12) [52]. Of course, still these battery systems are in the process of technological development, especially Zn metal anode one. Several issues that should be overcome are remained.

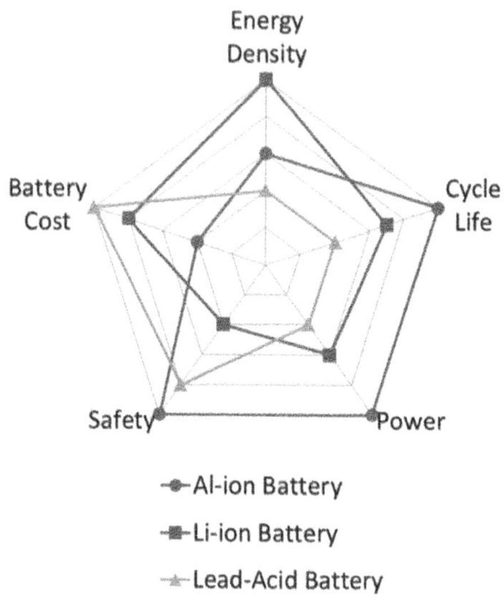

Fig. 12 Comparison of the performances of Al-ion, Li-ion, and lead-acid secondary batteries based on the 5 parameters: energy density, specific power, cost, cycle life, and safety Reproduced from Ref. [52] by permission of The Royal Society of Chemistry

For further development of these battery systems, it is vital to increase the scientists and engineers involved in the battery systems. I hope this article will be a good opener for the readers to initiate the studies on these fascinating battery systems.

References

1. https://en.wikipedia.org/wiki/Abundance_of_elements_in_Earth%27s_crust.
2. Lin, M.-C., Gong, M., Lu, B., Wu, Y., Wang, D.-Y., Guan, M., et al. (2015). *Nature, 520,* 324.
3. Tsuda, T., Stafford, G. R., & Hussey, C. L. (2017). *Journal of the Electrochemical Society, 164,* H5007. and references therein.
4. Wilkes, J. S., Levisky, J. A., Wilson, R. A., & Hussey, C. L. (1982). *Inorganic Chemistry, 21,* 1263.
5. Fannin, A. A., Jr., King, L. A., Levisky, J. A., & Wilkes, J. S. (1984). *Journal of Physical Chemistry, 88,* 2609.
6. Fannin, A. A., Jr., Floreani, D. A., King, L. A., Landers, J. S., Piersma, B. J., Stech, D. J., et al. (1984). *Journal of Physical Chemistry, 88,* 2614.
7. Abood, H. M. A., Abbott, A. P., Ballantyne, A. D., & Ryder, K. S. (2011). *Chemical Communications, 47,* 3523.
8. Abbott, A. P., Harris, R. C., Hsieh, Y.-T., Ryder, K. S., & Sun, I.-W. (2014). *Physical Chemistry Chemical Physics: PCCP, 16,* 14675.
9. Legrand, L., Heintz, M., Tranchant, A., & Messina, R. (1995). *Electrochemica Acta, 40,* 1711.
10. Hirato, T., Fransaer, J., & Celis, J.-P. (2001). *Journal of the Electrochemical Society, 148,* C280.
11. Kitada, A., Nakamura, K., Fukami, K., & Murase, K. (2014). *Electrochemistry, 82,* 946.
12. Endo, A., Miyake, M., & Hirato, T. (2014). *Electrochimica Acta, 137,* 470.
13. Fang, Y., Yoshii, K., Jiang, X., Sun, X.-G., Tsuda, T., Mehio, N., et al. (2015). *Electrochimica Acta, 160,* 82.
14. Lin, Y.-F., & Sun, I.-W. (1999). *Electrochimica Acta, 44,* 2771.
15. Hsiu, S.-I., Huang, J.-F., Sun, I.-W., Yuan, C.-H., & Shiea, J. (2002). *Electrochimica Acta, 47,* 4367.

16. Pitner, W. R., & Hussey, C. L. (1997). *Journal of the Electrochemical Society, 144,* 3095.
17. Chunyan, L., Nishikawa, K., Moon, J., & Kanamura, K. (2019). *Journal of Electroanalytical Chemistry, 832,* 467.
18. Xu, M., Ivey, D. G., Xie, Z., & Qu, W. (2013). *Electrochimica Acta, 89,* 756.
19. Liu, Z., Cui, T., Pulletikurthi, G., Lahiri, A., Carstens, T., Olschewski, M., et al. (2016). *Angewandte Chemie International Edition, 55,* 2889.
20. Fan, J., Xiao, Q., Fang, Y., Li, L., & Yuan, W. (2019). *Ionics, 25,* 1303.
21. Simons, T. J., MacFarlane, D. R., Forsyth, M., & Howlett, P. C. (2014). *Chemelectrochem, 1,* 1688.
22. Liu, Z., Pulletikurthi, G., & Endres, F. (2016). *ACS Applied Materials and Interfaces, 8,* 12158.
23. Liu, Z., Bertram, P., & Endres, F. (2017). *Journal of Solid State Electrochemistry, 21,* 2021.
24. Hussey, C. L. (1983). In G. Mamantov (Ed.), *Advances in Molten Salt Chemistry* (Vol. 5, pp. 185–230). New York: Elsevier.
25. Hussey, C. L. (1994). In G. Mamantov & A. I. Popov, (Eds.), *Chemistry of Nonaqueous Solutions, Current Progress* (pp. 227–275). New York: VCH Publisher.
26. Gifford, P. R., & Palmisano, J. B. (1988). *Journal of the Electrochemical Society, 135,* 650.
27. Takami, N., & Koura, N. (1989). *Journal of the Electrochemical Society, 136,* 730.
28. Tsuda, T., Kokubo, I., Kawabata, M., Yamagata, M., Ishikawa, M., Kusumoto, S., et al. (2014). *Journal of the Electrochemical Society, 161,* A908.
29. Chen, C.-Y., Tsuda, T., Kuwabata, S., & Hussey, C. L. (2018). *Chemical Communications, 54,* 4164.
30. Carlin, R. T., De Long, H. C., Fuller, J., & Trulove, P. C. (1994). *Journal of the Electrochemical Society, 141,* L73.
31. Pan, C.-J., Yuan, C., Zhu, G., Zhang, Q., Huang, C.-J., Lin, M.-C., et al. (2018). *Proceedings of the National Academy of Sciences of the United States of America, 115,* 5670.
32. Tsuda, T., Uemura, Y., Chen, C.-Y., Matsumoto, H., & Kuwabata, S. (2018). *Electrochemistry, 86,* 72.
33. Uemura, Y., Chen, C.-Y., Hashimoto, Y., Tsuda, T., Matsumoto, H., & Kuwabata, S. (2018). *ACS Applied Energy Materials, 1,* 2269.
34. Tsuda, T., Uemura, Y., Chen, C.-Y., Hashimoto, Y., Kokubo, I., Sutani, K., et al. (2017). *Journal of the Electrochemical Society, 164,* A2468.
35. Sun, H. B., Wang, W., Yu, Z. J., Yuan, Y., Wang, S., & Jiao, S. Q. (2015). *Chemical Communications, 51,* 11892.
36. Wang, D.-Y., Wei, C.-Y., Lin, M.-C., Pan, C.-J., Chou, H.-L., Chen, H.-A., et al. (2017). *Nature Communications, 8,* 14283.
37. Ui, K., Kuma, Y., & Koura, N. (2006). *Electrochemistry, 74,* 536.
38. Hudak, N. S. (2014). *Journal of Physical Chemistry C, 118,* 5203.
39. Mori, T., Orikasa, Y., Nakanishi, K., Chen, K. Z., Hattori, M., Ohta, T., et al. (2016). *Journal of Power Sources, 313,* 9.
40. Yu, Z. J., Kang, Z. P., Hu, Z. Q., Lu, J. H., Zhou, Z., & Jiao, S. Q. (2016). *Chemical Communications, 52,* 10427.
41. Wang, S., Yu, Z. J., Tu, J. G., Wang, J. X., Tian, D. H., Liu, Y. J., et al. (2016). *Advanced Energy Materials, 6,* 1600137.
42. Cohn, G., Ma, L., & Archer, L. A. (2015). *Journal of Power Sources, 283,* 416.
43. Gao, T., Li, X., Wang, X., Hu, J., Han, F., Fan, X., et al. (2016). *Angewandte Chemie International Edition, 55,* 9898.
44. Yu, X., Boyer, M. J., Hwang, G. S., & Manthiram, A. (2018). *Chem, 4,* 586.
45. Yu, X., Wang, B., Gong, D., Xu, Z., & Lu, B. (2017). *Advanced Materials, 29,* 1604118.
46. Thomas, R., Patole, S. P., & Costa, P. M. F. J. (2019). *Journal of Power Sources, 435,* 226780.
47. Hu, Y., Debnath, S., Hu, H., Luo, B., Zhu, X., Wang, S., et al. (2019). *Journal of Materials Chemistry A, 7,* 15123.
48. Schoetz, T., de Leon, C. P., Ueda, M., & Bund, A. (2017). *Journal of the Electrochemical Society, 164,* A3499.
49. Zhang, Y., Liu, S., Ji, Y., Ma, J., & Yu, H. (2018) *Advanced Materials*, 1706310.

50. Ru, Y., Zheng, S., Xue, H., & Pang, H. (2019). *Journal of Materials Chemistry A, 7,* 14391.
51. Tafur, J. P., Abad, J., Román, E., & Romero, A. J. F. (2015). *Electrochemistry Communications, 60,* 190.
52. Muñoz-Torrero, D., Palma, J., Marcilla, R., & Ventosa, E. (2019). *Dalton Transactions, 48,* 9906–9911.

The manufacturer's authorised representative in the EU is Springer Nature Customer Service Centre GmbH, Europaplatz 3, 69115 Heidelberg, Germany. If you have any concerns regarding our products, please contact ProductSafety@springernature.com

Printed and bound by CPI Group (UK) Ltd, Croydon, CR0 4YY
30/09/2026
02225296-0020